THE
BACTERIAL
SPORE

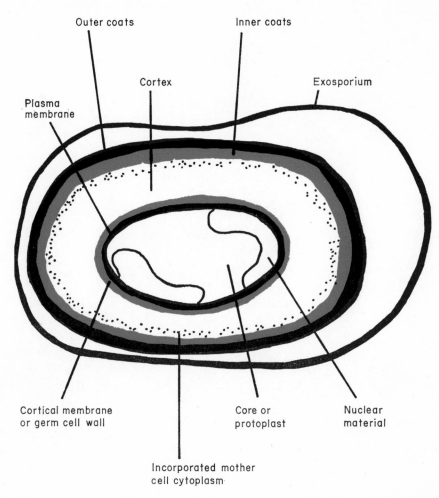

Outer coats

Inner coats

Cortex

Exosporium

Plasma membrane

Cortical membrane or germ cell wall

Core or protoplast

Nuclear material

Incorporated mother cell cytoplasm

Frontispiece: Diagrammatic representation of the structure of the bacterial spore

THE
BACTERIAL
SPORE

Edited by

G. W. GOULD

Unilever Research Laboratory,
Sharnbrook, Bedford, England
and

A. HURST

Microbiology Division,
Food and Drug Directorate,
Ottawa, Canada

1969

Academic Press
London and New York

ACADEMIC PRESS INC. (LONDON) LTD
Berkeley Square House
Berkeley Square
London, W1X 6BA

U.S. Edition published by
ACADEMIC PRESS INC.
111 Fifth Avenue
New York, New York 10003

Library of Congress Catalog Card Number: 69–16499
Standard Book Number: 12–293650–7

PRINTED IN GREAT BRITAIN BY
BUTLER AND TANNER LTD
FROME, SOMERSET, ENGLAND

List of Contributors

A. C. BAIRD-PARKER Unilever Research Laboratory, Sharnbrook, Bedford, England

R. W. BERNLOHR Department of Microbiology and Biochemistry, University of Minnesota, Minneapolis, Minnesota, U.S.A.

R. H. DOI Department of Biochemistry and Biophysics, University of California, Davis, California, U.S.A.

Z. EVENCHIK Institute for Biological Research, Ness Ziona, Israel

P. FITZ-JAMES Department of Bacteriology and Immunology, University of Western Ontario, London, Canada

G. W. GOULD Unilever Research Laboratory, Sharnbrook, Bedford, England

A. D. HITCHINS Department of Microbiology and Public Health, Michigan State University, East Lansing, Michigan, U.S.A.

J. R. HUNTER Microbiological Research Establishment, Porton, Salisbury, Wiltshire, England

A. HURST Microbiology Division, Food and Drug Directorate, Ottawa, Canada

M. INGRAM Agricultural Research Council, Meat Research Institute, Langford, Bristol, England

A. KEYNAN Institute of Life Sciences, Hebrew University, Jerusalem

C. LEITZMANN Departments of Microbiology and Biochemistry, University of Minnesota, Minneapolis, Minnesota, U.S.A.

J. C. LEWIS Western Regional Research Laboratory, Agricultural Research Station, U.S. Department of Agriculture, Albany, California, U.S.A.

W. G. MURRELL Commonwealth Scientific and Industrial Research Organization, Division of Food Preservation, Ryde, New South Wales, Australia

J. R. NORRIS "Shell" Research Ltd, Milstead Laboratory of Chemical Enzymology, Sittingbourne, Kent, England

T. A. ROBERTS Agricultural Research Council, Meat Research Institute, Langford, Bristol, England

H. L. SADOFF Department of Microbiology and Public Health, Michigan State University, East Lansing, Michigan, U.S.A.

R. E. STRANGE Microbiological Research Establishment, Porton, Salisbury, Wiltshire, England

A. S. SUSSMAN Department of Botany, University of Michigan, Ann Arbor, Michigan, U.S.A.

V. VINTER Department of General Microbiology, Institute of Microbiology, Czechoslovak Academy of Sciences, Prague, Czechoslovakia

E. YOUNG Department of Biology, York University, Toronto, Canada

Preface

THERE are numerous reasons why bacterial spores have claimed the attention of technologists, research workers and students of the biological sciences. Perhaps the most challenging reason is their ability to resist not only heat but also radiation, other physical agents, chemicals and enzymes. Spores are not devoid of proteins, nucleic acids, subcellular particles and structures typical of less resistant cells; however, even after a considerable research effort, the way in which the spore confers resistance on these various components is still neither understood nor reproducible *in vitro*. Dormancy and longevity of spores are as little understood as resistance. But although spores are exceedingly dormant, that same dormancy is lost dramatically during germination by "trigger" reactions which are so rapid as to be probably unique in biological systems. Germinated spores outgrow to form new vegetative cells by passing through a sequence of changes, the control of which is proving to be an important model for studies in developmental biology. Following growth and cell division vegetative cells must again form new spores to complete the cycle of growth of the spore-forming bacterium. During sporulation the vegetative genome is suppressed and the spore genome, controlling synthesis of new spore-specific enzymes, antigens and structures, is expressed within a very short period of time. In synchronously sporulating cultures this process is an excellent experimental system for studying the control of cellular differentiation.

There are of course equally good practical reasons for studying spores. Spore-forming bacteria include organisms of direct value to man as producers of antibiotics and insecticides as well as some of the most lethal toxins known. Spores are the life forms most resistant to almost any process designed to kill micro-organisms, be it sterilization of foodstuffs by heat, of space probes by ionizing radiation or of surgical instruments by chemicals. The experimentalist hopes that research may sufficiently reveal the secrets of resistance, dormancy, germination, outgrowth and sporulation that eventually novel ways of using, destroying, or controlling spores may be discovered.

The expansion of spore research in recent years is illustrated by the success of such ventures as the Illinois *Spores* I to *Spores* IV meetings and the wide distribution of the symposium papers. In Great Britain a more modest Spore Group has held annual meetings with steadily increasing attendance. A striking testimony to expansion of interest has been the success of *Spore Newsletter*, edited in Australia by W. G. Murrell, which contains abstracts of recently submitted papers and is distributed throughout the world.

Spore research is so wide in scope nowadays, and its expansion is increasing in impetus so rapidly, that adequate review of the subject in one volume will be difficult in the future. We therefore considered that it was an opportune time to bring together authoritative descriptions of the major phases of spore research, each written by specialists in their chosen areas of interest. We would like to thank the various authors and all those who assisted us in progressing the book. In particular we would like to thank Joan Gould for a great amount of help with referencing and proof reading.

It is hoped that this book will stimulate further research on bacterial spores by being a source of information about particular areas in depth. We also have very much in mind the young research worker and student, who, we hope, will be provided with a useful broad picture. Without doubt there is ample scope for further research.

May 1969   G. W. GOULD
                                                              A. HURST

Contents

1. The Prevalence and Role of Dormancy
A. S. SUSSMAN

2. Morphology of Sporulation
P. FITZ-JAMES AND E. YOUNG

3. Physiology and Biochemistry of Sporulation
V. VINTER

4. Changes in Nucleic Acids During Sporulation

R. H. DOI

5. Biosynthesis of Polypeptide Antibiotics

A. HURST

6. Control of Sporulation

R. W. BERNLOHR AND C. LEITZMANN

7. Chemical Composition of Spores and Spore Structures

W. G. MURRELL

13. Sporeformers as Insecticides

J. R. NORRIS

14. Medical and Veterinary Significance of Spore-forming Bacteria and Their Spores

A. C. BAIRD-PARKER

15. Sporeformers as Food Spoilage Organisms

M. INGRAM

16. Resistance of Spores

T. A. ROBERTS AND A. D. HITCHINS

Glossary

Activation of spores results from some treatment which does not itself initiate germination, but afterwards allows spores to germinate more rapidly or more completely or both.

Ångström (Å) is 10^{-7} mm or 10^{-1} mμ.

Anaplerotic sequences. In metabolic cycles there is a balance between the quantities of carbon which enter and those which leave the cycle, for example to be tapped off for the building of intermediates. Ancillary routes to that of the main cycle may operate to maintain the correct level of intermediates. These ancillary routes are collectively described as anaplerotic sequences.

Base ratio refers to the ratio of GC to AT or both in chromosomal material.

Cistron. Synonym of gene.

Commitment to sporulation is the stage of sporulation at which the sporal part of the sporangium can no longer return to vegetative growth on transfer to complete medium.

Core. Trivial name for the central part of a spore, i.e. that part bounded by the protoplast membrane or core membrane. Synonymous with spore protoplast.

Dalton. A unit of mass equal to one sixteenth the mass of an atom of oxygen. (One Dalton equals $1·6498 \times 10^{-24}$ g.)

Dormancy is the state of inertness of a cell towards a variety of exogeneous substrates. Spores may differ quantitatively in their dormancy: spores which are exceptionally unresponsive to germinants are termed "superdormant". (For a fuller discussion, see Chapter 9.)

Endotrophic sporulation occurs independently of external nutrients and is supported exclusively by the pre-existing make-up of the vegetative cell.

Episomes are extra-chromosomal repositories of genetic information. Episomes may exist autonomously or integrated into the bacterial chromosome.

Germination encompasses the rapid degradative changes in spores which are generally initiated by specific agents ("germinants") and which are accompanied by well-defined cytological and chemical changes. Germination leads to loss of dormancy and resistance and in normal growth germination is followed by "outgrowth".

The following terms have been used in the spore literature: "Physiological germination"; germination caused by the use of naturally occurring substances, e.g. L-alanine, inosine, etc.: "chemical germination"; germination caused by chemicals, mostly surface active compounds, e.g. n-dodecylamine: "mechanical germination"; germination caused mechanically, e.g. when an objective is suddenly pressed on a coverslip over a spore suspension, or when spores are cracked or abraded.

Integuments is a general term for the various coat layers of spores.

Maturation consists of the gradual development of the highest degree of resistance of the fully formed spore during sporulation.

Micron (μ) is 10^{-3} mm.

Microcycle sporogenesis is the resporulation which can occur within the first cell formed following germination and outgrowth, without the necessity for intervening cell division.

Murein is the trivial name for the unique type of polymer found only in bacterial cell walls. Synonymous with mucopeptide, glycosaminopeptide, peptidoglycan.

Muropeptide is the repeating unit of a murein.

Outgrowth is the process of development whereby a germinated spore grows to become a new vegetative cell prior to the first cell division.

Photoreactivation is the regain of viability of cells apparently killed by U.V. radiation when exposed to visible light.

Rejuvenation is the return to vegetative growth of the sporal part of the sporangium which occurs when sporulating cells are transferred to complete medium before the stage of commitment has been reached.

Replacement consists of the transfer of cells from a complete medium to a deficient one (step down). The deficient medium permits sporulation but not vegetative growth.

Sporangium is the whole cell in which the spore is being formed. It consists of the spore or forespore part and the parasporal part.

Torr is a measurement of pressure equivalent to mm of mercury.

Transduction is the transfer of genetic material from one cell to another by temperate bacteriophage.

Transfection is the production of a complete virus by cells infected with the nucleic acid isolated from that virus; it can be considered a type of bacterial transformation.

Transformation is the transfer of small segments of genetic material from one cell to another by free DNA.

Water activity (a_w) is the ratio of the vapour pressure of the solution to that of water. Equilibrium relative humidity (ERH) equals $a_w \times 100$.

Whitening period is a cytological term describing the gradual brightening observed in the phase contrast microscope of spores during sporulation.

Conventions

ADP, CDP, GDP, IDP, UDP: 5′-pyrophosphates of adenosine, cytidine, guanosine, inosine and uridine.

AMP, etc.: adenosine 5′-phosphate etc.

AT: adenine plus thymine.

ATP, etc.: adenosine 5′-triphosphate etc.

CAP: chloramphenicol.

DAP or DMP: diaminopimelic acid.

DNA: deoxyribonucleic acid.

DPA: dipicolinic acid.

EDTA: ethylenediaminetetra-acetate.

GC: guanine plus cytosine.

MAK: methylated albumin kieselguhr.

NAD: nicotinamide-adenine dinucleotide.

NADP: nicotinamide-adenine dinucleotide phosphate.

PHB: poly-β-hydroxybutyrate.

PTA: phosphotungstic acid.

RNA: ribonucleic acid; s-RNA, soluble RNA (synonymous with transfer or t-RNA); m-RNA, messenger RNA; r-RNA, ribosomal RNA.

SLS: sodium lauryl sulphate.

Tris: 2-amino-2-hydroxymethylpropane-1,3-diol.

CHAPTER 1

The Prevalence and Role of Dormancy

A. S. SUSSMAN

Department of Botany, University of Michigan, Ann Arbor, Michigan U.S.A.

I. Definitions and Terminology

A. CRYPTOBIOSIS

THE intensity of the dormant state has been expressed in terms of the organism's metabolic capacity, the time necessary for it to return to an active state, or the ease with which the latter condition can be achieved. Inasmuch as a lowered metabolic rate characterizes most, if not all, dormant organisms Keilin[1] adopted this measure of dormancy. He used the term *cryptobiosis* to set the lower limit of this condition, namely the state that is imposed by temperatures low enough to vitrify cells. In this state organisms show a minimum of activity and are presumed to be in the deepest state of dormancy. Other names have been used to describe this state, including *latent life* and *abiosis*[2] and *anabiosis*[3]. The latter term has been widely used for this purpose but originally it meant the resurrection of completely life-less organisms. However, the simultaneous use of abiosis in matters relating to the origin of life to describe the formation of a neobiont from

B

chemical mixtures induced Keilin to propose the term cryptobiosis which has been adopted here.

B. TYPES OF DORMANCY AND GERMINATION

Two types of phenomena involved in dormancy were clearly delineated by Doran[4] who described them as the "internal" and "external" factors. Included in the former were the maturity of the spore, its longevity and "animation" and "a poorly understood factor which may be called vitality . . ." The "external" factors included temperature, light, water, oxygen, nutrients, toxic substances and the length of the period during which these factors exerted their influence on the spore. Although this dichotomy is valid its application in terms of subjective factors like "animation" and "vitality" and the inclusion of time as one of the elements dissipates some of its usefulness. Thus, many instances exist in which the duration of treatment often is dependent upon the maturity of the dormant cell, which is an "internal factor"[5-7]. Consequently, the primary nature of the "internal factors" blurs the distinction between the two factors. However, any attempt to categorize these phenomena must be arbitrary and there are compelling arguments for distinguishing between "internal" and "external" factors[8]. Thus, although it is obvious that these factors are interrelated to the extent that any environmental stimulus is received and interpreted only through the apparatus of the cell, or its internal constituents, there is heuristic value in maintaining a dichotomy. However, the terminology and definitions suggested by Doran and later workers have been modified in order to ameliorate some of the difficulties alluded to above. The definitions given below are taken from Sussman[6], except as otherwise noted.

Dormancy. Any rest period or reversible interruption of the phenotypic development of an organism.

Constitutional (constitutive) dormancy. A condition wherein development is delayed due to an innate property of the dormant stage such as a barrier to the penetration of nutrients, a metabolic block or the production of a self-inhibitor.

Exogenous dormancy. A condition wherein development is delayed because of unfavorable chemical or physical conditions of the environment.

Maturation. The complex of changes associated with the development of the resting stage of dormant organisms or of the germinable stage in those without a dormant period.

Activation. As was pointed out by Gibbs et al.,[9] this term has been used to describe two phenomena. The first describes a treatment which leads to an increase in the *viable count*, as in the heat-activation treatments first described by Curran and Evans.[10] In addition, activation often is used to describe treatments that lead to an increase in the *rate of germination*, independent of the viable count. Inasmuch as the viable count and the rate of germination can vary independently of each other Gibbs *et al.* suggest that the term activation be qualified to distinguish between the two types of end-points of the germination process that are described above. Therefore, they suggest terms like "activation for viable count" and "activation for germination", "activation for rate of germination" or "kinetic activation"; and combinations of these are proposed for activation treatments in which both viability and rates are measured. Perhaps it is worth making a terminological distinction whenever activation is used but even greater difficulties exist in using the term germination which may be measured in at least a dozen different ways.[7] For this reason, comparative studies of germination *per se* require that the end-point measured be known, and this caveat holds for activation as well. But, having so cogently been warned by Gibbs *et al.*, I doubt whether the nomenclatorial complexities they introduce are necessary.

Germination. The first irreversible stage which is recognizably different from the dormant organism, as judged by morphological, cytological, physiological or biochemical criteria.

The intensity of the dormant period of individual spores cannot be expressed in absolute terms because it is a function of the conditions during spore formation, its genetics and the treatment that was administered during storage. Moreover, in the case of certain spores which have never been germinated in artificial culture,[11] it is impossible to distinguish between non-viability and dormancy. Theoretically, the upper limit of the duration of the dormant period is the time when non-viability sets in, and the lower limit is the time required to activate.

II. The Dormant State in Nature

A. DISTRIBUTION

Dormancy is widespread in nature and serves a number of purposes which have provided selective advantage during the course of evolution. The ubiquity of this phenomenon has led to several independent lines of study and to the adoption of distinctive terminologies in the separate disciplines. This diversity of terms is reflected in Table I which lists the nomenclatorial equivalents used in the study of dormancy.

B. CHARACTERISTICS OF DORMANT STAGES

1. *Morphological*

"Since during diapause (dormancy) the organism is subjected to widely differing and harsh conditions, adaptations of the diapause stages to climatic factors show greater specificity, and are in general more varied and complex than adaptations in the active stages."[19] And nowhere is this diversity of adaptation more evident than in the morphology of dormant stages.

There are many types of dormant organisms, ranging from protistan spores and cysts to multicellular seeds, eggs and hibernating animals of

TABLE I

Nomenclatorial equivalents used in the study of dormancy[a]

Term	Organisms	Reference
Constitutional (constitutive)		
dormancy		
True dormancy, physiological		
rest	Higher plants	12, 13
Rest	Higher plants	14
"*Wahre*" or "*Wirkliche*" *Ruhe*	Higher plants	13, 15
Hibernation[b], aestivation	Vertebrates	16
Diapause	Arthropods	17
Exogenous dormancy		
Imposed rest	Higher plants	12
Quiescence	Higher plants	14
"*Zwangsruhe*", "*Aufgezwungene*		
Wachstumsruhe"[c]	Higher plants	15
Hypothermia, torpor	Vertebrates	18
Quiescence	Arthropods	17

[a] Modified from Sussman and Halvorson.[7]

[b] According to Danilevskii[19], hibernation in insects is induced and controlled only by external conditions. Therefore, it would seem to belong among the exogenous types of dormancy. However, he gives reasons for considering it to be constitutional but the use of these terms in the insect literature differs from that in the literature on mammals.

[c] Dormancy induced by permeability barriers in the cell wall is considered by Vegis[15] to be of the exogenous type, whereas I have included such restraints among the constitutional factors.

diverse size. Even the vegetative cells of many protistans can survive extremes of temperature and atmospheric and osmotic pressure so that specially formed structures like spores are not the only means through which survivability is enhanced under these conditions.[20-24] However, it is usually the spores and cysts of protistans which survive under the most drastic circumstances, although modified cells like the "*dauerzellen*" of yeasts, and sclerotia, rhizomorphs and modified hyphae of other fungi can be very resistant to environmental upsets.[7] Animals also exhibit great morphological diversity in dormant stages, as in the case of protozoans[25] and insects, wherein diapause can occur in the egg, larval and pupal stages.[19] Among higher animals hibernation usually results in few anatomical changes other than increased growth of hair and widespread lipid deposits.[16] Moreover, surprising tolerance to temperature extremes is exhibited even by non-hibernating animals under certain conditions.[26]

2. Physiological

(i) *Respiration.* An almost universal concomitant of dormant systems is reduced respiratory capacity, as is revealed in the examples given in Table II. A more complete list of the respiratory characteristics of dormant organisms of several kinds is provided in the book by Altman and Dittmer[27]; bacteria and fungi are dealt with in the book by Sussman and Halvorson.[7] An exception is the uredospore of *Puccinia graminis*, the Qo_2 of which has been reported not to change significantly during germination.[28] However, the respiration of these spores was studied only 72 hr after germination was induced so transitory changes in rate may have been missed. Another exception is the macroconidium of *Fusarium solani*[29, 30] when provided with ethanol or mannose; but increased respiration does occur when glucose, fructose or trehalose are the substrates. Both of these exceptions are exogenously dormant spores and it is possible that respiratory increases during germination are not as characteristic of such systems as of constitutionally dormant ones.

In the case of mammalian hibernation it is often noted that respiratory rates are very irregular in deep hibernation. Thus, there may be times when no respiration is evident, followed by intervals of strengthened respiration. And, as hibernation becomes less deep, the irregularities become more frequent and the respiration rate greater.[31]

A surprisingly pervasive feature of dormancy is the use of lipid as an endogenous substrate during dormancy. Several fungus spores have been shown to metabolize principally fat when dormant, as have insects[17] and mammals.[16] In fact, in those mammals which have short phases of uninterrupted hibernation and which feed during arousals, such as hamsters, hibernation may be prevented by hyperglycemia. As for those which have a deep hibernation, a marked hypoglycemia is exhibited in comparison to the normothermic state.[16] On the other hand, the very low respiratory capacity of dormant bacterial spores[32] does not seem to be based upon lipids but, as in the cases mentioned above, their ability to utilize exogenous sugars is impaired until they are activated.

Another aspect of the ubiquitous use of lipid by dormant systems is its necessity, along with carbohydrates, for germination and arousal. Why this should be so is not known, but it may be that an important enzymatic step in glucose metabolism must be induced before the dependence upon fat metabolism is lost.

(ii) *Water content.* Another frequent concomitant of dormancy is a decrease in the total amount of water, often accompanied by its transfer, at least in part, to the colloidal state, or to a physical condition other than liquid. In fact, the "anhydrous core" theory of dormancy attempts to account for the characteristics of bacterial spores on the basis of the changed properties of water in the "bound" form. Although there are difficulties in determining the exact amount of water in each of its states in living cells,

TABLE II

Oxygen uptake of dormant and active stages of various organisms

Organism	Q_{O_2} (mm³/mg wet wt/hr)		Reference
	Dormant	Active	
Protistans			
Bacillus cereus	0·005[a]	6–10[a]	[33]
B. subtilis	1·0[a]	17[a]	[27]
Neurospora tetrasperma (ascospores)	0·03[a]	2·5[a]	[7]
Myrothecium verrucaria (conidia)	0·1	7·5[a]	[34]
Nematodes			
Eophila dollfusi	0·110	0·203	[35]
Insects			
Anatolica eremita	0·058–0·437	0·523–1·236	[36]
Deilephila euphorbia	0·025	0·275	[37]
Amphibians			
Rana esculenta	0·085	0·437	[27]
Mammals			
Mesocricetus auratus	0·070	2·90	[27]
Nictalus noctula	0·0004[b]	0·0035[c]	[27]
Higher plants			
Medicago sativa	0·0038	0·0106	[27]

[a] Data converted to mm³/mg (wet wt)/hr, assuming 90% H_2O
[b] At 20°C
[c] At 12·5°C

the total amount of water in many dormant protistans is reduced in comparison to active ones.[7] This is true for the seeds of higher plants[38] and diapausing insects[17] as well, and hibernating mammals must find special means of eliminating water which accumulates as a metabolic waste product.[39]

That the reduced metabolic capacity and dehydrated condition of dormant systems may be related is suggested by the data in Table III. These data suggest that as certain organisms become progressively dehydrated, there is a reduction, sometimes proportional, in oxygen consumption. Many such data exist for plants,[27, 44] including angiosperms, bryophytes and lichens. In the case of many of these, it has been shown that the process is reversible for rehydration results in the recovery of higher respiratory rates.[45] Therefore, the lowered respiratory rates which characterize dormancy could, in some cases, be engendered by dehydration alone which, in turn, might be caused by changed environmental conditions, such as in ponds which dry periodically, or in terrestrial habitats with limited water.

TABLE III

Relation between water content of dormant systems and respiratory rate

Plant and part	Unit	Water availability	Unit	Respiratory rate	Reference
Pisum sativum (Seeds)	% initial fresh wt	100 90 80 70 60 50	CO_2 Production % control	100 88 76 64 52 40	40
Secale cereale (Seed kernels)	% fresh wt	75·5 71·5 66·5 55·4 50·2 31·7	$mm^3 O_2/$ g fresh wt/hr	601 495 403 296 246 82	41
Funaria hygrometrica (Moss leaves)	water deficit, %	0 20 40 60 80 90	$mm^3 O_2/$ 100 g dry wt/10 hr	1370 1130 1020 700 380 260	42
Aspergillus niger (Conidia)	relative humidity, %	100 60	$mm^3 O_2/$ mg dry wt/hr	5 1	43

Although the mechanism of this effect is not yet understood completely, it is known that the functioning of the respiratory apparatus of cells, including mitochondria, is markedly affected by changes in tonicity. Moreover, isolated enzymes have been studied on soils and other surfaces, under different conditions of humidity, and inhibitory effects shown by desiccation.[46,47] However, much more needs to be learned in these *in vitro* systems, especially about respiratory enzymes which seem not to have been studied very intensively in this way.

Often associated with the low water content of dormant systems is a rigid and resistant wall. Spores, cysts, seeds and diapausing pupae are all characterized by the possession of such a covering. Only hibernating higher animals appear to lack special structures of this kind although increased hair growth sometimes is an accompaniment of this process. The removal of these heavy walls sometimes is necessary before development can occur, as in the case of some insects,[17] where a barrier to water entry exists in dormant eggs of *Melanoplus*. In this case the barrier must be eliminated before diapause can be broken and sufficiently long exposure to cold will accomplish this, as will immersion in xylol and other fat solvents. Seeds of many plants require scarification before germination will occur;[47] abrasion, or treatment with enzymes which soften wall layers, activate spores of bacteria and fungi,[7] and cysts of protozoa,[48] and nematodes.[49] However, it is still uncertain in all these cases whether access to water limits activation, for limitation of oxygen,[15] and release of endogenous inhibitors[50] have also been invoked to explain the need to breach the heavy walls of dormant organisms.

(*iii*) *Resistance.* The ability to survive environmental conditions that would kill other stages of an organism is a precondition of the dormant state. It also follows that constitutional dormancy will not exist in ephemeral organisms because those which initiate growth rapidly are favored by selection.[51, 52]

The heightened resistance of dormant systems to deleterious environmental conditions has been documented in detail elsewhere.[7, 26, 27] These data reveal that dehydration enhances resistance to injury from extremes of temperature as in the well-known superiority of wet sterilization over dry. Moreover, other data suggest that organisms with low water content survive for longer times in nature than do those with more water. In fact, the low water content of many dormant stages is evidence of a significant increase in the resistance of organisms to desiccation, for many organisms are sensitive to dehydration in their vegetative phases. Consequently spores and other dormant stages must have evolved means of mitigating the detrimental effects of drying, such as increased concentrations of protective chemicals like dipicolinic acid in bacteria, and trehalose, mannitol and glycerol in fungi, insects and other invertebrate animals like *Artemia*.[7, 53, 54] And it is worth noting that desiccation-resistant nonspore-forming bacteria like certain staphylococci may use unusual conjugates of sugars, nucleic acids and proteins for this purpose.[55] Thus, desiccation could lead to both enhanced resistance and reduced respiratory capacity in dormant organisms.

C. Induction of Dormancy

The means whereby dormancy may be induced are of two types, extrinsic (environmental) and intrinsic. Extrinsic factors predominate in Table IV, in which many of the inducers of dormancy in a variety of organisms are assembled. On the other hand, rhythmic determinants of dormancy also are indicated in the case of fungi, protozoa, seed plants, insects and mammals, and most of these are considered to be intrinsically controlled.[56] However, some examples of environmentally controlled rhythms do exist and there is still controversy concerning the extent of extrinsic control over other rhythms, including circadian types.[57] Another type of intrinsic control is represented by "sporogens", that is, substances produced by microorganisms which induce sporulation.[59] These have been described for bacteria and fungi and may be compared to the hormones of animals which also effect intrinsic control over development.

1. Extrinsic controls

It is clear from the data in Table IV that, except for the few reports of sporogens, nutritional factors are the principal inducers of sporulation and dormancy in bacteria and, perhaps, in protozoa. This is in contrast to other protistans, arthropods and plants wherein light and temperature influence the induction of dormant stages. In fact, only hibernating mammals appear to be as dependent as bacteria upon a single environmental inducer, in their case, temperature. Why is nutrition the prime determinant of sporulation in bacteria?

The answer must be sought in terms of the peculiarities of bacterial populations which have led to adaptive responses such as dormancy. Thus, their short generation time leads to the accumulation of enormous numbers of individuals relatively quickly. Consequently, the ebb and flow of bacterial populations must be measured in time periods which are shorter than those of diurnal environmental cycles of light, humidity, temperature, etc., and which are minute compared to the lunar, sidereal and annual cycles of many animals and plants. In fact, visible light which is an almost ubiquitous morphogenetic effector in green plants and insects actually is lethal to some bacteria.[71, 72] Therefore, the environmental cues which trigger developmental processes like dormancy in animals and plants, such as light and temperature, have restricted adaptive significance for bacteria. Furthermore, soils and water, which are the primary sources of the microenvironments of bacteria undergo fluctuations of temperature and humidity which are more damped than are those of the surrounding air, which is an environment that must often be frequented by higher organisms, even if only temporarily. But, selection pressure usually favors organisms that can adapt to periodicities in their surroundings,[73] so to which cues do bacteria respond?

As Stanier[74] has suggested, the type of food in a microbial ecosystem is one of the most important determinants of the kind of community that results. Moreover, this food usually is available *discontinuously* and in small

TABLE IV

References to means through which dormant stages are formed

Organisms	Temperature		Light		Nutrition	Other
	Low	High	Photoperiod	Other		
Protista						
Bacteria	—[32, 58]	—[32, 58]	—[32, 58]	—[32, 58]	Many[32, 58]	"Sporogen(s)"[59, 60]
Fungi	Some[11]	Some[11]	—[11, 61]	—[11, 61]	Many[11]	Desiccation[11] Rhythms[62]
Algae	Some[64]	Some[64]	Some[64]	?	Some[64]	"Sporogen(s)"[63] Desiccation[64]
Protozoa	Some[65]	—[65]	—[65]	—[65]	Many[65]	Desiccation, pH change[65] Rhythms
Plants						
Mosses and Liverworts	Some[67]	—[67]	—[67]	—[67]	Some[67]	Rhythms[69]
Seed plants	Some[68]	Some[68]	Many[68]	—[68]	?	
Animals						
Insects	Some[19]	—[19]	Many[19]	—[19]	Some[19]	Rhythms[19]
Mammals	Many[16]	—[16]	?[31]	—[16]	—[16]	Rhythms[31, 70]

Dashes indicate that the references cited, or others available to the author, did not show examples where such treatments were effective in inducing dormancy.

amounts which, taken together with the capacity of bacteria to exploit such opportunities rapidly, results in explosive growth and rapid transformation of the food to gaseous and other products. So the nature of the bacterial microenvironment is such that the major changes which occur over their time-span are nutritional ones, and responsiveness to these irregular, but recurring, stimuli is of great survival value and is one of the most important determinants of the community that develops.[75] That bacteria in soil almost invariably respond much more to the presence of plant roots, and their secreted nutrients, than do their neighboring actinomycetes, fungi, algae and protozoa[76] argues for the effectiveness of nutritional stimuli for bacteria in nature. Thus, nutrition is the factor in their surroundings to which bacteria most often respond, and a frequently encountered selection pressure. Of course, the food supply of certain bacterial environments is continuous, as in the rumen and other habitats of bacterial hosts[75] so that other factors must be limiting. Relatively little is known about the cues which induce vegetative dormancy in bacteria and it may be found that a wider diversity of environmental cues may influence these stages as compared to presporulation ones. For example, the role of water in the induction of dormancy in bacteria is not worked out whereas it is said to be the most important ecological determinant of protozoa[77] and is important to fungi as well.[52]

In contrast to the bacteria, fungi respond to a much more varied group of environmental variables, including light, temperature, nutrition and humidity. This may stem, in part, from the longer generation time of most fungi, which places a premium upon synchronization with rhythmically fluctuating environmental cycles such as the diurnal ones of light, temperature and humidity. Furthermore, the structural features that have been evolved by fungi to ameliorate desiccation, such as the cuticle of rhizomorphs, and the waxy coat and other walls of fruiting bodies, favor survival for sustained periods in environments that would be deleterious to bacteria. Finally, fungal pathogens gain selective advantage by coordinating their activities with those of their hosts; accordingly, light and temperature strongly influence dormancy in plant pathogens.[7]

Similar arguments can be advanced to explain the effectors of dormancy in other organisms listed in Table IV. Thus, the role of photoperiodism in the induction of flowering in plants[3] and diapause in insects[19] is too well documented to require further comment. However, it is worth stressing that the separation of the environmental factors in Table IV is arbitrary because they may interact. For example, although photoperiodic stimulation of diapause is of undoubted importance in insects, the response may be strongly affected by the temperature[19, 78] or nutritional factors (D. Shappirio, personal communication).[79] Not only can extrinsic factors interact with other environmental cues, but they may also do so with intrinsic factors like endogenous rhythms.[56] And, "sporogens" may be released into the environment so that they become, in effect, nutrients.

2. Intrinsic controls

Much has been said above about the interaction of intrinsic and extrinsic controls and it may be that these factors *always* interact in nature. Moreover, it should be stressed that the induction of dormant states may be of selective advantage because of biotic factors, as well as environmental ones. Thus, it is easy to see that internal rhythms would be of advantage to organisms which are nocturnal, such as some anthropods and mammals, in order that desiccation, or intense heat could be avoided. However, biotic reasons also exist such as the advantage to *Paramecium* of having its sexual phases in synchrony so that the different sexes have a better chance of meeting.[73] In fact, sexual activity often is linked to the formation of dormant stages, thereby providing a connection between the two phenomena.[80]

Asexual spores of fungi often are produced rhythmically and frequently the rhythm is very subject to external controls and cannot be considered to be intrinsic.[62, 81] However, circadian sporulation rhythms which appear to be truly intrinsic have been described in Neurospora[82] but their selective advantage is difficult to assess at this time.

D. DISRUPTION OF DORMANCY

Whereas return of the conditions which permit vegetative development leads to the breaking of exogenous dormancy, constitutional dormancy can only be overcome by a treatment that is not required by vegetative stages. Such treatments lead to activation and the kinds of activators and the organisms for which they are required will be explored next.

1. Physical activators

High temperature is the most common physical factor which activates bacterial spores, except for abrasion (Table V). The treatment usually given is $75–100°C$[7] for 5 to 60 min, and it is reasonable to question the significance of such a requirement in nature for there are few natural habitats where temperatures this high are reached. However, the data of Keynan et al.[83] reveal that lower temperatures suffice with spores of *Bacillus cereus* T if incubation times as long as 45 hr are provided. Thus, even 30° will activate maximally under these circumstances. Furthermore, the effectiveness of heat-activation is affected by the components of the medium in which the spores are suspended. For example, certain amino acids, reducing agents, as well as hydrogen ions, accelerate heat-activation, whereas others (potassium) inhibit.[7] So, activation in nature probably is accomplished by a combination of physical and chemical treatments which interact with

TABLE V

Physical factors required to break dormancy in organisms

Organisms	Temperature		Light		Other
	Low	High	Photoperiod	Continuous	
Protistans					
Bacteria	—[7]	Some[7]	—[7]	—[7]	Abrasion
Fungi[1]	Many[7,5]	Coprophiles[7,5]	—[7,5]	Many[7,5] (basidiomycetes)	Wetting and drying Soaking[7,5]
Algae	Some[5]	Some[7,5]	Some[5]	Few[5]	Change in osmoticum[5]
Protozoa	Woodruffia metabolica[84]	Ascaris[85]	—[85]	?	Change in osmoticum[5]
Plants					
Mosses, Liverwort	Cryptothallus mirabilis	Riella americana[5]	Some[5]	Some[5]	
Ferns	—[5]	Several[7]	—[5]	?	
Seed plants	Many[13,87]	Some[7,87]	Many[7,88]	Some[88]	Rhythms, scarification, etc.[47]
Animals					
Nematodes	Heterodera avenae[89]	Ascaris lumbricoides[7]	—[90]	—[90]	
Insects	Some[17,91]	Some[7,17]	Many[19]	?	Eclosion rhythm[92] Dehydration, injury, etc.[17] Rhythm[16]
Mammals	—[7]	Many[31,93]	?	?	

Dashes indicate that the references cited, or others available to the author, do not reveal examples where such treatments were effective in breaking dormancy.

varying degrees of effectiveness. In any event, nutritional factors are of great importance in the activation of bacterial spores (Table VI), as well as in their induction (Table IV), and probably for similar reasons.

Another case which emphasizes the duality of the treatment required for activation of dormant spores in nature is that of ascospores of *Neurospora*. Characteristically, these spores, along with those of other coprophilous and soil-inhabiting ascomycetes, require a heat-shock for germination (Table V). However, the temperature required is reduced from 60–70° to 42–50°C by the presence of chemicals like furfural.[7] Moreover, furfural and other heterocyclic activators are widely distributed breakdown products of pentosans which are present in most higher plant materials, thereby making more likely the significance of such activation in nature. Habitats in which these conditions probably exist include the digestive tracts of herbivores and their dung, compost and piles of rotting vegetation, etc. In fact, Welsford[95] found that cow dung at 30°C was an excellent germinating medium for dormant ascospores of *Ascobolus furfuraceus* and his work seems to have led to the first use of temperatures in excess of body temperature for activation.[96]

Dormancy of organisms other than bacteria and fungi is broken by high temperatures, including some algae, ferns, seed plants, insects and mammals. The latter group in particular depends upon such treatment for arousal from hibernation although it is possible that light may influence them as well.[31]

Cold temperatures are of the greatest importance to organisms of temperate environments since "overwintering" often is necessary in nature for the resumption of activity in the spring. Thus, more than a hundred species of phanerogams are listed by Crocker and Barton[38] as requiring such a treatment and many insects[17] and mammals do so as well.[18] Some fungi, algae, and a few other types of organisms are activated by cold temperatures (Table V). Frequently, alternate wetting and drying substitute for cold treatment, especially in the activation of seeds.[38] On the other hand, bacterial spores generally are unresponsive to activation by this means, but the fact that spores of both *Bacillus*[207, 208] and *Clostridium*[97] can germinate at 5°C suggests that more work is needed. Although the ecological correlates of cold-temperature requirements are clear in a number of cases because of seasonal changes which are characteristic of certain latitudes, they are less so in the case of the fungi, except for environmental dormancy. However, it is probably not just coincidence that the large majority of the fungi requiring cold-treatment for activation are pathogens for it is likely that synchronization of the fungus with the renewal of the activities of their host plants explains these data.[7]

A similar correlation exists in the case of the requirement for light-activation by fungi because, as in the case of cold-temperature, it is predominantly pathogens which have this need.[7] It is interesting to note that root and mycorrhizal fungi more often require cold treatment than light, as would be expected from the habitats they occupy.

Photosynthetic plants and insects rely heavily upon photoperiodic activation but usually in combination with cold- and chemical-treatment, as in the case of the induction of dormancy. These interactions are discussed in more detail elsewhere and will not be considered further here.[19, 13, 7]

A number of other physical activators have been described, including the abrasion of bacterial spores, which may break a permeability barrier or release a self-inhibitor. However, many substances including activators would be released by this treatment (Table VI), and the mechanism is still not clear. This treatment probably is analogous to scarification which frequently is used for activating seeds, most likely by releasing self-inhibitors.[98]

Recent work suggests that endogenous rhythms may control the arousal from dormancy, as well as its induction in insects,[19, 92] seed plants[99] and mammals.[70] Temperature, light and hormonal influences upon such rhythms are known and must be evaluated before their significance can be judged.

It should not be assumed from the foregoing that the ecological correlates of the induction of dormancy and its disruption always are exact or unambiguous. Indeed, they may be very difficult to make, or misleading, on occasion. A case in point are the spores of *Fomes igniarius* which can germinate at almost any temperature likely to be encountered during the growing season. By contrast, they germinate only very slowly below 15°C, so they prefer conditions such as those found during the summer. As the authors conclude (Good and Spanis)[100] "This is a surprising temperature adaptation in view of the extensive sporulation which takes place in early spring and fall, the somewhat greater rapidity of germination of spores produced in the autumn, and the frequency of adverse weather conditions during the warmer and drier months of the year."

2. Chemical activators

This class of activators has been divided into two types; the intrinsic and extrinsic ones. Among the former are hormones whose effects are widespread in the activation of multicellular animals and plants (Table VI). These substances often can induce the germination of dormant seeds by substituting for the action of physical factors like those listed in Table V.[98, 50] A few other cases exist where this is possible, such as in the ferns, bryophytes and in isolated instances among the algae and nematodes. However, in the latter cases and in the insects and mammals, where increased hormone production is necessary before activation can occur, it is likely that other events, possibly enzymatic ones, precede enhanced hormone flow.

TABLE VI

Chemical factors required to break dormancy in organisms

Organisms	Intrinsic		Extrinsic		
	Hormones	Other	Diffusates, leachates, etc.	Known factors	Other
Protistans Bacteria	—[7]	?	?	Many (amino acids, organic acids, nucleotides, sugar, salts, detergents, hydrocarbons, urea)[7]	
Fungi	Some[7] endogenous stimulators	Some[7, 6] (amino and organic acids, ethanol)	Many[7, 6] (soil, plant, dung extracts and others)	Amino and organic acids, ketones, aldehydes, alcohols, detergents, salts and others[7, 52]	Passage through animals and pH[7]
Algae	*Hydrodictyon reticulatum*[7] (indoleacetic acid?)	—[7]	Several[7, 52] (soil extracts)	A few[7, 52]	Distilled water[52]
Protozoans	—[7]	Some[101] (CO_2, amino acids, nucleotides, carbohydrates, organic acids)	Many[7] (animal, plant, bacterial extracts)	Some[7, 85, 101] (enzymes, methanol, amino acids, nucleotides, carbohydrates, organic acids)	Increase in O_2, distilled water, change in pH[85]

Plants					
Mosses and liverworts	Some[52, 94]	?	Some[7, 52] (Soil, pond and fungal extracts, peptone)	Marchantia[7, 52] (colchicine)	Distilled water[52]
Ferns	Some[52] (gibberellin)	?	Some[7, 52] (soil extracts)	Some[52] (salts)	
Seed plants	Many[38, 98] (gibberellin and others)	Some[50, 87] (CO_2 and others)	Striga, Orobanche	Many[50, 38] (CO_2, thiourea, nitrates, glutathione, uncouplers, uracil, etc.)	
Animals					
Nematodes	Some[102] (exsheathment factor)	—[90]	Many[7, 90] (leachates from hosts and soil)	Many[90, 103] (enzymes, salts, picolinic acid, anhydrotetronic acid, flavianic acid)	
Insects	Hypera postica[104] Platysamia cecropia[105]	?	—[17]	NH_4^+, urea, mineral acids, fat solvents[17]	
Mammals	Many[16]	?	—[16]	—[16]	

Carbon dioxide, because it is a metabolic product, is considered an intrinsic factor and, according to some,[101] excystment of many protozoans may be induced by this substance. They argue that such activation treatments as the addition of bacteria, organic acids, etc., could release molecular CO_2 which, therefore, would be the primary activator. This may be true for other organisms that respond to CO_2 as an activator, but defining the *primary* factor involved may be difficult.

Other naturally occurring substances like amino and organic acids, etc., are included as both intrinsic and extrinsic factors in Table VI. This is in recognition of the arbitrariness of the distinction and the difficulty in evaluating the most appropriate way of classifying these materials. Thus, although they are produced within organisms they also are released into soil and other natural environments where they become extrinsic factors.

In any event, a large number of natural products activate dormant stages. These include many complex materials such as soil and plant extracts, and diffusates from growing organisms. The ecological niche occupied by an organism frequently is a guide to the function of the activation treatment. For example, saprophytes like myxomycetes are stimulated to germinate by extracts of wood and leaves which are the substrata upon which the plasmodium feeds.[7] Puffballs and some other fleshy basidiomycetes cannot germinate in the absence of living yeast, or of other fungi, presumably organisms which co-exist with them in soil.[106] The relation between activator and ecology is particularly clear in the case of pathogens like nematodes of the genus *Heterodera*. In this case the source of the leachate usually serves as the host for the nematode whose cysts respond to the factor. However, the correlation is not complete for the pea-, cereal- and soybean-root eelworms are not stimulated to emerge by exudates from hosts.[107] Other correlations of this kind are provided by the parasitic plants, *Striga* and *Orobanche*, whose seeds are activated by leachates from their hosts (Table VI), and by sclerotia of *Sclerotium* whose germination is stimulated specifically by species of *Allium*.[108] But such specificity is by no means universal as Schroth and Hildebrand indicate.[109] Thus, they list a number of examples where resting spores of fungi germinate in exudates of non-susceptible plants. However, it is possible, as Coley-Smith and Holt[108] point out, that specificity obtains in natural soils whereas non-specific responses could be an artifact of the use of sterilized soil.

Many types of compounds serve as activators and some correlations of the kind made above for *Neurospora* ascospores and furfural can be made but these are exceptional. For the most part, the exact function of these chemicals in relation to the habitat of the organism is not known. This applies to chemical activators of bacterial spores as well and experiments to answer this question are badly needed. For example, what role do chemical and physical activators of bacterial spores serve in nature? And what is their source? Thus, is L-alanine, for instance, provided exogenously in bacterial habitats, or is it the product of the spore itself through the action of peptidases of the kind described as activators by Sierra?[110] And, if peptidases play such a role, are they of exogenous or endogenous origin?[111] Moreover, which of the several activating treatments are of significance in nature, including the combinations of metabolites reported by Wax and Freese?[112]

III. Mechanisms of Dormancy

Under this heading will be discussed only those mechanisms whereby dormancy is established and maintained. Thus, the related questions concerning the nature of the activation process and the steps involved in germination will not be considered directly.

A. CONSTITUTIONAL (CONSTITUTIVE) DORMANCY

All restraints upon development which originate directly from an innate property of the organism are included in this category. It is a state which, most often, is not entered or left suddenly in nature. Thus, complex structural and metabolic changes may occur before dormancy is imposed and such changes sometimes can take considerable time. In fact, Vegis[13] outlines several phases of the dormant condition in seeds and considers that the most intense state of dormancy exists in "middle "or "main rest". Varying depth of diapause in insects has been described by Danilevskii,[19] and both he and Vegis defend the adaptive significance of this phenomenon. And, it is clear from the descriptions of hibernating mammals[16] that the frequency of periodic arousals and subsidence changes during the course of dormancy, and is characteristic of the animal.

Moreover, the duration, as well as the intensity, of dormancy is subject to considerable variation in animals and plants. That this is the case for micro-organisms as well is frequently observed when a small proportion of dormant spores germinates without activation treatment. The selective advantage of this "background" germination could lie in the continued readiness of a few individuals to exploit environmental conditions which would sustain vegetative growth during times when activation stimuli are absent.

In discussing the mechanisms of dormancy it is useful to distinguish between effectors, or signals, and the targets affected by them. Thus, self-inhibitors, hormones and the means through which rhythmic stimuli are transmitted are considered to be effectors. Their targets include permeability barriers, enzymes and their precursors, nucleic acids and intracellular organelles like ribosomes and mitochondria.

1. Effectors

These include self-inhibitors and hormones, which may be overlapping categories because hormone action often involves the interaction of inhibitory and stimulatory substances. Nevertheless, these substances will be considered separately because, in practice, it is difficult to know whether self-inhibitors originate in tissues away from those they affect.

(*i*) *Self-inhibitors.* Frequently it is observed among fungi, bryophytes and vascular cryptogams that high concentrations of spores germinate less well than low ones.[5, 7] This has led to the finding of self-inhibitors in several fungi[113, 6] including alkaloids[114] and volatile materials.[115, 116] Bacterial spores may show similar effects.[117] Thus 10^8 spores per ml of *Bacillus cereus* yielded only 10% germination whereas 90% was obtained with 10^6 spores per ml. By contrast, the germination of spores of *B. megaterium* were unaffected by concentration. This mechanism is perhaps not universal in bacteria.

By far the most varied, and largest number of self-inhibitors has been described for seed plants where such substances appear to be ubiquitous. But the role of most of these in nature is not clear because they do not impose a dormant period upon seeds when used exogenously, except in a few cases like the dormins.[13, 98]

In multicellular organisms it is often difficult to make the distinction between self-inhibitors and hormones. It should be noted that the interplay of mutually antagonistic hormones, as in the case of insects may be relevant to this discussion.[19]

That the release or neutralization, of self-inhibitors can explain the effect of some activation treatments is suggested by the data in Table V. Thus, scarification of seeds and freezing and thawing may induce the loss of such substances from seeds, thereby permitting germination to occur.[98] According to Allen[115] floating rust spores on water and aeration dissipates a volatile inhibitor which is responsible for the inability of these spores to germinate. Furthermore, certain chemical treatments, such as the use of paraffin, may result in the binding of volatile materials and the subsequent activation of spores.[52] These are the only cases where the connection seems clear but it is likely that other correlations are possible.

(*ii*) *Hormonal effectors.* Multicellular organisms, by definition, are the only organisms which would be expected to have hormones. It might be argued that the release of self-inhibitors from unicells into the medium would be analogous to hormonal action but it hardly seems worth much debate. Therefore, it will only be noted that insect diapause and mammalian hibernation are probably induced by hormonal changes. In the case of the former, an attractive theory to explain the induction and termination of diapause through effects on the cholinergic system[118] has been shown not to be tenable but no good substitute has been proposed.[206] Nor are we closer to an understanding of the targets of mammalian hormones and plant hormones and self-inhibitors although some understanding of their interactions and properties has been reached.

(*iii*) *Rhythmic effectors*. Again we may be dealing with hormonal effectors, but their nature is not clearly understood, nor have they been well defined in the organisms in which rhythms may control dormancy.[56, 19, 99]

2. Targets

(*i*) *Permeability barriers*. Barriers to the passage, in either direction, of water and other nutrients have been invoked to explain dormancy in a number of organisms. Thus, Vegis[13, 15] seeks to explain seed dormancy in many cases on the basis of the inability of oxygen to penetrate the pericarp or other structures of the wall. However, dormant ascospores of *Neurospora* have been shown not to be affected by increased concentrations of oxygen but the penetration of oxygen was not studied.[119] Moreover, it is clear that dormant pupae of insects are not anaerobic.[120] Among other fungi several instances have been reported where the dormant period has been circumvented, or markedly reduced, under conditions where thin-walled spores have been produced.[7] Moreover, certain phycomycete spores germinate only after much of the outer wall is autodigested[121] or "cracked".[122] It will be noted in Table V that enzymes aid in the germination of cysts of protozoans and nematodes and it is possible that this treatment alters their permeability in such as way as to permit germination.

However, the relation to permeability of most of the phenomena discussed above is still not clear. Thus, although the breakage of the walls of seeds by scarification does change their permeability to water,[123] the loss of self-inhibitors also is encouraged so that the primary effect is difficult to pinpoint.

Anhydrobiosis is an aspect of permeability that has been used frequently to explain dormancy, especially in bacteria, but neither impermeability to water nor an anhydrous protoplast are very likely.[58] On the other hand, the "contractile cortex" theory[124] has much to recommend it and postulates the release of water from the cortex by the exertion of hydrostatic pressure and subsequent contraction. This theory does not require that the protoplast be impermeable to water, nor is there any evidence of extensive impermeability to dissolved materials.[125] Moreover, it helps to explain the heat resistance of bacterial spores despite their permeability to water.[58]

However, the existence of a "contractile core" must derive from other changes so that it cannot, by itself, explain the origin of dormancy, otherwise access of water alone should immediately activate spores even in the absence of other treatments.

(*ii*) *Respiratory enzymes*. The ubiquity of reduced metabolic gas exchange in dormant systems frequently has led to the suggestion that respiratory metabolism is the primary locus of a block which causes dormancy. This possibility is supported by the fact that compounds which overcome the self-inhibition of fungus spores, such as some of those listed in Table V as activators, also stimulate respiration. Also, uncouplers of oxidative phosphorylation break the dormancy of some seeds[126] and markedly enhance the respiration of certain fungus spores.[34] That hormonal controls can be influenced by uncouplers has been shown by Ritosa,[127] in insects wherein the puffing pattern of salivary gland chromosomes is changed by this means. Interestingly enough, the changed puffing pattern is identical to that which is induced by a heat-shock or exposure to anaerobiosis for 10 min. So, is dormancy of certain types maintained because of a lesion in respiratory metabolism, and does activation by the agents discussed above involve the repair of such a lesion?

A detailed examination of this question has been performed in dormant ascospores of *Neurospora*[7] whose respiratory rate is greatly reduced in comparison to germinating and vegetative stages (Table II). Immediately upon activation the release of fermentation products such as ethanol, acetaldehyde and pyruvate is greatly enhanced, followed by a levelling off in their production. The full complement of acids of the Krebs cycle is not found until an hour or longer after activation although the enzymes in this pathway appear to be present.[128] Experiments with ^{14}C-glucose reveal no impairment of anabolic and catabolic pathways, except in the ability to utilize a precursor of wall materials which accumulates only in dormant ascospores. As a result of these studies it has been concluded that the locus of the metabolic block is in the steps between trehalose and the phosphorylated hexoses in the Embden-Meyerhoff sequence of enzymes. Support for this hypothesis is provided by F. I. Eilers (personal communication),[129] who finds a burst in the formation of fructose 6-phosphate and glucose 6-phosphate within minutes after activation. The failure of exogenously added glucose to activate is ascribed to the notorious impermeability of this spore which limits the amount of sugar which enters. The respiratory metabolism of other fungus spores during germination has been studied,[113] as has the inability of certain spores to germinate without the provision of specific nutrients,[6] but the primary step in the activation step in these cases remains to be defined.

Extensive studies of the relation between respiratory metabolism and dormant insects reveal that the levels of inorganic phosphate may play a role in regulating the oxidation of dormant pupae but that phosphate acceptors are more controlled than controllers.[120]

Intense carbohydrate catabolism upon the arousal of hibernating mammals also has been reported along with increased activities of the associated enzymes.[130] But recent work suggests that the capacity for gluconeogenesis also must be increased at this time.[131] Although it has been suggested that a block exists in dormant seeds, somewhere in respiratory metabolism, Bradbeer and Colman[132] argue that low temperatures permit synthetic pathways to predominate over respiratory ones and remark that a wide range of metabolic activity is shown by the seeds they have studied. In any event, the exact sites of metabolic blocks in the respiratory metabolism of higher animals and plants have not been identified.

The effectiveness of L-alanine, glucose and other metabolites as activators of bacterial spores has led to several hypotheses involving respiratory metabolism. These include effects upon $NADH_2$ and glucose oxidation wherein DPA may substitute for FMN, regulation through the Krebs cycle and other mechanisms,[7] but none of these has gained complete acceptance.

(*iii*) *Protein synthesis and nucleic acid metabolism.* Once the locus of a metabolic block is recognized the distinction between an effect upon a pre-existing enzyme or precursor, and *de novo* synthesis must be made. And, if *de novo* synthesis is involved, the step in this process must be defined.

These questions have been studied in considerable detail in animal eggs, such as those of sea urchins, wherein many parallels exist with dormant systems. Thus, unfertilized eggs when activated by fertilization, or by artificial means, show increased respiration, substrate uptake, coenzyme content and proteolytic activity[133] and these changes are probably related to enzymic changes leading to processes required for cell division and differentiation. A number of workers have argued that renewal of protein synthesis is a key to the activation process and the lesion has been ascribed to the inability of otherwise normal ribosomes to translate maternal messages because these are in a "masked" state on other particles.[134] According to this hypothesis, messenger RNA is released upon fertilization to combine with free ribosomes which are present in excess. That "stable messages" are not uncommon in differentiating systems is claimed by Spirin on the basis of work with several organisms.[135]

This contrasts with the view of Monroy, Maggio and Rinaldi[136] that an inhibitory protein is removed from ribosomes as a result of fertilization or that they are inactive for other reasons.[137] But recent work has shown that there is a lag between the time when structural events occur following fertilization and increased protein synthesis.[133] This suggests the involvement of energy-linked processes and mRNA-ribosome attachment or allosteric effectors of enzymes, not involved in protein synthesis directly.

Dormin has been mentioned before as a widely distributed self-inhibitor of seeds of higher plants.[88] Therefore, it is of great interest that this substance appears to inhibit DNA synthesis in cultures of *Lemna* (duckweed) after just one day of treatment.[138] RNA synthesis also is decreased, but much later, and plant growth regulators like benzyladenine reverse the action of dormin.

The exact role of the synthesis of the components of the protein synthesis apparatus in activated seeds and animal embryos is uncertain. Thus, Marcus and Feeley,[139] Rieber[140] and others claim that there is a limiting concentration or "masking" of endogenous messenger RNA but that the rest of the apparatus is functional. That preformed mRNA and ribosomes occur in cotton seeds has been shown by Waters and Dure[141] so that differences may exist in different seeds. Thus, inhibition changes the amount of sRNA and rRNA in hazel-nut cotyledons[142] and similar changes may occur in peanut and wheat embryo rRNA and mRNA.[139] By contrast unimbibed peanut cotyledons have active sRNA and unimbibed cotton embryos contain an active mRNA.[143] The need for further extension of this work is obvious.

Evidence has been presented that mRNA is not stored in dormant ascospores of *Neurospora* and that they do not contain polyribosomes, although ribosomes and sRNA are present.[144, 145] Presumably germination is accompanied by mRNA synthesis and subsequent banding of ribosomes (polysomes), but the kinetics of respiratory changes in activated ascospores makes it unlikely that these are primary events. Other work suggests that mRNA synthesis occurs soon after the germination of spores of *Aspergillus niger*.[146] An interesting mechanism through which protein synthesis is restrained may occur in zoospores of *Blastocladiella* in which a nuclear cap, which consists of ribosomes, surrounds the nucleus.[147] The cap may serve as a source of RNA and protein, or of programmed ribosomes, that might initiate protein synthesis after its dispersal when germination begins. Other changes in nucleic acid metabolism upon germination of fungus spores are reviewed by Allen.[113]

Some of the uncertainties in such work are mirrored in the results on bacterial spores where no satisfactory explanation of the activating effect of inosine is available, for example. Furthermore, it is not clear whether the ribosomes of endospores are normal and whether the accompanying "soluble system" is deficient.[148, 149] Although polysomes and mRNA have been reported to be absent, work in progress suggests that these may, in fact, be present.

(*iv*) *Other targets*. Intracellular organelles, like nuclei, have been considered to be the target of hormonal effectors like ecdysone in insects by influencing puffing patterns.[120] Electron microscope observations of fungus spores reveal that mitochondria undergo distinct changes upon activation, but these appear to lag behind certain biochemical changes.[6] Changes in polyribosomes, or other particulate fractions which mask messenger RNA, may be characteristic of the dormant state (see above), but their primacy also is still conjectural. Mechanical constraints by the embryo of certain plants is considered to explain dormancy in some seeds[150] and a curious organelle which may be a cytolysome has been found in other embryos.[151] Finally, compartmentation is a possible mechanism through which dormancy might be imposed by the intracellular separation of enzymes and substrates, or of ribosomes from other elements of the protein synthetic system, as in the nuclear cap of *Blastocladiella*.[147] A compartment which might separate trehalose from its hydrolase has been described in *Neurospora* ascospores and has been related to dormancy in these spores.[128] "Channeling" is a possible means whereby multi-enzyme sequences might be separated from other systems, even in procaryotes.

B. Exogenous Dormancy

Environmental conditions of two types, including physical and chemical, can impose a state of rest, quiescence or torpor (Table I) upon organisms. Among the physical conditions are unfavorable levels of temperature, moisture, osmoticums, and light, or photoperiods which induce dormancy. Chemical factors like the gaseous atmosphere and inhibitors also can impose this state.

Exogenous dormancy is most important for organisms that have not evolved the special structural and physiological mechanisms that character-ize constitutionally dormant organisms. Vegetative stages and those which are slightly modified from these, like involution forms, some types of conidia, sclerotia, etc., can become dormant but only if they are resistant to the conditions in which they are placed.

1. Physical Factors

Many animals and plants are reversibly inactivated by extremes of temperature. Thus, cold serves this purpose in the temperate species of insects[19] and plants[15] and photoperiod plays a similar role in these organ-isms. Such periods of rest are less likely to be induced by temperatures above the maximum for growth, except under dry conditions or with highly resistant forms such as certain fish, insects, rotifers, protozoans, and algae which inhabit tide pools and other wet places which become dry periodically. Lichens and some desert plants are examples of organisms with vegetative stages that can survive in habitats which are surprisingly low in water most of the year. In these cases, the decreased levels of respira-tion engendered by reduced water content (Table III) may contribute to survivability.[45]

"Secondary dormancy" can be imposed by low temperatures in spores of *Neurospora*,[152] Phycomyces[153] and certain bacteria,[10] thereby extending the period of rest.

2. *Chemical factors*

(*i*) *Respiratory gases.* The variations in the level of O_2 and CO_2 that occur in soil and water habitats can impose reversible periods of inactivity upon micro-organisms. Thus, the germination of spores of certain mucors is arrested by high concentrations of CO_2[154] as is that of condia of *Erysiphe graminis.*[155] That this gas exerts a strong influence in nature was suggested by Burges and Fenton,[156] who showed that tolerant organisms populate regions of soils with high concentrations of CO_2. More recently, Stotzky and Goos[157, 158] have shown that of soil biota some fungi were the most tolerant to CO_2 and that segments of the actinomycete population were the most sensitive, with bacteria being intermediate. This work revealed that the sensitive organisms reacted to CO_2, for many survived complete anaerobiosis for some time. However, it was not established whether the new population was a new one selected for its resistance or whether adaptation was physiological. Protozoa often are inhibited strongly by high concentrations of CO_2 and the ability to develop under these conditions may determine which species survives in polluted habitats.[77]

Low oxygen tensions may, however, be inhibitory, as Griffin[159] has pointed out. Thus, he claims that the deleterious effects of high soil moisture content are due primarily to deficits of O_2 and excesses of CO_2. Moreover, low O_2 concentrations may induce "secondary dormancy" by reversibly inhibiting the germination of some fungus spores[6] and seeds.[47] It is likely that similar effects are engendered in other micro-organisms, for a a number of morphogenetic phenomena in fungi and animals are triggered by even small amounts of CO_2.

(*ii*) *Inhibitory factors in soil and water.* Spores of many fungi do not germinate in natural soil, except near undecomposed organic matter or in the rhizosphere.[160] Glucose and a variety of other substances can reverse this effect, as can autoclaving of the soil. Several theories have been offered to explain the failure of these spores to germinate, including:

(1) Diffusible inhibitors in soil
(2) Localized formation of antibiotics due to enhanced activity of soil micro-organisms
(3) Nutrient deficiency in spores or their environment

Although inhibitors have been separated from soils as sterile extracts[161, 162] none of these match the characteristics of soil fungistasis, such as strong inhibition, ubiquity and broad inhibitory spectrum. Moreover, numerous inconclusive results, or outright failures, have been recorded[160] so alternative mechanisms must be considered.

When spores were placed on soil they were found to create in their vicinity increased microbial activity. It was suggested that increased antiobiotic production exerted an influence in preventing spore germination.[163] This was supported by the observation that certain fungus spores would not germinate in the presence of washed cells of a variety of bacteria. However, no inhibitors could be isolated, nor is it easy to explain long term fungistasis by this means because spores do not secrete nutrients indefinitely.

Therefore, the nutrient status of soil and the necessity of exogenously provided nutrients for the germination of spores of various fungi were studied.[164] Most of those tested required added nutrients for germination and all of those that did were inhibited on soil. The reciprocal also is true, for the most part, in that nutritionally independent fungi were not sensitive to the fungistatic effect. A few exceptional fungi among those that did not require exogenous nutrients failed to germinate in soil or when exposed to dripping water. These results suggest that there is an irreversible sink for nutrients in soil fed by a strong diffusion gradient and that, in some soils at least, this leads to deprivation of factors required for the germination of many spores.

But there may be other types of fungistatic principles in soil and one that is not reversed by sugars or autoclaving has been described by Dobbs and Gash[165]. These workers also have described a thermosensitive type of fungistasis which is less sensitive to the addition of glucose than is the more widespread one, so differences may exist between soils.

Antibiotics and their effects in natural environments upon bacteria have been discussed in great detail elsewhere[166-168] so they will not be covered herein. Although antibiotics seem not to be involved in most cases of fungistasis studied, they cannot be ruled out entirely because of the diversity of soils, many of which show differences in their effects upon microorganisms.[165, 169]

Inhibitory principles also are demonstrable in aquatic environments. Thus, *Chlorella* produces antibacterial substances which are recoverable from the culture medium and cells.[170] However, antibacterial action was induced in extracts only upon exposure to light and air and further work revealed that unsaturated fatty acids were the substrate for photo-oxidation. This is of interest in view of the toxicity of certain lipids for bacterial spores and other stages. Other inhibitory materials of algal origin have been discovered and are reviewed by Hartman,[171] but their relevance to bacterial growth in these environments is uncertain. Antibacterial substances do exist in sea water[172] and it would be interesting to know whether they are related to the substances described from *Chlorella*.

Estuarine sediments have been reported to be fungistatic.[173] The effect was specific to fungus spores and was reversed by the addition of small amounts of nutrients. Thus, the data do not distinguish between the presence of toxic materials and the lack of nutrients required for germination.

In very few of the cases discussed above has work been done specifically with bacterial spores. However, bacteriostatic material has been found so that a form of exogenous dormancy can be induced. Moreover, the action of some activation treatments in natural habitats may be ascribable to the reversal of the effect of inhibitory principles such as those described in this section; treatments such as heating and addition of nutrients have been explained in this way.[7]

3. Lytic factors

Spores of many fungi are soon destroyed when mixed with raw soil but survive for a long time in autoclaved soil,[174, 175] and mycelium is subject to lysis as well.[176] These effects have been ascribed to enzymes[168] or to the combined action of starvation and antibiotics.[177] Therefore, it has been suggested[178] that spore formation protects bacilli against their own autolytic enzymes because some asporogenous mutants are known to be unstable and to lyse readily.

Another aspect that may relate to the growth of micro-organisms in soil is the release of nutrients from lysing fungal hyphae and plant and animal matter. After such lysis the growth of bacteria and actinomycetes is stimulated in a regular sequence, presumably in response to the nutrients released by the fungi.[179]

IV. The Role of Dormancy in Nature

Three roles have been ascribed to spores, including enhancement of survivability, disseminability and coordination of development with favorable environmental conditions.[7]

A. SURVIVABILITY

A detailed examination of the factors in survivability in micro-organisms has been made by Garrett[51] and Sussman,[52] so only those that impinge on bacterial spores will be discussed herein.

According to Park,[180] the modes of survival of fungi in soil can be divided into the active types whereby the organism continues to grow as a parasite, commensal or saprophyte, or inactive types, as in exogenous or constitutional dormancy. These divisions probably are relevant to bacteria as well. Of the factors that contribute to survivability, resistance to deleterious agents, nutritional capacity and mutability are influenced by dormancy.

1. Resistance to deleterious agents

Bacterial spores have been demonstrated to be more resistant to toxic chemicals than are vegetative stages although the results with fungal spores are ambiguous.[7] In nature, enrichment of spore-forming bacteria in soil has resulted from chemical treatment of several kinds,[181-183] and fumigation shifts the balance in favor of spore-forming fungi as well.[184] Lytic factors in the environment that destroy micro-organisms have been described above and it has been suggested that spore formation protects bacilli against their own autolytic enzymes.[178] Whether spores protect against the lytic factors of other micro-organisms is not well understood. Such lytic factors include murein-destroying enzymes, antibiotics which interfere with cell wall synthesis and metabolic products, whose action differs from that of antibiotics, which induce lysis.[185]

Bacterial spores also are more resistant to extremes of temperature than are their vegetative stages. However, examples of very resistant vegetative stages of procaryotes exist as in thermophilic bacteria lacking spores, as well as the blue-green algae of hot springs and snow fields.[24] Moreover, the flora of pasteurized milk often consists of a majority of nonspore-forming species (Hammer, 1948). Low temperatures also can be borne by vegetative stages with equal success; thus *Streptococcus lactis* survived 111 days at $-191°C$, or 45 min at $-253°C$.[187] On the other hand, *Bacillus subtilis*, *B. licheniformis* and *B. megaterium* predominate in the Egyptian desert during times of drought, whereas nonspore-formers are recovered only occasionally.[188] In addition, spore-formers account for a large majority of the microbial flora of desert soils[209] and around desert plants, in contrast to the flora of fertile Egyptian soils.[189] It is a frequent observation that a high percentage of dark-spored fungi occur in desert soils;[190] perhaps melanized bacterial spores enjoy selective advantages in similar environments.

Because oceanic water, which usually is below 5°C covers about three-quarters of the earth's surface, and large land areas are maintained at low temperatures as well, the survival and growth of micro-organisms under psychrophilic conditions is of great importance. Yet, relatively little is known about this subject. Although psychrophily has been observed most often among Gram-negative organisms (i.e. nonspore-formers),[24] it is known that spores of some obligate anaerobes germinate at 5°C[97, 210] and develop well at 0°C.[191] Larkin and Stokes[192] have reported on spore formation and germination at 0°C by aerobic spore-formers from soil, mud and water, but the role of spores in these environments remains largely unstudied, as does the phenology of psychrophilic bacteria.

Studies of the spectral distribution of sunlight at the surface of the earth reveal that the lower limit in the ultraviolet is at about 290nm.[193, 194] Although considerable infrared radiation reaches the earth, its effects are probably due to high temperature. As in the case of the other environmental variables reviewed in this section, bacterial spores are most resistant to radiation, although several very resistant nonspore-formers are known.[7] The recovery of fungi after irradiation has been studied in greater detail than that of bacteria. It has been found that the resistance of certain fungi to radiation damage is related directly to their rate of growth in pure culture under sublethal doses of continuous irradiation.[195] Fungi with melanized walls were the most resistant and, frequently, species were recovered which were not encountered in untreated soils. These phenomena, as well as the role of bacterial spores are worth study.

2. Nutritional status and capacity

The strong dependence of bacteria upon nutrient supply for the initiation of growth in natural environments and for the induction of spore formation has already been discussed. Therefore, it is clear that nutrition must play a large role in determining their survival.

Inasmuch as endospores of some bacteria have a greater requirement than vegetative cells for adenosine, serine and other activators (Table VI) germination may be delayed until enough metabolites are available to permit the continued development of the vegetative organism. If this were the case selective advantage would derive from the prevention of the precocious germination of spores. Also, multiple auxotrophy is a distinct advantage in cases where auxotrophic strains "commit suicide" in the absence of the required nutrient. Moreover, in the case of such "suicide" strains bacteriostatic materials like some antibiotics, might enhance survival by delaying germination until more propitious circumstances obtain, such as those wherein sufficient nutrients are available.[7] In fact, multiple requirements for activation, as a substitute for L-alanine alone, have been described,[112] and may represent a means through which activation is accomplished in nature.

Antibiotic production by bacteria, such as bacitracin[196] or the one found in *Bacillus subtilis*,[197] appears in some cases, at least, to be a necessary concomitant of sporulation. Perhaps one role of these substances is to increase the "inoculum potential" of germinating spores[51] by suppressing the growth of competitors.

3. Genetic factors in survivability

Although the spores of fungi and other protistans often distribute the products of meiotic recombination[80] bacterial spores may not function in this way. But Anagnostopoulos and Spizizen[198] have reported that transformability in *Bacillus subtilis* reaches its peak during presporulation and that nonsporulating strains never become competent. In this respect, at least, the ability to sporulate confers a definite selective advantage.

A genetic role for spores may derive from the fact that bacilli are the only known exceptions to the generalization that "the temperature optima of organisms are quite stable genetic properties that are not subject to change by selection".[199] Thus, mesophilic aerobic spore-formers can be adapted to grow at higher temperatures and the converse is also true in that thermophilic bacilli can be adapted to lower temperatures.[200, 201] Brock suggests that the ability to form spores is in some way related to the great adaptability of bacilli toward temperature. I believe that this is correct and related to the great survivability of spores at temperature extremes. Under temperature stress mutations that may occur in spores have a greater probability of surviving and responding favourably to the environmental pressure which results in evolutionary selection. In fact, mutability at high temperatures is enhanced so that organisms able to survive under these conditions would have great selective advantage. Conversely, it would be much harder for non-sporeformers to respond to such environmental pressures because, despite equivalent ability to mutate, the phenotypic changes which lead to adaptation lag behind the genotypic ones. Therefore, although the adapted genome may arise, the rest of the cellular machinery that is needed to ensure its replication and survival is inadequate to the task. This is not to say that small extensions of temperature range cannot occur over long periods of time, for nonspore-formers abound in hot springs and other extreme habitats. However, the time involved in such adaptations must be appreciably greater than for those observed in the laboratory. In summary, it can be said that spore-formation confers selective advantage in permitting rapid adaptation to temperature shifts. However, given enough time, even nonspore-formers have demonstrated their ability to respond to such environmental pressures. In fact, their predominance in certain extreme environments in nature suggests that they enjoy selective advantages over sporeformers in some situations. Still, even these conclusions must be qualified for, as Brock contends, it is likely that the upper limits of temperature at which organisms will grow in nature may yet be unknown. Moreover, for various reasons, it is often difficult to establish the number of sporeformers in an environment so precise data on this subject are lacking.

B. Disseminability

Whereas large spores, like those of rusts, lichens and bryophytes frequently are deposited by impaction, those of bacteria and some other protistans are too small to be deposited by this method. Thus, other means have been evolved for their deposition, including sifting through vegetation during rains.[202]

As for the role of bacterial spores in dispersal, it has been pointed out by Lamanna[203] that vegetative cells probably become air-borne as readily. The unique attribute of spores, vital to the organisms that form them, is their survivability. Thus, the greater resistance of spores to radiations, temperature extremes and desiccation enhances the likelihood of their survival over that of vegetative cells, in air and the other media of transport.

Studies of natural populations from air and aquatic environments in many parts of the world reveal the widespread presence of spores and spore-formers.[202, 7] However, other bacteria occur frequently and attest to the survivability of some nonspore-formers.

C. Spores as Timing Devices

The important role of dormancy in enhancing survivability and disseminability often obscures another important role. "No less important is the fact that diapause (dormancy) determines the constancy of life cycles and the synchronization of developmental stages with periods of the year to which they are adapted. Thus the eco-physiological features of diapause form the basis of the entire life cycles of insects".[19] The data in Table IV reveal the ecological factors which cause periodic cessation of development, including extremes of temperature, photoperiod and desiccation. Special thermal adaptations, like dormant stages, are not required in those tropical and sub-tropical conditions where the annual range of temperature encompasses that effective for development. But in habitats with dry and hot seasons insects undergo aestivation and some fungi form sclerotia. Dormancy also may evolve as an adaptation to scarcity of food, especially in the case of micro-organisms and insects which live in close association with plants.

Although many of the phenomena which induce dormancy are periodic, they are not invariably so. This is clearly the case in the micro-environments of bacteria and other micro-organisms. Other examples include insects like *Rhodnius* and many ticks which undergo dormant periods, due to temporary shortages of food, as they lie in wait for hosts.[19] Similarly, irregular dormant periods occur in aquatic forms like certain mosquitoes, fishes and amphibians whose habitats dry up occasionally.

It is likely that there has been selection for strains of micro-organisms which produce spores in response to deprivation of nutrients and other environmental upsets. So, to the extent that this generalization is valid, spore formation may represent a timing device which serves to maintain viability during unfavorable periods. It has been proposed[7] that activation treatments like inducers of dormancy can be viewed as serving to synchronize development with environmental factors. Thus, extreme temperatures may be cues which anticipate seasonal changes, and inhibitors probably restrain development until propitious times. Moreover, the role of chemical activators, in some cases at least, may be to reverse inhibitors and serve as indicators of periods of nutritional plenty. But, many of these suppositions are highly speculative, expecially where bacteria are concerned, and can only serve to set directions for future ecological and physiological studies.

Finally, activators, "sporogens", antibiotics and inhibitors of other kinds may represent a class of substances labeled "ecological ectocrines".[204] They are biosynthesized by a species which exerts an effect on another via the external medium. Of course, these may be produced by the same species they affect but, when different species are involved, such phenomena as ecological exclusion and succession in microbial populations may be involved.[205]

ACKNOWLEDGEMENTS

I should like to acknowledge the support received from National Science Foundation Grant No. GB–2620. The stimulating discussions with my colleague Dr. Harry Douthit were of great help in preparation of this manuscript.

REFERENCES

1. Keilin, D. (1959). *Proc. R. Soc.* B. **150**, 150.
2. Schmidt, P. (1955). "Anabioz". Izd. AN SSR. Moscow and Leningrad.
3. Preyer, W. (1891). *Biol. Zbl.* **11**, 1.
4. Doran, W. L. (1922). *Bull. Torrey Bot. Club* **49**, 313.
5. Sussman, A. S. (1965). *In* "Encyclopedia of Plant Physiology" (A. Lang, ed.), Vol. 15, Part 2, p. 933, Springer-Verlag, Berlin.
6. Sussman, A. S. (1966). *In* "The Fungi" (G. C. Ainsworth and A. S. Sussman, eds), Vol. 2, p. 733. Academic Press, New York and London.
7. Sussman, A. S. and Halvorson, H. O. (1966). "Spores: Their Dormancy and Germination". Harper and Row, New York.
8. Mandels, G. R. and Norton, A. B. (1948). *Q. gen. Lab. Res. Rept. Microbiol* Ser. 11, 1.
9. Gibbs, P. A., Gould, G. W., Hamilton, W. A., Hitchins, A. D., Hurst, A., King, W. L., Roberts, T. A. and Wolf, J. (1967). *Spore Newsletter* **2**, 152.
10. Curran, H. R. and Evans, F. R. (1945). *J. Bact.* **49**, 335.
11. Hawker, L. E. (1957). "The Physiology of Reproduction in Fungi". Cambridge Univ. Press, London.
12. Hill, A. G. G. and Campbell, G. K. G. (1949). *J. exp. Agric.* **17**, 259.
13. Vegis, A. (1964). *A. Rev. Pl. Physiol.* **15**, 185.
14. Samish, R. M. (1954). *A. Rev. Pl. Physiol.* **5**, 183.
15. Vegis, A. (1965). *In* "Encyclopedia of Plant Physiology" (A. Lang, ed.), Vol. 15, Part 2, p. 183. Springer-Verlag, Berlin.
16. Kayser, C. (1965). *In* "Physiological Mammalogy", Vol. 2, p. 179. Academic Press, New York and London.
17. Lees, A. D. (1955). "The Physiology of Diapause in Arthropods". Cambridge Univ. Press, London.
18. Lyman, C. P. and Dawe, A. R. (1960). *Bull. Mus. comp. Zool. Harv.* 124, 1.
19. Danilevskii, A. S. (1965). "Photoperiodism and Seasonal Development of Insects". Oliver and Boyd, Edinburgh.
20. Gaughran, E. R. L. (1947). *Bact. Rev.* **11**, 189.
21. Waksman, S. A. and Corke, C. T. (1953). *J. Bact.* **66**, 377.

c

22. Johnson, F. H. (1957). *Symp. Soc. gen. Microbiol.* **7**, 134.
23. Hawker, L. E. (1957). *Symp. Soc. gen. Microbiol.* **7**, 238.
24. Farrell, J. and Rose, A. (1967). *A. Rev. Microbiol.* **21**, 101.
25. Tartar, V. (1967). *In* "Research in Protozoology" Vol. 2, p. 1. Pergamon Press, New York.
26. Altman, P. L. and Dittmer, D. S. (1966). "Environmental Biology". *Fedn. Am. Soc. Exp. Biol.*, Bethesda, Maryland, U.S.A.
27. Altman, P. L. and Dittmer, D. S. (1964). "Biology Data Book". *Fedn. Am. Soc. Exp. Biol.*, Bethesda, Maryland, U.S.A.
28. Shu, P. K., Tanner, G. and Ledingham, G. A. (1954). *Can. J. Bot.* **32**, 16.
29. Cochrane, V. W., Berry, S. J., Simon, F. G., Cochrane, J. C., Collins, C. B., Levy, J. A. and Holmes, P. K. (1963). *Pl. Physiol.*, *Lancaster* **38**, 533.
30. Cochrane, V. W., Cochrane, J. C., Collins, C. B. and Serafin, F. G. (1963). *Amr. J. Bot.* **50**, 806.
31. Hoffman, R. A. (1964). *Ann. Acad. Sci. Fenn. A.* IV 71/14, 201.
32. Halvorson, H. O. (1962). *In* "The Bacteria" (I. C. Gunsalus and R. Y. Stanier, eds), Vol. 4, p. 223. Academic Press, New York and London.
33. Church, B. D. and Halvorson, H. O. (1957). *J. Bact.* **73**, 470.
34. Mandels, G. R. (1963). *Ann. N.Y. Acad. Sci.* **102**, 724.
35. Gallisian, A. (1967). *C.R. hebd. séanc. Acad. Sci., Paris* **264**, 1190.
36. Edelman, I. M. (1951). *Ent. Oboz.* **31**, 374.
37. Heller, J. (1926). *Biochem Z.* **169**, 208.
38. Crocker, W. and Barton, L. V. (1953). "Physiology of Seeds". *Chronica Botanica*, Amsterdam.
39. Fisher, K. C. and Manery, J. F. (1965). "Water and Electrolyte Metabolism in Heterotherms". *Third International Symposium in Natural Mammalian Hibernation*. Toronto (unpublished).
40. Wager, H. G. (1954). *New Phytol.* **53**, 354.
41. Shirk, H. G. (1942). *Am. J. Bot.* **29**, 105.
42. Kernbach, B. (1960). *Z. Bot.* **48**, 415.
43. Terui, G. and Mochizuki, T. (1955). *Techn. Rep. Osaka Univ.* **5**, 219.
44. Stiles, W. (1956). *In* "Encyclopedia of Plant Physiology", Vol. 3, p. 652. Springer-Verlag, Berlin.
45. Stocker, O. (1956). *In* "Encyclopedia of Plant Physiology", Vol. 3, p. 696. Springer-Verlag, Berlin.
46. Skujins, J. J. and McLaren, A. D. (1967). *Science*, N.Y. **158**, 1570.
47. Barton, L. V. (1965). *In* "Encyclopedia of Plant Physiology" (A. Lang, ed.), Vol. 15, Part 2, p. 726. Springer-Verlag, Berlin.
48. Kozloff, E. N. and Brown, H. E. (1963). *J. Protozool.* **10**, 377.
49. Ritterson, A. L. (1966). *J. Parasit.* **52**, 157.
50. Hemberg, T. (1965). *In* "Encyclopedia of Plant Physiology" (A. Lang, ed.), 15, Part 2, 669. Springer-Verlag, Berlin.
51. Garrett, S. D. (1956). "Biology of Root Infecting Fungi". Cambridge Univ. Press, London.
52. Sussman, A. S. (1965). *In* "Ecology of Soil-borne Plant Pathogens—Prelude to Biological Control" (K. F. Baker and W. C. Snyder, eds), p. 99. Univ. of California Press, Berkeley, California, U.S.A.
53. Dutrieu, J. and Chrestia-Blanchine, D. (1967) *C.R. hebd. séanc. Acad. Sci., Paris* **264**, 2941.
54. Asahina, E. and Tanno, K. (1964). *Nature, Lond.* **204**, 1222.
55. Webb, S. J. (1960). *Can. J. Microbiol.* **6**, 89.

56. Bünning, E. (1964). "The Physiological Clock". Academic Press, New York and London.
57. Brown, F. A., Jr. (1962). *Ann. N. Y. Acad. Sci.* **98**, 775.
58. Murrell, W. G. (1967). *Adv. microbial Physiol.* **1**, 133.
59. Srinivasan, V. R. (1966). *Nature, Lond.* **209**, 537.
60. Wooley, B. C. and Collier, R. E. (1966). *Bact. Proc.* 16.
61. Marsh, P. B., Taylor, E. E. and Bassler, L. M. (1959). *Plant. Dis. Rep. Suppl.* **261**, 251.
62. Jerebzoff, S. (1965). *In* "The Fungi" (G. C. Ainsworth and A. S. Sussman, eds), Vol. 2, p. 625, Academic Press, New York and London.
63. Trione, E. J., Leach, C. M. and Mutch, J. T. (1966). *Nature, Lond.* **212**, 163.
64. Lang, A. (1965). *In* "Encyclopedia of Plant Physiology" (A. Lang, ed.), Vol. 15, Part I, p. 690. Springer-Verlag, Berlin.
65. Van Wagtendonk, W. J. (1955). *In* "Biochemistry and Physiology of Protozoa" (S. H. Hutner and A. Lwoff, eds), Vol. 2, p. 57. Academic Press, New York and London.
66. Ehret, C. (1959). *Fedn. Proc.* **18**, 1232.
67. Bopp, M. (1965). *In* "Encyclopedia of Plant Physiology" (A. Lang, ed.), Vol. 15, Part I, p. 802. Springer-Verlag, Berlin.
68. Lang, A. (1965). *In* "Encyclopedia of Plant Physiology" (A. Lang, ed.), Vol. 15, Part I, p. 1380. Springer-Verlag, Berlin.
69. Bünning, E. (1961). *Can. J. Bot.* **39**, 461.
70. Pengelley, E. T. and Fisher K. C. (1961). *Can. J. Zool.* **39**, 105.
71. Futter, B. V. and Richardson, G. (1967). *J. appl. Bact.* **30**, 347.
72. Ashwood-Smith, M. J., Copeland, J. and Wilcockson, J. (1967). *Nature, Lond.* **214**, 33.
73. Cloudsley-Thompson, J. L. (1961). "Rhythmic Activity in Animal Physiology and Behaviour". Academic Press, New York.
74. Stanier, R. Y. (1953). *Symp. Soc. gen. Microbiol.*, **3**, 1.
75. Hungate, R. E. (1962). *In* "The Bacteria" (I. C. Gunsalus and R. Y. Stanier, eds), Vol. 4, p. 95, Academic Press, New York and London.
76. Rovira, A. D. (1965). *A. Rev. Microbiol.* **19**, 241.
77. Noland, L. E. and Gojics, M. (1967). *In* "Research in Protozoology" (T-T Chen, ed.), Vol. 2, p. 215. Pergamon Press, New York and Oxford.
78. Saunders, D. S. (1967). *Science, N. Y.* **156**, 1126.
79. Shappirio, D. (1967). Personal communication.
80. Bonner, J. T. (1958). *Am. Nat.* **92**, 193.
81. Sussman, A. S., Durkee, T. L. and Lowry, R. J. (1965). *Mycopath. Mycol. Appl.* **25**, 381.
82. Sargent, M. L., Briggs, W. R. and Woodward, D. O. (1966). *Pl. Physiol., Lancaster* **41**, 1343.
83. Keynan, A., Evenchik, Z., Halvorson, H. O. and Hastings, J. W. (1964). *J. Bact.* **88**, 313.
84. Johnson, W. H. and Evans, F. R. (1941). *Physiol. Zool.* **14**, 227.
85. Kudo, R. R. (1966). "Protozoology". 5th Ed., Charles C. Thomas, Springfield, Illinois, U.S.A.
86. Fairbairn, D. (1961). *Can. J. Zool.* **39**, 153.
87. Stokes, P. (1965). *In* "Encyclopedia of Plant Physiology" (A. Lang, ed.), Vol. 15, Part 2, p. 747. Springer-Verlag, Berlin.
88. Evenari, M. (1965). *In* "Encyclopedia of Plant Physiology". (A. Lang, ed.), Vol. 15, Part 2, p. 804. Springer-Verlag, Berlin.

89. Fushtey, S. G. and Johnson, P. W. (1966). *Nematologica* **12**, 313.
90. Thorne, G. (1961). "Principles of Nematology". McGraw-Hill, New York.
91. Danilevskii, A. S. (1949). *Entomol. Oboz.* **30**, 194.
92. Pittendrigh, C. (1965). *In* "Circadian Clocks" (J. Aschoff, ed.), p. 277. North-Holland Publishing Co., Amsterdam.
93. Twente, J. W. and Twente, J. A. (1965). *Proc. natn. Acad. Sci. U.S.A.* **54**, 1058.
94. Benson-Evans, K. (1960). *Br. Bryol. Soc., Trans* **3**, 729.
95. Welsford, E. (1907). *New Phytol.* **6**, 156.
96. Dodge, B. O. (1912). *Bull. Torrey Bot. Club* **39**, 139.
97. Roberts, T. A. and Hobbs, G. (1968). *J. appl. Bact.* **31**, 75.
98. Wareing, P. F. (1965). *In* "Encyclopedia of Plant Physiology" (A. Lang, ed.), Vol. 15, Part 2, p. 909. Springer-Verlag, Berlin.
99. Kummerow, J. (1965). *In* "Encyclopedia of Plant Physiology" (A. Lang, ed.), Vol. 15, Part 2, p. 271. Springer-Verlag, Berlin.
100. Good, H. M. and Spanis, W. (1958). *Can. J. Bot.* **36**, 421.
101. Averner, M. and Fulton, C. (1966). *J. gen. Microbiol.* **42**, 245.
102. Rogers, W. P. and Sommerville, R. I. (1957). *Nature, Lond.* **179**, 619.
103. Clarke, A. J. and Shepherd, A. M. (1967). *Nature, Lond.* **213**, 419.
104. Bowers, W. S. and Bleckenstaff, C. C. (1966). *Science, N.Y.* **154**, 1673.
105. Williams, C. M. and Adkisson, P. L. (1964). *Biol. Bull.* **127**, 511.
106. Fries, N. (1941). *Archs. Mikrobiol.* **12**, 266.
107. Curtis, G. J. (1965). *Nematologica* **11**, 213.
108. Coley-Smith, J. R. and Holt, R. W. (1966). *A. appl. Biol.* **58**, 273.
109. Schroth, M. N. and Hildebrand, D. C. (1964). *A. Rev. Phytopathol.* **2**, 101.
110. Sierra, G. (1967). *Can. J. Microbiol.* **13**, 489.
111. Levinson, H. S. (1957). *In* "Spores" (H. O. Halvorson, ed.), p. 120, Am. Inst. Biol. Sci., Washington, D.C., U.S.A.
112. Wax, R. and Freese, E. (1968). *Spore Newsletter* **2**, 171.
113. Allen, P. J. (1965). *A. Rev. Phytopath.* **3**, 313.
114. Lingappa, B. T. and Lingappa, Y. (1967). *Nature, Lond.* **214**, 516.
115. Allen, P. J. (1955). *Phytopathology* **45**, 259.
116. Robinson, P. M. and Park, D. (1966). *Trans. Br. Mycol. Soc.* **49**, 639.
117. Halvorson, H. O. (1959). *Bact. Rev.* **23**, 267.
118. Van der Kloot, W. G. (1955). *Biol. Bull.* **109**, 272.
119. Goddard, D. R. (1939). *Cold Spr. Harb. Symp. quant. Biol.* 7, 362.
120. Harvey, W. R. and Haskell, J. A. (1966). *Adv. Insect Physiol.* **3**, 133.
121. McKay, R. (1939). *J. R. Hort. Soc.* **64**, 272.
122. Cantino, E. C. (1951). *Antonie van Leeuwenhoek* **17**, 59.
123. Evenari, M., Koller, D. and Gutterman, Y. (1966). *Aust. J. Biol. Sci.* **19**, 1007.
124. Lewis, J. C. Snell, N. S. and Burr, H. K. (1960). *Science, N.Y.* **132**, 544.
125. Black, S. H. and Gerhardt, P. (1961). *J. Bact.* **82**, 743.
126. Ballard, L. A. T. and Lipp, A. E. G. (1967). *Science, N.Y.* **156**, 398.
127. Ritossa, F. M. (1964). *Expl. Cell. Res.* **35**, 601.
128. Budd, K., Sussman, A. S. and Eilers, F. I. (1966). *J. Bact.* **91**, 551.
129. Eilers, F. I. (1968). Personal communication.
130. Daudova, G. M. and Stepanova, M. (1966). *Fedn. Proc. Transl. suppl.* **25**, Part II, T 273.
131. Burlington, R. F. and Klain, G. J. (1967). *Comp. Biochem. Physiol.* **22**, 701.
132. Bradbeer, J. W. and Colman, B. (1967). *New Phytol.* **66**, 5.
133. Epel, D. (1967). *Proc. natn. Acad. Sci., U.S.A.* **57**, 899.

134. Stavy, L. and Gross, P. R. (1967). *Proc. natn. Acad. Sci., U.S.A.* **57,** 735.
135. Spirin, A. S. (1966). *Curr. Topics Develop. Biol.* **1,** 1.
136. Monroy, A., Maggio, R. and Rinaldi, A. M. (1965). *Proc. natn. Acad. Sci., U.S.A.* **54,** 107.
137. Bell, E. and Reeder, R. (1967). *Biochem. biophys. Acta* **142,** 500.
138. van Overbeek, J., Loeffler, J. E. and Mason, M. I. R. (1967). *Science, N.Y.* **156,** 1497.
139. Marcus, A. and Feeley, J. (1965). *J. biol. Chem.* **240,** 1675.
140. Rieber, M. (1967). *Biochem. J.* **103,** 64.
141. Waters, L. C. and Dure, L. S. (1966). *J. molec. Biol.* **19,** 1.
142. Wood, A. and Bradbeer, J. W. (1967). *New Phytol.* **66,** 17.
143. Dure, L. and Waters, L. (1965). *Science, N.Y.* **147,** 410.
144. Henney, H. and Storck, R. (1963). *J. Bact.* **85,** 822.
145. Henney, H. and Storck, R. (1964). *Proc. natn. Acad. Sci., U.S.A.* **51,** 1050.
146. Hoshino, J. Nishi, A. and Yanagita, T. (1962). *J. gen. Appl. Microbiol., Tokyo* **8,** 233.
147. Lovett, J. S. (1963). *J. Bact.* **85,** 1235.
148. Bishop, H. L. and Doi, R. H. (1966). J. Bact. **91,** 695.
149. Kobayashi, Y., Steinberg, W., Higa, A., Halvorson, H. O. and Levinthal, C. (1965). *In* "Spores III" (H. O. Halvorson, ed.), p. 200. *Am. Soc. Microbiol.* Ann Arbor, Michigan, U.S.A.
150. Chen, S. and Thimann, K. V. (1966). *Science, N.Y.* **153,** 1537.
151. Villiers, T. A. (1967). *Nature, Lond.* **214,** 1356.
152. Sun, C. Y. and Sussman, A. S. (1960). *Am. J. Bot.* **47,** 589.
153. Halbsguth, W. and Rudolph, H. (1959). *Arch. Mikrobiol.* **32,** 296.
154. Lopriore, G. (1895). *Jh. Wiss. Bot.* **28,** 531.
155. Brodie, H. J. and Neufeld, C. C. (1942). *Can. J. Res.* **124,** 318.
156. Burges, A. and Fenton, E. (1953). *Trans. Br. Mycol. Soc.* **36,** 104.
157. Stotzky, G. and Goos, R. D. (1965). *Can. J. Microbiol.* **11,** 853.
158. Stotzky, G. and Goos, R. D. (1966). *Can. J. Microbiol.* **12,** 849.
159. Griffin, D. M. (1963). *Biol. Rev.* **38,** 141.
160. Lockwood, J. L. (1964). *A. Rev. Phytopathol.* **2,** 341.
161. Dobbs, C. G. (1963). *Recent Prog. Microbiol.* **8,** 235.
162. Vaartaja, O. and Agnihotri, V. P. (1966). *Phytopathology* **56,** 905 (Abstr.).
163. Lingappa, B. T. and Lockwood, J. L. (1964). *J. gen. Microbiol.* **35,** 215.
164. Ko, W-h. and Lockwood, J. L. (1967). *Phytopathology* **57,** 894.
165. Dobbs, C. G. and Gash, M. J. (1965). *Nature, Lond.* **207,** 1354.
166. Brian, P. W. (1957). *Symp. Soc. gen. Microbiol.* **7,** 168.
167. Waksman, S. A. (1961). *Perspect. Biol. Med.* **4,** 271.
168. Brock, T. D. (1966). "Principles of Microbial Ecology". Prentice-Hall, Englewood Cliffs, New Jersey, U.S.A.
169. Baker, K. F., Flentje, N. T., Olsen, C. M. and Stretton, H. M. (1967). *Phytopathology* **57,** 591.
170. Spoehr, H. A., Smith, J. H. C., Strain, H. H., Milner, H. W., and Hardin, G. J. (1949). "Fatty Acid Anti-bacterials from Plants". Carnegie Institution Wash., Publ. No. 586.
171. Hartman, R. T. (1960). *In* "The Ecology of Algae" (C. A. Tryon, Jr. and R. T. Hartman, eds), p. 38. Special Publication No. 2, Pymatuning Lab. of Field Biol., Univ. of Pittsburgh.
172. Saz, A. K., Watson, S., Brown, S. R. and Lowery, D. L. (1963). *Limnol. Oceanogr.* **8,** 63.

173. Borut, S. and Johnson, T. W. (1962). *Mycologia* **54**, 181.
174. Park, D. (1955). *Br. Mycol. Soc. Trans.* **38**, 130.
175. Lloyd, A. B., Noveroske, R. L. and Lockwood, J. L. (1965). *Phytopathology* **55**, 871.
176. Acha, I. G. Leal, J. A. and Villanueva, J. (1965). *Phytopathology* **55**, 40.
177. Lloyd, A. B. and Lockwood, J. L. (1966). *Phytopathology* **56**, 595.
178. Iichinska, E. (1960). *Microbiology (Mikrobiologiya).* Translation, A.I.B.S. **29**, 147.
179. Bumbieris, M. and Lloyd, A. B. (1967). *Aust. J. biol. Sci.* 20, 103.
180. Park, D. (1965). *In* "Ecology of Soil-Borne Pathogens" (K. F. Baker and W. C. Synder, eds), p. 82. Univ. California Press, Berkeley.
181. Allison, L. E. (1951). *Soil. Sci.* **72**, 341.
182. Crump, L. M. (1953). *Rothamsted Expl. Sta. Rept.* No. 1952, p. 58.
183. McKeen, C. D. (1954). *Can. J. Bot.* **32**, 107.
184. Wensley, R. N. (1953). *Can. J. Bot.* **31**, 277.
185. Welsch, M. (1958). *J. gen. Microbiol.* **18**, 491.
186. Hammer, B. W. (1948). "Dairy Bacteriology". 3rd Ed., John Wiley, New York.
187. Beijerinck, M. W. and Jacobson, M. H. (1908). p. 9. *Proc. 1ᵉʳ Congress Int. ou Froid, Paris.*
188. Elwan, S. H. and Mahmoud, S. A. Z. (1960). *Arch. Mikrobiol.* **36**, 360.
189. Mahmoud, S. A. Z., Elfadl, M. A. and Elmofty, M. K. (1964). *Folia microbiol., Praha* **9**, 1.
190. Nicot, J. (1960). *In* "The Ecology of Soil Fungi" (D. Parkinson and J. S. Waid, eds), p. 94. University of Liverpool Press, Liverpool.
191. Sinclair, N. A. and Stokes, J. L. (1964). *J. Bact.* **87**, 562.
192. Larkin, J. M. and Stokes, J. L. (1966). *J. Bact.* **91**, 1667.
193. Blum, H. F. (1961). *Q. Rev. Biol.* **36**, 50.
194. Gates, D. M. (1966). *Science, N.Y.* **151**, 523.
195. Johnson, L. F. and Osborne, T. S. (1964). *Can. J. Bot.* **42**, 105.
196. Bernlohr, R. W. and Novelli, G. D. (1963). *Archs Biochem. Biophys.* **103**, 94.
197. Balassa, G., Ionesco, H. and Schaeffer, P. (1963). *C.R. hebd. séanc. Acad. Sci., Paris* **257**, 986.
198. Anagnostopoulos, C. and Spizizen, J. (1961). *J. Bact.* **81**, 741.
199. Brock, T. D. (1967). *Science, N.Y.* **158**, 1012.
200. Allen, M. B. (1953). *Bact. Rev.* **17**, 125.
201. Dowben, R. M. and Weidenmüller, R. (1967). *Fedn Proc.* **76**, Abstr. No. 1920.
202. Gregory, P. H. (1961). "The Microbiology of the Atmosphere". Interscience Publ., New York.
203. Lamanna, C. (1952). *Bact. Rev.* **16**, 90.
204. Lucas, C. E. (1947). *Biol. Rev.* **22**, 270.
205. Schneider, H. A. (1967). *Science, N.Y.* **158**, 597.
206. Shappirio, D. G., Eichenbaum, D. M. and Locke, B. R. (1967). *Biol. Bull.* **132**, 108.
207. Wolf, Z. J. and Mahmoud, S. A. Z. (1957). *J. appl. Bact.* **20**, 373.
208. Halvorson, H. O., Wolf, J. Z. and Srinivasan, V. R. (1961). *In* "Low Temperature Microbiology", p. 27, Campbell Soup Company, Camden, N.J.
209. Mahmoud, S. A. Z., Barker, A. N. and Wolf, J. Z. (1956). *6th Inter. Congr. Soil Sci., Paris* **3**, 34.
210. Segner, W. P., Schmidt, C. I. and Boltz, J. K. (1966). *Appl. Microbiol.* **14**, 49.

CHAPTER 2

Morphology of Sporulation

PHILIP FITZ-JAMES AND ELIZABETH YOUNG

*Department of Bacteriology and Immunology, University of Western Ontario,
London, Canada, and Department of Biology, York University,
Toronto, Canada*

I. Introduction

THE essential characteristics of the differentiation which converts vegetative cells of *Bacillus* and *Clostridium* species into dormant spores (Figs 1 and 2) may be recognized by microscopists limited to the use of visible light. Early workers recognized that there was a series of morphological changes (for early review see [1]) but Bayne-Jones and Petrilli [2] in 1933, were the first to present photographic evidence demonstrating the correct sequence of these changes in the living cell. Their evidence indicated these essentials: one cell gives rise to one spore; the spore arises within the cytoplasm at one end of the cell; the developing spore is a more dense area of protoplasm incapable of penetration by granules elsewhere in the cell; the spore area increases in size; the development of refractility is a rapid terminal process in the formation of the mature spore only to be followed by its liberation from the disintegrating remains of the cell.

The use of phase contrast microscopy confirms all these features[3] and hints that the area which will eventually become the spore is, at an early stage, surrounded by a limiting membrane.[3, 4]

Further definition of the process was made possible with the development of a reliable method for demonstrating bacterial chromatin. Thus it was revealed that the arrangement and distribution of the chromatin in the cell prior to the appearance of the spore forms a predictable sequence.[5]

Two factors have allowed a more intimate understanding of the sporulation process: the development of cultural methods encouraging synchronous sporulation of a population of cells and the refinement of techniques for use with the electron microscope. These aids permit a more detailed knowledge of the fine structure of the various spore parts and of the mechanics of their evolution. This in turn makes meaningful, as it relates to the developmental process, the increasing amount of chemical and genetical data. It is the aim of this chapter to present the current understanding of the anatomy of the sporulation process and to provide a structural framework for the chemical and physiological studies which are discussed elsewhere in this book.

II. Overall Summary of the Sporulation Process

Bacterial spore formation is, in the first instance, a process of cellular division (Figs 1 and 2) by which a protoplast is formed within the cytoplasm of a cell. Secondly, it is an ordered developmental process in which the two main layers, cortex (Fig. 1, stage 5; Fig. 2, stage 6) and coats (Fig. 1, stages 6 and 7; Fig. 2, stages 7 and 8), are laid down about the internal or spore protoplast.

Conditions necessary and favourable to the initiation of this morphological sequence are discussed elsewhere but, for the purpose of the discussion to follow, it is necessary to repeat that normally in the growth cycle sporulation occurs after an extended period of logarithmic growth and division, although it is possible to circumvent this period of increase in cell numbers and proceed directly from a spore to a spore or to "microcycle" (Fig. 1, paths A, B and C; refs 6, 7 and 8, see Section IX). In either case, normal or microcycle, the time elapsed between germination and the appearance of the crop of spores is the same for any one species. Sporulation proceeds at the same rate and through the same set of morphological changes under both conditions.

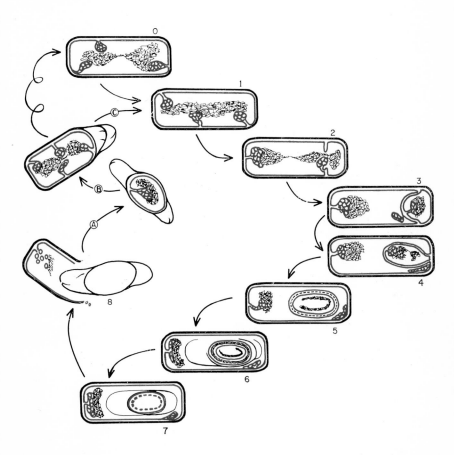

FIG. 1. Diagrammatic summary of sporulation in a *Bacillus* species. Membrane and associated mesosomes are shown in red. 0–1, transition from replicating cell to axial stage; 2, 3 and 4, stages in forespore development. At stage 4 the cell becomes "committed" to proceed to 8. Thus a cell at stage 3 can be returned to the vegetative form by fresh medium ([58] and E. Young, unpublished). Stage 5, cortex development commences (dotted line) and continues through 6 when the coat protein is deposited. Stage 7 is characterized by a dehydration of the spore protoplast and an accumulation of DPA and calcium in the spore. Stage 8, complete refractility, a lytic enzyme acts to release the spore. Also shown are germination A; out-growth to a primary cell B; from which the cell may, under special conditions, enter sporulation by a shortcut C, "the microcycle", but normally undergoes logarithmic growth (spiral arrow).

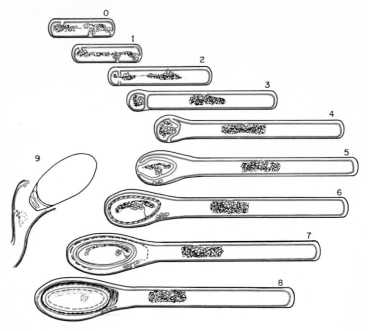

FIG. 2. Diagrammatic summary of sporulation in a *Clostridium*. The processes, although basically those shown in Fig. 1 for a *Bacillus*, are accompanied by considerable growth in cell length and a clubbing to accommodate first the forespore (steps 4 and 5), then the cortex (steps 6 and 7) and finally coats. Refractility (step 8) glazes the electron microscopic detail and lysis (step 9) liberates the completed spore. Membranes are in red. Storage substance fills the stem of the cell but is not drawn in. Taken from a study of *Cl. pectinovorum*.[18]

III. Transitional Cell

A. CHANGES IN SHAPE AND DISTRIBUTION OF CHROMATIN

Conventional cytology tells us that the vegetative cell of the spore-formers contains two chromatin bodies in stages of division, whereas the spore contains only one in a state of rest.[3] This, expressed in chemical terms, means that the average vegetative cell contains approximately twice the quantity of deoxyribonucleic acid (DNA) as does the spore of the same species.[9] Electron microscopy of thin-section of such cells has given no insight into the significance of the bizarre shapes of chromatin character-istic of these duplicating and dividing bodies but has revealed that at all times they are in close association with membranous invaginations of the plasma membrane mesosomes.[10, 11, 12] Subsequent evidence has suggested that the equal division and positioning of the chromatin bodies is effected by the mesosomes[12] and that the synthesis of DNA, duplication of the chromatin bodies and synthesis of membrane are inextricably involved.[13, 14]

B. The Axial Filament

With the shift from logarithmic to linear growth which just precedes the beginning of spore formation, the chromatin bodies in stained preparations appear compact, discrete, with two in each cell.[3] In electron micrographs no fine textural change is evident to explain the cytologically condensed bodies; they remain suspended to the peripheral membrane through associated mesosomes (Fig. 1, stage 0). The synthesis of DNA has now ceased[3, 15, 16] but the chromatin bodies continue to change in shape and distribution, flowing one into the other, until cytologically there is a single rope or axial thread within each cell (Fig. 1, stage 1; Fig. 2, stage 2). Again no further insight is gained from electron micrographs (Fig. 3a), the fine structure remaining unchanged, attachment with the mesosomes being intact and usually to the two poles of the cell.

The significance of the axial thread of chromatin is not clear. Such a pattern is known to precede the appearance of refractile spores by some 6 hr under the standard cultural conditions in the laboratory but it may not be a necessary pattern for the devolution of the process. Its appearance has been reported in several varieties of *B. cereus*[3, 15], *B. subtilis*[17], *Cl. pectinovorum*[18] and *Cl. histolyticum*[19]. However Flewett[20] has remarked on the wide distribution in the percentage of cells showing this form in *B. anthracis*. In some strains, only 1% of the cells had axial filaments whereas others had 95%. Furthermore, some strains with a naturally low frequency could be induced to a higher frequency by growth in increased levels of calcium. Although a potentially significant observation, these results might merely represent degrees of asynchrony of the population (i.e. of those with low frequency) which can be reduced by growth in calcium.

A comparable distribution of the cell's chromatin has been described in aged cells,[21] cells exposed to antibiotics[22] and cells exposed to cold and high salt concentrations.[23] In all of these situations, including spore formation, a cessation or reduced rate of DNA synthesis has occurred. Thus it may be found that this peculiar formation is the result of a lack of modulation between DNA and lipid synthesis—the continuing synthesis of the latter and involution with or without cell elongation could be imagined to drag or stretch the chromatin via the membrane or mesosomal attachments. It is clear, however, that this deformation of the chromatin is reversible; axial filaments associated with incipient sporulation rapidly disperse into conventional patterns if the sporulation medium is replaced by fresh (I. E. Young, unpublished observations). DNA synthesis resumes and the chromatin bodies and cells divide in the normal way. Thus, contrary to our[3] and others[24] earlier suggestions, the axial filament does not represent an irreversible stage in, or commitment to, the sporulation process.

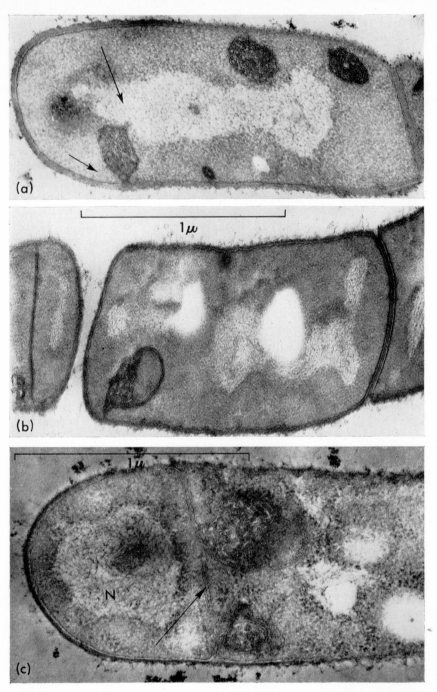

FIG. 3. a. Axial filament stage of transition from a growing to a sporulating cell in a thin section of a *B. cereus* variant. The nucleoid, as a central mass, appears suspended or supported by three mesosomes. The attachment of one of these to both chromatin and to peripheral membranes is marked by arrows. Magnification as indicated in 3b and c, The formation of the forespore septum as seen in two different variants of *B. cereus*. Mesosomes are prominent during both the early "budding" of double septum, 3b, and after the septum (arrow 3c) is complete. The future spore nucleoid (N) is surrounded by ribosomes suggesting continued function.

IV. The Sporangial Cell

A. DEVELOPMENT OF THE FORESPORE

1. Membranous division

The development of the forespore septum is the first definite sign of a cell entering sporulation. Many adequate descriptions of this internal membranous division have already appeared.[10, 18, 25] What is not understood clearly is the relation of the mesosomes anchoring the axial filament to those seen at the site of the forespore fold. Morphological evidence indicates that one pair of DNA-attached mesosomes a short distance from the pole, such as shown in Fig 3a, acts as a site of forespore membrane synthesis as indicated in Fig. 3b. After the membrane induced pocketing of the future spore chromatin and its mesosome, some two mesosomes, outside the forespore and not associated with DNA, develop (Fig 3c) and these continue membrane synthesis as the structure proliferates around the cell end (Fig 5a; Fig 1, steps 3 and 4).

Further studies employing serial sections of cells through stages 0–1–2 of Fig 1 are needed to establish properly the membrane-mesosome-DNA actions at this significant moment in cell differentiation.

2. The site of synthesis of forespore membrane

The appearance of the forespore development does not differentiate between a local membrane synthesis or the shifting of membrane into forespore folds from synthesis elsewhere in the cell. Yet it does seem reasonable to predict, on the basis of the suggested role of the mesosome in membrane synthesis of vegetative cells,[26, 27] that the mesosomes and membranes seen folding into the forespore are being newly synthesized and not simply shifted into folds from membrane formed elsewhere along the sporangium. This fundamental question concerning the site of new membrane synthesis has been studied using a series of 20 min pulse labellings of lipids with $1-^{14}C$ acetate. The results with a reasonably synchronous culture of *Bacillus cereus* strain A$^-$ are summarized in Fig 4. The stages of sporulation at each of the 4 pulse periods are indicated by the line drawings. After each exposure to label, cells from a portion of the culture were disrupted in a carrier wash and their lipid extracted and analysed for specific activity. Cells from another portion of each pulse labelled culture were washed into unlabelled medium and allowed to complete sporulation. These spores were disrupted and the specific activity of their phospholipids also determined. In Fig. 4 the maximum spore phospholipid labelling is associated with the maximum cell phospholipid labelling and the maximum of development of the forespore membrane. Thus the membrane of the forespore is formed *de novo* at the sites of development indicated by electron microscopy.

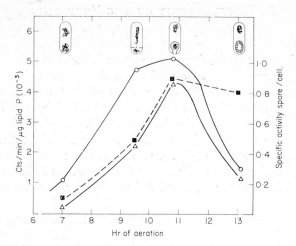

FIG. 4. ^{14}C-Acetate labelling of lipids of sporulating cells. ($\triangle$----$\triangle$) Specific activity (cts/min/μg lipid P) of sporangial cell lipid after 30 minutes' exposure of a sporulating culture to 1–C[14]-acetate (1 μc/ml) at the times indicated. ($\bigcirc$----$\bigcirc$) Specific activity of lipids from cleaned spores derived from above labelled sporangia. ($\blacksquare$----$\blacksquare$) Ratio specific activity sporangial cell: spore from above curves. Stages of membrane development at the end of the labelling periods are indicated diagrammatically.

B. CORTICAL DEVELOPMENT

Even the earliest micrographs of sectioned spores[28] clearly showed a unique cortical layer. Further studies with drugs and coat-free or deficient spores have confirmed that the cortex is a critical structure in the maintenance of the sporulating state (Fig. 1, stages 5 and 6). The development of the cortex is first noted in sections as a widening of the already complete forespore profile (Fig. 5b) by a dense narrow band. During early development in competent variants this zone widens out into an undifferentiated band (Fig. 5c). Then when fully formed (in the mature spore) an inner band appears, slightly more dense than the cortex and lying on the surface of the plasma or inner forespore membrane. Its location and subsequent persistence suggests a role as a primordial cell wall (Figs 6a and b and 12).

Mutants of *B. cereus* and *B. megaterium* which fail to complete a cortex also form a first clearly dense separation of the membranes but go no further. Whether this initially formed cortical primordium, so evident in such mutants, is the same cell wall precursor seen in mature normal spores is still open to question.

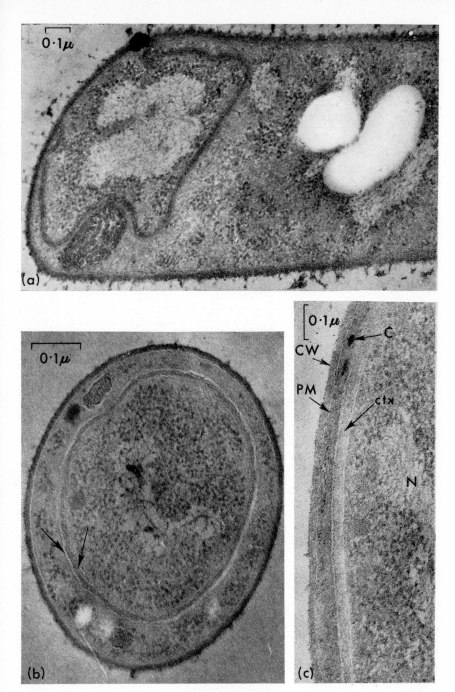

FIG. 5. a. A final stage in the development of the forespore in a large variant of *B. cereus* showing a mesosome attached at its base to the plasma membrane and at its apex to the growing double membrane of the forespore. b, A tangential section through the head of a sporangium of *Clostridium pectinovorum* at stage 5 indicated in Fig. 2. The inner and outer membrane of the forespore are now separated by the beginnings of cortical material. c, A later stage of development shown in 5b. The cortex between the forespore membranes is now wider and the deposition of spore coat (C) is beginning. The spore nucleoid (N) and cytoplasm are still stainable. (CW = cell wall, PM = plasma membrane).

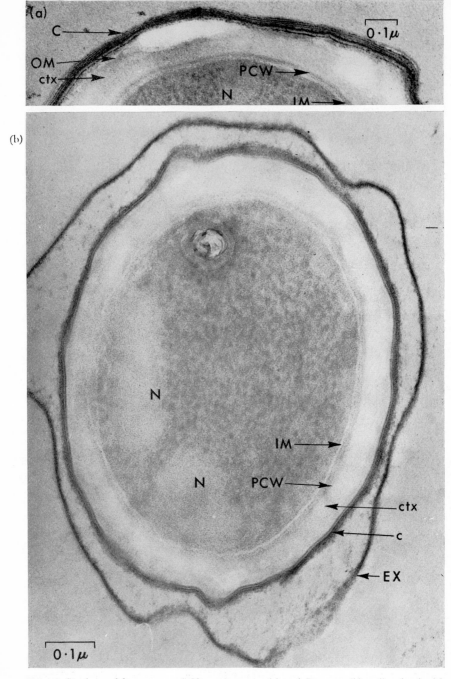

F IG. 6. Sections of free spores of *Cl. pectinovorum* (a) and *B. cereus* (b) well stained with lead after uranyl acetate in order to display the spore parts obscured by the refractility reaction. The multi-layered spore coat (C) of the clostridial spore and the exosporium (EX) and coat layer (c) of the *B. cereus* spore are both seen external to partly hidden outer membrane (OM). The cortex (ctx), primordial cell wall (PCW), inner membrane (IM) and spore cytoplasm are similar in both species. Of particular interest is the granular appearance of the poorly stained peripherally located nucleoids (N) in resting spores.

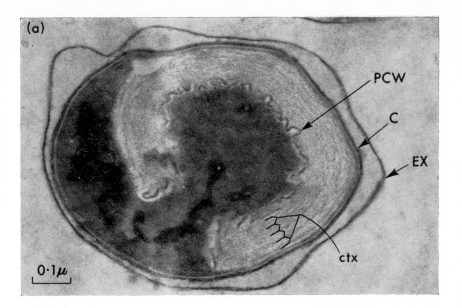

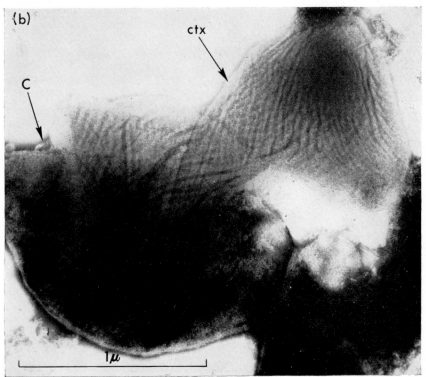

FIG. 7. a. A section of a spore treated with nitric acid prior to fixation and embedding as a means of displaying spore structures. The broken cortex (ctx) is markedly swollen and made up of some four layers. The flexible but less elastic primordial cell wall (PCW) presents a wrinkled profile attached to the inner surface of the cortex. The exosporium (EX) and coat (C) are little changed by the treatment but the extruded cytoplasm and nucleoid are severely damaged. b, Negatively stained (1% PTA) whole mount of a coat-cortex preparation of *Bacillus megaterium*. The sticky cortical (ctx) mass shows a striate structure distinct from the rigid spore coat (C).

D

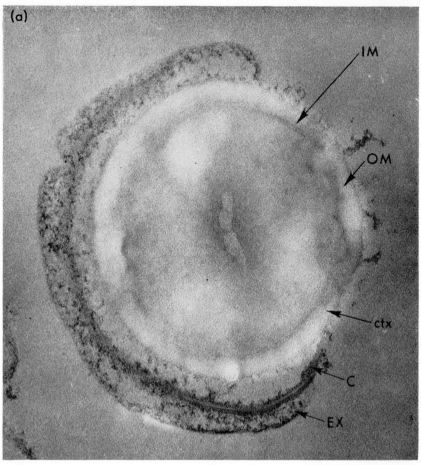

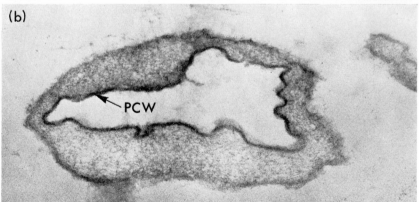

FIG. 8. a. Addition of chloramphenicol at 30 μg/ml during coat formation (the phase whitening stage) in a variant of *B. cereus* has blocked the synthesis of a large segment of spore coat. Thus the free spore has exposed part of its cortex (ctx) covered by remnants of the outer membrane (OM). b, A thin section of a cortical preparation obtained by disruption of boiled spores of *B. cereus* from which the coats had been previously extracted by Cleland's reagent in SLS at pH 11·0.[33] The fibrous nature of the cortical mass and the intimately attached primordial cell wall partly folded by the cortical swelling are evident.

The undifferentiated appearance of the cortical layer can be altered by a number of manœuvres and so reveal some structure in this important spore layer. With acid treatment of resting spores,[29] a severe swelling of the cortex occurs along with a rupturing and escape of the internal spore contents. This method was used earlier by Mayall and Robinow[30] and by us[31] to study the cortical structure and variations in refractility. The swollen cortical remnant of spores so treated is composed of some four major bands. The inner layer, again suggestive of a cell wall primordium, fails to adjust with the cortical swelling and appears as a much folded lining (Fig. 7a). Likewise, the studies of coat-cortex preparations by Warth et al.[32] display the striated nature of the cortical layer and the difference in contractility between it and the inner "cortical membrane"[32] layer.

The striated and plastic nature of the liberated cortical remnant can also be displayed by negative staining of "coat-cortex" preparations from spores previously boiled to destroy lytic enzymes (Fig. 7b). The cortex can also be seen as the major covering on spores whose coat has been partially deleted by late additions of chloramphenicol (Fig. 8a).

Sections of cortices isolated from spores already freed of their coats by extraction with Cleland's reagent and SLS[33] show, in contrast to the intact cortex, a spongy, fibrous matrix but still lined on their inner surface by the narrow, dense, wall-like primordium (Fig. 8b).

In spite of its sensitivity to germination and to lysis by acid and spore lytic enzyme, the spore cortex does not appear dependent on an intact coat for stability. When spore coat completion is prevented by the late addition of chloramphenicol, leaving an exposed segment of cortex, the final refractility of the spore core or its cortex is not altered (Fig. 8a).

C. COMMITMENT

With the completion of the forespore and the beginning of cortical synthesis, the sporulating cell reaches a stage of commitment such that it can no longer withdraw from completing a spore before entering vegetative growth. Prior to this, a culture of cells at, for example, stage 2–3, Fig. 1, will revert to vegetative growth upon the addition of fresh medium. In such an event, the forespore septum appears to convert to a transverse septum with the deposition of wall material between the membrane layers. Thus, two cells result, one long (non-spore part) and one short (potential spore). With resumption of vegetative growth, they soon divide—the long ones first, then the shorter ones.

After completion of the forespore, the addition of nutrients leads to a prolonged delay of sporulation. The forespore, usually with a variable amount of cortical development, persists in the now nondividing sporangium which (in B. cereus) becomes lipid-loaded. After some 3–6 hr, depending on the degree of dilution with new medium, the cell continues with sporulation but the spores, when freed, usually germinate.

D. Coat Formation

The deposition of the spore coat is first seen in many sporulating bacteria shortly after the beginning of the cortical structure. In *Cl. pectinovorum* it first appears as discontinuous masses of dense coat material (Fig. 5c) which then link together into a continuous layer (Fig. 6a).[18] This discontinuous and multi-regional deposition of coat protein is different from the spreading appearance of the coat in many of the *Bacillus* spores. However, it would seem from kinetic studies of coat synthesis that the deposition into a visible coat structure is a relatively late, cystine-dependent assembly of protein units whose synthesis can be traced back to the fore-spore stages.[33] Hence the electron microscope is revealing differences in mode of assembly rather than differences in the number of sites of protein synthesis. Coat protein synthesis does not appear unusual in its basic mechanism. It can be demonstrated as an activity dependent on ribosomes and with the expected sensitivity to inhibitors[34] (Section VI).

Perisporal mesosomes *appear* to play some role in the deposition of coats. However, this role is far from proven and is indeed open to some doubt, since coat formation can proceed in protoplasts of *B. megaterium* which appear to be mesosome free or, at least, to have these in an everted state well removed from the spore surface.[4]

E. Refractility

With coat development, the majority of sporulating cells, when viewed in phase contrast, undergo a whitening which ultimately clearly displays the forming spore as a dull white ovoid. This initial whitening should not be confused with the terminal refractility state of sporulation which, beginning during late coat formation, is associated chemically with the synthesis of DPA and the major uptake of calcium both of which are entirely located in the finished spore.

Although the "refractility" state can be differentiated from the "phase-white" state in a normal sporulating culture by phase microscopy, the most reliable method of observing it is by the use of the bright-field objective. If a well-dried nigrosin smear of a water suspension of cells on a coverslip is mounted, smear side down, on a slide and observed by oil immersion through the back, the air mounted spores glow with a refractility readily distinguishable from those in the phase-white state. This method, devised by Professor C. Robinow, is most useful for studies of the late stages of sporulation.[31]

The refractility state is associated with a difficulty in obtaining good fixation and hence proper embedding for thin-section electron microscopy. Thus, fixation and embedding times suitable for vegetative cells must be extended for the spore state. Refractility can be related to a glazing of the detail of the spore core (Fig. 8a). However, staining with uranyl or lead will faintly reveal the spore inner membrane, ribosomal and nuclear material. Even in well-stained sections the latter never appears in its usual fibrous state but rather as a compact mass of low density material showing a fine granular structure (Fig. 6b). This change in the character of the nuclear material can be used to detect the early development of refractility.

V. The Use of Variants (Mutants) to Study Sporulation

A. The Degrees of Asporogeny

In many laboratories, a number of variant, or mutant, cultures of *Bacillus* species have been isolated, showing arrests at a variety of stages of sporulation. In some, these blocks are apparently complete and the culture is incapable of producing any normal spores (i.e. no spores in 10^8 cells) as determined by absence of growth following heat treatment and are considered asporogenous (Sp$^-$). In others, a rare cell (one in 10^5–10^6) succeeds in producing a normal heat-resistant spore which, on recycling, produces a similarly blocked culture with the same low frequency of sporulation. Such cultures are now referred to as oligosporogenous (Osp.).[35]

Thus, from a morphological vantage point, one may describe three general types of sporulation mutants. First, there are cultures competent in all the synthetic events of sporulation but unable to regulate some aspect of the proper sequence of sporulation from initiation on down to terminal lysis. Hence, although a few cells may be able to achieve stable refractility in the proper order of synthesis, the majority of spores are aberrant in structure and aborted. The morphological pictures will depend on the location and extent of the disorder of control in the particular mutant. Many of the spore structures may be victims of whatever terminal lysis still does occur and most of the Osp-mutants would be the control-aberrant type.

Secondly, there are mutants completely blocked at a definite step of sporulation and unable to proceed beyond it. Such blocks, if occurring early, would give the culture the appearance in phase contrast of being completely asporogenous.

Thirdly, mutants have been isolated which show complete absence of one specific spore structure. Thus, the cells are capable of synthesis of structures formed before and after the expected time of the deleted structural change.

B. MORPHOLOGICAL MUTANTS OF SPORULATION

1. Mutants arrested at axial filament stage (Stage 1, Fig. 1)

Many completely asporogenous cultures, when aerated in a complete sporulation medium, remain in the axial filament stage of sporulation.[36] In our own studies, these mutants appear with high frequency following acridine orange or acriflavine treatment (Fig 3a).

2. Mutants arrested at forespore stage (Stage 3, Fig. 1)

Ryter et al.[36] have reported mutants of B. subtilis which do not proceed beyond the formation of the forespore septum. Such mutants resemble more closely division mutants since cell-wall material is deposited between the two layers of the septum, as would occur during normal vegetative division. In this same category, and consistent with this interpretation, is a larger group of mutants which form a "typical" forespore septum but at both poles of the cell[37] (Fig. 9). Cell wall material is likewise laid down between both septa to produce, following lysis of the central sporangium, two coccoids or pseudo-forespores, both of which are viable, though not heat resistant. In B. cereus such mutants fall into the oligosporogenous group. That is, some cells proceed to make a single refractile spore. Cells which do so usually appear at a frequency of $1/100-1/500$ and occur with the mutant form in the same chain of cells. The frequency of appearance of refractile spores in such cultures is constant for any one isolate and remains the same if propagated, after heating, from the spores only.

Many asporogenous mutants apparently blocked at this stage when seen in the light microscope are, when examined in the electron microscope, found to have multiple membranous septa with cell wall deposited between many of the membrane layers (Fig. 14a and b).[38] Thus, such "monsters"[39] appear to be mutants blocked in the sporulation process to which some previously suppressed vegetative activity has returned. This ready return of vegetative activity in this group of mutants is consistent with the ease with which cell division and growth can be brought back by the addition of nutrients during normal forespore development (see p. 74).

3. Cortexless mutants

These most interesting mutants appear to be completely lacking the ability to form a normal cortex. To date, we have encountered one of these variants in the B. cereus group, designated A-1.[34, 37, 40] This mutant (Sp-) has not yielded revertants to normal cortex function but does constantly yield a completely asporogenous variant. In sporulation medium, it synthesizes cortical peptides which appear normal, takes up the normal amount of calcium, forms the usual complement of DPA, synthesizes a

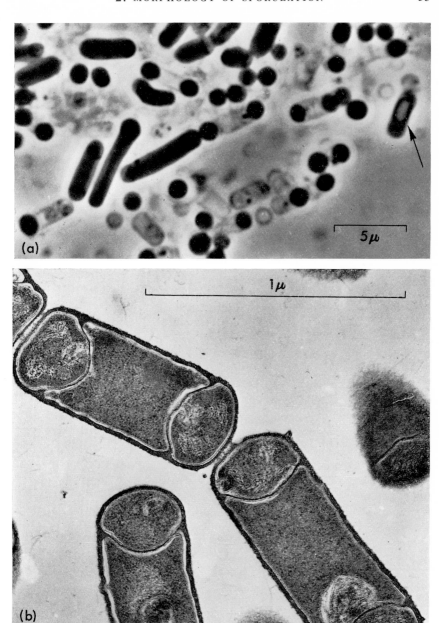

FIG. 9. a. A mutant of *B. cereus* derived from the cortexless mutant illustrated in Fig. 10 in which two spherical cell masses develop from blocked forespores af either cell end. A rare cell does complete the cortexless sporulation of the original strain (arrow). Phase contrast micrograph. b, A thin section of a similar mutant of *B. subtilis*. (Micrograph prepared by Dr Ryter, Institut Pasteur).

normal spore coat. However, following completion of the forespore it is unable to deposit properly the cortical structure (Fig 10) suggesting either that some acceptor is absent or that the system for transporting cortical substance between the forespore layers is faulty.

The synthesis of spore coat, as seen by electron microscopy (Figs 10b, c, d) and measured by cysteine uptake, is normal but advanced in completion by approximately one hour. Some of the spore coats show the alteration in nuclear detail usually associated with the beginning of refractility (Fig. 10c). Although the spore coats are usually completed, full refractility is not seen. Terminal lysis destroys the core and all spore layers except the resistant spore coat which survives (Fig. 10d).

More recently, a cortexless mutant of B. megaterium has been found which, like the cereus organism, is unable to survive the final lysis of the sporulation process. This culture, however, also has a defect in its calcium uptake and dipicolinic acid synthesizing machinery. Before being released by lytic enzyme, the calcium level reaches only 34% of the wild strain level and the DPA 23%.

It is interesting to compare these cortexless mutants, one with a normal capacity and one with a much reduced capacity of DPA synthesis, with a B. cereus mutant "21" which has been found by Dr A. Aronson (personal communication) to have a partial block of DPA synthesis such that only 17–21% of that found in normal B. cereus is formed. Nevertheless, mutant "21" takes up and holds a normal amount of calcium and, in thin section, has a normal cortex. These incomplete comparisons suggest the DPA synthesis can be independent of calcium uptake and that cortical development may be independent of both Ca and DPA levels (see p. 101).

Cortexless mutants have also been noted in electron microscopic examinations of Osp. mutants. In some of these, the defect is apparently a loss of control in the sequence of events. Hence premature refractility reaction begins before most spores have completed cortex or coat (Fig. 11a). A similar response can also be effected by the addition of chloramphenicol to a synchronous culture in late forespore or early cortical stage of development (Fig. 11b).

4. Aberrant coat mutants

Few mutants of this type have been described. Interesting physico-chemical investigations of resting spores will be possible if and when true coatless spore mutants are isolated.

Mutants with aberrant coats, however, have been described. One of these, presumably an Osp. Clostridium histolyticum appears to form normal coat, but is unable, in many instances, to control the proper deposition of the coat on the spore and instead forms a multiple layered mass of coat nearby.[41]

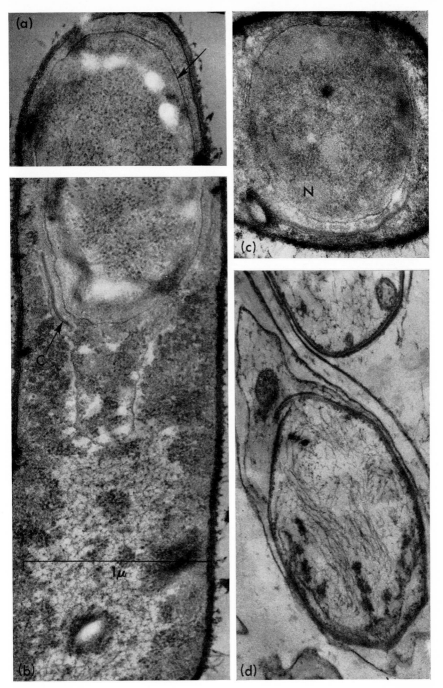

FIG. 10. Sporulation in a cortexless variant of *B. cereus*. (a) The completed forespore forms only a narrow dense profile (arrow) between its inner and outer membranes. (b) Instead of a normal period of cortical development coat deposition commences (arrow). (c) In many spore protoplasts the condensation of the nucleoids (N) into poorly defined granular masses, suggesting the beginning of a refractility reaction, is seen. (d) With the onset of terminal lysis the spore interior is also lysed leaving a spore coat. Magnification as indicated in (b).

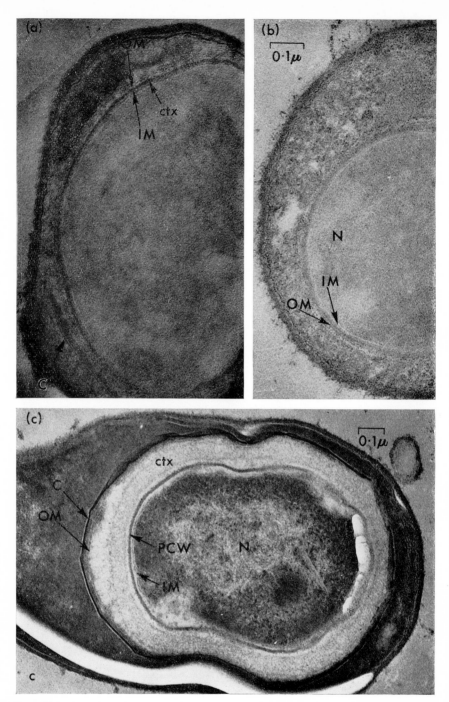

FIG. 11. a. A section of a mutant of *B. cereus* in which refractility appears to be completed
before the cortex (ctx) is properly formed. Likewise in most profiles, only a small segment
of spore coat was formed before refractility and lysis terminated sporulation. Magnification
as indicated in 11b. b, The effect of adding chloramphenicol (30 μg/ml) to a sporulating

VI. Modification of Sporulation Morphology by Inhibitors

Antibiotics of known prime sites or modes of action are useful as tools to test concepts of sporulation. Likewise known mechanisms of morphological variation can be used to test or confirm modes of action of inhibitors. The absence of cell enlargement and division in most sporulating species should in many instances permit a more precise morphological and biochemical demonstration of the primary site of antibiotic action than that possible in rapidly dividing cultures.

A. CHLORAMPHENICOL

A synchronously sporulating culture of *B. cereus* treated with 30–100 μg/ml of chloramphenicol (CAP) will show a variety of interesting and informative effects depending on the stage of development at which the drug is added. Addition during forespore development will block further development and no evidence of cortex, coat or refractility will be found. Addition of CAP at the completion of the forespore will prevent the formation of most of the cortex and all of the coat; yet in many cells, the completed forespores will become refractile. Additions of CAP after cortical development has begun, but before spore coats are laid down, leads to the formation of a fully refractile, cortex-covered but coatless spore.

In these last two types of CAP treated cultures the terminal lysis is also delayed. Indeed, many of the cortex and coat deficient refractile forespores failed to survive the eventual terminal lysis. However, before lysis begins it is possible to show that refractility, even in the absence of cortex and spore coat, is associated with a normal heat resistance.

Finally, treatment with CAP when coat formation, as indicated by the uptake of cysteine, is some four-fifths complete produces spores showing segmented defects in the protein spore coat (Fig. 8a). Such spores have a normal cortex. Moreover, even though terminal calcium uptake and DPA synthesis were only beginning when CAP was added, these activities were little affected. These coat-deficient spores appeared normal in structure and heat resistance but had a marked tendency to auto-germinate in aqueous suspensions. Their survival as dried spores has not yet been studied.

---

culture of *B. cereus*. At early cortex stage of sporulation the cortical zone between the inner (IM) and outer membrane (OM) is incomplete, spore coats are absent, yet refractility as indicated by the loss of detail in the spore protoplast is complete. Control culture did not reach this stage of refractility until a further hour of aeration. (Section; uranyl, but not lead stained.) c, Section of a sporulation mutant of *Cl. histolyticum* unable to achieve cortical rigidity or to ripen its spores. The section is dominated by a wide pliable cortical profile. (Micrograph from the work of Dr A. Ryter, Institut Pasteur.)

Such inhibitor studies indicate that the machinery for the refractility reaction is established soon after the completion of the forespore and is unaffected by late additions of CAP although the synthesis of coat protein is still sensitive.[34]

These studies also reveal that the establishment of refractility appears to be independent of cortex formation but that this layer does protect the refractile DPA-containing core from lytic attack and that the spore coat, in addition to other possible protective functions, prevents premature germination.

B. ACTINOMYCIN D

The time of onset of refractility can be readily pinpointed with this drug. Additions of Actinomycin D (10–30 μg/ml) at the forespore stage, or earlier, blocks further sporulation changes. Additions during early cortex formation block the development of full heat resistance and refractility as well as coat deposition[34]. However, some of the morphological beginnings of refractility can be seen in thin section (Fig 12a). Some time following the establishment of spore cortex, yet before the coat protein has begun to appear on the spore, actinomycin fails, in spite of its continued effect on the cortex and coat, to prevent heat resistance and refractility from developing. Presumably at this time enough of the mechanism for refractility is established and since this terminal reaction does not involve protein synthesis, it is refractory.

C. PENICILLIN

Once growth ceases and spore formation is established, penicillin has no apparent effect on the cell until cortex formation begins. Then sporulating cells in the presence of penicillin (or one of its derivatives) show a defect of development which mimics in appearance in the light microscope that shown by the cortexless mutants. Electron microscopy of such treated cells reveals a peculiar and specific lesion in the spore cortex (Fig 12b). Although material is deposited in the cortical zone, it is of variable width and lacks the usual apparent rigidity. Hence considerable distortion of the spore profile is seen. Diaminopimelic acid uptake is normal or even above normal,[42] but that taken up is lost along with the cortical structure, with terminal lysis (P. Fitz-James, unpublished observations). Proper full refractility is never achieved, only a phase-whitening or partial refractility. The retention of DPA in the spores is impossible in the absence of a rigid cortex[34] and, like the spores of the cortexless mutant A⁻1, those formed in the presence of penicillin are similarly sensitive to lytic enzymes, leaving only the spore coat whose synthesis is not sensitive to this drug (see also p. 97).

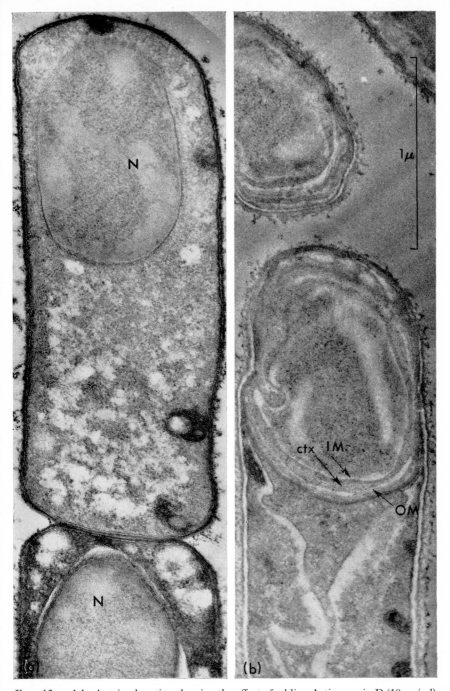

FIG. 12. a. A lead-stained section showing the effect of adding Actinomycin D (10 μg/ml) to a synchronously sporulating culture of *B. cereus* just prior to the synthesis of cortex. Partial refractility, as indicated by nucleoid (N) faintness, has developed in spores essentially devoid of cortex and spore coat. An untreated control by this time had well-developed coats and cortices but similar refractility. b, The effect of adding penicillin (Methicillin, 1 mg/ml) just after forespore completion to sporulating *B. cereus*. Spore coat formation is complete, but the cortex (ctx) between the inner (IM) and outer (OM) membranes of the forespore is uneven and non-rigid.

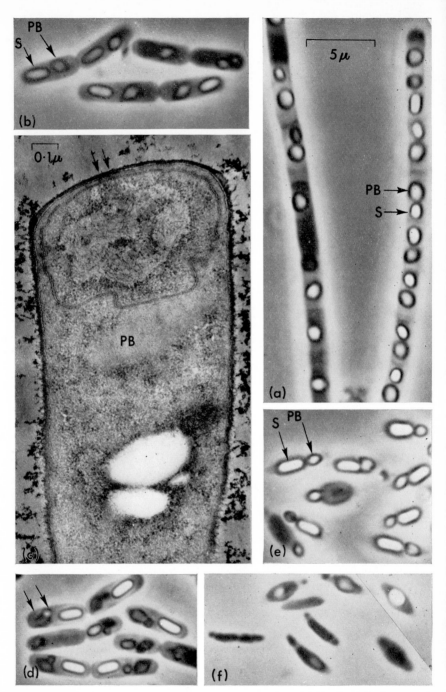

FIG. 13. a. The filamentous sporeformer *B. cereus* var *medusa* which at the end of log growth commences synthesis of a parasporal body some 3 hr prior to forespore formation and so ultimately produces a large parasporal body and spore (S) in each cell. Phase contrast micrograph. b, Late sporulation in *B. cereus* var *alesti* showing spores and parasporal crystals (PB) nearing completion. Phase contrast magnification as indicated in 13a. c, The

VII. Formation of Parasporal Bodies

The present interest in parasporal bodies originates with the redirection of microbiological attention to the "Rest-korper" of Berliner[43] by Hannay in 1953.[44] The demonstration that these were crystalline proteins and insect toxins has widened the interest from that of a microbiological curiosity to models for protein biochemistry and to practical methods for combating insects (see Chapter 13).

These structures appear in many shapes. Even two distinct structures can form inside the sporulating cell alongside a typical spore. Based on when they first appear in the sporulating cell, they can be separated into the two categories described below.

A. PARASPORAL BODIES INITIATED PRIOR TO FORESPORE APPEARANCE

The only representative of this group is a filamentous sporeformer originally isolated from cow dung and called *Bacillus medusa* by Dr C. F. Robinow. Its growth and spore morphology suggest that it may belong to the *B. cereus* group.[9] Although a formal description has not been made, the formation of its forespore[10] and the membrane-associated origin of its parasporal protein have been noted.[45] Following the end of log growth and some three hours prior to the appearance of the forespore, a protein inclusion develops against the membrane in one corner of each cell. These can be seen in phase contrast as bright dots which grow in size to ovoid inclusions which move into the centre of the cells. When these "bods" are about one-half of their ultimate size, the sporulation process commences. Feulgen stains reveal the axial strand of chromatin stretched around the parasporal body prior to the development of the forespore. Four or five hours later, the fully developed spores and "bods" fill the filamentous cells (Fig 13a).

The well-developed parasporal synthesis preceding sporulation appears to make spore formation itself highly sensitive to inhibitors. The addition of amino acids after the commencement of synthesis, for example, so favours the continued synthesis of protein that sporulation never occurs. When

---

initial stage of parasporal body (PB) formation in *B. cereus* var *alesti* adjacent to the original face of the forespore septum. The forespore membranes are almost complete (arrow). d, *B. cereus* var *schwetzova* forms besides a spore a diamond-shaped and a spherical parasporal body. Magnification as in 13a. e, The parasporal body (PB) attached to *Bacillus popilliae* spores (S) from infected "June bug" larvae. Magnification as in 13a. f, The parasporal body is much less prominent when *B. popilliae* sporulates on artificial medium[50]. Magnification as in 13a.

penicillin is added to prevent the development of the cortex, the protein "bod" appears to develop with such vigour that the cell is not able to proceed with the synthesis of structures, such as spore coats, which are normally insensitive to penicillin. Likewise in microcycle sporogenesis (see Section IX) "bod" formation appears favoured over sporulation.

Several mutants of this strain have been isolated. One still capable of synthesizing the "bod" at the end of log growth but unable to commence sporulation is obtained by a variety of mutagens. In thin section, these "bod-only" mutants reveal repeated abortive attempts at forespore development (Fig 14a). No toxic action to insects has yet been shown for the parasporal protein of this organism.

B. Parasporal Bodies Initiated during Forespore Development

Most parasporal crystalliferous bacilli are in this group. Many, like the "medusa" bacillus, appear from spore morphology and egg yolk reaction[46] to be variants of *B. cereus*. However, insect microbiologists prefer to classify them as variants of *Bacillus thuringiensis*. The development of the crystalline protein inclusion which are toxic to many *Lepidoptera* larvae is readily followed by phase contrast microscopy (Fig 13b). It is initially seen as a small refractile dot on the just completed forespore septum. Thin sections reveal the highly intimate relationship of the octahedral protein crystal and membrane.[45] One of the eight faces appears firmly bound to a flattened part of the forespore membrane (Fig 13c). In well-stained (uranyl-lead) sections the cross striations of the crystal structure are readily seen at high magnifications. Carbon replicas are also useful for displaying the surface crystalline characteristics of these inclusions[47]. When the crystals are about one half their ultimate size, they appear to become detached from the forespore and continue to grow in size while the spore is completed. As the spore develops refractility, the crystals also appear to become more refractile.

A number of organisms which form peculiar parasporal bodies have also been collected by microbiologists. Several double crystal-forming cultures have been studied. These are mainly from the collection of the late Dr C. Toumanoff, Institut Pasteur. For example, one designated B. 30–1 forms the typical, slightly elongated octahedron as well as a protein inclusion of different solubility in the form of a flattened trihedron. Another peculiar crystal former, *B. cereus* var *schwetzova* (Fig. 13d), forms both a crystal and ovoid inclusion besides a normal spore in each cell. A bacillus usually referred to as Fowler's bacillus[48] forms a multi-crystalline aggregate within the exosporium.[49] This parasporal protein can be extracted and readily recrystallized (P. Fitz-James, unpublished work).

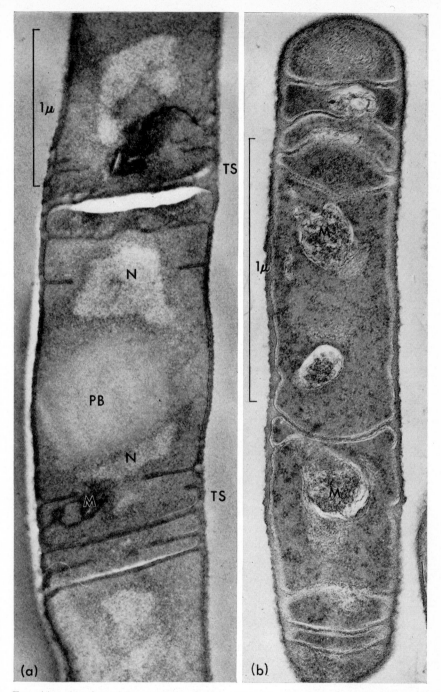

FIG. 14. a. A variant of *B. cereus* var *medusa* capable of forming only parasporal bodies. In thin section multiple forespore membranous septa are found. The transverse septa (TS) are slightly wider than the abortive forespore membranes. b, A variant of *B. subtilis* which is unable properly to control sporulation produces instead numerous membranous septa associated with mesosomes (M). Preparation and electron micrograph by Dr A. Ryter.

Bacillus popilliae, one of the most interesting and, until recently, difficult to propagate insect pathogens, forms a large refractile parasporal inclusion attached to the spore when cultivated in the hemolymph of its host (Fig. 13c), but when grown in artificial sporulation medium[50] this inclusion is very small or absent (Fig. 13f) (P. Fitz-James, unpublished studies). For a fuller treatment of insect pathogens, see Chapter 13.

C. EXTRACELLULAR APPENDAGE ASSOCIATED WITH SPORULATION

During a survey of sporulation morphology in the smaller highly motile bacilli aerated in shake culture, we observed that some of these species such as *B. laterosporus*, *B. larvae* are inhibited from sporulation by excess shaking. Apparently these small motile bacilli lose their flagella as soon as sporulation approaches and form pellicles on the surface of fluid medium. The discharged flagella will, in some species, coalesce into large spirochaete-like masses referred to as "whips".[51] This apparent requirement for a surface, presumably for aeration, is met by one of these motile sporeformers, *B. brevis*, by the development of a highly tactile polar secretion which appears in shake cultures to function as a "hold-fast". During the beginning or early steps of sporulation, shake cultures of *B. brevis* show rims of cells on the flask wall. By rotating the flask as the culture enters sporulation, fresh rims of cells become visible in 5–10 min. Vegetative cultures do not show this phenomenon nor does an asporogenous culture. Cells scraped from the walls show a faint amorphous structure attached at one pole of most sporangia (Fig. 15a and b). In thin sections this appears as a fibrous mass apparently attached to a small local region of the cell wall (Fig. 15c). A more complete description of this sporulation appendage on *B. brevis* is now in preparation.

Appendages on spores are also well documented. As examples, the reader may be directed to earlier studies of the canoe structure under the coats of *B. laterosporus*[52] and the recent description of fascinating appendages on the spores of *Cl. botulinum E*[53] and *Cl. bifermentans*.[54]

VIII. Sporulation in Gram-negative and Atypical Bacilli

A number of the small motile spore-forming bacilli react negatively in the gram reaction (e.g. *B. circulans*, *B. laterosporus*, *B. brevis*) and also display a multi-layered cell-wall structure. Another gram-negative bacillus, *Lineola longa*, capable of forming endospores is shown in Fig. 15d. The cell wall of this bacillus also shows the multilayers typical of gram-negative organisms. The early processes of sporulation are not greatly different from

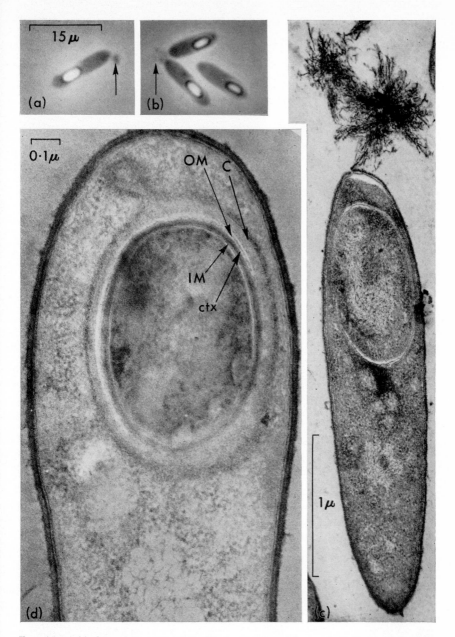

FIG. 15a and b. Sporulating cells of *Bacillus brevis* scraped from a shake flask wall showing one single (a) and (b) two interlocked "hold-fast" structures. c, A thin section of a sporulating *Bacillus brevis* cell showing the densely stained "hold-fast" structure attached to a displaced part of the cell wall. d, Sporulation in a Gram-negative bacillus, *Lineola longa*. In this species the spore coat (c) and cortex (ctx) appear to be formed concurrently. The inner and outer membranes outlining the cortex are also marked.

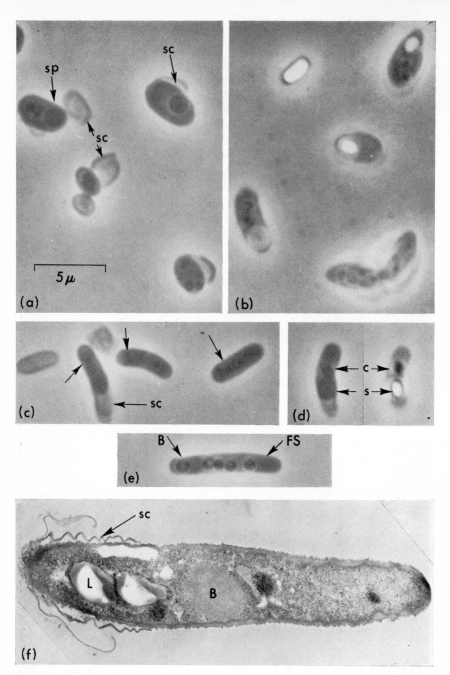

FIG. 16. a, b and c. Phase contrast microphotographs of microcycle sporulation in various *Bacillus* species. Magnification is indicated in 16a. a, Primary cells of *B. megaterium* aerated for 12 hr after germination in the medium of Freer and Levinson[7]. The majority of the cells have attached spore coats (sc) and well-formed forespores (sp). b, At 15 hr the primary cells show well-developed spores. The two rounded cells (upper right) are still partly contained by spore coats. One primary cell has managed to achieve one division. c, Primary cells of *B. cereus* var *alesti* from spores germinated in the medium of Mac-Kechnie and Hanson[56] and aerated for 10 hr. Some of the cells have the spore coats (sc) still adherent. A forespore septal membrane is seen in each cell (arrows). d, In the mature

that of gram-positive bacilli as outlined in Fig. 1. However, in thin section, the cortex and spore coat appear to be formed concurrently (Fig. 15c).

The sporulating coccus *Sporosarcina ureae* and a sporulating lacto-bacillus *Sporolactobacillus inulinus*[55] have been studied by thin-section electron microscopy and found to follow generally the steps outlined in Fig. 1. The initial steps of membrane development in *Sp. urea* are difficult to distinguish from cell division peculiar to this organism.

IX. Microcycle Sporulation

In 1965 Vinter and Slepecky[6] described the occurrence of sporulation in short, partly elongated cells obtained from germinated spores. This short cut to sporulation, eliminating the vegetative proliferation usually occurring, is termed microcycle sporogenesis and is indicated as pathway C in Fig. 1. The nutritional requirements for microcycle sporogenesis have been studied by Holmes and Levinson[7] for *B. megaterium* and by MacKechnie and Hanson[56] for *B. cereus* var T.

The fine structure of *B. megaterium* during this form of sporulation is well depicted by Freer and Levinson[7]. After 1–2 hr in a chemically defined medium lacking an energy source and limited in phosphorus, the germinated spore forms a primary cell. Then, following some six hours of further aeration, this cell begins the process of sporulation following the usual stages of morphogenesis. Thus the germinated spore or the primary cell, faced with insufficient nutrients for division, develops an extended generation time and becomes re-depressed into sporulation.[57] During this period the level of DNA increases some 3-fold before the cell proceeds to sporulate.[8]

Our interest in the ability of cells to form parasporal protein inclusions both prior to and along with spore synthesis prompted a morphological comparison of microcycle sporulation in several *Bacillus* species, including the *medusa* and *alesti* variants of *B. cereus*. *B. megaterium* was germinated with the medium of Freer and Levinson[7] and *B. cereus* variants with the medium of MacKechnie and Hanson.[56] Micrographs of some of these studies are gathered in Fig. 16.

The size of inoculum for a given medium appeared to be critical to the

---

spores (s) the developing parasporal protein crystal (c), although smaller than normal (cf. Fig. 13b), can be made out. e, A long primary cell from the spore of the filamentous *medusa* variant. A parasporal body (B) is forming at the distal end and possibly forespore (FS) at the proximal end of each robust cell. Lipid granules are also plentiful (cf. Fig 13a). f, A thin section electron micrograph of a long primary cell of the medusa bacillus. The parasporal body (B) in most cells was initiated at the end distal to the spore coat (sc) and showed the usual intimate association with the membrane[45]. Forespores were occasionally seen. ($\times$ 22,500).

proper yield of undivided primary cells. Too small an inoculum would permit one or more divisions of the primary cells before sporulation began. At the proper spore concentration the crystal-forming strains formed primary cells which developed typical forespore membranes and parasporal inclusions usually difficult to see in phase contrast micrographs (Fig. 16c). With continued aeration the primary cell completed spore and crystal formation. However, the crystal size was seldom that achieved by the normal process (Fig. 13b). When the inoculum is reduced, many of the primary cells make one division following germination, then each cell of the pair develops a typical spore and crystal.

The effect of spore density on the number of divisions could be most readily demonstrated with the filamentous *medusa* strain. Regardless of the number of divisions occurring, the cells remained attached and the number per chain can be readily noted. With a very small inoculum for a given volume of medium, chains of 9–12 cells formed from each primary cell. As the inoculum was increased, the number of divisions decreased until an optimum was reached where each spore formed a single, somewhat longer primary cell. Following the usual indefinite generation time, a protein inclusion formed at the distal tip of the cell. Then some 3 hr later a fore-spore began to appear at the other extremity of some of these same cells, usually proximal to the still attached spore coat. In most of the cells, the synthesis of the already initiated parasporal body was favoured and only rarely did the spore proceed beyond the forespore stage. The detection of the "bod" was made difficult by lipid inclusions (Fig. 16e). However, thin sections revealed a growing inclusion and sometimes a forespore-containing primary cell (Fig. 16f.) Inclusion formation also occurred in each of the cells of a short chain formed by division of a primary cell. However, in most of these cells spore formation seldom proceeded beyond the forespore stage.

X. Future Work

Now that much of the basic structural processes of sporulation have been analysed at the level of the electron microscope, what should the future offer in new, rather than repetitious data? Electron microscopes have now become so plentiful and the techniques so reasonably standardized that one would hope that, henceforth, genetic and biochemical studies on sporulation would be closely related, where necessary, to the fine structure. The future should thus bring more fine structure analyses of spore coats and other layers in relation to function and composition. Extracellular synthesis of spore layers and components, when documented by well-resolved data of fine structure, should inform our successors on the nature of the assembly processes of sporulation. Thus, the hope also rises that future works on the bacterial spore would not include an isolated chapter on morphology.

REFERENCES

1. Knaysi, G. (1948). *Bact. Rev.* **12,** 19.
2. Bayne-Jones, S. and Petrilli, A. (1933). *J. Bact.* **25,** 261.
3. Young, I. E. and Fitz-James, P. C. (1959). *J. biophys. biochem. Cytol.* **6,** 467.
4. Fitz-James, P. C. (1964). *J. Bact.* **87,** 667.
5. Robinow, C. F. (1960). *In* "The Bacteria" (I. C. Gunsalus and R. Y. Stanier, eds) Vol. I, p. 207. Academic Press, New York and London.
6. Vinter, V. and Slepecky, R. A. (1965). *J. Bact.* **90,** 803.
7. Freer, J. H. and Levinson, H. S. (1967). *J. Bact.* **94,** 441.
8. Holmes, P. K. and Levinson, H. S. (1967). *J. Bact.* **94,** 434.
9. Fitz-James, P. C. and Young, I. E. (1959). *J. Bact.* **78,** 743.
10. Fitz-James, P. C. (1960). *J. biophys. biochem. Cytol.* **8,** 507.
11. Van Iterson, W. (1961). *J. biophys. biochem. Cytol.* **9,** 183.
12. Ryter, A. (1967). *Folia microbiol. Praha* **12,** 283.
13. Ryter, A. and Jacob, F. (1964). *Annls Inst. Pasteur, Paris* **107,** 384.
14. Fitz-James, P. C. (1965). *Bact. Rev.* **29,** 294.
15. Young, I. E. and Fitz-James, P. C. (1959). *J. biophys. biochem. Cytol.* **6,** 843.
16. Canfield, R. E. and Szulmajster, J. (1964). *Nature, Lond.* **203,** 596.
17. Ryter, A. (1964). *Annls Inst. Pasteur, Paris* **10,** 40.
18. Fitz-James, P. C. (1962). *J. Bact.* **84,** 104.
19. Kawata, T., Inoue, T. and Takagi, A. (1963). *Jap. J. Microbiol.* **7,** 23.
20. Flewett, T. H. (1948). *J. gen. Microbiol.* **2,** 325.
21. Delaporte, B. (1950). *Genetics* **3,** 1.
22. Kellenberger, E. (1953). *Symp. on Bacterial Cytology, VIth Intern. Congr. Microbiol. Rome,* p. 45.
23. Whitfield, J. F. and Murray, R. G. E. (1956). *Can. J. Microbiol.* **2,** 245.
24. Halvorson, H. O. (1965). *Symp. Soc. gen. Microbiol.* **15,** 343.
25. Ohye, D. F. and Murrell, W. G. (1963). *J. Cell Biol.* **14,** 111.
26. Fitz-James, P. C. (1967). *In* "Microbiol Protoplasts, Spheroplasts and L-forms" (L. B. Guze, ed.), 1966 Brook Lodge Conference. Williams and Wilkins, Baltimore, Md, U.S.A.
27. Fitz-James, P. C. (1967). *In* "Protides of the Biological Fluids" (H. Peeters, ed.), XV Ann Colloq. Elsevier, Amsterdam.
28. Robinow, C. F. (1953). *J. Bact.* **66,** 300.
29. Robinow, C. F. (1951). *J. gen. Microbiol.* **5,** 439.
30. Mayall, B. H. and Robinow, C. F. (1957). *J. appl. Bact.* **20,** 333.
31. Young, I. E. and Fitz-James, P. C. (1962). *J. Cell. Biol.* **12,** 115.
32. Warth, A. D., Ohye, D. F. and Murrell, W. G. (1963). *J. Cell. Biol.* **16,** 593.
33. Aronson, A. I. and Fitz-James, P. C. (1966). *Bact. Proc.* 31, and (1968). *J. molec. Biol.* **33,** 199.
34. Fitz-James, P. C. (1965). *In* "Regulations chez les Microorganismes" (M. J-C Senez, ed.), p. 529. Centre Nat. Recher. Sci. Paris.
35. Schaeffer, P., Ionesco, H., Ryter, A. and Balassa, G. (1965). *In* "Regulations chez les Microorganismes" (M. J-C Senez, ed.), p. 353. Centre Nat. Recher. Sci. Paris.
36. Ryter, A. Schaeffer, P. and Ionesco, H. (1966). *Annls Inst. Pasteur, Paris* **110,** 305.

37. Young, I. E. (1964). *J. Bact.* **88**, 242.
38. Remsen, C. C. and Lundgren, D. G. (1965). *J. Bact.* **90**, 1426.
39. Ryter, A., Ionesco, H. and Schaeffer, P. (1961). *C. hebd. séanc Acad. Sci.*, *Paris* **252**, 2675.
40. Fitz-James, P. C. (1962). *Proc. VIII Int. Cong. of Microbiol., Montreal.* A 2.1.
41. Bayen, H., Frehel, C., Ryter, A. and Sebald, M. (1967). *Annls Inst. Pasteur, Paris* **113**, 163.
42. Vinter, V. (1963). *Experimentia* **19**, 307.
43. Berliner, E. (1911). *Z. ges. Getreidew.* **3**, 63.
44. Hannay, C. L. (1953) *Nature, Lond.* **172**, 1004.
45. Fitz-James, P. C. (1962). *Proc. Fifth Int. Cong. of Electronmicroscopy z*, RR 10.
46. Colmer, A. R. (1948). *J. Bact.* **55**, 777.
47. Labaw, L. W. (1964). *J. Ultrastruct Res.* **10**, 66.
48. Fowler, E. H. and Harrison, J. A. (1952). *Bact. Proc.* **1952**, 30.
49. Hannay, C. L. (1961). *J. biophys. biochem. Cytol.* **9**, 285.
50. St. Julian, G., Pridham, T. G. and Hall, H. H. (1963). *J. Insect. Path.* **5**, 440.
51. White, G. F. (1920). *In* "American Foulbrood". U.S. Dept. of Agric. Bull. 800.
52. Hannay, C. L. (1956). *Symp. Soc. gen. Microbiol.* **6**, 318.
53. Hodgkiss, W. and Ordal, Z. J. (1966). *J. Bact.* **91**, 2031.
54. Pope, L., Yalton, D. P. and Rode, L. J. (1967). *J. Bact.* **94**, 1206.
55. Kitahara, K. and Suzuki, J. (1963). *J. gen. appl. Microbiol., Tokyo.* **9**, 59.
56. MacKechnie, I. and Hanson, R. S. (1967). *J. Bact.* **95**, 355.
57. Grelet, N., (1957). *J. appl. Bact.* **20**, 315.
58. Vinter, V. (1967). *Folia microbiol., Praha* **12**, 89.

CHAPTER 3

Physiology and Biochemistry of Sporulation

VLADIMÍR VINTER

Department of General Microbiology, Institute of Microbiology, Czechoslovak Academy of Sciences, Prague, Czechoslovakia

I. Determinative Role of Environment in Sporogenesis

A. Induction of Sporogenesis

SPOROGENESIS in bacteria is a special case of cell adaptation to the periodic accessibility of growth nutrients in the environment. A spore is not a necessary step in the developmental cycle of spore-formers. Each cellular form of this cycle may be induced or maintained under certain conditions of environment. Spore formation normally occurs after the logarithmic growth phase when the generation time increases due to the limitation of nutrients. However, spore formation may be induced also in the primary cell which develops after germination and outgrowth of the spore without intervening cell division (see Section I F). The classical definition by Knaysi[1] that "spores are formed by healthy cells facing starvation" is the best expression of the determinative role of environment in the induction of spore formation. The term, "healthy cells" points out that the conditions of the environment suitable for metabolic activities of the cells involving multiple factors are the necessary prerequisite for spore formation. The term "starvation" includes the deficiency of factor or factors required for optimal growth but not of factor(s) specifically needed for sporogenesis.

When the level of nutrients has decreased and growth ceases, the following changes are induced in sporogenic bacteria. After a typical rearrangement of nuclear material a specialized cell division requiring extremely long period as compared to normal generation time is started by the formation of the forespore septum. This separates the sporal area from the parasporal part of the sporangium (see previous chapter). These changes are part of a series of sequential processes leading to sporogenesis and may eventually result in production of mature spores. Changes in the level of nutrients may affect this relatively long process. In characterizing this process it is wrong to use as the main or only criterion the appearance of the end cellular products such as the sporangium with more or less refractile endospore, or free mature spore. It is more pertinent to study the induction of sporogenesis, i.e. the derepression of the spore genome, detectable for instance by the changes of nuclear material, production of proteinase, turnover of proteins, formation of antibiotic, invagination of cytoplasmic membrane, etc., but these steps do not always lead to the completion of cellular differentiation. The problem then arises in which period of the development the factors of environment, mainly the level of nutrients, can actually still play the determinative role in morphogenesis. In the critical period of forespore septum formation the cell (sporangium) becomes a compartmentalized system with the capacity of developing either into a spore or into a vegetative cell. Depleted medium allows the completion of the development of the spore. Replacement of the depleted medium by fresh medium, before or at the beginning of this critical period, restores the activity of the vegetative genome and growth and division may be resumed in the parasporal compartment of the cell. This phenomenon of "rejuvenation" or "rejuvenescence" was described in some strains of B. cereus.[2, 3, 4, 5] Unfortunately, no data are available on this phenomenon in other species. Complete separation of the sporal part by the growing septum and formation of the forespore obviously prevents exogenous nutrients from having a determinative role in the forespore part. In this period of progressive changes the sporal part is being further developed and the stage is reached of the so-called "irreversible commitment to sporogenesis." The parasporal compartment may retain an ability for autonomous response to changes of nutrient level and this complicates a precise definition of inductory conditions for sporogenesis. The phenomenon of rejuvenation further complicates the picture and emphasizes the need to distinguish between conditions inducing the primary changes preceding forespore formation and changes occurring after forespore formation, i.e. the morphogenetic processes leading to the irreversible alteration of the sporal compartment into mature spore. In the classical studies which introduced the term "irreversible commitment to sporogenesis" into spore literature[6,7,8] the main criterion used was the ability of sporangia to complete spore formation after transfer into fresh medium. No detailed data are available on the fate of vegetative cells and sporangia in various stages of development immediately after transfer to fresh medium. The term "commitment to sporogenesis" is often misused when describing populations which show some very early signs of sporogenesis. It may be relatively simple to characterize the conditions inducing sporogenesis in those batch cultures in which growth stops because of the exhaustion of nutrients. But sporogenesis can also occur in logarithmically growing cultures (batch or continuous). The probability of any one vegetative cell reaching the stage of irreversible commitment to sporulation depends on the growth rate; lower growth rates favour spore formation. A comparison of the morphogenetic stages of sporeformers using appropriate batch and continuous cultivation techniques may contribute to the clarification and definition of both inductory and regulatory mechanisms involved in morphogenesis of sporangium.

Two main external factors are usually considered as likely inducers of sporogenesis: reduction of the level of growth-supporting substrates which repress sporulation, and accumulation of some catabolites. A number of reviews have been written summarizing both these types of control systems.[1, 9, 10, 11, 12, 13, 14, 15, 16] The first comprehensive investigations of factors whose limitation may induce sporogenesis in aerobic and anaerobic sporeformers were done by Manteifel[17, 18] and Grelet.[19] Among these were listed carbon or nitrogen sources, growth factors and inorganic constituents. The effect of these factors on growth could be distinguished from that on sporogenesis. Sporulation occurred in the presence of an excess of every medium constituent which was not the limiting factor under investigation. Differences in induction conditions, as far as components of the medium and oxygen demand are concerned, were found for individual species and even for different strains of the same species. Similarly the optimal conditions for growth and sporulation differed in individual species. Many sporulation enzymes are under glucose catabolite repression and the exhaustion of glucose was a necessary prerequisite for these enzymes to be derepressed, e.g. in *B. cereus*. In some other bacilli, under certain conditions, sporogenesis may occur even in the presence of glucose. Clostridia may sporulate after completion of growth in relatively high concentrations of glucose. However, the critical limiting factor which may determine sporogenesis is difficult to detect in the very complex media which are often used for the cultivation of clostridia. Some examples of the changes in the medium composition and of the conditions optimal for sporogenesis will be mentioned in Sections I B, C, and II A, B, of this chapter.

There may be external factors already present in the sporulation medium or released from the cells during growth which induce or allow spore formation. Two sporogenic peptide components of trypticase are specifically required for sporogenesis of *Cl. roseum*.[20] A low-molecular ninhydrin-negative sporulation factor was isolated and purified from *B. cereus T* and *B. subtilis* produced in the presporulation period. The addition of this factor to vegetative cultures suspended in media not supporting growth may prevent lysis and induce sporogenesis.[21, 22, 23] An unknown factor was excreted into the medium in exponentially growing *Cl. butyricum* which diminished the growth rate and induced sporulation.[24] The decrease of growth rate may be considered as one factor inducing sporogenesis. A steady formation of sporangia was reported in *B. megaterium* or *B. subtilis*[25, 26, 27, 28] growing exponentially in minimal media. In these experiments the ratio: spores/vegetative growing cells depended on the composition of the medium, mainly on the nature of carbon and especially nitrogen sources. The probability of cells of *B. subtilis* in a growth-supporting medium becoming committed to sporulation was explained by Schaeffer et al.[29] as a reflection of the intracellular concentration of some nitrogen-containing catabolite(s) which may act as repressors of at least one sporulation-specific enzyme. Growth and sporulation of *B. subtilis* is known to occur better in a medium rich in amino acids.[30] The regulatory role of glucose in the induction of sporogenesis, i.e. derepression of spore genes after glucose depletion, pH changes and induction of the tricarboxylic acid cycle (see Section II A), is similar to that exerted on other aerobic sporeformers.

To sum up, competent non-growing cells may sporulate with or without a previous history of cell divisions. In cultures growing on minimal media sporulation occurs in the relatively small fraction of committed cells. Lastly, in cultures growing on complete media almost all the cells may become competent to sporulate as the generation time increases. Decrease of the level of one or more factors in the medium required for growth induces sporogenesis sometimes regardless of the presence of excess of other nutrients.

B. Sporogenesis in Batch and Continuous Cultures

Most of the data about the induction of sporogenesis and the sequential differentiation of sporangia have been obtained from classical batch cultivations. Under these conditions the composition of the medium is being gradually changed by the culture itself. These changes and the response of the culture to environmental conditions will be summarized in Sections I C and II A of this chapter. As was mentioned above, the replacement of depleted medium by fresh medium may result in resumption of proliferation and further development up to the early stages of sporogenesis. Repeated transfer of growing culture of sporeformers into fresh medium completely prevented sporogenesis and showed the importance of the nutrient level in determination of sporogenesis.[31] A few repeated transfers of heavy inocula of logarithmically growing cells into fresh medium followed by the completion of growth after the last transfer was successfully used to achieve a high degree of synchrony of sporogenesis of bacilli and clostridia.[32] Continuous cultivation techniques gave similar results. The cultivation of sporeformers in a chemostat showed that under appropriate conditions the culture can be maintained indefinitely in the phase of vegetative growth without spore formation.[33, 34] A multi-stage continuous cultivation method was used for the characterization of growth and sporogenesis of B. pumilus in casamino acids medium.[35] Under suitable dilution rate a constant fraction of sporangia was formed. The ratio of vegetative cells to sporangia decreased in the second and third vessel of the cultivation system. The cells grew in chains; the individual cells within the chain showed high variability in their ability to sporulate. The equilibrium between the number of vegetative, granular and sporulating cells was extremely sensitive to the dilution rate. In this species the presence of glucose in the medium did not prevent sporogenesis.

A simple equipment was developed for cultivation of clostridia in collodion sacs allowing the continuous exchange of nutrients.[36, 37] Based on this principle, microchambers were designed which enabled the continuous microscopic examination of sporulating cells. The microchamber was used to study the effect of the medium on different stages of growth and sporogenesis.[38] The results showed that sporangia with forespores either attain the stage of mature spores and thus terminate the development cycle or die but cannot return to vegetative growth.

Sporogenesis of B. subtilis and B. cereus T was studied by continuous culture technique.[27] Synchrony was increased in both these cultures, but B. cereus T could be maintained as a steady-state population of vegetative cells for a longer period than B. subtilis. The latter was able to sporulate after prolonged incubation even in the presence of glucose or glutamate. The inability or ability of various species to form spores in the logarithmic phase were related to hypothetical differences in the rate of synthesis of the endogenous "sporulation factor".[21,22,23] Some differences may apparently exist between individual strains of the same species. For example, another strain of B. subtilis in continuous culture could be maintained indefinitely in the phase of vegetative growth.[37] Continuous system of cultivation can be successfully used for the study of the sporulation induction in relation to generation time. In *B. megaterium* sporogenesis in a chemostat was induced by diminishing the growth rate from 0·7 to 0·5 divisions per hour.[39]

Clostridia often require very complex media, and as a special case of continuous process of cultivation a biphasic system was successfully used for production of concentrated suspensions of relatively clean spores.[40] In its simplest form the biphasic system consists of a layer of nutrient agar overlayered with a liquid medium; the liquid layer is continuously enriched by diffusion of fresh nutrients from the agar. *B. stearothermophilus* spores could not be produced in a two-stage continuous cultivation system due to extraordinary lysis of the culture during cultivation.[41]

Application of continuous cultivation methods to sporology offers many further possibilities. Continuous systems may be helpful in the further study of the relationship between growth rate and sporulation ability, of the pre- and post-commitment period of cell development and of rejuvenation phenomena.

C. Factors Influencing the Quality and Quantity of Spores

Environmental factors determine to a high degree the ability of cells to form spores, the rate of sporogenesis, the yield of spores and their quality. The conditions required for growth and spore formation are different. The medium is being changed during growth and many influencing factors are interdependent. The altered quality of spores may be detected by their resistance to deleterious factors, ability to remain in the dormant state, activation requirements and germination response. However, the quality of the spores is not easy to determine by these tests because individual spores within a spore population may show considerable heterogeneity. Much information has been accumulated in the literature on the effect of environmental conditions on growth, sporogenesis and spore yields. They are summarized in a number of comprehensive review articles.[1, 9,11,12,13,19,42–45]

The most important environmental factors which determine the ability of growing cells to sporulate, or the rate and completion of spore formation and the quality of the resulting spores, are the level and nature of nutrients, mineral composition of the medium, pH, temperature and aeration.

1. The level and nature of nutrients

High concentrations of available nutrients may be often suitable for optimal growth but may retard the induction of sporogenesis and reduce the spore yield. This effect is often interdependent with pH changes occurring in the medium and oxygen requirements. A similar interrelationship was observed between the level and nature of carbon and nitrogen sources (for details see [19,13]). A close interdependence of glucose concentration, pH changes and aeration rate was a main determining factor in sporulation ability of B. licheniformis.[46] Some carbon sources leading to overproduction of poly-β-hydroxybutyric acid in B. megaterium markedly decreased the yield of sporangia and spores unless aeration was sufficiently increased.[47] A few examples of the interrelation between nutrient levels, pH changes and biochemical pathways determining the sporulation ability of bacilli will be discussed later in this chapter (Section II A, B). The replacement of the complex media used for growth and sporogenesis of aerobic bacilli[48,49, 50,51] by maximally defined media [52-60] simplified the study of the interdependence of individual factors in determining sporogenesis and enabled more precise characterization of sporulation requirements. In some of these media suitable concentrations of carbon and nitrogen sources optimal for growth and/or spore formation of B. subtilis[59] or B. cereus [58] were defined. In another study with another strain of B. subtilis increase in the concentration of nitrogen source (glutamic acid or alanine) resulted in enhancement of total viable count without any effect on the yield of sporangia.[30] The growth and production of spores in Cl. butyricum were not influenced by an increase of glucose or complex nitrogen source above a certain level and different strains of this species could sporulate in the relatively rich medium.[61,62] In continuous culture another strain of Cl. butyricum was not able to form spores in rich media. Sporogenesis was dependent on the exhaustion of nitrogen sources and vitamins.[36, 38] Growth and sporogenesis of anaerobic sporeformers usually requires complex media which are neither defined nor constant in composition. The media and the cultivation techniques found to be optimal for given species and strains has revealed further interdependent factors.[63-69] One of the methods by which the interrelationship between environmental conditions and triggering of sporogenesis could be studied in Cl. botulinum and Cl. perfringens was based on cultivation in dialysis sacs. Under these cultivation conditions the yield of spores was increased but their resistance to γ-radiations was diminished.[70]

Various components of the sporulation medium are capable of greatly modifying individual stages of culture development and spore properties. In a synthetic medium arginine has been reported to play an important role in sporulation of Cl. botulinum.[71] In complex media high concentration of glucose and L-cysteine exerted an inhibitory effect on sporogenesis of Cl. botulinum and markedly altered the response to heat activation of the spores.[72] Different monosaccharides may exert stimulatory or repressory effects on sporogenesis of Cl. thermosaccharolyticum.[73]

2. Mineral composition of the medium

Different cations and anions have been reported to influence growth, sporogenesis and quality of spores in bacilli and clostridia (see reviewing articles [13,45,74]). Manganese is specifically required in the presporulation and sporulation period of many bacilli[50, 59, 75-78] and possibly plays a role in the activation of some enzymes involved in spore formation. In *B. megaterium* both viable count and spore yield are markedly increased in glucose-salts medium by the addition of 5×10^{-6}M manganese at zero time.[79] A few other metals are required for sporogenesis, e.g. zinc for *B. coagulans var thermoacidurans*[80] and for *B. cereus*,[55, 81] trace amounts of copper for some variants of *B. cereus* and *B. megaterium*[82, 83] and molybdenum for *B. megaterium*.[83] Calcium is the most important metal determining the quality of the spores.[59, 74, 77, 84-90] Some other ions seem to be antagonistic to calcium in sporulation medium, e.g. sodium[91] or phosphates;[90,92,93] the effect of partial or complete calcium replacement by other metals on sporogenesis, heat resistance and other properties of spores was also studied in various species of bacilli.[54,94-97] Different levels of calcium in sporulation medium may influence also the germination properties of aerobic spores.[59,90]

3. pH

The tolerance of growing cells to pH changes is usually higher than that of sporulating cells. Some of the data on pH changes and pH optima for sporogenesis have been reviewed.[1, 9, 13, 44, 45] Although the pH optimum for sporogenesis is mostly close to neutral, the metabolic activity of growing cultures bring about a more or less pronounced drop of pH in unbuffered media. The accumulation of pyruvic, acetic and/or other acids in the medium reported for bacilli[32,46,55,98] and clostridia[99] is followed by their utilization and pH rise (for further details see Section II A of this chapter).

4. Temperature

The temperature of cultivation affects growth, the yield of spores and their properties. Some older data on these effects may be found in review articles.[13, 45] Elevated temperature during growth and sporulation resulted in increased heat resistance of spores of *B. subtilis* and *B. coagulans*,[95] but the reverse was found in *B. stearothermophilus*.[100] The optimal sporulation temperature allowing maximal thermoresistance of *B. cereus* T spores and optimal spore yield was 30°C. Shifts in sporulation temperature to either side lowered both heat resistance and the number of spores formed.[96] A few mutants prepared from *B. cereus* T were able to sporulate at 28° but unable to form spores at 37°C.[55] The pattern of zinc, potassium, sodium and calcium uptake, but not of iron and manganese uptake, were affected in temperature-sensitive asporogenic mutants.[81,89,101]

5. Aeration

An increased rate of aeration was often reported to be required for the presporulation period and for optimal sporogenesis of aerobic bacilli.[32,102-104]

In some species this higher oxygen demand coincides with the drop of pH and subsequent development of terminal respiratory system (see Section II A, B). A higher aeration rate is needed for the utilization of the intracellular reserve of poly-β-hydroxybutyric acid (PHB) accumulated before sporogenesis. In *B. megaterium*, the degradation of PHB was directly related to the sporulation ability.[47] In *B. stearothermophilus* grown from spores in complex media a higher rate of aeration promoted growth and spore formation. When vegetative cells were used as inoculum, aeration at higher temperature allowed vigorous growth but inhibited spore formation.[51] Although oxygen inhibits growth and sporulation of anaerobic sporeformers its presence at the end of sporogenesis has no effect on the cells and, moreover, promotes the liberation of spores.[64,105]

6. Miscellaneous factors

A few miscellaneous factors affecting the rate of sporogenesis in bacilli and clostridia and the resistance of spores were reported in complex or synthetic media, e.g. fatty acids,[8, 84] amino acids[59, 90, 99, 106] or growth factors.[44, 107, 108] A few so far unknown factors of biological origin may enhance sporulation in aerobic bacilli.[109, 110] *B. popilliae* and *B. lentimorbus* are very special in that until recently they only sporulated in the haemolymph of specific larvae. Now, however, complex sporulation media and suitable techniques have been described enabling sporogenesis in a fraction of the culture (for details see Chapter 13).

We may conclude that numerous often interdependent environmental factors influence the phenotypic expression of sporulation genes affecting the number and the quality of the spores produced. These factors are more or less defined, but the actual mechanism of their action is mostly unknown. The differences among individual species should be respected in order to avoid misleading generalizations.

D. ENDOTROPHIC SPOROGENESIS

The term "endotrophic sporogenesis" has been used to describe sporulation occurring under extreme conditions of starvation, after the vegetative cells were transferred to water or saline. Endotrophic sporogenesis was characterized as "occurring independently of exogenous nutrition and supported exclusively by the pre-existing makeup of the vegetative cell".[111] The ability of vegetative cells of sporeformers to sporulate in water was first reported by Buchner in 1890[31] and Schreiber in 1896.[112] The technique of replacement of complex media by water or by very simple media was later used by many authors for the study of sporogenesis. The method has the following advantages. Firstly, the medium for sporogenesis is maximally simplified and, secondly, it is possible to distinguish which

F

requirements for spore formation may be covered by the pre-existing material of the cell and which additional components of the medium determine the quality of spores produced.

A sudden limitation of exogenous nutrients in the different phases of the developmental cycle may enable the detection of the basic regulatory abilities of cells at particular stages of development. The culture may respond by two alternatives to such treatment: sporogenesis and/or lysis. The autolysis of a minor or a major part of the population was reported in some species.[5, 12, 21, 44, 86, 94, 113-115] The yield of spores after replacement depends on the species or strain, the age of vegetative cells at the moment of transfer and the composition of the replacement medium. In some cases sporulation takes place in the absence of lysis[8, 54, 111, 116-118] and, moreover, a few methods were developed to prevent lysis.[114] The hypothesis that sporogenesis is strictly an endogenous process[8] may be valid only if there is no contamination of the replacement medium by lytic products during the "endotrophic sporulation". This hypothesis assumes that the synthesis of proteins and nucleic acids during sporogenesis occurs *de novo* from low-molecular precursors originated by previous degradation of pre-existing macro-molecules. Many findings confirm that during sporogenesis in the absence as well as presence of exogenous nutrients, or low-molecular precursors a significant degradation of pre-existing macromolecules, enrichment of the intracellular pool with low-molecular precursors and fast resynthesis of macromolecules takes place.[119-125] On the other hand, some typical spore constituents can be hardly derived only from pre-existing intracellular reserves of progenitor cells. Calcium is a classical example. In its absence defective relatively heat-sensitive and DPA-deficient spores are produced.[86, 94, 117, 126] The high content of cystine found in spores of aerobic-organisms[127,128] can be at least partially derived from pre-existing methionine.[129,130] There may be, however, some other constituents required for sporogenesis in higher quantity. For instance, among amino acids requiring mutants of *B. subtilis* the lysine-dependent mutant did not sporulate in minimal media enriched by different amounts of lysine, unless the concentration of this amino acid exceeded that sufficient for complete growth.[131] An increased requirement for lysine during early stage of sporogenesis was observed also in *B. cereus*.[132] Glutamic acid was required for sporogenesis by endotrophically developing cells of *B. cereus*, which were transferred to salts-replacement medium before nuclear rearrangement. Cells with forespores were able to sporulate in its absence.[133] The deficiency of some pre-existing components may be brought about by lower rate of presporulation degradation of macromolecules or by the degradation of precursors produced due to changes of enzymatic pathways. A few examples of such changes in amino acid metabolism were described in *B. licheniformis*.[124, 134, 135]

There are almost no data on the endotrophic sporogenesis of clostridia. However, experiments with *Cl. botulinum* indicate that the cells of this anaerobe, although committed to sporulation, are not able to complete this process endotrophically. A few amino acids are needed for forespore maturation.[99]

The method of replacement of nutrient medium by water or saline was also used for the study of resistance development during sporogenesis,[136-138] of the properties of spores produced endotrophically in the presence of different cations,[54, 86, 94, 116, 117] of sensitivity to antibiotics[139] and of the effect of amino acids analogues.[140]

The techniques and findings obtained by this method have been reviewed.[12-15, 42, 114]

E. DEGREE OF SYNCHRONY OF SPOROGENESIS

Under conventional cultivation techniques the microbial population may exist in all phases of development. A number of methods have been developed for the synchronization of nuclear and cellular division of cultures of micro-organisms using single or multiple temperature shifts, medium changes, mechanical selection, etc.[141-143] Study of these cultures may enable one to deduce the sequence of events occurring in a single cell undergoing division. Normal division requires a relatively short generation time but sporogenesis, as a special type of cell division, is characterized by an extremely long period. Cyto-differentiation occurs involving nuclear changes, compartmentalization, development of the sporal part in the sporangium and subsequent destruction of the parasporal part. A temporal scale often used for classification of individual morphological stages of sporogenesis[28] starts at the end of growth ($= t_o$, see page 174). However, the rate of sporogenesis proper may vary in individual species or strains, especially under different cultivation conditions. Additional characterization of stages (I–VII) has been developed[2, 15] which is based on morphological stages defined by electron microscopy of thin sections.

In suitable media permitting the fast growth and simultaneous production of forespores and refractile spores in the major part of the culture, the time sequence of morphological, biochemical and physiological changes of cells could be clearly distinguished.[14, 28, 128, 132, 144, 145] In these cases the relatively good synchrony of the culture permitted the interfering effect of various antibiotics[2,146-152] and of different analogues[153,154] to be observed on the sequential synthetic activities in presporulation and sporulation period. The main event used as the criterion for characterization of the degree of synchrony was the appearance of forespores. Septation can be detected by negative staining with phosphotungstate.[155] In addition to the more laborious and time-consuming method of electron microscopy a fast staining procedure was developed for detection of septa. Mounting the live cells in dilute crystal violet (0·03% W/V) results in an immediate uptake of stain by the invaginated forespore septum.[156]

In bacilli the degree of synchrony during sporogenesis may be relatively high. A significant heterogeneity of clostridial cells in all phases of the development stimulated the elaboration of the method for synchronization of the culture in presporulation period. This method is based on a few sequential transfers of actively growing culture to fresh medium. For instance after subculturing three times, a high degree of sporulation synchrony was achieved with *Cl. roseum*. This allowed the study of individual steps of sporogenesis: formation of refractile spores, DPA and acquisition of heat resistance.[32] The same technique of cultivation was later used also for synchronization of sporogenesis in bacilli,[21, 32, 133] of a putrefactive anaerobe[64] and for detailed study of development of heat resistance of sporangia of *Cl. roseum* in relation to the intracellular content of calcium and DPA.[157] Sequential transfer of heavy inocula of *Cl. thermosaccharolyticum* in relatively simple media highly increased the degree of synchrony of sporulating cells and allowed the study of the effect of different monosaccharides on sporulation, accumulation and metabolism of labelled acetate.[73]

Another method based on filtration procedure was developed for synchronization of growth and sporulation of *B. megaterium* grown in synthetic medium.[158, 159] The high degree of synchrony achieved by this method was evident by the stepwise pattern of growth, by the doubling of cell numbers at each division and the high division index. Subsequently rapid sporulation occurred due to the homogeneity of the cell types in the filtered cultures and contrasted with asynchronous cultures. Division and sporulation patterns of filtered population of *B. megaterium* grown in synthetic medium are illustrated in Fig. 1. Designation of cell types in this figure: forespores and sporangia, corresponds approximately to the stages II–III and IV, respectively. The changes in activity of a few enzymes could be distinguished in these synchronized cells during division(s) preceding sporulation and in individual stages of sporogenesis.[160]

The synchronization of the last division(s) and sporogenesis may help in elucidating the sequential processes accompanying spore formation and especially the key intracellular events occurring at the end of growth and sporogenesis (but note, growth and sporogenesis are not incompatible phenomena).

F. MICROCYCLE SPOROGENESIS

In the presence of utilizable nutrients sporogenesis as a specialized cell division takes place only after a few non-specialized, normal divisions.

Microbial populations can be rather heterogeneous at the beginning of sporulation. In order to simplify the whole development cycle a simple method was developed by Vinter and Slepecky[115] for the induction of sporogenesis in the primary cell, i.e. in the cell formed immediately after

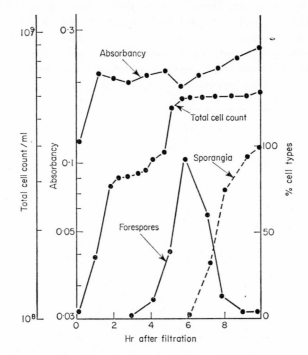

FIG. 1. Growth and sporulation of a filtered culture of *Bacillus megaterium* grown in a sucrose-salts-glutamic acid medium. (Reproduced with permission from Imanaka *et al.*[159])

outgrowth of the spore. Replacement of a complex bactopeptone medium by a diluted one during swelling of spores of *B. cereus* or *B. megaterium* prevented cell division. The primary cells were capable of resporulating without intervening cell division. This shortening of the development to the simplest cycle: spore → vegetative cell → sporangium → free spore was called "microcycle sporogenesis".[115] In *B. cereus* the envelopes of the original spore (coat + exosporium) remain attached to the new vegetative cell and the new sporangium. Microcycle sporogenesis is diagrammatically outlined in Chapter 2 (p. 41). The ability of outgrowing spores to change to new sporangia depended on the stage of development of cells at the moment of replacement. The primary cells transferred during the first DNA replication gave the highest yield of new sporangia. The cells transferred before this stage were not able to complete the elongation and resporulation. "Shift-down" treatment of the cells during first division(s) led to the lysis of the major portion of the culture.[5, 115] During microcycle sporogenesis at least one replication of DNA is completed in *B. cereus*.[5]

Microcycle sporogenesis could be induced in *B. megaterium* in chemically defined minimal media without dilution or replacement.[161] The DNA content of sporangia was found to be tripled. Some data on the formation of the second-stage spores in *B. megaterium* are presented in Fig. 2. In this species acetate and a small amount of a tricarboxylic acid cycle intermediate were the minimum organic requirements for microcycle sporogenesis. The ability of the primary cell to oxidize acetate via tricarboxylic acid cycle develops only during outgrowth. A functioning tricarboxylic acid cycle

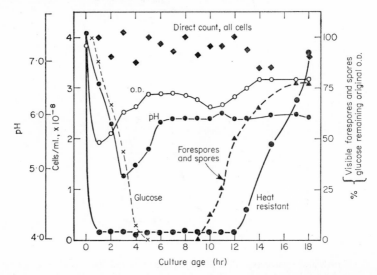

FIG. 2. Events occurring during microcycle sporogenesis of *Bacillus megaterium* QM B1551 in glucose-ammonia medium. (Reproduced with permission from Holmes and Levinson.[161])

seems to be a prerequisite for microcycle sporulation which is similar to normal sporogenesis of bacilli after the multiplication interphase (see Section II A of this chapter). The regulatory role of other nutrients was observed also in microcycle sporogenesis. The addition of glucose suppressed resporulation of *B. megaterium* without promoting cell division which could be resumed only after addition of complex media. A temporary shift-down treatment of primary cells of *B. megaterium* cultured in the glucose-ammonia medium also prevented resporulation.[161] The sequence of morphological events during microcycle sporogenesis of *B. megaterium*[162] was very similar to that observed in macrocycle sporogenesis of bacilli.

A similar system allowing the microcycle sporogenesis to occur without replacement of medium was developed also for *B. cereus*.[163] The critical factors determining the new morphogenetic cycle without intervening cell division were the concentrations of yeast extract and phosphate.

Microcycle sporogenesis may offer a very good and simplified system for the study of the direct and relatively synchronous transition of the primary cell into new sporangium.

II. Changes in Biochemical Pathways During Growth and Sporulation

A. OXIDATIVE METABOLISM

One of the most important determining factors of sporogenesis is the level of available carbohydrates, mainly glucose. Investigations of the metabolic and regulatory role of glucose before and during sporogenesis[8, 98, 106, 164] stimulated the interest of sporologists in quantitative modifications in the pathway of glucose catabolism in relation to the separate stages of development of aerobic sporeformers. Vegetative cells contain enzymes of the glycolytic pathway (Embden-Meyerhof, EM) as well as enzymes of the hexosemonophosphate pathway (HMP).[165-167] The utilization of glucose by these simultaneously operating pathways throughout the whole cycle of development of *B. cereus* T is summarized in Fig. 3. These data from *in*

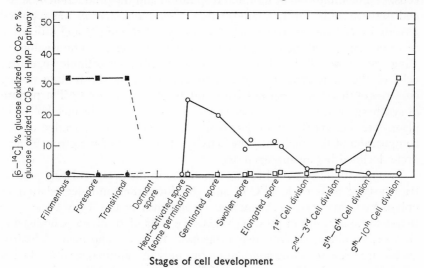

FIG. 3. The relation between the % of glucose oxidized by the hexose monophosphate pathway or the % of 6-^{14}C glucose oxidized to CO_2, a measure of the terminal respiratory pathway, and the stage of development of *B. cereus* T. ● % Glucose via the HMP-cells grown in modified G medium; ○ % Glucose via the HMP-cells grown in trypticase-soy medium; ■% 6-[14]C Glucose oxidized to CO_2 cells grown in modified G-medium; □% 6-[14]C Glucose oxidized to CO_2 cells grown in trypticase-soy medium. (Reproduced with permission from Blumenthal.[167])

vivo studies are based on the yield of ^{14}C from $[1-^{14}C]$ glucose and $[6-^{14}C]$ glucose. The amount of glucose metabolized by the HMP reached 10–20% in germinated and outgrowing spores[167] arrested at defined stages of development.[168] Vegetative cells of *B. cereus* operate almost exclusively via the EM pathway. The EM pathway is also the most important one in stationary phase cells of *B. subtilis*, only a minor part of the available glucose being catabolized by the HMP pathway.[169] Studies of glucose catabolism involving the EM pathway in spores have yielded some contradictory results. Presence of traces of enzymes involved in the glycolytic pathway of *B. cereus* T was reported by Goldman and Blumenthal,[170] but Doi *et al.* could not confirm this.[171] According to Doi *et al.* glucose is initially oxidized to gluconate by a soluble $NADH_2$-linked glucose dehydrogenase and converted to 2-ketogluconate. This is phosphorylated to 2-keto-6-phosphogluconate, which in turn is reduced in part to 6-phosphogluconate by a $NADH_2$-requiring 2-keto-6-phosphogluconate reductase and in part to pyruvate by an as yet unidentified pathway.[171–174] A particulate system was reported for terminal oxidation of pyruvate, acetate and of some tricarboxylic acid cycle intermediates in spore extracts of *B. cereus* T and for other strains of this species.[175] However, very low terminal respiratory activity was observed from germination up to first division(s) of *B. cereus* cells.[176] The inhibition of terminal respiration and the production of acetic and pyruvic acids during late logarithmic phase of *B. cereus* was reported already by Nakata and Halvorson.[98] Subsequent rise of pH and utilization of acetate suggested the induction of a glyoxylate cycle.[177] The predominating importance of the glyoxylic acid cycle in late logarithmic vegetative cells of the same strain[178, 179] was substantiated by the finding of key enzymes of this cycle, isocitrate lyase and malate synthetase.[180] The pattern of synthesis of these enzymes as well as the pathway for the metabolism of glyoxylate, i.e. via the glyoxylate or the glycerate cycle, depended on the composition of the medium. The activity of enzymes of the glyoxylic acid cycle declined before sporogenesis.

Two metabolic systems become functional in the presporulation period: tricarboxylic acid cycle (TCA) and the formation and degradation of poly-β-hydroxybutyrate. A number of enzymes of the TCA cycle were found to be limiting in growing vegetative cells of *B. cereus*. Condensing enzyme, aconitase, succinic dehydrogenase, fumarase and malic dehydrogenase are present during sporulation but only in minute quantities during vegetative growth. The formation of these enzymes and the oxidation of glucose C-6 to CO_2 were detected only immediately before or during sporogenesis.[170, 176, 178, 179] The importance of Krebs cycle enzymes for sporogenesis was confirmed also for other species of bacilli.[181–184] The involvement of the tricarboxylic acid cycle in sporogenesis was first suggested by Martin and Foster[185] and other investigations have emphasized the role of this cycle during the presporulation and sporulation periods. Genetic block of aconitase formation or presence of repressors of its synthesis, e.g. catabolites of glucose, glycerol or glutamate, resulted in inhibition of sporogenesis.[183] A sporogenic strain (Sp$^+$) of *B. subtilis* showed high NADH oxidase, NADH dehydrogenase, diaphorase and cytochrome *c*-reductase activity during sporogenesis, while these enzymes were low in activity or inactive in asporogenous mutants.[186] The activity of particulate NADH oxidase increased during sporogenesis of a Sp$^+$ strain, being low in asporogenic mutants.[187]

The accumulation of acetate and pyruvate in the medium during growth was observed in a number of *Bacillus* species. In unbuffered media it resulted in a significant drop of pH,[30, 46, 55, 98, 178, 179, 188, 189] followed by the rise of pH before and during sporulation. In *B. cereus* T grown in glucose-yeast extract-minerals medium the rise of pH and sporulation were inhibited by α-picolinic acid. This inhibition could be reversed by increased amounts of yeast extract,[190] by a few amino acids and by some organic acids, some of them being intermediates of the TCA cycle.[191, 192] α-Picolinic acid was reported to inhibit the synthesis of aconitase.[179] Fluoroacetic acid and several ethyl esters of organic acids reversing the inhibitory effect of α-picolinic acid interfered also with the formation and function of acetate utilizing system and with sporogenesis and/or synthesis of dipicolinic acid.[192] The effect of these inhibitors depended on the time of addition. The inhibitory effect of ethyl malonate on sporulation and antibiotic production was shown also in *B. licheniformis*.[46] *B. cereus* T can hydrolyse ethyl malonate and ethyl succinate to their respective monoethyl esters. They possess therefore no specific inhibitory capabilities against sporulation.[193]

Growth of *B. cereus* T in glucose-containing but well-buffered media resulted in accumulation of much higher amount of acetic and lactic acid than in unbuffered media and no significant drop of pH was observed. Production of poly-β-hydroxybutyric acid was much lower at pH 7·0 and above. As no differences in sporulating ability of culture and in heat resistance of the resulting spores could be detected, the high polymer content of cells is not imperative for the completion of sporulation.[194] In fact, in *B. megaterium* the yield of spores did not depend on the amount of PHB but on the rate of its utilization. If the bacteria were too rich in polymer or contained normal amounts, but conditions of cultivation were unfavourable for its utilization, the production of spores was limited.[47] PHB occurring in sporeformers in the form of inclusions is usually completely utilized during sporogenesis,[47, 119, 195] and obviously serves as an endogenous carbon and energy source to fuel the sporulation process. Occasionally it may be trapped in the spore integuments.[196, 197] Biosynthesis of PHB depends on adaptive formation of acetoacetyl coenzyme A reductase. Inhibition studies confirm the involvement of tricarboxylic acid cycle in polymer synthesis. An outline of some metabolic transformations which occur during the growth and sporogenesis of *B. cereus* T is presented in Fig. 4. The areas which are affected by pH are designed by dashed arrows. The major part of radioactive carbon from $[2-^{14}C]$ acetate is incorporated into PHB in *B. cereus* sporulating in chemically defined medium. After polymer degradation its breakdown products appear mainly in spore constituents.[198] The utilization of intracellular PHB, as well as of acidic products in the medium during sporogenesis of bacilli, can be markedly enhanced by increased

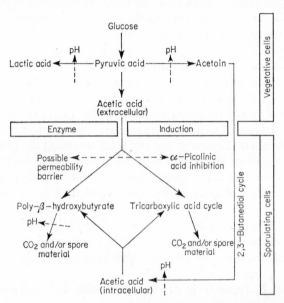

FIG. 4. A schematic summary of some metabolic transitions occurring during the growth and sporulation of *B. cereus* T and the areas which are affected by pH. (Reproduced with permission from Kominek and Halvorson.[228])

oxygen supply,[47, 199] and higher requirement of oxygen during some early stage of sporogenesis is often reported in some species and strains of bacilli.[8, 32, 102, 103, 104, 195, 200] Cells grown under nutritional conditions not permitting a higher accumulation of organic acids do not require higher oxygen supply for sporogenesis.[30, 46] On the other hand, in media rich in glucose the increased aeration markedly enhanced the production of organic acids and prevented sporogenesis.[46] Acetate itself does not induce the formation of tricarboxylic acid cycle enzymes. Exhaustion of glucose is a necessary prerequisite for their derepression.[178, 179, 183]

The metabolic processes preceding sporogenesis are naturally interdependent. For instance, glutamic acid plays a very important role. Simultaneous presence of glutamic acid and glucose represses aconitase formation in *B. subtilis*.[183] Glutamic acid was found to affect the oxidative pathway of glucose in the same species,[165] to serve as a precursor for DPA synthesis[185] as a carbon and nitrogen source for the synthesis of other amino acids during sporogenesis.[124, 201] Numerous other data confirm the importance of this amino acid in metabolic activities of the cell before and during sporogenesis.[133, 135, 184, 202, 203]

The data presented in this part emphasize the significance of the quantitative and qualitative changes in oxidative metabolism during the whole development cycle of bacilli. Sporogenesis is directly dependent on the formation of a terminal oxidative system, some metabolites playing a key role in the ability of cells to sporulate.

B. ENZYME PATTERN OF SPORULATING CELLS

Sporogenesis in bacteria represents the formation of a cell with new qualities and is associated with quantitative and qualitative changes of enzyme pattern. The process involves an overall redistribution of intracellular precursors and macromolecules. An extensive degradation of pre-existing proteins[111, 123, 125, 204, 205] and of all three major classes of RNA[122, 125, 204, 206] occurs before and during sporogenesis, followed by the utilization of precursors for the synthesis of new macromolecules. An interdependence was suggested by Balassa[207, 208] between amino acid supply derived from protein degradation on the one hand and stimulation of new mRNA and protein synthesis on the other. The sensitivity of the early stages of sporogenesis to analogues of nucleic acid bases,[111, 149, 153, 206] amino acids[56, 120, 209] and to antibiotics interfering with RNA and protein synthesis[125, 139, 147, 149, 151, 206, 210] clearly show the dependence of this process on *de novo* syntheses. Turnover studies showed that 80–100% of spore proteins were formed *de novo* during sporulation.[14] It was likely that at least one hundred structural and regulatory genes participated.[14, 121]

Sporogenesis, being intimately related to metabolic phenomena, is an excellent system for studying the sequence of syntheses and activities of enzymes involved in the formation of new molecular species and cell structures. In bacilli and clostridia a large number of spore-specific antigens were detected in spores[212–218] or sporangium.[219–222] Antigenic differences were found between sporogenous and asporogenous strains.[223] Some of the low- and high-molecular species are sporangium- or spore-specific and during the further development of the sporangium they may be released, metabolized or integrated into spore structures. The timing of the production of these components, e.g. sporogenic factor[21, 23] proteases,[224–226] antibiotic,[227] poly-β-hydroxybutyric acid,[47, 194, 228] N-succinyl-L-glutamic acid,[229, 230] muropeptides, cystine-rich structures, dipicolinic acid, etc., may more or less precisely define the individual stages of sporogenesis. A great number of pre-existing, *ad hoc* formed or activated enzymes may be involved in the synthesis of these sporangium or spore constituents. Much information is available on enzymic constitution of vegetative cells and spores or on changes in enzyme syntheses, activities and properties during sporogenesis.[14, 15, 45, 231, 232]

The following properties, often overlapping one another, can be used for the characterization of changing enzyme patterns before and during sporogenesis:

(a) Time sequence of formation or increase in activity of enzymes in relation to known biochemical, physiological and morphological changes occurring after and up to the completion of sporogenesis. Many of these data were summarized by Murrell,[15] the appearance or changes in activity of individual enzymes being related to individual stages of sporogenesis (I–VI). Halvorson *et al.* reviewed the sequential expression of biochemical events during sporogenesis and the possible mechanisms for the regulation of the irreversible sequence leading to spore formation.[14, 232] The elaboration of methods for synchronization of spore formation in the whole population[32, 158, 159] enabled the more precise determination of changing enzyme pattern during this process. Periodic changes of activity of different enzymes were reported during the last division(s) and sporogenesis of synchronously grown *B. megaterium*.[160]

(b) The way in which individual enzyme systems participate in spore formation. Some of the enzymes involved in sporogenesis are the necessary prerequisite for this process, providing the cell with sources of energy and precursors of macromolecules. Among these systems the enzymes of tricarboxylic acid and glyoxylic acid cycle,[98, 167, 178–183] protease[225] and enzymes involved in the metabolism of glutamate[133, 135, 184, 203] and poly-β-hydroxybutyric acid[228] are of prime importance (see Section IIA of this chapter, and Chapter 6). Much less is known, however, of the enzymes directly involved in the formation of spore constituents e.g. keratin-like polymers, cortical muropeptides and dipicolinic acid. Inhibition studies showed that the enzymes involved in the synthesis or polymerization of spore muropeptides in *B. cereus* are formed *de novo* in the forespore stage.[147, 233] The decrease of diaminopimelic acid (DAP) decarboxylase activity during sporogenesis was observed in *B. sphaericus*[234] and *B. cereus*.[132] In the former this decrease coincided with the appearance of DAP-containing material which was absent in vegetative cells. A particular role in the final stages of sporogenesis is played by lytic enzymes which dissolve the sporangial wall thus liberating the spores from the mother cell.[235, 236]

(c) Synthesis of spore-specific enzymes detectable in sporangia or spores but not in growing vegetative cells. In *B. cereus* T a heat-resistant glucose dehydrogenase is produced in the early stages of sporogenesis.[237] In another strain of *B. cereus*, ribosidase is formed during the later stage of sporogenesis. Its activity is very high in preparations from disintegrated free spores, being absent in vegetative cells or early in sporogenesis.[238] Spore lytic enzymes formed during sporogenesis of *B. cereus*[235, 236] are another example of spore-specific enzymic component. Production of different molecular forms of enzymes and their incorporation into developing spores has been often reported during sporogenesis of bacteria. Heat-resistant catalase was found in sporulating cells[239] and spores[240] of *B. cereus* T. A similar enzyme was detected in *B. subtilis* spores.[12] Thermo-resistant alanine racemase was reported in sporulating cells and spores of *B. cereus* T.[241, 242] Formation of a heat-stable form of $NADH_2$ oxidase was found during sporogenesis and in spores of *Cl. botulinum*.[243] Comparison of spore and vegetative cell enzyme showed that these two $NADH_2$ oxidases are distinctly different proteins.[244] By means of starch gel electrophoresis Baillie and Norris[220] were able to distinguish the original heat-labile from the heat-resistant form of catalase newly formed during sporogenesis of *B. cereus*. In the same species, immunoelectrophoresis enabled the monitoring of the increasing ratio of heat resistant to thermolabile antigens during sporogenesis.[245]

(d) Enzyme activities paralleling sporogenesis but not directly involved in this process. They occur in some species in the parasporal part of the sporangium. The formation of large crystal-like proteinaceous inclusions toxic for larvae of *Lepidoptera* was observed in postcommitment phase of sporogenesis of certain bacilli.[246–249] The synthesis of these inclusions was studied chemically[122, 153] and serologically,[123] but no data are available on the enzymes involved in their synthesis. A few extracellular enzymes may be produced in postlogarithmic phase of growth, e.g. amylase and protease by *B. subtilis*.[250] The formation of these enzymes in *B. licheniformis* as well as of bacitracin is prevented when subsequent sporulation is inhibited by ethyl malonate.[224] However, the overproduction of bacitracin paralleling the appearance of forespores[46, 189] is a special case of parasporal syntheses. This antibiotic is partly incorporated into the spores, thus serving as a precursor for the synthesis of some spore component, probably coat.[251] The regulatory mechanism resulting in its overproduction is not yet understood. Enzymes involved in the synthesis of circulin in *B. circulans*[252] α-toxin in *Cl. histolyticum*,[253] or antibiotic in *B. subtilis*[28] represent further systems paralleling sporulation process. Their relation to sporogenesis proper is not clear (also see p. 172).

(e) Qualitative or quantitative changes in enzyme pattern. This aspect of the study of the changing enzyme pattern is rather limited by the degree of sensitivity of enzymological methods, by the heterogeneity of populations and by the masking of some enzymic activities in spores. The changes in metabolic pathways during growth and sporogenesis are mostly quantitative. Among enzymes formed during sporulation but completely absent in log phase cells are, for instance, glucose dehydrogenase,[237] acetoacetyl coenzyme A reductase[228] protease in *B. licheniformis*[224, 225] or α-ε-diketopimelic acid oxidation system in *B. subtilis*.[254] The function and importance of the terminal oxidative pathway was already discussed in Section II A. Another example of quantitative difference in enzymic machinery may be the significantly increased rate of decarboxylation of malonate in *B. cereus* T in early stages of sporogenesis.[255] The production of arginase by sporulating but not by vegetative cells was reported in *B. licheniformis*. The appearance of this enzyme is connected with the change in arginine metabolism early in sporogenesis from biosynthetic to degradative.[256] However, arginase can be induced also in growing cells unless glucose is present.[135, 257] Also in the same organism aspartokinase activity completely disappears during sporogenesis.[258] Extracellular protease was found to be produced before sporogenesis but not during growth in *B. cereus* T in the absence of amino acids.[259]

Typical quantitative changes of enzyme pattern during growth and sporogenesis can be observed in glucose catabolism (Section II A) and in electron transport. While in activated spores of bacilli the electron transport proceeds through soluble flavoprotein oxidase, the vegetative cells use almost exclusively cytochrome system for terminal oxidation. The corresponding shifts in respiratory enzyme systems were observed in individual stages of development.[175, 260-263]

Biosynthesis of L-leucyl-β-naphthylamide hydrolase in *B. megaterium* is much more repressed during growth than during sporulation.[264] Some data are also available on the enzymes of glucose and pyruvate catabolism[243] and arginine metabolism[71, 265, 266] in vegetative cells and resting or germinating spores of clostridia. However, detailed information on the changes in enzyme pattern in individual stages of sporogenesis is lacking for this genus.

(f) Distribution of enzyme systems in the compartmentalized sporangium: their synthesis and function in the sporal or parasporal parts. It is difficult to localize exactly the synthetic activities occurring in the sporal or parasporal compartments. In sporulation stage III in *B. subtilis* autoradiographic methods showed that tritiated uracil was being incorporated into both parts of sporangium.[267] The enzymes of the tricarboxylic acid cycle in *B. subtilis* are synthesized in the parasporal part of the sporangium and are not incorporated into the spore.[183] Similarly, the α-ε-diketopimelic acid oxidation system synthesized during sporogenesis of *B. subtilis* is located exclusively in the sporangial part.[254] β-galactosidase may be induced in sporulating cells of *B. megaterium* being located in the parasporal part.[268] The ability of sporangia to resume the synthesis of typical

growth enzymes after induction may be closely related to the phenomenon of rejuvenation.[2, 5]

The coat proteins-synthesizing system in *B. cereus* was reported to be
bound to membranes,[269-271] but no further data on its location in the
sporangium are available. Similarly, an extremely interesting problem in
sporogenesis remains unsolved: which activities directly involved in the
synthesis of spore components and structures are located in the spore
compartment proper and which of them occurs outside the cortical barrier.
Cytoplasmic enzymes may be trapped between cortex and coat, in the
subexosporial cytoplasm or in the exogenous parasporal part of the sporangium. The presence of the major part of important spore enzymes e.g.
alanine racemase and adenosine deaminase was found in the exosporium or
in subexosporial space of *B. cereus*.[272] This clearly shows the site of their
formation during sporogenesis.

In summary the sporulation process is associated with quantitative and
qualitative changes in enzyme pattern and metabolic pathways. These
events may be characterized by the time sequence typical of sporogenesis;
they may be specific and directly involved in sporogenesis; their location
in the compartmentalized sporangium remains to be studied further.

III. Changes in Spore Constituents During Sporulation

A. FORMATION OF SPORE MUROPEPTIDES

The synthesis of the vegetative cell wall continues up to the period of
nuclear rearrangement and ceases during invagination of the cytoplasmic
membrane. Formation of new spore-specific muropeptidic structures is
detectable only after the closure of the forespore septum. Two periods of
increased incorporation of exogenous radioactive diaminopimelic acid into
the hot trichloroacetic acid (TCA)-insoluble fraction were observed in
B. cereus during sporogenesis.[132] The first peak of incorporation was paralleled by the septation, the second by the synthesis of DPA and accumulation of calcium. These results are illustrated in Fig. 5. Early labelling of
vegetative cell walls with ^{14}C-DAP resulted in practically no label being
detected in spore muropeptides after the lysis of sporangia. Both muro-
peptide-containing structures of spores are thus formed *de novo* without
utilization of pre-existing DAP-containing material of the cell. Only free
external ^{14}C-DAP could be incorporated into newly formed spore murein.
Radioactive fragments originating from vegetative cell wall turnover
during growth and then partially released into the medium[273-275] are not
utilized for the synthesis of spore muropeptides neither in *B. cereus*,[132]
nor in *B. megaterium*.[276] However, the addition of a mixture of spore wall

peptide components to the culture of *B. cereus* grown in glucose-glycine-salts medium accelerates sporogenesis.[277] Non-participation of pre-existing vegetative cell wall components in the synthesis of spore muropeptides was confirmed also by sporulation of protoplasts of *B. megaterium*.[278-280]

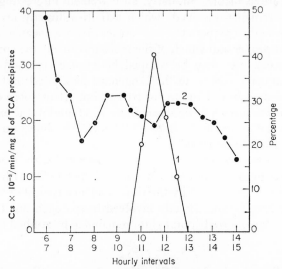

F IG. 5. Sequential labelling (length of incorporation = 1 hr) of sporulating cells of *B. cereus* with ^{14}C-DAP (excess of ^{12}C-lysine added). 1, increase in percentage of weakly refractile forespores, 2, radioactivity of trichloroacetic acid-insoluble fraction.[132]

During the formation and ripening of endospores in *B. cereus* no degradation of pre-existing cell wall can be observed. It occurs only after the intrasporangial development of spores is complete.[132] The transition of muropeptides of sporangial cell wall from the insoluble into a soluble fraction reported in *B. cereus* occurred only in the late stage of sporogenesis.[238]

Diaminopimelic acid is located mainly in the cortex. Sequential formation of two different murein structures detected in *B. cereus* by ^{14}C-DAP incorporation[132] may reflect the heterogeneity of this structure. In germination studies of pre-labelled *B. cereus* spores the second murein structure (originally formed during "whitening" period) is completely depolymerized within a few minutes. The first murein layer formed in the septation period is relatively stable during germination. Athough it undergoes a slow turnover during outgrowth it represents the germ cell wall.[281-283] Besides this case of functional heterogeneity some chemical differences of individual murein layers may exist in spores. They were demonstrated for instance by different sensitivity of cortex, residual cortex or cortical membrane to lytic enzymes[284, 285] or by fractionation of fluoro-benzene derivatives of muropeptides isolated from *B. coagulans*.[96] The role of murein layers in morphogenetic processes during spore formation of *B. cereus* was partially elucidated by inhibition studies. Pencillin does not prevent incorporation of ^{14}C-DAP into insoluble fraction of the first murein structure (stage II-III) and the development of the sporangium appears to proceed approximately to the stage IV. The final polymerization and cross-linking of cortical material is drastically inhibited by this antibiotic.[286] A similar inhibition of cortex formation by penicillin was observed in different variants of *B. cereus*.[2] Inhibition of muropeptide polymer formation by penicillin in *B. cereus* always prevented the retention of calcium and dipicolinic acid in the intrasporal structure. The addition of penicillin during calcium accumulation and DPA synthesis brought immediate release of cation but permitted a short continuation of DPA synthesis.[141, 286] Later on both calcium and DPA were rapidly released, further cytodifferentiation being irreversibly interrupted. Sporangia then lysed and only empty spore coats were present in the culture. A similar failure to complete spore formation was found in *B. cereus* when cortex polymerization was prevented by cycloserine.[96] Spores formed in the presence of this antibiotic were unstable and unable to retain typical spore constituents. Cortex-less mutants of *B. cereus* are able to accumulate calcium and synthesize dipicolinic acid, but after the completion of the coat formation both DPA and Ca^{2+} were released into the medium.[2, 296]

What is the main morphogenetic role of cortex and by what mechanism can it participate in the accumulation and retention of spore components? The chemical composition and structural arrangement of this endogenous wall is described in Chapters 2 and F. Lewis, Snell and Burr[287] suggested that this layer was contractile. According to this hypothesis the mechanism for the contraction may be based on the reduction in the electrostatic repulsion of the carboxyl groups by the introduction of Ca^{2+} or Ca-DPA. During sporulation the contraction of this layer may dehydrate the protoplast. Morphological data on the cortical layer or on residual muropeptides adhering to spore protoplast obtained from the studies of disintegrated spores support this hypothesis.[284, 285, 288] Additional support was obtained from studies of the reversion of germinated and outgrowing spores to refractile bodies caused by basic peptides and proteins.[289] A high affinity and stable binding of cortical muropeptides to neighbouring structures was found. This may explain the difficulty of analysing spore integuments. For example, hexosamine-containing peptides are associated with coat preparations and released during enzymic digestion of spore integuments.[235, 236, 284, 285] Polymerization of muropeptides in the cortical region and of cystine-rich proteins in the coat layer(s) (see below) during sporogenesis, closely followed by contraction and stabilization by electrostatic forces and/or by crosslinking of the protein with SS-bridges, is a unique and effective morphogenetic mechanism of the bacterial cyto-differentiation.

B. Cystine-rich Proteins

One of the differences in the chemical composition of free spores and corresponding vegetative cells of bacilli is the markedly higher content of cystine-rich proteins of spores. The higher requirement of sulphur amino acids for sporogenesis was reported already by Blair[290] for *Cl. botulinum* and by Ordal for *B. coagulans*.[44] Chromatographic analysis of cell hydrolysates showed a higher level of cyst(e)ine in spores of *B. subtilis* and *B. mycoides* than in vegetative cells.[291] Using a cytochemical method Widra[292] demonstrated an increase of SH-groups in proteins of *B. megaterium* during formation of forespores. By different chemical methods the cystine content in proteins of spores of *B. cereus* and *B. megaterium* was found to be about five times higher than that in homologous vegetative cells.[127, 128] Similar data were obtained for several other species and strains of bacilli.[128, 144] More recently the marked increase in the cyst(e)ine content of spores compared to vegetative cells was found also in *Cl. botulinum*.[65] By analysing the distribution of ^{35}S after pre-labelling of vegetative cells of *B. megaterium* with different ^{35}S-sources it was shown that most cystine sulphur found in spores was derived from methionine sulphur.[129] Recently the metabolic conversion of methionine into cysteine during sporogenesis was found also in *B. subtilis*.[130] The stage of formation of cystine-rich proteins during cytodifferentiation of sporangia was identified in synchronously sporulating *B. megaterium* and *B. cereus* by chemical analysis[129] or by pulse-labelling with ^{35}S-cysteine during different stages of sporogenesis.[293, 294] Cystine-rich proteins are synthesized relatively early during spore formation. In free spores the cystine-rich proteins are located in a coat fraction.[128, 144] There is almost no information on the site of their synthesis during sporogenesis or on the mechanism of their polymerization into rigid coat layers. Aronson suggested that coat proteins are synthesized by polysomes which are bound, together with a stable mRNA, to the cytoplasmic membrane.[269-271] Some facts are available on the sequence and interdependence of biochemical and morphogenetic processes in relation to the formation of cystine-rich proteins. Inhibition of the synthesis of these proteins and obviously of the formation of disulfide links, during so-called "cystine inhibition" results in the retardation of further development of sporangia of *B. megaterium* and *B. cereus*.[85, 294] Defective spores are produced with abnormal morphology, calcium content and heat resistance. On the other hand, the agents which cause a relatively specific blockage of later stages of sporogenesis, e.g. tetracyclines[146, 295] or penicillin,[147, 286] have no injurious effect on the formation of cystine-rich proteins or coats. Recent studies have shed further light on the location of cystine-rich proteins in coats of bacilli.[2, 128, 144, 296, 297] X-ray diffraction patterns showed the presence of crystalline keratin-like structures in free spores and enzymatically purified coats of *B. subtilis*. The lattice distance in these structures was calculated to be 9·8 Å and they were absent in vegetative cells. The coat components were found to be similar to α- and β-keratin of wool.[298, 299] Hexagonal single crystal pattern was detected in spore coat fragments by electron diffraction investigation.[300] The analysis of crystal type by this method showed that the structure of the spore coat protein is similar to that described for polyglycine II.[301]

An important process which parallels the formation of cyst(e)ine-rich proteins during sporogenesis of bacilli is the conversion of SH-groups to SS-bridges. Although some cysteine was found in vegetative cells or in sporangia with forespores, only cystine was detected in free spores of *B. cereus*.[128] In agreement with these findings during sporogenesis of *B. cereus* an increasing ratio of SS/SH occurred and -SH groups could not be detected in spores.[302] A negligible content of sulphydryl as compared to disulphide groups was reported also in spores of *B. subtilis* var. *niger*.[303]

The formation of cyst(e)ine-rich proteins as well as the oxidation of sulphydryl to disulphide groups crosslinking the coat proteins may be an important morphogenetic mechanism in sporogenesis and in other biological systems.[304] The ratio of SS/SH during sporogenesis may affect morphological and other properties of spores.[85, 305] The disulphides of spore coat apparently play a prominent role in regulation of permeability of resting spores of bacilli. The artificial reduction of SS to SH groups alters some spore properties[306–309] and the conversion of disulphides to sulphydryl groups was observed during physiological germination.[128, 310]

A full understanding of the role of cyst(e)ine-rich proteins in sporogenesis also requires a knowledge of the fate of these compounds during germination. In addition, attempts have been made, so far only with *B. cereus*, to integrate the events observed with these proteins with events occurring with other compounds e.g. the murein structures described in the previous part of this chapter. Although these studies are limited to one strain, it is thought that they may have general validity.

The so-called "labile murein",[132, 281] DPA and calcium[128, 311] formed or taken up during maturation, are released within the first few minutes of germination.[128, 287] The structures or their precursors which are formed earlier, i.e. before maturation, are more stable and resist the depolymerization which occurs early in germination. This second group of compounds are the "cyst(e)ine-rich proteins"[128, 129] and the stable murein structure"[132, 281] which then protect the primary cell for almost the whole period of outgrowth.[128, 281–283] It is possible, as outlined above, that these compounds, operating together, provide the mechanism for the contraction of the spore protoplast during sporogenesis. The chemical composition predetermines both latter structures to exert at least potentially the changes in elasticity which may occur during sporogenesis as well as during outgrowth. Mechanisms of contraction of murein layers postulated by several authors,[96, 284, 285, 281] and crosslinking of SS-proteins by SS-bridges to form a compact keratin layer of coat, may well operate together during sporogenesis in mechanical contraction of spore protoplast. On the other hand, the architecture of these structures now encasing the spore protoplast may, during germinative degradation of cortex, provide an easily adaptable structure needed for subsequent rapid changes of cell volume brought about by swelling. The turnover of the "stable murein layer" occurring in outgrowing *B. cereus* spores[281, 282] may be one of these mechanisms. The possibility of its recontraction during outgrowth by adsorption of basic peptides and proteins[289] underlines its role in elasticity changes. Similarly, the "opening" of disulphides in the keratin-like coat layer may offer an efficient mechanism of flexibility. These changes may be similar to those described in the division of some yeasts.[312]

C. ACCUMULATION OF CALCIUM

Spectrophotometric measurements done 25 years ago already showed that spores of bacilli contain much more calcium than the homologous vegetative cells.[313] The attention of sporologists was drawn to this cation by its possible role in the heat resistance of proteins.[84] The precise stage of calcium accumulation during the differentiation of the sporangium was defined much later by chemical analysis during sporogenesis of *B. cereus* and *B. subtilis*,[238] by spectral analysis in *B. megaterium*[85, 314] and by incorporation of radioactive calcium in *B. cereus*[128] or *B. cereus* var *alesti*.[88] Recent investigation defined the period of calcium accumulation in synchronously sporulating *Cl. roseum*.[157] In some of these studies calcium uptake was related to the degree of heat resistance of sporangia.[128, 144, 157, 311] The incorporation of $^{45}Ca^{2+}$ during sporogenesis of *B. cereus* starts during the development of forespores. This is "labile calcium" partially removable by washing procedures and it is not clear whether it is bound only to the surface or is partly incorporated into the sporangium.[146] An increased affinity of cell surface to calcium during forespore formation of *B. cereus* and *B. megaterium* was demonstrated by a fluorometric method and by calcium-mediated binding of tetracycline antibiotics to the cell surface.[146, 295] Binding of calcium in bacilli usually precedes the synthesis of DPA[88, 146, 311] and the same phenomenon was observed in *Cl. roseum*.[157] Some results suggest that calcium has a significant role in the induction of DPA synthesis. Spores of *B. cereus* T formed endotrophically were found to be deficient in dipicolinic acid. Presence of calcium during endotrophic sporogenesis increased the amount of DPA in spores, and there was a direct dependence of DPA content on the time of addition of calcium. Calcium added after beginning of cortex formation was not capable to stimulate higher DPA synthesis.[86] In *B. cereus* var *alesti* grown in complex but calcium-deficient medium this cation was able to induce maximum DPA synthesis when added even at the beginning of "whitening" of forespores. Similarly, later addition had no stimulatory effect on DPA synthesis.[88] The effect of calcium on DPA synthesis is demonstrated also by the fact, that the final DPA content of some species or strains depends on the calcium level of the sporulation medium.[86, 88, 89, 305, 311, 315, 316] In other strains or species, however, the final content of DPA does not seem to be influenced by the level of calcium in the medium.[317] Some cations can replace calcium in spores[54] and can also stimulate the synthesis of DPA, e.g. strontium is very effective in DPA stimulation.[94, 97] Spores of *B. cereus* or *B. megaterium* developing in low levels of calcium[76, 88, 314] or produced endotrophically[86] have lower refractility. The same is true when calcium accumulation during sporulation of *B. megaterium* or *B. cereus* was diminished by the presence of excess cysteine[85, 318] or its analogue, thioproline.[305] In addition to refractility changes these spores were smaller and rounder. Spores of *B. megaterium* endotrophically produced in the medium supplemented with barium did not reach the normal refractility of calcium- or strontium-spores.[97] Since calcium-barium-spores were fully refractile, the co-presence of calcium seems to be necessary for normal refractility. In these experiments strontium or barium plus calcium were able to stimulate nearly maximum DPA synthesis, but barium alone was not. Despite the specific effect of calcium in stimulation of DPA synthesis, this cation was not able to induce DPA formation in an "abortively (at 37°C) sporulating"[55,81] mutant of *B. cereus*.[89] However, the DPA level in spores of this mutant produced at 28°C was fully influenced by the concentration of calcium in the medium. As was mentioned above, an inductory level of calcium must be available during a very early period of endotrophic sporogenesis[86] or, in complex medium, early in the "whitening period".[88] Calcium can serve to trap intracellular DPA, thus "pulling" the reaction sequence. Unfortunately no convincing data are available on the intrasporangial distribution of calcium during sporogenesis or on the site of DPA synthesis, mobility and interaction with calcium. The mechanisms of the final integration of DPA with intrasporal structure also remains obscure.

D. BIOGENESIS OF DIPICOLINIC ACID (DPA)

Since the discovery in 1953 of dipicolinic acid in spores and spore exudates during germination[319] a number of further studies have been carried out on the biosynthesis and function of this compound. Despite imperfect synchrony of sporulating cells of bacilli and clostridia it was possible to correlate roughly the period of DPA synthesis with some events occurring in sporangia. In the early studies DPA synthesis was found to be paralleled by the appearance of refractile spores in sporangia,[53, 195, 238] accumulation of calcium, iron and by changes in distribution of hexosamine-containing material.[238] DPA was estimated by u.v. spectrophotometry of spore extracts[53, 238, 320-323] or by spectrophotometry of ferrous-DPA complex.[324] DPA appears in bacilli during cortex formation and maturation of sporangia (i.e. stages IV–VI); curves of the time course of its biosynthesis and of firm binding of calcium in sporangia are often superimposable.[88, 128, 137, 311] Similar results were obtained in clostridia.[99, 157, 325] Thus DPA synthesis takes place at a relatively late stage of intrasporangial development of spores. In this period the spore protoplast is enclosed by the forespore septum and is being gradually encased by the cortical zone and rigid layer of coat(s). The site of DPA synthesis is not known but, in general, three areas may be considered: a) spore protoplast, b) region of spore integuments (cortex, coat, exosporium), c) parasporal (sporangial) part. The possible synthetic role of the mother cytoplasm, which may be engulfed by the coat[15, 326] or of cytoplasmic material encased by exosporium and partially resisting lysis,[285, 327] is not clear. According to Kondo et al.[254] the primary site of DPA biosynthesis in B. subtilis may be the sporangial cytoplasm. This assumption originated from the finding that DPA liberation takes place when walls of cells with forespores are destroyed by lysozyme and osmotic shock. Such treatment when applied in later stages of sporogenesis does not have this effect. However, the function of forespore envelopes in retention of spore constituents was not studied.

Although the site of DPA synthesis remains unclear the possible pathways of DPA biosynthesis are better understood. A greater portion of radioactive carbon from ^{14}C-diaminopimelic acid (DAP) incorporated during endotrophic sporogenesis of B. cereus var mycoides was found in pyridine-2, 6-dicarboxylic acid than in 12 other amino acids in spores.[53] Breakdown of the carbon chain of DAP and deamination were involved in the DPA biosynthetic reactions. In fact, some shorter chain compounds were most efficient precursors of the carbon skeleton of DPA which could be synthesized apparently by condensation of C-4 and C-3 compounds.[320] The disintegrated cells of sporogenous and asporogenous strains of bacilli and homogenates of E. coli were able to form DPA from α-ε-diketopimelic acid and ammonia.[328] No diketopimelic acid was found in sporulating cells of B. cereus in these studies. However, this precursor was detected in Penicillium citreo-viride capable of synthesizing DPA.[329] In B. subtilis the α-ε-diketopimelic acid oxidative activity paralleled the formation of spores.[254] A hypothetical scheme for the biosynthesis of DPA proposed by Benger ascribes key roles to glutamic acid and glycolate. According to this hypothesis the DPA is formed by C-2 plus C-5 condensation, ring closure to 3-oxy-4,5-dihydropyridine-2, 6-dicarboxylic acid and by reduction and dehydrogenation of the latter.[330, 331] Recent evidence suggests that there is a similarity in the biosynthetic pathways of bacteria and moulds and supports the original proposal of Martin and Foster[185] who thought that C-3 + C-4 compounds were the most probable precursors of DPA. Aspartic semialdehyde and pyruvic acid were recently reported to serve for DPA synthesis.[332] 2-Keto, 4-hydroxy, 6-aminopimelic acid seems to be the key intermediate giving, by ring closure and desaturation, dipicolinic acid.[333,334] These data (Fig. 6) are based on experiments using different ^{14}C-labelled precursors and on studies of degradation of labelled DPA. Some of the

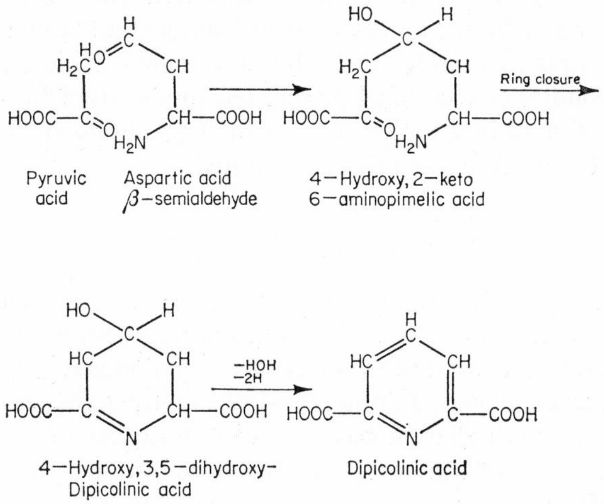

Pyruvic   Aspartic acid   4−Hydroxy, 2−keto
acid   β−semialdehyde   6−aminopimelic acid

4−Hydroxy, 3,5−dihydroxy−   Dipicolinic acid
Dipicolinic acid

FIG. 6. Hypothetical pathway of dipicolinic acid biosynthesis (According to Kanie *et al.*[334]).

labelled precursors used in these studies are known to be involved in the lysine pathway. The participation of lysine precursors in DPA synthesis was found in extracts from sporulating *B. megaterium*.[335] Two types of lysine auxotrophs of *B. cereus* T were prepared which had blocks at different steps of the lysine pathway and which were producing different amounts of DPA. A conversion of dihydrodipicolinic acid, an intermediate in this pathway to DPA, can be assumed from these experiments. These mutants, blocked prior to the proposed shunt to DPA, produced low amounts of DPA, the synthesis in mutants which were blocked after the shunt being normal.[336]

IV. Development of Resistance of Sporulating Cells

It is apparent, from what has been written so far, that different spore constituents, structures and tertiary arrangements change during the whole course of sporogenesis. A hopeful line of research is provided by the correlation of these changes in the sporangium with changes in resistance to deleterious physical agents, which is characteristic of the spore.

Such studies, in turn, may shed new light on the mechanism and function of the differentiation process. In this section we shall consider development of resistance of the sporangium to radiation, heat and other miscellaneous factors.

Most of these studies have, however, been marred by inadequate techniques. Chemical and morphological characteristics of whole populations have been studied in some cases without adequate synchrony of the culture. Precise deduction from samples derived from a whole population cannot be made to apply to an individual cell unless there is a high degree of synchrony. Secondly, conventional plating techniques used for the detection of survivors should be supplemented with steps designed to diminish the likelihood of clumps or chains being counted.

A. RADIATION RESISTANCE

Ultraviolet light was first used to test the radiation resistance of sporulating cells. Romig and Wyss[136] showed that endotrophically sporulating *B. cereus* develops resistance to ultraviolet radiation somewhat in advance of heat resistance. A period of increased sensitivity to the lethal and mutagenic effects of u.v. radiation was observed before the appearance of forespores. The cells were photoreactivable before but not after the period of sensitivity.[136, 137]

A similar increase in ultraviolet resistance about 60–90 min before increase of heat sensitivity was demonstrated in *B. cereus* sporulating synchronously in complex medium.[144] This stage corresponded to the formation and completion of forespore septum and to the beginning of synthesis of cystine-rich proteins in sporangia. Some shielding effect could be expected during this compartmentalization of sporangia. However, heat-sensitive forespores of *B. subtilis* released from sporangia by lysozyme treatment also pose a higher resistance to ultraviolet radiation than vegetative cells.[338] During the forespore stage the resistance of cells of *B. cereus* increase not only to ultraviolet light but also to penetrating type of radiation, e.g. X-rays.[144, 311, 339] Special cultivation equipment, with an efficient vibration stirrer for splitting of chains of bacteria was used in these experiments.[144] Resistance to X-rays developed in the stage corresponding to the formation of cyst(e)ine rich proteins in sporangia. Some evidence from the literature supports the hypothesis that these proteins may contribute to the radioprotection of sporangia towards ionizing radiation. Sulphydryl compounds have been known to protect various biological objects against radiation injury[340–342] including spores.[343, 344] The radioprotective efficacy of some of these agents was directly dependent on their ability to form disulphides.[345] The results of the study of paramagnetic resonance of X-irradiated proteins[346, 347] indicated that the disulphide bonds could serve as donor sources of electrons for repairing radiation injury. Whenever in a protein molecule an electron is knocked out by ionizing radiation the vacancy can be filled by an electron borrowed from a disulphide linkage. Radioprotectors containing sulphur evidently change the paramagnetic resonance of X-irradiated proteins[348, 349] or yeasts.[350]

The sequence of development of radiation and heat resistance was confirmed also in endotrophically sporulating cells of *B. subtilis*. The X-ray resistance increased about 2 hr before heat resistance and DPA synthesis.[138] The data obtained with *B. subtilis* are illustrated in Fig. 7. The radiation

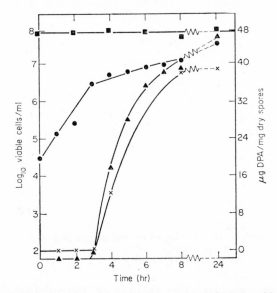

FIG. 7. Changes in X-ray resistance, heat resistance and dipicolinic acid content during endotrophic sporulation of *B. subtilis*. Symbols: ■ total viable cells; ● cells resistant to X-ray dose of 50 kr; ▲ cells resistant to heating at 80°, for 15 min; × dipicolinic acid. (Reproduced with permission from Rowley and Newcomb.[138])

resistance of transaminases isolated from sporulating cells of *B. subtilis* remained essentially unchanged, whereas the dehydrogenases exhibited a decreased resistance in early stage of sporogenesis.[138] Bott and Lundgren,[302] studying the development of resistance to X-rays in sporulating prototrophic strains and a sulphur-requiring auxotrophic mutant of *B. cereus* found a direct correlation between radiation resistance and SS/SH ratio in cell proteins (Fig. 8).

The resistance of sporulating cells of *B. cereus* grown in bactopeptone medium continued to increase even after completion of synthesis of cys-t(e)ine-rich proteins.[311] The crosslinking of these proteins by SS-bridges and especially some other mechanisms, e.g. dehydration, changes occurring in the DNA molecule, development of repair mechanisms, etc., may contribute to the final high radioresistance of spores. Unfortunately, nothing is known about these processes during sporogenesis proper.

A problem arises whether the SS-groups as such present in the proteins may contribute to the radiation protection of spores. As was mentioned above, the sulphydryl constituents of cells are commonly believed to be

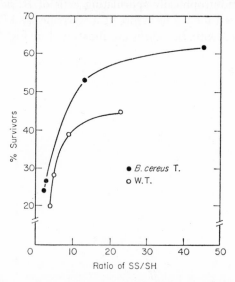

FIG. 8. Relationship of the fraction of control populations of wild type of *B. cereus* (W.T.) and *B. cereus* T surviving 160 kr of X-rays to the ratio of SS/SH. (Reproduced with permission from Bott and Lundgren.[302])

effective radioprotectors. The simplest of the sulphydryl compounds, hydrogen sulphide, was reported to decrease the radiation sensitivity of dry spores of *B. megaterium*.[343, 344] These results were supported by the observation that the rupture of part of spore disulphide bonds in *B. cereus* by thioglycolate did not decrease the resistance of spores to γ-radiation.[351] The role of the cystine-rich structure alone in the radioprotective mechanisms of mature spore can only be surmised. Definite evidence will become available when mutants of the same species markedly differing in the content of these proteins are prepared. The evidence that the damage caused by ionizing radiation may be taken over by disulphides was obtained, however, in many cases. Beside the above-mentioned detection of unpaired electrons in X-irradiated proteins by paramagnetic resonance, the opening of SS-bridges by ionizing radiation was observed in simple disulphides,[352] in proteins[353] and also in spores.[354] The keratin-like structures of spores may operate in this way by non-destructively dissipating radiation energy, thus contributing to the other radioprotective mechanisms of spores.

B. Heat Resistance

The hypothesis that calcium and dipicolinic acid are specifically associated with the acquisition of heat resistance originated immediately after the finding of these components in spores.[313, 319, 355] The first studies correlating DPA synthesis and development of heat resistance were done with synchronously grown *Cl. roseum*[32] and *B. cereus* T.[137] Formation of heat-resistant sporangia has also been related to the uptake of radioactive calcium in *B. cereus*.[128, 144] An attempt to integrate the development of thermoprotective mechanism in differentiating sporangia in relation to both Ca^{2+} and DPA was done in *B. cereus*[311] and *Cl. roseum*.[157] The data obtained from these experiments are illustrated in Figs 9 and 10. A time

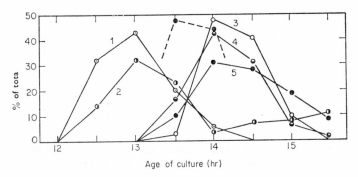

Age of culture (hr)

FIG. 9. Development of radioresistance and thermoresistance and the synthesis of typical constituents in sporulating cells of *B. cereus* (From Vinter.[311])
   1, increase in percentage of cells surviving X-irradiation with 120 kr.
   2, increase in cystine content,
   3, increase in percentage of cells surviving heating at 65°/15 min,
   4, increase in incorporated $^{45}Ca^{2+}$,
   5, increase in dipicolinic acid content.
The broken line marks the peak of the curve showing the percentage of increase in mildly refractile forespores.

lag was sometimes observed between the appearance of the two spore components and acquisition of thermoresistance. This lag is especially striking in *Cl. roseum*.[32, 157] In *Cl. botulinum* refractility developed 3 to 5 hr in advance of heat resistance, but the biosynthesis of dipicolinic acid was closely paralleled by the development of thermoresistant cells.[99, 325] In synchronously sporulating *B. cereus* T no increased heat resistance of sporangia was observed unless the DPA content reached one-third of the maximum level.[137] A correlation between heat resistance of sporangia and DPA content seems to occur only above a threshold level of this spore constituent. A very interesting phenomenon was observed in *B. cereus* grown in modified G-medium. The addition of gluconate, 0·2%, did not affect exponential growth, but refractile spores occurred much earlier and simultaneously with DPA synthesis. However, these spores became heat resistant in the same period as those formed in the control culture.[356] This

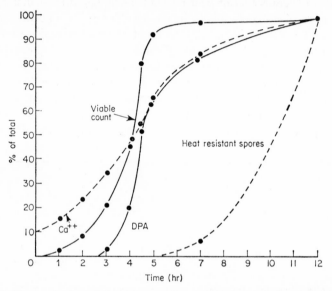

FIG. 10. Changes in the cellular concentration of calcium and dipicolinic acid during the sporulation of *Cl. roseum*. (Reproduced with permission from Wooley and Collier.[157])

is another example of the separation of refractile spore formation and DPA synthesis from acquisition of heat resistance.

For sporulating cells the term "thermoresistance" has evidently a relative meaning. A gradual decrease in susceptibility to different heat treatments lethal for vegetative cells but harmless to mature spores was observed in sporulating cells of *B. cereus*.[311] In this case the resistance of sporangia to different "pasteurization" temperatures developed in accordance with the amount of DPA synthesized. However, the resistance to higher, spore-inactivating temperatures was not acquired simultaneously with achieve-ment of maximum level of DPA in sporangia. This delay, observed in this and other aforementioned studies, may be due to the subsequent development of fully functional thermoprotective mechanisms, possibly by the final interaction of calcium, DPA and other spore constituents.

This period of maturation is characterized by the formation of cortex which itself may play, directly or indirectly, an important role in thermoprotection. A direct correlation was found between diaminopimelic acid content and heat resistance of spores.[96] The possible participation of Ca^{2+} or Ca-DPA in mechanism of cortex contraction postulated by several authors[96, 284, 285, 287] accentuates the interrelationship of cortical muropeptides, calcium and DPA in acquisition of thermoresistance during maturation of sporangia. The cortex evidently contributes to the retention of calcium and DPA in spores. Penicillin prevents the formation of cortex in *B. cereus* and brings about rapid release of calcium from sporangia. Less affected sporangia may complete sporogenesis, but the resulting spores are more heat sensitive than normal spores.[147, 286] In *B. cereus* T celbenin or cycloserine also interfered with cortex formation. DPA-containing but heat-labile spores were produced in certain concentrations of cycloserine.[96] The evidence that not so much the quantity of DPA alone, but rather the quality of its complex with other spore constituents, is responsible for thermoprotection is confirmed also by some other data. Prolongation of ^{45}Ca uptake, DPA biosynthesis and of refractility development by addition of thioproline to *B. cereus* in forespore stage resulted in the production of spores with almost normal DPA content but they were heat sensitive. Their calcium content was lower and the extractibility of this cation from spores was increased.[305, 316] Typical heat inactivation curves of this organism have an initial shoulder during which rate of killing is diminished. Spores of some species which contain normal amounts of DPA can be produced endotrophically in presence of β-thienylalanine (a cysteine analogue) and in adequate level of calcium. Heat survival curves obtained with these spores, however, lack the shoulder typical for calcium-rich spores.[140]

Lower DPA content but higher Ca : DPA ratio and higher heat resistance was observed in spores of *B. coagulans* produced at 45°C compared to those formed at 30°C.[95] The same effect of higher incubation temperature on Ca : DPA ratio was observed in *B. subtilis* but, in contrast to *B. coagulans*, some increase of DPA was also detected.[95] The cells of *B. cereus* T growing at 30°C but sporulating at various temperatures produced spores differing in heat resistance and in the content of spore constituents. In the most resistant spores (30° and 37°C crops) calcium content as well as the Ca : DPA ratio were higher. Further elevation of sporulation temperature lowered heat resistance in this strain.[96] Conditions of sporogenesis allowing, in addition to maximum DPA synthesis the higher saturation of sporangia with calcium increase not only the heat resistance of the spores produced as calculated from the final slope of inactivation curves but prolong also the initial shoulder of the thermal inactivation curve,[95, 305, 315, 316] (also see Chapter 16). On the other hand, factors which retard or prevent calcium accumulation[85, 293, 305] lower the heat resistance and shorten or completely eliminate the thermoinactivation shoulder.

The complex interaction of calcium and DPA with cortical muropeptides and other spore constituents in the establishment of a more or less efficient thermoprotective mechanism may be influenced in many ways during sporogenesis. No common scheme for the optimal interaction of these components resulting in maximum resistance valid for all species of bacilli and clostridia can be made. The delicate chemical and structural variations of cortical muropeptides and their affinity to cations and DPA determining the mechanical properties of cortex and the different degrees of solute packing are obviously highly specific for each species. In addition to this hypothetical complex some other changes in spore constituents evidently participate in the overall integrated mechanisms of thermoprotection during sporogenesis, e.g. stabilization of proteins with calcium alone, changes in the physico-chemical state of proteins and nucleic acids, changes in the tertiary structure of cystine-rich proteins in the spore periphery, etc. The importance of these mechanisms will be further discussed in the chapters on composition and resistance of bacterial spores.

C. RESISTANCE TO OTHER FACTORS

The difference in resistance of spores and homologous vegetative cells to various other deleterious factors: metabolic inhibitors, disinfectants, desiccation, etc., is well documented (see Chapter 16), but our knowledge on the development of these types of resistance during sporogenesis is rather limited. We may predict that the mechanisms of resistance to different harmful factors develop separately from each other. Studies of heat- and radiation-resistance already described illustrate this point. For example, spores produced under conditions interfering with optimal development of heat resistance remain resistant to ultraviolet or ionizing radiation or to phenol.[54, 137, 140, 305, 315, 316] The increasing resistance of sporulating cells of bacilli to cyanide reflected the shift from cytochromes- to flavin-based systems for terminal oxidation.[175, 261, 262] The high sensitivity of cells in presporulation period and in early stages of sporogenesis to analogues of amino acids or nucleic acid bases[56, 111, 120, 149, 153, 209] was due to extensive syntheses of macromolecules *de novo*. In *B. cereus* var *alesti* a continuous decrease of the inhibitory effect of 8-azaguanine on sporogenesis and crystal formation was observed. Although 8-azaguanine was incorporated into RNA, in later stages it may be incorporated into a position in, or a portion of, the RNA which is not involved in spore protein formation.[153] A similar increase in the resistance to base analogues and actinomycin D was observed at the beginning of sporulation of *B. cereus* T.[150] A membrane-bound stable mRNA apparently involved in the synthesis of spore coats was found in these cells.[269, 271] In the same species a similar development of the resistance to 8-azaguanine was reported during transition from vegetative growth to presporulation stage. When subinhibitory concentration of radioactive 8-azaguanine was used for pre-labelling the cells, in the period of acquisition of resistance to this analogue, a release of label was observed and its replacement by normal base derived from the intracellular pool.[154] In *B. subtilis* resistance to actinomycin developed only in the final stages of sporogenesis, no stable mRNA being detectable throughout this process.[152] In the same species a sequential increase of the resistance to octanol, chloroform and heat was observed. Chloramphenicol inhibited all stages of spore development except the final phase in which the octanol resistant sporangia became heat-resistant.[28, 152] The development of forespores of *B. subtilis* which have not yet reached stage V (coat formation) was stopped by chloramphenicol. In later stages this antibiotic blocked coat formation but did not impair cortex formation. Synthesis of dipicolinic acid stopped only about 1 hr after addition of chloramphenicol.[151] A similar inhibition of coat synthesis measured by ^{35}S-cystine incorporation was caused by chloramphenicol in *B. cereus* var *alesti*. Subsequent lysis and loss of refractility was a result of autogermination apparently brought about by the incomplete spore coat. Calcium uptake and DPA synthesis were not greatly affected by the late addition of chloramphenicol.[2] Penicillin inhibited the cortex formation and retention of calcium and DPA in sporulating variants of *B. cereus*.[2, 147, 286] Sporangia become resistant to this antibiotic at the end of DPA synthesis and calcium accumulation.[147]

Some data are available on the development of resistance to various factors during sporulation of *Cl. welchii*. The encystment of forespores was associated with increased resistance of sporangia to hydrolysis[357] and a sequential development of resistance to different heat treatments, desiccation and toxic concentration of thiomersalate was observed in the same species.[358] The final steps of forespore maturation in *Cl. botulinum* were completely resistant to high levels of chloramphenicol, 8-azaguanine or mitomycin C.[325] Sporulating cells gradually acquire resistance to agents which inhibit synthetic activities, but the immediate effect can be established with some certainty. However, the end result of these agents as estimated by survival studies may be complicated by numerous factors. For example it may be difficult to determine the precise sequence of development of resistance due to asynchrony of the culture, adsorption of inhibitors on the surface of spores, altered response of spores to germination media, etc. In characterization of resistance development some specific changes occurring in sporangium must be taken into account. A decreased penetration of inhibitory substances to their site of action can be predicted due to the increasing isolation of spore protoplast by the complex envelope system. Formation of forespore septum, cortical layer, coat(s) and exosporium represents the development of an efficient secondary intracellular barrier which is additional to the permeability regulation exerted by the mother cell. Simultaneous cessation of synthetic activities involved in spore formation and dehydration of spore protoplast naturally synchronizes the acquisition of resistance to both metabolic inhibitors and desiccation.

V. Formation of Lytic Enzymes and Release of the Spore

The spore may be regarded as an "intracellular cell" and the parasporal part of the sporangium has many functions in its development. Early in sporogenesis calcium receptors appear on the sporangial surface,[146, 295] metabolic pathways alter so as to provide the forespore with sources of energy and precursors (see Section II A, B). Later in the development, the parasporal part presumably contains enzymes involved in the formation of peripheral spore polymers. The precise stage in which the spore formation is completed with respect to all spore properties is difficult to define. Some signs that forespores can be completed independently of sporangial wall were observed in studies of sporulating protoplasts.[278, 280] Forespores of B. subtilis isolated from sporangia by lysozyme treatment were able to grow and form colonies.[338] Spores of bacilli usually do not germinate inside the sporangium unless they are fully mature and the external level of nutrients is readjusted to become high. In some species of clostridia sporangial components or structures may persist around mature spores, e.g. part of cytoplasm and/or sporangial wall;[326] in other species no remnants of sporangium or exosporium were observed.[359] Liberation of clean spores from vegetative sporangia of Cl. botulinum may be accomplished by the use of lytic enzymes and sonic oscillation.[360] In bacilli the sporangial wall and cytoplasm are usually degraded and free spores are liberated into the medium. The production of cell wall lytic systems in growing or sporulating cultures[361–365] and the turnover of cell walls[273, 274, 366] were observed in bacilli. Some of these lytic systems were isolated and partially characterized.[367–369] A transition of DAP and hexosamine from insoluble to a soluble fraction of disintegrated cells of B. cereus was found in late stages of sporogenesis.[238] An enzyme was isolated and characterized from spores of B. cereus and B. anthracis which released non-dialysable components from wall preparations of vegetative cells of the same or other species. α-ε-Diaminopimelic and glutamic acid, alanine, amino sugars and glucose were detected in these components.[235] Two types of lytic enzymes were found in sporulating cells of B. cereus differing in substrate specificity, pH optima and activation requirements.[236] These results showed that the lytic system called enzyme V may be mainly concerned with the release of free spores from sporangia and the so-called enzyme S with the lytic processes which accompany spore germination. The products of hydrolysis after chemical or enzymic degradation and the mode of action of different lytic principles, their effects on vegetative cell walls and spore integuments of bacilli and clostridia were mentioned or reviewed in some articles.[370–375] The cytological effects of sporangiolytic enzymes of the V-type during sporogenesis of various strains of B. cereus, B. thuringiensis and B. anthracis were characterized by Tomcsik and Bouille.[376] The effect of these sporangiolytic enzymes was retarded or blocked by cell-wall anti-polysaccharide antibodies, whereas antipolypeptide antibodies had almost no influence. Three vegetative cell-wall lytic enzymes were isolated and purified from sporulating cells of B. thuringiensis. They differ by the site of action, pH optima and mode of activation.[377] During sporogenesis of B. cereus no release was observed of ^{14}C-DAP-containing material from vegetative cell walls pre-labelled during growth. After the completion of intrasporangial development practically the whole amount of label was released from sporangia, only exogenous free ^{14}C-DAP being incorporated into spore muropeptides.[132] All factors retarding sporogenesis of B. megaterium and B. cereus also retard or prevent liberation of completed spores.[85, 146, 147] The sporangial wall of B. cereus is broken down gradually and the penetration of fluorescent antibodies to the surface of the endospore may take place before actual liberation of spores from the chains of cells.[222] Free spores of B. cereus and of some other bacilli are extremely hydrophobic and in aerated liquid media may accumulate in the foam immediately after release from sporangia.[378] Flotation ability of spores of bacilli may be successfully used for collecting free and clean spores[379] but can often result in serious difficulties in cell counting, estimation of heat resistance and germination rate.[128, 378]

Maturation of spores inside sporangia is also connected with the completion of its autonomous surface layers. These layers determine some of the new qualities of spores, e.g. altered permeability, specific response to nutrient substrates and germination stimulators and increased resistance to lytic enzymes of all kinds.

REFERENCES

1. Knaysi, G. (1948). Bact. Rev. 12, 19.
2. Fitz-James, P. C. (1963). In "Mécanismes de régulation des activités cellulaires chez les microorganismes" (J. Senez, ed.), p. 529. Gordon and Breach, New York.
3. Young, I. E. (1966). Personal communication.
4. Vinter, V. (1967). Folia. microbiol., Praha 12, 87.
5. Vinter, V. and Chaloupka, J. (1967). In "Cell Differentiation" (O. Nečas and M. Dvořák, eds), p. 29, Acta Fac. Medic. Univ. Brun.
6. Bayne-Jones, S. and Petrilli, A. (1933). J. Bact. 25, 251.
7. Grelet, N. (1946). C. r. hebd. Séanc. Acad. Sci., Paris 222, 418.
8. Hardwick, W. A. and Foster, J. W. (1952). J. gen. Physiol. 35, 907.
9. Leifson, E. (1931). J. Bact. 21, 331.
10. Cook, R. P. (1932). Biol. Rev. 7, 1.
11. "Symposium on the Biology of Bacterial Spores" (1952). Bact. Rev. 16, 89.
12. Murrell, W. G. (1955). "The Bacterial Endospore", University Sydney Press, Sydney.

G

13. Murrell, W. G. (1961). *Symp. Soc. gen. Microbiol.* **11**, 100
14. Halvorson, H. O. (1965). *Symp. Soc. gen. Microbiol.* **15**, 343.
15. Murrell, W. G. (1967). *In* "Advances in Microbial Physiology" (A. H. Rose and J. F. Wilkinson, eds), Vol. 1, p. 133. Academic Press Inc., New York and London.
16. Szulmajster, J. (1964). *Bull. Soc. Chim. biol. Paris* **46**, 443.
17. Manteifel, A. J. (1940). *Mikrobiologiya* **9**, 89.
18. Manteifel, A. J. (1941). *Mikrobiologiya* **10**, 419.
19. Grelet, N. (1957). *J. appl. Bact.* **20**, 315.
20. Wooley, B. C. and Collier, R. E. (1966). Bact. Proc. 16
21. Srinivasan, V. R. and Halvorson, H. O. (1963). *Nature, Lond.* **197**, 100.
22. Srinivasan, V. R. (1965). *In* "Spores III" (L. L. Campbell and H. O. Halvorson, eds), p. 64. Am. Soc. Microbiol., Ann Arbor, Michigan, U.S.A.
23. Srinivasan, V. R. (1966). *Nature, Lond.* **309**, 537.
24. Bergère, J. L. and Hermier, J. (1965). *Annls Inst. Pasteur, Paris,* **109**, 391.
25. Aubert, J.-P., Millet, J. and Castoriadis-May, C. (1961). *C.r. hebd. séanc. Acad. Sci., Paris* **253**, 1731.
26. Szulmajster, J., Blicharska, J. and Spotts, C. C. (1962) *C.r. hebd. séanc. Acad. Sci., Paris* **254**, 4533.
27. Kerravala, Z. J., Srinivasan, V. R. and Halvorson, H. O. (1964). *J. Bact.* **88**, 374.
28. Schaeffer, P., Ionesco, H., Ryter, A. and Balassa, G. (1963). *In* "Mécanismes de régulation des activités cellulaires chez microorganismes" (J. Senez ed.), p. 553. Gordon and Breach, New York.
29. Schaeffer, P., Millet, J. and Aubert, J.-P. (1965). *Proc. natn. Acad. Sci., U.S.A.* **54**, 704.
30. Bergère, J. L. and Hermier, J. (1964). *Annls Inst. Pasteur, Paris* **106**, 214.
31. Buchner, H. (1890). *Zentbl. Bakt. ParasitKde* **8**, 1.
32. Halvorson, H. O. (1957). *J. appl. Bact.* **20**, 305.
33. Monod, J. (1950). *Annls Inst. Pasteur, Paris* **79**, 390.
34. Perret, C. J. (1953). *Atti VI, Congr. Intern. Microbiol. Roma* I, 287.
35. Málek, I., Chaloupka, J., Vosyková, L and Vinter, V. (1953). *Čs. biol.* **2**, 323.
36. Jerusalimskij, N. D. and Rukina, E. A. (1956). *Mikrobiologiya* **25**, 649.
37. Jerusalimskij, N. D. (1958). *In* "Continuous Cultivation of Microorganisms", p. 62. Symp. Czech. Acad. Sci. Prague.
38. Jerusalimskij, N. D. and Rukina, E. A. (1959). *Mikrobiologiya* **28**, 801.
39. Aubert, J. P. and Millet, J. (1961). *C.r. hebd. séanc. Acad. Sci., Paris* **253**, 1880.
40. Tyrrell, E. A. (1962). Ph.D. Thesis, University of Michigan.
41. Humphrey, A. E., Kitai, A. and Cooney, C. L. (1966). *J. Fermt. Technol. Osaka* **44**, 283.
42. Foster, J. W. (1956). *Q. Rev. Biol.* **31**, 102.
43. Stedman, R. L. (1956). *Am. J. Pharm.* **128**, 84.
44. Ordal, Z. J. (1957). *In* "Spores" (H. O. Halvorson, ed.), p. 18. Amer. Inst. Biol. Sci., Washington, U.S.A.
45. Halvorson, H. (1962). *In* "Bacteria IV. Physiology of Growth", (I. C. Gunsalus and R. Y. Stanier, eds), p. 223. Academic Press Inc., New York and London.
46. Bernlohr, R. W. and Novelli, G. D. (1960). *Archs Biochem. Biophys.* **87**, 232.
47. Slepecky, R. A. and Law, J. H. (1961). *J. Bact.* **82**, 37.
48. Foster, J. W. and Heiligman, F. (1949). *J. Bact.* **57**, 613.

49. Levinson, H. S. and Sevag, M. G. (1953). *J. gen. Physiol.* **36**, 617.
50. Amaha, M., Ordal, Z. J. and Touba, A. (1956). *J. Bact.* **72**, 34.
51. Long, S. K. and Williams, O. B. (1960). *J. Bact.* **79**, 625.
52. Grelet, N. (1946). *Annls Inst. Pasteur, Paris* **72**, 153.
53. Perry, J. J. and Foster, J. W. (1955). *J. Bact.* **69**, 337.
54. Slepecky, R. A. and Foster, J. W. (1959). *J. Bact.* **78**, 117.
55. Lundgren, D. G. and Beskid, G. (1960). *Can. J. Microbiol.* **6**, 135.
56. Perkins, J. P., Louie, D. D. and Aronson, J. N. (1963). *Can. J. Microbiol.* **9**, 791.
57. Nakata, H. M. (1964). *J. Bact.* **88**, 1522.
58. Johnson, R. M. and Gilda, G. (1964). *Trans. Am. Microsc. Soc.* **83**, 226.
59. Donnellan, J. E. Jr., Nags, E. H. and Levinson, H. S. (1964). *J. Bact.* **87**, 332.
60. Aubert, J.-P. and Millet, J. (1963). *In* "Mécanismes de régulation des activités cellulaires chez les microorganismes" (J. Senez, ed.), p. 545. Gordon and Breach, New York.
61. Nasuno, S. and Asai, T. (1960). *J. gen. appl. Microbiol, Tokyo* **6**, 71.
62. Bergère, J. L. and Hermier, J. (1965). *Annls. Inst. Pasteur, Paris* **109**, 80.
63. Hitzman, D. O., Halvorson, H. O. and Ukita, T. (1957). *J. Bact.* **74**, 1.
64. Zoha, S. M. S. and Sadoff, H. L. (1958). *J. Bact.* **76**, 203.
65. Tsuji, K. and Perkins, W. E. (1962). *J. Bact.* **84**, 81.
66. Uehara, M., Fujioka, R. S. and Frank, H. A. (1965). *J. Bact.* **89**, 929.
67. Perkins, W. E. (1965). *J. appl. Bact.* **28**, 1.
68. Roberts, T. A. (1965). *J. appl. Bact.* **28**, 142.
69. Roberts, T. A. (1967). *J. appl. Bact.* **30**, 430.
70. Schneider, M. D., Grecz, N. and Anellis, A. (1963). *J. Bact.* **85**, 126.
71. Perkins, W. E. and Tsuji, K. (1962). *J. Bact.* **84**, 86.
72. Roberts, T. A. and Ingram, M. (1966). *In* "Botulism, 1966" (M. Ingram and T. A. Roberts, eds), p. 169. Chapman and Hall, London.
73. Pheil, Ch. G. and Ordal, Z. J. (1967). *Appl. Microbiol.* **15**, 893.
74. Curran, H. R. (1957). *In* "Spores" (H. O. Halvorson, ed.), p. 1. Am. Inst. Biol. Sci., Washington, U.S.A.
75. Charney, J., Fisher, W. P. and Hegarty, C. P. (1951). *J. Bact.* **62**, 145.
76. Grelet, N. (1952). *Annls Inst. Pasteur, Paris* **82**, 66.
77. Grelet, N. (1952). *Annls Inst. Pasteur, Paris* **83**, 71.
78. Weinberg, E. D. (1955). *J. Bact.* **70**, 289.
79. Weinberg, E. D. (1964). *Appl. Microbiol.* **12**, 436.
80. Ward, B. Q. (1947). Ph. D. Thesis, Univ. of Texas.
81. Lundgren, D. G. and Cooney, J. J. (1962). *J. Bact.* **83**, 1287.
82. Kolodziej, B. J. and Slepecky, R. A. (1962). *Bact. Proc.*, 48.
83. Kolodziej, B. J. and Slepecky, R. A. (1964). *J. Bact.* **88**, 821.
84. Sugiyama, H. (1951). *J. Bact.* **62**, 81.
85. Vinter, V. (1957). *J. appl. Bact.* **20**, 325.
86. Black, S. H., Hashimoto, T. and Gerhardt, P. (1960). *Can. J. Microbiol.* **6**, 213.
87. Halvorson, H. and Howitt, C. (1961). *In* "Spores II", (H. O. Halvorson, ed.), p. 149. Burgess Publishing Co., Minneapolis, Minn., U.S.A.
88. Young, I. E. and Fitz-James, P. C. (1962). *J. Cell. Biol.* **12**, 115.
89. Cooney, J. J. and Lundgren, D. G. (1962). *Can. J. Microbiol.* **8**, 823.
90. Levinson, H. S. and Hyatt, M. T. (1964). *J. Bact.* **87**, 876.
91. Fleming, H. P. and Ordal, Z. J. (1964). *J. Bact.* **88**, 1529.
92. El-Bisi, H. M. and Ordal, Z. J. (1956). *J. Bact.* **71**, 1.

93. Amaha, M. and Ordal, Z. J. (1957). *J. Bact.* **74**, 596.
94. Pelcher, E. A., Fleming, H. P. and Ordal, Z. J. (1963). *Can. J. Microbiol.* **9**, 251.
95. Lechowich, R. V. and Ordal, Z. J. (1962). *Can. J. Microbiol* **8**, 287.
96. Murrell, W. G. and Warth, A. D. (1965). *In* "Spores III", (L. L. Campbell and H. O. Halvorson, eds), p. 1. Am. Soc. Microbiol., Ann Arbor, Michigan, U.S.A.
97. Foerster, H. F. and Foster, J. W. (1966). *J. Bact.* **91**, 1333.
98. Nakata, H. M. and Halvorson, H. O. (1960). *J. Bact.* **80**, 801.
99. Day, L. E. and Costilow, R. N. (1964). *J. Bact.* **88**, 690.
100. Falcone, G., Armani, G., Previtera, A. and Bertini, V. (1966). *Boll. Ist. Sieroter. Milanese* **45**, 11.
101. Beskid, G. and Lundgren, D. G. (1961). *Can. J. Microbiol.* **7**, 543.
102. Tinelli, R. (1955). *Annls Inst. Pasteur, Paris* **88**, 1.
103. Tinelli, R. (1955). *Annls Inst. Pasteur, Paris* **88**, 642.
104. Roth, N. G., Lively, D. H. and Hodge, H. M. (1955). *J. Bact.* **69**, 455.
105. Collier, R. E. (1957). *In* "Spores" (H. O. Halvorson, ed.), p. 10. Am. Inst. Biol. Sci., Washington, U.S.A.
106. Foster, J. W. and Heiligman, F. (1949). *J. Bact.* **57**, 639.
107. Williams, O. B. and Harper, O. F., Fr. (1951). *J. Bact.* **61**, 551.
108. Lund, A. J., Janssen, F. W. and Anderson, L. E. (1957). *J. Bact.* **74**, 577.
109. Brady, R. J., Chan, E. C. S. and Pelczar, M. J. Jr. (1961). *J. Bact.* **81**, 725.
110. Borowski, E. and Krynski, S. (1963). *Chemotherapia* **6**, 289.
111. Foster, J. W. and Perry, J. J. (1954). *J. Bact.* **67**, 295.
112. Schreiber, O. (1896). *Zentbl. Bakt. ParasitKde.* **20**, 353, 429.
113. Powell, J. F. and Hunter, J. R. (1953). *J. gen. Physiol.* **36**, 601.
114. Black, S. H. and Gerhardt, P. (1963). *Ann. N.Y. Acad. Sci.* **102**, 755.
115. Vinter, V. and Slepecky, R. A. (1965). *J. Bact.* **90**, 803.
116. Portellada, P. C. L. (1959). *Anais Microbiol.* **7**, 71.
117. Portellada, P. C. L. (1959). *Anais Microbiol.* **7**, 77.
118. Portellada, P. C. L. (1960). *Anais Microbiol.* **8**, 65.
119. Tinelli, R. (1955). *Annls Inst. Pasteur, Paris* **88**, 364.
120. Nakada, D., Matsushiro, A. and Miwatani, T. (1956). *Med. J. Osaka Univ.* **6**, 1047.
121. Young, I. E. and Fitz-James, P. C. (1959). *J. biophys. biochem. Cytol.* **6**, 467.
122. Young, I. E. and Fitz-James, P. C. (1959), *J. biophys. biochem. Cytol.* **6**, 483.
123. Monro, R. E. (1961). *Biochem. J.* **81**, 225.
124. Bernlohr, R. W. (1965). *In* "Spores III", (L. L. Campbell and H. O. Halvorson, eds), p. 75. Am. Soc. Microbiol., Ann Arbor, Michigan, U.S.A.
125. Balassa, G. (1966). *Annls Inst. Pasteur, Paris* **110**, 316.
126. Keynan, A., Murrell, W. G. and Halvorson, H. O. (1962). *J. Bact.* **83**, 395.
127. Vinter, V. (1959). *Nature, Lond.* **183**, 998.
128. Vinter, V. (1960). *Folia microbiol., Praha* **5**, 217.
129. Vinter, V. (1959). *Folia microbiol., Praha* **4**, 216.
130. Kadota, H. and Uchida, A. (1967). *J. Agric. chem. Soc. Japan* (In press.).
131. Jičínská, E. (1964). *Folia microbiol., Praha* **9**, 73.
132. Vinter, V. (1963). *Folia microbiol., Praha* **8**, 147.
133. Buono, F., Testa, R. and Lundgren, D. G. (1966). *J. Bact.* **91**, 2291.
134. Ramaley, R. F. and Bernlohr, R. W. (1966). *Archs. Biochem. Biophys.* **117**, 34.
135. Ramaley, R. F. and Bernlohr, R. W. (1966). *J. molec. Biol.* **11**, 842.
136. Romig, W. R. and Wyss, O. (1957). *J. Bact.* **74**, 386.

137. Hashimoto, T., Black, S. H. and Gerhardt, P. (1960). *Can. J. Microbiol.* **6**, 203.
138. Rowley, D. B. and Newcomb, H. R. (1964). *J. Bact.* **87**, 701.
139. Brock, T. D. (1962). *Nature, Lond.* **195**, 309.
140. Weisová, H., Vinter, V. and Stárka, J. (1966). *Folia microbiol., Praha* **11**, 387.
141. Maaløe, O. (1963). *In* "The Bacteria" (I. C. Gunsalus and R. Y. Stanier, eds), Vol. IV, p. 1. Academic Press Inc., New York and London.
142. Zeuthen, E. (ed.) (1964). "Synchrony in Cell Division and Growth", Interscience Publishers, Inc., New York.
143. Cameron, I. L. and Padilla, G. M. (eds), (1966). "Cell Synchrony Studies in Biosynthetic Regulation". Academic Press, Inc., New York and London.
144. Vinter, V. (1961). *In* "Spores II" (H. O. Halvorson, ed.), p. 127. Burgess Publishing Co., Minneapolis, Minn. U.S.A.
145. Szulmajster, J. and Canfield, R. E. (1963). *In* "Mécanismes de régulation des activités cellulaires chez les microorganismes". (Senez J., ed.), p. 587. Gordon and Breach, New York.
146. Vinter, V. (1962). *Folia microbiol., Praha* **7**, 275.
147. Vinter, V. (1964). *Folia microbiol., Praha* **9**, 58.
148. Szulmajster, J., Canfield, R. E. and Blicharska, J. (1963). *C.r. hebd. séanc. Acad. Sci., Paris* **256**, 2057.
149. Balassa, G. (1963). *In* "Mécanismes de regulation des activités cellulaires chez les microorganismes" (J.S enez, ed.), p. 6. Gordon and Breach, New York.
150. Aronson, A. I. and Rosas del Valle M. (1964). *Biochim. biophys. Acta.* **87**, 267.
151. Ryter, A. and Szulmajster, J. (1965). *Annls Inst. Pasteur, Paris* **108**, 640.
152. Balassa, G. (1966). *Annls Inst. Pasteur, Paris* **110**, 175.
153. Young, I. E. and Fitz-James, P. C. (1959). *J. biophys. biochem, Cytol.* **6**, 499.
154. Stahly, D. P., Srinivasan, V. R. and Halvorson, H. O. (1966). *J. bact.* **91**, 1875.
155. Ryter, A., Schaeffer, P. and Ionesco, H. (1966). *Annls. Inst. Pasteur, Paris* **110**, 305.
156. Gordon, R. A. and Murrell, W. G. (1967). *J. Bact.* **93**, 495.
157. Wooley, B. C. and Collier, R. E. (1965). *Can. J. Microbiol.* **11**, 279.
158. Gillis, J. R., Rosas del Valle, M., Vinter, V. and Slepecky, R. A. (1965). *Bact. Proc.* 37.
159. Imanaka, H., Gillis, J. R. and Slepecky, R. A. (1967). *J. Bact.* **93**, 1624.
160. Imanaka, H. and Slepecky, R. A. (1967). *Bact. Proc.*, 28.
161. Holmes, P. K. and Levinson, H. S. (1967). *J. Bact.* **94**, 434.
162. Freer, J. H. and Levinson, H. S. (1967). *J. Bact.* **94**, 441.
163. MacKechnie, I. and Hanson, R. S. (1967). *Bact. Proc.* 29.
164. Grelet, N. (1951). *Annls Inst. Pasteur, Paris* **81**, 430.
165. Keynan, A., Strecker, H. J. and Waelsch, H. (1954). *J. biol. Chem.* **211**, 883
166. Pepper, R. E. and Costilow, R. N. (1964). *J. Bact.* **87**, 303.
167. Blumenthal, H. J. (1965). *In* "Spores III" (L. L. Campbell and H. O. Halvorson, eds), p. 222. Am. Soc. Microbiol., Ann Arbor, Michigan, U.S.A.
168. Goldman, M. and Blumenthal, H. J. (1961). *Can. J. Microbiol.* **7**, 677.
169. Goldman, M. and Blumenthal, H. J. (1963). *J. Bact.* **86**, 303.
170. Goldman, M. and Blumenthal, H. J. (1964). *J. Bact.* **87**, 377.
171. Doi, R., Halvorson, H. and Church, B. D. (1959). *J. Bact.* **77**, 43.
172. Halvorson, H. and Church, B. D. (1957). *J. appl. Bact.* **20**, 359.

173. Halvorson, H., O'Connor, R. and Doi, R. (1961). *In* "Cryptobiotic Stages in Biological Systems" (N. Grossowicz, S. Hestrin and A. Keynan, eds), p. 71. Elsevier Publishing Co., Amsterdam.

174. Church, B. D. and Halvorson, H. (1957). *J. Bact.* **73**, 470.

175. Nakada, D., Matsushiro, A., Kondo, M., Suga, K. and Konishi, K. (1957). *Med. J. Osaka Univ.* **7**, 809.

176. Goldman, M. and Blumenthal, H. J. (1964). *J. Bact.* **87**, 387.

177. Halvorson, H. O. (1961). *In* "Growth in Living Systems" (M. X. Zarrow, ed.), p. 107. Basic Books, Inc., New York.

178. Hanson, R. S., Srinivasan, V. R. and Halvorson, H. (1963). *J. Bact,* **85**, 451.

179. Hanson, R. S., Srinivasan, V. R. and Halvorson, H. O. (1963). *J. Bact.* **86**, 45.

180. Megraw, R. E. and Beers, R. J. (1964). *J. Bact.* **87**, 1087.

181. Hanson, R. S., Blicharska, J. and Szulmajster, J. (1964). *Biochem. biophys. Res. Commun.* **17**, 1.

182. Hanson, R. S., Blicharska, J., Arnaud, M. and Szulmajster, J. (1964). *Biochem. biophys. Res. Commun.* **17**, 690.

183. Szulmajster, J. and Hanson, R. S. (1965). *In* "Spores III" (L. L. Campbell and H. O. Halvorson, eds), p. 163. Am. Soc. Microbiol., Ann Arbor, Michigan, U.S.A.

184. Hanson, R. S. and Cox, D. P. (1967). *J. Bact.* **93**, 1777.

185. Martin, H. H. and Foster, J. W. (1958). *J. Bact.* **76**, 167.

186. Szulmajster, J. and Schaeffer, P. (1961). *Biochem. biophys. Res. Commun.* **6**, 217.

187. Szulmajster, J. and Schaeffer, P. (1961). *C.r. hebd. séanc. Acad. Sci., Paris* **252**, 220.

188. Bernlohr, R. W. and Novelli, G. D. (1959). *Nature, Lond.* **184**, 1256.

189. Bernlohr, R. W. and Novelli, G. D. (1960). *Biochim. biophys. Acta* **41**, 541.

190. Gollakota, G. K. and Halvorson, H. O. (1960). *J. Bact.* **79**, 1.

191. Gollakota, G. K. and Halvorson, H. O. (1961). *In* "Spores II" (H. O. Halvorson, ed.), p. 113. Burgess Publishing Co., Minneapolis, Minn. U.S.A.

192. Gollakota, G. K. and Halvorson, H. O. (1963). *J. Bact.* **85**, 1386.

193. Kominek, L. A. and Halvorson, H. O. (1964). *J. Bact.* **88**, 263.

194. Nakata, H. M. (1963). *J. Bact.* **86**, 577.

195. Tinelli, R. (1955). *Annls Inst. Pasteur, Paris* **88**, 212.

196. Yoneda, M. and Kondo, K. (1959). *Biken's J.* **2**, 247.

197. Kondo, M., Yoneda, M., Nishi, Y. and Fukai, K. (1961). *Biken's J.* **4**, 41.

198. Nakata, H. M. (1966). *J. Bact.* **91**, 784.

199. Stevenson, J., Miller, K., Strothman, R. and Slepecky, R. A. (1962). *Bact. Proc.*, 47.

200. Knaysi, G. (1945). *J. Bact.* **49**, 473.

201. Millet, J. and Aubert, J. -P. (1960). *Annls Inst. Pasteur, Paris* **98**, 282.

202. Grelet, N. (1955). *Annls Inst. Pasteur, Paris* **88**, 60.

203. Buono, F., Nass, J. and Lundgren, D. G. (1967). *Bact. Proc.*, 28.

204. Spotts, C. R. and Szulmajster, J. (1962). *Biochim. biophys. Acta* **61**, 635

205. Aubert, J. -P. and Millet, J. (1963). *C.r. hebd. séanc. Acad. Sci., Paris* **256**, 5442.

206. Balassa, G. (1963). *Biochim. biophys. Acta* **76**, 410.

207. Balassa, G. (1964). *Biochem. biophys. Res. Commun.* **15**, 236.

208. Balassa, G. (1964). *Biochem. biophys. Res. Commun.* **15**, 240.

209. Aronson, J. N. and Wermus, G. R. (1965). *J. Bact.* **90**, 38.

210. Leitzmann, C. and Bernlohr, R. W. (1965). *J. Bact.* **89**, 1506.
211. Schaeffer, P. (1965). *Symp. Biol. Hung.* **6**, 147.
212. Lamanna, C. (1940). *J. infect. Dis.* **67**, 193.
213. Lamanna, C. (1940). *J. infect. Dis.* **67**, 205.
214. Tomcsik, J. and Baumann-Grace, J. B. (1959). *J. gen. Microbiol.* **21**, 666.
215. Norris, J. R. and Wolf, J. (1961). *J. appl. Bact.* **24**, 42.
216. Norris, J. R. (1962). *J. gen. Microbiol.* **28**, 393.
217. Cavallo, G., Falcone, G. and Imperato. S. (1963). *Bact. Proc.*, 25.
218. Walker, P. D., Baillie, A., Thomson, R. O. and Batty, I. (1966). *J. appl. Bact.* **29**, 512.
219. Tomcsik, J. and Baumann-Grace, J. B. (1958). *Schweiz. Z. Path. Bakt.* **21**, 914.
220. Baillie, A. and Norris, J. R. (1963). *J. appl. Bact.* **26**, 102.
221. Walker, P. D. and Batty, I. (1964). *J. appl. Bact.* **27**, 137.
222. Walker, P. D. and Batty, I. (1965). *J. appl. Bact.* **28**, 194.
223. Falcone, G. and Imperato, S. (1966). *G. Microbiol.* **14**, 7.
224. Bernlohr, R. W. and Novelli, G. D. (1963). *Archs Biochem. Biophys.* **103**, 94.
225. Bernlohr, R. W. (1964). *J. biol. Chem.* **239**, 538.
226. Michel, J. F. (1967). *Folia microbiol.*, Praha **12**, 297.
227. Balassa, G., Ionesco, H. and Schaeffer, P. (1963). *C.r. hebd. séanc. Acad. Sci.*, Paris **257**, 986.
228. Kominek, L. A. and Halvorson, H. O. (1966). *J. Bact.* **90**, 1251.
229. Millet, J. and Pineau, E. (1960). *C.r. hebd. séanc. acad. Sci.*, Paris **250**, 1362.
230. Aubert, J.-P., Millet, J., Pineau, E. and Milhaud, G. (1961). *Biochim. biophys. Acta* **51**, 529.
231. Sussman, A. S. and Halvorson, H. O. (1966). "Spores: Their Dormancy and Germination". Harper and Row, New York.
232. Halvorson, H. O., Vary, J. C. and Steinberg, W. (1966). *A. Rev. Microbiol.* **20**, 169.
233. Vinter, V. (1963). *Experientia* **19**, 307.
234. Powell, J. F. and Strange, R. E. (1957). *Biochem. J.* **65**, 700.
235. Strange, R. E. and Dark, F. A. (1957). *J. gen. Microbiol.* **16**, 236.
236. Strange, R. E. and Dark, F. A. (1957). *J. gen. Microbiol.* **17**, 525.
237. Bach, J. A. and Sadoff, H. L. (1962). *J. Bact.* **83**, 699.
238. Powell, J. F. and Strange, R. E. (1956). *Biochem. J.* **63**, 661.
239. Sadoff, H. L. (1961). *In* "Spores II" (H. O. Halvorson, ed.), p. 180. Burgess Publishing Co., Minneapolis, Minn. U.S.A.
240. Lawrence, N. L. and Halvorson, H. O. (1954). *J. Bact.* **68**, 334.
241. Stewart, B. T. and Halvorson, H. O. (1953). *J. Bact.* **65**, 160.
242. Stewart, B. T. and H. O. Halvorson (1954). *Archs Biochem. Biophys.* **49**, 168.
243. Simmons, R. F. and Costilow, R. N. (1962). *J. Bact.* **84**, 1274.
244. Green, J. H. and Sadoff, H. L. (1965). *J. Bact.* **89**, 1499.
245. Baillie, A. and Norris, J. R. (1964). *J. Bact.* **87**, 1221.
246. Hannay, C. L. and Fitz-James, P. C. (1955). *Can. J. Microbiol.* **1**, 694.
247. Angus, T. A. (1956). *Can. J. Microbiol.* **2**, 122.
248. Fitz-James, P. C. and Young, I. E. (1958). *J. biophys. biochem. Cytol.* **4**, 639.
249. Heimpel, A. M. and Angus, T. A. (1960). *Bact. Rev.* **24**, 266.
250. Nomura, H. and Yoshikawa, H. (1959). *Biochim. biophys. Acta* **31**, 125.
251. Bernlohr, R. W. and Sievert, C. (1962). *Biochem. biophys. Res. Commun.* **9**, 32.
252. Jann, G. J. and Eichhorn, H. H. (1964). *Bact. Proc.*, 13.

253. Sebald, M. and Schaeffer, P. (1965). *C.r. hebd. séanc. Acad. Sci.*, Paris **260**, 5398.
254. Kondo, M., Takeda, Y. and Yoneda, M. (1964). *Biken's J.* **7**, 153.
255. Lee, G. L. C., Srinivasan, V. R. and Halvorson, H. O. (1964). *Bact. Proc.*, 13.
256. Ramaley, R. F. and Bernlohr, R. W. (1966). *J. biol. Chem.* **241**, 620.
257. Laishley, E. J. and Bernlohr, R. W. (1966). *Biochem. biophys. Res. Commun.* **24**, 85.
258. Stahly, D. P. and Bernlohr, R. W. (1966). *Bact. Proc.*, 80.
259. Levisohn, S. and Aronson, A. I. (1967). *J. Bact.* **94**. (In press.)
260. Keilin, D. and Hartree, E. F. (1949). *Nature, Lond.* **164**, 254.
261. Spencer, E. J. and Powell, J. F. (1952). *Biochem. J.* **51**, 239.
262. Hachisuka, Y., Asano, N., Kaneko, M. and Kanbe, T. (1956). *Science, N.Y.* **124**, 174.
263. Doi, R. H. and Halvorson, H. (1961). *J. Bact.* **81**, 51.
264. Aubert, J.-P. and Millet, J. (1965). *C.r. hebd. séanc. Acad. Sci.*, Paris **261**, 4274.
265. Costilow, R. N. (1962). *J. Bact.* **84**, 1268.
266. Mitruka, B. M. and Costilow, R. N. (1967). *J. Bact.* **93**, 295.
267. Ryter, A., Bloom, B. and Aubert J.-P. (1966). *C.r. hebd. séanc. Acad. Sci.*, Paris **262**, 1305.
268. Aubert, J.-P. and Millet, J. (1963). *C.r. hebd. séanc. Acad. Sci.*, Paris **256**, 1866.
269. Aronson, A. I. (1965). *J. molec. Biol.* **11**, 576.
270. Aronson, A. I. (1965). *J. molec. Biol.* **13**, 92.
271. Aronson, A. I. (1965). *Fedn Proc.* **24**, 282.
272. Berger, J. A. and Marr, A. G. (1960). *J. gen. Microbiol.* **22**, 147.
273. Chaloupka, J. Křečková, P. and Říhová, L. (1962). *Folia microbiol.*, Praha **7**, 269.
274. Chaloupka, J. Křečková, P. and Říhová, L. (1962). *Experientia* **18**, 362.
275. Chaloupka, J., Říhová, L. and Křečková, P. (1964). *Folia microbiol.*, Praha **9**, 9.
276. Chaloupka, J. (1966). Personal communication.
277. Ellar, D. J. and Lundgren, D. G. (1966). *J. Bact.* **92**, 1748.
278. Salton, M. R. J. (1955). *J. gen. Microbiol.* **13**, iv.
279. Stárka, J. and Čáslavská, J. (1964). *Folia microbiol.*, Praha **9**, 21.
280. Fitz-James, P. C. (1964). *J. Bact.* **87**, 667.
281. Vinter, V. (1965). *In* "Spores III" (L. L. Campbell and H. O. Halvorson, eds), p. 25. Am. Soc. Microbiol., Ann Arbor, Michigan, U.S.A.
282. Vinter, V. (1965). *Folia microbiol.*, Praha **10**, 230.
283. Vinter, V. (1965). *Folia microbiol.*, Praha **10**, 288.
284. Warth, A. D., Ohye, D. F. and Murrell, W. G. (1963). *J. Cell. Biol.* **16**, 579.
285. Warth, A. D., Ohye, D. F. and Murrell, W. G. (1963). *J. Cell Biol.* **16**, 593.
286. Vinter, V. (1962). *Experientia* **18**, 409.
287. Lewis, J. C., Snell, N. S. and Burr, H. K. (1960). *Science, N.Y.* **132**, 544.
288. Hitchins, A. D. and Gould, G. W. (1964). *Nature, Lond.* **203**, 895.
289. Vinter, V. and Šťastná, J. (1967). *Folia microbiol.*, Praha **12**, 301.
290. Blair, E. B. (1950). *Tex. Rep. Biol. Med.* **8**, 361.
291. Pfennig, N. (1957). *Arch. Mikrobiol.* **26**, 345.
292. Widra, A. (1956). *J. Bact.* **71**, 689.
293. Vinter, V. (1958). IV Int. Congr. Biochem., Vienna, *Abstracts*, p. 129.
294. Vinter, V. (1959). *Folia microbiol.*, Praha **4**, 1.

295. Vinter, V. (1962). *Nature, Lond.* **196,** 1336.
296. Fitz-James, P. C. (1962). *Proc. VIIIth Congr. of Microbiology, Montreal,* A 2, 1.
297. Aronson, A. I. and Fitz-James, P. C. (1966). *Bact. Proc.,* 31.
298. Kadota, H. and Iijima, K. (1965). *Agric. biol. Chem.* **29,** 80.
299. Kadota, H., Iijima, K. and Uchida, A. (1965). *Agric. biol. Chem.* **29,** 870.
300. Hiragi, Y., Iijima, K. and Kadota, H. (1967). *Nature, Lond.* **215,** 154.
301. Crick, F. H. C. and Rich, A. (1955). *Nature, Lond.* **176,** 780.
302. Bott, K. F. and Lundgren, D. G. (1964). *Radiat. Res.* **21,** 195.
303. Mortenson, J. E. and Beinert, H. (1953). *J. Bact.* **66,** 101.
304. Brachet, J. (1961). *In* "Growth in Living Systems" (M. Zarrow, ed.), p. 241, Basic Books, New York.
305. Vinter, V. and Věchet, B. (1964). *Folia microbiol., Praha* **9,** 238.
306. Gould, G. W. and Hitchins, A. D. (1963). *Nature, Lond.* **197,** 622.
307. Gould, G. W. and Hitchins, A. D. (1963). *J. gen. Microbiol.* **33,** 413.
308. Gould, G. W. and Hitchins, A. D. (1965). *In* "Spores III" (L. L. Campbell and H. O. Halvorson, eds), p. 213. Am. Soc. Microbiol, Ann Arbor, Michigan, U.S.A.
309. Keynan, A., Evenchik, Z., Halvorson, H. O. and Hastings, J. W. (1964). *J. Bact.* **88,** 313.
310. Blankenship, L. C. and Pallansch, M. J. (1966). *J. Bact.* **92,** 1615.
311. Vinter, V. (1962). *Folia microbiol., Praha* **7,** 115.
312. Nickerson, W. F. and Falcone, G. (1959). *In* "Sulphur in Proteins" (R. Benesch, R. E. Benesch, P. D. Boyer, I. M. Klotz, W. R. Middlebrook, A. G. Szent-Gyögyi and D. R. Schwarz, eds), p. 409. Academic Press Inc., New York and London.
313. Curran, H. R., Brunstetter, B. C. and Myers, A. T. (1943). *J. Bact.* **45,** 485.
314. Vinter, V. (1956). *Folia biol., Praha* **2,** 216.
315. Vinter, V. and Vechet, B. (1964). *Folia microbiol., Praha* **9,** 352.
316. Vinter, V. (1965). Conference of IAEA, *Publ. No.* 653207, F.A.O., Vienna.
317. Levinson, H. S., Hyatt, M. T. and More, F. E. (1961). *Biochem. biophys. Res. Commun.* **5,** 417.
318. Vinter, V. (1957). *Folia biol., Praha,* 3, 193.
319. Powell, J. F. (1953). *Biochem. J.* **54,** 210.
320. Martin, H. H. and Foster, J. W. (1958). *Arch. Mikrobiol.* **31,** 171.
321. Slepecky, R. A. (1961). *In* "Spores II" (H. O. Halvorson, ed.), p. 171. Burgess Publishing Co., Minneapolis, Minn., U.S.A.
322. Holsinger, V. H., Blankenship, L. C. and Pallansch, M. J. (1967). *Archs Biochem. Biophys.* **119,** 282.
323. Lewis, J. C. (1967). *Analyt. Biochem.,* **19,** 327
324. Janssen, F. W., Lund, A. J. and Anderson, L. E. (1958). *Science, N.Y.* **127,** 26.
325. Day, L. E. and Costilow, R. N. (1964). *J. Bact.* **88,** 695.
326. Takagi, A., Kawata, T., Yamamoto, S., Kubo, T. and Okita, S. (1960). *Jap. J. Microbiol.* **4,** 137.
327. Gerhardt, P. and Ribi, E. (1964). *J. Bact.* **88,** 1774.
328. Powell, J. F. and Strange, R. E. (1959). *Nature, Lond.* **184,** 878.
329. Tanenbaum, S. W. and Kaneko, K. (1964). *Biochemistry.* **3,** 1314.
330. Benger, H. (1962). *Z. Hyg. Infektionskr.* **148,** 318.
331. Benger, H. (1967). *Zentbl. Bakt. ParasitKde* (In press.)
332. Bach, M. and Gilvarg, C. (1964). *Fedn. Proc.* **23,** 313.

333. Hodson, P. H. and Foster, J. W. (1966). *J. Bact.* **91**, 562.
334. Kanie, M., Fujimoto, K. S. and Foster, J. W. (1966). *J. Bact.* **91**, 570.
335. Bach, M. L. and Gilvarg, C. (1966). *J. biol. Chem.* **241**, 4563.
336. Aronson, A. I., Henderson, E. and Tincher, A. (1967). *Biochem. biophys. Res. Commun.* **26**, 454.
337. McDonald, W. C. and Wyss, O. (1959). *Radiat Res.* **11**, 409.
338. Kondo, M., Teshima, K. and Kawasaki, C. (1967). *Jap. J. Bact.* **22**. (In press.)
339. Vinter, V. (1961). *Nature, Lond.* **189**, 589.
340. Eldjarn, L. and Pihl, A. (1957). *J. biol. Chem.* **225**, 499.
341. Alexander, P. (1962). *Trans. N.Y. Acad. Sci.* **24**, 966.
342. Ormerod, M. G. and Alexander, P. (1962). *Nature, Lond.* **193**, 290.
343. Powers, E. L. and Kaleta, B. F. (1960). *Science, N.Y.* **132**, 959.
344. Powers, E. L. (1961). *J. cell. comp. Physiol.* **58**, 13.
345. Eldjarn, L. and Pihl, A. (1958). *Radiat. Res.* **9**, 110.
346. Gordy, W., Ard, W. B. and Shields, H. (1955). *Proc. natn. Acad. Sci., U.S.A.* **41**, 983.
347. Gordy, W. and Shields, H. (1958). *Radiat. Res.* **9**, 611.
348. Norman, A. and Ginoza, W. (1958). *Radiat. Res.* **9**, 77.
349. Gordy, W. and Miyagawa, I. (1960). *Radiat. Res.* **12**, 211.
350. Smaller, B. and Avery, E. C. (1959). *Nature, Lond.* **183**, 539.
351. Hitchins, A. D., King, W. L. and Gould, G. W. (1966). *J. appl. Bact.* **29**, 505.
352. Cavallini, D., Mondovi, B., Giovanella, B. and De Marco, C. (1960). *Science, N.Y.* **131**, 1441.
353. Ray, D. K., Hutchinson, F. and Morowitz, H. J. (1960). *Nature, Lond.* **186**, 312.
354. Gould, G. W. and Ordal, Z. J. (1967). *J. gen. Microbiol.*, **50**, 77.
355. Powell, J. F. and Strange, R. E. (1953). *Biochem. J.* **54**, 205.
356. Sadoff, H. L. (1966). *Biochem. biophys. Res. Commun.* **24**, 691.
357. Smith, A. G. and Ellner, P. D. (1957). *J. Bact.* **73**, 1.
358. Cash, J. D. and Collee, J. G. (1962). *J. appl. Bact.* **25**, 225.
359. Takagi, A., Kawata, T. and Yamamoto, S. (1960). *J. Bact.* **80**, 37.
360. Grecz, N., Anellis, A. and Schneider, M. D. (1962). *Bact. Proc.*, 29.
361. Ivanovics, G. and Alfoldi, L. (1955). *Acta microbiol., Polon.* **2**, 275.
362. Greenberg, R. A. and Halvorson, H. O. (1955). *J. Bact.*, **69**, 45.
363. Nomura, M. and Hosoda, J. (1956). *J. Bact.* **72**, 573.
364. Dark, F. A. and Strange, R. E. (1957). *Nature, Lond.* **180**, 759.
365. Norris, J. R. (1957). *J. gen. Microbiol.* **16**, 1.
366. Chaloupka, J. (1967). *Folia microbiol., Praha* **12**, 264.
367. Young, F. E. (1964). *Bact. Proc.*, 31.
368. Singer, H. J. and Church, B. D. (1964). *Bact. Proc.*, 32.
369. Brown, W. C. and Fraser, D. K. (1967). *Bact. Proc.*, 29.
370. Strange, R. E. (1959). *Bact. Rev.* **23**, 1.
371. Work, E. (1961). *J. gen. Microbiol.* **25**, 167.
372. Young, F. E. (1966). *J. Bact.* **92**, 839.
373. Pickering, B. T. (1966). *Biochem. J.* **100**, 430.
374. Novotný, P. and Čáslavská, J. (1967). *Folia microbiol., Praha* **12**, 274.
375. Rogers, H. J. and Perkins, H. R. (eds). "Cell Walls and Membranes", E. & F. N. Spon, Ltd, London. (In press.)
376. Tomcsik, J. and Bouille, M. (1961). *Annls Inst. Pasteur, Paris*, **100**, 25.

377. Kingan, S. L. and Ensign, J. C. (1967). *Bact. Proc.*, 30.
378. Dobiáš, B. and Vinter, V. (1966). *Folia. microbiol, Praha* **11,** 314.
379. Gaudin, A. M., Mular, A. L. and O'Connor, R. F. (1960). *Appl. Microbiol.*, **8,** 91.

CHAPTER 4

Changes in Nucleic Acids During Sporulation

ROY H. DOI

Department of Biochemistry and Biophysics, University of California, Davis, California, U.S.A.

I. Introduction

SPORULATION in bacteria is limited to species which have the genetic potential to undergo this morphological change. Moreover, this conversion of an active vegetative cell to a dormant spore is dependent on proper growth conditions and a regulated sequence of biochemical events.[1] The central role of nucleic acids in the storage, transcription, and translation of genetic information is well documented. It is likely, therefore, that sporulation results from an effect of the cellular environment on the bacterial genome, expressed through the process of differential protein synthesis. The specific factor(s) which initiates sporulation is unknown. Various metabolic and environmental changes which occur prior to sporulation are discussed in Chapters 3, 5, and 6. The elucidation of the manner in which these changes affect gene expression in sporulating cells is necessary in order to obtain a clear understanding of the total process. In the absence of precise knowledge on the regulation of gene expression in sporulating cells, investigations have been directed towards understanding the biochemical relationships of DNA and the various RNA fractions.

There are several questions concerning the properties and functions of these fractions: (1) What is the replicative state, the origin, the amount and the chemical nature of the DNA incorporated into spores; (2) What are the properties, functions and relationships of the various RNA species during sporulation; (3) Is the complete machinery for transcription and translation of the genome present in dormant spores; (4) Is there a genetic relationship between spore-forming bacilli? The purpose of this chapter is to present modern information related to these questions.

Although much information is available on these topics, a general picture of the sporulating cell is hampered by practical problems. First, most studies present at best an average picture of a population of cells which are in different stages of morphological development, i.e. there is a lack of cellular homogeneity in sporulating cultures. Conditions have been derived to maximize the synchrony of sporulation, at any given time, but there are still a number of different forms, since the appearance of endospores occurs over a period of a few hours. Second, the studies on dormant spores, which appear to be a relatively homogeneous population, present technical problems such as difficulty in breakage of the spores and in quantitative recovery of various cellular components. The high level of nucleases in sporulating cells and spores which are liberated on breakage is a third problem.

Aside from the technical aspects the other major factor which makes integration of cellular events during sporulating somewhat tenuous, is the differences between the diverse *Bacillus* species studied. Surprisingly even strains of the same species may display large differences. The variations observed between related species may or may not be important, however, in the actual process of sporulation. The composite picture obtained from studies on different organisms is qualified therefore by several exceptions. In spite of these difficulties the general properties and functions of nucleic acid fractions of sporulating cells have been analysed and a rather consistent pattern has emerged which will be discussed in this chapter.

Most of the results were derived from studies on *Bacillus* species. Very little information is available about the nucleic acid fractions of *Clostridium* species and *Sporosarcina ureae*. The nucleic acid fractions of the following *Bacillus* species have been studied most extensively: *B. cereus*, *B. subtilis* and *B. megaterium*. Much of the data is true for only a single species or even a strain and this will be noted in most cases.

II. General Pattern of Nucleic Acids During Sporulation

The RNA and DNA contents of sporulating cells generally follow the patterns illustrated in Fig. 1. The rapid synthesis of RNA and DNA during

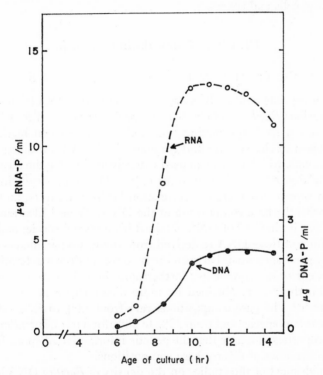

FIG. 1. The RNA and DNA content of sporulating cells. Sporulation begins at 10 hr and endospores begin to appear at 14 hr. (Reproduced in part by permission of The Rockefeller University Press from Ref. 2).

the exponential phase of growth ceases near the end of the log phase.[2] Usually no net synthesis of RNA or DNA occurs during the sporulation phase which covers a period of several hours after the end of growth. In some cases a slow synthesis of DNA occurs during sporulation, but a constant level is reached before endospore formation.[3] This level appears to be a multiple of the basic content of a genome. In all cases RNA synthesis occurs actively during sporulation, but the total RNA of sporulating cells remains constant and then decreases as the endospore matures and lysis of the sporangium commences. Both DNA and RNA are present in the metabolically inert spore.[4] They are present in lesser amounts than that found in vegetative cells, although the actual concentration of the two forms in the compact spore protoplasts may be similar. The ratio of RNA to DNA is reduced in spores when compared to the ratio found in vegetative cells and this is consistent with the relationship found between RNA content and growth rates of bacterial cells. The general pattern of nucleic acid synthesis and content is correlated therefore with the metabolic status of sporulating cells and spores.

III. DNA of Sporulating Bacteria

A. Properties of DNA from *Bacillus* Species

It is appropriate to describe a few properties of the DNA from vegetative cells of various *Bacillus* and *Clostridium* species prior to a discussion of the DNA fraction from sporulating cells and spores. Several methods have been devised to characterize and compare the DNA from diverse species. Two methods, which have been used extensively, involve the measurement of thermal denaturation temperature (T_m)[5] and buoyant density in CsCl[6]. It has been shown that there is a correlation between both T_m and buoyant density with the base composition of the DNA. Table I illustrates the T_m and the density in CsCl of DNA obtained from several species and the base composition of these DNA as derived from these two parameters. There is relatively good agreement between the base compositions determined from T_m and density measurements; furthermore these data are in good agreement with the results obtained by more direct chemical analysis of the base ratio.[7, 8] The base compositions vary from 32% to 52% GC for the various *Bacillus* species and from 32% to 48% for a few *Clostridium* species. These properties illustrate that the genetic information of spore-formers is stored in genomes of diverse base compositions.

Although most of the studies on the density of *Bacillus* DNA have been determined with CsCl solutions, a few studies have been done with Cs_2SO_4 which may be preferable for certain types of experiments. The density of *B. subtilis*[11] and *B. cereus* T[12] DNA have been reported to be 1·424 and 1·425, respectively in Cs_2SO_4 solutions.

The chemical composition of DNA from various species shows the usual complementarity between purine and pyrimidine bases. A small amount of 5-methylcytosine and 6-methylamino purine is present in the DNA from *B. subtilis*, *B. subtilis* var. *aterrimus*, and *B. niger*.[13] The DNA from several species of *Bacillus* and *Clostridium* also show methyl group acceptor activity to *B. subtilis*[14] and *E. coli*[15, 16] methylating enzymes. The biological function of this modification which occurs at the polynucleotide

TABLE I

Properties of DNA from spore-forming bacteria[a]

Organism	T_m (°C)	Density (g/cc)	Mole %GC from T_m	Density
B. cereus	83	1·696	33	37
B. anthracis[b]	83	—	34	—
B. thuringiensis	83·5	1·695	34	36
B. circulans	84	—	35	—
B. megaterium	85	1·697	37	38
B. lentus	85	—	37	—
B. sphaericus	85	—	39	—
B. pumilus	85·5	—	39	—
B. laterosporus	86	—	40	—
B. firmus	86·5	—	41	—
B. subtilis var. atterimus	87·5	—	43	—
B. brevis	87·5	1·704	43	45
B. subtilis	87·5	1·703	43	44
B. natto	87·5	1·703	43	44
B. niger	87·5	—	43	—
B. polymyxa	88	—	44	—
B. licheniformis	88·5	1·705	46	46
B. macerans	90·5	1·713	50	45
B. stearothermophilus[c]	89	1·713	49	50
Cl. perfringens	80·5	1·691	26·5	32
Cl. tetani	80·5	1·693	26·5	34
Cl. chauvei	80·5	1·691	26·5	32
Cl. madisonii	80·5	1·693	26·5	34
Cl. kluyveri	—	1·694	—	35
Cl. butylicum	—	1·697	—	38
Cl. nigrificans	—	1·707	—	48

[a] Data from Marmur and Doty[5] and Schildkraut *et al.*[6];
[b] McDonald *et al.*[9];
[c] Welker and Campbell.[10]

level is still unknown, but may be involved in the final conformation of the molecule. In general the chemical composition and gross physical properties of the DNA from spore-formers are similar to that of other bacteria.

B. DNA CONTENT OF VEGETATIVE CELLS AND SPORES

Bacillus species synthesize DNA rapidly during the exponential phase of growth and usually contain two chromatin bodies.[4] As Fig. 1 illustrates, the net synthesis of DNA ceases during early sporulation. The cellular DNA level remains constant until lysis of the sporangium occurs. At this time the DNA in the sporangium is released into the medium resulting in a decrease in cellular DNA. The mature spore which is released contains a discrete amount of DNA depending on the species.

1. DNA content of different species

The relative content of DNA has been determined for vegetative cells and spores of several *Bacillus* species. It was observed that the cells of *B. cereus* and *B. megaterium* commencing sporulation contained approximately twice as much DNA as their spores.[2, 3] Spores from a number of different batches contained a constant amount of DNA and only a single nuclear structure compared to the binucleate vegetative cells. Changes in the composition of the medium or in the growth rate did not affect the amount of DNA ultimately incorporated into the spores.

Significant differences were noted in the relative amounts of DNA found in spores of *B. cereus* and *B. megaterium*. When this analysis of species difference was extended to the spores of a variety of *Bacillus* species, it was found that the DNA content per spore varied over a four-fold range and was a multiple of the amount of DNA found in the smallest spores.[17] This is illustrated in Fig. 2. The average amount of spore DNA remained constant for a particular specie under different conditions of sporulation, e.g. various growth media and different aeration and agitation conditions, and this DNA content was characteristic of the specie. The results with X-ray inactivation on spores showed "single hit" inactivation with spores with the minimum amount of DNA and "multiple hit" inactivation with spores with multiple levels of DNA.[18] Fitz-James and Young[17] interpreted these results as illustrating that a single chromosome set was present in cells with the minimal amount of DNA and that spores with multiple levels of DNA contained 2, 3, and 4 similar chromosome units.

In general, spores incorporate only one half the DNA present in vegetative cells. At least one complete nuclear unit enters the spore. The actual amount of DNA present in spores is unique to each specie since it appears to depend on the spore size or volume. This is shown in Table II in which the variation of spore volume ranges from 17 to $71.8 \times 10^{-2} \mu^3$. The actual DNA content based on DNA-P ranges from 5.0 to 20.8×10^{-16}g P per spore. The larger spores usually contain more DNA although an absolute relationship is not seen.

2. The DNA content of the B. subtilis chromosome

According to several investigators the molecular weight of the *B. subtilis* chromosome has been estimated to be 1.3 to 10×10^9.[19-24] The DNA content of a *B. subtilis* spore is 5.4×10^{-15}g which sets an upper limit to

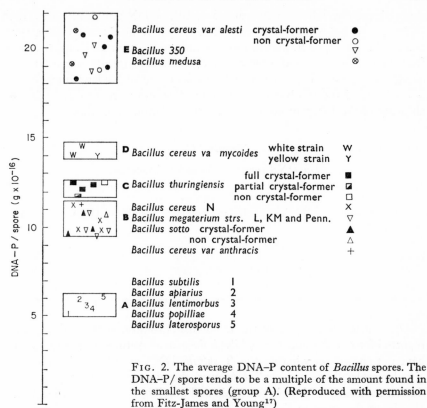

DNA–P/ spore (g x 10⁻¹⁶)

Bacillus cereus var alesti   crystal-former ●
                             non crystal-former ○
E Bacillus 350   ▽
Bacillus medusa   ⊗

D Bacillus cereus va mycoides   white strain W
                                 yellow strain Y

                                 full crystal-former   ■
C Bacillus thuringiensis partial crystal-former ◪
                                 non crystal-former   □
Bacillus cereus N   X
B Bacillus megaterium strs. L, KM and Penn.   ▽
Bacillus sotto   crystal-former   ▲
                 non crystal-former   △
Bacillus cereus var anthracis   +

Bacillus subtilis   1
Bacillus apiarius   2
A Bacillus lentimorbus 3
Bacillus popilliae   4
Bacillus laterosporus   5

FIG. 2. The average DNA–P content of *Bacillus* spores. The DNA–P/ spore tends to be a multiple of the amount found in the smallest spores (group A). (Reproduced with permission from Fitz-James and Young[17])

TABLE II

The RNA and DNA content of spores[a]

Organism	Spore volume $\mu^3 \times 10^2$	RNA-P-content 10^{-16} g P/spore	DNA-P-content 10^{-16} g P/spore	RNA/DNA
B. cereus	23·9	49·6	9·8	5·0
B. sotto	17·0	41·6	10·7	3·9
B. thuringiensis	24·8	46·5	12·1	3·8
B. cereus var. alesti	49·3	65·9	20·8	3·1
B. medusa	40·2	100·2	19·2	5·2
B. cereus var. mycoides	71·5	94	14·6	6·6
B. megaterium L	71·8	40·9	9·8	4·1
B. megaterium KM	46·1	62·6	10·8	5·7
B. subtilis	16·5	21·0	5·0	4·2
B. apiarius	38·6	12·1	5·9	2·0

[a] From Fitz-James and Young.[17]

the molecular weight of DNA at 3.3×10^9. A DNA molecule with this molecular weight would have a length of 1700 μ in the B form. The longest continuous length of B. *subtilis* DNA identified by autoradiography ranged from 800 to 900 μ[21] and 1200 to 1300 μ.[22] These results suggest that a single chromosome is present in a B. *subtilis* spore with a length of 1700 μ. The shorter lengths observed by autoradiography are probably due to shearing of the DNA molecule during the isolation procedure.

TABLE III

The DNA content of bacterial chromosomes

Organism	DNA per cell	Mol wt. per genome	Length of DNA	Reference
	(μg)		(μ)	
B subtilis	5.4×10^{-9}	3.0×10^9	1300	22
E. coli	4.6×10^{-9}	2.8×10^9	1100	25
H. influenzae	2.6×10^{-9}	1.6×10^9	830	26

The content and length of the B. *subtilis* chromosome is similar to that reported for E. *coli*[25] and *Hemophilus influenzae*.[26] Autoradiographic studies have not shown yet whether the B. *subtilis* DNA exists as a single closed circular molecule as in E. *coli*. However, the pictures do show long linear duplex molecules with only two ends and possibly replication forks. Table III shows the comparative value for the molecular weight and length of chromosomes from three bacterial species.

In the discussion to follow it will be assumed that the B. *subtilis* chromosome or genome is a single DNA molecule.

C. STATE OF CHROMOSOME AT VARIOUS GROWTH PHASES

After a period of rapid DNA synthesis in the culture during the log phase of growth, DNA synthesis decreases at the beginning of the stationary phase and usually little or no DNA synthesis is observed during the sporulation phase.[2] Once started, DNA synthesis appears to proceed to completion as long as the precursors are present. Therefore, one might expect that this aspect of cell metabolism would be rather constant among all *Bacillus* species. However, the state of the chromosome at the sporulation stage appears to vary depending on the organism. In analysis of the sequential replication of the B. *subtilis* chromosome, it was observed that the chromosome of stationary phase B. *subtilis* W23 cells was in a completed form, i.e. the genetic markers at the distal end of the chromosome occurred with equal frequencies as those markers near the proximal end of the replicative cycle.[27, 28] In this species the replication of the chromosome is completed as the cell reaches the stationary phase. In molecular terms all the DNA molecules were completely replicated in these stationary phase cells.

On the other hand, in *B. subtilis* W168 cells most of the chromosomes are not in a completed state in the stationary phase.[29] No polarity of DNA replication was observed after transfer of this organism from the stationary phase into fresh growth medium. This indicated that the chromosomes in the stationary phase were in different stages of replication and completion. However, before the DNA is incorporated into the spore, chromosome replication is completed. This is indicated by the synchronous and sequential replication of the *B. subtilis* W168 spore genome during germination.[30] It appears that DNA replication is completed during sporulation since low levels of DNA synthesis are detectable in some species and the genetic evidence strongly suggests that spore DNA is in a fully replicated state. Furthermore, analysis of the germinating spore has shown that the *B. subtilis* genome can replicate in a sequential manner with either one or three growing points depending on the growth rate.[30]

Cytological studies have demonstrated rather discrete nuclear areas in several *Bacillis* species. Is the DNA in a "nuclear body" or is the DNA in a free state? Several investigators have shown that "nuclear bodies" can be isolated from *B. megaterium*[31, 32] and *B. subtilis*[33-35] vegetative cells. Usually the method for obtaining these nuclear bodies includes a step in which protoplasts are produced. This is followed by gentle lysis of the protoplast and isolation of the nuclear body by differential centrifugation. The nuclear body consists chiefly of DNA, RNA and protein. In *B. megaterium* nuclear bodies have been obtained containing DNA, RNA and protein in the following ratios: $1 : 1 : 3$[31] and $7 : 1 : 10$[32]. A *B. subtilis* deoxynucleoprotein with 43% DNA, 5% RNA and 52% protein has been isolated.[33] The ratio of these macromolecules varies, depending on the isolation procedure; however, the amount of DNA in this fraction represents a major fraction (up to 95%) of the total cellular DNA. Up to 30% of the total RNA and 45% of the total protein of a cell may be found in this "nuclear" fraction.

In the chromosomes of higher forms basic proteins such as histones are associated with the DNA. When the protein in the nuclear bodies from *B. megaterium* was analysed, no basic proteins were found.[32] However, 40% of the protein from the *B. subtilis* deoxynucleoprotein complex was soluble in 0·25 N HCl and two basic protein fractions were isolated which contained no cysteine or tryptophan.[33] The question arises whether these basic proteins have a function similar to that of histones or whether these proteins have some other functions. The nuclear fraction contains RNA polymerase and probably DNA polymerase. Since the fraction is rather viscous, it is possible that some of the protein is only fortuitously associated with it. Most of the RNA associated with the nuclear bodies are in "nuclear ribosomes" which appear to associate much more rapidly than cytoplasmic ribosomes with newly synthesized RNA, presumably mRNA.[36] That this newly made RNA in the nuclear fraction is mRNA has been demonstrated by pulse labelling experiments[37] and by the stimulation of an *in vitro* protein synthesizing system by purified nuclear RNA[35].

D. Origin of the Spore DNA

Is the DNA present in spores made during sporulation or is it derived from vegetative cells? It has been demonstrated by various radioisotopic labeling experiments that spore DNA is derived directly from vegetative cell DNA. *B. cereus* var *alesti* vegetative cells were labeled with ^{32}P during growth and transferred to unlabeled medium after growth had stopped. The specific activity of the resulting spore DNA was compared with the specific actitivy of the DNA of vegetative cells and sporulating cells. Very little or no difference was found in their specific activities, indicating that there was an absence of DNA turnover during spore formation.[2] In this case DNA had to be derived directly from vegetative cell DNA.

In the case of *B. cereus* some synthesis of DNA occurred during sporulation, but this synthesis was confined to the non-spore region of the sporangium.[3] The small amount of DNA synthesis which occurs gradually during sporulation of *B. subtilis* probably reflects the completion of replication of the vegetative cell DNA which is incorporated as completely replicated chromosomes into spores.

In *B. megaterium* cells very little or no DNA synthesis occurred during endotrophic sporulation since less than 1% of the spore DNA contained radioisotopic label from several radioactive precursors for DNA.[38]

(i) *Presence of phage DNA in spores.* A spore can contain, in addition to its own genome, the DNA of an infective phage. If vegetative cells of some *Bacillus* species are infected with phage prior to sporulation, the cell may continue to function and actually sporulate.[11, 39-41] The infected spores lyse upon germination and release newly synthesized phage. The phage genome must be in the spore since it is resistant to heat and other conditions which ordinarily kill the phage. When DNA was isolated from *B. subtilis* spores infected with phage PBS1 and subjected to equilibrium density centrifugation in CsCl, DNA bands at two densities were observed.[42] One band was at the same density as *B. subtilis* vegetative cell DNA and the other band was at the density of PBS1 DNA. The proportion of the two DNAs indicated that several phage genomes were present per spore. The physical separation of the two bands and the complete absence of DNA at an intermediate density demonstrated that the two DNAs were not covalently linked and that they existed separately within the spore. The presence of phage genomes in spores is observed in nature, since one of the sources of *Bacillus* phages is heat treated soil samples.[41, 43]

E. Properties of Spore DNA

The differential synthesis of mRNA during sporulation and germination and the dormant nature of spores indicate that gene expression is well regulated during these different phases of growth. Furthermore, during germination DNA synthesis is delayed for an hour or longer depending on the germination medium.[44-46] These facts indicate a precise regulation of gene replication and transcription during sporulation and germination. The absence of RNA synthesis in the spore and the delay in DNA synthesis during germination suggests that the DNA may be in a physical state which precludes its participation in these reactions. Is there any evidence that such may be the case? To answer this question several types of analyses have been made on the DNA isolated from spores.

The observation that approximately one half of the vegetative cell DNA is incorporated into spores[2] can be interpreted in two ways. It could mean that only half of the fully replicated DNA molecules is incorporated into spores or that only one of two complementary DNA strands is enclosed. If the first case is true, a double stranded DNA molecule would be present in spores, whereas in the second case, a single stranded DNA molecule would be present. Since single stranded DNA molecules do not appear to be capable of directing mRNA synthesis *in vivo*, this interpretation could explain the absence of mRNA synthesis in spores.

A direct examination of this question showed that spore DNA was in fact a double stranded molecule.[47] The DNA from *B. subtilis* var *niger* spores exhibited a sharp melting curve during heat denaturation indicating a separation of two complementary strands. When the heated DNA was cooled slowly, hypochromicity occurred, indicating a renaturation of the two separated strands. The DNA had a T_m of 88·1°C which is essentially identical to that of native DNA from vegetative cells. The double helical nature of DNA from *B. subtilis* spores was illustrated by equilibrium density centrifugation in CsCl. The density of spore DNA was identical to that of native vegetative cell DNA.[42] Single stranded DNA would have had a higher density. DNA isolated from *B. cereus* T. spores was also double stranded since alkaline denaturation of the DNA resulted in an increased density as expected if single stranded DNA were formed.[12]

From these physico-chemical data it is concluded that spore DNA is double stranded; it supports the idea that one half of the completely replicated DNA molecules present in the sporangium is incorporated into the spore. Further support for these conclusions comes from genetic and autoradiographic studies. The transformation studies performed with DNA from *B. subtilis* spores suggest that completely replicated molecules are present since the frequency of transformation of all markers is unity.[30] This would only be true if the DNA molecule had completed its replication cycle and all markers were present in equal amounts. The autoradiographic analyses of sporulating cells show that both the sporangial DNA and the forespore DNA are actively engaged in the synthesis of RNA.[48] Since the RNA polymerase utilizes double stranded DNA, the active synthesis of RNA in forespores indicates the presence of double stranded DNA which presumably is maintained throughout the sporulation process and is present in the mature spore.

The differences between the DNAs from vegetative cells and spores may not be obvious from physico-chemical studies if the changes are small and do not alter the gross structure of the DNA. It is possible that a specific conformation of the DNA molecule plays an important role in the control of gene transcription. Any small alteration in conformation may allow transcription or repression of specific genes. One of the minor elements in DNA structure is the methylated bases which occur at low frequency in *Bacillus* DNA.[13] Methylation of DNA appears to depend in part on specific base sequences, GC composition and conformation. The methyl acceptor ability of DNA preparations can be tested by use of a methylase from a heterologous bacterial strain, since DNA cannot be methylated *in vitro* by enzymes prepared from a homologous strain. If two DNA preparations differ in conformation, then their methyl accepting ability may be altered.

To test for any differences between *B. subtilis* spore and vegetative cell DNA preparations, the acceptor ability for methyl groups was tested using methyl transferase from *E. coli* infected with T2 phage.[49] DNA from both spores and vegetative cells were methylated to the same extent of 0·40% of the total base residues. It appears from these results that the two DNA preparations contain similar amounts of the specific sequences required for methylation. Since this method measures only the extent and not the distribution of the methylated bases, it is difficult to conclude whether the DNA preparations are identical. However, according to the specific parameter of methyl accepting capacity the two DNAs are identical. These data suggest that the DNA from spores and vegetative cells have similar secondary structure and conformation.

Two types of study have indicated that the physical state of DNA in spores may be quite different from that in vegetative cells. The physical state of DNA in spores and vegetative cells of *B. megaterium* has been investigated by ultraviolet irradiation of DNA preparations labeled with tritiated thymidine.[50] The spores were irradiated at 2650 Å with an average radiation intensity of 5×10^4 erg/mm² while the vegetative cells received 2×10^3 ergs/mm². Paper chromatography of the hydrolyzed DNA from both morphological forms indicated the presence of uracil-thymine dimers and thymine-thymine dimers in vegetative cell DNA, but very little thymine-thymine dimers in spore DNA. However, two major unidentified photoproducts were observed in the hydrolyzed DNA from irradiated spores.

Similar results have been obtained when spores and vegetative cells of *B. subtilis* were irradiated.[51] Although thymine-thymine dimers were found in DNA isolated from vegetative cells, very little was found in DNA from irradiated spores. A new photoproduct was produced in irradiated spores which had properties similar to but not identical with 5-hydroxyuracil and 5-hydroxymethyluracil. The irradiation of dry DNA preparations gave photoproducts similar to that obtained from the DNA of irradiated spores. These results suggest that the spore DNA is in a less hydrated state than the vegetative cell DNA.

The presence of different photoproducts indicate that the physical state of the DNA *per se* or the environment of the DNA within the spore cytoplasm is quite different from that found in vegetative cells.

The other type of evidence that suggests that spore DNA is physically different from vegetative cell DNA, is the fact that some spore DNA preparations contain two components of differing densities in CsCl and Cs_2SO_4 density gradients.

Two types of DNA occur in *B. cereus* T spores. DNA extracted from *B. cereus* T spores bands at densities of 1·696 g cm⁻³ and 1·725 g cm⁻³ in a CsCl gradiant.[12, 52] The lighter DNA component has a density identical to that of vegetative cell DNA. The heavier DNA component comprises up to 20% of the total DNA and, according to the relationship between density in CsCl and base composition, it would have a GC composition of 66·4% compared to the GC content of 36·7% for the lighter component. Both DNA bands contained double stranded DNA since alkaline denaturation resulted in a shift of both bands to a heavier density as expected if single stranded DNA were formed. The material in the heavy band was insensitive to ribonucleases and pronase, but was completely hydrolysed by treatment with snake venom phophodiesterase and pancreatic deoxyribonuclease.

The heavy DNA was present maximally in spores and shortly after germination, but gradually disappeared during cell division. The heavy DNA was completely absent from vegetative cells and was not detected in the sporangium later at sporulation.

The function and properties of this heavy DNA are not understood. Although it had been proposed earlier that episomes may be involved in controlling sporulation,[53, 54] the high percentage (up to 21%) of the total DNA in this satellite fraction appears to preclude this genetic function. Furthermore, an incorporation of this heavy DNA into the major light DNA fraction would have caused a shift in density which was not detected. It has been proposed that sporulation and dormancy may be due to a modified structure of the DNA, which results in differential transcription. The further clarification of this phenomenon should lead to an understanding of the physico-chemical state of this heavy DNA and its functional role in sporulation.

F. Enzymes Concerned with DNA Synthesis

During sporulation little or no DNA synthesis occurs; however, RNA synthesis continues until the final stages of sporulation. During germination an almost immediate synthesis of RNA ensues, but DNA synthesis begins only late in outgrowth. From these observations the enzymes associated with RNA synthesis appear to be present at all stages of bacterial morphogenesis. On the other hand, the enzymes associated with DNA synthesis appear to be either absent or reversibly inactivated in spores. The long interval of protein synthesis preceding DNA synthesis during germination could be interpreted as a period in which enzymes or structural proteins necessary for DNA replication are being synthesized. Cell-free extracts of spores have been examined to see directly whether DNA polymerase and other enzymes required for producing DNA precursors were present.

DNA polymerase activity was found in extracts of *B. subtilis* spores and was purified 86-fold.[55, 56] The spore DNA polymerase had many properties in common with the enzyme from vegetative cells.[57] They had the same pH optima (pH 8·2), primer, deoxyribonucleotide triphosphate, and cation (Mg^{++}, Mn^{++}) requirements, $S_{20,w}$ of 4·0, molecular weight of 46,000, electrophoretic mobility, and stability *in vitro*. Both enzymes were unstable *in vitro* although whole spores retained 40% of their polymerase activity after heating at 80°C for 15 min. The level of polymerase activity was measured in the parasporal parts of the sporangium during various stages of sporulation. A dramatic decrease in enzyme levels was noted during sporulation and only 30% of the activity found in vegetative cells was recovered in spores. This 70% decrease resulted from the destruction of the enzyme and not from the selective incorporation of the polymerase into the spore, since the specific activity of the enzyme was similar in spore and vegetative cell extracts. However, the total amount of enzyme per spore was three times less than that found in vegetative cells. It has not been determined whether the polymerase is newly synthesized during sporulation or is encapsulated by the forepore membrane along with the nuclear body. In any case the lag of DNA synthesis during germination is not due to absence of DNA polymerase, but to some other regulatory process.

Many other enzymes related to the synthesis of nucleic acids are found in spores and vegetative cells of *B. subtilis*.[55] (Table IV). These enzymes include the nucleotide kinases, a nucleotide phosphatase, and an inorganic pyrophosphatase. The inorganic pyrophosphatase from spores and vegetative cells had similar amino acid compositions, molecular weights of about 68,000, $S_{20,w}$ of 4.4, electrophoretic mobility, heat lability, substrate specificities, metal requirements, and kinetic properties.[58] During sporulation inorganic pyrophosphatase levels declined following a pattern observed with DNA polymerase[56] and DNA methylase.[14]

The purine nucleoside phosphorylase (PNPase) from spores and vegetative cells of *B. cereus* T, had similar electrophoretic mobilities, pH optima,

TABLE IV

Enzymes associated with DNA synthesis found in spores

Enzymes	Reference
Adenosine monophosphate kinase	55
Deoxyadenosine monophosphate kinase	55
Deoxycytidine monophosphate kinase	55
Deoxythymidine monophosphate kinase	55
Deoxyguanosine monophosphate kinase	55
Adenosine triphosphatase	55
Deoxyribonucleic acid polymerase	55, 56
Purine nucleoside phosphorylase	59
Inorganic pyrophosphatase	58, 60

$S_{20,w}$ of 4·7–4·8, molecular weights of 80,000 and heat inactivation.[59] They had the same substrate specificities although the K_m for inosine was $1·1 \times 10^{-4}$M for vegetative cell enzyme and $1·4 \times 10^{-4}$M for spore enzyme. PNPase activity, in contrast to some of the other enzymes associated with nucleic acid metabolism, continued to increase during sporulation. The spore in fact had higher levels than that found in vegetative cells.

Inorganic pyrophosphatase activity was observed also in extracts from spores and vegetative cells of *B. megaterium*.[60] The enzyme level reached its maximum in sporulating cells and remained constant until the appearance of endospores. The amount of enzyme incorporated into spores was only 18% of the peak value found in sporulating cells. The enzyme from both morphological forms were identical in electrophoretic mobility, pH optimum of 7·0, response to inhibitors and Mn^{++} requirement.

In all cases tested no significant differences have been observed between enzymes from spores and from vegetative cells. The stability of spore enzymes *in situ* appears to be a function of the spore cytoplasm and not of specific properties of the enzymes. So far no enzymes unique to the spore have been found. An unanswered question is whether enzymes found in spores were derived directly from vegetative cells or whether they were synthesized *de novo* during the sporulation period. Since protein turnover occurs actively during sporulation, it does not seem unreasonable that proteins found in spores could be synthesized during the sporulation phase. (See also Chapter 4.)

IV. RNA of Sporulating Cells

A. Properties of RNA from *Bacillus* Species

The RNA fractions of cells have been studied extensively since they perform major roles during protein synthesis. Besides carrying the information from the DNA template to the site of peptide bond synthesis as mRNA, it is an important component of ribosomes (rRNA), acts as carrier of amino acids (tRNA) and also as adaptor molecules during the translation process. Since sporulation occurs as a regulated sequence of biochemical events dependent on protein synthesis, it is natural that a great interest has been taken in the RNA of sporulating cells. The mRNA and tRNA populations are of particular interest, since several current models of regulation suggest that differential gene expression depends on their functional concentration and properties.[61-65]

1. The general classes of RNA

The *Bacillus* species contain the usual types of RNA found in bacteria. They include the 5s, 16s, and 23s rRNAs, the 4s tRNAs, and the mRNAs. The latter two fractions contain a heterogeneous population of RNA in contrast to the relatively homogeneous rRNA fraction. One of the properties characteristic of these RNA fractions is their base composition. The base ratios of various RNA fractions from several *Bacillus* species are listed in Table V. The composition of rRNAs is characterized by their high guanine content and their high GC content. The tRNAs are identified by the complementarity between adenine and uracil and between guanine and cytosine. Furthermore, they usually have a high GC content, even higher than that of rRNA. The base compositions of rRNA and tRNA are closely related to the RNA from many other distant bacterial, plant, and animal species. This suggests that these RNA molecules have been highly conserved during the course of evolution.[7] On the other hand, the mRNA base compositions are generally complementary to the base composition of the homologous DNAs.

A comparison of the RNA from spores and vegetative cells of *B. subtilis* indicates that they have the same elution profiles from a methylated albumin *kieselguhr* (MAK) column (Fig. 3). This demonstrates that the size and conformation of the tRNA and rRNA from the two forms are very

TABLE V

The Base Composition of RNA from Bacillus *species*

| Organism | Moles % | | | | | Reference |
	C	A	U	G	GC	
1. Bulk RNA						
B. subtilis	22·1	25·5	21·0	31·4	53·5	[66]
	22·8	25·2	20·7	31·2	54·0	[67]
B. cereus	21·5	25·5	22·6	31·4	52·9	[68]
	21·3	25·5	22·5	30·7	52·0	[69]
	20·1	24·8	23·8	31·2	51·3	[7]
B. stearothermophilus	23·5	23·0	18·2	34·0	57·5	[70]
B. megaterium	21·9	22·4	23·6	32·6	53·9	[71]
Cl. perfringens	24·5	26·1	19·2	30·3	54·8	[7]
2. 23s rRNA						
B. subtilis	22·5	26·5	19·3	32·0	54·5	[66]
	22·8	26·4	20·7	30·0	52·8	[67]
B. cereus	21·4	27·9	21·5	29·2	51·6	[72]
B. stearothermophilus	23·1	25·0	18·8	33·1	56·2	[70]
3. 16s rRNA						
B. subtilis	22·3	26·5	21·6	29·6	51·9	[66]
	23·4	24·8	21·1	30·8	54·2	[67]
B. cereus	21·7	26·8	22·6	28·9	50·6	[72]
B. stearothermophilus	29·4	21·0	17·7	32·0	61·4	[70]
4. tRNA						
B. subtilis	26·9	21·9	20·3	30·8	57·7	[73]
	28·3	20·2	17·6	33·9	62·2	[66]
	28·5	19·1	19·4	32·9	61·4	[67]
B. cereus	28·0	20·4	20·4	31·1	59·1	[73]
	28·3	20·9	20·6	30·2	58·5	[72]
B. stearothermophilus	30·5	18·7	18·9	31·8	62·3	[70]
5. 5s rRNA						
B. subtilis	27·1	22·5	19·9	30·3	57·4	[74]
6. mRNA						
B. subtilis						
log phase	22·0	29·0	25·7	23·3	45·3	[75]
sporulation	20·4	28·6	28·3	22·5	42·9	[75]
germination	20·8	29·0	24·2	25·9	46·7	[75]
B. cereus						
log phase	20·4	28·6	30·0	21·0	41·4	[72]
sporulation	19·8	32·7	28·5	20·8	40·6	[68]

TABLE VI

Base compositions of RNA from spores and vegetative cells of B. subtilis[a]

RNA	C	A	U	G	%GC
			Moles %		
1. tRNA					
Spores	29·4	18·9	18·2	33·5	62·9
Vegetative cells	28·5	19·1	19·4	32·9	61·4
2. 16s rRNA					
Spore	24·0	24·4	20·6	30·9	54·9
Vegetative cells	23·4	24·8	21·1	30·8	54·2
3. 23s rRNA					
Spore	22·6	26·2	19·8	31·4	54·0
Vegetative cells	22·8	26·4	20·7	30·0	52·8

[a] From Doi and Igarashi[67]

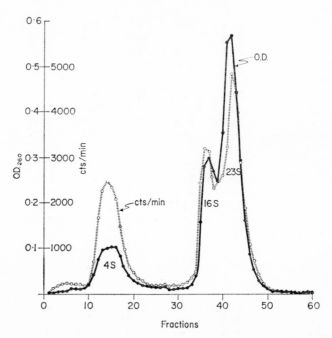

FIG. 3. The elution profiles of RNA from vegetative cells and spores of *B. subtilis* from a MAK column. ○, spore RNA as measured by radioactivity. ●, vegetative cell RNA as measured by optical density at 260 mμ. (Reproduced with permission from Doi and Igarashi.[67])

similar. This is further supported by the similarity in base composition of these RNA fractions (Table VI). Although 5s rRNA has been characterized only from vegetative cells, it appears to be present in spores also. The presence of 5s rRNA is suggested by the MAK column elution profile of rRNA isolated from spore ribosomes.[76] In addition to the 16s and 23s rRNA peaks, a small peak is eluted at the distal end of the tRNA elution profile. The position is consistent with that expected with 5s rRNA, since its molecular weight is larger than that of 4s tRNA. As with the other rRNA species, the exact function of 5s rRNA is still unknown. Presumably, they serve as structural components of the ribosomes which serve to bring the various components of the protein synthesizing system into the proper conformation for peptide bond synthesis.

2. RNA content of vegetative cells and spores

The RNA content of several *Bacillus* species has been analysed. In contrast to the DNA content which remained constant for a given species, the RNA content of vegetative cells and spores depended on the species, the cell or spore volume, and the growth condition. Bacterial cells contain more RNA when they are growing rapidly than slowly and it is not surprising that *Bacillus* species conform to this pattern. Figures for RNA content per cell are meaningful only when the exact conditions for growth are also noted. In all cases more RNA than DNA was found in vegetative cells and spores. The RNA/DNA ratio in spores ranges from 1·9 to 7·2 depending on the species.[17] Within the same species this ratio could vary by 60% depending on the growth condition.

The RNA content per cell is always much higher for vegetative cells than spores. For *B. cereus* and *B. megaterium* the vegetative cells contain approximately twice as much RNA as the spores.[4] The higher rate of protein synthesis in vegetative cells and the turnover of RNA during sporulation probably accounts for the higher level of RNA in vegetative cells. The actual concentration of RNA in spores may be equal to or greater than that found in vegetative cells, since a large fraction of the spore mass is occupied by the complex cortex, spore coat and exosporium.

Example of the range of RNA concentration found in spores expressed as RNA-phosphorus (RNA-P) per spore was from 12×10^{-16} g in *B. apiarus* to 100×10^{-16} g in *B. cereus* var *mycoides*.[17] The content of RNA-P for several other species are given in Table II. These values must be considered as values for spores obtained under specific conditions, since considerable variation occurred depending on the growth condition. The size usually determined the total RNA content of the spore. The largest spores (e.g. from *B. megatermiu*) had a volume of approximately $93 \times 10^{-2} \mu^3$ which was some sevenfold greater than the smallest spores (e.g. from *B. sotto*) which had a volume of $12·8 \times 10^{-2} \mu^3$. Depending on the growth conditions, spores of the same species could vary in volume by 75%. In these cases it was found that larger spores usually contained more RNA.

B. RNA Pattern During Sporulation

Following the rapid synthesis of RNA during exponential growth, the net synthesis of RNA ceases at the end of the log phase of growth (Fig. 1). The intracellular level of RNA stays constant throughout most of the sporulation phase and a decrease in the total RNA content occurs only late in spore formation. Although there is no net synthesis of RNA during sporulation, active turnover of RNA is observed.[2, 3, 77] The turnover continues until quite late in sporulation since radioactive precursors are incorporated into RNA after the appearance of forespores. All the major RNA species are synthesized at this time since mRNA was identified after short term incorporation of radioactivity, and long term incorporation studies indicate that radioactive precursors are incorporated into tRNA and rRNA. The latter two fractions from dormant spores contain radioactive label which is added late in sporulation.[67]

The ratio of tRNA to rRNA increases during the transition of a vegetative cell to a spore.[78] The tRNA/rRNA ratio of 0·20 in vegetative cells increases during sporulation to 0·36 and to 0·43 in dormant spores of *B. subtilis*. This change in the ratio of the two RNA species is caused by an increasing degradation of the ribosomes which were present in the cells at the end of the log phase.[79] The ribosomes which are present in the sporulating cells are active since pulse-labeling with ^{35}S-sulfate showed the presence of nascent proteins on ribosomes. These nascent proteins were chased into the soluble protein pool by unlabeled sulfate. Furthermore, the incorporation of radioactive amino acids into proteins during sporulation is indicative of normally functioning ribosomes.[68, 79, 80]

The synthesis of RNA has been analysed by use of inhibitors and base analogs. The addition of the analogs 8-azaguanine and 5-fluorouracil during any part of the log phase prevents the formation of *B. cereus* spores. Both of these analogs are readily incorporated into the RNA fractions and presumably form faulty RNA. However, once sporulation has commenced the cells become increasingly resistant to the presence of these analogs and spore formation occurs.[81–83] The cells of *B. cereus* are not inhibited by analogs added some 4 hr after the end of the exponential phase and 5 hr before the appearance of refractile endospores, although incorporation of the analog into RNA is still occurring. This suggests that RNA for sporulation was synthesized prior to the addition of the analogs.

Actinomycin D, which prevents the synthesis of all RNA,[84] inhibits sporulation when added to vegetative cells and sporulating cells of *B. subtilis* before a critical point is reached. After this time the cells become less sensitive to actinomycin D.[85] Actinomycin D inhibits growth of vegetative cells of *B. cereus* T, but if it is added to the culture shortly after the end of the log phase at a level of $10\mu g/ml$, then virtually all the cells can sporulate. The exact level of actinomycin D appears to be critical in these inhibitor studies. Therefore, in *B. cereus* T neither analog incorporation into RNA nor the absence of RNA synthesis is inhibitory to the sporulation process once a certain amount of mRNA has been synthesized. From these types of results it has been proposed that a stable mRNA is required for the sporulation of *B. cereus* T cells.[82] In *B. subtilis* cells, which are much more sensitive to actinomycin D, this inhibitor prevents sporulation at all stages except at the final phase when those endospores which are octanol resistant become heat resistant (see Fig. 3, Chapter 5). In this case the continued synthesis of mRNA appears to be essential during all phases of sporulation except for the final maturation phases.[85, 86]

C. Messenger RNA (mRNA)

Considerable attention has been focused on the mRNA fraction of sporulating cells to determine whether there is a correlation between differential protein and mRNA synthesis. If specific proteins are made for sporulation, then it is reasonable to expect to find mRNA from genes which are unexpressed during the log phase of growth. Furthermore, it was of interest to study and compare other properties of sporulation mRNA such as half-life, size, base composition, and homogeneity.

1. General properties of mRNA

The mRNA formed during sporulation has been identified by pulse-labeling cells with radioactive RNA precursors at various times of sporulation. Most of the RNA that is formed during short periods of labeling sediments slightly slower than 16s rRNA and faster than tRNA. There is heterogeneity in the sedimentation pattern, suggesting that a population of mRNAs is being synthesized. The mRNA fractions were identified in cells at the forespore, cortex and coat-forming stages of *B. cereus* var *alesti* showing a rather continuous synthesis of mRNA during spore formation.[79] When the incorporation of radioactive uracil was compared between a sporogenous and an asporogenous strain, a difference in the total uptake of radioactivity and in the sedimentation pattern was noted.[87, 88] The asporogenous mutants synthesized far less mRNA and the 8s–14s mRNA peak was depressed.

The half-life of the mRNA fraction in *B. subtilis* sporulating cells was approximately 2 min[86] which is similar to that for mRNA of log phase cells.[84] The half-life was similar for mRNA made at all stages of sporulation and germination,[89] thus indicating a continuous need for mRNA synthesis

H

at all growth phases. The only mRNA with long half-life has been reported for *B. cereus* T and this phenomenon will be discussed later.

The base composition of *B. subtilis* and *B. cereus* mRNA formed during sporulation is similar to that of the homologous DNAs (Table V). Discrete differences in over-all base composition were noted between the mRNAs formed during log phase, sporulation and germination.[89] These results suggest that either different genes are being expressed during sporulation or the same genes are being expressed at different rates during these different growth phases. In *B. cereus* the total pulse-labeled RNA had a composition somewhat between DNA and bulk RNA, which suggested that mRNA, rRNA and tRNA were all being synthesized during spore formation. The composition of the RNA which hybridized with high efficiency with the DNA was very similar to that of *B. cereus* DNA.[91]

2. Differential synthesis of mRNA during sporulation

To demonstrate that the mRNA formed during sporulation was different from the log phase mRNA, DNA-RNA hybrid competition studies were performed. The rationale for these experiments is that, if RNAs are made for sporulation-specific proteins, then the mRNA population from log phase cells would not compete for the same sites on the DNA during competitive hybridization procedures. Therefore, labeled sporulation mRNA would be able to hybridize in the presence of a great excess of unlabeled log phase mRNA. It was shown in such an experiment that a 15–30% fraction of sporulation mRNA from *B. subtilis* was able to hybridize with DNA in the presence of an 150 fold excess of log phase or germination phase RNA.[75] These results illustrate that genes are transcribed during spore formation which are not expressed during growth or germination.

Another way to illustrate the differential expression of the genome was to fractionate the DNA by use of a MAK column and then hybridize mRNA from different growth phases with the fractionated DNA. If the mRNA are similar, the degree of hybridization with each of the DNA fractions would be similar. However, if the mRNA fractions are different either quantitatively or qualitatively, a difference in the hybridization pattern would arise. Distinct patterns of hybrid formation were noted when log phase, germination and sporulation mRNA fractions from *B. subtilis* were hybridized with the fractionated DNA.[90] The results show that the relative amounts of mRNA for specific sites on the DNA are quite different for the mRNA from the three stages investigated.

3. Stable sporulation mRNA

Although continuous synthesis of RNA is required for the sporulation of *B. subtilis* and *B. megaterium*, the use of base analogs and inhibitors of RNA synthesis revealed that stable mRNA was present during spore formation in *B. cereus*.[68, 82] These cells were able to sporulate in the presence of 8-azaguanine, 5-fluorouracil, and actinomycin D, if they were added after the end of the log phase of growth. After this critical period the incorporation of the analogs into RNA did not affect the final level of spores formed. The complete inhibition of RNA synthesis by actinomycin D also did not hinder sporulation. Since sporulation-specific proteins are made after this critical period, it is evident that stable mRNA for these proteins had been made earlier in the sporulation stage. However, the necessity of continued protein synthesis is shown by inhibition of sporulation by addition of chloramphenicol at any stage. The functions of genetic transcription and peptide bond synthesis are clearly separated by these inhibitors.

In a critical analysis of the stable mRNA,[91] the DNA-RNA hybridization studies revealed that sporulation mRNA was less heterogenous than log phase mRNA. By hybrid competition studies it was shown that a similar population of mRNA existed from the earliest to the latest stages of sporulation. The results suggested that this common mRNA population could be synthesized continuously or could be synthesized for a short time early in sporulation and persist throughout spore formation. Further studies revealed that a population of mRNA was made in the initial stages of sporulation and persisted throughout the subsequent synthetic processes; this RNA population was not continuously synthesized. These data complement the studies with the inhibitors of RNA synthesis. As the actively growing cell approaches the end of the log phase, a deficiency in the medium must trigger the synthesis of a mRNA population which is stabilized and participates throughout the synthetic processes of spore formation.

4. Membrane-bound mRNA

A possible mechanism to prevent the degradation of mRNA is to complex it with either ribosomes and tRNAs or with the membrane of the sporangium. In inhibitor studies it was revealed that mRNA could be protected by processes which prevented protein synthesis but did not disrupt the ribosome-mRNA-tRNA complex.[92] Since protein synthesis occurs throughout sporulation, this mechanism seems unlikely. However, a stable membrane bound mRNA fraction has been identified in vegetative cells of *B. cereus* and *B. megaterium*.[93-96] At the end of the log phase of growth, the cells were pulse labeled with radioactive uracil for 4 min. The labeled cells were then chased for 3 hr with excess unlabeled uracil and cytidine in the presence of actinomycin D. At the end of 3 hr, hybridization with DNA showed that 20% of the labeled RNA made at 4 min still remained. This RNA made during the 4 min pulse survived not only protein synthesis, but the presence of actinomycin D which stops all RNA synthesis. All the stable mRNA remaining after this chase period was membrane bound RNA and was found actually in membrane-bound polysomes, but not in soluble polysomes. Membrane-bound polysomes were more stable than polysomes from log phase cells when subjected to repeated washing and ribonuclease treatment. At least part of the RNA persisting in the presence of actinomycin D was involved in the synthesis of spore coat proteins. Lysine incorporation into spore coat protein was observed after 45 min pretreatment of the sporulating cells with actinomycin D before the addition of radioactive lysine.

It has been suggested that the affinity of the polysomes for the cell membrane could be facilitated by the nascent polypeptides associated with the polysome.[94] The protein being made may have a high affinity for membranes and bind the polysome complex to the membrane. This complex would stabilize the mRNA and prevent it from being degraded. *In vitro* studies have shown that polysomes with nascent polypeptides tend to stick to membranes whereas the removal of nascent proteins decreases the adsorption of polysomes to membranes. The synthesis of structural proteins which have a high affinity for membranes *in vivo* may result in the formation of stable mRNA and polysomes during sporulation. To extend this thought would also suggest that structural spore proteins may be synthesized early in the sporulation process.[97]

5. Do spores contain mRNA?

The synthesis of stable mRNA during sporulation of *B. cereus* T[82] and the presence of stable mRNA in non-growing and differentiated cells suggest that spores could contain mRNA. The mRNA fraction which usually comprises 2–5% of the total RNA of bacterial cells is difficult to demonstrate except by radioactive pulse-labeling techniques. When bulk spore RNA is extracted and examined by the usual physico-chemical methods, no evidence for a mRNA fraction is observed.[67]

Two approaches have been used to test for the presence of mRNA in dormant spores. Spores of *B. cereus* T were germinated in the presence and absence of actinomycin and radioactive amino acids. Since actinomycin D inhibits all RNA synthesis, any incorporation of radioactive amino acids would be dependent on the presence of stable mRNA in the spore. In these experiments no significant incorporation of radioactive amino acids into proteins was observed unless RNA synthesis was permitted, i.e. in the absence of actinomycin D.[89] In fact, even without the inhibitor protein synthesis lagged behind RNA synthesis by 3 to 5 min.[98] These results suggest that spores contain no stable mRNA which could be utilized during germination. Similar conclusions were reached concerning *B. subtilis* spores.[99]

The other test for the presence of stable mRNA in dormant spores was to determine the ability of a spore extract to incorporate labeled amino acids into proteins. In experiments of this type a very low level of phenylalanine incorporation was observed with *B. cereus* T spore extracts; however, the amount was so small that one could conclude that little or no endogenous mRNA was present in dormant spore extracts.[89] A definite interpretation of these results is difficult for the following reasons: (a) a deficiency in the cell-free synthesizing system derived from *B. cereus* T spores has been reported;[89, 100] (b) spores contain a high level of RNase[76] which is released upon disruption and which may destroy endogenous mRNA; and (c) low levels of protein may be missed even by *in vivo* experiments. More definitive information is required before a firm conclusion can be reached regarding the presence of mRNA in dormant spores.

D. RIBOSOMES AND RIBOSOMAL RNA (rRNA)

1. General properties

The vegetative cells of *Bacillus* species contain the usual ribosomal components.[70, 76, 101–103] The presence of ribosomes in *B. subtilis* spores was revealed by sedimentation analyses of spore extracts. Ribosomes with s values of 50s and 68s were detected after extraction of crushed spores with citrate buffer.[101] When a *tris*-$MgCl_2$ buffer was used for extraction, ribosomes with sedimentation values of 69s, 52s, and 30s were obtained.[76] The absence of 30s ribosomes in extracts prepared with citrate buffer could be attributed to the high degree of instability of 30s ribosomes of *B. subtilis*. A direct comparison of ribosomes from vegetative cells and spores of *B. subtilis* showed that they had similar sedimentation coefficients and were alike in RNA and protein content.[76] Ribosomes from both morphological forms contained approximately 60% RNA and 40% protein. The ribosomes also contained 23s and 16s rRNA which were similar in base composition, in elution pattern from a MAK column, and in sedimentation pattern in sucrose gradients. An instability of ribosomes from *B. subtilis* spores was noted possibly due to the high levels of protease and nuclease in the spore extracts. The gross physical and chemical properties of spore ribosomes are indistinguishable from those of vegetative cell ribosomes. However, it is still possible that the difficulty experienced in extracting ribosomes from spores may be due to the absence of some minor component of the ribosome.

Although polysomes have been identified readily in vegetative cells of several *Bacillus* species,[104–106] no evidence has been obtained for their presence in spores. The difficulty of breaking spores gently, the presence of ribonucleases, and the low level or absence of mRNA make it difficult to isolate polysomes from spores. The best chance of observing polysomes may be by electron microscopy of thin sections. To date, however, even these examinations give no clear evidence.

2. Ribosomal RNA sites in the B. subtilis genome

The DNA-RNA hybridization experiments between rRNA and DNA of *B. subtilis* have indicated that there may be two or more cistrons for both 16s rRNA and for 23s rRNA. This may be significant since the possibility exists that ribosomes are not merely structural sites for the synthesis of proteins, but are active in directing protein synthesis. In this connection it has been proposed that mRNA synthesis may be facilitated by the action of ribosomes,[63, 107] and chemical analyses have shown that there are at least two species each of 16s and 23s rRNAs in *E. coli* the ratios of which are altered by growth conditions.[108, 109]

The percent of the *B. subtilis* genome which hybridizes with each of the rRNA species is listed in Table VII. The average values for 16s and 23s

TABLE VII

Percent of B. subtilis *genone hybridizing with ribosomal RNA*

16s rRNA	23s rRNA	16s + 23s rRNA	Reference
0·26	0·36	0·62	[110]
0·13	0·25	0·38	[111]
0·22	0·26	0·48	[112]
Ave. 0·20	0·29	0·49	

rRNAs are 0·20% and 0·29% respectively. From the molecular weight of a *B. subtilis* chromosome (3×10^9), the percent hybridization, and the molecular weights of 16s and 23s rRNA (assuming they are the same as that for *E. coli* rRNA, i.e. 0·56 and $1·1 \times 10^6$, respectively [113]), there appear to be 7 to 10 sites for 16s rRNA and 5 to 8 sites for 23s rRNA in the *B. subtilis* genome. The lower numbers of sites have been obtained by correcting for the location of rRNA cistrons near the origin of replication and the state of the DNA used for the hybridization studies. The possibility of more than one cistron for each of the two rRNAs of *B. subtilis* was proposed from results obtained with hybrid competition studies.[112] In these experiments it was shown that *B. subtilis* 16s and 23s rRNAs hybridized with *B. cereus* and *B. stearothermophilus* DNAs. Furthermore, it was shown that different sequences of each of the two rRNAs were hybridizing with the two heterologous DNAs. This suggested that either different sequences of the same RNA molecule were hybridizing with the two heterologous DNAs or that different cistrons existed for each of the two rRNAs. Further studies are needed to clarify this situation; however, from the number of likely sites on the *B. subtilis* genomes, it appears that several cistrons exist for each of the rRNA species.

E. TRANSFER RNA (tRNA)

The *Bacillus* species appear to have the usual array of tRNA species.[114] The amount of tRNA relative to rRNA increases during the transition of a vegetative cell to a spore.[78, 115] In actively growing *B. subtilis* cells tRNA comprises about 17% of the bulk RNA. This percentage rises to 26·5% during sporulation and finally approaches 30% of the total RNA in spores. This change in ratio is caused by degradation of ribosomes which results in breakdown of rRNA. Although no net synthesis of RNA occurs during sporulation, the synthesis of tRNA continues until late in spore formation. This was demonstrated by the addition of radioactive RNA precursors to cultures during various stages of sporulation. Radioactivity was found in the tRNA of dormant spores.[67] It has been suggested that tRNA degradation occurs during sporulation;[116] more definitive evidence is required to clarify this point.

Evidence has been obtained for a change in the tRNA fraction of sporulating cells of *B. subtilis*. The tRNA for all the amino acids except proline, asparagine, glutamine and cysteine were examined by MAK column chromatography.[114] All the aminoacyl-tRNAs from vegetative cells and sporulating cells eluted in a similar fashion except for valyl-tRNA. This aminoacyl-tRNA eluted as two peaks. The ratio of the two peaks was 1·4 in vegetative cells, but shifted to 0·6 in sporulating cells. This ratio was maintained until late in sporulation when the ratio returned to 1·1. The ratio of the two valyl-tRNAs was 1·6 in dormant spores. Thus, the relative amounts of the two valyl-tRNAs varied during early and mid-sporulation, but by late sporulation and in the dormant spore the ratio was similar to that of the log phase. Furthermore, the addition of glucose in the initial stages of sporulation reversed the process and growth resumed. During this period there was a rapid return of the ratio from 0·6 to 1·3 (within 20 min.[117]) There is either a differential synthesis of the two valine tRNAs or there is a modification of the amino acid accepting function which alters the apparent concentration of the two valine tRNA species. Whether there is a direct correlation between the valyl-tRNA pattern changes and sporulation has not been determined. Furthermore, it is possible that other aminoacyl-tRNA patterns are changing but not being detected by MAK column chromatography. At present it appears that the pattern changes are a function of the change in the growth medium or growth rate.

The elution profile of total *B. subtilis* spore tRNA from a MAK column is different from that of the vegetative cell.[67] To test whether this change could be attributed to changes in a specific tRNA, the aminoacyl-tRNA patterns of vegetative cells and spores were compared by double isotope labeling and MAK column chromatography. The patterns of leucyl-, lysyl-, phenylalanyl-, arginyl-, valyl- and glutamyl-tRNAs were compared and all had similar patterns, except lysyl-tRNA.[118] Furthermore, the amount of lysine tRNA was increased fourfold in the spore relative to that found in vegetative cells. A study of the relative increase in lysine tRNA revealed that spores contained a greater amount of a second lysine tRNA fraction which was absent from vegetative cell tRNA. The increase of this lysine tRNA fraction occurred during the formation of the spore wall and the appearance of refractility. This implies that the synthesis or appearance of this tRNA occurred in the endospore and was not the result of enclosure of this lysine tRNA from the sporangium.

The presence of this second lysine tRNA in *B. subtilis* spores is dependent on the sporulation medium.[119] Spores obtained from cells grown on complex medium containing tryptone and yeast extract contained the new lysine tRNA, whereas spores from cells grown on synthetic or dilute complex media contained little or none of the new component. The difference in lysine tRNA pattern cannot be attributed directly to the growth rate since the doubling times of the cells in complex and diluted complex media were quite similar. It appears that the concentration of specific nutrients in the growth medium determines the pattern of lysine tRNA in spores. This alteration does not appear to be correlated directly with sporulation, since an asporogenous mutant blocked in the initial stages of sporulation exhibited the change in lysine tRNA pattern identical to that of wild type spore-formers. Also spores could be obtained by growth under special conditions which did not contain any of this component. At present there is no direct evidence that any of the tRNA changes are associated with regulation or initiation of spore formation.

F. REGULATION OF THE RATE OF RNA SYNTHESIS

The rate of RNA synthesis during sporulation may be regulated by the available supply of amino acids. The synthesis of RNA was compared in a wild type sporulating cell (SP+) and an asporogenous mutant (SP−) which had an early block in spore formation. The rate of RNA synthesis was 5 times greater in the SP+ cells than in the SP− mutant in the presence or absence of chloramphenicol.[120] Chloramphenicol stimulates RNA synthesis by increasing the supply of available amino acids or aminoacyl-tRNA.[121, 122] When chloramphenicol and twenty different amino acids were given simultaneously to the SP− strain, then its rate of RNA synthesis was similar to that of the SP+ cells. Therefore, the rate appears to be controlled by the availability of amino acids. Asporogenous mutants of *B. subtilis* were divided into two classes called TO+ and TO− depending on whether the addition of chloramphenicol alone stimulated RNA synthesis.[123] All the TO+ strains had an increased rate of RNA synthesis after the addition of only the inhibitor. All the TO− mutants were early block mutants (t_0 and t_1 of Ryter et al[124]; see Fig. 3, p. 174). It was proposed that an intracellular protease which would provide amino acids during sporulation was missing from the TO⁻ mutants.

The turnover of protein has been demonstrated for *Bacillus* species during sporulation and resting states indicating the presence of proteolytic activity.[125-127] It is possible that the amino acid pool is replenished by protein turnover which allows RNA synthesis to occur even though the medium is deficient in amino acids. The presence of extra-cellular proteases has been reported for several species. There appears to be some correlation between protease activity and sporulation. Exceptions have been reported in that some asporogenous *B. subtilis* mutants were able to produce protease[128] and sporulation of *B. cereus* T occurred in the absence of an extracellular protease.[129] The identification of an intracellular protease would be valuable in order to determine whether RNA synthesis during sporulation was directly correlated with its activity.

G. SITE OF RNA SYNTHESIS

1. Nuclear fractions

Nuclear fractions containing DNA, RNA and protein have been isolated from vegetative cells of *B. subtilis*[33-35] and *B. megaterium*.[31, 32] These fractions have five times the RNA polymerase activity of cytoplasmic fractions.[35] The DNA in the nuclear fraction is more active than purified DNA as the *in vitro* template for RNA polymerase.[34] A higher rate and final level of RNA production is achieved when nuclear DNA is used as the template. The sedimentation coefficients of the RNA produced with purified DNA and nuclear DNA are 4s and 14s, respectively. This suggests that the nuclear template is able to provide longer stretches of DNA template for the RNA polymerase. Since RNA made *in vivo* can compete with the RNA made *in vitro* during DNA/RNA hybridization experiments, it shows that the RNA polymerase is transcribing the genetic sequences which are ordinarily expressed during growth of the cell. Part of the RNA in the nuclear fraction is mRNA, since the nuclear RNA was more effective than rRNA in promoting the incorporation of amino acids into proteins in an *in vitro* system.

In *in vivo* experiments it was shown that RNA accumulated in the nuclear body when *B. megaterium* cells were treated with chloramphenicol, 8-azaguanine and tetracyclines.[37] The inhibition of protein synthesis appears to block the newly made RNA in the nuclear fraction. Rapidly labeled RNA was also shown to be associated with the nuclear fraction. This RNA fraction could be chased from the nuclear fraction into the cytoplasm. All of these results suggest that the nuclear fraction is the initial site of RNA synthesis. The association with ribosomes may occur here before the complex appears in the cytoplasmic fraction, since nuclear ribosomes active in the attachment of newly made mRNA have been isolated from *B. megaterium*[36].

2. RNA synthesis in spores

The rapid synthesis of RNA during germination[98] suggested that dormant spores contained DNA dependent RNA polymerase. Earlier studies did not give unequivocal evidence for its existence in spores,[55] but the presence of RNA polymerase has been demonstrated in *B. subtilis* spore extracts.[130, 131] The system shows an almost complete dependence on Mg^{++} for optimal activity. The reaction is inhibited by DNase, RNase, or actinomycin D. The spore RNA polymerase differs in at least two respects from the vegetative cell enzyme. It is more heat resistant and has a different sedimentation behaviour in glycerol gradients.

The existence of RNA and DNA polymerase in spore extracts suggests that the enzymatic machinery for DNA replication and transcription were incorporated into spores during their formation. Other factors must be involved in the regulation of these enzymatic activities during dormancy. The almost immediate resumption of RNA synthesis during the initial stages of germination[98] shows that the spore RNA polymerase is in a state in which activity can be readily induced. Perhaps the hydration of the spore core is sufficient to activate all the spore enzymes.

Autoradiography revealed that the genomes in the parasporal and forespore parts of the sporangium were both active sites of RNA synthesis during sporulation.[48] The implications of this observation are exciting, since it is conceivable that the two genomes are being expressed differentially. For instance, the synthesis of dipicolinic acid, which never appears in the parasporal part and is localized in the forespore, could be controlled by the genome of the forespore. An interesting possibility is that the parasporal genome is involved in synthesizing the outer spore coat while the forespore genome contributes to the synthesis of the cortex regions. In any case it is not difficult to imagine that the microenvironment of the forespore is different from that of the sporangium. Since the environment of the cell controls the expression of its genome, it follows that the two genomes can have totally different functions.

V. Cell-free Protein Synthesis in *Bacillus* Systems

Cell-free systems from *Bacillus* species have been developed for the synthesis of specific enzymes,[130] for the analysis of protein synthesis at varying temperatures,[131, 132] for comparison with the *E. coli* system,[133] for comparison of vegetative cell and spore components involved in peptide bond synthesis,[89, 100, 134] and for correlation of tRNA changes with the initiation of protein synthesis.[135] In most cases cells harvested in early or mid-log phase provide the most active extracts. Cells from late log phase tend to be less active, perhaps because they contain more ribonuclease and protease which are released at that time. The synthesis of polypeptides in the presence of endogenous mRNA is normal, but with synthetic polynucleotides it is usually less than that in *E. coli* systems. The reasons for this difference are unknown; however, the instability of amino-acyl synthetases may be involved since the incorporation of amino acids from aminoacyl-tRNAs was comparable to that of the *E. coli* system.[133]

A. THE *B. SUBTILIS* SYSTEM

The usual macromolecular components (ribosomes, tRNAs, endogenous or synthetic mRNAs, and enzymes), amino acids, cations and energy sources are required[130, 133] in the system developed from vegetative cells. The addition of methanol, ethanol, isopropanol, di-isopropylfluorophosphate (DFP), and spermidine stimulated phenylalanine incorporation when polyuridylic acid (poly U) was used as the mRNA.[130] The organic solvents cause structural changes in the nucleic acid components of the *in vitro* system resulting in increased efficiency. DFP is an inhibitor of esterases, and spermidine stabilizes ribosomes and nucleic acids against nucleases. The amino acid activating enzymes have been used extensively for the analysis of aminoacyl-tRNAs.[114, 118]

An instability of ribosomes of *B. subtilis* has been noted.[133] The particular lability of 30s ribosomes was revealed during studies of polyphenylalanine synthesis with poly U templates. A rapid decrease in synthetic capacity was observed when ribosomes were stored at 0–4°C instead of at −15°C. It was found that these ribosomes could be reactivated by the addition of *E. coli* 30s ribosomes. The more stable *B. subtilis* 50s ribosomes combine with the *E. coli* 30s ribosomes to form heterologous 70s ribosomes. These hybrid 70s ribosomes are active in polypeptide synthesis. The degradation of 30s ribosomes is stimulated markedly by low Mg^{++} concentrations. This may explain the extreme lability of *B. subtilis* spore ribosomes which are extracted in the presence of endogenous dipicolinic acid, a chelating agent.

The *in vitro* synthesis of large polypeptides has been shown by using radioactive alanine and methionine which labeled the NH_2-terminal amino acids of the newly synthesized polypeptides. In contrast the newly synthesized smaller peptides (containing up to 10 amino acids) have no free NH_2-terminal amino acids.[135] *B. subtilis* contains the enzymatic system to synthesize N-formylmethionyl-tRNA,[156] which may be masking the NH_2-terminus.

B. The *B. cereus* T System

Cell-free protein synthesis has been examined with extracts from vegetative cells and spores of *B. cereus* T.[89, 134] One of the primary reasons for developing these systems has been to detect any deficiencies in the synthetic machinery of the spore. The vegetative cell extracts have the usual requirements for amino acid incorporation.[134] Spermidine increased amino acid incorporation by 30%. The system was inhibited by puromycin, chloramphenicol and RNase, but not by DNase. The complete system is stable for 1 month when kept in liquid nitrogen, but at $-15°C$ the soluble fractions lose 60% of their activity. The ribosomes are stable under these conditions in contrast to the ribosomes from *B. subtilis*. The system is quite stable during polypeptide synthesis, since polyphenylalanine incorporation in the presence of poly U continues for more than 1 hr. This is comparable to the rabbit reticulocyte[137] and *E. coli*[138] systems.

The cell-free system from *B. cereus* T spores incorporates amino acids very poorly in the absence of exogenous mRNA, indicating that the spore contains little or no endogenous mRNA.[89] When poly U was added to the spore system for the synthesis of polyphenylalanine, the spore extract had only 2·5% of the synthetic capacity of comparable vegetative cell extracts. Various components of the vegetative cell extract were added to the spore extract to determine whether the spore system was deficient in a particular component. The supernatant fraction from vegetative cells stimulated incorporation fivefold, while the addition of ribosomes from vegetative cells gave the incorporation expected from the sum of the ribosomes from the two systems. Although spores contained a normal level of phenylalanyl-tRNA (phe-tRNA) synthetase, they were deficient in the transfer enzymes for peptide bond synthesis.[100] In addition to the enzyme deficiencies, the ribosomes from dormant spores failed to bind phe-tRNA in the presence of poly U; ribosomes from germinating spores and vegetative cells were able to bind phe-tRNA normally. Both the ribosome and transfer enzymes appear to be deficient or altered in dormant spores of *B. cereus* T.

C. The *B. stearothermophilus* System

This system from a thermophilic spore-former offers some advantages over the other systems because it is active from 30° to 70°C.[131, 132] When protein synthesis was studied with endogenous mRNA, the optimal temperature was found to be 55–60°C. When poly U was used as the mRNA, the optimum temperature depended on the Mg^{++} concentration. With 0·01 M Mg^{++} incorporation was maximal at 37°C; with 0·018 M Mg^{++} at 65°C. Polyadenylic acid stimulation of lysine incorporation was better at 65° than at 37°C because of its loss of secondary structure and the greater ease of attachment to ribosomes at the higher temperature. This system appears to be ideally suited to the analysis of reactions at high temperatures which may affect the secondary structures of tRNA, mRNA and ribosomes.

The cell-free systems from *Bacilli* appear to have requirements and properties very similar to that reported for *E. coli*. It appears that minor differences among the *Bacillus* systems depend on species specificities, the presence or absence of ribonucleases and proteases, the inherent instability of one or more of the many components required for peptide bond synthesis, and methods of extract preparation. Although certain deficiencies were noted in spore extracts from *B. cereus* T, it is not yet certain that the optimum conditions have been obtained for deriving the spore components. The rapid synthesis of proteins during germination suggest that only mRNA synthesis is required. Although the level of some of the components may be low in spores, e.g. ribosomes, the complete machinery appears to be present and capable of replenishing the components required for the first cell division.

VI. Relationship between Bacilli at the Molecular Level

The relationship between *Bacillus* species has been investigated by genetic and physico-chemical techniques. The base composition of the DNA from various *Bacillus* species illustrates their heterogeneous nature (Table I). Transformation[139] and transduction[140] have been utilized to test the degree of genetic interaction possible between two *Bacillus* species. The technique of DNA–DNA and DNA–RNA hybridization has been employed to determine the homologous sequence of nucleotides in the DNA from various *Bacilli*. Both types of analyses have revealed that a precise relationship exists between various species but that a considerable alteration of nucleotide sequences has occurred in their DNA during evolution.

A. TRANSFORMATION AND TRANSDUCTION

Transformation in *Bacillus* species was first demonstrated with *B. subtilis*[139] and has been extended to *B. natto*, *B. subtilis* var *aterrimus*, and *B. licheniformis*.[141–143] The test for interspecies transformation has revealed that only species with closely related DNA compositions can be transformed for auxotrophic markers.[110, 141, 142, 144]

However, it is not sufficient for the DNA compositions to be similar, since many species with similar DNA compositions cannot transform each other. The DNA from the following organisms were able to transform *B. subtilis* auxotrophic mutants: *B. natto*, *B. subtilis* var *aterrimus*, *B. subtilis* var *niger* and *B. polymyxa*.[141, 142] Reciprocal transformation of the *try* marker was shown between *B. subtilis* and *B. natto*.[142] In one case transformation of *B. subtilis* by *B. megaterium* DNA was observed.[144] This was a rather unique situation since their DNA compositions differed by 3% GC. However, another possibility is that the *B. megaterium* was classified incorrectly since its DNA has a 45% GC content whereas most *B. megaterium* DNAs have a 37–39% GC content.

Interspecies transformation has been demonstrated for a greater number of species when transformation to antiobiotic resistance was used as the marker.[110] DNA from *B. pumilus* (40–41% GC) and *B. licheniformis* (45–46% GC) were able to transform *B. subtilis* (43% GC) cells to streptomycin and micrococcin resistance; however, DNA from *B. megaterium* (39–40% GC), *B. macerans* (52–53% GC) and *B. polymyxa* (47–48% GC) did not transform these markers. Transformation of *B. licheniformis* cells by *B. subtilis* DNA to streptomycin resistance has been accomplished; thus reciprocal transformation of this marker occurred between these two species.[145] The transformation of antibiotic markers among heterologous species occurs much more frequently than for auxotrophic markers. The antibiotic markers appear to be involved in determination of ribosome structure. It may mean that cistrons for ribosome proteins and RNA have been conserved more highly than other cistrons.

Transduction of genetic markers has been shown to occur with phage systems from *B. subtilis* and *B. licheniformis*.[140] Interspecies transduction occurred between *B. subtilis* and *B. natto*,[142] but not between *B. subtilis* and *B. licheniformis*.[146] Although the same phage could infect and produce progeny in two *Bacillus* species, the process of transduction occurred very infrequently between two infectable species. A close correlation was observed between transformation and transduction, indicating that homology between DNA sequences was essential for both types of genetic information transfer. Except for antibiotic markers, transformation and transduction between *Bacillus* species occurs only with very closely related species or strains.

B. HYBRID FORMATION BETWEEN HETEROLOGOUS NUCLEIC ACIDS

1. DNA-DNA hybridization

The direct homology between DNA from different *Bacillus* species was examined by testing for DNA–DNA hybrid formation. This method depends on the existence of complementary nucleotide sequences between the DNA from the two test species. Hybrid formation has been observed between DNA from *B. subtilis* and the DNAs from *B. subtilis* var *aterrimus* *B. natto* and *B. megaterium* 203, but not with DNA from many other species including *B. laterosporus*, *B. megaterium*, *B. polymyxa*, *B. cereus*, *B. macerans*, *B. circulans* and *B. mycoides*.[141, 147] Although this method is sensitive and would give the best information concerning the direct relationship between two species, it has not been investigated extensively. As in most of these analyses, B. subtilis has been compared with the other Bacillus species; the other species have not been studied as thoroughly, since most of the genetic analysis of Bacilli is limited to B. subtilis and B. licheniformis.

2. DNA-mRNA hybridization

The molecular homology between different species can be determined by the degree of hybridization between RNA from one species and the DNAs from heterologous species. All RNA fractions can be tested in this manner since all RNA are transcribed from DNA.[148] Messenger RNA from sporulating cells and log phase cells of B. cereus and B. subtilis hybridized well with only their homologous DNAs and very poorly with the DNA from several other species.[110, 149] Even mRNA from one strain of B. subtilis hybridized less efficiently with the DNA from another strain. The relative degree of hybridization between RNA and heterologous DNA was determined by the base composition of the heterologous DNA; the closer its composition to that of the homologous DNA, the better it hybridized with the RNA. The poor homology between sporulation mRNAs and heterologous DNA indicates that the sequence of nucleotides for cistrons involved in sporulation have not been highly conserved.

3. DNA-rRNA hybridization

When the hybridization of rRNA of B. subtilis was tested with DNA from various species, a surprisingly high degree of heterologous DNA–rRNA complementarity was observed.[110, 149, 150] Table VIII lists the hybridization efficiency of B. subtilis rRNA with heterologous DNAs. The results indicate that 36 to 100% of the nucleotide sequences for rRNA have been conserved among species whose DNA compositions varied greatly. This high degree of interspecies homology existed only among Bacillus species. A much smaller amount of hybridization occurred with species from other genera. However, the close relationship of Bacillus species at the molecular level is indicated most clearly by these results. Both 16S and 23s rRNA cistrons were conserved to an equal extent.[110, 112] This high degree of interspecies homology for rRNA suggest that they have been highly conserved during evolution.

The conservation of rRNA nucleotide sequences and ribosome structure among bacterial species has been shown further by the formation of hybrid 70s ribosomes in which the 50s and 30s subunits are derived from two different species. The formation of 70s ribosomes from 50s and 30s subunits from E. coli and B. subtilis[143, 151] and from E. coli and B. stearothermophilus[152] has been demonstrated. The sedimentation coefficients and functions of the hybrid 70s ribosomes were almost identical to the 70s ribosomes made from homologous subunits. Thus, two species which are very different in overall genetic make-up still have conserved genetic information to allow formation of a complex morphological entity such as

TABLE VIII

Hybridization, as a percentage of the homologous reaction of B. subtilis *ribosomal RNA, with DNA of different species of bacilli*

Organism	Percent Hybridization			
	23s + 16s[a]	16s[b]	23s[b]	23s[c]
B. subtilis	100	100	100	100
B. cereus	63	—	—	—
B. megaterium	54	121	118	75–93
B. stearothermophilus	55	—	—	—
B. macerans	48	78	78	—
B. niger	—	52	56	—
B. pumilus	—	75	71	—
B. licheniformis	—	118	113	—
B. polymyxa	—	43	36	—
B. circulans	—	—	—	43

[a] From Doi and Igarashi[149]
[b] From Dubnau, Smith, Morell and Marmur[110]
[c] From Takahashi, Saito and Ikeda[150]

the ribosome. While the rRNA and ribosomal proteins are not identical, the functional conformation of the final structure has been conserved. This may be related to the specific functions of ribosomes which require a particular conformation in order to interact with mRNA and tRNA during protein synthesis. Perhaps the early evolution of this efficient system precluded subsequent evolutionary changes in which several simultaneous changes were required.

4. DNA-tRNA hybridization

The tRNA of *B. subtilis* has been tested for interspecies homology. As in the case with rRNA, a high degree of complementarity was found between *B. subtilis* tRNA and DNA from various species ranging from 26 to 87% of the hybrid formed in the homologous situation.[110] Thus, it appears that the cistrons involved in the synthesis of tRNA, 16s rRNA and 23s rRNA have been highly conserved in sporulating bacteria with DNA containing 32–52% GC. Once a favourable overall structure is attained the multiple functions of tRNA with several specific sites, e.g. amino acid accepting end, site for recognition by aminoacyl synthetase, ribosome binding site, and anticodon, also probably precludes much alteration during evolution.

The hybridization technique in conjunction with equilibrium density centrifugation methods has been used to map the chromosomal locations of rRNA and tRNA in *B. subtilis*.[111, 153, 154] The order of the genes from the point of origin is *antibiotic markers*, 5s rRNA, tRNA, 16s rRNA and 23s rRNA.[74] These markers, which are all highly conserved in *Bacillus* species, are located close to the origin of the chromosomal replication site.

VII. Conclusion

Nucleic acid changes of sporulating cells have a general pattern, which allows the cell to carry out the necessary functions for spore formation and ensures both a complete genome for the dormant spore and the synthetic machinery necessary for its germination. At the same time, sporulation involves the synthesis of spore components and the controlled degradation of some RNA and protein fractions, since sporulation generally takes place when the medium is deficient for growth. The ribosome fraction which is present in excess at the end of the log phase of growth, is degraded and serves as a source for amino acids and nucleotides. These amino acids and nucleotides are used for the formation of mRNA and proteins required for the sporulation process. The use of the *excess* ribosomes for this purpose fits into the current concept of the direct relationship between the rate of protein synthesis and ribosome concentration.

The differential synthesis of mRNA emphasizes the change in cellular physiology which is occurring in sporulating cells. Furthermore, genetic studies indicate that many functions are involved. As in most systems concerned with cellular differentiation, the regulation of gene expression at the transcription and translation level requires further elucidation. The information obtained to date is only a beginning and the analysis between mRNA synthesis and tRNA functions may be one fruitful area of investigation.

The accumulating evidence suggests that the dormant spore contains the complete machinery for gene transcription and protein synthesis. The difference between vegetative cells and spores in this respect may be only quantitative and not qualitative; only an increased concentration of cellular components may be necessary for outgrowth and division to occur. In this regard the analysis of microcycle sporogenesis in which sporangia are obtained directly from outgrowing spores without cell division may be particularly useful.[155] The very low level of mRNA in dormant spores suggests that protein synthesis is completed during spore maturation and that mRNA is degraded during this period. The synthetic machinery lies dormant but in readiness for germination since RNA and protein synthesis begins without delay.

Many questions related to nucleic acids during sporulation are concerned with regulatory mechanisms. Of major interest are the factors which initiate differential mRNA synthesis, the process of sequential gene expression, the mechanism of chromosome and cytoplasmic enclosure into the forespore, the controlled turnover of RNA and protein during sporulation, the differential expression of the parasporal and forespore genomes, and the nature and activation of the synthetic machinery in dormant spores. These and many other intriguing problems will be answered most probably in the general context of cellular regulatory mechanisms. The study of sporulation offers an opportunity for the analysis of morphogenesis at a level of complexity above that of protein-nucleic acid interactions, but below that of cellular interactions. Although the added complexity has technical disadvantages, its major contribution may lie in serving as a model for the study of biological systems enclosed in membranes.

REFERENCES

1. Halvorson, H. O., Vary, J. C. and Steinberg, W. (1966). *A. Rev. Microbiol.* **20**, 169.
2. Young, I. E. and Fitz-James, P. C. (1959). *J. biophys. biochem. Cytol.* **6**, 483.
3. Young, I. E. and Fitz-James, P. C. (1959). *J. biophys. biochem. Cytol.* **6**, 467.
4. Fitz-James, P. C. (1955). *Can. J. Microbiol.* **1**, 502.
5. Marmur, J. and Doty, P. (1962). *J. molec. Biol.* **5**, 109.
6. Schildkraut, C. L., Marmur, J. and Doty, P. (1962). *J. molec. Biol.* **4**, 430.
7. Belozersky, A. N. and Spirin, A. S. (1960). *In* "The Nucleic Acids" (E. Chargaff and J. N. Davidson, eds), Vol. III, pp. 147–185, Academic Press, New York and London.
8. Sueoka, N. (1961). *J. molec. Biol.* **3**, 31.
9. McDonald, W. C., Felkner, I. C., Turetsky, A. and Matney, T. S. (1963). *J. Bact.* **85**, 1071.
10. Welker, N. E. and Campbell, L. L. (1965). *J. Bact.* **89**, 175.
11. Bott, K. and Strauss, B. (1965). *Virology* **25**, 212.
12. Douthit, H. A. and Halvorson, H. O. (1966). *Science N.Y.* **153**, 182.
13. Doskocil, J. and Sormova, Z. (1965). *Biochim. biophys. Acta* **95**, 513.
14. Oda, K. and Marmur, J. (1966). *Biochemistry* **5**, 761.
15. Fujimoto, D., Srinivasan, P. R. and Borek, E. (1965). *Biochemistry* **4**, 2849.
16. Gold, M., Hurwitz, J. and Anders, M. (1963). *Proc. natn. Acad. Sci. U.S.A.* **50**, 164.
17. Fitz-James, P. C. and Young, I. E. (1959). *J. Bact.* **78**, 743.
18. Woese, C. R. (1958). *J. Bact.* **75**, 5.
19. Ganesan, A. T. and Lederberg, J. (1965). *Biochem. biophys. Res. Commun.* **18**, 824.
20. Massie, H. R. and Zimm, B. H. (1965). *Proc. natn. Acad. Sci. U.S.A.* **54**, 1636.
21. Dennis, E. S. and Wake, R. G. (1966). *J. molec. Biol.* **15**, 435.

22. Dennis, E. S. and Wake, R. G. (1967). Abstr. Intern. Congr. "Replication and Recombination of Genetic Material", Canberra.
23. Eberle, H. and Lark, K. G. (1967). *Proc. natn. Acad. Sci. U.S.A.* **57**, 95.
24. Kelly, M. S. and Pritchard, R. H. (1965). *J. Bact.* **89**, 1314.
25. Cairns, J. (1963). *Cold Spr. Harb. Symp. quant. Biol.* **28**, 43.
26. MacHattie, L. A., Berns, K. I. and Thomas, C. A. Jr. (1965). *J. molec. Biol.* **11**, 648.
27. Yoshikawa, H. and Sueoka, N. (1963). *Proc. natn. Acad. Sci. U.S.A.* **49**, 559.
28. Yoshikawa, H. and Sueoka, N. (1963). *Proc. natn. Acad. Sci. U.S.A.* **49**, 806.
29. Yoshikawa, H, O'Sullivan, A. and Sueoka, N. (1964). *Proc. natn. Acad. Sci. U.S.A.* **52**, 973.
30. Oishi, M., Yoshikawa, H. and Sueoka, N. (1964). *Nature, Lond.* **204**, 1069.
31. Spiegelman, S., Aronson, A. I. and Fitz-James, P. C. (1958). *J. Bact.* **75**, 102.
32. Butler, J. A. V. and Godson, G. N. (1963). *Biochem. J.* **88**, 176.
33. Bhagavan, N. V. and Atchley, W. A. (1965). *Biochemistry* **4**, 234.
34. Mizuno, S. and Whiteley, H. R. (1967). *Bact. Proc.*, 116.
35. Oishi, M., Kitayama, S., Takahashi, H. and Maruo, B. (1964). *J. Biochem., Tokyo* **56**, 108.
36. Butler, J. A. V. and Godson, G. N. (1964). *Nature, Lond.* **201**, 876.
37. Ezekial, D. (1960). *J. Bact.* **80**, 119.
38. Hodson, P. H. and Beck, J. V. (1960). *J. Bact.* **79**, 661.
39. McCloy, E. (1953). *VI Congr. Intern. Microbiol. Rome* **2**, 210.
40. Goldberg, I. and Gollakota, K. (1960). *In* "Spores II" (H. O. Halvorson, ed.), pp. 276–296, Burgess Publishing Co., Minneapolis.
41. Yehle, C. O. and Doi, R. H. (1967). *J. Virology* **1**, 935.
42. Takahashi, I. (1964). *J. Bact.* **87**, 1499.
43. Romig, W. R. and Brodetsky, A. M. (1961). *J. Bact.* **82**, 135.
44. Fitz-James, P. C. (1955). *Can. J. Microbiol.* **1**, 525.
45. Young, I. E. and Fitz-James, P. C. (1959). *Nature, Lond.* **183**, 372.
46. Woese, C. R. and Forro, J. R. (1960). *J. Bact.* **80**, 811.
47. Mandel, M. and Rowley, D. B. (1963). *J. Bact.* **85**, 1445.
48. Ryter, A., Bloom, B. and Aubert, J. P. (1966). *C.r. hebd. séanc. Acad. Sci. Paris* **262**, 1305.
49. Falaschi, A. and Kornberg, A. (1965). *Proc. natn. Acad. Sci. U.S.A.* **54**, 1713.
50. Donnellan, J. E., Jr. and Setlow, R. B. (1965). *Science, N.Y.* **149**, 308.
51. Smith, K. C. and Yoshikawa, H. (1966). *Photochem. Photobiol.* **5**, 777.
52. Douthit, H. A. and Halvorson, H. O. (1966). *Fedn. Proc.* **25**, 707.
53. Jacob, F., Schaeffer, P. and Wollman, E. L. (1960). *Symp. Soc. gen. Microbiol.* **10**, 67.
54. Rogolsky, M. and Slepecky, R. A. (1964). *Biochem. biophys. Res. Commun.* **16**, 204.
55. Falaschi, A., Spudich, J. and Kornberg, A. (1965). *In* "Spores III" (L. L. Campbell and H. O. Halvorson, eds), p. 88. Am. Soc. Microbiol., Ann Arbor, Michigan, U.S.A.
56. Falaschi, A. and Kornberg, A. (1966). *J. biol. Chem.* **241**, 1478.
57. Okazaki, T. and Kornberg, A. (1964). *J. biol. Chem.* **239**, 259.
58. Tono, H. and Kornberg, A. (1967). *J. biol. Chem.* **242**, 2375.
59. Gardner, R. and Kornberg, A. (1967). *J. biol. Chem.* **242**, 2383.
60. Tono, H. and Kornberg, A. (1967). *J. Bact.* **93**, 1819.
61. Jacob, F. and Monod, J. (1961). *J. molec. Biol.* **3**, 318.

62. Ames, B. N. and Hartman, P. E. (1963). *Cold Spring Harb. Symp. quant. Biol.* **28**, 349.
63. Stent, G. S. (1964). *Science, N.Y.* **144**, 816.
64. Imamoto, R., Ito, J. and Yanofsky, C. (1966). *Cold Spring Harb. Symp. quant. Biol.* **31**, 235.
65. Newton, W. A., Beckwith, J. R., Zipser, D. and Brenner, S. (1965). *J. molec. Biol.* **14**, 290.
66. Midgley, J. E. M. (1962). *Biochem. biophys. Acta* **61**, 513.
67. Doi. R. H. and Igarashi, R. T. (1964). *J. Bact.* **87**, 323.
68. Aronson, A. I. and Rosas del Valle, M. (1964). *Biochim. biophys. Acta* **87**, 267.
69. Stuy, J. H. (1958). *J. Bact.* **76**, 179.
70. Saunders, G. F. and Campbell, L. L. (1966). *J. Bact.* **91**, 332.
71. Hayashi, M. and Spiegelman, S. (1961). *Proc. natn. Acad. Sci. U.S.A.* **47**, 1564.
72. Levin, D. H. (1966). *Biochemistry* **5**, 1618.
73. Miura, K. I. (1962). *Biochim. biophys. Acta* **55**, 62.
74. Morell, P., Smith, I., Dubnau, D. and Marmur, J. (1967). *Biochemistry* **6**, 258.
75. Doi, R. H. and Igarashi, R. T. (1964). *Proc. natn. Acad. Sci. U.S.A.* **52**, 755.
76. Bishop, H. L. and Doi, R. H. (1966). *J. Bact.* **91**, 695.
77. Balassa, G. (1963). *Biochim. biophys. Acta* **76**, 410.
78. Doi, R. H. and Igarashi, R. T. (1964). *Nature, Lond.* **203**, 1092.
79. Fitz-James, P. C. (1965). *In* "Mecanismes de regulation chez les microorganismes". Collq. Intern. Centre Natl. Rech. Sci. (Paris), No. 124, p. 529.
80. Canfield, R. E. and Szulmajster, J. (1964). *Nature, Lond.* **203**, 596.
81. Young, I. E. and Fitz-James, P. C. (1959). *J. biophys. biochem. Cytol.* **6**, 499.
82. Rosas del Valle, M. and Aronson, A. I. (1962). *Biochem. biophys. Res. Commun.* **9**, 421.
83. Stahly, D. P., Srinivasan, V. R. and Halvorson, H. O. (1966). *J. Bact.* **91**, 1875.
84. Levinthal, C., Keynan, A. and Higa, K. (1962). *Proc. natn. Acad. Sci. U.S.A.* **48**, 1631.
85. Szulmajster, J., Canfield, R. E. and Blicharska, J. (1963). *C.r. hebd. séanc. Acad. Sci., Paris* **256**, 2057.
86. Balassa, G. (1966). *Annls Inst. Pasteur, Paris* **110**, 175.
87. Spotts, C. R. and Szulmajster, J. (1962). *Biochim biophys. Acta* **61**, 635.
88. Yamagishi, H. and Takahashi, I. (1967). *Bact. Proc.*, 28.
89. Kobayashi, Y., Steinberg, W., Higa, A., Halvorson, H. O. and Levinthal, C. (1965). *In* "Spores III" (L. L. Campbell and H. O. Halvorson, eds), p. 200. Am. Soc. Microbiol., Ann Arbor, Michigan, U.S.A.
90. Doi, R. H. (1965). *In* "Spores III" (L. L. Campbell and H. O. Halvorson, eds), p. 111. Am. Soc. Microbiol., Ann Arbor, Michigan, U.S.A.
91. Aronson, A. I. (1965). *J. molec. Biol.* 11, 576.
92. Levinthal, C., Fan, D. P., Higa, A. and Zimmerman, R. A. (1963). *Cold Spr. Harb. Symp. quant. Biol.* **28**, 183.
93. Aronson, A. I. (1965). *J. molec. Biol.* **13**, 92.
94. Aronson, A. I. (1966). *J. molec. Biol.* **15**, 505.
95. Kennell, D. (1964). *J. molec. Biol.* **9**, 789.
96. Schlessinger, D., Marchesi, V. T. and Kwan, B. C. K. (1965). *J. Bact.* **90**, 456.
97. Aronson, A. I. and Fitz-James, P. C. (1966). *Bact. Proc.*, 31.
98. Torriani, A. and Levinthal, C. (1967). *J. Bact.* **94**, 176.

99. Balassa, G. and Contesse, G. (1965). *Annls Inst. Pasteur, Paris* **109**, 683.
100. Kobayashi, Y. and Halvorson, H. O. (1967). *Bact. Proc.* 22.
101. Woese, C. R., Langridge, R. and Morowitz, H. J. (1960). *J. Bact.* **79**, 777.
102. Woese, C. R. (1961). *J. Bact.* 82, 695.
103. Taylor, M. M. and Storck, R. (1964). *Proc. natn. Acad. Sci. U.S.A.* **52**, 958.
104. Schlessinger, D. (1963). *J. molec. Biol.* **7**, 569.
105. Pfister, R. M. and Lundgren, D. G. (1964). *J. Bact.* **88**, 1119.
106. Fitz-James, P. C. (1964). *Can. J. Microbiol.* **10**, 92.
107. Bremer, H. and Konrad, M. W. (1964). *Proc. natn. Acad. Sci. U.S.A.* **51**, 801.
108. McIlreavy, D. J. and Midgley, J. E. M. (1967). *Biochim. biophys Acta* **142**, 47.
109. Nichols, J. L. and Lane, B. G. (1967). *Can. J. Biochem.* **45**, 937.
110. Dubnau, D., Smith, I., Morell, P. and Marmur, J. (1965). *Proc. natn. Acad. Sci. U.S.A.* **54**, 491.
111. Oishi, M. and Sueoka, N. (1965). *Proc. natn. Acad. Sci. U.S.A.* **54**, 483.
112. Doi, R. H. and Igarashi, R. T. (1966). *J. Bact.* **92**, 88.
113. Kurland, C. G. (1960). *J. molec. Biol.* **2**, 83.
114. Kaneko, I. and Doi, R. H. (1966). *Proc. natn. Acad. Sci. U.S.A.* **55**, 564.
115. Balassa, G. (1966). *Annls Inst. Pasteur, Paris* **110**, 17.
116. Balassa, G. (1966). *Annls Inst. Pasteur, Paris* **110**, 316.
117. Doi, R. H. and Kaneko, I. (1966). *Cold Spr. Harb. Symp. quant. Biol.* **31**, 581.
118. Lazzarini, R. A. (1966). *Proc. natn. Acad. Sci. U.S.A.* **56**, 185.
119. Lazzarini, R. A. and Santangelo, E. (1967). *J. Bact.* **94**, 125.
120. Balassa, G. (1964). *Biochem. biophys. Res. Commun.* **15**, 236.
121. Kurland, C. G. and Maaloe, O. (1962). *J. molec. Biol.* **4**, 193.
122. Stent, G. S. and Brenner, S. (1961). *Proc. natn. Acad. Sci. U.S.A.* **47**, 2005.
123. Balassa, G. (1964). *Biochem. biophys. Res. Commun.* **15**, 240.
124. Ryter, A., Schaeffer, P. and Ionesco, H. (1966). *Annls Inst. Pasteur, Paris* **110**, 305.
125. Monro, R. E. (1961). *Biochem. J.* **81**, 225.
126. Urba, R. C. (1959). *Biochem. J.* **71**, 513.
127. Bernlohr, R. W. (1967). *J. Bact.* **93**, 1031.
128. Spizizen, J. (1965). *In* "Spores III" (L. L. Campbell and H. O. Halvorson eds), p. 125. Am. Soc. Microbiol., Ann Arbor, Michigan, U.S.A.
129. Angelo, N. and Aronson, A. (1967). *Bact. Proc.*, 29.
130. Hirashima, A., Asano, K. and Tsugita, A. (1967). *Biochem. biophys. Acta* **134** 165.
131. Friedman, S. M. and Weinstein, I. B. (1966). *Biochim. biophys. Acta* **114**, 593
132. Bubela, B. and Holdsworth, E. S. (1966). *Biochim. biophys. Acta* **123**, 376.
133. Takeda, M. and Lipmann, F. (1966). *Proc. natn. Acad. Sci. U.S.A.* **56**, 1875.
134. Kobayashi, Y. and Halvorson, H. O. (1966). *Biochim. biophys. Acta* **119**, 160.
135. Horikoshi, K. and Doi, R. H. (1967). *Fedn Proc.* **26**, 457.
136. Clark, B. F. C. and Marcker, K. A. (1966). *Nature, Lond.* **211**, 378.
137. Allen, E. H. and Schweet, R. S. (1962). *J. biol. Chem.* **237**, 760.
138. Matthaei, J. H. and Nirenberg, M. W. (1961). *Proc. natn. Acad. Sci. U.S.A.* **47**, 1580.
139. Spizizen, J. (1958). *Proc. natn. Acad. Sci. U.S.A.* **44**, 1072.
140. Thorne, C. B. (1962). *J. Bact.* **83**, 106.
141. Marmur, J., Seaman, E. and Levine, J. (1963). *J. Bact.* **85**, 461.
142. Aoki, H., Saito, H. and Ikeda, Y. (1963). *J. gen. appl. Microbiol., Tokyo* **9**, 307.
143. Gwinn, D. D. and Thorne, C. B. (1964). *J. Bact.* **87**, 519.

144. Ikeda, Y., Saito, H., Miura, K., Takagi, J. and Aoki, H. (1965). *J. gen. appl. Microbiol., Tokyo* **11**, 181.
145. Goldberg, I. D., Gwinn, D. D. and Thorne, C. B. (1966). *Biochem. biophys. Res. Commun.* **23**, 543.
146. Taylor, M. J. and Thorne, C. B. (1963). *J. Bact.* **86**, 452.
147. Takahashi, H., Saito, H. and Ikeda, Y. (1966). *J. gen. appl. Microbiol., Tokyo* **12**, 113.
148. Spiegelman, S. and Hayashi, M. (1963). *Cold Spr. Harb. Symp. quant. Biol.* **28**, 161.
149. Doi, R. H. and Igarashi, R. T. (1965). *J. Bact.* **90**, 384.
150. Takahashi, H., Saito, H. and Ikeda, Y. (1967). *Biochim. biophys. Acta* **134**, 124.
151. Lederberg, S. and Lederberg, V. (1961). *Expl. Cell Res.* **25**, 198.
152. Chang, F. N., Sih, C. J. and Weisblum, B. (1966). *Proc. natn. Acad. Sci. U.S.A.* **55**, 431.
153. Dubnau, D., Smith, I. and Marmur, J. (1965). *Proc. natn. Acad. Sci. U.S.A.* **54**, 724.
154. Oishi, M., Oishi, A. and Sueoka, N. (1966). *Proc. natn. Acad. Sci. U.S.A.* **55**, 1095.
155. Vinter, V. and Slepecky, R. A. (1965). *J. Bact.* **90**, 803.
156. Horikoshi, K. and Doi, R. H. (1967). *Archs. Biochem. Biophys.* **122**, 685.

CHAPTER 5

Biosynthesis of Polypeptide Antibiotics

A. HURST

*Unilever Research Laboratory, Sharnbrook, Bedford, England**

I. Introduction

POLYPEPTIDE antibiotics are synthesized by a wide variety of micro-organisms. Penicillin made by *Penicillium chrysogenum* contains heterocyclic structures which could have arisen from condensations in a peptide chain. The actinomycins made by members of the genus *Streptomyces* are a family of antibiotics in which two peptides are attached by peptide linkage to a chromophore moiety of the molecule; the different actinomycins differ solely in the peptide portion of the molecule. The discussion of this chapter is, however, restricted to the principal antibiotics synthesized by members of the genus *Bacillus*. For a discussion of polypeptide antibiotics irrespective of their source the reader should see ref. 1. A wide variety of poly-peptides, animal, plant, fungal and bacterial in origin are discussed in ref. 2. A detailed account of many antibiotics, including some of the bacilliary antibiotics is in ref. 3.

The term polypeptide is open to some ambiguity. Some authorities use it to describe a defined portion of a protein macromolecule which may be dissociated from the molecule by urea or other H bond breaking reagents.

* Present address: Microbiology Division, Food and Drug Directorate, Ottawa, Canada

For example, lactic dehydrogenase (but the same applies to *Escherichia coli* tryptophanase or haemoglobins[4, 5]) contains two distinct polypeptides, A and B, four polypeptide chains constituting an enzyme macromolecule.[6] The synthesis of these polypeptides is under direct genetic control; the evidence about polypeptide antibiotics suggests that their synthesis is enzymic (see below) and therefore the genetic control is indirect being applied first to the synthesis of the enzymes necessary for the synthesis of the antibiotic. It is not clear just at what point in the complexity of a molecule the precise control of protein synthesis is needed and how far is enzyme synthesis satisfactory in making a functional molecule. In general, bacillus polypeptide antibiotics have a molecular weight of only 1400 (i.e. about 10 amino acids per molecule) and enzyme synthesis is conceivable; however, subtilin, made by *Bacillus subtilis* has a molecular weight of 3200 and nisin made by *Streptococcus lactis* 7000.[7, 8] This should be contrasted with insulin, a protein which has a molecular weight of 6000.[9]

In addition to their low molecular weights polypeptide antibiotics share the property of containing amino acids not normally found in proteins e.g. ornithine, di-aminobutyric acid. Also the amino acids may be in the "unnatural" D-configuration and it is not yet clear how these are formed and incorporated into the peptide chain. In addition, they may be completely cyclic as in tyrocidine or gramicidin S or contain an important cyclic moiety as in the polymyxins.

The facts listed have added to our difficulties in understanding the physiological role of these substances. Antibiotics arise late in the growth cycle, after the phase of exponential growth, but they are not the only substances made at this stage. For example several species of *Bacillus* excrete an extra-cellular amylase[10] or protease.[11] In principle, polypeptide antibiotics could be synthesized in one of three ways:

(1) By the cleavage of proteins already formed during preceding growth
(2) By the normal mechanism of protein synthesis
(3) By the sequential building up of amino acids by specific enzymes.

These possibilities were reviewed by Bodansky and Perlman.[12] There appears to be no experimental evidence in support of the first. As regards the second possibility, the genetic code is fully occupied by the usual amino acids found in proteins.[13, 14] This would imply that the genetic code which is thought to be valid for all living things does not apply to *Bacillus*—an unattractive proposition. Furthermore, the triplet code permits at maximum only 64 combinations, and virtually this number of unnatural amino acids has already been described in the antibiotic literature. Most of the evidence therefore favours the third view of enzymic synthesis which will be discussed in some detail.

TABLE I

Composition of gramicidins

Compound	1	2	3	4	5	6	7	8	9	10	11	12	13	14	15
I	Valine-gramicidin A:														
	HCO-L-Val-Gly-L-Ala-D-Leu-L-Ala-D-Val-L-Val-D-Val-L-Try-D-Leu-L-*Try*-D-Leu-L-Try-D-Leu-L-Try-NHCH$_2$CH$_2$OH														
II	Isoleucine-gramicidin A:														
	HCO-L-Ileu-Gly-L-Ala-D-Leu-L-Ala-D-Val-L-Val-D-Val-L-Try-D-Leu-L-*Try*-D-Leu-L-Try-D-Leu-L-Try-NHCH$_2$CH$_2$OH														
III	Valine-gramicidin B:														
	HCO-L-Val-Gly-L-Ala-D-Leu-L-Ala-D-Val-L-Val-D-Val-L-Try-D-Leu-L-*Phe*-D-Leu-L-Try-D-Leu-L-Try-NHCH$_2$CH$_2$OH														
IV	Isoleucine-gramicidin B:														
	HCO-L-Ileu-Gly-L-Ala-D-Leu-L-Ala-D-Val-L-Val-D-Val-L-Try-D-Leu-L-*Phe*-D-Leu-L-Try-D-Leu-L-Try-NHCH$_2$CH$_2$OH														
V	Valine-gramicidin C:														
	HCO-L-Val-Gly-L-Ala-D-Leu-L-Ala-D-Val-L-Val-D-Val-L-*Try*-D-Leu-L-*Try*-D-Leu-L-Try-D-Leu-L-Try-NHCH$_2$CH$_2$OH														
VI	Isoleucine-gramicidin C:														
	HCO-L-Ileu-Gly-L-Ala-D-Leu-L-Ala-D-Val-L-Val-D-Val-L-*Try*-D-Leu-L-*Try*-D-Leu-L-Try-D-Leu-L-Try-NHCH$_2$CH$_2$OH														

II. Brief History, List and Composition of the
Principal Bacillary Antibiotics

Antagonism among microbes was recognized and its practical significance in medicine appreciated since the last decades of the 19th century.
However, the preparations were either toxic or contained virulent bacteria
and the danger to the patient from such treatment was nearly as great as the
disease itself. For an early review see Florey *et al.*[15] Dubos[16] in 1939
described a soil bacillus, products isolated from its culture fluids being
capable of destroying pneumococci. These substances were identified as
gramicidin and tyrocidine and the producer organism as *B. brevis.*[17] Subsequent chemical work has shown that neither gramicidin nor tyrocidine
are single entities. According to the older literature the most common form
of gramicidin is gramicidin D;[18] it differs from the other gramicidins by
containing isoleucine. Gramicidin A constitutes the major part of commercial gramicidin.[19] Ramachandran[20] recognizes the following gramicidins:

Gramicidin A Gly, Ala$_2$, Val$_4$, Leu$_4$, Try$_6$ + Ethanolamine
Gramicidin B Gly$_1$, Ala$_2$, Val$_4$, Leu$_4$, Phe, Try$_{3-4}$ + Ethanolamine
Gramicidin C Gly$_1$, Ala$_2$, Val$_4$, Leu$_4$, Try$_1$, Try$_6$ + Ethanolamine
Gramicidin D Gly$_1$, Ala$_2$, Val$_3$, Ileu$_1$, Leu$_4$, Try$_6$ + Ethanolamine

This composition is very close to that described by Witkop and his
collaborators.[19] The sequence of amino acids in gramicidins A, B, C is shown
in Table I, and is taken from the work of Sarges and Witkop.[21]

Tyrocidine is a cyclic decapeptide[22] and the forms A, B, C, D, shown
in Fig. 1, I, are taken from the work of Mach and Tatum;[23] these forms
differ in the kinds of aromatic amino acids they contain.

As a result of the impetus given by the pioneering work of Dubos[16] other
antibiotics were re-examined, the first practical and successful application
being penicillin. Gramicidin S, also produced by strains of *B. brevis*, was
described in 1944 and was apparently used clinically during World War
II.[24, 25] It is another cyclic decapeptide (Fig. 1, II). In addition to a number
of D amino acids it contains ornithine, a non-protein amino acid. Note
that gramicidin S is structurally related to tyrocidine but tyrocidine and
gramicidin, although synthesized concurrently, are not related.

Polymyxin and bacitracin are two other polypeptide antibiotics which
have clinical use. Polymyxin and aerosporin were discovered at about the
same time by three groups of independent workers, two in America[26, 27]
working with *B. polymyxa* and a group in England[28] working with
B. aerosporus. Later, the two microbes were shown to be similar and the
antibiotics to be closely related, as closely as the tyrocidines. The name
aerosporin is no longer used; altogether 5 polymyxins (A–E) were recognized and their relationship was reviewed in 1956.[29] The circulins made by

B. circulans are also closely related chemically to the polymyxins. Little has been published about circulins (but see ref. 30). Polymyxin B appears to possess the least toxicity; its amino acid sequence was studied by

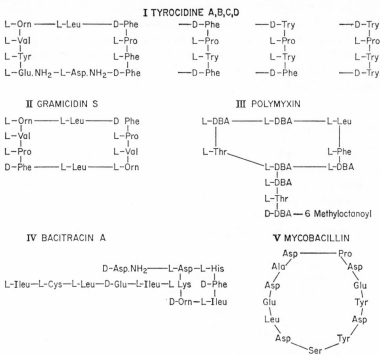

FIG. 1. Structures of some polypeptide antibiotics synthesized by *Bacilli*.

Hausmann[31] and its composition was unequivocally established by the synthetic studies of Vogler *et al.*[32] (see Fig. 1, III). The non-protein amino acid of polymyxin is α-diaminobutyric acid (DBA) which occurs as both the D and L isomer.

Bacitracin was first described in 1945 as being formed by a member of the *B. subtilis* group.[33] The organism was later classified as *B. licheniformis* which was already known to produce several antibiotics, the most important being ayfivin. Ayfivin and bacitracin were shown to be so closely related that the name ayfivin was abandoned in 1950[34] and the active components were referred to as bacitracin A, B, C. The possible structure of bacitracin A based on reference 35 is shown in Fig. 1, IV. As with gramicidin S, the non-protein amino acid is ornithine, but in the D-configuration.

Mycobacillin, so named on account of its wide antifungal spectrum, was first described in 1955 and its structure is shown in Fig. 1, V.[36, 37] With the possible exception of glutamic acid all the amino acids appear to be in the L-form.[38]

Edeine produced by a strain of *B. brevis* was described by Borowska and Tatum.[39] Bacilysin, a dipeptide of alanine and possibly tyrosine, is produced by *B. subtilis* and has been studied by Roscoe and Abraham.[40]

Other bacillary peptide antibiotics have been described and are listed by Bodansky and Perlman.[12] Perhaps the most interesting of these is subtilin made by *B. subtilis*[41, 42] and considered at one time as a possible preservative in food.[43] It contains the unusual S amino acid lanthionine and another unidentified S amino acid,[7] the empirical formula of which corresponds with methylanthionine also occurring in nisin.[8]

III. The Growth Cycle, Sporulation and Antibiotic Synthesis

It is widely accepted that sporulation is characteristic of stationary phase cells. Antibiotic synthesis precedes sporulation as illustrated with *B. licheniformis* in Fig. 2. It will be seen that bacitracin synthesis started at 7 hr, by which time more than half the vegetative growth was completed. Free spores did not appear until 20 hr, by which time the culture was well into the stationary phase. Similar findings are reported for polymyxin;

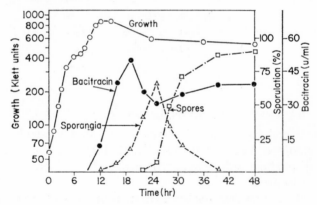

F I G. 2. The time course of growth, bacitracin concentration and the percentage of sporangia and free spores in the culture

Paulus and Gray[44] and Daniels[45] both found that release of bioassayable polymyxin into the medium occurred most rapidly during the second half of the exponential growth phase and continued into the stationary phase.

Gramicidin S also is synthesized during late exponential and stationary phases.[46, 47] In the experiments of Winnick *et al.*[47] growth rate was exponential for about 40 hr although active protein synthesis occurred for 70 hr, by which time the antibiotic also reached the very high level of 150 mg/1; The gramicidin/protein ratio was 0·04 at 24 hr and 0·14 at 72 hr.

In contrast to polymyxin, which is largely found in the medium, gramicidin S is largely intracellular.

Although antibiotic synthesis occurs at the end of logarithmic phase and precedes sporulation, the evidence that antibiotic synthesis is related to the process of spore formation is still controversial. According to Woodruff,[48] the change in metabolism which occurs in cultures of *Bacillus* during antibiotic synthesis is due to "secondary metabolism". This is a means of using fully formed enzymes and substrates when, in the stationary phase, an essential metabolite is already exhausted. Antibiotic synthesis thus can be regarded as an alternative to a breakdown of cellular control mechanisms; it only occurs in "undifferentiated" cells such as microbes, higher forms of life having acquired more sophisticated methods.[48] However, this view does not seem to be applicable to polypeptide biosynthesis which is an energy-requiring process. It seems unlikely that during stationary phase when energy resources could also become limiting the cell should turn to the production of apparently useless compounds.

Alternative explanations have been suggested.[49] For instance it is not necessary to assume that sporulation can only occur in deficient media. Strains of *B. megaterium* and *B. subtilis* have been isolated which form spores while growing exponentially in a mineral-salts/glucose medium[50] (see Chapter 3). The number of spores and viable cells increase at the same rate but the spore-to-cell ratio depends on the composition of the medium. In any one medium, this ratio can be used to predict the probability of a cell continuing to multiply or to form a spore, and from this the internal biochemical environment of the cell can be deduced. The presence of glucose breakdown products which act as genetic repressors (catabolite repression) leads to a sequence of biochemical events which, probably indirectly, control the expression of the sporulation genes. One of the biochemical events in this sequence could be the synthesis of antibiotic polypeptides and hence the apparent or real correlation between sporulation and antibiotic synthesis.

Electron-microscopic studies of sporulating *B. subtilis* (strain Marbug) revealed a sequence of structural changes summarized in Fig. 3.[51, 52] Mutants may be selected which are blocked at almost any of the stages and Schaeffer[53] has examined a total of about 7500 mutant colonies to relate biochemical properties with sporulation. The results suggested that the production of antibiotic, protease and wall lytic activity was required for sporulation to reach stage I. Production of amylase, DNase and RNase were not required for sporulation to occur.

Independent studies using different genetic techniques led to similar conclusions. An association or linkage was found between a gene controlling sporulation and that of an antibacterial or antibiotic factor.[54] Spizizen[55] went on to select mutants after irradiation of *B. subtilis* with ultra-violet which killed 90–95% of the individuals. He isolated asporogenous mutants in which this genetic defect was in the same locus as defects for two proteolytic enzymes and antibiotic production. The evidence favoured the

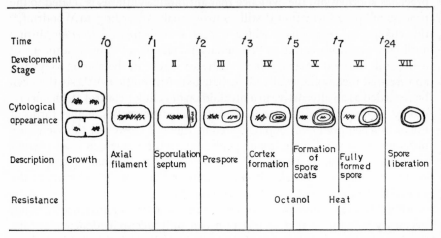

Time	t_0	t_1	t_2	t_3	t_5	t_7	t_{24}
Development Stage	0	I	II	III	IV	V	VI	VII
Cytological appearance								
Description	Growth	Axial filament	Sporulation septum	Prespore	Cortex formation	Formation of spore coats	Fully formed spore	Spore liberation
Resistance						Octanol	Heat	

FIG. 3. Cytological stages during sporulation and the times when changes are observed.

view that the loci for antibiotic production and sporulation though close were separate. The ability of these mutants to sporulate could be restored when they became competent to take up DNA of wild strains; this also restored antibiotic production.[55]

Paulus (in ref. 3) lists the following 5 points which support the notion that polymyxin or circulin are involved in the sporulation process. These points are as follows:

(1) Mutants of *B. polymyxa* which do not produce polymyxin do not form spores

(2) Specific sporulation inhibitors also inhibit polymyxin synthesis

(3) Unaerated cultures of *B. circulans* do not sporulate and do not form circulin whereas aerated cultures produce both

(4) Temperature shifts during the growth cycle (i.e. 30° to 37°C) when they affect polymyxin synthesis also affect sporulation

(5) At the time of sporulation in the growth cycle, polymyxin becomes bound to spores.

IV. Mechanism of Antibiotic Biosynthesis

A. GENERAL

Most investigators aim to devise systems which synthesize protein and antibiotic. In most cases growing cultures are used but washed cell suspensions have also been employed. Attempts are then made to separate the processes of protein synthesis or growth from antibiotic synthesis by means of amino acid analogues or inhibitors of protein synthesis (e.g. certain antibiotics). The antibiotics principally used as "tools of biochemical research" are chloramphenicol and puromycin to inhibit protein synthesis, and actinomycin D to inhibit RNA synthesis (for a review on the mechanism of mode of action of antibiotics see reference 56).

Further analysis generally involves centrifugal fractionation of broken cells; ribosomal and particulate fractions sediment below 100,000 g, while enzymes and other soluble fractions are considered to remain in the supernatant.

B. WHOLE CELL SYSTEMS

1. Gramicidin S

Barry and Ichihara[46] were the first to use a washed suspension of cells which incorporated labelled amino acids into protein and into the newly synthesized decapeptide. Amino acid analogues (p-fluorophenylalanine, nor-leucine, nor-valine) inhibited antibiotic production but growth and protein synthesis were less sensitive.[57] Chloramphenicol and puromycin gave results contradicting those of the amino acid analogues: shortly after their addition growth and protein synthesis were inhibited but gramacidin S production remained unimpaired for several hours.[58] Actinomycin D gave similar results, inhibiting growth but not antibiotic synthesis.[59] These results suggested that gramicidin S was synthesized enzymically rather than ribosomally; this was confirmed by locating newly synthesized gramicidin S in the "soluble fraction" (supernatant after centrifuging broken cells at 100,000 g). The method used consisted in allowing cells to incorporate radio-active phenylalanine but only for a short time. The cells were then broken, differentially centrifuged, the gramicidin S isolated from the different fractions and the highest specific activity determined.[59] (Also p. 177, Section C).

2. Tyrocidine

The biosynthesis of tyrocidine was studied by Mach and Tatum[23, 60] in growing cultures of B. brevis. Chloramphenicol, puromycin and chlortetracycline prevented the incorporation of labelled amino acids into protein without affecting the synthesis of the antibiotic; protein and antibiotic synthesis could also be dissociated by the use of amino acid analogues. When cultures were grown on minimal media tyrocidine A, B, and C were formed (see Fig. 1) which differed in aromatic amino acid content and could be separated by chromatography on Sephadex G25. Addition of

L-phenylalanine resulted in the production of almost pure tyrocidine A, L-tryptophane resulted in the synthesis of almost pure tyrocidine D. This has not been previously described.

These observations indicate that tyrocidine is not synthesized ribosomally but enzymically, the enzymes having low specificity for aromatic amino acids.

3. Bacitracin

The synthesis of this antibiotic also differs from protein synthesis in not being sensitive to chloramphenicol, puromycin and tetracycline.[35, 61] D-Phenylalanine, a residue of which occurs in the antibiotic, inhibits antibiotic synthesis but not growth. Post-logarithmic but presporulation cells incorporate radio-active bacitracin, the label appearing in the spores. Bernlohr and Novelli[35] prepared bacitracin labelled both with ^{14}C-isoleucine and ^{3}H-ornithine. This was added at different times to growing cultures of *B. licheniformis*; the spores formed were harvested at 48 hr, hydrolyzed and the ratio of [14]C : [3]H determined. The original ratio of the isotopes was retained by those spores to which bacitracin had been added at 16–22 hr, ie. at a time when the cells were forming their sporangia. The incorporated label could not be washed out of the spores by a variety of solvents, suggesting that it became incorporated into the spore structure. Retention of the label in the original [14]C : [3]H ratio suggested that the whole undegraded molecule was utilized. Comparison of the amino acid analysis of bacitracin and spore coats suggested that bacitracin was an important structural component of the coat.[62]

4. Polymyxin

The evidence from three separate laboratories clearly indicates that polymyxin is not synthesized ribosomally. One of the early steps in protein biosynthesis is the activation of amino acids with ATP to give an aminoacyladenilate which is transferred by soluble RNA to ribosomes. Enzymes which activate all the L protein amino acids are known, but the D amino acids which occur in antibiotic peptides are not activated in this way.[63] It appears therefore that the D amino acids are not added directly to the antibiotic chain growing on ribosomes. D-Leucine which occurs in polymyxin D (not shown in Fig. 1) when added to the growth medium inhibited the production of polymyxin and the inhibition was reversed by L-leucine.[64] Paulus and Gray[44] and Daniels[45] were also able to separate growth and protein synthesis on the one hand from polymyxin synthesis on the other hand by using chloramphenicol, puromycin and actinomycin. Both authors used radioactive DAB, the unique amino acid of polymyxin and showed that intact cells incorporated it into polymyxin.

5. Mycobacillin, edeine and bacilysin

The use of inhibitors, amino acid analogues and certain genetic techniques, clearly dissociates the biosynthesis of these antibiotics from protein synthesis.[65, 66, 39, 40]

C. CELL-FREE SYSTEMS

1. Gramicidin S

The synthesis of the complete molecule by a cell-free system was first accomplished by Japanese workers in 1965.[67] Earlier reports with cell-free systems, derived from *B. brevis*, described the synthesis of peptides having amino acid sequences closely related to gramicidin S, e.g. the tri-peptide D-phenylalanyl-L-prolyl-L-valine[68] or of the tetrapeptide phenylalanyl-prolyl-valyl-ornithine.[69, 70]. Holm *et al.*[70] considered that the tetrapeptide was conjugated to an unknown residue possibly linked to the carboxyl group of ornithine. This link was quite labile and was readily broken with acid or alkali. The tetrapeptide + residue therefore had a different charge from the tetrapeptide alone and the change in mobility suggested that the residue was not of the nucleotidyl-peptide type. An analogue of phenylalanine competed for the incorporation of valine ^{14}C much more successfully than with the incorporation of leucine [14]C. This suggests that phenylalanine could be incorporated into a chain prior to valine—as in the above tetrapeptide. This makes it possible for the tetrapeptide to be an intermediate in gramicidin S biosynthesis which might start with phenylalanine.

Later reports by the same workers describe the complete synthesis of gramicidin S by a cell-free system.[71, 72] Again valine ^{14}C was used as the labelled amino acid. The age of the cells and the conditions for breaking them were found to be critical. For example cell-free systems made from young cells incorporate valine ^{14}C into protein and only negligible amounts went into gramicidin S. This incorporation is sensitive to inhibitors of protein synthesis such as chloramphenicol and puromycin. Cell-free systems made from old cells (stationary phase) incorporate negligible amount of valine into protein, but incorporation into gramicidin S is vigorous and chloramphenicol decreased incorporation by only 20%. Cells were best broken by ultrasonic treatment for as short a time as 1 min. The incorporating ability decreased when ultrasonic treatment was prolonged. The system required ATP or ATP generation, Mg^{2+} (optimum 0·05 M), pH optimum of 8·2–8·4, and K^+ (optimum 0·01 M). Later, an enzyme system was isolated and partially purified by precipitation with 40% ammonium sulphate; it appeared to take part in gramicidin S synthesis.[73]

Further purification and resolution of the enzyme system into two fractions which were insensitive to inhibitors of protein synthesis was reported by Tomino *et al.*[74] After precipitation with 50% ammonium sulphate and ion exchange chromatography the enzyme system was resolved into two fractions: fraction I synthesized a compound containing phenylalanine, proline, valine and ornithine and could have been the tetrapeptide described above, possibly incomplete gramicidin S; fraction II was concerned with the activation and isomerization of phenylalanine. Neither fraction alone could synthesize gramicidin S, but the combined fractions did, providing that all the constituent amino acids were present, as well as ATP, Mg^{2+} and β-mercaptoethanol; contrary to the results of Holm *et al.*[70] K was not required. The enzyme system appears to be absent in young cultures; it was synthesized suddenly in late exponential phase and disappeared again in stationary phase cells. It appears that the enzyme system can be degraded or inactivated.

Two other groups of workers have confirmed the cell-free synthesis of gramicidin S.[67, 75] The fraction retaining the gramicidin biosynthetic ability was the ribosome-free supernatant from the 105,000 g centrifugation. This system also was insensitive to chloramphenicol, puromycin, penicillin, ribonuclease or deoxyribonuclease.

2. Bacitracin, edeine, polymyxin

Preliminary studies with these antibiotics have broadly suggested properties similar to gramicidin S. Thus Shimura *et al*[76] produced bacitracin from sub-cellular fractions incubated with the constituent amino acids. The 20,000 g precipitate of broken cells was most active but the material synthesized did not chromatographically appear to be bacitracin. Material which chromatographed with authentic bacitracin was only synthesized when the 20,000 g supernatant fraction was returned; this system also appeared to require ATP.

An edeine synthesizing system was described by Borrowska and Tatum.[39] The 105,000 g supernatant from disrupted cells of *Bacillus brevis* Vm_4 was active; the system similarly required ATP and was insensitive to inhibitors of protein synthesis.

A cell-free system for the synthesis of polymyxin was described by Daniels.[45] Radio-active DAB was incorporated into the antibiotic and the labelled antibiotic was identified by its position after high voltage electrophoresis.

The most complete studies of the biosynthesis of polypeptide antibiotics by cell-free systems are those done with gramicidin S. In these the incorporation of one or two of the amino acids of gramicidin, e.g. ^{14}C-valine, into subsequently isolated products is taken as evidence of *de novo* biosynthesis, which is a reasonable but not a conclusive assumption. This type of approach will have to be extended to cover the other gramicidin S amino acids and the fate of these amino acids in the cell has to be followed.

For example, amino acids might be incorporated into gramicidin, protein or nucleic acid or even after de-amination, could serve as energy sources. Some amino acids might also be incorporated by specific enzyme reactions into peptides and small proteins and serve as cell wall or ribosome constituents. Eventually the isotope technique will have to be supplemented with sensitive biological assays.

D. INCORPORATION OF NON-PROTEIN AMINO-ACIDS

A general discussion of the mechanism for the incorporation of D or other unnatural amino acids into polypeptide chains would be out of place in this chapter, but some areas of uncertainty should be indicated.

The biosynthesis of polymyxin D (which contains D-leucine) can be inhibited by adding D-leucine to growing cultures of B. polymyxa; cell-free extracts of the microbe did not activate D- but did activate L-leucine. It is not known how or at what point in the biosynthesis of polymyxin isomerization occurs since the cells appear to lack a leucine racemase.[64] L-DAB has been reported to be 'activated' by B. polymyxa extracts.[77] An ATP-^{32}P-Pi exchange reaction was catalysed by the extract and could be used for the quantitative estimation of DAB in hydrolysates. Using this technique it could be estimated that B. polymyxa spores contain less than 0·2% polymyxin. In contrast, hydrolysates of B. licheniformis spore coats contain amino acids including ornithine in the same ratio as bacitracin. Based on this evidence and on that derived from the incorporation of doubly labelled bacitracin, it has been suggested that bacitracin was an important structural component of the spore coat.[35, 62] Extracts of B. brevis (a strain synthesizing gramicidin S) activated L-ornithine in an ATP exchange reaction, but the activated ornithine could not be transferred to sRNA,[73] other amino acids which are constituents of gramicidin S are also involved in ATP-PPi exchange activities, e.g. L-leucine and L-valine.[74] It is evident that cell-free extracts of microbes which synthesize antibiotics are capable of carrying out this reaction, but whether the reaction is essential for the peptide synthesis is still unknown.

A soluble enzyme from B. brevis could convert L-phenylalanine to the D-form.[78] It has been purified and further characterized. Supernatants after high-speed centrifugation of disintegrated cells were precipitated with 50% ammonium sulphate, redissolved and then finally resolved into two fractions by molecular sieving on Sephadex G200.[74] When fraction II was incubated with all the necessary components for gramicidin S synthesis only one unknown compound could be isolated. It was thought to be phenylalanine amide and its accumulation was interpreted as follows: D-phenylalanine is formed from L-phenylalanine and ATP and is then used for the synthesis of gramicidin S. In the absence of suitable acceptor (normally supplied by fraction I reaction products) the activated D-phenylalanine combines with NH^{4+}.

These results suggest that L- rather than D-phenylalanine is the normal precursor of the D-phenylalanine in gramicidin S. Isotope dilution experiments confirmed this. It seems that generally microbes prefer the L isomer even when the final product contains the D isomer. The precise details of activation and racemization remain to be discovered.

V. Conclusions

The history and biosynthesis of about six polypeptide antibiotics made by different species of *Bacillus* have been reviewed and discussed. Some of the synonyms are listed; most of the names probably represent families of closely related antibiotics which differ in amino acid composition. All the antibiotics described in some detail have molecular weights of about 1400, are probably synthesized enzymically, contain non-protein amino acids and amino acids in the D-configuration. In most cases the biosynthesis requires ATP, but it is not known how the non-protein amino acids are introduced into the molecule. The antibiotics may play a part in the biochemical differentiation processes connected with sporulation.

ACKNOWLEDGEMENTS

Table I is reproduced by permission of the American Chemical Society. The structure of the different forms of tyrocidine shown in Fig. 1 are reproduced by permission of the National Academy of Sciences (U.S.A.). The structure of bacitracin A (Fig. 1) and Fig. 2 are reproduced by permission of Academic Press Inc., New York and Fig. 3 has been modified by permission of Prof. P. Schaeffer, Institut Pasteur, Paris.

REFERENCES

1. Abraham, E. P., Newton, G. G. F. and Warren, S. C. (1965). *In* "Biogenesis of antibiotic substances" (Z. Vanek and Z. Hostalek, eds), p. 169. Academic Press, New York and London.
2. Waley, S. G. (1966). *Adv. Protein. Chem.* **21**, 1.
3. Gottlieb, D. and Shaw, P. D. (eds). (1967). "Antibiotics", Vol. II. Springer-Verlag, New York, U.S.A.
4. Yanofsky, C. (1967). *Scient. Am.* **216**, 80.
5. Perutz, M. F. (1962). "Proteins and Nucleic Acids". Elsevier Publ. Co., Amsterdam.
6. Markert, C. L. (1963). *In* "Cytodifferentiation and Macromolecular synthesis" (M. Locke, ed.), p. 65. Academic Press, New York and London.
7. Lewis, J. C. and Snell, N. S. (1951). *J. Am. chem. Soc.* **73**, 4812.
8. Hurst, A. (1966). *J. gen. Microbiol.* **44**, 209.
9. Sanger, F. (1959). *Science*, N.Y. **129**, 1340.

10. Collman, G. and Elliott, W. H. (1962). *Biochem. J.* **83**, 256.
11. Bernlohr, R. W. (1964). *J. biol. Chem.* **239**, 538.
12. Bodansky, M. and Perlman, D. (1964). *Nature, Lond.* **204**, 840.
13. Cohen, N. R. (1966). *Biol. Rev.* **41**, 503.
14. Soll, D., Ohtsuka, E., Jones, D. S., Lohrman, R., Hayatsu, H., Nishimura, S. and Khorana, H. G. (1965). *Proc. natn. Acad. Sci., U.S.A.* **54**, 1378.
15. Florey, H. W., Chain, E., Heatley, N. G., Jennings, M. A., Sanders, A. G., Abraham, E. P. and Florey, M. E. (1949). "Antibiotics". Oxford University Press, England.
16. Dubos, R. J. (1939). *J. exp. Med.* **70**, 1, **70**, 11.
17. Hotchkiss, R. D. and Dubos, R. J. (1941). *J. biol. Chem.* **141**, 155.
18. Work, T. S. (1948). *Biochem. Soc. Symp.* **1**, 61.
19. Cross, E. and Witkop, B. (1965). *Biochemistry* **4**, 2495.
20. Ramachandran, L. K. (1963). *Biochemistry* **2**, 1138.
21. Sarges, R. and Witkop, B. (1965). *Biochemistry* **4**, 2491.
22. King, T. P. and Craig, L. C. (1955). *J. Am. chem. Soc.* **77**, 6627.
23. Mach, B. and Tatum, E. L. (1964). *Proc. natn. Acad. Sci., U.S.A.* **52**, 876.
24. Gause, G. F. and Braznikova, M. G. (1944). *Nature, Lond.* **154**, 703.
25. Gause, G. F. (1946). *Lancet* (ii) 46.
26. Stansky, P. G., Shepherd, R. G. and White, H. J. (1947). *Bull. Johns Hopkins Hosp.* **81**, 43.
27. Benedict, R. G. and Langlykke, A. F. (1947). *J. Bact.* **54**, 24.
28. Ainsworth, G. C., Brown, A. M. and Brownlee, G. (1947). *Nature, Lond.* **160**, 263.
29. Newton, B. A. (1956). *Bact. Rev.* **20**, 14.
30. Koffler, H. (1959). *Science, N.Y.* **130**, 1419.
31. Hausmann, W. (1956). *J. Am. chem. Soc.* **78**, 3663.
32. Vogler, K., Studer, R. O., Lanz, P., Lergier, W. and Böhni, E. (1961), *Experientia* **17**, 223.
33. Johnson, B. A., Auker, H. and Meleney, F. L. (1945). *Science, N.Y.* **102**, 376.
34. Newton, G. G. F. and Abraham, E. P. (1950). *Biochem. J.* **47**, 257.
35. Bernlohr, R. W. and Novelli, G. D. (1963). *Archs Biochem. Biophys.* **103**, 94.
36. Majumdar, S. K. and Bose, S. K. (1958). *Nature, Lond.* **181**, 134.
37. Majumdar, S. K. and Bose, S. K. (1960). *Biochem. J.* **74**, 596.
38. Banerjee, A. B. and Bose, S. K. (1963). *Nature, Lond.* **200**, 471.
39. Borowska, Z. K. and Tatum, E. L. (1966). *Biochim. biophys. Acta* **114**, 206.
40. Roscoe, J. and Abraham, E. P. (1966). *Biochem. J.* **99**, 793.
41. Humfeld, H. and Feustel, I. C. (1943). *Proc. Soc. Exp. Biol. Med.* **54**, 232.
42. Lewis, J. C., Feeney, R. E., Garibaldi, J. A., Michener, H. D., Hirschmann, D. J., Traufler, D. H., Langlykke, A. F., Lightbody, H. D., Stubbs, J. J. and Humfeld, H. (1947). *Arch. Biochem.* **14**, 415.
43. Le Blanc, F. R., Devlin, K. A. and Stumbo, C. R. (1953). *Fd Technol.* **7**, 181.
44. Paulus, H. and Gray, E. (1964). *J. biol. Chem.* **239**, 865.
45. Daniels, M. (1966). PhD. Thesis. Cambridge University, England; (1968) *Biochim. biophys. A.ta.* **158**, 119.
46. Barry, J. M. and Ichihara, E. (1958). *Nature, Lond.* **181**, 1274.
47. Winnick, R. E., Lis, H. and Winnick, T. (1961). *Biochim. biophys. Acta* **49**, 451.
48. Woodruff, H. B. (1966). *Symp. Soc. gen. Microbiol.* **16**, 22.
49. Weinberg, E. D. and Tonnis, S. M. (1966). *Appl. Microbiol.* **14**, 850.
50. Schaeffer, P., Millet, J. and Aubert, J. P. (1965). *Proc. natn. Acad. Sci., U.S.A.* **54**, 704.

51. Ryter, A., Schaeffer, P. and Ionesco, H. (1966). *Annls Inst. Pasteur, Paris* **110**, 305.
52. Schaeffer, P., Ionesco, H., Ryter, A. and Balassa, G. (1963). *In* "Mechanismes de Regulation des Activites Cellulaires". (Ed. Centre Natl. Recherche Scientifique, Paris), p. 553.
53. Schaeffer, P. (1967). *Folia Microbiol., Praha* **12**, 291.
54. Spizizen, J., Reilly, B. E. and Dahl, B. (1963). *Proc.* 11th *Int. Congr. Genet.* **1.**
55. Spizizen, J. (1965). *In* "Spores" III. (L. L. Campbell and H. O. Halvorson, eds), p. 125. *Amer. Soc. Microbiol. Ann Arbor, Mich.*
56. Gale, E. F. (1963). *Pharm. Rev.* **15**, 481.
57. Winnick, R. E. and Winnick, T. (1961). *Biochim. biophys. Acta.* **53**, 461.
58. Eikhom, T. S., Jonsen, J., Laland, S. and Refsvik, T. (1963). *Biochim. biophys. Acta* **76**, 465; (1964). *Biochim. biophys. Acta* **80**, 648.
59. Eikhom, T. S. and Laland, S. (1965). *Biochim. biophys. Acta* **100**, 451.
60. Mach, B., Reich, E. and Tatum, E. L. (1963). *Proc. natn. Acad. Sci., U.S.A.* **50**, 175.
61. Snoke, J. E. and Cornel, N. (1964). *Biochim. biophys. Acta* **91**, 533.
62. Bernlohr, R. W. and Sievert, C. (1962). *Biochim. biophys. Res. Commun.* **9**, 32.
63. Cifferri, O., DiGirolamo, M. and DiGirolamo, A. B. (1961). *Nature, Lond.* **191**, 411.
64. DiGirolamo, M., Cifferri, O., DiGirolamo, A. B. and Albertini, A. (1964). *J. biol. Chem.* **239**, 502.
65. Banerjee, A. B. and Bose, S. K. (1964). *J. Bact.* **87**, 1397.
66. Banerjee, A. B. and Bose, S. K. (1964). *J. Bact.* **87**, 1402.
67. Yukioka, M., Tsukamoto, Y, Saito, Y., Tsuji, T., Otani, S. Jr., and Otani, S. (1965). *Biochem. biophys. Res. Commun.* **19**, 204.
68. Tomino, S. and Kurahashi, K. (1964). *Biochem. biophys. Res. Commun.* **17**, 288.
69. Tsuji, T. (1966). *J. Osaka City Med. Center* **15**, 1.
70. Holm, H., Froholm, L. O. and Laland, S. G. (1966). *Biochim. biophys. Acta* **115**, 361.
71. Berg. T. L., Froholm, L. O. and Laland, S. G. (1965). *Biochem. J.* **96**, 43.
72. Spaeren, U., Froholm, L. O. and Laland, S. G. (1967). *Biochem. J.* **102**, 586.
73. Sand, T., Vaage, O., Bredesen, J. Froholm, L. O., Siebke, J. C. and Laland, S. G. (1967). *FEBS (Abstr.)* 573.
74. Tomino, S., Yamada, M., Itoh, H. and Kurahashi, K. (1967). *Biochemistry* **6**, 2552.
75. Bhagavan, N. V., Rao, P. M., Pollard, L. W., Rao, R. K., Winnick, T. and Hall, J. B. (1966). *Biochemistry* **5**, 3844.
76. Shimura, K., Sasaki, T. and Sugawara, K. (1964). *Biochim. biophys. Acta* **86**, 46.
77. Brenner, M., Gray, E. and Paulus, H. (1964). *Biochim. biophys. Acta* **90**, 401.
78. Yamada, M., Tomino, S. and Kurahashi, K. (1964). *J. Biochem., Tokyo* **56**, 616.

CHAPTER 6

Control of Sporulation

ROBERT W. BERNLOHR AND CLAUS LEITZMANN

Departments of Microbiology and Biochemistry, University of Minnesota
Minneapolis, Minnesota, U.S.A.

I. Intracellular Differentiation

In the course of describing the physiological and biochemical events occurring during the formation of spores one observes that the whole process is regulated. The cell is faced with the task of controlling vegetative growth while excluding sporulation or controlling sporulation and excluding vegetative growth. The two processes are on different paths with differing energy-producing pathways, enzyme complements and nutritional supplies. In its simplest form, this regulation expresses itself in the timed synthesis of selected microbial enzymes and products. Even if the control were no more complex than this, the process would, in fact, be differentiation on a time scale, and has attracted workers in large numbers who use the system as a model of developmental control.

But the regulation of sporulation is not simple, nor is the whole process typical of the total problem of development. One spore is formed, potentially, in every cell and the mother cell is lost in the process. In eucaryotes, however, development leads either to a conversion of some cells of a population or to the production of multiple similar forms from one cell or an aggregate of cells. Morphogenesis in *Azotobacter* and the myxobacteria involves the retention of the entire DNA content of the cell, which is not the case for *Bacillus*. Thus, sporulation in *Bacillus* and *Clostridium* is a special case and our goal is to elucidate and understand how this system is controlled. If generalities prove feasible, so much the better.

A number of excellent reviews on sporulation have become available recently, and will be cited throughout. Many have sections on control and most discuss the physiological events during sporulation. In addition to the information contained in this volume, the text and references in the works by Fitz-James,[1] Foster,[2] Grelet,[3] Halvorson,[4] Halvorson,[5, 6] Knaysi,[7] Murrell,[8] Schaeffer *et al.*,[9, 10] Szulmajster,[11] Tomscik[12] and Vinter[13] should be considered. An excellent definition of differentiation, as it applies to sporulation is given by Baldwin and Rusch.[14]

II. Genetic Control

It is assumed throughout that the genus *Bacillus* is a homogeneous group and investigators strive to resolve differences reported from various laboratories. However, the DNA base composition of various species extends from 32 to 54% GC[15] and metabolic differences are well known (also see Table I, Chapter 4). Should we expect a *B. cereus* strain (32% GC) to control its metabolism and regulate sporulation in the same manner as a *B. macerans* strain (54% GC)? Obviously the genus is combined because of one property, sporulation. We are attempting to construct one hypothesis to satisfy all the data and this may not prove to be possible. However, for orientation, ignorance, and the sake of simplicity, this chapter will continue the practice.

A. CHROMOSOME REPLICATION

Autoradiography of *E. coli* has shown that the chromosome comprises a single piece of DNA and indicates duplication at a single growing point.[16] By lysing tritiated thymine-labeled cells with lysozyme, intact and replicating circles of DNA were found.[17] The autoradiography of the *B. subtilis* chromosome has not met with the same success. So far, continuous chromosomal lengths of 800–900μ have been found in vegetative cells,[18] but considering the fragility of the bacterial chromosome, it is possible that the total DNA of the genome is 1700μ in length. This is in agreement with the fact that the amount of DNA of *B. subtilis* spores corresponds to a molecular weight of 3×10^9 daltons.[19] In the same organism, the DNA may occur in an assembly of sub-units of about 2.5×10^8 daltons.[20]

Sueoka and Yoshikawa[21] measured the frequency of markers of the genome in cells in the exponential growth phase by genetic transformation and found that the chromosome of *B. subtilis* replicated sequentially as a unit from the origin to the terminus. The replicating chromosome has one replication point when cells are grown in a minimal medium and the genome exists as a single chromosome.[22, 23] The single fork replication is, however, not true when cells are dividing rapidly in an enriched medium,[24] where one finds multifork replication. Using the technique of synchronously germinating spores,[25] Oishi *et al.*[24] found that the chromosome of the spore is in a completed form and initiates DNA replication in the same definite sequential order as in the vegetative cell. In an enriched medium, a dichotomous replication occurs during outgrowth of spores in a manner similar to that observed during exponential growth. That the genome of *B. subtilis* is made up of a number of chromosomes or of several independently replicating sub-units was suggested from transformation studies.[26] Auto-radiography indicates, however, that the genome exists as a single chromosome[18] Although it was suggested that the spore might contain two chromosomal units, the fragility of the chromosome could account for this result. From morphological studies it had been suggested earlier that the spore contains only a single chromosome.[19] Also, endotrophically formed spores had less than one-half of the DNA content of the vegetative cell,[27] indicating that the vegetative cell has more than one morphologically identifiable chromosomal region. A different model of a replicating chromosome of *B. subtilis* was proposed recently which assumes the replicating chromosome to be circular and covalently linked at the initiation point.[28] The separation of the two daughter chromosomes requires the breakage of two phosphodiester bonds which may automatically initiate a new replication cycle. However, circularity of the *B. subtilis* chromosome has not been shown by transformation and transduction.[29]

B. GENE TRANSFER

Bacillus species lack a conjugation system, but the discovery of transformation, transduction and transfection has been used for genetic analysis and in the establishment of the genetic loci controlling sporulation. It seems that transformation and transduction share a common mechanism for genetic recombination.[30]

1. Transformation

Transformation is a process whereby purified DNA is accepted by cells and incorporated into their genetic make-up. The subject has been adequately reviewed by Spizizen et al.,[31] Schaeffer,[32] Hayes,[33] and Ravin.[34] A number of bacterial genera are capable of transformation[34] including strains of B. subtilis,[35] B. natto, B. niger, B. polymyxa,[36] and B. licheniformis.[37] Similarity of overall DNA base composition is a necessary but not sufficient requirement for the transfer and incorporation of genetic information among Bacillus species.[36] Studies on the mechanism of transformation have been discussed.[31, 38]

The transformation system requires "competent" cells, but the nature of this competence phenomenon still remains to be defined. Competence occurs under special growth conditions and is often of short duration.[39] In B. licheniformis, competence developed at a time at which the cells did not divide.[40] It was suggested that the competent state in B. subtilis might be associated with certain phases of the process leading to sporulation.[41, 42] Asporogenous mutants[43] and cells treated with the sporulation inhibitor α-picolinic acid were found to be non-competent.[44] In some asporogenous strains, the genetic locus controlling competence was separable from the locus for spore formation, but these loci are linked with genes controlling the formation of a protease and an antibiotic.[41] In other asporogenous mutants blocked at later stages, competence was normal.[32, 43] The cell wall, cytoplasmic membrane, or forespore membrane might be involved in competence.[45-48] It has been demonstrated that transforming DNA is bound to the surface of the cell in a DNase and lysozyme-insensitive form most frequently in association with structures resembling forespores.[48] From these results, it is believed that the occurrence of competence is related to the sporulation cycle as one of the early expressions.

2. Transduction

Transduction is a process whereby genetic information is conveyed from one cell to another by bacteriophages. Transduction has been reviewed[33, 49] and shown to occur in different genera including Bacillus.[50, 51] Transduction is a useful tool for mapping mutations and has been used to introduce sporogenesis into asporogenous strains.[50] Transduction has been shown to occur in non-competent cells, and the mechanism does not seem to be related to sporulation physiology. The size of the donor chromosome is much larger than that incorporated by recipient cells in transformation[52] allowing an analysis of linkage over longer distances.

Buoyant density profiles of DNA of spores derived from B. subtilis cultures infected with phage PBS-1 revealed the presence of a minor band that corresponded to that of the phage DNA.[53] The relative amount of the phage DNA present in the spores was estimated to be 11%, suggesting that spores of this organism may incorporate several copies of the phage genome.[53]

3. Transfection

Transfection is the production of a complete virus by cells infected with the nucleic acid isolated from that virus,[31] and can be considered a type of bacterial transformation. In *Bacillus*, it was found that transfection occurred only in cells competent for transformation,[54, 55] which in turn might be related to bacterial sporulation (see II B, 1). The kinetics of transfection suggest infection by a single DNA molecule and indicate that the characteristics of the initial stages are similar to the initial stages of transformation.[56] The use of transfection for studies on sporulation control has been limited, and has shown only that the state of competence may be related to the sporulation process.

C. GENETIC MAP

Since the genetic material of bacteria is functionally arranged in a linear manner, the recombination frequencies among a series of linked markers allow one to locate the position of these markers in their correct order. Genetic analysis of *Bacillus* has been performed mainly by transformation of mutants blocked in the biosynthesis of a number of amino acids. Linkage has been observed for loci controlling tryptophan synthesis[57] and for loci controlling the biosynthesis of shikimic acid, tyrosine, histidine and regulation genes for aromatic amino acid synthesis.[58, 59] Loci for histidine[60] and arginine[61] were located on different transforming particles and the genetic loci controlling isoleucine, valine and leucine biosynthesis[62] were not clustered but located close to loci for lysine and arginine. Transformation and transduction gave identical recombination frequencies between markers of some linkage groups.[63] The location of genes concerned with the synthesis of ribosomal and soluble RNA can be assigned to the starting region of the chromosome.[64-66]

Using asporogenous mutants of *B. subtilis*, the presence of the first sporulation marker was demonstrated by transformation[67] and transduction.[50] Transduction studies showed that three distinct sporulation loci were linked to amino acid markers or to antibiotic markers.[68] More recently, reciprocal transduction experiments suggested that there are numerous genetic loci at which a mutation can affect sporulation and that these are distributed randomly along the chromosome.[69, 70] Transformation studies indicated that there are at least three unlinked loci for sporulation[71] one of which is a group of linked genes which control the synthesis of a proteolytic enzyme, a wall-lytic enzyme[41, 72] and a sporulation control marker.[73] The size of the donor chromosome transferred by transduction is much larger than that incorporated by transformation and could be used to extend the transformation studies.[52]

The genetic map of *B. subtilis* strain 168 shown in Fig. 1, is that of Oishi *et al.*[24] and Oishi and Sueoka.[64] The positions are based on marker frequency analysis. The positions of ade$_{16}$–ade$_6$ represent sequence, not map distance. The spore markers that have been obtained by linkage with other markers by genetic transfer are indicated by arrows, and are those determined by Takahashi,[70] indicated by (*), and Rogolsky and Spizizen,[74] indicated by (+). Takahashi's positions on the map do not correspond to the

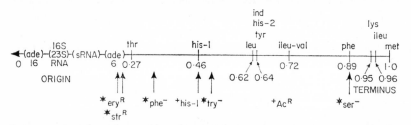

FIG. 1. A map of selected markers of the *B. subtilis* 168 chromosome, indicating the probable location of six (↑) markers for asporogeny. For details, see the text.

marker positions indicated, but since he obtained his own mutants, that difference may be due to separate mutation for the same end product on an unlinked position on the chromosome.

Although genes controlling related functions may be closely associated to form a cluster, the spore genes are scattered over the entire bacterial chromosome. Rogolsky and Spizizen[74] concluded that there are at least 5 separate regions with spore genes, and Schaeffer[73] suggested that there are at least several tens of genes at which mutation can affect sporulation. Halvorson[6] has suggested that the cell may contain 100 or more structural and regulatory "sporulation" genes, a figure that now seems conservative. The absence of genetic linkage between the sporulation genes demands some co-ordination in their control, which is likely to be physiological (see Section III).

D. Episome

The suggestion has been made that sporulation may be controlled by genetic elements of an episomal nature.[75] Episomes are genetic elements of an accessory nature which may occur in an autonomous or integrated form. In the light of new genetic data (see II C) Schaeffer *et al.*[9] did not favor the episomal control of sporulation. The idea has, however, persisted for two reasons. First,[76] acridine dyes can eliminate the episomal sex factor in

E. coli F[+] and can produce asporogenic mutants in *B. subtilis.*[77] Recent work, however, has shown that acridine orange influences the respiration of the cell and consequently prevents the increased metabolic activity that may be necessary for spore synthesis.[78] It is well known that acridine produces respiratory (petite) mutants in yeast. The chromosomal location of the genetic elements controlling penicillinase production in *Staphylococcus* were also believed to be episomal since they could be eliminated by acriflavin treatment[79] but now are postulated to be chromosomal.[80] That this element may have a different location in different strains[80, 81] and that the DNA of these episomes may be circular was also suggested.[82]

The other reason for the continued interest in the episomal control of sporulation is the finding of a DNA of different density upon germination of spores.[83] Since this satellite DNA is probably not a phage, its function may be related to spores, but the occurrence of multiple density DNA preparations is not restricted to developing organisms. Sueoka[84] has listed all known cases of satellite DNA. Rolfe,[85] working with *B. subtilis*, has given a plausible explanation for the occurrence of satellite DNA in bacteria during exponential growth. He believes that the heavy band does not have an ordered helical structure and collapses. This may also be the case for the heavy band seen upon germination.

E. Asporogenous Mutants

Asporogenous mutants have been obtained by the use of different agents, mechanisms and selection methods.[5, 86, 87] Mutants have been helpful in establishing the genetic map by genetic transfer experiments (see II C) and in determining sporulation specific reactions. Sporulation mutants have been divided into two groups,[73] the first of which will not sporulate under any conditions. This group is called asporogenous and designated by Sp-. The second group will sporulate with a low frequency under optimal standard conditions and is called oligosporogenous, designated Osp.[71] The sporulation frequency of Osp mutants varies with cultural conditions, so it is assumed that they have all the information required to sporulate, but this information is rarely expressed.[88]

Restoration to sporulation (at the same frequencies as other mutational characters) has been accomplished in both groups of mutants by transformation[71] and transduction.[50] The data discussed for the genetic map (II C) suggest that many genes can be affected by mutation to lead to asporogeny. Asporogenous mutants grow and behave during vegetative growth like the wild type strains from which they were derived and show their mutation only when differentiation would normally be initiated.

During sporulation the bacteria undergo a number of successive structural changes that have been described by Fitz-James and Young (see

Chapter 2) and by Ryter[89] and Ryter *et al.*[90] An asporogenous mutant may be blocked in any one of these stages.[10] By testing for the accumulation of sporulation-specific products and correlating this with morphology, different classes of mutants were identified (Table I). Mutants (Sp_0^-) have been found that are deficient in a protease, an antibiotic, competence and in a cell-wall lytic activity, but not all combinations of positive and negative responses of these characters are found.[10] One class of mutants lost at least five apparently independent characters by a single step mutation of a

TABLE I

Characteristics of asporogenous mutants

Mutant class[a]		Linked group[b]				Turnover[c]	Unlinked markers[b] (Amy, RNase DNase)
Stage	Time of block	Ab	Prot	Lz	Cpt		
Sp_0^-							
Type a	before	—	—	—	—	—	+
b	or	—	±	±	±		
c	at t_0	+	+	+	+		
Sp_I^-	t_0–$_1$	+	+	+	+	+	+
Sp_{II}^-	t_1–t_2	+	+	+	+	+	+
Sp_{III}^-	t_2–t_3	+	+	+	+	+	+
S_{IV}^-	t_3–t_5	+	+	+	+	+	+
Sp +		—		+		+	+

[a] Mutants are subdivided into stages on a morphological basis according to Ryter.[89] See also Fig. 2, p. 198, and Fig. 3, p. 174.
[b] Schaeffer.[10] Cpt = competence; Prot = protease; Ab = antibiotic; Lz = wall-lytic enzyme; Amy = amylase; RNase = ribonuclease; DNase = deoxyribonuclease.
[c] Balassa.[112]

pleiotropic gene controlling the triggering of the sporulation process.[10] It seems now that the production of the antibiotic, protease and wall lytic activity is required for sporulation to reach t_0–t_1 (stage I of Ryter).[89] (see Chapter 5, Fig. 3). Isolation of specific mutants indicated that activities that appear in the late log phase of growth (e.g. amylase) are not required for sporulation.

Slightly different but not necessarily anomalous results have been obtained by Bott and Davidoff-Abelson.[78] Table II shows that in Osp mutants prepared by acriflavin treatment, markers shown to be linked in Table I are separable in some cases. Thus, the protease, antibiotic and autolysis genes are not always lost together. However, it should not be

TABLE II

Selected characteristics of oligosporogenous mutants[a]

Character	I	II	III	IV
Frequency of Sp+	10^{-8}	10^{-5}	10^{-4}	10^{-2}
Autolysis	+	−	−	−
Antibiotic	±	+	+	+
Protease	+	+	+	+

[a] From Bott and Davidoff-Abelson.[78]

assumed that different laboratories are examining the same markers, as a number of antibiotics or proteases may be synthesized by some *Bacilli*.[91]

F. CONCLUSIONS

Once the genetic constitution of sporeforming organisms is characterized and, more importantly, the genes that are involved in sporulation are identified, we come to the question of how the cell controls the expressions of these genes. Why are the markers of the entire genome not transcribed at all times? Although it has been suggested[92] that whole segments of the yeast genome are not available for reading, normal control mechanisms (e.g. catabolite repression) may be responsible. In the lambda phage systems, a repressor(s) for the vegetative state of the phage is produced that prevents the expression of the lysogen.[93] This is superficially similar to the process in which the sporulation genome is repressed during the vegetative growth of *Bacilli*.

We also have to consider sporulation control at the level of translation. The changes in quantities of polymers in the vegetative or sporulating cell probably reflect control mechanisms for synthesis at the translation level.[92] It is the purpose of the following section to examine the roles of physiological control on the expression of all of the genes available to the cell. From this it is hoped that a relatively clear picture will emerge which would explain the total control system of sporulation. It is important to note, however, that few definitive experiments have been performed so far, and that the proposal is, although plausible, only a working hypothesis.

III. Physiological Regulation

A. GROWING CELLS

In the past decade a tremendous amount of work has been done on mechanisms of physiological and genetic regulation of enzymatic activities in a whole spectrum of living cells. This field has progressed to the state that current writers have rarely attempted to discuss the whole area[94-96] but have restricted themselves to the sub-topics of induction and repression,[97, 98] feedback inhibition,[99-102] or catabolite repression.[103] Thus, it is obvious that this section can do no more than to orient the reader to the concepts and send the serious student into the library.

This part has been divided so that a clear picture of controls during growth or during development can be described. We want a cell to grow efficiently, at a maximal rate and with a significant yield per unit energy used. Conversely, we know that the formation of a spore involves no net increase in cell mass and that, via turnover, a resistant cell is formed in an environment of endogenous nutrients.

1. Repression

The steady state repression of enzyme (generally anabolic) synthesis by an end product of a reaction sequence allows a *growing* cell to synthesize only the amount of enzyme protein needed to provide a pool of the end product. To overproduce the enzyme is extravagant from the viewpoint of the energy required to make the enzyme. Since end product inhibition can be assumed to be operative (see III A, 3), the cell is probably not synthesizing an excess of end product. Underproduction of the enzyme leads to obvious consequences. Thus, the overall damping effect of repression results in the efficient use of the energy that is expended in the synthesis of enzymes.

2. Induction–catabolite repression

The induced disproportionate synthesis of enzymes (generally catabolic) allows the cell to metabolize an unusually large number of organic compounds, most commonly glycosides, aromatics and amino acids. Since the overall elemental composition of bacterial cells is different from most inducers, the utilization of the above compounds as primary substrates often forces the cell into inefficient oxidation-reduction reactions that lead to lowered growth yields. Thus, nature seems to "prefer" glucose or some other easily metabolizable substrate such as glycerol for efficient growth. Glucose is ubiquitous and microbes apparently are adapted to growth on glucose to such an extent that they do not synthesize the many other catabolic enzymes. This must mean that this is most efficient for *growth*.

However, when one of the preferred substrates is not present, enzymes are induced in response to the exogenous presence of many nutrients. Microbes have the capability to grow on different compounds, but their enzymatic capacity is controlled so that they use the preferred substrate first. This phenomenon of induction-catabolite repression[103] functions to allow the cell to synthesize only those enzymes necessary for growth on the available milieu.

The mechanism of catabolite repression is still unclear[104, 105] and does not really concern us here. However, the stringency of repression of induced enzyme synthesis by catabolites of glucose (and other compounds) is directly related to growth rate, a fact that may be very important in the control of sporulation.

3. Feedback inhibition

In addition to the repressive effect mentioned above, end products are able to inhibit the activity of one or more enzymes (usually the initial reaction) that function in the biosynthesis of the end product. As this effect does not require significant periods of time, it is probable that the pool sizes of metabolites are controlled during growth through this mechanism, and that this is the primary function of feedback inhibition.

All of these control mechanisms have been amply described in *Bacillus*. Although variations in the details of the mechanism of feedback inhibition occur in many micro-organisms, there appears to be nothing unique about the types of control found in the aerobic sporeformers during vegetative growth.

B. DEVELOPING CELLS

Many reviewers have treated this area in some detail,[6, 8, 73, 106–108] but often with chosen restriction. A complete summary seems appropriate.

In bacterial sporulation and in other developmental systems the metabolism of the cell changes drastically at the time of initiation of development. Growth ceases for a number of reasons, but in general, turnover increases, pool sizes rise and the net flow of carbon compounds changes in all cases. In most studies the carbon source, glucose, is exhausted at the end of growth and the cells change from a glycolysis-biosynthesis metabolism to one of gluconeogenesis and acid oxidation. It is probable that the level of ATP is low during sporulation and that, energetically, the cells can be considered to be on the verge of starvation.[109]

1. Repression

With this background, repression of enzyme synthesis would seem to be of some use to the cell, even though the enzymes already formed are not diluted in the absence of an increase in cell mass. As mentioned above, repression would prevent the synthesis of enzymes which are no longer needed and relieve the energy requirements of the cell. Since pool sizes rise during sporulation and since the steady state levels of many biosynthetic enzymes are probably in the first order range with respect to the co-repressor in *Bacilli*, this control phenomenon would be expected to be functioning in an important way during development.

2. Induction–catabolite repression

In its most significant sense, catabolite repression is related to growth rate. When glucose or some other "preferred" carbon source is used, generation times of *Bacilli* in minimal media are in the 50–75 min range and catabolite repression is probably functioning. Generation times of 80–120 min are observed when glutamate or acetate are used and catabolite repression is not seen.[110] Since the effect of catabolite repression[103] requires an accumulation of a catabolite(s) of glucose plus a high amount of an energy related molecule (ATP?), the use of this control mechanism fits well with data on the control of sporulation. Thus, when nutrients are exhausted from the medium, catabolite repression control would cease and sporulation specific enzymes would be induced. A scheme of catabolite repression control of sporulation could very clearly delineate growth from development and allow growing cells to continue growth as long as nutrients are available. It should be indicated that the mechanism of catabolite repression control does not preclude the regulation of biosynthetic enzymes by this procedure.

There are two primary problems left open in this description. First, the induction of selected enzymes would have to be triggered by endogenous molecules. Although it can be shown that the exogenous addition of the same molecule will stimulate the synthesis, alternate hypotheses on the cause of appearance of these enzymes cannot be ruled out. The second problem involves a decision as to which induced enzymes are important for sporulation. A growing *Bacillus* cell, after exhausting glucose from the medium, will induce enzymes that will degrade exogenous inducers[110] and can undoubtedly induce enzymes in response to high levels of endogenous metabolites. Under these conditions enzyme induction is correlated with sporulation. Since this does not prove a direct relationship, it is important to use supplementary data in relating the function of an enzyme induced during sporulation with sporulation-related metabolism.

3. Feedback inhibition

The role of feedback inhibition of allosteric enzymes in the control of development has not been emphasized in the past. It is known that pools of some metabolites rise after growth, but the reason for this increase is unclear. Growth stops because of the exhaustion of nutrients and this should lead to lower pool sizes. However, hydrolases are known to increase at this time[111, 112] and it is assumed that the pools increase as a response to increased turnover of cell polymers.[113] This would have two effects. Increases in the intracellular concentration of amino acids and nucleotides would both repress the formation of biosynthetic enzymes and inhibit the further synthesis of these end products by feedback inhibition. This would allow the cell to construct a spore using the necessary induced enzymes plus those uninhibited ones that are derived from the growing cell. The main benefit to the sporulating cell would again be an efficient use of the total energy of the cell.

A new aspect of the utility of allosteric enzymes in development should be considered. The basal activity of many enzymes of carbohydrate metabolism is affected by metabolites that could not be strictly considered to be end products but which are theoretical signposts to the energetic state of the cell. The inhibition of phosphofructokinase by ATP is a good example. The rationale for the control of glycolysis and gluconeogenesis is dependent on this link to the energy-yielding reactions of the cell. *Bacilli*, sporulating in an exhausted minimal medium or under endotrophic conditions, must have an effective allosteric control to allow amino sugar synthesis (cortex) in a cell that is starving.

Several insights into this problem have been offered recently. A number of biosynthetic enzymes have been observed to decrease drastically in *activity* at a time when the flow of metabolites through pathways is reversing.[114, 115] This is not a function of simple feedback inhibition but a relatively stable inactivation correlated with the energy balance of the cell. This type of phenomenon has been observed so far only in yeast and *Bacilli*. Phosphofructokinase (PFK) is one such enzyme studied by the authors. During growth on glucose, PFK is required and is present in an active form which can be inhibited by ATP. Glycolysis occurs at the rate demanded by the cell and the overall control appears to be very similar to that of any microbial cell. However, upon the initiation of sporulation, PFK activity decreases many-fold and this occurs coincident with a *decrease* in the ATP pool of the cell. Thus, the PFK activity is lost by a mechanism that is not obviously feedback inhibition and at a rate much greater than expected by protein turnover. A working hypothesis would have to include an allosteric transition controlled by a metabolite that would fluctuate in response to the needs of sporulation metabolism. With PFK in an inactive form, gluconeogenesis can proceed during sporulation in a starving cell. This may be the only type of control unique to development.

4. Messenger RNA

The problem of the effect of actinomycin-D on development in general and on the role of stable mRNA in particular is still unresolved. The occurrence of stable messages has been suggested,[116, 117] but the idea is opposed by others.[118–120] Since sporulating cells can synthesize RNA, there is no *a priori* reason for depending on stable messages. (See page 146 for further details.)

5. Summary

From the foregoing it is clear that developing cells could use enzyme repression, induction-catabolite repression and several forms of allosteric activation-inhibition to great advantage. These mechanisms are sufficient to provide the physiological control needed for sporulation but additional unknown mechanisms may exist.

C. INFLUENCE OF NUTRIENTS

The belief that bacterial sporulation may be initiated by a limitation of nutrients has received extensive experimental consideration. It is evident now that there is a correlation in time between these two phenomena.[8] The exhaustion of selected nutrients, which in this context could include carbon, nitrogen and energy sources, inorganic ions, oxygen and also conditions like pH and temperature, are needed to bring about the physiological changes to initiate the sporulation process. In the evaluation of nutrient effects, concentration is a critical factor, since it has been shown that low substrate concentrations may allow sporulation to occur in a certain percentage of organisms during vegetative growth.[121]

Once the carbon source in the medium is nearly depleted, the cells induce enzymes to utilize some of the metabolic by-products which in the case of glucose are mainly acetate, lactate and pyruvate.[122, 123] This shift in metabolism is accompanied by a drastic change in the pH value of the medium coincident with the initiation of sporulation. The change in the pH value, however, is not required for sporulation since organisms grown on yeast extract or glutamate without glucose do not go through this pH change and sporulate normally.[110] If the phosphate buffer concentration is increased to prevent the drop in pH, the acetate that is produced is still oxidized and sporulation ensues normally.[124, 125]

A protease and the enzymes involved in arginine metabolism in *B. licheniformis* are derepressed during vegetative growth with glutamate as the sole carbon and nitrogen source.[110] These enzymes are low in activity or not detectable in vegetative cells grown on glucose or glycerol. It was suggested, therefore, that enzymatic activities normally associated with sporulation are merely under catabolic repression during growth, and while they may have some function during sporulation, they are not necessarily sporulation-specific.

Glutamate seems to be important in the sporulation process when added to a glucose medium. Even though little glutamate is metabolized under these conditions, a reduction in glutamate concentration slowed the growth rate and delayed spore formation.[126] In *B. licheniformis*, glutamate, together with alanine, comprise 75% of the amino acid pool during sporulation and their concentration increases initially and then decreases rapidly.[127] The relation of nutrients to sporulation in *Clostridia* is much less studied but seems to follow a similar pattern as that observed for *Bacillus* species.[8]

The absence of nutrients as a condition for sporulation has been studied under endotrophic conditions.[128] It was shown that protein degradation and utilization of low molecular weight solutes occurred during sporulation. This interpretation has received some criticism[129] but in general has been confirmed in many *Bacillus* species.[130] *B. licheniformis* sporulates endotrophically after the induction of a proteolytic enzyme.[131] The degradation of enzymes no longer required by the cell must be a selective process since cells are not committed to sporulation irreversibly until late during spore formation. Some *Bacilli* synthesize reserve materials in the form of polysaccharides or lipids during the later stages of growth and utilize these in the initial stages of sporulation. The accumulation of special storage compounds like poly-beta-hydroxybutyric acid (PHB) is not required, however, since sporulation is normal in the absence of PHB production.[124, 132, 133]

Thus, nutrients play a role during growth only to the extent of forcing the cell into a metabolic pattern. If this pattern is drastically different from that of sporulating cells, an inhibition of sporulation during the logarithmic phase is generally observed, and an exhaustion of the nutrient(s) is required before the sporulating phenotype can be expressed. In view of the wide variability of the genus *Bacillus*, it is not surprising that there are significant differences in the effects of individual nutrients on different sporulating species. (For a further discussion of the effects of nutrients see Chapter 3.)

D. SPORULATION-SPECIFIC ENZYMES

Changes in the enzymatic constitution of the cell during sporulation are quantitative and qualitative and have been tabulated by Halvorson[5] and Murrell.[8] While some of these changes may be coincident, others are functionally associated with the sporulating process. Synchronized cultures are not easily obtained, but give the best information about enzymes during development. As the medium becomes depleted of its carbon and energy sources, the metabolic pattern of the cell changes to the synthesis of new enzymes needed for spore formation. Figure 2 schematically presents the general picture of these changes. During growth the enzyme complement of the cell is what we would consider normal for growth. At t_0, sporulation is initiated and the protease, antibiotic and wall-lytic systems are formed (open circle), and concurrently, many vegetative cell enzymes decrease in activity. This decrease can be divided into two types, the first apparently involving inactivation of selected enzymes (closed triangle). These include lysine decarboxylase,[134] isocitric dehydrogenase,[11] aspartokinase,[114] threonine dehydratase[135] and phosphofructokinase (author's unpublished work).

Other enzymes decrease slightly or remain almost constant in activity during sporulation (open triangle).

At about t_1 to t_4 many other enzymes appear in the sporulating cells, some of which are listed by Halvorson[5] and Murrell.[8] The early group in Fig. 2 (t_1–t_2) includes some of the TCA-cycle enzymes and acid oxidizing

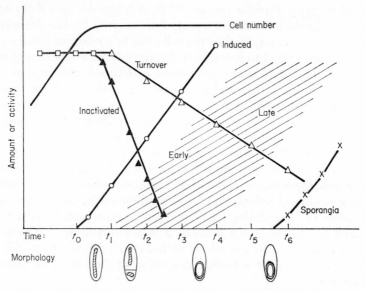

FIG. 2. Hypothetical time course to illustrate the changes in levels of enzymes during sporulation. For details see the text. The symbols represent inactivated enzymes (▲); induced enzymes (○); the profile of the loss of enzyme activity due to turnover (△); and heat stable sporangia (×). The morphological forms at the bottom are redrawn from Ryter.[89] The shaded area represents the whole spectrum of enzyme activities increasing in a parallel fashion with the induced (○) enzymes but 1–2½ hrs later (early) or 2½–4 hrs later (late).

enzymes. The later group (t_3–t_4) includes many of the enzymes responsible for the synthesis of spore components, such as DPA. Finally, approximately six hours after growth has ceased, sporangia appear.

The following sections amplify the results obtained on some of these enzymes as they apply to the control of sporulation. For more details on the enzymes themselves see Halvorson,[5] Murrell[8] and other chapters in this volume.

1. Protease and turnover

Extracellular proteolytic enzymes are characteristically produced by *Bacillus* species, especially after growth has been completed (at t_0 in Fig. 2). The functions of these proteases is not established, but an association between sporulation, turnover and the enzyme has been shown in

B. licheniformis,[113] *B. subtilis,*[43, 10, 91] *B. megaterium*[136] and *B. cereus.*[137] Genetic studies of asporogenous mutants indicate a functional relation or genetic linkage between the capacity to form the enzyme and the ability to sporulate. The formation of the protease in *B. megaterium* seems to be repressed by a group of amino acids of the aspartic acid family.[138] A similar observation was made for the enzyme from *B. cereus*[137] with a different group of amino acids. *B. subtilis* produces several extracellular proteases and attempts were made to establish to what extent these proteases were actually required for sporulation.[91] At least two proteases were found, one of which was involved in an early stage of the sporulation process. It was not determined whether the production or the activity of the enzyme was associated with the sporulation process.[91] From the effect of amino acids on protease production, the hypothesis has been adopted[137] that the enzyme is controlled by the level of a repressor which is directly or indirectly an intermediate in many pathways, and which may be the same compound that functions in the regulation of the synthesis of spore structural compounds.[121] Since all protease-deficient mutants are asporogenous,[43, 10] it seems clear that the protease is required as one of the early biochemical steps in sporulation.[111, 10] Unpublished data indicate that some mutants producing very low levels of extracellular protease are still able to sporulate.[137]

Spizizen[43] mapped the genes of two proteolytic enzymes of *B. subtilis* and found them clustered with the genes controlling spore formation. Balassa proposed that the co-ordination between protein and RNA turnover during sporulation occurs at the level of a protease.[112]

There has been a fair amount of speculation about the role of the proteolytic activity during sporulation,[10, 111, 139, 140] but it is not clear whether the activity that is genetically related to the initiation of sporulation can function as the agent catalyzing protein turnover. Complete protein hydrolysis by a purified extracellular protease has not been observed. However, recent studies by the author have demonstrated that the total intracellular proteolytic activity of sporulating *B. licheniformis* cells can completely degrade several different proteins to amino acids, a fact consistent with the proposition that complete turnover of cell protein can be accomplished during sporulation.

2. Wall-lytic enzymes

The physiological significance of autolytic enzymes in *Bacillus* species is not understood, since they occur during growth, in the stationary phase or in association with spore germination.[141] It was suggested that one function of these enzymes might be in differentiation[142] or transformation.[141] Wall-lytic enzymes have also been detected by Schaeffer[10] and mutants of *B. subtilis* that have lost this ability are asporogenous. From this it was concluded that wall-lytic activity is required for sporulation (Table I, p. 190).

3. TCA cycle enzymes

The involvement of the tricarboxylic acid (TCA) cycle in sporulation was postulated by Martin and Foster.[143] Hanson and colleagues have investigated the relationship between the TCA cycle enzymes and sporulation in several *Bacillus* species.[144, 145] They came to the conclusion that the TCA cycle enzymes are repressed during vegetative growth in a complex medium containing glucose, and they postulated that derepression of these enzymes is necessary for sporulation to occur.[146] The authors noted that this repression of the TCA cycle enzymes during growth is not found in *B. licheniformis* or in *B. subtilis*[147] when grown on a minimal salts-glucose medium. From these studies it appears that the presence of the TCA cycle enzymes is indispensable for the sporulation process in *Bacillus* although they may be present during the entire life cycle of the organism.

These data are consistent with the finding that sporulation is inhibited when the culture medium drops to a pH below 5·0, caused by a high initial carbohydrate concentration. This condition is believed to be unfavorable for induced enzyme formation and for the action of the TCA cycle enzymes and prevents sporulation.[122, 148]

Genetic evidence for the involvement of the TCA cycle in sporogenesis comes from the isolation of asporogenous mutants, several of which showed a low activity of a particulate NADH oxidase during the time of sporogenesis. Transformation of the mutant restored both the enzyme activity and the ability to sporulate.[149]

4. Amino acid-degrading enzymes

Several *Bacillus* species form spores from vegetative cells at the expense of endogenous nutrients which may be lipid or carbohydrate storage compounds. *B. licheniformis* does not contain any detectable storage compounds and it is believed that the cell derives its energy and building blocks from amino acid degradation.[113, 150] This protein turnover followed by amino acid degradation probably occurs in all sporeformers and must be performed by specific enzymes synthesized during sporulation and absent during growth. Several examples of these degradative enzymes have been studied in *Bacillus*.[113, 151] In other developing systems amino acid oxidizing enzymes have also been observed.[108]

E. SPORULATION-SPECIFIC PRODUCTS

The synthesis of sporulation-specific products is dependent on new enzyme synthesis. However, since it is the product that is studied rather than the enzymes that synthesize the product, they will be discussed separately in this section. Several products change quantitatively during sporulation and others appear *de novo*. Only the products that have been related to control mechanisms will be discussed. One compound, N-succinyl-L-glutamic acid, appears only for a short period during sporulation and its physiological function is not yet understood.[152] In reference to Fig. 2, the antibiotics will appear from t_0 (open circle), whereas DPA is found in cultures at later times (t_4–t_6).

1. Antibiotics

Several *Bacillus* species produce antibiotic substances that are released into the medium and appear after the end of growth. The production of an antibiotic in *B. subtilis* was linked with an early step in sporulation by genetic studies.[72, 10, 153] In addition, a number of mutants have been found that are poor antibiotic producers but normal sporulators.[10] From preliminary work it appears that in the crude filtrates of sporulating cultures of *B. subtilis* three distinct antibiotics are present. The possibility has to be considered that not all but at least one of the antibiotic substances is required for the initiation of sporulation.

It was suggested that the antibiotic produced early in sporulation is a possible candidate for one of the first products of transcription of the spore genome and the antibiotic is not produced by the vegetative cell genome.[112] This hypothesis was extended by Halvorson,[6] who postulated that the production of vegetative cell mRNA is selectively blocked by the antibiotic substance. The work of Bernlohr and Novelli[122, 131] and Bernlohr and Sievert[154] indicated that bacitracin may be a primary structural unit of the spore coat of *B. licheniformis*, but this could not be confirmed by Snoke.[155]

2. Nucleic acids

The changes occurring in the nucleic acids of the cell during sporulation are discussed in Chapter 4. In relation to the control of sporulation it is significant that only part of the DNA of the cell is incorporated into the spore[19, 156] and that there seems to be no stable mRNA synthesized during sporulation.[6, 11, 109, 112, 157] The mechanism of action of actinomycin D is not clear[120] and any interpretation of results on development using this inhibitor should be drawn with caution.

During sporulation, a nuclease activity increases in the culture medium of *Bacillus* species. This may have the same significance as the appearance of proteolytic activity discussed earlier.

3. Dipicolinic acid

Dipicolinic acid (DPA) is not found in vegetative cells of *Bacilli* but is found in spores and constitutes 5–15% of the dry weight.[8, 158] That the synthesis of this component is intimately associated with the sporulation process is indicated by the lack of heat resistance of spores of mutants that are poor DPA producers and by the appearance of DPA during growth in mutants that sporulate during growth.[159] The mechanism of the biosynthesis of DPA is discussed in Chapter 3 and it is fairly clear that it is produced by enzymes synthesized during sporulation. The physiological control of this process is unknown.

4. Factors

The initiation of the sporulation process has fascinated microbiologists for as long as the phenomenon has been known. It was only natural that a factor or a specific enzymic reaction responsible for the initiation of this process was anticipated and looked for by Foster and Heiligman.[160] A factor called "Sporogen" has indeed been reported in cultures of *B. cereus* and is present in the non-protein fraction of broken cells.[161, 162] It is believed that this specific endogenous substance has a hormone-like activity as speculated by Murrell.[163] The assay for "Sporogen" is performed with cells destined to lyse; after the addition of "Sporogen", lysis is prevented and sporulation takes place. So far, "Sporogen" has not been shown to play a primary role in cultures sporulating normally, and it is possible that its activity may be secondary.

Two peptide factors isolated from *Clostridium roseum*[164] and the above-mentioned competence factor (II B) may be similar to the "Sporogen". A growth-retarding factor from *C. roseum* has been reported[165, 166] that allows sporulation to occur in growing cells. This factor may simply have the indirect effect of releasing catabolite repression, a phenomenon that should not be ignored in cases where other "factors" could not be found.

F. OTHER ENZYMES

Enzymes which appear *de novo* or increase in activity or amount during sporulation may be functionally related to this process. Although some of these enzymes have not as yet been shown to be required for sporulation, it has not been shown that they are not required for this process. Most enzymes studied have been discussed and tabulated by Halvorson[5] and Murrell.[8] A good example of this group of enzymes is α-amylase[167] which is produced in increased amounts in the post-logarithmic phase of growth in *B. subtilis*. A higher rate of enzyme synthesis was observed during growth when starch and maltose were used as carbon sources, rather than glucose or glycerol. This we interpret as catabolic repression. Mutants unable to synthesize amlyase were found to be normal sporulators and it was concluded that amylase production is not needed for sporulation.[10] The relationship of other enzymes to sporulation is still unclear, although many authors point out possible roles. The enzymes inorganic pyrophosphatase,[168] DNA polymerase,[169] and nucleoside phosphorylase[170] all appear to be similar when isolated from the vegetative form or isolated from spores, even though they have different stabilities *in situ*. Whether these enzymes function during sporulation or are simply placed in the spore for subsequent use remains to be determined.

In other developing systems similar changes in the levels of enzymes have been observed and have been related to similar types of control. The UDP-galactose transferase[171] and pyrophosphorylase[172] from a cellular slime mould, the glucose-6-P dehydrogenase of *Neurospora*,[173] the glutamate dehydrogenase of *Schizophyllum*[174] and several systems in *Blastocladiella*[175] should be mentioned. A great deal of work has been done on the control of enzyme levels in yeast and fungi. Thus, there are enough examples of similar phenomena in many lower organisms to warrant optimism that a general or unified picture will emerge in the future.

IV. Models

A. PROCARYOTIC AND EUCARYOTIC REGULATION

The relationship between nuclear structure and the complexity of organization of cells has been emphasized.[176] Thus, bacteria and blue green algae (procaryotes) probably contain only a single DNA molecule (excepting episomes) which represents the entire genome, whereas higher systems (eucaryotes) contain two or more sub-units (chromosomes) complexed with histones and enclosed by a membrane. We tend to believe, without justification, that cells are more able to control a segment of their activity if the information for that segment is physically segregated on the genome. Among the procaryotes, *Bacillus*, *Clostridia*, *Azotobacter*, the myxobacteria, the actinomycetes and blue green algae exhibit a stage in their life cycle that could be called development. These micro-organisms appear to contain typical single genomic units, although in most cases definitive experiments have not been performed. At the present time, it must be assumed that *Bacilli* can sporulate without a possible advantage (?) of a separate piece of DNA and can control the process of sporulation by differentially translating specific segments of the genome.

B. PLAUSIBLE MODEL

Excellent and relatively consistent hypotheses concerning the control of sporulation have been presented recently by Schaeffer[73] and Halvorson.[6] Because of the considerable amount of new data on the subject, an extension is now possible.

The DNA of the cell can be transcribed at all times during growth and sporulation.[92] Thus, mRNA can be produced when required, and the amount of mRNA is controlled via the feedback type mechanisms proposed by Jacob and Monod[177, 178] and Stent.[179] The control of the phenotype of the cell is at the translation level. Stable mRNA does not appear to be involved in the process of sporulation.

The genome of *Bacillus* is one piece of DNA and the "spore" genes (more than 100) are distributed throughout (Fig. 3). Although some close linkages are known for some of the genes, it is probable that there are many spore gene sites on the whole genome and thus many controlling operator sites must exist. The mechanisms of control operating throughout all stages

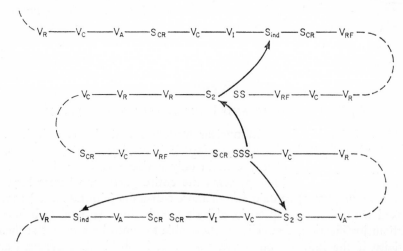

FIG. 3. A segment of the *Bacilllus* genome representing the model scheme described in the text.

of the cycle are: (1) repression of enzyme synthesis, (2) feedback inhibition of enzyme activity, (3) induction-catabolite repression of spore enzymes and catabolic enzymes. The model presents a picture consistent with the controls operating when cells are growing either on a relatively high level of easily utilizable carbohydrate, such as glucose, or on a glucose-free rich medium that also produces a high growth rate, such as nutrient broth. If neither of these conditions are applied, sporulation at variable rates can occur during growth and the controls which are operating would have to be inferred in each case.

All the necessary biosynthetic enzymes and anaplerotic enzymes[180] are produced during growth (Fig. 3; V_C and V_A respectively). The level of the biosynthetic group would be regulated by repression (V_R) and the activity by feedback inhibition (V_{RF}) so that the most efficient steady state will operate. All of the "sporulation" enzymes would be inducible or linked to control genes so that they would be absent or at extremely low levels as a result of catabolite repression (S_{ind} or S_{CR}). The amounts of mRNA in the growing cell would reflect this state since messages for the vegetative enzymes would be present in large quantities, whereas messages for "sporulation" enzymes would be repressed.[92, 181]

Growth would cease upon the exhaustion of nutrients and the cell would convert to sporulation metabolism. This change could take many forms depending on the original nutrient array, but it is inferred that there will be a decrease in the level of the intracellular catabolite(s) which was responsible for catabolite repression during growth. The products of the early sporulation genes (Sp_0) would be induced endogenously, in a manner similar to the increase in the levels of β-galactosidase in cells grown in the absence of a catabolite repressor. The protease, antibiotic, and wall-lytic enzymes are logical candidates, as pointed out by Schaeffer.[73] In addition, latent nuclease activity becomes apparent[112] and the rate of polymer turnover increases several-fold. Intracellular pools of nucleotides and amino acids increase. If a storage material exists (PHB), it is oxidized quite rapidly.[132, 182]

The cell now has two alternatives: to sporulate or to lyse. If efficient utilization of internal monomers can be managed, sporulation is favored. Thus, the production of amino acid and nucleotide biosynthetic enzymes (V) would be repressed and the resident enzymes inhibited as a result of the increase in the pools of the end products. Certain other enzymes, e.g. phosphofructokinase, threonine dehydratase and isocitric dehydrogenase, would be allosterically inactivated (V_I), probably by metabolites that are related to the energetics of the cell. Finally, a whole host of catabolic enzymes (S_{CR}) would arise, their main function being the degradation of endogenous nutrients to easily oxidizable substrates. A good example of the latter is the group of enzymes which degrade arginine to glutamic acid.[151] This whole complex process would be initiated and functioning during the initial stages of "sporulation" metabolism.

The cell is now set to express its sporulation phenotype efficiently. Enzymes specific for the synthesis of spore products (S) are now formed at varying times from t_1 to t_{5-8}. There has been sufficient work on the expression of a number of enzymes (see Halvorson, ref. 6, for a more complete description) to warrant the conclusion that these enzymes are produced in a specific order and at a specific time during sporulation. Little is known about the control of the timing of the synthesis of these enzymes but the similarity to the control of the lambda genome is striking[93] and will be described.

The phage lambda can be maintained as a stable prophage within the chromosome of lysogenic strains of *E. coli*. The lambda prophage produces a diffusible repressor molecule which inhibits the expression of the phage vegetative state. Upon prophage induction the repressor is inactivated, allowing the formation of a group of early gene products, one of which is an inducer for later gene products. Thus, a sequential synthesis of gene products is observed resulting from a sequential formation of inducers.

An hypothesis of the sequential nature of the synthesis of spore products involves similar ideas. The initial spore products are formed as a result of the release of catabolite repression. One or more of these products (S_1) would be an inducer molecule that would allow the synthesis of later spore enzymes (S_{ind}). This sequence could repeat as required (S_2) or the induction would be dependent on the concentration of the inducer. Some products would be formed at low inducer concentration (earlier enzymes) and others at higher concentrations. It has been postulated that the sporulation factors, mentioned earlier, could be such inducers. Also, the antibiotic produced during the initiation phase may be related to this phenomenon.

The control of the process of commitment to sporulation can not be adequately described. Committed cells will sporulate after being removed from a nutrient medium, washed and suspended in a non-nutrient medium. Logarithmic phase cells growing under catabolite repression conditions are not committed but assume the committed state at the time of exhaustion of the catabolite. This time is about t_0 and is coincident with the initiation of the first septum formation[183] and the production of the Sp_0 phenotypes. Cells grown on a minimal medium in the absence of glucose produce the Sp_0 phenotypes.[110] Some confusion arises because many *Bacilli* lyse suspended in a salts mixture[73] making it impossible to determine if the cells were or were not committed.

Thus, it is assumed that the commitment is no more than the initiation of septum formation and the production of the Sp_0 phenotypes, both events probably being under catabolite repression control. If repressed cells are washed and suspended in the complete absence of nutrients, they may be energetically unable to proceed to this first stage and can not continue sporulation. However, if cells exhaust their nutrient via their own metabolism, catabolite repression would be expected to be released before the complete loss of nutrient, allowing the synthesis of the relatively few initiation products.

There is also a stage of sporulation after which cells that are provided with growth nutrients will not grow but will complete sporulation even in the presence of high levels of glucose. This irreversible stage occurs much later than the time of commitments, from t_4–t_6 By this late time all of the Sp^+ products have been synthesized, so even if new synthesis is repressed the cell is equipped to complete the construction of the spore. We do not know why the vegetative genome cannot be expressed, but several possibilities come to mind. The sporulating cell may produce a repressor as one of the last expressions of sporulation metabolism that prohibits the translation of vegetative enzymes. This repressor could not be any of the factors mentioned earlier since they are produced at earlier stages. The second possibility involves permeation. Uptake systems for amino acids have been shown to become relatively inactive during the later stages of sporulation.[127] Thus, if nutrients cannot get into developing cells, then the process would appear to be irreversibly controlled. Finally, the degradative activity of the nuclease may have irreversibly cleaved the nucleic acids of the sporangium and conversion to growth would be impossible.

This total picture of the control of sporulation is consistent with almost all of the available data and is considered by the authors as being quite plausible. It is far from being proven in many areas, one of which involves the identification of the *in vivo* catabolites that repress in different *Bacilli* grown under different conditions. Some conflicting data exist here. Thus, it should go without saying that the model is no more than a rather detailed working hypothesis.

V. Biological Role

The biological role of endospores is still unknown but speculations have not been lacking. Many of the earlier suggestions have been reviewed by Knaysi,[7] Foster[2] and Murrell.[163] Henrici[184] was satisfied to call spores essentially resting forms, since only one spore is formed by each cell and each spore gives rise to only one cell. The close relationship between spore formation and the exhaustion of nutrients has led to the interpretation that the cell recognizes an unfavorable environment and in order to survive it forms an endospore. The organisms would die if it did not sporulate, and since the dormant spore completely lacks biosynthetic activity, it preserves the species. The conclusion has been, therefore, that it is extremely likely that spores represent a stage in the life cycle that allows certain bacteria to bypass unfavorable conditions.[185] Objections to this particular role of spores have been raised and modifications have been presented[7, 186] and discussed.[163, 187] However, none of these proposals have been accepted unanimously because they do not appear to endow the process with factors that would allow a significant selective advantage. As an alternate line of thought, the idea has always existed that the formation of this resting body has a significance in the bacterial life cycle that is greater than its obviously advantageous production of a temporarily resistant form.[188] Since spores in higher plants have a role in sexual processes which result in nuclear re-arrangements, a corresponding role for endospores has been sought. The pertinent literature on this subject has been reviewed by Knaysi.[7] Endospore formation was suggested to be an autogamic reproductive process and this notion was supported by cytological studies, but (i) the necessary diploid phase was never observed, (ii) it was established that nuclear phenomena are vegetative, and (iii) no evidence for sexuality or autogamy was observed.[189] That the ability to sporulate in *Bacillus* may confer a genetic advantage upon the organism via the development of transformation has also been suggested.[190]

Although it has been argued that sporulation cannot be a reproductive process because no multiplication occurs, it may represent a primitive form of recombination and segregation. It is obviously no advantage to reproduce a great number of cells that cannot compete with the environment, but it is an advantage to produce offspring that have new genetic potential for survival and multiplication.[191] The discovery of transformation in *Bacillus* by Spizizen[35] lends support to this latter biological role of endospore formation. The development of sensitivity to transformation (competency) occurs toward the end of the logarithmic phase.[31, 47, 48] Pieces of DNA are readily incorporated into the genome of the cell, showing that the recombination mechanisms are active. In strains which sporulate poorly, DNA uptake and transformation are markedly reduced. Thus, competency in *Bacillus* could be correlated with the presence of the ability to sporulate.[45]

Bacilli are soil micro-organisms and individual cells are restricted to limited areas where there is no constant supply of nutrients, whereas enteric bacteria (all non-sporulators) are continually provided with a wide variety of carbon and nitrogen compounds. Recombinants of the latter have the opportunity to express their genomes quickly, whereas mixing of genes of *Bacilli* would not be an advantage to a starving cell.

Because all extant hypotheses do not completely satisfy many of those who are concerned with the problem, we would like to present an additional scheme. It is likely that sporulation evolved under physical and nutritional conditions that were quite different from those that exist today. Sporulation, as a mechanism of producing a resistant and resting form, may have been a much greater advantage two billion years ago. In the intervening years, however, this has ceased to be a significant advantage as evidenced by the occurrence of the large majority of non-sporulating genera; nevertheless, mutation to asporogeny in *Bacillus* is still a disadvantage. At the same time, living systems were experimenting with more efficient mechanisms of recombination. Thus, *Bacilli* may have adapted the mechanics of sporulation to a method for segregating a recombined portion of the total DNA of the cell, while other primitive microbes chose alternate processes.

It is our suggestion, therefore, that the formation of spores is a relatively primitive genetic process and functions to integrate recombined nuclear material by transformation and to segregate it in a separate package, the spore. Since growth cannot take place, this haploid recombined is put into a resting form to wait for more favorable conditions, or to be transported to them, as emphasized by Bisset.[186] The current selective advantage of sporulation would then involve transformation followed by a mechanism for putting the cell into a new environment, that mechanism being sporulation.

ACKNOWLEDGEMENTS

Research performed in the authors' laboratories was supported in part by grants AI-05096 and GM-K3-7709 from the United States Public Health Service and GB-4670 from the National Science Foundation. Helpful discussions with many persons, especially Dr Martin Dworkin and Dr S. G. Bradley, are gratefully acknowledged.

REFERENCES

1. Fitz-James, P. C. (1965). *Symp. Soc. gen. Microbiol.* **15**, 369.
2. Foster, J. W. (1956). *Quart. Rev. Biol.* **31**, 102.
3. Grelet, N. (1957). *J. appl. Bact.* **20**, 315.
4. Halvorson, H. O. (1961). In "Growth in Living Systems" (M. X. Zarrow ed.), p. 107. Basic Books, Inc., New York.
5. Halvorson, H. (1962). In "The Bacteria" (I. C. Gunsalus and R. Y. Stanier, eds), Vol. 4, p. 233, Academic Press Inc., New York.
6. Halvorson, H. (1965). *Symp. Soc. gen. Microbiol.* **15**, 343.
7. Knaysi, G. (1948). *Bact. Rev.* **12**, 19.
8. Murrell, W. G. (1967). *Adv. microbial Physiol.* **1**, 133.
9. Schaeffer, P., Ionesco, H., Ryter, A. and Balassa, G. (1965). In "Mechanisms de regulation des activites celluaires chez les micro-organismes" (J. Senez, ed.), p. 553. Gordon and Breach, New York.
10. Schaffer, P. (1967). *Folia microbiol., Praha* **12**, 291.
11. Szulmajster, J. (1964). *Bull. Soc. Chim. biol.* **46**, 443.
12. Tomscik, J. (1962). *Ergebn. Mikrobiol. Immun. Forsch. exp. Ther.* **35**, 39.
13. Vinter, V. (1967). *Folia microbiol., Praha* **12**, 89.
14. Baldwin, H. H. and Rusch, H. P. (1965). *A. Rev. Biochem.* **34**, 565.
15. Hill, L. R. (1966). *J. gen. Microbiol.* **44**, 419.
16. Cairns, J. (1963). *J. molec. Biol.* **6**, 208.
17. Cairns, J. (1963). *Cold Spring Harb. Symp. quant. Biol.* **28**, 43.
18. Dennis, E. S. and Wake, R. G. (1966). *J. molec. Biol.* **15**, 435.
19. Fitz-James, P. C. and Young, I. E. (1959). *J. Bact.* **78**, 743.
20. Massie, H. R. and Zimm, B. H. (1965). *Proc. natn. Acad. Sci., U.S.A.* **54**, 1636.
21. Sueoka, N. and Yoshikawa, H. (1963). *Cold Spring Harb. Symp. quant. Biol.* **28**, 47.
22. Yoshikawa, H. and Sueoka, N. (1963). *Proc. natn. Acad. Sci., U.S.A.* **49**, 559.
23. Yoshikawa, H. and Sueoka, N. (1963). *Proc. natn. Acad. Sci., U.S.A.* **49**, 806.
24. Oishi, M., Yoshikawa, H. and Sueoka, N. (1964). *Nature, Lond.* **204**, 1069.
25. Wake, R. G. (1963). *Biochem. biophys. Res. Commun.* **13**, 67.
26. Kelly, M. S. and Pritchard, R. H. (1965). *J. Bact.* **89**, 1314.
27. Mandel, M. and Rowley, D. B. (1963). *J. Bact.* **85**, 1445.
28. Yoshikawa, H. (1967). *Proc. natn. Acad. Sci., U.S.A.* **58**, 312.
29. Yoshikawa, H. (1967). *Genetics* **54**, 1201.
30. Hoch, J. A., Barat, M. and Anagnostopoulos, C. (1967). *J. Bact.* **93**, 1925.
31. Spizizen, J., Reilly, B. E. and Evans, A. H. (1966). *A. Rev. Microbiol.* **20**, 371.

K

32. Schaeffer, P. (1964). *In* "The Bacteria' (I. C. Gunsalus and R. Y. Stanier, eds), Vol. 5, p. 87. Academic Press Inc., New York.
33. Hayes, W. (1964). "The Genetics of Bacteria and their Viruses". J. Wiley, New York.
34. Ravin, A. W. (1961). *Adv. Genet.* **10**, 61.
35. Spizizen, J. (1958). *Proc. natn. Acad. Sci., U.S.A.* **44**, 1072.
36. Marmur, J., Seaman, E. and Levine, J. (1963). *J. Bact.* **85**, 461.
37. Gwinn, D. D. and Thorne, C. B. (1964). *J. Bact.* **87**, 519.
38. Chilton, M. (1967). *Science* **157**, 817.
39. Anagnostopoulos, C. and Spizizen, J. (1961). *J. Bact.* **81**, 741.
40. Leonard, C. G., Mattheis, D. K., Mattheis, M. J. and Housewright, R. D. (1964). *J. Bact.* **88**, 220.
41. Spizizen, J. (1961). *In* "Spores II" (H. O. Halvorson, ed.), p. 142. Burgess Publ. Co., Minneapolis, Minn.
42. Nester, E. W. (1964). *J. Bact.* **87**, 867.
43. Spizizen, J. (1965). *In* "Spores III" (L. L. Campbell and H. O. Halvorson, eds), p. 125. Am. Soc. Microbiol. Ann Arbor, Michigan.
44. Gollakota, G. K. and Halvorson, H. O. (1960). *J. Bact.* **79**, 1.
45. Young, F. E. and Spizizen, J. (1961). *J. Bact.* **81**, 823.
46. Young, F. E., Spizizen, J. and Crawford, I. P. (1963). *J. biol. Chem.* **238**, 3119.
47. Young, F. E. (1965). *Nature, Lond.* **207**, 104.
48. Young, F. E. (1967). *Nature, Lond.* **213**, 773.
49. Campbell, A. (1964). *In* "The Bacteria" (I. C. Gunsalus and R. Y. Stanier, eds), Vol. 5, p. 49. Academic Press, Inc., New York.
50. Takahashi, I. (1961). *Biochem. biophys. Res. Commun.* **5**, 171.
51. Thorne, C. B. (1962). *J. Bact.* **83**, 106.
52. Takahashi, I. (1966). *J. Bact.* **91**, 101.
53. Takahashi, I. (1964). *J. Bact.* **87**, 1499.
54. Romig, W. R. (1962). *Virology* **16**, 452.
55. Földes, J. and Trautner, T. A. (1964). *Z. VererbLehre* **95**, 57.
56. Reilly, B. E. and Spizizen, J. (1965). *J. Bact.* **89**, 782.
57. Anagnostopoulous, C. and Crawford, I. P. (1961). *Proc. natn. Acad. Sci., U.S.A.* **47**, 378.
58. Nester, E. W. and Lederberg, J. (1961). *Proc. natn. Acad. Sci., U.S.A.,* **47**, 52.
59. Nester, E. W., Schafer, M. and Lederberg, J. (1963). *Genetics* **48**, 529.
60. Ephrati-Elizur, E., Srinivasan, P. R. and Zamenhof, S. (1961). *Proc. natn. Acad. Sci., U.S.A.* **47**, 56.
61. Mahler, I., Neymann, J. and Marmur, J. (1963). *Biochim. biophys. Acta* **72**, 69.
62. Barat, M., Anagnostopoulos, C. and Schneider, A. M. (1965). *J. Bact.* **90**, 357.
63. Ephrati-Elizur, E. and Fox, M. S. (1961). *Nature, Lond.* **192**, 433.
64. Oishi, M. and Sueoka, N. (1965). *Proc. natn. Acad. Sci., U.S.A.* **54**, 483.
65. Dubnau, D., Smith, I. and Marmur, J. (1965). *Proc. natn. Acad. Sci., U.S.A.* **54**, 724.
66. Oishi, M., Oishi, A. and Sueoka, N. (1966). *Proc. natn. Acad. Sci., U.S.A.* **55**, 1095.
67. Schaeffer, P., Ionesco, H. and Jacob, F. (1959). *C.r. hebd. séanc. Acad. Sci., Paris* **249**, 577.
68. Takahashi, I. (1963). *Proc. XI. Intern. Congr. Genetics.* The Hague, Vol. 1, 31.
69. Takahashi, I. (1965). *J. Bact.* **89**, 294.

70. Takahashi, I. (1965). *J. Bact.* **89,** 1065.
71. Schaeffer, P. and Ionesco, H. (1959). *C.r. hebd. séanc. Acad. Sci., Paris* **249,** 481.
72. Spizizen, J., Reilly, B. and Dahl, B. (1963). *Proc. XI. Intern. Congr. Genetics.* The Hague, Vol. 1, 31.
73. Schaeffer, P. (1965). *Sym. Biol. Hungary* **6,** 147.
74. Rogolsky, M. and Spizizen. J. (1967). *Bact. Proc.* 22.
75. Jacob, F., Schaeffer, P. and Wollman, E. L. (1960). *Symp. Soc. gen. Microbiol.* **10,** 67.
76. Hirota, Y. (1960). *Proc. natn. Acad. Sci., U.S.A.* **46,** 57.
77. Rogolsky, M. and Slepecky, R. A. (1964). *Biochem. biophys. Res. Commun.* **16,** 204.
78. Bott, K. F. and Davidoff-Abelson, R. (1966). *J. Bact.* **92,** 229.
79. Harmon, S. A. and Baldwin, J. N. (1964). *J. Bact.* **87,** 593.
80. Asheshov, E. H. (1966). *Nature, Lond.* **210,** 804.
81. Poston, S. M. (1966). *Nature, Lond.* **210,** 802.
82. Richmond, M. H. (1967). *J. gen. Microbiol.* **46,** 85.
83. Douthit, H. A. and Halvorson, H. O. (1966). *Science* **153,** 182.
84. Sueoka, N. (1964). *In* "The Bacteria" (I. C. Gunsalus and R. Y. Stanier, eds), Vol. 5, 419. Academic Press Inc., New York.
85. Rolfe, R. (1963). *Proc. natn. Acad. Sci., U.S.A.* **49,** 386.
86. Chiasson, L. P. and Zamenhof, S. (1966). *Can. J. Microbiol.* **12,** 43.
87. Northrop, J. and Slepecky, R. A. (1967). *Science* **155,** 838.
88. Aubert, J. P. and Millet, J. (1961). *C.r. hebd. séanc. Acad. Sci., Paris* **253,** 1880.
89. Ryter, A. (1965). *Annls Inst. Pasteur, Paris* **108,** 40.
90. Ryter, A., Schaeffer, P. and Ionesco, H. (1966). *Annls Inst. Pasteur, Paris* **110,** 305.
91. Michel, J. F. (1967). *Folia microbiol., Praha* **12,** 297.
92. Cline, A. L. and Bock, R. M. (1966). *Cold Spring Harb. Symp. quant. Biol.* **31,** 321.
93. Naono, S. and Gros, F. (1966). *Cold Spring Harb. Symp. quant. Biol.* **31,** 363.
94. Holzer, H. (1963). *Naturwiss.* **50,** 260.
95. Umbarger, H. E. (1964). *Science N.Y.* **145,** 674.
96. Moyed, H. S. and Umbarger, H. E. (1962). *Physiol. Rev.* **42,** 444.
97. Vogel, H. J. (1961). *In* "Control Mechanism in Cellular Processes" (D. M. Bonner, ed.), p. 23. The Ronald Press Co., New York.
98. Maas, W. K. and McFall, E. (1964). *A. Rev. Microbiol.* **18,** 95.
99. Atkinson, D. E. (1966). *A. Rev. Biochem.* **35,** 85.
100. Umbarger, H. E. (1961). *Cold Spring Harb. Symp. quant. Biol.* **26,** 301.
101. Cohen, G. N. (1965). *A. Rev. Microbiol.* **19,** 105.
102. Stadtman, E. R. (1966), *Adv. Enzymol.* **28,** 41.
103. Magasanik, B. (1961). *Cold Spring Harb. Symp. quant. Biol.* **26,** 249.
104. Loomis, Jr., W. F. and Magasanik, B. (1966). *J. Bact.* **92,** 170.
105. Palmer, J. and Moses, V. (1967). *Biochem. J.* **103,** 358.
106. Umbarger, H. E. (1965). *Brookhaven Symp. Biol.* **18,** 14.
107. Sussman, M. (1965). *Brookhaven Symp. Biol.* **18,** 66.
108. Wright B. (1966). *Science N.Y.* **153,** 830.
109. Leitzmann, C. and Bernlohr, R. W. (1965). *J. Bact.* **89,** 1506.
110. Laishley, E. J. and Bernlohr, R. W. (1966). *Biochem. biophys. Res. Commun.* **24,** 85.

111. Bernlohr, R. W. (1964). *J. biol. Chem.* **239**, 538.
112. Balassa, G. (1964). *Biochem. biophys. Res. Commun.* **15**, 240.
113. Bernlohr, R. W. (1965). *In* "Spores III" (L. L. Campbell and H. O. Halvorson, eds), p. 75. Am. Soc. Microbiol., Ann Arbor, Michigan.
114. Stahly, D. P. and Bernlohr, R. W. (1967). *Biochim. biophys. Acta* **146**, 467.
115. Ferguson, Jr., J. J., Boll, M. and Holzer, H. (1967). *European J. Biochem.* **1**, 21.
116. Aronson, A. I. (1965). *J. molec. Biol.* **13**, 92.
117. Del Valle, M. R. and Aronson, A. I. (1962). *Biochem. biophys. Res. Commun.* **9**, 421.
118. Balassa, G. (1966). *Annls Inst. Pasteur, Paris* **110**, 175.
119. Pannbacker, R. G. and Wright, B. E. (1966). *Biochem. biophys. Res. Commun.* **24**, 334.
120. Wright, B. and Pannbacker, R. (1967). *J. Bact.* **93**, 1762.
121. Schaeffer, P., Millet, J. and Aubert, J. P. (1965). *Proc. natn. Acad. Sci., U.S.A.* **54**, 704.
122. Bernlohr, R. W. and Novelli, G. D. (1960). *Archs. Biochem. Biophys.* **87**, 232.
123. Nakata, H. M. and Halvorson, H. O. (1960). *J. Bact.* **80**, 801.
124. Nakata, H. M. (1963). *J. Bact.* **86**, 577.
125. Hanson, R. S., Srinivasan, V. R. and Halvorson, H. O. (1963). *J. Bact.* **85**, 451.
126. Buono, F., Testa, R. and Lundgren, D. G. (1966). *J. Bact.* **91**, 2291.
127. Bernlohr, R. W. (1967). *J. Bact.* **93**, 1031.
128. Hardwick, W. A. and Foster, J. W. (1952). *J. gen. Physiol.* **35**, 907.
129. Powell, J. F. and Hunter, J. R. (1953). *J. gen. Physiol.* **36**, 601.
130. Perry, J. J. and Foster, J. W. (1954). *J. gen. Physiol.* **37**, 401.
131. Bernlohr, R. W. and Novelli, G. D. (1963). *Archs. Biochem. Biophys.* **103**, 94.
132. Slepecky, R. A. and Law, J. H. (1961). *J. Bact.* **82**, 37.
133. Nakata, H. M. (1966). *J. Bact.* **91**, 784.
134. Mandelstam, J. (1954). *J. gen. Microbiol.* **11**, 426.
135. Leitzmann, C. and Bernlohr, R. W. (1968). *Biochim. biophys. Acta.* **151**, 461.
136. Vinter, V. (1956). *Cs. Mikrobiol.* **1**, 63.
137. Levisohn, S. and Aronson, A. I. (1967). *J. Bact.* **93**, 1023.
138. Chaloupka, J. and Kreckova, P. (1966). *Folia microbiol., Praha* **11**, 82.
139. Millet, J. and Aubert, J. P. (1964). *C.r. hebd. séanc. Acad. Sci., Paris* **259**, 2555.
140. Mandelstam, J. (1960). *Bact. Rev.* **24**, 289.
141. Young, F. E. (1966). *J. biol. Chem.* **241**, 3462.
142. Shockman, G. D. (1965). *Bact. Rev.* **29**, 345.
143. Martin, H. H. and Foster, J. W. (1958). *J. Bact.* **76**, 167.
144. Szulmajster, J. and Hanson, R. S. (1965). *In* "Spores III" (L. L. Campbell and H. O. Halvorson, eds), p. 162. Amer. Soc. Microbiol., Ann Arbor, Michigan.
145. Hanson, R. S. and Cox, D. P. (1967). *J. Bact.* **93**, 1777.
146. Hanson, R. S., Blicharska, J. and Szulmajster, J. (1964). *Biochem. biophys. Res. Commun.* **17**, 1.
147. Hanson, R. S., Blicharska, J., Arnaud, M. and Szulmajster, J. (1964). *Biochem. biophys. Res. Commun.* **17**, 690.
148. Bergere, J. L. and Hermier, J. (1964). *Annls Inst. Pasteur, Paris* **106**, 214.
149. Szulmajster, J. and Schaeffer, P. (1961). *C.r. hebd. séanc. Acad. Sci., Paris* **252**, 220.

150. Foster, J. W. and Perry, J. J. (1954). *J. Bact.* **67**, 295.
151. Ramaley, R. F. and Bernlohr, R. W. (1966). *Archs. Biochem. Biophys.* **117**, 34.
152. Aubert, J. P., Millet, J. and Castoriadis-May, C. (1961). *C.r. hebd. séanc. Acad. Sci., Paris* **253**, 1731.
153. Balassa, G., Ionesco, H. and Schaeffer, P. (1963). *C.r. hebd. séanc. Acad. Sci., Paris* **257**, 986.
154. Bernlohr, R. W. and Sievert, C. (1962). *Biochem. biophys. Res. Commun.* **9**, 32.
155. Snoke, J. E. (1964). *Biochem. biophys. Res. Commun.* **14**, 571.
156. Hodson, H. H. and Beck, J. V. (1960). *J. Bact.* **79**, 661.
157. Doi, R. H. and Igarashi, R. I. (1964). *Proc. natn. Acad. Sci., U.S.A.* **52**, 755.
158. Powell, J. F. (1953). *Biochem. J.* **54**, 210.
159. Szulmajster, J., Blicharska, J. and Spotts, C. C. (1962). *C.r. hebd. séanc. Sci., Paris* **254**, 4533.
160. Foster, J. W. and Heiligman, F. (1949). *J. Bact.* **57**, 639.
161. Srinivasan, V. R. and Halvorson, H. O. (1963). *Nature, Lond.* **197**, 100.
162. Srinivasan, V. R. (1965). *In* "Spores III", (L. L. Campbell and H. O. Halvorson, eds), p. 64. Am. Soc. Microbiol., Ann Arbor, Michigan.
163. Murrell, W. G. (1961). *Symp. Soc. Gen. Microbiol.* **11**, 100.
164. Woley, B. C. and Collier, R. E. (1966). *Bact. Proc.* 16.
165. Bergere, J. L. and Hermier, J. (1965). *Annls Inst. Pasteur, Paris* **109**, 80.
166. Bergere, J. L. and Hermier, J. (1965). *Annls Inst. Pasteur, Paris* **109**, 391.
167. Coleman, G. and Grant, M. A. (1966). *Nature, Lond.* **211**, 306.
168. Tono, H. and Kornberg, A. (1967). *J. Bact.* **93**, 1819.
169. Falaschi, A. and Kornberg, A. (1966). *J. biol. Chem.* **241**, 1478.
170. Gardner, R. and Kornberg, A. (1967). *J. biol. Chem.* **242**, 2383.
171. Loomis, Jr., W. F. and Sussman, M. (1966). *J. molec. Biol.* **22**, 401.
172. Ashworth, J. M. and Sussman, M. (1967). *J. biol. Chem.* **242**, 1696.
173. Brody, S. and Tatum, E. L. (1966). *Proc. natn. Acad. Sci., U.S.A.* **56**, 1290.
174. Dennen, D. W. and Niederpruem, D. J. (1967). *J. Bact.* **93**, 904.
175. Cantino, E. C. and Lovett, J. S. (1964). *Adv. Morphogen.* **3**, 33.
176. Ris, H. and Chandler, B. L. (1963). *Cold Spring Harb. Symp. quant. Biol.* **28**, 1.
177. Jacob, F. and Monod, J. (1961). *J. molec. Biol.* **3**, 318.
178. Jacob, F. and Monod, J. (1961). *Cold Spring Harb. Symp. quant. Biol.* **26**, 193.
179. Stent, G. (1964). *Science* **114**, 816.
180. Kornberg, H. L. (1965). *Symp. Soc. gen. Microbiol.* **15**, 8.
181. Doi, R. T. (1965). *In* "Spores III" (L. L. Campbell and H. O. Halvorson, eds), p. 111, Am. Soc. Microbiol., Ann Arbor, Michigan.
182. Kominek, L. A. and Halvorson, H. O. (1965). *J. Bact.* **90**, 1251.
183. Gordon, R. A. and Murrell, W. G. (1967). *J. Bact.* **93**, 495.
184. Henrici, A. T. (1934). "The Biology of Bacteria". D. C. Heath and Co., New York.
185. Cook, R. P. (1932). *Biol. Rev.* **7**, 1.
186. Bisset, K. A. (1950). *Nature, Lond.* **166**, 431.
187. Lamanna, C. (1952). *Bact. Rev.* **16**, 90.
188. Zinsser, H. and Bayne-Jones, S. (1934). "A Textbook of Bacteriology". 7th ed. D. Appleton-Century Co., New York.
189. Hunter, M. E. and DeLamater, E. D. (1952). *J. Bact.* **63**, 23.
190. Sussman, A. S. and Halvorson, H. O. (1966). "Spores, their Dormancy and Germination". Harper and Row, New York.
191. Anderson, E. S. (1966). *Nature, Lond.* **209**, 637.

CHAPTER 7

Chemical Composition of Spores and Spore Structures

W. G. MURRELL

Commonwealth Scientific and Industrial Research Organization
Division of Food Preservation, Ryde, New South Wales, Australia

I. Introduction

SINCE 1950 studies have shown that spores are complex in both structure
and chemical composition. They differ from vegetative cells in the composition of the various new layers, in having unique substances such as
dipicolinic acid (DPA), in various new antigens, and in having marked

quantitative differences, e.g. calcium at over 1000 times the concentration present in vegetative cells.[1]

The data on the chemical composition of the spore are however still far from complete, and in many ways uncritical and perhaps not always reliable. This is largely the result of the increasingly higher standards demanded in the preparation of spores for analysis. Vegetative cells can be grown rapidly in crystal clear chemically well-defined media and harvested within minutes of their formation. On the other hand, spores are formed slowly (6–10 hr) in the accumulating metabolic by-products in cultures 16 hr or much more in age, often containing unlysed vegetative cells and mother cells, complex lytic products and debris; they have to be freed of all these. Other reasons for the sometimes unsatisfactory state of the chemical composition of the spore are the methods of chemical analysis, and because the composition of the spore is perhaps not such a fixed characteristic as previously accepted. Metal and lipid contents in particular are affected greatly by the method of preparation of the spores, and apparent lipid contents by the extraction procedures.

A. Preparation of Spore Crops for Chemical Analysis

The ultimate aim of the preparation procedures is to obtain in gramme lots spore crops having:

(i) 100% fully refractile spores
(ii) no cellular debris, or cells other than spores, and
(iii) surface-clean spores free of external contamination by the particular substances or chemical fraction to be analysed.

Most of these standards can only be approached and are unnecessarily strict if a major component (>5%) is being determined. When a substance, however, occurs at levels of 1% or less, extreme precautions must be taken to avoid contamination of a constituent with surface-adsorbed or interstitial contaminating material. The surface area of a gramme of dry spores is approximately three square metres; this indicates the enormous adsorption problems possible. Likewise the elimination of particulate debris of similar dimensions from 10^{12} spores (c. 1 g) requires exhaustive use of physical separation procedures.

1. Removal of vegetative and mother cells

When possible it is preferable to start with a spore crop in which most of the mother cells have lysed and which contains a minimum of vegetative cells and stainable or non-refractile spores. Unlysed mother cells and vegetative cells should be degraded enzymically. The following enzymes have proved useful—lysozyme (0·2 mg/ml),[2,3] trypsin (0·1 mg/ml),[3] papain (1 mg/ml)[4] sometimes coupled with sonication.[3] Lytic enzymes from sporulated cultures of other organisms may also be useful. The enzyme treatments need to be carried out under conditions which avoid or minimize germination of the spores in the presence of breakdown products, e.g. low temperature (0–5°); inhibitors (0·1% 1:1 chloroform-toluene).[4]

Free spores may also be selectively harvested from crops containing vegetative cells by repeated use of one of the following methods

(i) the two-phase treatment of Sacks and Alderton;[5] unfortunately this causes germination of spores of some species[6]
(ii) flotation methods[7, 8, 9]
(iii) density gradient methods.[10, 11, 12, 13]

Spores of some strains also germinate as a result of osmotic shock or the ionic conditions present during density gradient separation.

With a number of strains of the *Bacillus* species commonly studied and a few strains of *Clostridium* the above enzyme treatments are unnecessary as these strains possess very active lytic systems which completely autolyse the mother cells and any vegetative cells.

2. Spore cleaning

The spores, now free of other cell types, should be water-washed several times to remove products of cellular degradation. They now have to be enzyme cleaned and freed of precipitates and particulate debris.

Enzyme cleaning involves treating the spores sequentially with a number of enzymes to remove protein, fat, DNA and RNA contamination. This is possible because the coats are resistant to most known enzymes, e.g. proteolytic enzymes,[4, 14, 15, 16, 17] lipases,[14] lyzozyme,[4, 14] phosphatase,[14] bacterial proteinase.[17]

The spores also appear impermeable to molecules of these enzymes as they remain refractile for long periods in their presence,[18, 19] unless germination occurs in response to breakdown products.

Normally, a sequence of at least a proteolytic enzyme, ribonuclease and lipase is advisable, with, in addition, any enzyme to remove the particular constituent of analytical interest, e.g. DNase if DNA is to be analysed. The soluble enzymes and their products should then be removed by centrifugation.

The spore suspensions may still contain organic aggregates, cellular particles, inorganic or organometal precipitates of similar density and sedimentation rate to the spores. This contamination is not removed by ordinary centrifugation. Metal precipitates are satisfactorily removed in some cases by acid-washes (0·03N) in the cold (0–2°).[6, 20, 21]

Physical methods are usually needed to remove material with similar sedimentation rates to spores. Molecular sieve methods using charged particles are probably impracticable with large quantities of spores. Density gradient centrifugation methods are satisfactory if the contaminating material differs sufficiently in density or in sedimentation rate and the solute for preparation of the gradient lacks harmful effects.[12] $CsCl_2$ and sucrose are unsatisfactory for some spores,[12] Urografin appears satisfactory but is expensive.[13] A long chain water-soluble chelate of lead $Pb[OOCCH_2N(C_2H_4OH)C_2H_4N(C_2H_4OH)CH_2COO]$ has been used[12] but is unavailable commercially.

The two-phase system of Sacks and Alderton[5] when this could be used proved useful for removing metal precipitates as the heavy inorganic precipitates and crystals do not appear to become entrapped in the upper spore layer.[6]

Electrophoretic separation should be very useful for separating, without harmful effects, particles with charges differing from spores; use of this method has not yet been reported. Use of both gradient and electrophoretic methods may be necessary under some conditions, e.g. thermophilic crops where undesirable precipitates form during growth and sporulation.

Ideally for comparison of spore crops produced under various conditions these harvesting and cleaning methods should not result in loss of lighter spores or spores of different composition from the main sample, i.e. there should be no selection of spores of particular properties or composition.

Some standard description of the degree of purity should be adopted, at least in regard to other cellular forms and debris, e.g. number of stained (phase-dark) spores, particles of debris, and rods per 1000 refractile spores. These data should be obtained by phase contrast microscopy or by observation of the suspension mounted with the stain under a cover slip. Often more debris is revealed if the lightly stain-mounted suspension is viewed by phase-contrast. The degree of purity should be reported in all papers on chemical composition.

II. Gross Composition

Very little difference was evident in early studies[22] in the gross composition of spores and vegetative cells. Spores contain about 10% nitrogen (Table I) and 6–11% ash (Table IV). Following the discovery of DPA in spores[23] more specific analyses have been undertaken. Tinelli's analyses[24] show that spores of *B. megaterium* are much richer in protein, and relatively low in carbohydrate and β-hydroxybutyrate compared with vegetative cells (Table II). The protein content of *B. megaterium* vegetative cells was very low compared with that for *B. subtilis* and *B. mycoides*; this ensued from the high β-hydroxybutyrate content (Table II) which probably resulted from growth of this organism in a glucose medium. Except for cystine, the mole proportions of the amino acids of spores and vegetative cells of three species are remarkably similar (Table III; Fig. 1); alanine, glycine, valine, lysine, leucine, aspartate and glutamate were more commonly in highest concentration (Table III; Fig. 1); *Bacillus* spores appear to have more tryptophan, tyrosine and threonine, and less asparagine, glutamate and lysine than the vegetative cells. The chief differences in amino acid

TABLE I

Nitrogen content of vegetative cells and spores

Species	Vegetative cells	Spores Range	Av.	Ref.
	Total Nitrogen			
Bacillus megaterium	6·3	9·23–13·86	10·36	33
			11·60	24
3 strains, different media			11·33	21
Bacillus cereus				
var *mycoides*	10·3–11·3	8·9–11·3	10·5	22
	10·6		10·7	34
var *anthracis*	6·8		12·24	35
3 strains		8·69–13·36	10·58	21
Bacillus subtilis				
Strain 1	10·64		10·43	33
Strain 11			9·37	33
	10·58		11·10	34
		9·04–12·30	11·07	21
Bacillus mesentericus			10·20	33
Bacillus polymyxa		10·61–13·72	12·05	21
	α-Amino Nitrogen			
Bacillus cereus var *mycoides*	7·7		7·9 ⎫	
Bacillus subtilis	7·7		8·2 ⎬ 34	
	7·5		7·6 ⎪	
	6·75		7·4 ⎭	

composition between the *Bacillus* and *Clostridium* species are greater mole contents of aspartate, glutamate and lysine in the *Clostridium* spores.

In *B. cereus* T vegetative cells 3·5% of the amino acid nitrogen was in the D-form while only 1% occurred in this form in spores.[25] Glutamate was present as an equal mixture of the L-and D-isomer in spores whereas 63% occurred in the D-form in vegetative cells. Spores of a smooth variant of *B. stearothermophilus* were generally 20–50% richer than the rough variant

TABLE II

Gross composition of spores and vegetative cells

	% dry weight	
	Vegetative cells	Spores
Bacillus megaterium (Data of Tinelli)[24]		
Protein (N × 6·25)	39·4	68
β-Hydroxybutyrate	28·5	0
Carbohydrate[a]	18·2	4·8
Calcium dipicolinate	Nil	15
Ash	9·1	8[b]
	95·2	95·8
Protein[c] (Data of Pfennig)[34]		
Bacillus subtilis	68·5	76
Bacillus mycoides	67·5	71·5

Carbohydrate (Data of Walker *et al.*)[21]	Range	Av.
Bacillus megaterium (3 strains)	1·30–4·01	2·39
Bacillus subtilis (1 strain)	1·46–3·80	2·27
Bacillus cereus (3 strains)	1·30–2·58	2·08
Bacillus polymyxa (2 media)	4·15–4·66	4·40

[a] Includes glucose, galactose and traces of uronic acid.
[b] Less calcium (3%).
[c] Amino acids less hexosamine; will include amino acids in muropeptide.

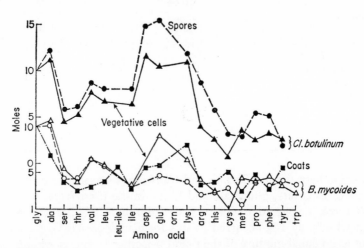

FIG. 1. Amino acid composition of vegetative cells and spores of *Bacillus mycoides* and *Clostridium botulinum* type A, and the average amino acid composition of spore coats of five *Bacillus* species, Data from Tables III and VII in moles relative to ten moles of glycine.

TABLE III

Amino acid composition of vegetative cells and spores of Bacillus subtilis,
Bacillus mycoides *and* Clostridium botulinum

Amino Acid	Data[a] of Pfennig[34] % dry weight				Data[b] of Tsuji and Perkins[36] μ moles/mg Kjeldahl N[c]		
	Bacillus subtilis		*Bacillus mycoides*		*Clostridium botulinum* 62A		
	Vegetative cells	Spores	Vegetative cells	Spores	Vegetative cells	Spores	
					Polypeptone medium		Trypticase medium
Glycine	4	5	4	5	4·2	3·1	3·6
α-Alanine	5	6	5	6	4·7	3·8	3·3
β-Alanine	1	+	1	—			
Serine	3	4	3	3	1·9	1·8	2·1
Threonine	3	5	2·5	3·5	2·2	1·9	2·2
Valine	4	5	4	5	3·2	2·7	3·1
Leucine	5	6	4	5	2·8	2·5	2·9
Isoleucine	2·5	3	2·5	3	2·7	2·5	2·9
Aspartic					4·8	4·6	5·4
Asparagine	4	3	4	3			
Glutamic	7·5	5	7	4·5	4·4	4·8	5·1
Lysine	5	4	5	4	4·6	3·7	4·2
Arginine	4	3	4	3	1·7	2·7	2·6
Histidine	2·5	3	2·5	3	1·1	1·8	1·5
½-Cystine	+	4	+	4	0·3	1·0	1·5
Methionine	1·5	+	3·5	1·5	1·5	0·9	1·2
Proline	2·5	3	2·5	3	1·1	1·7	1·8
Hydroxyproline	+	—	+	—			
Phenylalanine	4	5	4	4	1·4	1·6	1·8
Tyrosine	3·5	6	3·5	5	1·1	0·6	0·9
Tryptophan	3	5	3	5			
α-ε-Diaminopimelic	2·5	3	2·5	3	0·5	0·3	0·2
Hexosamine	5·5	3	5·5	6			
TOTAL	73·0	79·0	73·0	77·5			

[a] Semi-quantitative determination in the total hydrolysate (6N HCl, 17%v/v amyl alcohol, 100°).

[b] Automatic amino acid analyser, 6N HCl hydrolysates, 110°.

[c] Assuming N equals 10% dry wt. the approximate % dry wt. of amino acid may be obtained by multiplying by 1/100 mol. wt.; where the mol. wt. is about 100, the above figures are approximately the % dry wt.

in all the amino acids except ornithine; they contained over twice as much
$\frac{1}{2}$-cystine but only half as much ornithine.[179]

In addition to protein, carbohydrate and dipicolinic acid spores may
contain a variable amount of lipid and ash (Table IV). The composition of
the ash is discussed on p. 246.

TABLE IV

Ash content of spores and vegetative cells

Species	Vegetative cells	Spores % dry wt	Ref.
B. *anthracis*		1·15	35
	7·76		37
(rough)	13·57		38
(smooth)	12·7		38
B. *mycoides*			
Exp. 1	9·32	6·40, 9·55	22
2	5·6	5·8	22
3	5·2	7·0	22
B. *megaterium*	9·1	11·0	24
Bacillus (13 strains,		9·17	39
11 species)		(6·1–13·1)	

III. Composition of Spore Structures

A. Exosporium

A single exosporium preparation has been analysed (Table V). This was
prepared by stripping off the exosporia as the spores passed through the
Ribi press, and separating the exosporia by differential centrifugation.[26] A
phospholipoprotein composition similar to that of unit membranes is
suggested with, in addition 10·4% sugars and 11·2% glucosamine. Traces
only of diaminopimelic acid (0·2%) occurred and glucosamine was the
only amino sugar detected. Organic phosphate occurred mainly as teichoic
acid (2·0%). The non-hydrolysable fraction was reported to consist mostly
of degraded carbohydrates.[26]

The exosporium is therefore a complex structure (Chapter 2) and no
information exists on the composition of the nap, intermediate or basal
layer. The inner basal layer has a hexagonally perforate surface pattern
made up of lamellae which apparently fragment into crystal-like elements
giving an X-ray diffraction pattern.[27]

TABLE V

Composition of the exosporium of Bacillus cereus

Component	% dry wt
Amino acids (15)	37
Lipid, ether extractable	17·9
Sugars (glucose, galactose, rhamnose, ribose)	10·4
Glucosamine	11·2
Phosphate, organic	2·1
RNA	1·2
Non-hydrolysable fraction	18·0
	97·8

(Data from Matz and Gerhardt.[26])

B. Coats

The coats constitute a major part of the spore. They occupy about 50% of the spore volume and yields of water-washed integument preparations of 40–60% of the dry weight are common.[4, 17, 28, 29] In *Bacillus megaterium* this diminishes to about 30% after lysozyme treatment.[30]

The preparation of coats for analysis usually involves mechanical disruption of the spores by shaking with glass beads and separation of the insoluble integuments from the material that does not sediment at 10,000 g ("soluble fraction"). The integuments usually contain hexosamine-containing peptides which can be removed with lysozyme.[4, 14, 29] Most of the hexosamine-containing material is considered to be attached cortical material.[4, 29, 31] The coats are considered to be the insoluble fraction from which all the cortical material has been removed as determined by electron microscopy of thin sections and chemically by hexosamine and diamino-pimelate analyses. Whether the portion of diaminopimelate and hexosamine not removed in some species by lysozyme or by spore lytic enzymes that degrade cortical material constitutes part of the coat fraction, or is residue not accessible or susceptible to digestion by these enzymes, is not clear.

During disruption considerable contamination with other structures and membranes is possible. Precise methods of freeing coats from these have not been employed.

Other methods that can be used to make coat preparations are:

(i) germination with separation of the shed coats or enzymic removal of the outgrowing germinated cells[32]

(ii) autoclaving intact spores for 2 hr in 0·1 N HCl at 121° to hydrolyse non-coat material[16]

(iii) chemical degradation of spores with hypochlorite[40]

(iv) "acid-popping" of spores[41, 42] and enzymic removal of nuclear material, and cortical material.

Most methods of coat preparation allow possible changes in the composition of the coats during disruption of the spores and purification of the coats. A number of enzyme and chemical schedules have been used to clean the coats:

(i) 1 M NaHCO$_3$, water, 1 N HCl, water[14]

(ii) Trypsin, water[15]

(iii) Trypsin, pepsin, ribonuclease[16]

(iv) Lysozyme[4, 29, 30]

(v) Lysozyme, ribonuclease, trypsin + bacterial proteinase.[17]

As the coats are composed of several layers (Chapter 2) and the stability of all of these to proteolytic enzymes is unknown, it is inadvisable to subject the coat preparations to proteolytic enzymes. While the outer coats or coat surfaces of intact spores are resistant and coat preparations appear resistant to the following enzymes—lipases, lysozyme, phosphatase and various proteolytic enzymes—it is not known whether these or other enzymes can remove various groups or degrade some of the coat layers. The layer of mother cell cytoplasm incorporated between the outer forespore membrane and the inner coat[31] is quite likely degraded by some of the above enzymes. Lysozyme itself does not normally release much protein or peptide material during removal of cortical material from the integuments.[29] Enzymes as above, therefore, may be useful in determining the composition of the incorporated mother cell cytoplasm layer if the inner and outer coat are resistant to these.

The following analyses of coat layers are thus not altogether satisfactory because (i) the many coat layers have not been analysed separately, (ii) some of the preparations contained residual cortical material based on the criteria that diaminopimelic acid and glucosamine were present and, (iii) some preparations may have contained exosporium material.

The coats consist mainly of protein with 1–3% ash, 0–3% phosphorus and some lipid (Table VI). The protein content varies from 60% in B. coagulans spore coats[4] to 80% in B. licheniformis.[28] The coats are composed of a large number of amino acids, with all except B. licheniformis having glycine and lysine as the predominant amino acids (Table VII, Fig. 2). The coats of most species were high in tyrosine, aspartic acid, leucine-isoleucine and histidine. Except for cystine, valine, glutamate and

TABLE VI

Composition of spore integuments

			% dry wt				
	Total N	L-amino N	Total P	Ash	Total Lipid	Acid-insoluble residue	Ref.
Bacillus							
megaterium	12·8	5·6	0·3	1·4[a]	—		14
Bacillus cereus	13·2	9·5	1·2	2·8[a]	0·9[b]		14
	10·6	6·9	1·3				
Bacillus subtilis	13·1	8·2	1·6	2·8[a]	1·1[b]		4
	12·9		1·4		3·0(1·4)[c]		15
	11·5	7·5	1·3				4
Bacillus							
licheniformis		12·8[d]	+	Variable	2·0	9·1	28
Bacillus							
coagulans	9·7	6·2	2·8				4
Bacillus stearo-							
thermophilus	11·1	7·0	0·7				4

[a] Sulphated.
[b] After acid hydrolysis.
[c] Ether extractable.
[d] Amino acid content ÷ 6·25.

lysine the amino acid spectrum of the coats strongly resembles that of vegetative cells (Fig. 1) (see also [180]).

Figures for hexosamine and carbohydrate have been omitted from Table VI, and in some cases those for alanine, and glutamic acid from Table VII as the evidence of Warth *et al.*[4, 29] suggests that in many species all the diaminopimelate and hexosamine and a large part of the alanine and glutamate were derived from cortical material. Alanine and glutamate, however, are certainly not absent from coats.

In *Bacillus* strain 636 with a ridged coat, lysozyme removed most of the diaminopimelate but only about half of the glucosamine and it was suggested that the remainder occurred in the ridge structure. Coats of this strain also contained much glutamate and taurine. Taurine was not present in the soluble fraction of this strain or in whole spore hydrolysates of fourteen other *Bacillus* species.[29]

In *B. sphaericus* aspartic acid occurs as the L-form in spore coats but chiefly as the D-form in the vegetative cell walls. Amino-succinyl-lysine is formed as an artefact by hydrolysis of aspartyl-lysine in vegetative walls but appears to be a real constituent of the spore coats (Tipper, private communication).

TABLE VII

Amino acid composition of spore coat preparations

mole ratios

Amino acid	Bacillus megaterium			Bacillus cereus	Bacillus subtilis		Bacillus licheniformis	Bacillus coagulans			Bacillus stearothermophilus	
	Alkali-soluble fraction	Para-crystal fraction	Resistant residue					heavy fraction	light fraction		Unfractionated	Heavy fraction[a]
Ref.	30	30	30	4	15	4	43	4	4	16	4	4
Glycine	10.0	10.0	10.0	10.0	10	10.0	10.0	10.0	10.0	10.0	10.0	10.0
Alanine	4.6	2.9	7.6	b	8	b	5.0	b	b	9.3	b	5.4
Serine	4.2	4.1	9.8	4.5	8	4.3	3.7	4.3	4.8	3.6	2.5	1.9
Threonine	3.3	1.8	8.2	3.1	2	2.9	3.1	3.7	4.1	2.9	3.3	2.3
Valine	3.3	1.4	6.1	3.1	4	2.5	3.2	3.6	3.7	3.9	4.2	2.7
Leucine	4.2	0.2	tr.	5.3	4	4.1	3.2	6.5	6.8	5.1	6.7	4.4
Isoleucine	2.6	1.2	9.4		4		2.5			3.3		
Aspartic	8.1	9.4	17.7	4.4	2	4.2	10.5	5.0	5.7	10.1	5.1	4.2
Glutamic	6.2	5.2	16.1	b	6	b	5.0	b	b	8.4	b	4.0
Ornithine							0.8					
Lysine	6.2	1.6	30.0	7.4	12	6.0	7.9	5.4	6.5	19.7	5.1	2.3
Arginine	5.1	3.5	7.2	3.3	+	2.6	1.8	4.6	5.3	6.0	3.8	2.2
Histidine	4.0	1.5	5.0	3.4	+	3.0	4.8	3.4	2.5	10.4	2.6	1.5
½-Cystine	1.1	5.1	3.1	+	4	+	3.0	+	+	7.2	+	+
Methionine	0.9	0.6	5.6	+	+	+	1.5	+	+	3.6	+	+
Proline	5.4	0.8	7.7	+	+	+	4.0	+	+	5.6	+	+
Phenylalanine	3.9	4.2	8.7			2.7	4.6	2.6	4.0	5.8	2.3	2.3
Tyrosine	9.6	1.3	7.9	2.6	3	4.3	11.5	4.4	4.1	9.1	4.2	4.6
Ammonia				4.6			11.0					
Enzyme cleaning treatment	←————— lysozyme —————→			Nil	Trypsin	Nil	Nil	Nil	Nil	Trypsin Pepsin RNase	Nil	Nil

All spores except 16 were mechanically disrupted; 16, spores autoclaved 121°, 2 hr in 0·1 N HCl.

a Practically free of diaminopimelate and hexosamine.

b Omitted, see text.

Spores contain about 26 μg cystine sulphur/mg protein nitrogen, while the vegetative cells have only about 7 μg.[44] Most of the cystine occurs in the coat fraction[45] (Table VIII). Enzyme-purified coats of *B. subtilis* contained 292 μg cysteine + cystine sulphur per mg. N, similar to wool or hair keratin.[17] In *B. megaterium* most of this occurred in the paracrystal coat fraction[30] (see later) (Table VIII).

The coats are resistant to many chemicals, e.g. 8 M-urea, 8 M-LiBr, 2 M-CaCl$_2$ (90°), formic acid (98%), performic acid (98%), N-NaOH, phenol (80%) and surface-active compounds. This has been interpreted as indicating relatively stable covalent cross-linking and that coat resistance is not entirely dependent on disulphide cross-linking or hydrogen, ionic or hydrophobic bonding.[29] Resistance to alkali and surfactants also may

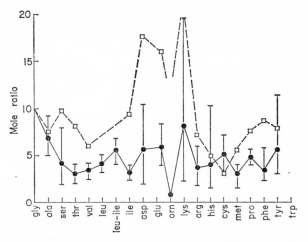

FIG. 2. Amino acid composition of spore coats of five *Bacillus* species (●) and of the resistant residue coat fraction of *Bacillus megaterium* (□). Data from Table VII in moles relative to ten moles of glycine.

suggest that normal lipid and lipoprotein are not very important to structural integrity.[29] Strong alkali causes coats to become gelatinous.[14] Disulphide bond-breaking reagents stimulate germination (Chapter 11) and probably change permeability by rupturing these bonds in the coats. Eighty to 90% of the coat proteins of *B. cereus* T spores have been solubilized in alkaline thioglycollate or dithiothreitol.[180]

Little information on coats of *Clostridium* species has been published. Stewart[47] prepared spore coat skeletons of *Cl. botulinum* 62A by autoclaving the spores in acid (0·1N HCl) for 90–105 min at 121°. These were free of stainable ninhydrin positive material, contained 56·7% C, 10·5% N, 8·2% H, 17·3% O, 2·1% S and 0·1% P, they were insoluble in 0·1N H$_2$SO$_4$, conc. HCl, 5 N NaOH and fat solvents and were unaffected by α- and

TABLE VIII

The (cystine + cysteine) sulphur content of vegetative cells, spores and cell fractions

Species	Cystine-S (μg S/mg protein N)	
Data from Vinter [45]	Vegetative cells	Spores
Bacillus megaterium B2	7·7	27·9
Bacillus megaterium NCI B8291	7	26·9
Bacillus cereus NCI B8122	6	29·8
Bacillus cereus var *mycoides* B6	9·8	27·7
Bacillus agrestis B3	4	27·6
Bacillus subtilis NCI B8060	8·1	15[c]
Data of Bott and Lundgren [46]	μg S/mg dry wt	
Bacillus cereus WT	3·70	7·14
Bacillus cereus T	5·63	7·91

Structures	Cysteine + cystine-S (μg S/mg Total N)	
Bacillus subtilis (Data of Kadota et al.[17])		
Intact cell	26	108
Wall	10	
Crude coat[a]		190
Coats—lysozyme treated		212
Coats—treated lysozyme + ribonuclease		252
Coats—purified[b]		292
Bacillus megaterium (Data of Kondo and Foster[30])		Cysteine (% dry wt)
Whole coat		1·7
Alkali-soluble coat fraction		Nil
Paracrystal coat fraction		6·6
Resistant coat residue		0·8

[a] N, 11·2% dry wt.
[b] N, 14·5% dry wt.
[c] Sporangia with spores.

β-amylases, α-glucosidases and cellulases. Qualitative tests and infra-red spectra indicated reducing compounds, the indole nucleus, p-unsubstituted phenol-, carboxyl-, amide-, aldehyde- and ketone-groups.

Kondo and Foster[30] have attempted to fractionate chemically and en-zymically the coats of *Bacillus megaterium*. They separated the lysozyme-treated coats into an alkali-soluble fraction, a paracrystal fraction (sonicate of the alkali-insoluble fraction), and a pronase-resistant residue. These formed 30, 30 and 40% respectively of the lysozyme-treated coats. The paracrystal fraction appeared to represent the middle particulate layer of the coat sandwiched or cemented between the alkali-soluble layer on one side and the resistant-residue layer on the other.

The alkali-soluble layer contained one or more proteins, insoluble at pH 7 but soluble at pH 9·5, composed of most of the amino acids (Table VII) and 0·9% P. It was rich in glycine and exceptionally rich in tyrosine (12%), and contained very little cystine.

The paracrystal fraction appeared to contain a keratin-like basal substance from its X-ray diffraction pattern. It was composed of amino acids, muramic acid and phosphorus in the mole ratio of 40 (alanine equivalents): 6 : 1. It contained 0·8% hexosamine (not acetyl glucosamine, glucosamine or galactosamine). Glycine, aspartate, glutamate and cysteine made up half the amino acid residues (Table VII). The half-cystine residue occurred once in every twelve amino acid residues, the dicarboxylic acids and glycine once in every four.

The resistant-residue coat fraction appeared to contain a phosphomuramyl polymer combined with peptide or protein. It contained 28% amino acids (alanine equivalents), 14% muramic acid, a trace of glucosamine and 26% ash (7% P), the remainder was unaccounted for. It was exceptionally high in lysine (equivalent to that of cytochrome C), low in cystine and contained three unknown ninhydrin-positive peaks. Three phosphorus fractions were present. Fraction (i), the organo-P (50%), contained muramic acid, lysine, ortho-P in a 1 : 1 : 4·8 mole ratio and appeared to be a polymer of phosphomuramic acid and lysine. Fraction (ii) contained 38% of the soluble-P and was orthophosphorus and fraction (iii), 12% of the soluble-P, contained orthophosphorus and muramic acid.

The mole ratios (compared to 10 moles of glycine) of the above three coat fractions obtained from B. megaterium resemble closely the amino acid spectrum for coats of other species (Fig. 2), except that the amino acid content relative to glycine of the resistant residue fraction is high throughout. This could simply mean that this fraction is low in glycine, compared to the other two fractions and coats generally. The similarity in the amino acid spectra suggests that the fractionation of the coats may not have fractionated or separated the protein component of the coats.

The biosynthesis of the alkali soluble and insoluble fractions of spore coats of B. cereus T, B. cereus var alesti (A-) and B. megaterium KM was recently studied by Aronson and Fitz-James.[180] The coat precursor proteins containing some of the cystine were synthesized during the early spore stages (I and II) several hours before the appearance of morphologically distinct coat laminae. Throughout sporulation 80–90% of the coat proteins (16–17% N) could be solubilized in alkaline thioglycollate or dithiothreitol; in the late stages when morphologically distinct coat structures were present a resistant coat profile and the exosporium remained after the treatment. The insoluble residue contained 7% N and was relatively rich in cystine. The amino acid spectra of the vegetative cell and coat proteins were very similar and resembled those in Fig. 2. Cystine was incorporated at a high rate in the late stages during the formation of the coat structural components and it was suggested that the incorporation occurred by disulphide exchange between the cystine and pre-existing protein disulphide or sulphydryl groups, this rapid exchange resulting in conformational changes in the polypeptides. The structural layers were dissolved by sodium lavyrl sulphate plus dithiothreitol, releasing free cystine. Aronson and Fitz-James concluded that the structural layers (laminae) consisted predominantly of polypeptides interacting by hydrophobic bonds with deep-seated cystine residues. The polypeptides appeared to be rather limited in kind.[180]

The electrophoretic mobility studies of Douglas and Parker[18, 48-50] suggest that the resistant-residue layer of B. megaterium is probably the outer surface layer of the coat. These studies suggest that the outermost, electrokinetically effective surface layer of spores is non-ionogenic, acquiring its charge by adsorption of ions (note p. 246). Spores differed, B. subtilis behaving as though the surface was polysaccharide and that of B. megaterium having surface lipid and protein which rendered the normal ionogenic groups inoperative.[51] The mobility of B. subtilis spores was altered by lysozyme, hyaluronidase and polyglutamylpeptidase treatment.[18] This could indicate a polysaccharide surface layer either constituent or derived from contaminating mother cell material. B. megaterium spore mobility was slightly increased by lipase but more so by glutamylpeptidase, suggesting a glutamyl-peptide and lipid surface. Spores of B. cereus were not affected by any of the enzyme treatments.[18] The differences in spores could relate to surface cleanliness resulting from the different activities of the autolytic enzymes of the mother cells, as no mention of previous enzymic cleaning of the spore surface was made.

The concentration of cations needed to reduce mobility to zero ("cation charge reversal spectra"), i.e. the net charge on the spores, showed that the effective charge group is—COOH, with some phosphate as well in B. megaterium spores. With the proviso above this evidence suggests that the surface matrix substance could well be polysaccharide-peptides with in addition in B. megaterium a phospholipid or phosphorus-containing component.

Coats are therefore composed of chemically very stable disulphide-rich proteins and 1–2% lipid, phosphorus and inorganic matter, and sometimes glycopeptide or hexosamine. X-ray diffractometry of B. subtilis coats indicates periodic (crystalline) structures with a lattice distance of 9·8 Å, very similar to wool keratin.[17, 52] The diffraction patterns showed 9 and 9·6° peak structures, and eighteen peaks corresponded to those in wool keratin, nine with α- and nine with β-keratin, suggesting that coats are composed of α- and β-forms like keratin. The lipid and phosphorus contents suggest phospholipoprotein components perhaps in between protein layers in the well-defined laminated layers of the inner coats, and perhaps in some (*B. megaterium*) on the outer surface.

C. INCORPORATED MOTHER CELL CYTOPLASMIC LAYER (OR INNERMOST COAT)

This layer of mother cell cytoplasm appears to be physically trapped between the coat layers and the outer forespore membrane and remains intact throughout sporulation[31, 178] and could be designated as the innermost or inner coat.[178] In species with an exosporium this should not strictly be called mother cell cytoplasm as this material would have to be synthesized *de novo* between the exosporium and the outer forespore membrane unless there is still a connection between the two regions. In electron micrographs it has the appearance of typical cytoplasm. No direct information has been obtained on its composition, but it may be deduced that at the time of incorporation it is probably similar to that of the mother cell cytoplasm at that stage of spore development; secondary, tertiary and other chemical changes must occur during maturation, and these may be involved in the bonding of the cortex and outer forespore membrane to the coats.

The studies of sections of spores of *B. cereus* T stained with ferritin-labelled antibody to vegetative cells show in plates 16–21 of Walker *et al.*[53] that this layer stains with the antibody, indicating (i) similarity of constituents of this layer to vegetative cell material, and (ii) that if the layer is synthesized *de novo* then it has components similar in composition to those of the mother cell.

D CORTEX

The insoluble structural material of the cortex consists largely of a muropeptide polymer (murein) resembling those found in vegetative cell walls; spore murein is not necessarily identical to vegetative murein. A structure consistent with the available data is shown in Fig. 3.

Disruption of spores under conditions of inhibited autolytic activity (pH 9–10, plus EDTA to bind Ca^{2+}) or after heating to inactivate the autolytic enzymes, yields integument preparations containing nearly all of the structural material of the cortex including the germ cell wall, and containing essentially all the hexosamine and diaminopimelate of the spore.[4, 29] The insoluble cortical material can then be removed by lysozyme or the spore lytic enzymes, and analysed. This provides no information about substances possibly occurring in the cortex of the intact spore and which are released during disruption. Enzyme digestion of the insoluble cortical material also usually degrades the germ cell wall; this will therefore probably result in errors in the analyses of cortical murein.

Lysozyme or autolytic degradation of the integument preparations containing the insoluble cortical structure yields non-dialysable muropeptide units and a small amount ($<4\%$) of low mol. wt solutes. These include amino acids and a variety of peptides composed of amino acids common to the side-chain peptide of the muropeptide. They may be derived from the polymer or from the coats as a result of enzymic impurities[29, 54] in the lysozyme preparations or from a mixture of enzymes in the spore autolytic enzyme complex.

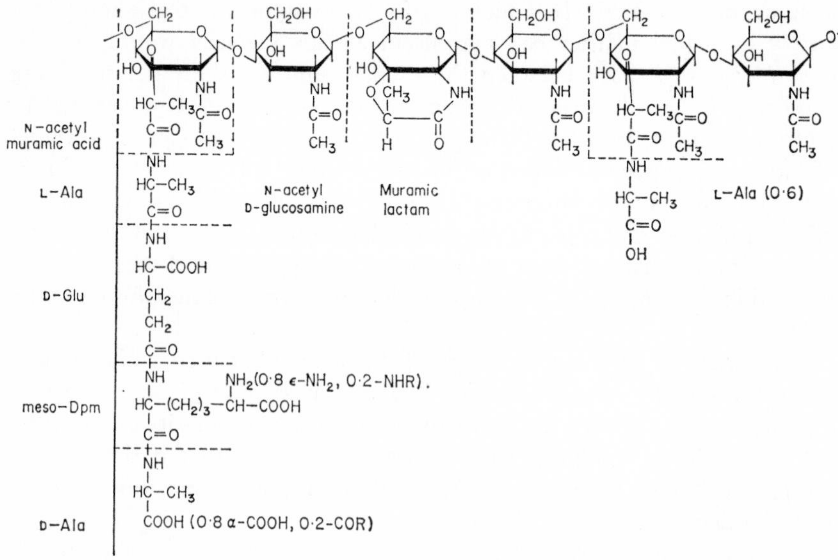

FIG. 3. Hypothetical average unit of the spore muropeptide polymer. Dpm indicates a,ϵ- diaminopimelic acid (after Warth,[54] and Warth, private communication). Numbers indicate relative amounts of each substituent in the polymer.

The first evidence of the composition of the cortical muropeptides came from studies of the "germination exudate",[55, 56] although it was not realized at the time that the muropeptides arose from degradation of the cortex during germination. The hydrolytic products of the non-dialysable muropeptide in the germination exudate and extracts from disrupted spores were very similar (Tables IX, X). As the spore peptides are released during germination by the spore lytic system this may be expected. The germination exudate murein from *B. megaterium* had a mol. wt of 15,300.[58] Some of the muropeptide from *B. megaterium* passed through the dialysis membrane, indicating a mol. wt of about 10,000. Strange and Powell[56] suggested that various preparations may contain units of different chain length, a suggestion supported by viscosity measurements.

TABLE IX

Composition of non-dialysable peptides (g/100 g) from spore cortexes

	B. megaterium		B. cereus		B. subtilis		B. coagulans
	Germination exudate	Extract	Germination exudate	Extract	Germination exudate	Extract	Extract
Ref.	56	56	56	56	56	56	54
Hexosamine	40·3	43·2	42·5	48·2	37·0	38·0	65·4[a]
Acetyl	10·9	12·0	9·9	10·1	10·4	9·9	8·2
Total N	7·37	7·35	7·3	7·3	6·6	7·3	
α-Amino N	2·91	2·75	2·9	2·5	2·6	3·1	4·2[b]
Reducing power of hydrolysate (as glucose)	1·24	1(5·0)[c]	50·0	51·7	37·5	36·1	
Total carbohydrate (as glucose)	5·8	7·0	4·7	2·1	4·4	5·7	2·06[d]
Sulphated ash (mainly Ca)	0·03		2·7	3·2	8·7	6·8	
Phosphorus	0·2						
Pentose							

[a] Glucosamine (25·4%) + muramic acid (40%).
[b] Calculated from amino acids present.
[c] Unfractionated.
[d] Not sulphated, mainly Ca, Na, P.

The average composition of the non-dialysable muropeptide from four species was about 45% amino sugars, amino acids (7% total N; 3% α-amino-N) and small amounts of ash (P and mainly Ca) (Table IX). The major products of the nondialysable peptide after acid-hydrolysis occurred in the mole ratios of α, ε-diaminopimelic acid (DAP, 1), glutamic acid (1), alanine (3), and acetylhexosamine (6-8) (Table X) In *B. coagulans* the products were the same from lysozyme or autolytic digestion.[29] Lower hexosamine values occur when muramic acid is not estimated. Only small amounts of other amino acids were present in acid hydrolysates of the muropeptide.[29, 54] This was found more often when the spores were autoclaved before disruption, suggesting retention of coagulated included protein in the non-dialysable fraction. A qualitative analysis of the cortical muropeptide of *B. sphaericus* indicates 3 glucosamine : 1 glutamate : 2 alanine resembling that of *B. subtilis* (D. J. Tipper, private communication). Detailed analysis[54] of a well-dialysed non-diffusible muropeptide preparation from *B. coagulans* gave glucosamine, muramic acid, L-alanine, D-alanine, D-glutamate and meso-DAP in the molecular proportions of 2·8 : 3 : 1·6 : 1·0 : 1·0 : 1·0.

Uronic acids, purines, pyrimidines,[56] hexoses, pentoses and methylpentose[54] were not found in muropeptide preparations. Traces of phosphorus (Table XIV) indicated the absence of significant amounts of teichoic acid. No teichoic acid was found in spores of *B. subtilis* or *B. licheniformis*.[183] The ash, mainly Ca^{2+}, probably arises from metal ions associated with the carboxyl groups of the polymer during disruption or germination.

The structure of the muropeptide of *B. coagulans* (Fig. 3) is suggested from titration, dinitrophenylation and degradation studies. About 80% of the amino groups of DAP and 3·8 carboxyl groups per polymer unit were free; alanine in trace amounts (1 in 300 residues) was the only N-terminal amino acid detected. The results are consistent with the cortical structure being a series of cross-linked polymer chains made up as in Fig. 3 to give a three-dimensional network with a considerable excess of carboxyl groups. One-fifth of the amino groups of DAP are probably peptide-bonded with carboxyl groups, possibly in the adjacent chain. Only one L-alanine residue possibly occurs in each peptide side-chain, but some muramic residues may be unsubstituted by peptide side-chains whilst others bear only a single alanine residue. About six units (mol. wt 11,000) probably occur for each non-dialysable muropeptide fragment released. Evidence of inhomogeneity and variable polymer length exists.[54, 56]

Autolysis of the cortex of *B. subtilis* involves splitting some of the muramyl-alanine amide bonds (A. D. Warth and J. L. Strominger[182], private communication). N-terminal alanine increased, and peptides I and II were split off during autolysis.

TABLE X

Composition of non-dialysable muropeptide of the spore cortexes of Bacillus *species*

Component	Contents (mole ratios): *Bacillus megaterium* Germination exudate	*Bacillus megaterium* spore extracts	*Bacillus megaterium* spore extracts	*Bacillus cereus* T spore extracts	*Bacillus subtilis* spore extracts	*Bacillus strain* 645 spore extracts	*Bacillus coagulans* spore extracts	*Bacillus coagulans* spore extracts	*Bacillus stearothermophilus* spore extracts
Ref.	56	56	6	57	6	6	6	54	6
α, ε-Diaminopimelic acid	1·00	1·00	1·00		1·00	1·00	1·00	1·00	1·00
D-Glutamic acid	1·35	1·18	0·8	1·0	0·80	0·88	0·95	0·95	0·78
L-Glutamic acid			0·11		0·20	0·22	0·03	0·03	0·04
L-Alanine / D-Alanine	3·8	3·1	2·2	2·0	2·4	2·6	2·5	2·5	2·6
Hexosamine	8·4	8·2	2·1[a]	3·3	2·5[a]	2·4[a]	4·7	5·6	1·9[a]
Glycine			0·20		0·26	0·22	0·03	0·05	0·08
Aspartic acid			0·17		0·21	0·18	0·03	0·03	0·07
Acetyl groups								3·6	

[a] This value does not include muramic acid.

I
$$(H_2N)—Ala—Glu—(COOH)$$
$$|$$
$$^—DAP—(COOH)$$
$$(H_2N)—|—(COOH)$$

II
$$(H_2N)—Ala—Glu—(COOH)$$
$$|$$
$$^—DAP—Ala—(COOH)$$
$$(H_2N)—^—(COOH)$$

The cortex, because of its morphological development and its chemical composition can be considered a large endogenous cell wall-like layer.[29, 31] It is of considerable interest to compare the structure of the spore and vegetative cell polymers to see in what ways the cortex may be specifically distinguished. Some insight may also be gained into the properties and possible function of the spore cortex. All spore mureins so far studied are fairly similar in composition and differ from known cell wall polymers in having muramic δ-lactam in place of every second or third N-acetyl-muramic acid residue[184] and a two- to three-fold excess of glucosamine or muramic acid residues over glutamic acid or DAP residues (Fig. 3). All spore mureins analysed and many species of spores contained DAP although vegetative cell walls commonly have lysine or other dibasic amino acids replacing DAP. Spore muropeptide and germination exudate also contain more than one L-alanine per glutamic acid, DAP or D-alanine. The spore polymer is only lightly crosslinked, 80% of the DAP ε-amino groups were free in a *B. coagulans* preparation,[54] whilst some cells wall, e.g. *Staphylococcus aureus* are very highly cross-linked. Finally the spore muropeptide does not seem to be intimately associated with other polymers such as the teichoic acid, teichuronic acid and complex polysaccharides which often form a major part of cell walls.

More specific information has been obtained by a direct comparison of the muropeptide from both vegetative cells and spores of *B. subtilis*. During autolysis of the cell wall, the muramyl-alanine bonds are broken and the cell wall peptides are freed from the polyaminosugar "backbone".[59]

The structure (Warth and Strominger[182]) of the predominant peptides III, IV, and V can be compared with the peptides released from spore muropeptide partially by autolysis or completely by the N-acetylmuramyl-L-alanine amidase of Ghuysen.[60]

$$(H_2N)—Ala—Glu—(COOH)$$
$$|$$
$$^—DAP—(COOH)$$
$$(H_2N)^—(CONH_2)$$

III

$$(H_2N)—Ala—Glu—(COOH)$$
$$(H_2N)—Ala—Glu—(COOH)^—DAP$$
$$|$$
$$^—DAP—D—Ala^—(CONH_2)$$
$$(H_2N)—COR$$

IV   R = NH₂
V   R = OH

The vegetative cell wall peptides are all amidated, the dimers IV and V predominate, monomer III and trimers exist in lesser amounts. No carboxyl terminal alanines are found. In the spore, however, the peptides were not amidated and are strongly acidic. Compound II comprises about 80% of the peptide units isolated, the crosslinked dimer or higher homologues being much less abundant. Notable also is the frequent occurrence of the C terminal alanine. Crosslinking in cell walls involved the ε-amino group of 35% to 45% of the DAP residues whilst in the spores only 12% to 19% of these residues were substituted. *B. sphaericus* spore murein also appears less crosslinked than the vegetative cell wall polymer (Tipper, private communication).

Nearly all the muramic acid residues of *B. subtilis* vegetative cell walls must be peptide substituted, but the major products of lysozyme digestion of spore murein (Warth and Strominger, private communication) are tetrasaccharide-tetrapeptide, $GlcNAc_2$ $NAMA_2$ Ala_2 Glu_1 DAP_1, and tetrasaccharide-alanine, $GlcNAc_2$ $NAMA_2$ Ala_1, with smaller amounts of hexasaccharide-tetrapeptide, hexasaccharide alanine and disaccharide tetrapeptide or alanine (NAMA = N-acetylmuramic acid, GlcNAc = N-acetylglucosamine).

All aspects of the differences in structure between the two muropeptides of *B. subtilis* may not turn out to be applicable to all species. Although spore muropeptides seem very similar, the cell wall muropeptide of *B. megaterium*, for example, is not amidated and is only slightly cross-linked.[61]

The cortical structure in integument preparations is resistant to a variety of enzymes[14, 29] and to many chemical treatments. Formic acid, performic acid, N-NaOH, 80% phenol, cetyltrimethyl ammonium bromide and polysept S (quaternary ammonium glycine) released less than 14% hexosamine; N-NaOH for 1 hr at 100° released 28% and 5% trichloroacetic acid (15 min, 90°) released 43% of the hexosamine.[29]

The cortical material is not simply material trapped between the protoplast membrane and the outer forespore membrane and coats; it remains attached to the coats on disruption.[29] During spore maturation the cortex must become cemented or bonded to the coats and to the incorporated mother cell cytoplasm by tertiary changes in chemical structure.[31] The type of linkage involved in this attachment is not known, but apparently it is not by peptide bonding as Strange and Dark[14] found that the hexosamine-containing peptides associated with their coat proteins were not removed by proteinases but were removed by lysozyme.

Little information has been published on the composition of cortexes of *Clostridium* species, but their formation is cytologically similar and germination exudates resemble in composition those of *Bacillus*,[47, 62] and the spores contain diaminopimelate and hexosamine,[36] suggesting that similar types of muropeptides are present.

Warth[184] has recently shown that the spore muropeptide of *Cl. sporogenes*, like that of *Bacillus* species, contains muramic lactam and L-alanine substituted muramic acids. Six compounds, isolated from *Cl. sporogenes* muropeptide lysozyme digest, were identical to those from the *B. subtilis* spore muropeptide. [182, 184]

E. GERM CELL WALL

Although this layer has not been isolated and analysed, three types of evidence suggest that its integrity is dependent on a murein composition. First, integument preparations of *B. subtilis, Bacillus* strain 636, *B. coagulans* and *B. stearothermophilus* lose this layer during lysozyme digestion.[29] In *B. cereus* T preparations it persisted. However, the presence of 36% of the hexosamine and 20% of the diaminopimelate in the lysozyme-treated integuments and the presence of extra hexosamine and only traces of diaminopimelate in the exosporia also suggests a diaminopimelate-containing muropeptide composition. Secondly, Vinter[63] has shown that labelled diaminopimelate incorporated into the first formed muropeptide during sporulation persists during germination. Thirdly, Walker *et al.*[53] have shown that ferritin-labelled antibody to the vegetative cell wall stains this layer.

F. PROTOPLAST

It is not possible to isolate and determine directly the chemical composition of the core, or more definitively the protoplast (cytoplasm + nucleoid(s) + plasma membrane) of the mature spore. The "soluble" fraction (10,000 g supernatant) after disruption of the spore and removal of the insoluble integuments includes the major part of the protoplast and any soluble material lost from the outer layers of the spore during disruption. The "soluble fraction" could be further subdivided into soluble constituents and insoluble macromolecular components.

The "soluble fraction" includes most of the free solutes, enzymes, ribosomes, RNA and DNA together with various membrane inclusions derived from the protoplast membrane, mesosomes or rudimentary mesosomes (Chapter 2) and membrane elements from just inside the protoplast membrane.[31, 64] The nucleic acid fraction contains sulphur and the greater part of this is present as methionine and cystine.[185]

Chemical analyses of the acid hydrolysates of the "soluble fraction" from four species show about 9% total nitrogen, 4–6% α-amino nitrogen, 1–2% phosphorus, 6–7% sugars and 10–26% DPA (Table XI). The DPA content was greater in the more heat-resistant species, but this resulted largely from the decrease in the amount of the "soluble fraction" from the more resistant species. The hexosamine present is most probably mainly derived from lytic degradation of the cortex as lytic enzymes were not inactivated before or during disruption in this study. The soluble fraction from the mesophiles contained more hexosamine as the lytic enzymes of these organisms are apparently more active at 1° than those of the thermophiles.[4]

The amino acid composition of the hydrolysed "soluble fractions" were in general similar to those for the coats (Table VII) except that cystine

TABLE XI

Composition of the "soluble fraction" from spores of Bacillus *species* [a]

	% dry wt			
	Bacillus cereus	*Bacillus subtilis*	*Bacillus coagulans*	*Bacillus stearothermophilus*
Total nitrogen	9·1	9·7	—	9·2
α-Amino nitrogen	—	—	5·6	4·3
Total phosphorus	2·3	0·7	2·2	0·8
Hexosamine (as glucosamine)	5·5	7·0	2·8	2·6
Hexose (as glucose)	2·3	0·5	2·1	3·9
Methyl pentose (as rhamnose)	1·09	0·75	0·0	0·0
Dipicolinic acid	9·4	12·3	17·6	26·4
Fraction (% whole spores)	52	48	34	40
Sugars:				
Glucose	+ +	+ +	+ + +	+ + +
Galactose	+	+	+ + +	+
Mannose	—	—	—	+ + + +
Rhamnose	+ + +	+ +	—	—
Ribose	+ +	+ +	+ + +	+ +

[a] The "soluble fraction" was the supernatant after disruption of the spores and centrifugation for 10,000 **g** for 15 min: (Data from Warth *et al.*[4])

was not detected. Glucose, galactose and ribose were the main sugars; in addition, rhammose occurred in *B. cereus* and *B. subtilis* and mannose in *B. stearothermophilus*.

Numerous active enzymes occur in this fraction[31] (see Chapter 8), but quantitatively it is not known how much of the protein is enzymic and how many soluble proteins are present.

DNA forms about 1% of the dry weight of the spore (Chapter 4), but as the protoplast forms only 30% or less of the spore volume, the DNA content of the protoplast will be at least 3%. Various types of RNA and microsomal particles are also present[31] (Chapter 4).

(*i*) *Free low mol. wt solutes.* The amount of free solutes is probably an important indicator of the physico-chemical state of the resting spore. Free

TABLE XII

Free amino acids in spores

Amino acid	Bacillus subtilis (% dry wt) Boiled 10 min 75% ethanol	Bacillus megaterium Disrupted, 100°, 15 min	Bacillus megaterium Boiled in water 90 min	Bacillus megaterium Broken in 50% ethanol	Bacillus megaterium Stored at 3° for 40 days, then boiled	Bacillus cereus 90°, 100 min
Ref.	34	66	65	65	65	67 cd
			(% dry wt)			
Cysteic		+a				+b
Aspartic	0·04	+				} +
Serine	0·04					++
Glycine	0·12	+			0·11	
Arginine			} 0·50	} 0·50	} 0·82	} ++
Histidine				0·02		++
Threonine				0·74		
Glutamic	0·4	+	1·03	0·74	0·74	+++
Tyrosine						
Leucine	0·04				} 0·39	} +
Isoleucine		+			0·13	
Alanine	0·04				0·22	
Lysine	0·04			0·07		
Asparagine	0·08					
Valine	0·04					
Glutamine		+				
TOTAL (% dry wt)	0·84		1·53	1·33	2·41	

a + = present.   b + = very faint, ++ = faint, +++ = medium.   c Paper chromatography.   d Thin layer.

amino acids comprise about 1% of the dry wt of spores[34, 65] (Table XII) compared with 5% or more in vegetative cells.[34] After hydrolysis of alcoholic extracts the amount of amino acids from spores was half that from vegetative cells. The alcohol extracted, in addition to free amino acids, some low mol. wt peptide material which contained proline, tyrosine, phenylalanine and threonine.[34] *B. megaterium* contained six free amino acids and five amino acids associated with DPA.[66] In *B. megaterium* glutamate was the major free amino acid; more amino acids became free on storage in the cold[65] (Table XII).

In *B. cereus* Farkas and Kiss[67] observed only two faint ninhydrin-positive spots on the chromatogram of a hot-water extract (90°, 100 min) from intact spores, and four on thin layer plates (Table XII).

Autoclaving (20 min 121°) released the following growth factors and coenzymes from *B. subtilis* spores:-folic acid, 2·8 mμ moles/g; folinic acid, 0·2; pyridoxal, 1·4; pantothenic acid, 0·7; nicotinic acid, 28·5; biotin, 3·3 (after hydrolysis of combined forms); purines, 800; serine, 2200; and methionine, 300 mμ moles per g. dry spores.[68]

In addition to such free solutes as amino acids, growth factors and purines spores may contain other readily extracted low molecular weight solutes. Spores of four species contain as the major acid-soluble phosphate 3-phospho-D-glyceric acid.[186] *B. subtilis* spores contain as much as 6% of their dry wt as L-sulpholactic acid, which is extractable by hot water or by sonication. It was not found in spores of three other species.[186]

(*ii*) *Cytochrome and haematin.* Spores of *B. subtilis* contain about 6% of the cytochrome of vegetative cells and about half (0·016% dry wt) the haematin content.[69] No information on the location of these constituents is available.

IV. Composition of Chemical Fractions

A. LIPID

Knowledge of the type and distribution of lipids in spores is meagre and not very satisfactory. Lipid contents vary from less than 1 to 13% dry wt in *Bacillus* to 38% in *Clostridium* spores (Table XIII). Surface contamination causes considerable errors[76] and disruption of the spores is necessary for good extraction.[73, 79] Extraction procedures not involving disruption are reported to have little or no effect on viability or heat resistance.[72, 80]

Poly-β-hydroxybutrate (PHB) was not found by several workers in spores (Table XIII), but Yoneda and Kondo[73] found 4·1% and Akashi,[75] 3·6% in spores of *B. subtilis*; much of the PHB occurred in the coat and exosporium fraction.[73] The lipid content of integument preparations is normally low (Table VI). Akashi[75] isolated "sporin" as the sole component

L

TABLE XIII

Lipid contents of spores

Species	Type of lipid	Amount (% dry wt)		Ref.
		Spores	Insoluble fraction (integuments)	
Bacillus megaterium	β-Hydroxybutyrate	0		24, 70, 71
Bacillus cereus	C₂–C₈ acids	4		72
	Acid hydrolysate		0·9	14
	Disrupted spores			
	β-Hydroxybutyrate	4·1	7–9ᵇ	73
	Lipid phosphorus	0·012		74
	Lipid phosphorus in acid hydrolysate	0·03		74
Bacillus subtilis	Acid hydrolysate		1·1	14
			3·0 (1·4 before hydrolysis)	15
	β-Hydroxybutyrate	3·6		75
	Fatᵃ	2·0		75
	Phospholipid	0·8		75
Bacillus polymyxa	C₂–C₈ acids	8·0		72
Bacillus licheniformis			about 2·0	28
Bacillus stearothermophilus	Surface	6·9–13·1		76
	Total constituent	1·3–1·9		
	Phospholipid	0·3–0·4		
Clostridium perfringens	C₁₂–C₂₂ acids (behenic, stearic, palmitic, linoleic, myristic, and lauric acids)	38·0		77
Clostridium thermosacchorolyticum	Acid hydrolysate			
	Strain 3814	16·3		78
	Strain TA-37	13·5		
Clostridium botulinum 62A		20·0		47

ᵃ Ethanol–ether (1 : 1) extract.
ᵇ Kondo, personal communication.

of the chloroform-soluble wax fraction and established its structure as a polyester of β-hydroxybutyric acid or PHB with a mol. wt of 63,000. No details of the spore cleaning procedures were given. In "water-washed" spores of *B. megaterium* Mastroeni, Contadini and Teti[187] found 4·4% free lipids, 4·1% bound lipids, 3·3 mg non-esterefied fatty acids and 17·5 mg. phosphatides per 100 mg total lipids; the figures for *B. subtilis* were 6·7, 6·3, 1·7 and 12·8 respectively. The vegetative cells of these species contained more non-esterified fatty acids than the spores but much less phospholipid.

In *B. cereus* T and *B. polymyxa* the lipid was composed of C_2–C_8 acids mainly, with large amounts of acetic acid, less butyric and propionic acids and some non-saponifiable waxy residue with a low melting point. Small amounts of C_{14}, C_{16}, C_{18} and C_{20} acids occurred in *B. polymyxa* spores.[72]. Much higher lipid contents were obtained when the spores were produced in a medium containing 5% glycerol.

Spores of two thermophilic anaerobes (TA-37 and *Clostridium thermosaccharolyticum* 3814) contained firmly bound lipid, extractable only after acid hydrolysis. The major acids in both had a carbon chain length to double bond ratio of 14 : 0; 14 : 1; 16 : 0; 16 : 1; 18 : 0; 18 : 1 and a hydroxy acid of 18 : 0. The position of the double bonds was not determined. Shorter chain fatty acids appeared to be present in trace amounts only.[78]

The following phospholipids were identified in both spores and vegetative cells of *B. polymyxa*: phosphatidylethanolamine, lysophosphatidylethanolamine, lysophosphatidylserine, lysolecithin, phosphatidic acid and phosphatidylglycerol. The first was the major phospholipid in both spores and vegetative cells.[81] Quantitative estimations were not made. The phospholipids in both cell forms of *B. megaterium* were phosphatidylethanolamine, phosphatidylglycerol and phosphatidylcholine. *B. subtilis* cells and spores contained in addition to these phosphatidylinositol.[187] On the other hand, Yamakawa *et al.*[82] found a progressive decrease in the amount of phospholipid per cell of *B. megaterium* during spore formation, the free spore having only 31% that of the stationary phase cell. Kornberg and Birch (personal communication) have observed cardiolipin as the major (30%) phospholipid component in *B. megaterium* spores: it occurred to only a minor extent in vegetative cells. The dominant phospholipid in vegetative cells was phosphatidylglycerol.

B. PHOSPHORUS

Extensive membrane synthesis occurs during sporulation, and many of the layers of the spore are either formed from membranes or result from membrane activity. The phospholipid composition of membranes and the general importance of phosphorus-containing constituents in cell metabolism make an understanding of these constituents in the various parts of the spore highly desirable.

Phosphorus occurs in all fractions of the spore and constitutes up to 7% dry wt in some (Table XIV). Its distribution varies greatly in different species.[79, 83]

TABLE XIV

Phosphorus contents of Bacillus *spores and spore fractions*

Fraction	Phosphorus (% dry wt of fraction)	Ref.
Whole spores (range in 27 strains)	0·45–2·70	Table XV
Morphological		
Exosporium (*B. cereus* T)	2·1	26
Cortical muropeptide		
B. megaterium	0·03	56
B. coagulans	0·04	54
Coats (range in 6 species)	0·3–2·8	Table VI
Soluble contents (range in 4 species)	0·7–2·3	Table XI

Coat fractions (Data from Kondo and Foster[30])

	Spores	Crude coat	Lysozyme insoluble coat	Alkali and sonic resistant insoluble residue	Pronase resistant residue
B. megaterium QMB1551	2·2	3·2	4·80	6·30	7·00
B. megaterium 19213	1·1	0·42	0·43	0·48	0·40
B. megaterium 13368	2·0	0·68	0·76	0·35	0·60
B. licheniformis 9259	0·83	0·23	0·28	0·05	0·14
B. subtilis 6633	2·39	3·17	3·83	4·00	4·54
B. polymyxa 842	1·72	1·54	1·56	2·10	3·80

Chemical (Data from Table VII, Fitz-James[79])

	(% total P fraction)		(% spore dry wt)	
	Bacillus cereus	*Bacillus megaterium*	*Bacillus cereus*	*Bacillus megaterium*
Cold acid-soluble	6	4	0·05	0·07
Lipid phosphorus	7	3	0·06	0·06
Hot 5% TCA-soluble				
Total phosphorus			0·57	0·51
RNA-phosphorus	50	20	0·45	0·35
DNA-phosphorus	10	4	0·09	0·07
Labile phosphorus	9	—	0·08	—
(7 min. in N—HCl at 100°)				
TCA residue (empty coats, lysozyme-resistant)				
Total phosphorus	22	60	0·20	1·1
Labile phosphorus	14	—	0·12	—
Sum total phosphorus			0·9	1·7

The lipid and cold acid soluble phosphorus (0.2 N $HClO_4$ or 10% TCA, $0-2°$) could be properly estimated only with disrupted spores.[79] Chemical and cytochemical studies with acid-treated spores ("acid-popped") which extrude the spore protoplast confirmed that most of the RNA-P and DNA-P was located in the protoplast.[74]

Removal of nucleic acid-phosphorus with hot 5% TCA left some phosphorus-containing compounds. The larger part of this was alkali-soluble and acid-labile, and resembled polymetaphosphate. The smaller part was stable to ribonuclease and acid- and alkali-insoluble; this was associated with the coats.[79] In some strains of B. megaterium, the spores had twice the phosphorus content of B. cereus spores, and the acid and alkali-insoluble residual phosphorus formed 60% of the total phosphorus; this was only 4% of the total in B. cereus.[79] In twenty-two strains (ten species and subspecies) of Bacillus, the amount of this fraction varied 90-fold.[83]

B. megaterium spores contained 17% more stable sugar-P and 317% more residual-P per cell than did the stationary phase vegetative organisms. In contrast, spores contained only 3% of the labile sugar-P, 17% of the nucleotide and 12% of the polyphosphate[82] compared with the stationary phase cells.

Spores of B. laterosporus, which have a canoe-shaped parasporal body closely attached or continuous with the coat layers, are unusually rich in phosphorus;[32] about 50% of this occurs in the canoe-like body. The phosphorus is mostly (80%) acid-soluble inorganic orthophosphate plus phosphoprotein and residual-P or spore coat fraction.[32]

C. INORGANIC MATTER

Metals, particularly calcium, which accumulate during sporulation account for most of the high ash content of spores. Metals play important functions in formation and in the final properties of the spore. Excess of one metal affects the content of others.[20] Any treatment that affects the Ca^{2+} content usually affects DPA content and heat resistance. Similarly, but in reverse, any condition affecting DPA content will probably affect or perhaps determine the Ca^{2+} content, and affect the content of other metals and heat resistance.

Spores of most species contain $2-3\%$ Ca^{2+} and $5-15\%$ dry wt DPA in about a 1 : 1 mole ratio. The DPA content may also be varied by use of various organic inhibitors during sporulation; in many of these experiments the Ca^{2+} and other metal contents were not determined (Table XIX). However, the general instability (readiness to germinate) of spores with reduced DPA makes it unlikely that a full complement of Ca^{2+} is retained. Likewise although spores formed in the presence of penicillin[84, 85] and cycloserine [6, 57] formed near normal amounts of DPA, the DPA readily leaked from the unstable spores formed and the Ca^{2+} content presumably varied similarly.

At Ca^{2+} concentrations in the medium of less than 25 $\mu g/ml$ the Ca^{2+} and DPA contents of the spore are directly related to the Ca^{2+} availability (Fig. 4).

The above points should be kept in mind when considering the inorganic composition and Ca–DPA–status of the spore.

1. Inorganic constituents

Spores formed in media with a reasonable metal balance contain a wide range of inorganic elements.[31, 86] The major elements are Ca, K, P, Mg, Mn and occasionally Cu and Si in some crops (Table XV). Calcium is accumulated preferentially, reaching ten times the concentration of Mg even though the concentration of Mg, Na and K in the medium is many times that of calcium. The Si may result from added antifoam or glassware. In addition many minor or trace elements occur.[31] Copper and some other metals probably have functional significance, but most are probably adsorbed by the cation-exchange system of the spore. Copper and Mn^{2+} are concentrated to a greater extent in spores than in vegetative cells.[86] Aluminium, copper and iron were usually present at higher concentrations in spores than in the medium indicating accumulation.[86]

2. Variation in metal content

The metal content of spores is greatly affected by the cultural conditions and the preparation of spores for analysis. Spores, or some of the peripheral layers, behave essentially as an ion-exchange system,[87, 88] and so can lose and gain metals during harvesting and cleaning procedures. Losses may occur from initiation of germination. Metal-containing precipitates formed in the medium during growth, especially at high temperatures, are very difficult to remove unless a procedure such as the two-phase system of Sacks and Alderton[5] is used.

Acid-washing procedures are essential to remove some precipitates;[6, 20] they are unlikely to remove constituent Ca^{2+} if the pH is held above 2 and the temperature near 0°.[6, 89] Acid (0·03–0·05 N HCl) at <5° removed <10% of the Ca^{2+}, but 50–80% of the Mg^{2+} and Mn^{2+} in the first wash; storage for over 17 hr failed to remove more unless the spores had been freeze-dried.[6] The initial rapidly removable fraction was considered to be contaminating metal precipitates in the spore crops. Some of the "acid-released" Ca^{2+} and PO_4^{3-} of spores in crops grown in the presence of high concentrations of Ca^{2+} phosphate and amino acids[90] was very likely derived from inorganic precipitates in the crops; the very high Ca/DPA ratios (2–5) emphasize the possibility.

The metal content depends on the types of metals and their availability

TABLE XV

Major inorganic elements in spores
% dry wt

	Ca	K	Mg	Mn	Na	Fe	Al	Si	P	Ref.
Bacillus species[a]										
Average	2·27	0·65	0·45	0·23	0·08	0·07	0·026	0·53	1·10	⎫
Range Min	1·00	0·05	0·07	0·003	0·03	0·008	0·009	0·20	0·45	⎬ 86, 21, 91, 92, 39
Max	5·00	3·80	1·30	1·50	0·22	0·40	0·066	1·20	2·70	⎭
Clostridium bifermentans	6·4	1·6	0·25	0·01						91

[a] Average of 13 species and 14 unidentified strains.

during growth and sporulation, temperature of growth and base exchange reactions. Apparently normal spore crops were grown when the metal concentrations in the medium were varied as much as 2000-fold (Ca^{2+} and Zn^{2+}) although in the spore only 3- to 600-fold variations in metal content occurred[20] (Table XVI). The higher the metal concentration in the medium

TABLE XVI

Metal contents of spores formed in media containing, respectively, minimum and maximum concentrations of individual metals

Metal ion	Metal ion concentration in medium ($\mu g/ml$)	Ca^{2+} in medium ($\mu g/ml$)	Ca^{2+} in spores (% dry wt)	Metal ion in spores (% dry wt)
Ca^{2+}				
Min	—	0·4	0·69	—
Max	—	722	3·04	—
Zn^{2+}				
Min	0·2	1·8	3·60	0·08
Max	454	1·8	0·62	0·46
Ni^{2+}				
Min	0	1·8	3·60	<0·003
Max	11·2	1·8	0·62	1·81
Cu^{2+}				
Min	0	1·8	0·65	<0·0001
Max	0·16	1·8	0·77	0·01
Co^{2+}				
Min	0	1·8	0·65	0·04
Max	40	1·8	0·78	0·14
Mn^{2+}				
Min	0·22	1·8	0·47	0·10
Max	2·28	1·8	0·10	0·78

Data from Slepecky and Foster.[20]

the more was incorporated. A significant concentration mechanism operated for Ca^{2+}, Ni^{2+}, Zn^{2+}, and Mn^{2+}. At high concentrations, Zn^{2+}, Ni^{2+} and Mn^{2+} competed for Ca^{2+} sites in the spore.[20, 90] The increased content of the substituent metal was less than the decrease in the amount of Ca^{2+}, i.e. it was not a quantitative substitution (Table 16). Slepecky and Foster[20] suggested that the low content of Ca^{2+} in the Co^{2+}, Cu^{2+} and Mn^{2+} treatments resulted possibly from interference by other metals in the medium (Mn^{2+}, Fe^{2+}, Zn^{2+}); these metals were not assayed in these crops. The Ni^{2+} content reached 1·8% when the Ni^{2+} concentration in the medium (toxic limit) was only 1/95 that of Ca^{2+} in the calcium treatment; this probably results from stronger adsorption of Ni^{2+} and the greater stability of the Ni-DPA-chelate.

Spores grown at or allowed to develop at higher temperatures may contain more Ca^{2+}.

A small portion of the metal content of spores can be titrated away at pH4 to give the H^+-form of spores; the H^+ can then be replaced by a variety of metals.[87] This exhangeable fraction has important effects on germination in *B. megaterium*[93] and heat resistance of a number of species.[87] In *B. megaterium* titration to pH4 removed only about 5%[89] of the Ca^{2+} and probably other metals in the surface region. No DPA leakage occurred at pH4.[87] When the H^+ of *B. megaterium* spores were replaced by Ca^{2+} at 60° and 25°, the Ca^{2+}-spores prepared at 60° were found not to be dependent on electrolytes for germination and retained about twice as much Ca^{2+} as those formed at 25°. At the higher temperature extra groups reactive with Ca^{2+} are apparently exposed which in respect to H^+ exchangeability are qualitatively different or inaccessible to H^+ at pH 1·8 and 25°.[89]

The coat fraction from disrupted spores may contain substantial amounts (*c.* 0·4%) of Ca^{2+} and Ba^{2+} (from spores formed in the presence of $BaCl_2$ instead of $CaCl_2$); these are readily removable by titration to pH4.[89, 93]

3. Location of metals

Knowledge of the site of the metals, particularly calcium, in spores is of considerable importance. Spodography (microincineration for 30 min at 500–525°) using the light microscope suggests that much of the ash is located in the protoplast.[94] Thomas,[92] using electron microscopy of low temperature-ashed sections, showed that the ash pattern defined the various layers of the spore. A considerable amount of ash occurred in the protoplast region and in the two coat layers. The innermost coat (incorporated mother cell cytoplasm) and cortex were missing as a result of fixation procedures. The amount of mineral matter lost with the cortex and during thin sectioning onto water is unknown. If these losses are small then the ash patterns suggest that a major part of the mineral matter may be located in the outer two coats of *B. megaterium*. How much of this pattern is due to phosphorus is not clear. *B. cereus*, with a much lower phosphorus content, had a less dense ash pattern.[92]

V. Dipicolinic Acid and Calcium

Dipicolinic acid as a cell constituent is unique to bacterial spores; it has been isolated from spores of all *Bacillus* and *Clostridium* species analysed and also from spores of *Sporosarcina ureae*,[95] and always in about a 1 : 1

mole ratio with Ca^{2+}. As the DPA content of spores of different species ranges from about 5 to 15% dry weight and the calcium 1 to 3%, these two constituents are of considerable quantitative as well as qualitative significance, especially if they are located together in one region of the spore. The role of these two constituents has tended therefore, to dominate much of the thinking and work on spores during the last fifteen years.

A. CONTENT

The range in content mentioned above is true of spores produced on media suitable for growing good crops of spores typical and characteristic of the species (Table XVII). Figures for *Clostridium* species are not very

TABLE XVII

Heat resistance and dipicolinic acid contents of spores of Bacillus *species[a]*

Organism	D_{100}[b] (min)	DPA (% dry wt)	Ca/DPA ratio
Bacillus cereus T	0·83	9·42	0·96
Bacillus 668	0·99	8·78	0·86
Bacillus 653	1·39	6·90	0·88
Bacillus subtilis var *niger* (1)	1·67	7·35	1·07
Bacillus 652	2·00	7·14	0·95
Bacillus megaterium	2·10	8·80	0·76
Bacillus 645	2·38	9·34	0·89
Bacillus apiarius	5·00	5·06	1·02
Bacillus subtilis var *niger* (2)	6·67	9·30	1·00
Bacillus 611	8·33	6·60	1·12
Bacillus cereus var *mycoides*	10·00	7·36	1·28
Bacillus cereus	14·2	6·14	1·47
Bacillus licheniformis	24·1	6·99	1·16
Bacillus 636	35·2	7·06	1·12
Bacillus 669	43·8	11·28	0·94
Bacillus 670	68·5	5·74	3·25
Bacillus 671	81·3	9·25	1·14
Bacillus coagulans	270	10·42	0·94
Bacillus stearothermophilus (1)	459	9·77	0·86
Bacillus stearothermophilus (2)	714	13·55	0·96

[a] Data from Warth and Murrell.[6]
[b] The decimal reduction time (D_{100}) is the time required for a 90% decrease in the number of viable spores at 100°.

common but the range is similar (Table XVIII). Although, in any one species or strain, the DPA content can be varied rather widely, rather drastic conditions have to be applied to achieve this. Spores produced on the usual media normally are very similar in composition and have similar properties and levels of heat resistance.[6, 24, 96, 97] Two factors probably favour this result. First, poor or very poor crops of spores on media unsatisfactory for spore formation are not usually studied and are very difficult to analyse. Secondly, harvesting and cleaning procedures tend to lose low density spores which are probably low in calcium and DPA. Mutants of *B. cereus* T have recently been isolated which form spores without any DPA; the spores are cytologically normal and refractile but low in heat resistance and dormancy (H. Orin Halvorson, private communication).[188]

Dipicolinic acid and calcium content and Ca/DPA mole ratios of spores of
Clostridium *species*

Species	DPA (% dry wt)	Ca (% dry wt)	Ca/DPA mole ratio	Ref.
Clostridium roseum	6·2	0·62	0·42 ⎫	98
	9·0	0·92	0·44 ⎭	
	8·7			99
Clostridium bifermentans	7·6	6·4	3·50	91
Clostridium perfringens				
Heat-resistant strain	11·3		⎫	100
Heat-labile strain	9·6		⎭	
Clostridium botulinum 62A	14·0	3·0	0·89	47
Clostridium sporogenes PA 3679	8·2	2·67	1·37 ⎫	62
Clostridium PAS₂ᵃ	7·2	2·23	1·28 ⎭	

ᵃ Average of 4 spore crops.

Conditions and factors known to affect the DPA content in a single strain are listed in Table XIX. Availability of Ca^{2+} has the greatest effect, and several of the other treatments probably directly or indirectly interfere with Ca^{2+} availability, e.g. chelation by phenylalanine, precipitation by oxamic acid (from hydrolysis of ethyl oxamate). Cysteine and tetracyclines possibly interfere with Ca^{2+} adsorption sites.[103, 104]

The Ca/DPA ratio is not always close to unity. The divergence may result from partial replacement of Ca^{2+} by other metal ions, analytical factors, or unknown factors. A plot (Fig. 11 in ref. 31) of many of the published analyses suggests that there does not exist a true 1 : 1 stoichiometrical relationship between the two constituents, unless the errors from extraneous Ca^{2+} and assay methods are sufficient to obscure it. The lack of an exact relationship is not surprising if calcium has more than one role and if the specificity of Ca^{2+} is not absolute.

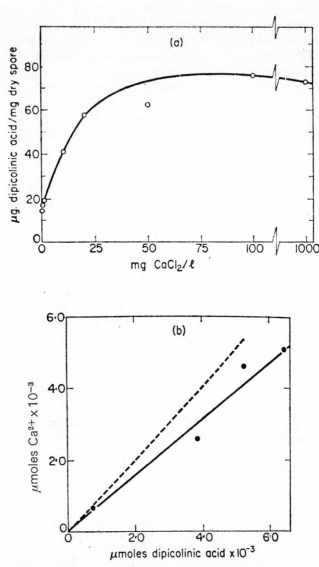

FIG. 4a. Relationship between calcium content of the medium and dipicolinic acid content of spores.

FIG. 4b. Relationship between calcium and dipicolinic acid contents of spores – – – equivolar ratio.

TABLE XIX

Factors affecting the dipicolinic acid contents of Bacillus *spores*

Factor	Species	Effect	Ref.
Nutritional			
Yeast extract (0–2 mg/ml)	*Bacillus cereus* T	Dependent on concentration	10
Pantothenic acid	*Bacillus cereus* T	Increases	
Thiamine	*Bacillus cereus* T	Increases	
Calcium (0–36 μg/ml)	*Bacillus cereus* T	Determines content (see Fig. 4a)	101
Physical			
Temperature	*Bacillus subtilis* *Bacillus coagulans* *Bacillus cereus* T	Increase with temperature to near maximum growth temperature	102, 6
Stimulatory			
Glycollic acid	*Bacillus megaterium*	84% increase	96
Glyoxylic acid	*Bacillus megaterium*	12% increase	
Inhibitory			
Cystine	*Bacillus megaterium*	Interferes with Ca^{2+} incorporation	103
Phenylalanine	*Bacillus cereus* T	Decreases	10
α-Picolinic acid	*Bacillus cereus* T	Decreases	105
Tetracyclines	*Bacillus cereus*	Inhibits Ca^{2+} incorporation	104
Ethyloxamate Ethylmalonate Ethylpyruvate Ethylpimelate Ethylsuccinate	*Bacillus cereus* T	Inhibits DPA formation indirectly	96, 105

B. SPECIFICITY OF CALCIUM

Apparently normal spores (refractile, non-staining) are formed in media with quite low concentrations of Ca^{2+} (1 μg/ml); the spores have low DPA contents and are unstable, becoming non-refractile and stainable during storage. [1,106] Slepecky and Foster[20] grew crops in media containing 1·8 μg Ca^{2+}/ml in the presence of low and high concentrations of other ions (Table XVI). At the high concentration, the Ca^{2+} content was decreased and the other ions (Zn^{2+}, Ni^{2+}, Mn^{2+}) accumulated in the spores instead of Ca^{2+}; but the replacement was not quantitative. The spores had less than 1% Ca^{2+}; their DPA content was not determined. The spores were

TABLE XX

Replacement of calcium by other metals in Bacillus spores

Organism	Replacement metal (present as the chloride; mM)	Metal content of spores (% dry wt)			Replacement metal (m atoms 100 g)	DPA (% dry wt)	Metal/DPA ratio	Ca/DPA	Heat resistance	Ref.
		Ca	Sr	Ba						
Bacillus cereus T										
Stage I[a]										
Ca²⁺	0·9					8·8			30[c]	101
Sr²⁺	0·9					7·5			30	
Ca²⁺	1·0	1·45			36·3	9·8		0·62	c. 450[d]	
Sr²⁺	1·0	0·003	2·28		26·0	8·2	0·53		c. 110	
Ba²⁺	1·0	0·002		6·2	45·1	5·0	1·60		c. 100	
Bacillus megaterium										
Stage I[a]										
Ca²⁺	1·0	1·74			43·5	10·8		0·67	c. 280	108
Sr²⁺	1·0	0·004	2·57		29·4	8·6	0·57		c. 240	
Ba²⁺	1·0	0·002		5·74	41·8	2·0	3·5		c. 20	
Ca²⁺	0·5				43·5	11·8		0·62		
Sr²⁺	0·5				34·3	10·3	0·56			
Ba²⁺	0·5				41·1	3·3	2·05			
Ca²⁺ + Sr²⁺ 0·5 ea					38·0 + 17·2	10·9	0·85	0·58		
Ca²⁺ + Ba²⁺ 0·5 ea					36·0 + 30·5	11·2	0·99	0·54		

Bacillus cereus T

		μg/spore ($\times 10^{-8}$)		
Stage II[b]				
Ca^{2+}	0.9	1·55		⎫
Sr^{2+}	0.6	1·60		⎬ 111
Ba^{2+}	0.5	0·32	0·92	⎭

Bacillus cereus[a]

		(% dry wt)			
Ca^{2+}	0	3·0	0·50	< 5[e]	⎫
	0·1	6·2	0·59	5	⎪
	1·0	9·9	0·96	60	⎪
	10·0	8·4	0·93	60	⎪
Sr^{2+}	0·1	6·1		22	⎪
Ba^{2+}	0	2·9	0·71	< 5	⎬ 112
	0·1	1·9	0·94	8	⎪
	1·0	2·2	1·00	23	⎪
	1·0	3·0	0·33	29	⎪
Mg^{2+}	0·1	3·8	0·26	5	⎪
Zn^{2+}	0·1	1·5	0·67	< 10	⎪
Co^{2+}	0·1	1·1	0·43	< 5	⎪
Ni^{2+}	0·1	2·4	0·41	23	⎭

[a] Cells grown to Stage I and then placed in the metal-containing solutions.
[b] Spores formed in the glucose-yeast extract-salts medium of Stewart and Halvorson.[113]
[c] Heat resistance as the percentage of survivors after 30 min at 80°.
[d] Heat resistance as the D value (min) calculated from the data of Foerster and Foster.[108]
[e] Heat resistance as the time (min) for 99% death at 85°.

not stable to 10 min at 60°, and so were probably low in DPA also. The high concentrations of Zn^{2+}, Ni^{2+} and Mn^{2+}, although apparently replacing Ca^{2+}, either were not as satisfactory as Ca^{2+} or were held at another site. On disruption of the spores, they were present in the extracts as the metal-DPA chelate.[107]

Studies in the replacement of Ca^{2+} by Sr^{2+} and Ba^{2+} are summarized in Table XX. Sr^{2+} approaches Ca^{2+} in functional equivalence for heat resistance and DPA formation, but Ba^{2+} and the other divalent metals are both quantitatively and qualitatively much less effective. The Sr^{2+}-spores usually had less DPA, were not as heat resistant, and germinated at a much slower rate.[108] Both Sr^{2+} and Ba^{2+} appeared to be co-accumulated with Ca^{2+}, and their uptake seemed to be independent of Ca^{2+}. They did not interfere with germination provided Ca^{2+} was present. Many of the barium-spores were non-viable. The integument fraction (18,000 g, 15 min sediment) from disrupted spores retained 60–80% of the barium, while the Ca^{2+} from calcium-spores was mostly soluble.[108] This suggests that a large part of the high Ba^{2+} content (5–6%) was bound in the coat regions, and was not replacing Ca^{2+} at the internal Ca^{2+} sites.

The evidence above indicates that Ca^{2+} ions have a very high specificity in DPA formation, heat resistance and germination (see Chapter 11).

C. Chemical State of DPA in the Spore

When DPA is isolated from spores it apparently is nearly always in the Ca–DPA chelate,[109] but sometimes as the chelate of other divalent metals[107] and perhaps as a DPA–Ca amino acid complex.[66, 110] Evidence of a precursor form such as dihydrodipicolinic acid is lacking. Perry and Foster[114] isolated and identified monoethyl-DPA in spore extracts of B. cereus var mycoides and B. megaterium, and methyl dipicolinate was isolated in trace quantities from spores of B. cereus var globigii by Hodson and Foster.[115] The ethyl ester accounted for only 1% of the DPA. Enough ethyl groups occurred in spores, however, to account for the possibility of the occurrence of DPA in the esterified form in the spore. The acid extraction method caused hydrolysis, and other mild extraction methods failed to give higher yields. Perry and Foster[114] suggested, therefore, that DPA may occur in the spore in a bound form, and that conditions causing its release cause hydrolysis of the ester linkage.

Spectrometry studies suggest that DPA occurs in the chelate form in the spore. Infra-red difference spectra between dry intact spores and germinated spores resembled those for DPA, but lacked sufficient resolution to differentiate DPA esters and chelates.[116] Electron paramagnetic resonance spectra of spores formed in a Mn^{2+}-enriched medium strongly suggested the presence of the Mn chelate in spores, while on the other hand, the Cu

(II)-spectra indicated a Cu-protein complex.[117] Ultraviolet spectra of dry spores embedded in KBr strongly suggested the presence of calcium and manganese chelates.[91] The spores, however, suffered damage during embedding, and although water was substantially absent from the spores in the operation preceding u.v. spectroscopy it is possible some rearrangement occurred. The slight differences in the u.v. spectra of spores and Ca–DPA in solution may result from the different environmental conditions, but also because the compound in the spores, e.g. a protein–DPA–Ca complex, is not the same as that in water.[91]

Young[66] found DPA in spore extracts in combination with amino acids. Chromatograms of hot- and cold-water extracts gave a DPA spot with considerable tailing. It contained DPA and six amino acids, tentatively identified as α, ε-diaminopimelate, glutamate, tyrosine, valine and isoleucine. The difference in tailing of chromatogram spots between the above combination and a synthetic mixture suggested the difference resulted from the degree of polymerization. Ca^{2+} was required for stability of the combination on paper. It was therefore suggested that DPA occurred as a DPA–amino acid or –peptide complex with calcium; hydrolysis or Ca^{2+} removal released the amino acids. The amino groups in the complex were ninhydrin-negative suggesting some protective stabilization by DPA. (Note also Ref. 110.) Lund[118] has reported a greater solubility of spore Ca–DPA than synthetic Ca–DPA also, suggesting a form other than the simple chelate.

DPA forms chelates with many divalent metals. The order of stability is $Cu > Ni > Zn > Co > Cd > Ca > Mn > Sr > Ba > Mg$, the stability constant (log K) decreasing from 10 for Cu–DPA to 4·4 for Ca–DPA to 2·4 for Mg–DPA.[62, 119] The stability constant for the Ca–DPA chelate decreases in the presence of high ammonium acetate concentrations, at high temperatures and at low pH.[119] The structure of the chelate was recently studied by Strahs and Dickerson.[120] It crystallized in a sesquihydrate and a trihydrate form. The X-ray diffraction pattern of the trihydrate indicated a dimer of planar units related by a centre of symmetry half-way between the calcium atoms (Fig. 5). They considered that the dimer could be quite stable because the Ca—O (2) bond length was the shortest in the structure. The crystal is held together by several intermolecular hydrogen bonds. Oxygen atoms 1 and 4 both accept two hydrogen bonds from water molecules, while O (3) accepts one hydrogen bond. The polymeric Ca—O chains and the extensive hydrogen bonding may contribute to spore stability.[120]

A wide variety of treatments release Ca–DPA from spores. It is doubtful whether these give information on the chemical bonding of Ca–DPA in the spore; most cause some physical or chemical change which induces permeability changes, physical disruption or chemical degradation. The following may be listed—moist heat, germination (Chapter 11) irradiation,[121, 67] sonication,[122] mechanical breakage,[66, 121] various solvents,[121] 1% thio-glycollic acid in 8M-urea with lysozyme,[123] long-chain alkylamines, and a variety of cationic surface-active compounds.[121, 124, 125] Disruption in cold water (2–4°) releases all the DPA;[66, 4] this may indicate that DPA is not peptide-linked to spore matter; on the other hand, release could result

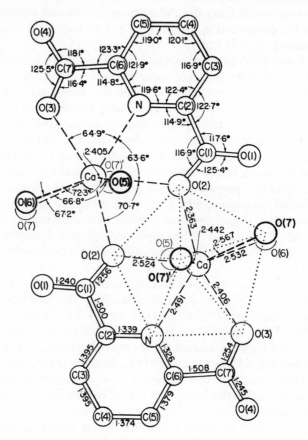

FIG. 5. Geometry, numbering, bond lengths, and bond angles in the Ca–DPA–3H$_2$O dimer. Hydrogen atoms omitted for clarity. (From Strahs and Dickerson.[120])

from mechanical germination and the activation of lytic enzymes. Vinter[84] observed that DPA was released more slowly than Ca^{2+} from spores formed in the presence of penicillin.

Electrodialysis,[121, 126] repeated freezing and thawing, many solvents at 25°, NaOH (pH 12·2)[121] and many anionic and non-ionic surface-active compounds[124] do not release DPA.

It is possible that the DPA or Ca–DPA chelate polymer is held loosely in the muropeptide polymer meshwork, perhaps as a transitory dihydro-form forming part of the meshwork, and it is released as soon as the spore structure is disturbed. This could explain its occurrence in a complex with α, ε-diaminopimelic acid under some circumstances and would indicate a location in the cortex. For instance, if the spore lytic enzymes had mura-mylamidase activity and split the side chain from the amino sugar backbone (see p. 232). On the other hand, if calcium and DPA are located in the protoplast or throughout the spore inside the coats a more general association with a gel structure would be required. The exact location of calcium and DPA are of paramount importance for a complete understanding of dormancy and the heat resistance of spores.

D. LOCATION OF DPA

There is no good direct evidence to indicate the location of DPA. Other u.v.-absorbing materials in spores prevent good definition by u.v. micro-scopy.[127] Indirect evidence is not convincing. The rapid release of Ca^{2+} and DPA under a wide variety of treatments, the cytological[42] and chemical evidence[55, 109] of the degradation of the cortex during germination, Ca^{2+} uptake and formation of DPA at the time of cortex formation (Chapter 3) have been suggested as evidence favouring the cortex location.[129, 128, 31] Hachisuka and Kuno[129] studied this possibility by examining the electron microscopic and stained appearance of spores of *B. subtilis* before and after DPA extraction by the Janssen *et al.* method.[130] The spores (73 mg) lost 11·2 mg of dry matter, of which 10·3 mg were DPA; the only obvious change was the apparent disappearance of the cortex. The cortex is difficult to resolve in intact mature spores, and is unlikely to be removed by the heat treatment;[29] perhaps the uranyl acetate staining of the thin sections removed the cortical material. It is more likely, however, that the cortex was still present, as the cortical muropeptide would account for much more than the 0·9 mg discrepancy. The change in appearance of the cortical region, moreover, does not prove that the DPA was extracted from this region. Carbol fuschin filled the electron transparent cortical area formed by the almost specific loss of DPA.

Donellan and Setlow[131] observed unidentified photo-products which were similar in hydrolysed DNA from u.v.-irradiated spores and in irradi-ated DNA previously dried in the presence of Ca-DPA or salts. As calcium dipicolinate is probably the major salt in spores, this suggests that the DPA may occur in the protoplast.

The approximate 1 : 1 mole ratio of Ca^{2+} and DPA in spores, and their close quantitative relationship during sporulation (Chapter 3) could be interpreted as indicating a molecular association and support their occur-rence in the chelate form. If these two constituents are not associated in the spore, then other cations would seem necessary to neutralize the DPA, unless the carboxyl groups of DPA or dihydro DPA are attached to amino or similar groups in spore structures. In an analysis of most of the *Bacillus* species the mole sum of all other inorganic cations determined was much less than Ca^{++31} (average <0.5; Table XV). Therefore, it would seem that much of the DPA is associated with Ca^{2+}. The location of calcium is, however, equally indefinite.

E. ROLE OF DPA

A variety of roles have been suggested for DPA.[111] These were grouped as non-specific (general or overall effect) or specific (related to a given reaction or process). These are listed in Table XXI. The reader is referred to a number of references and other parts of the book for detailed discussion of these. It is clear from the work on spore formation (in the presence of low concentrations of Ca^{++}, inhibitors (Table XIX)), relationship of biochemical and maturation changes (Chapter 3), that DPA is involved in the production of the heat resistant, resting state of spores.[31] The heat resistant state, however, is probably of secondary importance in the evolution of the spore as a "bacterial seed" for dissemination.[132] It is therefore not surprising that the DPA content is not the determining factor in the degree of heat resistance of each species[6, 31] (p. 250).

TABLE XXI

Roles in which DPA is implicated directly or indirectly[a]

Phenomenon	System	Mechanism proposed	Ref.
NON-SPECIFIC			
Heat resistance	Spores	Protein stabilization Water removal	177
Dormancy	Spores	Protein stabilization Water removal	Chap. 9
SPECIFIC			
Calcium accumulation	Stage IV cells	Chelation	111
Stimulation of electron transport	Extracts of spores and vegetative cells	Electron acceptor	161
Enzyme stability	Extracts:		161
	irradiation and		159
	heat resistance		162
Enzyme inhibition	Extracts		164
Stimulation of germination	Spores	Ionic, or change in chelation	163

[a] Modified from Halvorson and Howitt[111].

VI. Composition and Heat Resistance

The development of heat resistance and the related biochemical changes (Chapter 3) and the mechanism of heat resistance (Chapter 16) are discussed elsewhere. It is evident from these discussions that the mechanism of heat resistance is not fully understood. An adequate explanation of the chemical and biophysical basis of heat resistance needs to explain (i) the increase in heat resistance of the vegetative cell-to-spore transformation, and (ii) the large range in heat resistance of different species of spores.[31] Most bacteria form spores roughly similar in cytological composition and in DPA contents and yet the spores differ in their level of heat resistance by as much as 100,000 times. What is the basis of this difference? It is clear that there is no single chemical determinant that can explain both aspects of heat resistance. Rather the spore is a complex cytological and chemical structure whose properties, and in particular heat resistance, depend on the integrity and necessary complementation of many structural components and chemical constituents. Most of these are known to be involved both in the formation of a heat resistant spore and in inter-species differences in heat resistance. The purpose of this section is to examine the chemical characteristics of each of these in relation to heat resistance and if possible describe how they determine or affect heat resistance.

A. INTEGUMENTS

1. Coats

There is no evidence that the exosporium affects heat resistance or that its possession confers any additional heat resistance; most species possessing an exosporium are relatively low in heat resistance. Coats, on the other hand, are essential for the stability of the resting state and normal heat resistance.

Coatless mutants have been reported (Sp_{68},[133] $Osp_{iv}B$[134]) but their heat resistance was not given. However, spores have been produced without or deficient in coat layers by growth in the presence of chloramphenicol added at Stage V;[85, 128] these are refractile and resistant to mild heat treatments (10 min at 80°), but are unstable, germinating within a few days at 0–1°. Coats vary greatly in the complexity of the layers but the coats of the various species have about the same thickness. Knaysi[135] suggested that the spores with tougher coats which germinate by shedding their coats are more resistant than those wholly or partially absorbing the coats. As the majority of species shed their coats,[136,137] this relation between B. megaterium and B. subtilis cannot be general. Further, spore coats of B. stearothermophilus, the most resistant species, are more easily broken into fragments by shaking with glass beads (0·1 mm diam.) than many less resistant spores.[4] Chemical analysis of coats from five species suggest no marked differences in the amino acid composition of the coat preparations from the more heat-resistant species (Table VII). Indicative of the importance of chemical differences of the coat proteins in heat resistance is a significant relation between alkali resistance and heat resistance;[138] this is found under conditions in which the alkali does not appear to degrade the cortical muropeptide but could be disrupting disulphide and hydrogen bonds in the coats.

2. Cortex

Cortexless mutants,[139, 85] cortex-deficient spores and spores formed in the presence of penicillin[85] (a specific inhibitor of muropeptide polymer formation), indicate that a properly formed cortex is necessary for a stable resting spore with full heat resistance. Even though Ca^{2+} uptake and DPA formation during sporulation in the presence of penicillin are initially normal, they are lost during sporangial lysis and germination of the spore. Diaminopimelate incorporation occurred, but the cortical polymer was not fully polymerized as labelled diaminopimelate occurred in the diffusible peptide fraction. Less than full refractility occurred.

Assuming (i) that the cortexes of spores of all species are constituted in part of a muropeptide polymer of the type indicated in Fig. 3 as seems likely on present evidence (p. 231), (ii) that a fully formed cortex is essential for retaining Ca^{2+} and DPA, and for refractility and heat resistance, it becomes necessary to define which groups or characteristics of the cortical muropeptide may be of most importance in determining species differences in heat resistance. In a comparison of twenty spore preparations of strains of *Bacillus* with a 700-fold range in heat resistance, a highly significant relationship was found between diaminopimelate content and heat resistance, but not between hexosamine content and heat resistance.[6] Hexosamine content, however, was one factor in a multiple regression equation which predicted heat resistance with a multiple correlation coefficient of 0·96. This was because hexosamine was sufficiently well correlated with one or several of the other varieties to be included in the "best" equation.[6] These results suggest that while the amount of hexosamine backbone of the cortical polymer (*index of amount of cortex*) does not vary significantly in the spores of different species, the amount of diaminopimelate as an *index of the number of side chains* does and is significantly related to species heat resistance. As the amino group of diaminopimelate is involved in cross-linking, this could mean that the cortexes of more heat resistant spores are more cross-linked.

That the peptide side-chain is probably involved in the stability of the cortex is also suggested by experiments on spores produced in the presence of cycloserine (oxamycin; Murrell *et al.* unpublished). This antibiotic added at Stage IV (cortex formation) like penicillin interferes with cortex formation; Ca^{2+} uptake and DPA formation occur but the spores are unstable and leak Ca–DPA. Cycloserine is a competitive inhibitor of D-alanine for alanine racemase and for D-ala-D-ala ligase, which forms D-alanine-D-alanine for addition to the side-chain of the muropeptide. Its action on the spore is reversible by D-alanine. The muropeptide appears not to be fully polymerized (non-dialysable in extracts) and diaminopimelate-containing substances can be more readily extracted from maturing spores than from control spores. Electron microscopy indicates "apparently" normal cortex formation. Instability of the spore and cortex thus seems dependent on interference at the D-alanine group during peptide-side-chain formation and in the polymerization of the muropeptide units and perhaps cross-linking of polymer chains. If the situation is parallel to that in vegetative wall synthesis cycloserine inhibits transpeptidization of the penultimate D-alanine of the pentapeptide to the free amino group of the neighbouring muropeptide chain,[140] in spores most probably to the amino group of diaminopimelate.

B. Calcium and DPA

Calcium and DPA are involved in the development of the mature spore (Chapter 3) and substitution of Ca^{2+} by other metals reduces heat resistance (p. 254, Table XX). A deficiency of Ca^{2+} reduces DPA content and heat resistance[10] (Fig. 4). But does the content of these constituents determine the species level of heat resistance? Two groups of workers found no relation each with four species.[21, 141] Levinson *et al.*,[141] however, observed a relation between the Ca/DPA ratio and heat resistance, the higher the ratio the greater the resistance. In the analysis of the *Bacillus* species referred to above (Table XVII) the Ca/DPA ratio was about one for all the *Bacillus* species (except one suspected of contamination with metal precipitates) with a 700-fold range in heat resistance; Ca^{2+} content was related to heat resistance but DPA content was not. As the amount of DPA may not be expected to be related to species heat resistance (Table XVII), the significant relation of calcium to heat resistance may be deduced to result from the role of the pH4 exchangeable calcium[87-89] (p. 249). That is, the Ca^{2+} bound to DPA need not be expected to exert a significant effect on species heat resistance, but when this is supplemented by pH4 exchangeable Ca^{2+} the relation may become significant.

C. METALS

Metals acting in a number of ways have a considerable effect on heat resistance. First, a number, e.g. Mn, K, are required in spore formation having indirect effects on the composition and properties of spores. Secondly, many are able to replace Ca^{2+} under certain environmental conditions (p. 254, Table XX) reducing Ca^{2+} content and heat resistance. Thirdly, the metal composition may be changed before or during heating by base exchange,[87] the latter giving rise to the phenomenon of "heat adaptation".[88] Ca^{2+}, K^+, Na^+, Li^+, the ions giving the highest pH when added to the H^+-form of spores (p. 249) of *B. megaterium* all resulted in greatly increased heat resistance.[87] The effectiveness of these ions in increasing heat resistance is pH-dependent.[88] Ca^{2+} exchange apparently only affects a small part of the metal content of the spore (p. 249). The rate of cation uptake suggested two kinds of binding sites, different in accessibility by diffusion. An initial rapid uptake resulted in only a small increase in heat resistance, compared to that resulting from the later slow uptake; the slower rate was pH and temperature dependent. Fourthly, higher growth and sporulation temperatures for *B. subtilis* produced spores with more Ca^{2+}, Mg^{2+}, Mn^{2+} and DPA; with *B. coagulans* the metal content changed little but the DPA content was less at the higher temperatures. In both organisms, however, the cation/DPA ratio increased as the heat resistance increased.[102] Cells of *B. cereus* grown at 30° to Stage II of sporulation then placed at various temperatures to complete spore development formed spores with different Ca^{2+} and Mg^{2+} contents and heat resistance.[6] Ca^{2+} content increased as the temperature increased from 15 to 37°C but at 41°C, a super optimum temperature, the Ca^{2+} content decreased even though the DPA content was higher. Heat resistance ($D90°_2$) increased sevenfold from 4·8 min for spores grown at 15° to 36·1 min for spores grown at 30°, then decreased to 16·9 min for spores grown at 41°. Maximum heat resistance (30°- and 37°-developed spores) was associated with maximum Ca^{2+} and Mg^{2+}, high DPA and the maximum diaminopimelate/hexosamine mole ratio.[6]

The Ca/Mg ratio is an important determinant of species heat resistance. Walker *et al.*[21] observed three groups of ratios in seventeen strains of four species. The most heat-resistant group had ratios of 2·8 to 9·1, the medium resistant group 0·4 to 2·1 and the least resistant 2·3 to 5·2. They suggested that magnesium accumulation interferes in formation of the DPA–Ca complex, group 3 being an exception due to other factors. In a much larger group of *Bacillus* species a highly significant inverse relation was found between Mg^{2+} content and the Mg/Ca ratio and heat resistance.[6] This strengthens the idea that Mg^{2+}, which forms a weaker DPA-chelate, is interfering in one or more functions of calcium and that the more heat-resistant species can more effectively exclude Mg^{2+}, even though the Mg^{2+} concentration of the medium may be ten times or more greater than that of Ca^{2+}.

VII. Composition and Radiation Resistance

Spores are usually about 8–12 times more resistant to irradiation than their progenitor vegetative cells[142, 67, 143–145] and more resistant than cells of most other species. This resistance develops during spore formation in association with a number of biochemical and cytological changes (Chapter 3). The protective mechanisms developed may involve therefore cytological, biochemical or biophysical changes or all of these.

The chemical basis of radiation resistance should be considered in relation to (a) the change in degree of resistance during spore morphogenesis in a single strain and (b) the levels of radiation resistance of different species of spores. As the exact mechanism(s) of radiation resistance is not known (Chapter 16) and as the increase in radiation resistance developed during sporulation and between spores of different species is much less (over a hundredfold smaller increase) than that developed in the case of heat resistance it is more difficult to define the appropriate components or chemical constituents involved in radiation resistance. The more important of these will be considered.

(i) *Cytological components.* It is highly probable that the many membranous layers of the spore contribute to its radiation resistance. For instance, enzymes in particulate fractions are more resistant than in the non-sedimentable fraction.[146] Indirect effects of hits, electrons, or free radicals, generated in areas away from the vital protoplast of the spore are dissipated, neutralized or removed before damage to the vital areas occurs. Low voltage electrons (<200 V) have no effect on spores as they presumably do not penetrate the coat layers; 600–900 V electrons are believed to penetrate only regions of low sensitivity in the coat or cortical region of the spore, several hits being needed for inactivation[147] (presumably enzymes required in germination and outgrowth); higher energy electrons inactivate the most sensitive portion of the spore, units of large mol. wt dispersed throughout the interior of the spore.[147]

Development of radiation resistance occurs first during the engulfment and very early development of the forespore, before Ca^{2+} uptake and DPA formation[45, 143, 144, 148, 149] (Fig. 10 in chapter 3). However, a secondary increase in radiation resistance continues throughout the later stages of spore formation[143] associated with Ca^{2+} uptake, DPA formation, cortex and coat formation, therefore the additional layers may well be involved in enhancing radiation resistance. Evidence of whether spores with more complex layering, greater H-bonding and cross-linking or with differences in the chemical composition of these layers show a greater increase in resistance over that of their progenitor vegetative cells, and whether such spores are more resistant than cytologically less complex spores of other species, is lacking. Certainly, the spore coats appear to be able to act by isolating the spore protoplast from the environment.[150] Coats possibly provide some protection during the very early stages of germination (see later).

(ii) *Cystine- or disulphide-rich structures* (*CRS*). Vinter[149, 151] has shown an association between the development of radiation resistance and the increase in CRS (Fig. 10 in Chapter 3). Practically all of the increase in ^{35}S-uptake is found eventually in the spore coat fraction.[149, 151] However, the period of increase of the cystine or ^{35}S occurs before spore coat formation and therefore the ^{35}S-uptake must involve incorporation into some fraction, e.g. soluble proteins that are being assembled in the mother cell or forespore for imminent coat formation. Therefore if the CRS are involved in radiation resistance at this first stage in development of radiation resistance it is possible that the —S—S— rich soluble proteins may be acting as electron donors to replace electrons knocked out by the irradiation[143] or for non-destructive energy dissipation by the splitting of the disulphide bonds.[143]

Once the cystine or disulphide bonds are incorporated into the coats the CRS appear no longer to be involved in radiation resistance, as Hitchins et al.[152] have shown that spores are still as resistant to γ-radiation when 20–30% of the disulphide bonds are disrupted or when irradiated in the presence of reagents which block the formation of thiol groups. In a highly organized structure like the spore considerable breakage of disulphide bonds could occur without the loss of molecular ordering; many of these bonds may reform due to the fixed proximity of the broken bonds; this is probably the explanation of the reversibility of effects due to heat activation and the action of disulphide bond breaking reagents (see Chapters 10 and 11).

The complexity of the situation is indicated by the fact that during germination radiation resistance is lost (see later) with loss of Ca^{2+} and DPA, although the coats still retain their sulphur content,[151] presumably as disulphide.[151, 153]

Bott and Lundgren[46] confirmed the general relation between X-radiation resistance and disulphide content with two strains of *B. cereus* but not in an organic-S requiring auxotrophic mutant; they found, however, a most significant relation between radiation resistance of vegetative cells, sporulating cells and spores and the SS/SH ratio. The relationship was not linear, varying with the cultural conditions. The most radiation-resistant spores were formed in media with the highest concentrations of S-containing amino acids, but the disulphide content of the spores varied, indicating that other factors are involved in resting spores besides the CRS.[46]

The increase in disulphide content during sporulation and the final content (Table VIII) appears to be similar for a number of species,[45] suggesting that the disulphide content is not closely related to the radiation resistance of the different species.

(*iii*) *DPA and Ca*$^{2+}$. The amount of Ca^{2+} and DPA in spores is closely related (p. 249) and both play important roles in dormancy (Chapter 9) and heat resistance (p. 263), therefore any mechanism by which one confers radiation resistance may involve the other. There is little evidence involving Ca^{2+} in radiation resistance but the role of DPA is conflicting.

First, development of radiation resistance appears to occur before DPA formation,[143, 144] however, if there is a primary and secondary resistance mechanism[154] (see above) this would not invalidate involvement of Ca^{2+} and DPA in the secondary mechanism. Secondly, spores with low DPA contents are, however, still resistant to ionizing and u.v.-radiation[156] and were actually more resistant than normal to X- and γ-radiation.[101,155] This suggests that either only a small amount of DPA is involved in radiation resistance or it is not involved at all. Thirdly, during germination DPA and Ca^{2+} are lost rapidly before degradation of coats, and before DNA replication (Chapter 12). Radiation resistance decreases more or less at the same time,[67, 145, 157] although Stuy[145] and Irie *et al.*[158] observed that during the first few minutes of active germination the u.v.-radiation resistance was greater than that of the resting spores. At this time absorbance was decreasing at its maximum rate, and presumably the rate of loss of Ca^{2+} and DPA was near maximum. Fourthly, the Ca–DPA chelate is more resistant to radiolysis than DPA,[67] and so if it is bound to proteins and other material in the spore it may enhance radiation resistance. DPA has a radio-protective effect on enzymes[159] and vegetative cells.[160] Finally, spores with high DPA contents have multiple target inactivation curves,[157] suggesting that additional sites important in viability (e.g. enzymes involved in germination and outgrowth) are being protected by Ca^{2+} and DPA. The multiple target inactivation curves changed to the two-target form on germination, i.e. after the need for such enzymes was lost.

There is insufficient data to deduce whether a relation exists between Ca^{2+} and DPA content and species radiation resistance.[157]

(*iv*) *Enzyme resistance.* Enzymes in the intact spore are usually more resistant than when in the extracted state.[144] Little change was observed in the radio-sensitivity of aspartate aminotransferase and alanine aminotransferase during spore formation; however, aspartate aminotransferase in spore extracts was more sensitive to radiation than in vegetative cell extracts.[144] Dehydrogenase activity in intact spores was more resistant to radiation than viability.[144] The increased resistance of enzymes in spores could be the result of the structural arrangement in particulate matter or to the water content, enzymes being sometimes more resistant at lower water contents.[165, 166] In general the increased radiation resistance of spore enzymes appears to be connected with the physical organization of the resting spore rather than being due to modified enzymes.

(v) *Germination system.* This system is more resistant to radiation than the biochemical system involved in outgrowth, many spores able to germinate being unable to outgrow.[167, 168] Dry spores of *B. subtilis* irradiated with heavy charged particles cease outgrowing at various stages, e.g. the phase dark, swollen or elongated stage, or they develop into monster forms.[167] The sensitive units for the various stages appear to be inactivated independently and by the same mechanism.[167] Perhaps there may be a relation between radiation resistance (viability) and the number of enzyme molecules or enzymic steps involved in germination and the multiplicity of ways of germination.

(vi) *Reactivation and repair mechanisms.* u.v.-irradiated resting and germinated spores of *B. cereus* and spores germinated for 3 hr in adenosine, L-alanine and tyrosine were not photo-reactivatable, the spore and germinated spore apparently having too limited a metabolic apparatus to repair u.v.-damage.[169] Spores, however, show "thermorestoration",[170] i.e. greater recovery if they are heated after X· radiation; this results from a reduction in the number of free radicals.[171] The constituents of the spore and some of its layers are possibly involved in this phenomenon by combining with some of the free radicals. Studies, however, have chiefly concerned the effect of added chemicals on removal of free radicals from the spore environment after irradiation.

Negligible amounts of cyclobutane-type thymine dimers are produced during u.v.-irradiation of spores of *B. megaterium* but three unidentified photo-products were produced.[131, 172] These involved 40% of the spore thymine. The photoproducts were not substrates for the photoreactivating enzyme, but other evidence suggested that they were thymine dimers different from the cyclobutane type found in vegetative cells. The evidence indicated that the DNA in the spores was in a form different from that in vegetative cells. During germination and formation of vegetative cells 60% of the photoproducts disappeared by means of a dark-repair mechanism which differed from the dark-repair mechanism of vegetative cells. Spores formed in the absence of added Mn^{2+} were 2·32 times more sensitive than those formed in the presence of added Mn^{2+}; 2·38 times more photo-products were produced in the sensitive spores.[172]

Wake[173] recently concluded that germinating spores of *B. subtilis* have no or only a very low repair synthesis of DNA.

Similar photoproducts to the above were formed during irradiation of DNA (from *Escherichia coli*) dried in the presence of various salts including NaCl and CaDPA.[131] Such photoproducts may therefore result from the high concentrations of calcium and DPA in the spore, and if the photoproducts are important in radiation resistance then calcium and DPA contents of spores could affect radiation resistance by control of the amount and type of photoproduct.

(vii) DNA. Zamenhof et al.[174] found no evidence which suggested that differences in u.v. and X-ray resistance of cells or spores were due to differences in the radiation resistance of their DNA. Strains differing in sensitivity were cross-transformable, indicating gross chemical structural similarity of the DNA. The sensitive and more resistant strain each showed an eleven- to twelvefold increase in resistance as a result of the morphogenetic change, indicating that during sporulation the mechanism of radiation resistance amplifies the resistance of the vegetative cells allowing retention of genetic differences. Whether the degree of amplification by a method such as reduction in amount of thymine dimer formation varies from species to species remains to be shown. The dose (D_{37}) for inactivation of colony-forming ability of spores of B. subtilis is only 115 krad (^{60}Co γ-rays) and for inactivation of transforming DNA is 6300 krad; the former is restored by anoxic heating (80°, 15 min) but not the latter (perhaps due to free radical saturation at the high dose).[175] These facts indicate that the DNA in spores is probably not the main target of ionizing radiation. This is supported by calculations that 87 hits are required to inactivate spores of Clostridium botulinum with γ-rays[176] and that the target of low voltage electrons are high molecular weight units dispersed throughout the interior of the spore.[147]

REFERENCES

1. Young, I. E. and Fitz-James, P. C. (1962). J. Cell Biol. 12, 115.
2. Brown, W. L., Ordal, Z. J. and Halvorson, H. O. (1957). Appl. Microbiol. 5, 156.
3. Grecz, N., Anellis, A. and Schneider, M. D. (1962). J. Bact. 84, 552.
4. Warth, A. D., Ohye, D. F. and Murrell, W. G. (1963). J. Cell Biol. 16, 579.
5. Sacks, L. E. and Alderton, G. (1961). J. Bact. 82, 331.
6. Murrell, W. G. and Warth, A. D. (1965). In "Spores III" (L. L. Campbell and H. O. Halvorson, eds), p. 1, Am. Soc. Microbiol., Ann Arbor, Michigan, U.S.A.
7. Boyles, W. A. and Lincoln, R. E. (1958). Appl. Microbiol. 6, 327.
8. Dobiás, B. and Vinter, V. (1966). Folia microbiol. Praha 11, 314.
9. Gaudin, A. M., Mular, A. L. and O'Connor, R. F. (1960). Appl. Microbiol. 8, 84, 91.
10. Church, B. D. and Halvorson, H. (1959). Nature, Lond. 183, 124.
11. Krask, B. J. and Fulk, G. E. (1959). Archs Biochem. Biophys. 79, 86.
12. Lewis, J. C., Snell, N. S. and Alderton, G. (1965). In "Spores III" (L. L. Campbell and H. O. Halvorson, eds), p. 47. Am. Soc. Microbiol., Ann Arbor, Michigan, U.S.A.
13. Tamir, H. and Gilvarg, C. (1966). J. biol. Chem. 241, 1085.
14. Strange, R. E. and Dark, F. A. (1956). Biochem. J. 62, 459.

15. Salton, M. R. J. and Marshall, B. (1959). *J. gen. Microbiol.* **21,** 415.
16. Hunnell, J. W. and Ordal, Z. J. (1961). *In* "Spores II" (H. O. Halvorson, ed.), p. 101. Burgess Publ. Co., Minneapolis, Minn., U.S.A.
17. Kadota, H., Iijima, K. and Uchida, A. (1965). *Agric. biol. Chem.* **29,** 870.
18. Douglas, H. W. and Parker, F. (1958). *Biochem. J.* **68,** 94.
19. Guse, D. G. and Hartsell, S. E. (1959). *Fd Technol. Champaign* **13,** 474.
20. Slepecky, R. and Foster, J. W. (1959). *J. Bact.* **78,** 117.
21. Walker, H. W., Matches, J. R. and Ayres, J. C. (1961). *J. Bact.* **82,** 960.
22. Virtanen, A. I. and Pulkki, L. (1933). *Arch. Mikrobiol.* **5,** 99.
23. Powell, J. F. (1953). *Biochem. J.* **54,** 210.
24. Tinelli, R. (1955). *Annls Inst. Pasteur, Paris* **88,** 212.
25. Lawrence, N. L. and Halvorson, H. O. (1954). *J. Bact.* **67,** 585.
26. Matz, L. and Gerhardt, P. (1964). *Bact. Proc.* p. 14.
27. Gerhardt, P. and Ribi, E. (1964). *J. Bact.* **88,** 1774.
28. Bernlohr, R. W. and Sievert, C. (1962). *Biochem. biophys. Res. Commun.* **9,** 32.
29. Warth, A. D., Ohye, D. F. and Murrell, W. G. (1963). *J. Cell Biol.* **16,** 593.
30. Kondo, M. and Foster, J. W. (1967). *J. gen. Microbiol.* **47,** 257.
31. Murrell, W. G. (1967). *Adv. Microbial Physiol.* **1,** 133.
32. Fitz-James, P. C. and Young, I. E. (1958). *J. biophys. biochem. Cytol.* **4,** 639.
33. Tarr, H. L. A. (1933). *Biochem. J.* **27,** 136.
34. Pfennig, N. (1957). *Arch. Mikrobiol.* **26,** 345.
35. Dyrmont, A. (1886). *Arch. exp. Path. Pharmakol.* **21,** 309.
36. Tsuji, K. and Perkins, W. E. (1962). *J. Bact.* **84,** 81.
37. Wheeler, S. M. (1909). *J. biol. Chem.* **6,** 509.
38. Damboviceanu, A. and Barber, C. (1931). *Archs. roumaines Path. exp. Microbiol.* **4,** 5.
39. Murrell, W. G., Swaine, D. J. and Warth, A. D. (In preparation.)
40. Rode, L. J. and Williams, M. G. (1966). *J. Bact.* **92,** 1772.
41. Robinow, C. F. (1953). *J. Bact.* **66,** 300.
42. Mayall, B. H. and Robinow, C. F. (1957). *J. appl. Bact.* **20,** 333.
43. Snoke, J. E. (1964). *Biochem. biophys. Res. Commun.* **14,** 571.
44. Vinter, V. (1959). *Nature, Lond.* **183,** 998.
45. Vinter, V. (1960). *Folia microbiol., Praha* **5,** 217.
46. Bott, K. F. and Lundgren, D. G. (1964). *Radiat. Res.* **21,** 195.
47. Stewart, G. J. (1963). Thesis, West Virginia University, Morgantown.
48. Douglas, H. W. and Parker, F. (1958). *Biochem. J.* **68,** 99.
49. Douglas, H. W. and Parker, F. (1957). *Trans. Faraday Soc.* **53,** 1494.
50. Douglas, H. W. and Parker, F. (1959). *Trans. Faraday Soc.* **55,** 850.
51. Douglas, H. W. (1955). *Trans. Faraday Soc.* **51,** 146.
52. Kadota, H. and Iijima, K. (1965). *Agric. biol. Chem.* **29,** 80.
53. Walker, P. D., Baillie, A., Thomson, R. O. and Batty, I. (1966). *J. appl. Bact.* **29,** 512.
54. Warth, A. D. (1965). *Biochim. biophys. Acta* **101,** 315.
55. Powell, J. F. and Strange, R. E. (1953). *Biochem. J.* **54,** 205.
56. Strange, R. E. and Powell, J. F. (1954). *Biochem. J.* **58,** 80.
57. Murrell, W. G., unpublished.
58. Record, B. R. and Grinstead, K. H. (1954). *Biochem. J.* **58,** 85.
59. Young, F. E., Tipper, D. J. and Strominger, J. L. (1964). *J. biol. Chem.* **239,** PC 3600.
60. Ghuysen, J-M., Leyh-Bouille, M. and Dierickx, L. (1962). *Biochim. biophys. Acta* **63,** 286.

61. Bricas, E., Ghuysen, J-M., and Dezélée, P. (1967). *Biochemistry, N.Y.* **6**, 2598.
62. Riemann, H. (1963). Thesis, Copenhagen.
63. Vinter, V. (1965). *Folia microbiol., Praha* **10**, 280.
64. Black, S. H. and Arredondo, M. I. (1966). *Experientia* **22**, 77.
65. Lee, W. H. and Ordal, Z. J. (1963). *J. Bact.* **85**, 207.
66. Young, I. E. (1959). *Can. J. Microbiol.* **5**, 197.
67. Farkas, J. and Kiss, I. (1965). *Acta microbiol. Hung.* **12**, 15.
68. Murrell, W. G. (1955). "The Bacterial Endospore", Mimeo, University of Sydney.
69. Keilin, D. and Hartree, E. F. (1947). *Antonie van Leeuwenhoek* **12**, 115.
70. Grelet, N. (1957). *J. appl. Bact.* **20**, 315.
71. Slepecky, R. A. and Law, J. H. (1961). *J. Bact.* **82**, 37.
72. Church, B. D., Halvorson, H., Ramsey, D. S. and Hartman, R. S. (1956). *J. Bact.* **72**, 242.
73. Yoneda, M. and Kondo, M. (1959). *Biken's J.* **2**, 247.
74. Fitz-James, P. C., Robinow, C. F. and Bergold, G. H. (1954). *Biochim. biophys. Acta* **14**, 346.
75. Akashi, S. (1965). *Nagoya med. J.* **11**, 133.
76. Long, S. K. and Williams, O. B. (1960). *J. Bact.* **79**, 629.
77. Meisel-Mikolajczyk, F. (1965). *Bull. Acad. pol. Sci. Cl. II Ser. Sci. biol.* **13**, 7.
78. Pheil, C. G. and Ordal, Z. J. (1967). *J. Bact.* **93**, 1727.
79. Fitz-James, P. C. (1955). *Can. J. Microbiol.* **1**, 502.
80. Sugiyama, H. (1951). *J. Bact.* **62**, 81.
81. Matches, J. R., Walker, H. W. and Ayres, J. C. (1964). *J. Bact.* **87**, 16.
82. Yamakawa, T., Aida, K. and Uemura, T. (1966). *J. gen. appl. Microbiol., Tokyo* **12**, 337.
83. Fitz-James, P. C. and Young, I. E. (1959). *J. Bact.* **78**, 743.
84. Vinter, V. (1962). *Experientia* **18**, 409.
85. Fitz-James, P. C. (1963). In "Mécanismes de régulation des activités cellulaires chez les microorganismes", (J. Senez, ed.), p. 529. Gordon and Breach, New York.
86. Curran, H. R., Brunstetter, B. C. and Myers, A. T. (1943). *J. Bact.* **45**, 485.
87. Alderton, G. and Snell, N. (1963). *Biochem. biophys. Res. Commun.* **10**, 139.
88. Alderton, G., Thompson, P. A. and Snell, N. (1964). *Science, N.Y.* **143**, 141.
89. Rode, L. J. and Foster, J. W. (1966). *J. Bact.* **91**, 1589.
90. Levinson, H. S. and Hyatt, M. T. (1964). *J. Bact.* **87**, 876.
91. Bailey, G. F., Karp, S. and Sacks, L. E. (1965). *J. Bact.* **89**, 984.
92. Thomas, R. S. (1964). *J. Cell Biol.* **23**, 113.
93. Rode, L. J. and Foster, J. W. (1966). *J. Bact.* **91**, 1582.
94. Knaysi, G. (1965). *J. Bact.* **90**, 453.
95. Thompson, R. S. and Leadbetter, E. R. (1963). *Arch. Mikrobiol.* **45**, 27.
96. Benger, H. (1962). *Z. Hyg. Infectkrankh.* **148**, 318.
97. Bergère, J. L. and Hermier, J. (1964). *Annls Inst. Pasteur, Paris* **106**, 214.
98. Wooley, B. C. and Collier, R. E. (1965). *Can. J. Microbiol.* **11**, 279.
99. Byrne, A. F., Burton, T. H. and Koch, R. B. (1960). *J. Bact.* **80**, 139.
100. Weiss, K. F. and Strong, D. H. (1966). *Bact. Proc.* p. 32.
101. Black, S. H., Hashimoto, T. and Gerhardt, P. (1960). *Can. J. Microbiol.* **6**, 213.
102. Lechowich, R. V. and Ordal, Z. J. (1962). *Can. J. Microbiol.* **8**, 287.
103. Vinter, V. (1957). *Folia biol., Praha* **3**, 193.
104. Vinter, V. (1962). *Folia microbiol. Praha* **7**, 275.

105. Gollakota, K. G. and Halvorson, H. O. (1963). *J. Bact.* **85,** 1386.
106. Keynan, A., Murrell, W. G. and Halvorson, H. O. (1962). *J. Bact.* **83,** 395.
107. Slepecky, R. A. (1961). *In* "Spores II" (H. O. Halvorson, ed.), p. 171. Burgess Publishing Co., Minneapolis, Minn., U.S.A.
108. Foerster, H. F. and Foster, J. W. (1966). *J. Bact.* **91,** 1333.
109. Powell, J. F. and Strange, R. E. (1956). *Biochem. J.* **63,** 661.
110. Martin, H. H. and Foster, J. W. (1958). *Arch. Mikrobiol.* **31,** 171.
111. Halvorson, H. and Howitt, C. (1961). *In* "Spores II" (H. O. Halvorson, ed.), p. 149. Burgess Publishing Co., Minneapolis, Minn., U.S.A.
112. Pelcher, E. A., Fleming, H. P. and Ordal, Z. J. (1963). *Can. J. Microbiol.* **9,** 251.
113. Stewart, B. T. and Halvorson, H. O. (1953). *J. Bact.* **65,** 160.
114. Perry, J. J. and Foster, J. W. (1956). *J. Bact.* **72,** 295.
115. Hodson, P. H. and Foster, J. W. (1965). *J. Bact.* **90,** 1503.
116. Norris, K. P. and Greenstreet, J. E. S. (1958). *J. gen. Microbiol.* **19,** 566.
117. Windle, J. J. and Sacks, L. E. (1963). *Biochim. biophys. Acta* **66,** 173.
118. Lund, A. (1961). *Rep. Hormel Inst. Univ. Minn.,* p. 20.
119. Fleming, H. P. (1964). Thesis: University of Illinois, Urbana, Illinois. Abstract *Spore Newsletter* **1,** No. 6, p. 3.
120. Strahs, G. and Dickerson, R. E. (1965). *Spore Newsletter* **2,** 30.
121. Rode, L. J. and Foster, J. W. (1960). *J. Bact.* **79,** 650.
122. Berger, J. A. and Marr, A. G. (1960). *J. gen. Microbiol.* **22,** 147.
123. Gould, G. W. and Hitchins, A. D. (1963). *J. gen. Microbiol.* **33,** 413.
124. Rode, L. J. and Foster, J. W. (1960). *Arch. Mikrobiol.* **36,** 67.
125. Rode, L. J. and Foster, J. W. (1960). *Nature, Lond.* **188,** 1132.
126. Harper, M. K., Curran, H. R. and Pallansch, M. J. (1964). *J. Bact.* **88,** 1338.
127. Hashimoto, T. and Gerhardt, P. (1960). *J. biophys. biochem. Cytol.* **7,** 195.
128. Ryter, A. and Szulmajster, J. (1965). *Annls Inst. Pasteur, Paris* **108,** 640.
129. Hachisuka, Y. and Kuno, T. (1963). Int. Symp. Physiol. Ecol. Biochem. of Germination, Greifswald, Germany, Sept. 1963.
130. Janssen, F. W., Lund, A. J. and Anderson, L. E. (1958). *Science, N.Y.* **127,** 26.
131. Donnellan, J. E. and Setlow, R. B. (1965). *Science, N.Y.,* **149,** 308.
132. Murrell, W. G. (1961). *Symp. Soc. gen. Microbiol.* **11,** 100.
133. Ryter, A., Ionesco, H. and Schaeffer, P. (1961). *C.R. hebd. séanc. Acad. Sci., Paris* **252,** 3675.
134. Ryter, A., Schaeffer, P. and Ionesco, H. (1966). *Annls Inst. Pasteur, Paris* **110,** 305.
135. Knaysi, G. (1938). *Bot. Rev.* **4,** 83.
136. Lamanna, C. (1940). *J. Bact.* **40,** 347.
137. Gould, G. W. (1962). *J. appl. Bact.* **25,** 35.
138. Anand, J. C., Barr, J. E. and Murrell, W. G. (In preparation.)
139. Fitz-James, P. C. (1962). *Abstr. VIII Int. Cong. Microbiol.* A2.1, p. 16.
140. Park, J. T. (1966). *Symp. Soc. gen. Microbiol.* **16,** 70.
141. Levinson, H. S., Hyatt, M. T. and Moore, F. E. (1961). *Biochem. biophys. Res. Commun.* **5,** 417.
142. Duggar, B. M. and Hollaender, A. (1934). *J. Bact.* **27,** 241.
143. Vinter, V. (1962). *Folia microbiol., Praha* **7,** 115.
144. Rowley, D. B. and Newcomb, H. R. (1964). *J. Bact.* **87,** 701.
145. Stuy, J. H. (1956). *Biochim. biophys. Acta* **22,** 241.
146. Serlin, I. and Cotzias, G. G. (1957). *Radiat. Res.* **6,** 55.
147. Davis, M. (1954). *Archs. Biochem. Biophys.* **48,** 469.

148. Romig, W. R. and Wyss, O. (1957). *J. Bact.* **74**, 386.
149. Vinter, V. (1961). *Nature, Lond.* **189**, 589.
150. Hill, E. C. and Phillips, G. O. (1959). *J. appl. Bact.* **28**, 8.
151. Vinter, V. (1961). *In* "Spores II" (H. O. Halvorson, ed.), *p.* 127. Burgess Publ. Co., Minneapolis, Minn., U.S.A.
152. Hitchins, A. D., King, W. L. and Gould, G. W. (1966). *J. appl. Bact.* **29**, 505.
153. Mortenson, L. E. and Beinert, H. (1953). *J. Bact.* **66**, 101.
154. Vinter, V. and Vechet, B. (1964). *Folia microbiol., Praha* **9**, 238.
155. Vinter, V. and Vechet, B. (1964). *Folia microbiol., Praha* **9**, 352.
156. Woese, C. (1958). *Abstr. Biophys. Soc. Meeting*, R7, p. 42.
157. Woese, C. (1959). *J. Bact.* **77**, 38.
158. Irie, R., Yano, N., Morichi, T. and Kembo, H. (1965). *Biochem. biophys. Res. Commun.* **20**, 389.
159. Braams, R. (1960). *Radiat. Res.* **12**, 113.
160. Leif, W. R. and Herbert, J. E. (1960). *Am. J. Hyg.* **71**, 285.
161. Doi, R. H. and Halvorson, H. (1961). *J. Bact.* **81**, 642.
162. Hachisuka, Y., Tochikubo, K., Yokoi, Y. and Murachi, T. (1967). *J. Biochem., Tokyo* **61**, 659.
163. Riemann, H. and Ordal, Z. J. (1961). *Science, N.Y.* **133**, 1703.
164. Hachisuka, Y., Tochikubo, K. and Murachi, T. (1965). *Nature, Lond.* **207**, 220.
165. Braams, R., Hutchinson, F. and Ray, D. (1958). *Nature, Lond.* **182**, 1506.
166. Hutchinson, F., Preston, A. and Vogel, B. (1957). *Radiat. Res.* **7**, 465.
167. Donnellan, J. E. and Morowitz, H. J. (1960). *Radiat. Res.* **12**, 67.
168. Costilow, R. N. (1962). *J. Bact.* **84**, 1268.
169. Stuy, J. H. (1956). *Biochim. biophys. Acta* **22**, 238.
170. Webb, R. B., Powers, E. L. and Ehret, C. F. (1960). *Radiat. Res.* **12**, 682.
171. Powers, E. L., Ehret, C. F. and Smaller, B. (1961). *In* "Free Radicals in Biological Systems". Academic Press Inc., New York.
172. Donnellan, J. E. and Stafford, R. S. (1968). *Biophys. J.* (In press.)
173. Wake, R. G. (1967). *J. molec. Biol.* **25**, 217.
174. Zamenhof, S., Bursztyn, H., Ramachandra Reddy, T. K. and Zamenhof, P. J. (1965). *J. Bact.* **90**, 108.
175. Tanooka, H. and Hutchinson, F. (1965). *Biochim. biophys. Acta* **95**, 690.
176. Grecz, N. (1965). *J. appl. Bact.* **28**, 17.
177. Mishiro, Y. and Ochi, M. (1966). *Nature, Lond.* **211**, 1190.
178. Ohye, D. F. and Murrell, W. G. (1968). *J. Bact.* submitted.
179. Fields, M. L. and Rotman, Y. (1968). *Appl. Microbiol.* **16**, 960.
180. Aronson, A. I. and Fitz-James, P. C. (1968). *J. molec. Biol.* **33**, 199.
181. Blankenship, L. C. and Pallansch, M. J. (1966). *J. Bact.* **92**, 1615.
182. Warth, A. D. and Strominger, J. L. (1968). *Bact. Proc.* p. 64.
183. Chin, T., Younger, J. and Glazer, L. (1968). *J. Bact.* **95**, 2044.
184. Warth, A. D. (1968). Thesis, University of Wisconsin, Madison.
185. Kadota, H. and Uchida, A. (1968). *Agric. Biol. Chem.* **32**, 759.
186. Nelson, D., Spudich, J., Bonsen, P., Bertsch, L. and Kornberg, A. (1969). *In* "Spores IV" (L. L. Campbell, ed.), Am. Soc. Microbiol., Ann Arbor, Michigan, U.S.A. (In press.)
187. Mastroeni, P., Contadini, V. and Teti, D. (1968). *J. Bact.* **95**, 1961.
188. Wise, J., Swanson, A and Halvorson, H. O. (1967). *J. Bact.* **94**, 2075.

M

CHAPTER 8

Spore Enzymes

H. L. SADOFF

Department of Microbiology and Public Health, Michigan State University, East Lansing, Michigan, U.S.A.

I. Introduction

MICROBIOLOGISTS have been aware of the existence of bacterial endospores since they were described by Cohn approximately 100 years ago. Although metabolic dormancy is one of the chief spore attributes, a more striking but perhaps more variable property is their thermal stability. It was the effort to understand and control this property in relation to food processing which was probably the initiating force responsible for research concerning bacterial spores over the past 35 years. Current efforts tend to be directed toward the study of cellular differentiation using sporulation as a model. Genetic and biochemical techniques are being successfully used to probe the nature of controls, the sequence of events in sporulation and the properties of homologous proteins in cells and spores. At the present "state of the art", such approaches would be exceedingly difficult or impossible to apply to differentiating tissues of higher life forms. In this chapter an attempt will be made to present an historical development and rationale for the use of enzymes extracted from cells and spores as models for investigating various aspects of sporulation.

The dilemma facing the worker studying the heat resistance of spores was the development of a suitable experimental system. The proper tool was thought to be an active enzyme associated with dormant spores which could then be used to determine the effects of heat and other denaturants on "native" spore and vegetative cell proteins. Hopefully, any differences noted could then be related to a unique structure or other properties of the proteins. Thus, a search for spore enzymes began. This search was important in that it brought a variety of approaches to the study of sporulation in bacteria and opened new areas in which to probe structure-function relationships, metabolic dormancy, developmental biology, and cellular control systems. It was successful to the extent that active enzymes were found in otherwise dormant spores. However, to date we still have little or no control over the thermal stability of spores.

A. Enzymes Associated With Intact Spores

The resting cell technique originally employed for the study of spore metabolism was unsuccessful because of the very fact of the spores' dormancy. Virtanen and Pulkii[1] concluded that spores had no catalase or polypeptidase and carried out no glucose fermentation. It was only after techniques were developed to break dormancy in the absence of germination, or to prepare spore-free extracts, that the full range of the spores' enzymatic capability was discovered.

Many early investigators including Ruehle[2] and Cook[3] reported evidence for a variety of enzymatic activities for spores, but these were probably the result of vegetative cell enzymes in cell debris or adsorbed protein which had not been removed by adequate washing of their spore preparations. Holst and Sturtevant[4] demonstrated how such enzyme carryover occurred and how it could be eliminated by sub-lethal heating of spore suspensions. A partial explanation for the low oxygen consumption by spores was provided by Keilin and Hartree,[5] who measured cytochromes at very low temperatures and pointed out that the level of these electron carriers in several species of *Bacillus* was only 5% that of the "parent" vegetative cells. Crook[6] measured the Q_{O_2} of clean *Bacillus anthracis* and *B. subtilis* spores using a microrespirometer. The low oxygen uptake was probably due to the activity of a diaphorase subsequently observed by Spencer and Powell.[7]

1. Alanine racemase

The researches of Hills[8, 9, 10, 11] and Powell[12, 13] on the germination of *Bacillus* spores provided the necessary leads pointing to the existence of active spore enzymes. Hills showed how L-amino acids, particularly L-alanine, were involved in spore germination and that the active accessory germinating component in yeast extract was adenosine. Both investigators provided evidence that D-amino acids antagonized the effects of the L-isomers. The specificity of the germinating agents and the rapidity of the germination event indicated some enzymatic function. Thus, when Stewart and Halvorson[14] noted that the germination solutions for spores of *B. terminalis* (*B. cereus*, strain T) became depleted and actually inhibited the process, they were prepared to look for the possibility of amino acid racemization and found an active alanine racemase in otherwise dormant spores. They observed that the enzyme in *B. cereus* T was similar to the racemase of *Streptococcus faecalis*[15] and that its content in spores was approximately 16 times as great as in vegetative cells. With keen insight they commented on the function of a germination enzyme which generated a product inhibitory to the process and speculated on the possibility that "this negative feed-back type of regulation of spore germination is an effective survival mechanism, ensuring that the supply of viable spores will not all be used up by a single set of favorable circumstances."[14]

Stewart and Halvorson[16] compared the properties of the spore and vegetative cell enzymes. They noted that the vegetative cell racemase was heat labile but that the spore protein was stable with a half-life of 14 hr at 80°C. Extracts of spores which had been prepared by sonication also contained heat stable alanine racemase with a half-life of 8·5 hr at 80°C. It is significant that this activity was associated with particles which could be precipitated with ammonium sulfate or sedimented in an ultracentrifuge. Furthermore, the precipitates could be resolved into labile enzyme and nonactive particles by prolonged sonication. These studies and those of Militzer et al.[17] provided the experimental basis for their hypothesis that spore proteins are stabilized by their organization into aggregates. That is to say, the effects of thermal agitation on the denaturation of protein could be overcome by incorporating the labile molecules into a rigid matrix.

2. Other enzymes

The catalase content of intact spores and extracts of *B. subtilis* was studied by Murrell[18] and reinvestigated in *B. cereus* T by Lawrence and Halvorson.[19] The catalase in intact spores of *B. subtilis* had a half-life of 30 min at 100°C and was completely resistant to 0·02 M azide and 5×10^{-4} M cyanide. When spore extracts were tested under the same conditions, the catalase was no more stable than the vegetative cell enzyme. These findings led Murrell[20] to conclude that the spore catalase was vegetative enzyme "bound into the structural network of the intact spore in some way imbued with heat resistance, or protected by the spore coat". The catalase of *B. cereus* T appeared to be particle-bound and to exist in a heat labile and stable form similar to that of alanine racemase.

The role of germinating agents was studied further by Lawrence[21, 22] and Powell and Hunter[23] who demonstrated that spores of a number of species of *Bacillus* would cleave adenosine and other ribosides to produce free base and sugar and the specificity of the hydrolytic enzyme varied with the strain of organism. The spores of *B. cereus* T degraded adenosine, guanosine, inosine, xanthosine, adenylic acid, cytidine, and uridine,[21] but had no effect on deoxyribosides.[24] Powell and Hunter[23] found that other strains of *B. cereus* and *B. anthracis* hydrolyzed ribosides at variable rates, and that inosine was more effective than adenosine in promoting the germination of their spore strains. The proof that adenosine in germination systems was the precursor of the active germinating agent was obtained by spectrophotometric identification of inosine, the reaction product, and by the quantitation of the liberated ammonia. The nucleoside ribosidase and the adenosine deaminase were both heat stable in intact spores and labile when solubilized. The ribosidase was similar in behaviour to the racemase in that spore-free extracts were particulate and heat stable.[25]

3. Stabilization of enzymes in spores

The results of research reported thus far had established that the enzymatically active proteins associated with dormant spores were stabilized in some manner by the spore structure and were not intrinsically heat stable molecules. The binding of proteins to particles has already been mentioned. Waldhalm and Halvorson[26] suggested that stabilization was due to bound protein since spores had less bound water than vegetative cells. The function of the spore integuments in contributing to heat resistance was unknown, but the possibility that the spore coats were involved was suggested by Murrell.[20] Curran *et al.*[27] noted that spores contained high levels of calcium, and the discovery of dipicolinic acid (pyridine-2, 6-dicarboxylic acid, DPA) as a unique spore component[28] led to its implication in a mechanism of heat resistance. Most models of DPA function were based on its chelating capacity for divalent metals,[29] calcium in particular, but little attention was given to how such chelates might function at the molecular level. The calcium chelate of DPA was shown by Riemann and Ordal[30] to be an effective germinating agent for spores of Bacillaceae.

B. Rationale for Embarking on a Study of Spore Enzymes

1. As models for the study of heat resistance

Proteins as a class are among the most heat labile molecules of cells, and protein denaturation is the probable cause of cell death at high temperature. Bacterial endospores are highly resistant cell forms which are characterized both by their resistance to adverse environments and by the rapidity with which they lose this property on germination. If spores are able to maintain their viability after being subjected to temperatures which are lethal for normal cells, those spore proteins essential for viability must be able to resist thermal denaturation. Therefore, the principal problem in understanding the mechanism(s) of resistance is to deduce how spore protein is protected from the deleterious effects of heat. Whatever the means of stabilization, protein denaturation is an event which occurs at the molecular level. The factors affecting its rate should be amenable to study with isolated proteins. The principal advantage of this approach is the experimental separation of the heat-resistance property from spore viability. The use of enzymes as models permits the direct observation of the effects of modifications in the ionic environments and solvent systems on the protein's activity and heat resistance. Although a valid model of the spore need not be as heat stable *in vitro* as *in vivo*, the protein must be capable of existing in either the labile or stable form. An intrinsically stable protein such as ribonuclease does not necessarily mirror the state of proteins in spores. In effect, the model must be capable of *in vitro* germination.

Although model systems may yield valuable information, the results from such studies must be interpreted with caution. Each particular model may reveal only a fraction of biological reality. Curran *et al.*[27] were aware of this limitation and made the following guarded statement with regard to the relationship between calcium content and the heat stability of spores: "It seems unlikely that any single factor is responsible for the enhanced resistance of spores: rather it would appear that many interrelated and complex factors contribute to this end."

2. *Approaches to differentiation*

It is now known that bacterial spores and their corresponding vegetative cells share many enzymatic abilities. However, it is still uncertain whether these activities are due to *identical* proteins or to structurally and functionally very *similar* proteins in cells and spores. This information is crucial to the understanding of spore morphogenesis and heat resistance and can be obtained by the study of physical, chemical, and functional properties of those isolated enzymes which occur in both cell forms. This may be the most productive approach to the understanding of differentiation in the Bacillaceae.

II. Comparison of Vegetative Cell and Spore Enzymes

A. ENZYMES PRESENT IN SPORES

The diversity of the enzyme complement of spores has been shown by biochemical studies following the extended, sublethal heating of spores in the presence of an incomplete germination system (alanine or adenosine for many *Bacillus* species) in order to interrupt dormancy. Utilizing this procedure plus improved techniques of spore rupture, Halvorson and Church[31, 32] and Doi et al.[33] were able to demonstrate the existence of a complete hexose monophosphate pathway of glucose oxidation in spores of B. cereus T. Goldman and Blumenthal[34] and Blumenthal[35] found both glycolytic and oxidative pathways present in germinating spores of this organism. These same groups of enzymes had already been detected in cells of Bacillus species,[36, 37] and their presence was reconfirmed in vegetative cells of B. cereus T.[38] Glucose oxidation was observed in spores of B. megaterium which had been rendered nonviable by heat or ionizing radiation.[39] Costilow[40] reported both glucose and amino acid fermentations in spores of Clostridium botulinum 62-A which had been irradiated and germinated in the presence of chloramphenicol. Simmons and Costilow[41] measured the levels of enzymes of the Embden-Meyerhof pathway as well as coenzyme A transferase, acetokinase, and phosphotransacetylase in cells and spores of Cl. botulinum. They also noted a soluble, heat-resistant, reduced nicotinamide adenine dinucleotide oxidase (NADH oxidase) which was subsequently studied in detail by Green and Sadoff.[42] Acetokinase and phosphotransacetylase had been reported in spores of B. cereus T by Krask and Fulk[43] and in its vegetative cells by Hanson et al.[44]

Spores of B. cereus contain alanine dehydrogenase[45] which deaminates the amino acid to pyruvate even though the equilibrium favors alanine production. The keto acid is subsequently metabolized during germination and outgrowth. The reverse reaction occurs in vegetative cells where alanine synthesis is presumed to be catalyzed by the same enzyme.[46] This enzyme was extracted from cells of B. subtilis by Yoshida and Freese[47, 48] and crystallized. Its physical properties were similar to the alanine dehydrogenase of B. cereus.

A partial description now emerges of the role of adenosine and/or inosine in the spore germination of some Bacillus species. Krask and Fulk[49] reported a nucleoside phosphorylase in extracts of spores of B. cereus which was specific for adenosine or inosine and yielded ribose-1-phosphate in the presence of inorganic phosphate. They also identified a phosphoribomutase and a magnesium dependent, adenosine triphosphate specific ribokinase which yielded ribose-5-phosphate; these enzymes functioned during germination to synthesize phosphate esters with the minimal expenditure of energy. Gardner and Kornberg[50] note that the nucleoside-phosphorylase from spores of B. cereus is indistinguishable from that of the vegetative cells.

Murrell[18] must be credited with the preparation of the first spore-free enzyme, inorganic pyrophosphatase from B. subtilis. It is noteworthy that Tono and Kornberg[51, 52] recently studied this enzyme from cells and spores of B. megaterium and found that in this case, too, the proteins were indistinguishable. Detailed studies of deoxyribonucleic acid polymerase by Falaschi and Kornberg[53] indicate that the spore and cell enzymes may be identical.

These citations are not intended as an extensive recapitulation of the literature but merely to provide an historical approach and to indicate the broad spectrum of enzymes in spores. Enzymes of carbohydrate, nucleic acid and lipid metabolism have already been mentioned. One notable difference is the low level in spores of heme proteins associated with electron transport. Reviews of spore metabolism have been made by Murrell,[20] Halvorson,[54] and more recently by Halvorson,[55] Murrell[56] and Kornberg.[57]

These authors point out that more than 60 enzymes are now known to be common to cells and spores. It is therefore unlikely that all spore proteins are uniquely different from their vegetative cell homologues. This would be biologically inefficient and would require inordinately large amounts of non-functional DNA to be carried by both cell forms. A more likely and conservative hypothesis is that a single genome directs the synthesis of a given vegetative cell and spore protein. The modifications necessary to impart the potential for heat resistance may occur following synthesis of the specific protein in the developing spore.

B. FUNCTION OF SPECIFIC ENZYMES

1. Sporangial enzymes

Many proteins which first appear or increase significantly with the onset of sporulation may be specific for that process. Protease,[58, 59, 60] heat-resistant catalase,[61] glucose dehydrogenase[62] and adenosine deaminase[23] are enzymes in this category and appear in cultures soon after growth ceases. Other enzymes are those of poly-β-hydroxybutyrate synthesis and metabolism,[63] tricarboxylic acid cycle enzymes,[64, 65] enzymes for the metabolism of acetoin,[63] DPA synthesis,[66] transaminases,[67, 68, 69] enzymes cleaving phosphate esters[70, 71] and purine nucleoside phosphorylase.[50] Their synthesis may result from derepression due to changes in the medium composition or, alternatively, they may play some specific role in sporulation. Early in the sporulation process the bacterial cell becomes compartmentalized into the pre-spore and pre-sporangial structures.[72] The "vegetative cell" portion continues to function metabolically and synthetically, thus contributing to the synthesis of the developing spore. The enzyme complement which is external to the developing spore is probably not incorporated per se, but is degraded to form an amino acid pool or is released into the medium along with the protease.[59, 60] Genetic[58] and kinetic data[73, 74, 75, 76] implicate protease in the extensive protein turnover which is associated with sporulation. The nature of the participation of other sporangial enzymes (those occurring during sporulation but not found in mature spores) in sporulation is unknown but their role can be inferred.

They must provide structural components, energy, and ultimately substrate from which the developing spore is synthesized.

2. Spore enzymes

Sporulation-specific proteins which are found in mature spores have been extensively studied to detect structural properties which could be related to the mechanisms of their heat resistance or dormancy. These enzymes are stable in the intact spore. Little is known of their function in the sporulation process, although the activities of adenosine deaminase and ribosidase in germination have been pointed out. While studying the role of glucose dehydrogenase in sporulation, Sadoff[77] noted that cultures of B. cereus grown in media containing 0·2% sodium gluconate formed refractile spores sooner than in media without gluconate. The sporulating cells in gluconate contained dipicolinic acid, were heat labile, but acquired heat stability at the same time that refractile, heat-resistant spores were formed in the control cultures. 50% of the gluconate added to the medium was metabolized or incorporated during growth and sporulation (Sadoff and McKay, unpublished). Its distribution in the spore fractions, shown in Table I, appeared quite general.

TABLE I

The distribution of radioactive carbon in spores of B. cereus grown in G-medium which had been supplemented with uniformly labelled ^{14}C-gluconate

Fraction[a]	Counts/min	% of total
Spores	5240	100
Acid solubles	50	1
Phospholipid	1230	23
Nucleic acids	1470	28
Protein	1920	37
Mucopeptide	570	11
Autoclaved spores	4170	80

[a] Fractionation procedure of Park and Hancock.[78]

C. TECHNIQUES FOR COMPARING CELL AND SPORE COMPONENTS

1. Immunological

Sporulation is a well documented example of unicellular differentiation, and, considering the extent of the morphological changes involved, it is to be expected that some spore-specific molecular species would exist. On the other hand, even a limited expression of the DNA in the pre-spore compartment of the cell would lead to the synthesis of macromolecules whose immune response would be identical to those of vegetative cells. These *post facto* generalizations are supported by the results of immunochemical investigations of vegetative cell and spore antigens over the past 65 years. These studies have been directed toward an understanding of the differ- ences between spore and vegetative cell antigens and have been somewhat dominated by a search for uniqueness in spore components which could be related to the properties of spores.

The advantages of serological procedures lie in their extreme specificity and sensitivity. Relatively minor molecular structural differences in anti- gens have a measurable effect on the extent of the immunochemical reaction but similarities in antigens can also be detected.

The two-dimensional immunodiffusion procedure of Ouchterlony,[79] especially when coupled to electrophoresis, constitutes a keen tool for studying the relationships between structures of cell and spore antigens. The chemical nature of the antigens in question can often be determined by staining the precipitates, utilizing procedures which are specific for proteins, carbohydrates or conjugated proteins. Many antibody-enzyme precipitates maintain their activity and, by judicious coupling procedures, the end- product of catalysis can be used to localize the precipitation band of specific enzymes in the diffusion matrices. Sadoff[61] utilized immunodiffu- sion in cellulose acetate films in an effort to demonstrate the identity of spore catalase of *B. cereus* with that occurring in cells committed to sporu- lation. His results were confirmed and extended by Baillie and Norris[80] and Norris and Baillie,[81] who separated vegetative cell and spore catalases of *B. cereus* by immunoelectrophoresis in agar. The enzyme bands were located in the dried gels by flooding with 10% hydrogen peroxide. A similar approach was used to establish that glucose dehydrogenase was an "early" enzyme in the sporulation of *B. cereus* and that the enzyme persisted in germinated spores. The enzyme-antibody precipitate was located by over- laying the diffusion matrix with glucose and NAD and locating the fluorescent NADH-enzyme complex under u.v. light.

The Cohn and Torriani technique[82] is a convenient variation of the quantitative precipitation procedure and was devised for use with impure enzymes as antigens and polyvalent antisera. Using this procedure, Green and Sadoff[42] showed that the NADH oxidases of spores and cells of *Cl. bot- ulinum* reacted only with homologous antisera and were, indeed, different molecular species. On the other hand, the fructose 1, 6-diphosphate aldo- lases from cells and spores of *B. cereus* were completely cross reacting.[83] They were identical in kinetic properties and identical in reactions with a variety of inhibitors but had different electrophoretic mobilities and mole- cular size.

Modern developments in electronmicroscopic cytochemistry make possible the localization of specific cell antigens. Thomson *et al.*[84] and Walker *et al.*[85] have used the ferritin labelling procedures of Singer[86, 87] to produce electron dense stains from antisera directed against vegetative cells and spores of *B. cereus*. They have shown by the electron microscopy of stained thin sections that cell antigens exist along the core and cortical membranes in spores. Their data reaffirm the cross reactivity of some cell and spore antigens and indicate the possible sites of enzyme activity in spores. Undoubtedly, current advances in the development of specific stains for electron microscopy should make possible, in the future, the visualization of sporogenesis using enzymes or other macromolecules as markers.

2. *Heat resistance*

The principal difference between vegetative cell and spore proteins is the extreme thermal stability of the latter *in situ*. The elucidation of the basis of heat resistance of spores depends in part on resolving the mode of transformation of proteins from the labile to the stable state. When the proteins are heated in aqueous or non-aqueous media or subjected to extremes of hydrogen ion concentration, they become denatured. Their configuration changes and, hence, one or more of their characteristic properties change. The most sensitive, and thus the most easily measured modification, is that of catalytic capability of the protein. Thus, the loss of enzymatic activity has been most widely used as a measure of the thermal denaturation of spore proteins.

The denaturation may be reversible, in which case the activity of the protein is recovered on cooling or on incubation with the specific substrate. In the case of irreversible denaturation, the proteins are thought to assume a more randomly coiled structure than exists for the globular protein.

If the rates of denaturation are slow enough, the kinetics of the process can be followed in an effort to understand to what extent the intramolecular forces which stabilize the protein have been altered. These studies can only reveal the overall effects of certain factors (hydrogen ion concentration, temperature, ionic strrength, or ionic species) on the rate of inactivation. Little can be learned of the fundamental mechanism of inactivation because, as Tanford[88] points out, neither the initial nor the final state is clearly characterized.

The bonds which are generally conceded to stabilize the structure of globular proteins are probably operative in maintaining the heat resistance of spore proteins *insitu*. Peptide hydrogen bonds of the proteins' secondary structure must be involved to some extent because of the general tendency of polypeptides to form helical structures. The amount of helicity in aqueous systems ranges from approximately 10% in B-lactoglobulin[89] to approximately 65% in hemoglobin.[90] Hydrophobic bonding[88, 91, 92] is thought to be one of the principal forces which hold proteins in their globular configurations. They are formed by the contact of nonpolar side chains with water, a condition which then produces an unfavorable arrangement of water molecules. The globular forms of proteins have few such contacts. The hydrophobic forces, are, in effect, a measure of the energy required to transport the nonpolar or aromatic amino acid side chains from the anhydrous interior to the "wetted" exterior of the protein. Such displacement may occur in the initial stages of protein denaturation leading to structural changes. The free energy change for the transfer of appropriate side chains approaches 1000 cal/mole.[91, 92] Spores are known to contain cysteine-rich components[94] which could participate in protein stabilization by S—S bond formation. The S—S bonds function to restrict the protein structure and perhaps permit optimal conditions for other kinds of bonding.

Sidechain hydrogen bonding has been invoked as a means of maintaining protein structure as have intramolecular ion pairs. Some difference of opinion exists as to the magnitude of the contribution of these forces, and the reader is referred to works of Tanford,[88] Joly[91] and Scheraga[95] for their respective discussions.

The thermal denaturation of proteins can be viewed as an activation process and treated by the Eyring theory of absolute reaction rates.[96] In this treatment of inactivation data, the native protein is assumed to be held somewhat rigidly and to denature rapidly when the stabilizing bonds are broken. At any time in the denaturation process there exists a concentration of activated, unstable molecules, C, and

$$- \frac{dC_{\text{native protein}}}{dt} = \frac{kT}{h} C$$

where $C_{\text{native protein}}$ is the concentration of native protein, k is the Boltzmann constant, T the absolute temperature and h is Planck's constant. The thermal inactivation of proteins approximates a first-order reaction, and the measured rate constant k' in units of sec^{-1} is

$$k' = \frac{kT}{h} e^{-\frac{\Delta F^{\ddagger}}{RT}}$$

or

$$\Delta F^{\ddagger} = RT \ln \frac{kT}{k'h}$$

where $\Delta F^{\ddagger}$ is the free energy change due to the formation of the activated complex and R is the gas constant. The enthalpy change, $\Delta H^{\ddagger}$, due to formation of the activated complex is

$$\Delta H^{\ddagger} = E - RT$$

where E is the activation energy derived from an Arrhenius plot of the effect of temperature on the denaturation rate. The entropy change, $\Delta S^{\ddagger}$, can be calculated by the equation

$$\Delta F^{\ddagger} = \Delta H^{\ddagger} - T\Delta S^{\ddagger}.$$

The $\Delta F^{\ddagger}$ for thermal denaturation of almost all proteins lies in the range 25 ± 5 kcal/mole. Joly[91] concludes that those protein inactivations with small $\Delta H^{\ddagger}$ involve the rupture of a few strong and co-operative bonds and those with large $\Delta H^{\ddagger}$ are characteristic of the rupture of a large number of weak, nonco-operative bonds.

Table II contains thermodynamic parameters which have been calculated for the denaturation of a variety of spore and cell enzymes. Upon examination it is seen that the free energies of the activation process involved in denaturation are all characterized by an energy change of approximately 25 kcal/mole. Despite a 2500-fold difference in heat resistance between the labile and stable catalases of *B. cereus*, relatively little difference exists between the thermodynamic constants for the denaturation process. Both proteins appear to be stabilized by a multiplicity of weak bonds. On the other hand, the stability of glucose dehydrogenase increases 240-fold over the pH range 7·5–6·5 and appears to be the result of a few strong bonds. The calcium-free vegetative cell and spore aldolase of *B. cereus* have identical thermal stabilities as do the purine nucleoside phosphorylases. The aldolases differ in their ability to bind calcium which stabilizes the spore protein and accounts for a 32-fold difference in heat resistance at pH 6·4. There are three forms of alanine dehydrogenase which occur in *B. cereus*. One of the cellular forms has thermal stability of approximately 50% that of the spore enzyme. The high enthalpies of activation may indicate that both the stable cell enzyme and the spore dehydrogenase are stabilized by many weak bonds.

3. Catalytic properties

Basically, sporulation in the Bacillaceae can be said to consist of the encapsulation within the vegetative cell of one complete genomic unit and ancillary proteins by means of a series of spore coats.[56] Among the proteins which can be then extracted from the resulting mature spores are a variety of enzymes with activities homologous to those in cells. Recent studies have shown that these enzymes are indistinguishable from vegetative cell proteins by several physical and functional criteria. We must investigate further the physico-chemical nature of spore proteins in order to formulate rational theories for the mechanism of heat resistance and metabolic dormancy.

Enzymes of similar catalytic properties from cells and spores may be compared on the basis of their pH optima and their interactions with specific substrates and inhibitors. The usual approach in the latter case is to study the change in reaction velocity with substrate concentration for a

TABLE II

Thermodynamic constants for the heat denaturation of a variety of cell and spore enzymes

Enzyme	Temp °C	pH	Velocity constant sec⁻¹	$\Delta F^\ddagger$ kcal/mole	$\Delta H^\ddagger$ kcal/mole	$\Delta S^\ddagger$ e.u.	Ref.
Catalase, vegetative, B. cereus	70	6·9	$1·7 \times 10^{-1}$	22	83	178	61
Catalase, spore, B. cereus	70	6·9	$6·7 \times 10^{-5}$	27	73	134	61
Glucose dehydrogenase, spore, B. cereus	55	5·5	$3·2 \times 10^{-4}$	24·5	19·1	−17	97
		6·0	$8·4 \times 10^{-5}$	25·4	19·1	−19	
		6·5	$1·7 \times 10^{-5}$	26·5	19·1	−23	
		7·0	$2·0 \times 10^{-4}$	24·8	19·1	−18	
		7·5	$4·1 \times 10^{-3}$	22·9	19·1	−2	
Glucose dehydrogenase, spore, B. cereus in 4M NaCl	55	6·5	$6·1 \times 10^{-9}$	31·6	19·1	−38	97
Purine nucleoside phosphorylase, cell, B. cereus	60	7·8	$3·9 \times 10^{-3}$	23·1	—	—	50
Purine nucleoside phosphorylase, spore, B. cereus	60	7·8	$3·9 \times 10^{-3}$	23·1	—	—	50
Aldolase of B. cereus cells and spores—no Ca++	53	6·4	$2·9 \times 10^{-3}$	22·9	—	—	83
Aldolase, spore, B. cereus + 10⁻²M Ca++	53	6·4	$2·4 \times 10^{-4}$	24·5	—	—	83
Aldolase, cell, B. cereus + 10⁻²M Ca++	53	6·4	$7·7 \times 10^{-3}$	22·3	—	—	83
Alanine dehydrogenase, spore, B. cereus	58	6·8	$7·0 \times 10^{-5}$	25·7	14·1	350	45, 46
Alanine dehydrogenase, cell, B. cereus, stable form	58	6·8	$1·4 \times 10^{-4}$	25·2	14·1	350	46
Alanine dehydrogenase, cell, B. cereus, labile form	58	6·8	$2·9 \times 10^{-3}$	23·4	16·7	−20	46
NADH oxidase, spore, Cl. botulinum	70	7·7	$5·9 \times 10^{-6}$	28·3	21·6	−20	42
NADH oxidase, cell, Cl. botulinum	70	7·7	$5·0 \times 10^{-2}$	22·2	11·3	−32	42

fixed enzyme concentration and to assign a parameter, K_m (Michaelis constant) which is the substrate concentration for half-maximal velocity. These types of data are especially significant when the respective K_m values indicate major differences in some aspects of the catalytic process. They are subject to interpretation when the differences are only two- to threefold. Table III lists representative data on cell-spore enzymes and illustrates the preceding statement. The NADH oxidases of *Cl. botulinum* have been shown to be distinct molecular species by a variety of parameters[42] and have approximately an eight-fold difference in their Michaelis constants. On the

TABLE III

Michaelis constants of representative homologous enzymes in cells and spores of Bacillaceae

Organism	Enzyme	Substrate	Cell		Spore		Ref.
			pH	K_m	pH	K_m	
C. botulinum	NADH oxidase	NADH	7·5	$7·9 \times 10^{-6}$	7·5	$5·9 \times 10^{-5}$	42
B. cereus	Alanine dehydrogenase	L-alanine	9·0	$3·6 \times 10^{-3}$	9·2	$1·7 \times 10^{-3}$	45, 46
		NAD	9·0	$4·8 \times 10^{-5}$	9·8	$1·3 \times 10^{-4}$	
		pyruvate	9·0	$5·3 \times 10^{-4}$	8·0	$5·9 \times 10^{-4}$	
		NH_4^+	9·0	$1·5 \times 10^{-2}$	8·0	$4·7 \times 10^{-2}$	
		NADH	9·0	$4·3 \times 10^{-5}$	8·0	$7·8 \times 10^{-5}$	
B. subtilis	Alanine dehydrogenase	L-alanine	10	$1·7 \times 10^{-3}$			48
		NAD	10	$1·8 \times 10^{-4}$			
		pyruvate	8	$5·4 \times 10^{-4}$			
		NH_4^+	8	$3·8 \times 10^{-2}$			
		NADH	8	$2·3 \times 10^{-5}$			
B. cereus	Purine nucleoside phosphorylase	inosine	7·5	$1·1 \times 10^{-4}$	7·5	$1·4 \times 10^{-4}$	50

other hand, the alanine dehydrogenases of cells and spores are thought to be very similar if not identical proteins but have two- to fourfold differences in K_m values for the various substrates in the reaction. Of particular interest are the kinetic data of Yoshida and Freese[48] for crystalline alanine dehydrogenase from vegetative cells of *B. subtilis*. Their values are in close accord with those of O'Connor and Halvorson,[45] who studied the deamination reaction catalyzed by the enzyme derived from spores of *B. cereus*.

The hydrogen ion concentration has a marked influence on the rate of enzyme catalyzed reactions because it influences the molecular configuration of the catalyst, the ionic state of the substrate (in some cases), and the extent of dissociation of cofactors or activating ions. Thus the study of the effect of pH on the reaction rate affords a sensitive means for the comparison of cell and spore enzymes. The nature of the pH-activity response might be anticipated and, on the basis of functional and structural similarities, could fall into the following categories: (1) Enzymes of similar function but dissimilar structure which catalyze a given reaction by the same mechanism; these would have similar pH optima but could differ in their relative enzymatic activities over a pH range; (2) Enzymes of similar function which have identical structures would yield identical pH response curves; and (3) Enzymes of similar function and similar but not identical structure might yield partially congruent pH response curves. Class 3

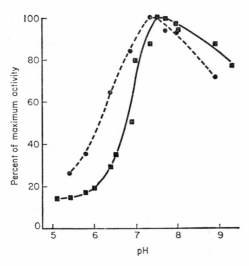

FIG. 1. Relative activity vs. pH for NADH oxidases derived from vegetative cells and spores of *Clostridium botulinum* 62-A.

results might occur if an enzyme derivative resulting from partial proteolysis, or oxidation, is compared with the native enzyme.

The first consideration is well illustrated in Fig. 1, where the pH-activity curves for cell and spore NADH oxidase of *Cl. botulinum* are shown.[42] These enzymes differ in their molecular size, kinetic parameters, heat resistance, and show no immunochemical relationship. They are, in fact, different molecular species.

The inorganic pyrophosphatases from cells and spores of *B. subtilis*[51] are identical by all parameters which have been utilized in their comparison. The enzymes were purified by preparative acrylamide gel electrophoresis and were identical in their amino acid composition, sedimentation properties, substrate specificities and pH response (Fig. 2).

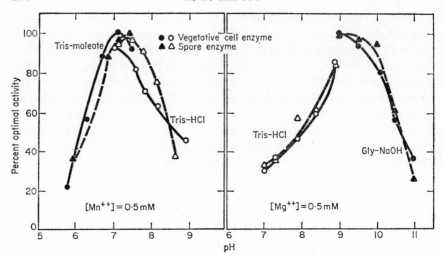

FIG. 2. Effect of pH on reaction rate of inorganic pyrophosphatase of *Bacillus subtilis* in the presence of the activating ions Mn^{2+} and Mg^{2+}. Reproduced by permission of Tono and Kornberg.[51]

The pH response of the enzymes alanine dehydrogenase,[45, 46, 48] DNA polymerase[53] and nucleoside phosphorylase[50] from cells and spores of *Bacillus* species corresponds to the third category presented. Fig. 3 presents the activity vs. pH data for the cell and spore purine nucleoside phosphorylase of *B. cereus*.[50] The response curves are congruent above pH 8 but differ markedly below this value. The enzymes tested had been prepared

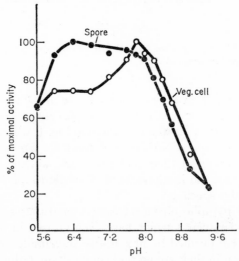

FIG. 3. Effect of pH on the enzyme activity of purine nucleoside phosphorylase of *Bacillus cereus*. Reproduced by permission of Gardner and Kornberg.[50]

by polyacrylamide gel electrophoresis and were approximately 70% pure. The proteins were very similar during sedimentation in sucrose density gradients and on electrophoresis in polyacrylamide gels. However, the spore enzyme was 30% more active in the arsenolysis of guanosine than its vegetative counterpart. The data suggest that the cell and spore nucleoside phosphorylases are similar but not identical proteins.

The purified DNA polymerases of cells and spores of B. subtilis[98, 99] are similar in many kinetic and functional properties and have a pH optimum of 8·2. However, the polymerase from cells is more active than the spore enzyme over the range 8·2–7·0.

Inorganic pyrophosphatase is the only enzyme of the Bacillaceae studied which has been shown to be identical in both cells and spores. As has been pointed out, this identity was verified by means of amino acid composition studies in addition to physico-chemical and kinetic parameters. All other protein pairs investigated thus far have exhibited some functional or physical property differences which most investigators have ascribed as being minor. The modifications in protein structure which produce these differences may in fact be the very means by which spores achieve heat resistance and dormancy.

4. Molecular size and weight

The molecular size of proteins can be determined by gel filtration using the calibration methods of Ackers.[100] He showed that gel filtration in most media can be treated by the Renkin equation for restricted diffusion [101] and that the elution volume for a given protein is dependent on the "pore" size of the gel and the Stokes' radius of the protein. Stokes' radii can be calculated from gel filtration data, and thus diffusion constants for proteins readily obtained.

The calibration of a gel column can be achieved by the passage of one protein of known Stokes' radius and the determination of the void volume and the internal volume of the gel. The unique advantage in the application of gel filtration to the determination of diffusion constants of enzymes is that relatively impure preparations of a particular protein can be used if a specific assay is available. By the application of sucrose density centrifugation[102] to obtain sedimentation constants, it becomes possible to determine the molecular weights and shape factors of partially purified cell and spore enzymes employing the usual relationships between sedimentation, diffusion, density and partial specific volume characteristics of the molecule.[103]

The results of a typical gel filtration comparison of vegetative cell and spore FDP aldolase are presented in Fig. 4. A column of Sephadex G-200 dextran, 2·5 × 30 cm, was prepared in 0·05 M phosphate buffer, pH 6·4. The void volume and the internal liquid volume were determined using high molecular weight dextran and $^{32}PO_4$ respectively. Cytochrome c,

Stokes' radius 1·42 mμ was employed to determine that the pore size of the gel was 18·8 mμ. The elution profiles of the vegetative cell (veg) and spore aldolases are presented in the two figures. It is apparent that the cellular form of the enzyme is an equilibrium mixture of two molecular sizes, each of which elutes in a lesser volume than the spore aldolases. The various physical constants for FDP aldolase are presented in Table IV.[104]

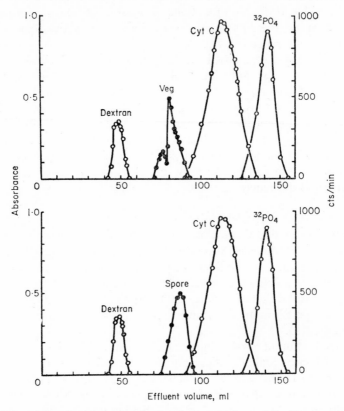

FIG. 4. The gel filtration properties of cell and spore aldolases of *Bacillus cereus* in 0·05 M phosphate buffer, pH 6·4. The Sephadex G-200 columns were calibrated with crystalline cytochrome C and had average pore radia of 18·8 mμ.

The ratio of molecular weights of the two forms of vegetative aldolase is 3 : 2. The spore enzyme has a weight slightly greater than one-half that of veg$_2$ and therefore it is of extreme interest that the vegetative and spore aldolases are identical in K_m, pH response, and immunochemical properties.[83] The spore enzyme may represent a stable, active subunit of the vegetative aldolase.

A similar study has been made of the alanine dehydrogenases from cells

TABLE IV

Properties of vegetative cell and spore FDP aldolases of B. cereus

	Veg$_1$[a]	Veg$_2$[a]	Spore
Stokes radius, mμ	4·6	4·1	3·2
Diffusion constant, $D_{20,w}$ (10^{-7} cm^2 sec^{-1})	4·7	5·3	6·7
Sedimentation constant, $s^{20}{}_{,w}$ (10^{-13} sec)	6·5	4·9	3·3
Molecular weight g/mole	120,000	79,000	43,000
Shape factor f/f_{min}	1·4	1·4	1·4

[a] The vegetative cell FDP aldolases in order of their elution from the column.

and spores of *B. cereus*,[105] where it was noted that the vegetative enzyme also existed in two equilibrium forms of 326,000 and 242,000 molecular weight and had sedimentation values of 9·4 and 8·6 s respectively. The larger molecular weight configuration was the predominant form at pH 6·4 in 0·05 M phosphate buffer. The spore enzyme possessed only one molecular configuration of 294,000 molecular weight and sedimentation of 10·1 s. The molecular weight of alanine dehydrogenase of *B. cereus* cells is 248,000 by sedimentation equilibrium[46] with a sedimentation of 10·2 s. The presence of a minor component was noted by these workers. Yoshida and Freese[47] determined that the molecular weight of crystalline enzyme from cells of *B. subtilis* was 228,000 with a sedimentation rate of 8·8 s. This rate corresponds to that of the minor component noted in alanine dehydrogenase from cells of *B. cereus* whose molecular weight is 242,000. It is significant that the molecular weight of the *B. subtilis* enzyme was 280,000 when measured

TABLE V

Physical parameters of alanine dehydrogenases

Organism	$s_{20,w}$[a]	$D_{20,w}$[b]	Molecular weight	Ref.
B. cereus cells	10·2	3·8	248,000	[46]
B. cereus cells	9·4	2·6	326,000	[105]
	8·6	3·2	242,000	
B. cereus spores	10·1	3·1	294,000	[105]
B. subtilis cells	8·8	—	228,000	[47]
B. subtilis cells	—	—	280,000 (light scattering)	[48]

[a] Units of 10^{-13} sec.
[b] Units of 10^{-7} cm^2 sec^{-1}.

by light scattering at pH 8·0.[48] The above results are summarized in Table V. The discrepancies in the results may be due to the inactivation (disaggregation) which occurs on dilution of alanine dehydrogenase.[48] Furthermore, the effect of hydrogen ion concentration on the aggregation or disaggregation of this protein is unknown and may be significant.

III. The *in vitro* Effect of Varying the Environment on the Heat Resistance of Spore Proteins

It is well known that a loss of heat resistance of proteins ensues when spores are disrupted, but the basis for stability in intact spores is unknown. There is obviously contained within the spore structure a unique environment which is capable of promoting the stability of molecules which have the potential for stabilization. Some hypotheses and experiments relating to the basis of heat resistance and its alteration will be described.

Lewis *et al.*[106] have attempted to explain thermal stability of spores by ascribing contractile properties to the spore cortex. Contraction of the structure would result in anhydrous conditions which might lead to heat resistance. Disruption of the spore or the modification of the cortex during germination might lead to labilization of the spores' macromolecules.

Spore enzymes were thought to be stabilized by adsorption to structural particles. This hypothesis was based on the observation that both alanine racemase[16] and catalase[19] from spores of *B. cereus* were stable and particulate when first isolated but became labile upon their dissociation from subspore particles.

1. Effect of ions

Disruption of spores in aqueous environments results in the dilution of ionic and organic materials such as dipicolinic acid, calcium, or manganese which might stabilize spore enzymes by direct physical or chemical interaction. The pyridine compound is almost unique to bacterial endospores and is an excellent chelating agent which may provide some unique "bridging" capacity when it is combined with a protein and a divalent ion. Ionic effects on the thermal resistance of spore proteins may be tested directly *in vitro*. Studies of purified glucose dehydrogenase[97] have shown that protonation of the enzyme in the pH range 8·0–6·5 produced a dimer to monomer interconversion with an attendant 260-fold increase in heat resistance. Upon increasing the ionic strength at pH 6·5 with NaCl, even further stability increases were noted. A million-fold increase was ultimately achieved, and the enzyme became as stable *in vitro* as in the intact spore. These modifications in stability were reversible and, upon dialysis, the glucose dehydrogenase was rendered labile. Similar but less extensive stabilization by calcium has been noted for fructose 1,6-diphosphate aldolase of *B. cereus*.[83] The purified spore and cell enzymes were of equal heat resistance with a half-life of four minutes at 53°C in 0·05 M phosphate buffer, pH 6·4. Upon adding calcium chloride to 4×10^{-2} M, the stability of the spore enzyme increased twelvefold progressively while the stability of the vegetative cell aldolase decreased 50%.

2. Chemical modifications

The amino acid side chains which determine the activity and stability of proteins can be modified by a variety of reactions including acetylation and oxidation. The use of semi-specific group reagents in the studies of active centers of enzymes has lead to a better understanding of catalytic processes. The same approaches should be applicable to the study of the kinds of side

TABLE VI

Thermal inactivation of acetylated glucose dehydrogenase of B. cereus
at 80°C in solutions of NaCl

	1M NaCl		2M NaCl	
	$k(\text{min}^{-1})$	$t_{\frac{1}{2}}(\text{min})$	$k(\text{min}^{-1})$	$t_{\frac{1}{2}}(\text{min})$
Control	0·029	23·9	0·0066	105
Acetylated enzyme				
No glucose present	0·51	1·4	0·035	19·8
Acetylated in presence of glucose	0·16	4·3	0·01	69

chains which are operative in the stabilization of spore enzymes. The reader is referred to various works[107][108] for reviews and critical evaluations of the methods of protein modification. Sadoff (unpublished) has studied the effects of acetylation on the heat resistance of glucose dehydrogenase of *B. cereus*. This protein was particularly suitable for the investigation because of the ease with which it could be stabilized. The enzymatic activity was maintained by carrying out the acetylation at pH 6·5 with excess acetic anhydride in the presence of 1M glucose and 1M NaCl. Under these conditions approximately 25% enzyme inactivation occurred. Full activity of the enzyme was restored by reacting the glucose dehydrogenase with 1·15 M neutral hydroxylamine. The procedure is known to selectively hydrolyze O-acetyl linkages formed in the acetylation of tyrosyl side chains. Omission of glucose during acetylation lead to extensive enzyme inactivation and a restoration of 75% activity by hydroxylamine. The enzymes were dialyzed and then inactivated at 80°C in 0·05 M imidazol buffer, pH 6·5, containing sodium chloride (Table VI).

It is apparent that acetylation of glucose dehydrogenase reduced its thermal stability even when its enzyme activity had been maintained. The enzyme whose active center had been covered with substrate was more stable than the protein which had been acetylated in the absence of glucose. The labilizing effects of acetylation were most apparent in low ionic strength buffers. The mechanisms of the stabilization of glucose dehydrogenase by high concentrations of Group IA cations may not involve those side chain groups on proteins which are susceptible to acetylation.

3. Urea inactivation

Urea inactivation studies of glucose dehydrogenase have been carried out to obtain an insight into the mechanism of stabilization of glucose dehydrogenase by high ionic strength.[109, 110] Urea is thought to denature proteins by modifying the arrangement of water molecules, thereby disrupting hydrophobic bonding. The denaturation kinetics are second order with respect to the urea concentration. Glucose, but not NAD, stabilizes the enzyme in the presence of urea. Other carbohydrates tested had no effect. It is significant that glucose also protects this enzyme from inactivation by a wide variety of chemical agents.[97] Glucose dehydrogenase was stable to 8 M urea under conditions which were also conducive to the promotion of heat resistance. An extension of the studies of Whitney and Tanford[93] showed that the precise conditions which stabilized glucose dehydrogenase were those which restricted from the solubilization of the side chains of aromatic and nonpolar amino acids.

IV. Conclusions

It is probable that the synthesis of a given enzyme in cells and spores is controlled by a single, common gene. The wide variety of identical or near identical enzymes in the two cell forms supports this view. Furthermore, mutations which limit or eliminate purine nucleoside phosphorylase[50] or alanine dehydrogenase[111] in vegetative cells are also expressed in the corresponding spores.

The dramatic differences in properties between cells and spores led early research workers to postulate that uniquely structured proteins were a requisite for spores. Recent findings have shown this postulate to be fallacious and we are, therefore, confronted by the same question which initiated the study of spores: "What is the basis of their heat resistance?"

Perhaps substantial progress in unravelling this problem may result from a reexamination of minor differences in the physico-chemical and functional properties between cell and spore enzymes. These differences may be a reflection of protein modifications which endow vegetative cell macromolecules with a potential for heat resistance and dormancy.

REFERENCES

1. Virtanen, A. I. and Pulkii, L. (1933). *Arch. Mikrobiol.* **4**, 99.
2. Ruehle, G. L. A. (1923). *J. Bact.* **8**, 487.
3. Cook, R. P. (1931). *Zentbl. Bakt. ParasitKde Abt. I, Orig.* **122**, 329.
4. Holst, E. C. and Sturtevant, A. P. (1940). *J. Bact.* **40**, 723.
5. Keilin, D. and Hartree, E. F. (1947). *Antonie van Leeuwenhoek. J. Microbiol. Serol.* **12**, 115.
6. Crook, P. G. (1952). *J. Bact.* **63**, 193.
7. Spencer, R. E. J. and Powell, J. (1952). *Biochem. J.* **51**, 239.
8. Hills, G. M. (1948). *Biochem. Soc. Symp.* **1**, 86.
9. Hills, G. M. (1949). *Biochem. J.* **45**, 353.
10. Hills, G. M. (1949). *Biochem. J.* **45**, 363.
11. Hills, G. M. (1950). *J. gen. Microbiol.* **4**, 38.
12. Powell, J. F. (1950). *J. gen. Microbiol.* **4**, 330.
13. Powell, J. F. (1951). *J. gen. Microbiol.* **5**, 993.
14. Stewart, B. T. and Halvorson, H. O. (1953). *J. Bact.* **65**, 160.
15. Wood, W. A. and Gunsalus, I. C. (1950). *J. biol. Chem.* **190**, 403.
16. Stewart, B. T. and Halvorson, H. O. (1954). *Archs Biochem. Biophys.* **49**, 168.
17. Militzer, W., Sonderegger, T. B., Tuttle, L. C. and Georgi, C. E. (1950). *Arch. Biochem.* **26**, 299.
18. Murrell, W. G. (1952). Ph.D. thesis, University of Oxford.
19. Lawrence, N. L. and Halvorson, H. O. (1954). *J. Bact.* **68**, 334.
20. Murrell, W. G. (1955). "The Bacterial Endospore", University of Sidney, Australia.
21. Lawrence, N. L. (1955). *J. Bact.* **70**, 577.
22. Lawrence, N. L. (1955). *J. Bact.* **70**, 583.
23. Powell, J. F. and Hunter, J. R. (1956). *Biochem. J.* **62**, 381.
24. Lawrence, N. L. (1957). *In* "Spores I" (H. O. Halvorson, ed.). Am. Inst. Biol. Sci., Washington, D.C., U.S.A.
25. Nakata, H. (1957). *In* "Spores I", (H. O. Halvorson, ed.). Am. Inst. Biol. Sci., Washington, D.C., U.S.A.
26. Waldhalm, D. G. and Halvorson, H. O. (1954). *Appl. Microbiol.* **2**, 333.
27. Curran, H. R., Brunstetter, B. C. and Myers, A. T. (1943). *J. Bact.* **43**, 25.
28. Powell, J. F. (1953). *Biochem. J.* **54**, 210.
29. Powell, J. F. and Strange, R. E. (1956). *Biochem. J.* **63**, 661.
30. Riemann, H. and Ordal, Z. J. (1961). *Science* **133**, 1703.
31. Halvorson, H. O. and Church, B. D. (1957). *J. appl. Bact.* **20**, 359.
32. Church, B. D. and Halvorson, H. O. (1957). *J. Bact.* **73**, 470.
33. Doi, R., Halvorson, H. and Church, B. D. (1959). *J. Bact.* **77**, 43.
34. Goldman, M. and Blumenthal, H. J. (1960). *Biochem. biophys. Res. Commun.* **3**, 164.
35. Blumenthal, H. J. (1961). *In* "Spores II" (H. O. Halvorson, ed.). Burgess Publishing Company, Minneapolis, Minn., U.S.A.
36. Dedonder, R. (1952). *Proc. Int. Cong. Biochem. 2nd Cong., Paris,* p. 77.
37. Keynan, A., Strecker, H. J. and Walsch, H. (1954). *J. biol. Chem.* **211**, 883.
38. Goldman, M. and Blumenthal, H. J. (1964). *J. Bact.* **87**, 377.
39. Levinson, H. S. and Hyatt, M. T. (1960) *J. Bact.* **80**, 441.
40. Costilow, R. N. (1962). *J. Bact.* **84**, 1268.
41. Simmons, R. J. and Costilow, R. N. (1962). *J. Bact.* **84**, 1274.
42. Green, J. H. and Sadoff, H. L. (1965). *J. Bact.* **89**, 1499.

43. Krask, B. J. and Fulk, G. E. (1960). *Archs Biochem. Biophys.* **90**, 304.
44. Hanson, R. S., Srinivasan, V. R., and Halvorson, H. O. (1963). *J. Bact.* **85**, 451.
45. O'Connor, R. J. and Halvorson, H. O. (1960). *Archs Biochem. Biophys.* **91**, 290.
46. McCormick, N. G. and Halvorson, H. O. (1964). *J. Bact.* **87**, 68.
47. Yoshida, A. and Freese, E. (1964). *Biochim. biophys. Acta* **92**, 33.
48. Yoshida, A. and Freese, E. (1965). *Biochim. biophys. Acta* **96**, 248.
49. Krask, B. J. and Fulk, G. E. (1959). *Archs Biochem. Biophys.* **79**, 86.
50. Gardner, R. and Kornberg, A. (1967). *J. biol. Chem.* **242**, 2383.
51. Tono, H. and Kornberg, A. (1967). *J. biol. Chem.* **242**, 2375.
52. Tono, H. and Kornberg, A. (1967). *J. Bact.* **93**, 1819.
53. Falaschi, A. and Kornberg, A. (1966). *J. biol. Chem.* **241**, 1478.
54. Halvorson, H. O. (1958). "The Physiology of the Bacterial Spore", Tronheim.
55. Halvorson, H. (1962). *In* "The Bacteria" (I. C. Gunsalus, and R. Y. Stanier, eds). Academic Press, New York, U.S.A.
56. Murrell, W. G. (1967). *In* "Advances in Microbial Physiology", Vol. I (A. H. Rose and J. R. Wilkinson, eds). Academic Press, London.
57. Kornberg, A., Spudich, J. A., Nelson, D. L. and Deutscher, M. P. (1968). *Ann. Rev. Biochem.* **37**, 51.
58. Spizizen, J., Reilly, B. and Dahl, B. (1963). *Abstr. Eleventh Int. Congr. Genetics, The Hague.*
59. Bernlohr, R. W. and Novelli, G. D. (1963). *Archs Biochem. Biophys.* **103**, 94.
60. Bernlohr, R. W. (1964). *J. biol. Chem.* **239**, 538.
61. Sadoff, H. L. (1961). *In* "Spores II" (H. O. Halvorson, ed.). Burgess Publishing Company, Minneapolis, Minn., U.S.A.
62. Bach, J. A. and Sadoff, H. L. (1962). *J. Bact.* **83**, 699.
63. Kominek, L. A. and Halvorson, H. O. (1965). *J. Bact.* **90**, 1251.
64. Hanson, R. S., Srinivasan, V. R. and Halvorson, H. O. (1961). *Biochem. biophys. Res. Commun.* **5**, 457.
65. Hanson, R. S., Srinivasan, V. R. and Halvorson, H. O. (1963). *J. Bact.* **86**, 45.
66. Bach, M. L. and Gilvarg, C. (1966). *J. biol. Chem.* **241**, 4563.
67. Bernlohr, R. W. (1965). *In* "Spores III" (L. L. Campbell and H. O. Halvorson, eds.), Am. Soc. Microbiol. Ann Arbor, Michigan, U.S.A.
68. Buono, F., Testa, R. and Lundgren, D. G. (1966). *J. Bact.* **91**, 2291.
69. Ramaley, R. F. and Bernlohr, R. W. (1966). *Archs Biochem. Biophys.* **117**, 34.
70. Millet, J. (1963). *C.r. hebd séanc. Acad. Sci., Paris* **257**, 784.
71. Taniguchi, K. and Tsugita, H. (1966). *J. Biochem.* **60**, 372.
72. Young, I. E. and Fitz-James, P. C. (1959). *J. biophys. biochem. Cytol.* **6**, 483.
73. Foster, J. W. and Perry, J. J. (1954). *J. Bact.* **67**, 295.
74. Urba, R. C. (1959). *Biochem. J.* **71**, 513.
75. Monro, R. E. (1961). *Biochem. J.* **81**, 225.
76. Aubert, J-P. and Millet, J. (1963). *C.r. hebd séanc. Acad. Sci., Paris,* **256**, 5442.
77. Sadoff, H. L. (1966). *Biochem. biophys. Res. Commun.* **24**, 691.
78. Park, J. T. and Hancock, R. (1960). *J. gen. Microbiol.* **22**, 249.
79. Ouchterlony, O. (1949). *Acta path. microbiol. scand.* **26**, 507.
80. Baillie, A. and Norris, J. R. (1963). *J. appl. Bact.* **26**, 102.
81. Norris, J. R. and Baillie, A. (1964). *J. Bact.* **88**, 264.
82. Cohn, M. and Torriani, A. M. (1952). *J. Immunol.* **69**, 471.

83. Sadoff, H. L., Hitchins, A. D. and Celikkol, E. (1967). *Bact. Proc.* 23.
84. Thomson, R. O., Walker, P. D. and Hardy, R. W. (1966). *Nature* **210**, 760.
85. Walker, P. D., Baillie, A., Thomson, R. O. and Batty, I. (1966). *J. appl. Bact.* **29**, 512.
86. Singer, S. J. (1959). *Nature* **183**, 1523.
87. Singer, S. J. and Schick, A. F. (1961). *J. biophys. biochem. Cytol.* **9**, 519.
88. Tanford, C. (1961). "Physical Chemistry of Macromolecules". John Wiley and Sons, Inc., New York, U.S.A.
89. Tanford, C., De, P. K. and Taggart, V. G. (1960). *J. Am. chem. Soc.* **82**, 6028.
90. Perutz, M. F., Rossmann, M. G., Cullis, A. F., Muirhead, G. W. and North, A. C. T. (1960). *Nature* **185**, 416.
91. Joly, M. (1965). "A Physico-Chemical Approach to the Denaturation of Proteins". Academic Press, New York, U.S.A.
92. Kauzman, W. (1959). *Adv. Protein Chem.* **14**, 1.
93. Whitney, P. L. and Tanford, C. (1962). *J. biol. Chem.* **237**, 1735.
94. Vinter, V. (1961). *In* "Spores II" (H. O. Halvorson, ed). Burgess Publishing Company, Minneapolis, Minn., U.S.A.
95. Scheraga, H. A. (1961). "Protein Structure", Academic Press, New York, U.S.A.
96. Eyring, H. (1935). *Chem. Rev.* **17**, 65.
97. Sadoff, H. L., Bach, J. A. and Kools, J. W. (1965). *In* "Spores III" (L. L. Campbell, and H. O. Halvorson eds). Am. Soc. Microbiol. Ann Arbor, Michigan, U.S.A.
98. Okazaki, T. and Kornberg, A. (1964). *J. biol. Chem.* **239**, 259.
99. Falaschi, A., Spudich, J. and Kornberg, A. (1965). *In* "Spores III" (L. L. Campbell, and H. O. Halvorson, eds.). Am. Soc. Microbiol., Ann Arbor, Michigan, U.S.A.
100. Ackers, G. K. (1964). *Biochemistry* **3**, 723.
101. Renkin, E. M. (1955). *J. gen. Physiol.* **38**, 225.
102. Martin, R. G. and Ames, B. N. (1961). *J. biol. Chem.* **236**, 1372.
103. Siegel, L. M. and Monty, K. J. (1965). *Biochem. biophys. Res. Commun.* **19**, 494.
104. Sadoff, H. L., Hitchins, A. D. and Celikkol, E. (1968). Submitted to *J. Bact.*
105. Sadoff, H. L. and Celikkol, E. (1968). Submitted to *J. Bact.*
106. Lewis, J. C., Snell, N. S. and Burr, H. K. (1960). *Science* **132**, 544.
107. Fraenkel-Conrat, H. (1957). *In* "Methods in Enzymology, Vol. IV" (S. P. Colowick and N. O. Kaplan, eds). Academic Press, New York, U.S.A.
108. Putnam, F. W. (1953). *In* "The Proteins Vol. I" (H. Neurath, and K. Bailey, eds). Academic Press, New York, U.S.A.
109. Sachar, K. and Sadoff, H. L. (1965). *Bact. Proc.* 79.
110. Sachar, K. and Sadoff, H. L. (1966). *Nature, Lond.* **211**, 983.
111. Freese, E., Park, S. W. and Cashel, M. (1964). *Proc. natn. Acad. Sci., U.S.* **51**, 1164.

CHAPTER 9

Dormancy

J. C. LEWIS

Western Regional Research Laboratory, Agricultural Research Service,
U.S. Dept. of Agriculture, Albany, California, U.S.A.

I. Introduction

DORMANCY is the central function, the essence, of sporulation. The formation of the bacterial endospore is a specialized response to the population explosions so common in the microbial world. It involves the orderly preparation of a cryptobiotic state in a time of plenty; of a durable cell able to persist with little or no metabolism; of an organelle designed to accomplish the rapid but controlled release of the dormant condition through the processes of activation and germination when subtle and specific messages from the environment determine the initiation of a new cycle of growth and multiplication.

The words *dormant* and *dormancy* express the general sense of a stage or condition of a living organism that is characterized by a lack of metabolism and developmental processes. The terms may be applied to an idealized spore, to a real spore, to a fraction of a population of spores, or to a characteristic unresponsiveness of the spores of a particular strain or species. The manifestations of dormancy and the available information and ideas on the functions of spore components in the development, maintenance and release of the state of dormancy will be treated in this chapter. Citations usually will be omitted when a more detailed discussion appears elsewhere in this book.

Unlike higher organisms with more fixed, more complex life cycles, the sporulating bacteria show a highly flexible alternation between growth and dormancy. They possess a remarkable facility of responding either to signals of optimism or of warning. The close linkage of environment and differentiation is manifest in various phenomena of sporulation; in control of the fraction of the vegetative population that becomes committed to sporulation; in limitation of the degree of dormancy achieved as is illustrated by *recycling*, the spontaneous germination of newly formed spores in the sporulating culture; in resporulation after only a few vegetative divisions; or—most dramatically—in *microcycle sporogenesis*, a resporulation of the germinating spore itself to give a functional spore. Sporulation without dormancy is an abortive process; a pathology, or a laboratory artifact.

The metabolic maintenance requirement of spores is very low, if one exists. Those enzymes that are most active in unbroken ungerminated spores seem not to play a respiratory role. The endogenous respiration rate may be only 10^{-4} of the maximal rate for vegetative cells metabolizing substrate. The exogenous rate for dormant spores is much less than that for activated spores or for spore extracts. Since activation and the early stages of germination precede protein synthesis it is clear that the respiratory apparatus of the dormant spore is only turned off, and lacks no essential protein component.

The natural release of spores from the dormant state is under precise control. The control function is a marvel of regulation by which the proportion of the population responding, the state of activation, and the speed of germination and outgrowth are determined by a variety of internal and environmental factors. It has adaptive values of spreading the spores in time to complement their random dispersal in space, of precognition of nutrient adequacy and the absence of deleterious conditions, and of recognition of a particular ecological niche so acute with some of the more specialized members (e.g., *B. popilliae*) that high germination rates have not yet been inducible under simulated physiological conditions. Among the absolute requirements of the germinative process is a coordination of the release of the restraining layers of the dormant spore with the biosynthesis of the new vegetative cell wall such that at no stage is the delicate vital core exposed to osmotic disruption. This involves a release of dipicolinate (DPA), a partial or complete lysis of the morphologically and compositionally unique cortex, and a well-timed splitting or dissolution of the spore coat, a proteinaceous structure notoriously unaffected by proteolytic enzymes in nature or in the laboratory.

II. Dormancy; Examples and Usage

In this chapter the unqualified words *dormant* and *dormancy* are reserved for their general sense. However, this sense often has been assumed implicitly in specialized discussions of spores and the terms have been used in more particular ways, as in comparing degrees of metabolic activity, or ease of germination, or for inhibitions that might better be called *sporostasis*. Examples of usage and of variables important for inducing or maintaining dormancy follow.

A. THE CRYPTOBIOTIC STATE

1. Measurement of metabolism

Dormancy implies a reduction of activity for an extended period. The extreme case characterized by no metabolic activity has been designated *cryptobiosis* (see Chapter 1). The dormant spore may be in this condition. Accurate assessment of the respiration of dormant spores requires crops free of vegetative cells and of activated, injured, or germinated spores which have much higher respiration rates than dormant spores. Crook[1] measured an endogenous rate by spores of *B. cereus* var *anthracis* equal to 0·3 μlitres O_2 per mg spores per hr before any heat treatment, and 6 and 20 times this rate for two extents of heat activation. Well-washed dormant spores of *B. cereus* T consumed less than 0·3 μlitres O_2 per mg spores per hr (the limit of detection by Warburg manometry) with or without 0·0025 M glucose, compared with 20 μlitres per mg per hr by heated aged spores with glucose.[2] A more sensitive test of dormant spores of the same strain grown on $^{14}CO_2$ photosynthate[3] gave less that 10^{-4} μlitres $^{14}CO_2$ per mg per hr by dry spores or by spores in water at 19°C; at 30°C in water or in buffer without glucose 0·012 to 0·04 μlitres per mg per hr was released, and at the optimal 40°C the rate was about nine-fold higher. Spores grown on radioactive glucose gave higher metabolic rates, which were increased by pretreatment at 70°C. The lower value gives a period of centuries for self-consumption by endogenous respiration. The total absence of free radicals in clean spores,[4] even after long storage in air, is consistent with an extremely low level of metabolic activity in the dormant spore.

2. Survival

The retention of viability for long extended periods by spores implies the cryptobiotic state. Examples of the persistence of spores under conditions that restrain germination (e.g., low water activity) have been compiled. Viable spores have been detected in the soil of old specimens from herbaria at frequencies that suggest decimal reduction times of the order of 50 to 100 years. (Self-sterilization of a kilogram of soil would require 1,000 years.) *B. anthracis* was obtained from soil and sealed vials that had been stored for 60 and 68 years. *Cl. sporogenes* survived in pathological specimens preserved in alcohol for 39 and 46 years. *Cl. tertium, sordelli* and *novyi* were isolated from the abdomens of mummies that had been at 8° to 10°C for 180 to 250 years. Moreover, viable spores have been reported from geological materials; for example, *B. circulans* from paleozoic salt beds. Precautions and difficulties in evaluating the presence of viable organisms in ancient materials have been discussed.[5]

Perhaps the most favorable circumstances for the preservation of spores are those obtained by sealing at an optimal water activity[6] in vacuum or inert gases after freeze-drying. Many vegetative cells survive for decades with such treatment. The ultimate viability of spores under such conditions may well be measured in centuries.

An instance of survival of spores under conditions where germination might have been expected is more interesting for a consideration of dormancy. Three types of Bacillus survived for 118 years in a can of veal that had been carried to the Arctic in 1824 by Peary, stored in a museum a few years later, and opened in 1938. The meat was found to be unspoiled microbiologically.[7] Sporostatic agents, as from the decomposition of fat, may have had the controlling influence.

3. Termination of the cryptobiotic state

After the last recognized morphological and compositional changes of sporulation have been completed (the perfection of the cortex and coats, the uptake of Ca, and the deposition of DPA), evidence of further maturation often is seen in enhancement of resistance and germinative dormancy. The cryptobiotically dormant spore is one that is more or less fully matured but not yet altered by the processes of activation and germination.

In the sequence of events that lead from the dormant spore to the vegetative cell three conceptually distinct processes have been postulated to occur in succession in individual spores.[8] These processes—*activation, germination* and *outgrowth*—are discussed in detail in the next three chapters. As the first of the sequence, activation bears an intimate relationship to dormancy. Operationally, it is the acquisition of a competence to initiate the irreversible steps of germination under a particular set of conditions. One may observe a change in the germination-triggering requirements, such as a simplification of the organic requirements or an extension of the tolerable physical conditions. With ageing, heating, reductants, low or high pH, or altered exchangeable cation load the optimally dormant state may be transformed into an activated state. Activation usually is considered not to require metabolic activity but rather to depend on changes in macromolecular configurations that arise in response to time, temperature, ionic environment, solvents, reductants, humidity and adsorbed gases. An activation phase may not be recognizable for all spore preparations or strains or for all germination conditions. The bypassing of requirements for specific germination signals as in mechanically, surfactant, or chelationally induced germination also bypasses the need for activation by sublethal heat treatment (the classical *heat shock*). If such treatments are interrupted before the germinative process becomes self-sustaining—as with a short exposure to 0·04 M Ca·DPA in the cold[9]—a deactivatable ('reversible') activation may be demonstrated.

The property of reversibility (or more precisely, the capability to undergo one or more cyclic transformations) that is shown in various circumstances promises to add to our knowledge of dormancy. A reversible phenomenon hints that a single reaction can be isolated, that significant thermodynamic calculations can be made, and that the reaction can be reconstructed with isolated structures and enzymes. The effects of the ionic environment within a physiological pH range usually are freely reversible. Some degree of reversibility has been demonstrated for low pH and exchangeable cation load,[10, 11] oxidation–reduction state,[12] heat activation,[13, 14] and activation by aqueous ethanol and other solvents.[15, 217]

The function, in an immediate sense, of activation may be an alteration of permeability or of access to a particular site; an unmasking of an enzyme or of a control site on an enzyme; or a loss of a spore component such as a superficial store of DPA or of an inhibitor.

The next phase in the breaking of cryptobiotic dormancy is *germination* (in a narrow sense); the early but irreversible processes of hydrolysis, depolymerization and exudation of low and high molecular weight components. The process ordinarily is observed as a substantial decrease in the optical density (O.D.) of a spore suspension, or as a darkening of individual spores under positive (dark) contrast phase microscopy. The requirements for germination customarily differ markedly from those for outgrowth or for multiplication. Germinants and inhibitors of germination are discussed in Chapter 11.

An apparent dormancy may reflect a *defective* spore, as with the inhibition of phase-darkening of *B. megaterium* spores produced in phage-infected industrial antibiotic fermentations.[16]

The retention of cryptobiotic dormancy thus depends on factors that

N

affect activation or germination or both processes; factors that Sussman and Halvorson[5] classify heuristically as *constitutive* (permeability barriers, metabolic blocks, autogenous inhibitors) and *exogenous* (physical or chemical conditions of the environment). Some of these effects will receive detailed attention below. In many cases the nature of the inhibition remains unknown, and for all cases elucidation of the molecular basis largely remains to be discovered.

B. Germinative Dormancy; Physiological Initiators of Germination

The term *dormancy* has been applied to a requirement for a special signal or trigger to initiate germination, beyond the usual heat activation and the nutritional and environmental conditions seemingly adequate for outgrowth and multiplication; according to Murrell[17], "viable spores failing to germinate in apparently favourable conditions are said to be 'dormant' ". Many idiosyncrasies have been observed in the response of spores to complex media. The basic parameter of this *germinative dormancy* is our state of ignorance. Hard-to-germinate strains often are found to germinate at familiar rates and high frequencies once the particular germination requirements have been identified and supplied, or inhibitions removed. Foerster and Foster[18] found that all but one of their *reluctant* strains responded either to lowered temperature (25° vs 40°C) or to the addition of 0·001 M L-leucine. Among the unusual physiological germinants discovered recently are fructose and adenine for spores of *B. macerans* strain 7 × 1; L-alanine, glucose and adenosine are ineffective.[19]

The general failure of spores to germinate in the medium in which they were formed has been attributed to exhaustion of nutrients.[20] A strain of *B. subtilis* that did not require heat activation germinated in "used" sporulation medium when L-alanine was added.[21]

Germinants act much like traffic signals. The positive or "go" signals include representatives of familiar nutrients such as glucose and L-alanine, and a multitude of others that accommodate the bewildering color-blindness of diverse strains. Negative or "don't-go" signals include extremes of pH and temperature, very low or very high ion concentrations, D-alanine, and many others including unidentified factors. Ordinary nutrients (for example, certain of the ordinary amino acids) often inhibit germination.

Another class of germinants do not show the strain specificities and activation requirements of the physiological germinants. They appear to bypass or short-circuit the mechanisms that respond to the physiological germinants. Such destabilizers of the organization or structure of the dormant spore include mechanical injury, and certain surfactants, chelates, gases and solvents (for example, n-dodecylamine, Ca·DPA, butane, chloral hydrate and water vapor). Their use has given considerable insight into the basis of dormancy of the bacterial spore.

Use of the term *dormancy* has been questioned for circumstances wherein the composition of the medium restrains the germination of spores: "The term 'dormancy' in the sense of delayed germination no longer has the degree of validity it formerly enjoyed, and there now appears to be some question as to the justification of its continued use, unless it occurs in cases in which the effect is shown to be independent of the germination medium" (Foster and Wynne).[22] On the other hand, Morrison and Rettger[23] wrote that "the dormancy of aerobic bacterial spores is largely, if not entirely, determined by conditions in the environment of the spores". It may be convenient to use the phrases *autogenous germinative dormancy* and *exogenous germinative dormancy* for circumstances appropriate to the opposing viewpoints.

It is often important with populations that do not germinate completely to make a total microscopic spore count. Use of a membrane filter instead of a counting chamber offers certain advantages.[24]

C. DORMANCY IN RELATION TO HEAT ACTIVATION

The important topic *heat activation* is discussed in the next chapter. It suffices here to state that exposure of most spores in aqueous suspension to a sub-lethal heating increases the proportion that will germinate under a particular condition, decreases the lag before germination becomes perceptible, decreases the concentrations of specific germinants required and the number of such components necessary for maximal germination rate, and broadens the variety of agents that will induce germination. Where it has been tested heat activation increases markedly both endogenous and exogenous metabolic rates. Keynan *et al.*[25] used the term *dormancy* "to describe the fact that spore suspensions respond poorly or not at all to germination agents under conditions that permit rapid germination of aged or heat-activated spores". They proposed to use the extent of time required for optimal heat activation to quantitate dormancy. The degree of dormancy by this measure was correlated with the composition of the spore (i.e., DPA content as varied by the conditions of growth and sporulation) and with the nature of the germinative agent.

D. VARIABLE GERMINATIVE DORMANCY IN SPORE POPULATIONS

The common experience that all members of a sample of spores do not germinate simultaneously implies that the individual spores vary in their susceptibility or that some local fluctuation of the environment affects the activation or the initiation of germination of only a part of the population. The former possibility has been demonstrated in particular cases, whereas the latter seems remote in many experiments. The proportion that remains dormant may be so large as to be characteristic of the species, or so small that it may be ignored. Population heterogeneity is expressed in *delayed germination*, a slow increase in colony counts or in "positive tubes" considered by Foster and Wynne[22] to be "the characteristic feature of classical dormancy". The small fraction of a spore population that remains dormant after exposure to a good germination condition has been termed *superdormant*[236] (see Chapter 11, Section II B).

Delayed germination has been a particular nuisance with *Cl. botulinum*. The absence of viable spores from canned foods must be assured without recourse to unacceptable additives or to quality-damaging treatments. However, Dickson[26] found that heated spores of *Cl. botulinum* were still germinating at a low frequency six years after sealing into tubes of broth. Over the years the germination rate of *Cl. botulinum* has been improved by the selection of favorable natural substrates such as pork and pea infusions, by the addition of 0.1% soluble starch to counteract certain inhibitors, by the addition of bicarbonate to supply an essential though low concentration of CO_2, by the provision of a low oxidation-reduction potential, and by the avoidance of the inhibitory effect of even short exposure to low concentrations of O_2. In recent years analogous problems of delayed germination have been encountered with *Cl. thermosaccharolyticum* and *B. popilliae*.

Observations of individual spores confirm a differential behavior toward germinative conditions. The relatively uniformly responding *B. cereus* T shows a variation in the times of passage through the microscopically distinguishable *microlag* and *microgermination* stages.[27] Microlag, the period between addition of the germinant and the first phase-microscopic change, is dependent on the extent of activation and the concentration of germinant. Microgermination, the period of phase-microscopically recognizable refractility change (a lytic stage), depends only on temperature.

The germinative response often varies continuously over a range of a factor, such as the concentration of a germinant. In other cases major fractions of the spore population differ qualitatively in their response. This has been called *fractional germination*.[28] Fig. 1 illustrates fractional germination of heat-activated *B. megaterium* QM B1551 induced by various salts. The population consisted of a mixture of phase-bright and dark spores after prolonged incubation. Similar effects have been seen in response to heat-activation, ageing, specific organic germinants, Ca·DPA, water activity, pH, inhibitors and other variables. Physical separations of germinatively dormant fractions of spore populations have not been made for lack of suitable methods. Sharp separation of spore populations into distinct classes by isopycnic density gradient centrifugation has been achieved recently,[10, 29, 30] but application to germinative behavior has not been reported.

Woese *et al.*[219] have proposed a model for the kinetics of germination that assumes a Poisson distribution of *germination units*, the product of

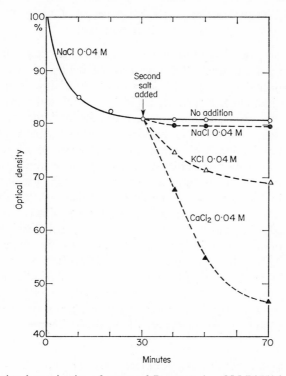

FIG. 1. Fractional germination of spores of *B. megaterium* QM B1551 in water after the addition of salts as shown. (From Rode and Foster.[28])

which is required in a critical concentration for germination to proceed. The *germination distribution* resulting is predicted to follow a step function.

Delayed germination and fractional germination probably have natural roles that are genetically determined. One need only postulate an assortment of distinct allosteric effector sites in the population. Attempts to isolate homogeneously responding components of the population by subculturing usually do not succeed. Sussman and Halvorson[5] have discussed the importance of spores as "a simple timing system in which an arresting device must be overcome by an activator". The relaxation of germination requirements with ageing fits well with this function. Variation of germinative behavior also would be expected from unavoidable variations of the sporulation microenvironment with position or time during spore formation. Reported experiments may greatly underrepresent the variation of germinative dormancy in nature through conscious and unconscious selection of strains. Of random isolates of Bacillus from soil 35% gave spores that did not germinate readily.[18]

E. GERMINATIVE DORMANCY OF CERTAIN STRAINS AND SPECIES

Spores of bacteria adapted to highly specialized environments might be expected to respond to unusual dormancy-breaking signals or, where these remain unidentified, to respond poorly in artificial situations. Such ecological niches include hot springs, frigid terrestrial and marine environments, anaerobic muds, alkaline soils, intestines and the larval hosts of insect pathogens. Thermophilic species tend to require a considerable degree of heat activation, which may favor restriction of germination to a thermal environment. Anaerobic species tend to germinate reluctantly, although an anaerobic environment is not necessary for the initial steps of germination in some obligate anaerobes. Spores of the thermophilic anaerobe *Cl. thermosaccharolyticum* respond very slowly to the usual germinants including the "universal germinant", n-dodecylamine.[31]

Reluctant germination in artificial culture also is characteristic of *B. popilliae*. Although large numbers of spores often are necessary for infection of the larval hosts, this does not necessarily indicate a low frequency of germination in the host. A selective advantage of a low rate of germination *in vitro* seems likely for a pathogen transmitted through the soil. Insect and animal pathogens are discussed in Chapters 13 and 14.

F. FACTORS THAT MAINTAIN, AUGMENT, OR INDUCE DORMANCY

Germinative dormancy often may be enhanced in spores that have completed a normal maturation, by a broad array of natural and artificial conditions. Disinfectants may act on the mechanisms essential for germination as well as on those essential for vegetative multiplication. Extreme dormancy is operationally indistinguishable from death.

1. Temperature

This pervasive variable affects both rates of reactions and states of equilibria, including cation loads and polymer configurations, and so may be expected to affect the dormancy of spores. Spore preparations often are chilled to restrain germination, during harvesting and cleaning or in the familiar and convenient mode of storage of aqueous suspensions under refrigeration. Inhibition by cold is reversed readily by restoration of a temperature suitable for germination.

Although various processes of germination are slowed or stopped at low temperatures, psychrophilic strains of both Bacillus and Clostridium germinate and grow near or below 0°C. Some other spores have more than one optimal temperature for germination. One may infer that different pathways of germination are dominant at the several optima or that certain pathways are inhibited beyond particular temperature ranges.

The effects of high temperatures upon germinative dormancy are complex and diverse. Heat activation has been summarized in Chapter 10. With some spores heat-shocking provokes spontaneous germination in water.[32, 33, 34] With others a relatively mild heating induces dormancy. *B. stearothermophilus* spores are heat activated at 105 to 110°C. A short exposure to a sub-activating temperature (80 to 100°C) repressed germination of up to 65% of the number that grew without heat.[35] The effects varied among strains and between smooth and rough colony variants. These variants were fairly stable upon isolation and subculture. Cleaned spores of the smooth phase grown at 50°C became less germinable on further exposure to 50°C than did spores of the rough phase. The spores of the smooth variant also were the more readily heat activated[36] and the more heat resistant.[37] The effects of cation load and heat adaptation (see p. 314) may well be important in these phenomena.

The survivors of severe heating are sensitive to outgrowth conditions. The numbers of colonies vary by as much as two or three orders of magnitude for variations of medium or other factors that give indistinguishable counts for healthy vegetative or spore inocula. The counts often are raised by supplementation of the outgrowth medium or by the inclusion of inhibitor adsorbents such as soluble starch. The low counts usually have been thought to arise from injuries but, until recently, the nature of the injuries had not been investigated. It now seems that the injury is, at least in part, to the germination mechanism(s). The initial heat-injury of spores of *B. subtilis* was overcome by germination with Ca·DPA.[38] The injury was more severe with the more rapid heating obtained by injection of steam than with heating in capillary tubes; thus, the injured germination system appeared to be protected partially by heat adaptation.[39] Sublethally heated spores of *B. subtilis* release glutamic acid. Glutamate or arginine permit germination and/or outgrowth.[40] Campbell *et al.*[41] isolated stable mutants of *B. stearothermophilus* 1518 with altered germination requirements (*germination mutants*) from severe heat treatments (about 10^{-4} survival on enriched medium, 10^{-7} on minimal medium). When the minimal medium was supplemented with L-glutamate the count rose about 40-fold to about 4% of that obtained with the enriched medium. With L-lysine a doubling of the count on the minimal medium was obtained. The amino acids were required for germination only, but not for outgrowth or for vegetative growth in subcultures. The results imply either a phenotypic expression, prior to transcription, of gene injury or a permanent gene repression in response to a phenotypic alteration. A more detailed investigation of these induced dormancies might have considerable interest.

2. Ionic environment, pH, and exchangeable cation load

The absence of ions and other solutes favors the retention of the dormant state. Extensive washing is commonly employed in the preparation of spore suspensions. Sediments of well-washed spores ordinarily do not germinate, especially when refrigerated. Germination occasionally observed in thick suspensions is attributed to the release of endogenous germinants from the spores. Anaerobes tend to germinate spontaneously in the medium in which they are produced, but thorough washing stabilizes them for storage in water or inorganic buffer solutions.[42]

Spores usually initiate germination most readily at intermediate concentrations of ions, in the broad range 0·01 to 0·1 M. [11, 18, 28, 43, 44, 45] The extraordinarily sensitive B. megaterium B1551 germinates with pH 3·0 HNO_3 or with 0·001 M KCl + 0·03 mM de-ionized glucose,[28] or with 0·025% n-hexanol or with n-pentane (1 atmosphere) in distilled water.[46] As will be discussed below, the ion concentration requirement also may be dispensed with if the spores carry a suitable exchangeable cation load.[11] Very high ion concentrations tend to inhibit germination. The germination of spores of Cl. botulinum is inhibited by 6% NaCl and of Cl. roseum, B. cereus T, and B. mycoides by 9% NaCl.[42] High concentrations of non-ionized solutes are relatively ineffective. Thus, osmotic effects seem not to be highly important in the retention of dormancy.

Many strains of spores show remarkable specificities in the ionic environments that permit or restrain germination. For example, spores of B. cereus T[47, 48] and Cl. bifermentans[14] require Na^+, whereas K^+ is not effective. A break between these ions of the alkali metal series in the ordering of water structure has been noted. Phosphate, acetate and other organic anions, and F′ were effective on spores of B. cereus T but Br′, Cl′ and I′ were not. L-Alanine, inosine, or glucose were effective in the presence of certain strong electrolytes, but not in the presence of others or in the absence of any.[47] Spore germination may be either stimulated (various Clostridia) or inhibited (B. subtilis[49]) by bicarbonate ion.

Spores of certain strains (for example, B. megaterium B1551,[28] cereus T,[47] and subtilis L[43]), germinate in particular ionic environments without the addition of organic germinants. This has been called ionic germination by Rode and Foster.[28] They considered that ions are "critical initiating agents of germination" and that the role of the "so-called organic germinative compounds . . . is an augmentation, in varying degrees, of the germinative potential of particular ionic systems". However, it should be noted that the most potent physiological organic germinants act at concentrations far below those required for the most effective inorganic ions.

Germination of heat-shocked spores of *B. licheniformis* NCTC 9945[50] is triggered by a few minutes' exposure to L-alanine at 37°C and pH 8. The triggering, but not the turbidity drop after triggering, is prevented below 20°C, below pH 6, or by D-alanine. Various salts slowed the triggering reaction at 0·03 to 0·08 M, and NH_4Cl, $MgCl_2$ and $CaCl_2$ did so at much lower concentrations. The effects were reversible if the salt solutions were washed out within 15 min but not after 2 hr.

Cations affect the rate of heat activation of freshly harvested highly dormant spores of *B. popilliae*.[51] Less than 1% germinated without activation, but the germination rate was increased greatly by ageing, by sonication, or by extended washing as well as by heating. All of these activations were prevented by low pH or by low concentrations of K^+. K^+ had no effect on germination and its effect in restraining heat activation was fully reversible. Ca^{2+} (less effectively: Mn^{2+}, Sr^{2+} or Mg^{2+}) increased the rate of heat activation whereas K^+ (less effectively: Li^+, Na^+, Rb^+ or Cs^+) inhibited the effect of Ca^{2+}. A plot of the reciprocals of the activation rate and the Ca^{2+} concentration (Fig. 4 of Ref. 51) indicates that Ca^{2+} and K^+ compete for the same site during the activation process.

Most spores germinate best near neutrality, although *thermoacidurans* strains of *B. coagulans* complete all phases of the spore cycle at pH 4·5[52] and *B. pasteurii* does so at pH 10.[53] Spores of strains of *B. cereus* and *B. circulans* selected for their alkali-resistance are able to germinate and grow at pH 10·0 and 11·0.[54] The germination of heat-activated spores of *B. licheniformis* by L-alanine is repressed below pH 6.[50] The pH range for ionic germination of spores of *B. megaterium* B1551 with KCl is 4·1 to 10·6.[28] Cationic surfactants are ineffective below about pH 4, anionic surfactants above this pH.[55] Ca·DPA germination is blocked below about pH 6 and Ca·DPA-activated spores are deactivated at pH 3.[9] Under various laboratory conditions all except the earliest steps of germination are restrained at the extremes of pH. The pH has a major role in the control of the lytic enzymes that are active in the solubilization phase of germination. Warth *et al.*[56] used 1% EDTA at pH 8·5 to restrain the disappearance of the cortex in mechanically disrupted spores.

The heat activation of *B. cereus* T at 65°C for germination with L-alanine + adenosine occurs over a very wide pH range but most rapidly at pH 2 to 3.[12, 13] At pH 1 activation for germination by L-alanine was lost on extended heating but the spores germinated in a rich nutrient broth. The Ca^{2+}-stimulated heat activation of *B. popilliae*[51] is highly sensitive to pH, with an optimum near pH 7 and half-maximal responses at pH 6 and 8·5. In tris Cl′ buffer maximal heat-activation was found at pH 9 to 10. The effects of K^+ and Na^+ were not markedly sensitive to pH. Heat activation of *Cl. bifermentans* CN 1617 at pH 10·5, as compared with pH 7·4, markedly simplified the germinant requirements and eliminated the spontaneous de-activation on storage.[14]

The exchangeable cation load of spores[57] has profound effects on heat

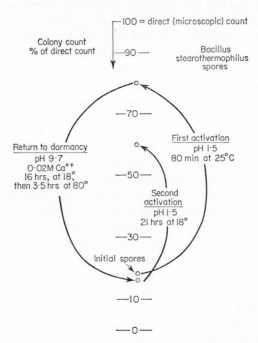

FIG. 2. Activation of spores of *B. stearothermophilus* FS1518 by exposure to acid, and restoration of germinative dormancy by exposure to Ca^{2+} at alkaline pH (Temp. °C). (From Lewis, Snell and Alderton.[10])

resistance, activation requirements and germinative response; pH plays a major role through its effects on the cation load. At a sufficiently low pH the spore becomes "stripped" of its exchangeable (usually metal) cations, which are replaced by protons. At intermediate pH values divalent cations are bound more effectively than monovalent cations at the exchange sites. At high pH values the proton cannot compete effectively. The kinetics of the uptake of Ca^{++} explain the phenomenon of *heat adaptation*.[39] The activation energy for the acquisition of a more heat-resistant condition by *B. megaterium* NRRL B938 is 18,000 to 20,000 calories as compared with about 100,000 calories for killing by moist heat.

Spores of *B. stearothermophilus* 1518 require a severe heat treatment (for example, 4 min at 115°C) to achieve a maximal colony count. This deep germinative dormancy is broken without exposure to elevated temperatures by treatment at a low pH. It is restored by exposure to Ca^{2+} at a much higher pH and at an elevated temperature to facilitate the uptake of Ca^{2+} or its penetration to the activation controlling sites. The transformations can be repeated cyclically (Fig. 2). Reloading the stripped spores with Ca^{2+} at pH 5·7 restores heat resistance but not germinative dormancy.

Spores of *B. megaterium* Texas lightly loaded with Ca^{2+} (*Ca-spores*)[11] dispense with the need for strong electrolytes for induction of germination by L-alanine + inosine. Ba-spores and Sr-spores, but not Mg-spores, responded like Ca-spores. Stripped spores (H-spores) and native spores

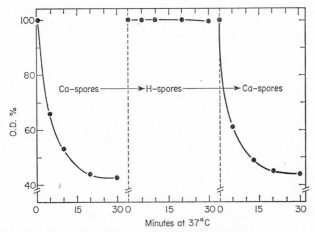

FIG. 3. Effect of cation load on germinative dormancy of spores of *B. megaterium* Texas. Ca-spores (left) were converted to H-spores (centre) by titration to pH 1·8 with 0·01 N HNO_3 and washing in water. They were then converted to Ca-spores (right) by exposure to calcium acetate and washing in water. (From Rode and Foster.[11])

required strong electrolytes. The germinability was reversible through cyclic shifts from Ca-spores to H-spores to Ca-spores (Fig. 3). A survey of 72 isolates of Bacillus disclosed a variety of responses to cation load. Minimal germinative dormancy of the Ca^{2+} loaded form was observed for a majority of the soil isolates and for several species from culture collections.

Ca-spores of *B. macerans* strain 7×1, which have unusual germination requirements for 0·02 mM adenine + 0·01 M fructose,[19] show an enhanced response to adenine or to adenine + fructose but a reduced response to fructose in comparison with native spores.

The germinative response of spores of *B. megaterium* B1551 to glucose[58] was eliminated almost completely by 5 min exposure at 30°C to 0·4 M thioglycolate in 0·05 M glycine Cl' at pH 2·5. Germinability was restored by the addition of Ca^{2+}, Mn^{2+}, Ba^{2+} and Mg^{2+} but not by K^+ or Na^+. Inasmuch as a similar effect was obtained with pH 2·6 glycine buffer alone after 2 hr, Rowley and Levinson[58] suggested that the thioglycolate had speeded the removal of exchangeable cations. The proposal is consistent with the induction of sensitivity to heat by the pH 2·6 thioglycolate treatment but not by 5 min exposure to the pH 2·6 glycine buffer alone. The interrelationships of cation load, heat activation, and specific germinants for these spores have been summarized recently.[220]

We do not know as yet the identity or even the localization of the cation exchange groups of spores. The exchange rates for the Ca^{2+} titration of spores are biphasic.[39, 59] The exchange capacity of isolated coats of *B. megaterium* Texas is similar to that of whole spores on an equal dry weight basis.[59] The coats take up Ca^{2+} rapidly, whereas the profound effects of Ca^{2+} loading on the germinability and heat resistance of dormant spores are exerted much less rapidly.

3. Water activity

The germinability of spores is restrained by marked reduction of the water activity (a_w) of suspensions, powders or isolated spores. The a_w of spores may be manipulated in two ways: (*a*) the concentration of ions or neutral solutes may be varied in water suspensions of spores; (*b*) the relative humidity of the atmosphere may be used to control the a_w of spores in powders or on surfaces. As was discussed in the preceding section, certain ions and range of concentrations are necessary for optimal germination of most spores, although germination in distilled water has been observed. Low ranges of a_w cannot be studied in water suspensions because high ion concentrations become inhibitive at relatively high values of a_w, whereas the viscosity and other physical effects of the more inert neutral solutes become limiting at concentrations for which a_w is still high. Germination of several species took place with physiological germinants in sucrose solutions with a_w as low as 0·96 to 0·98, depending on species and storage history.[60] However, the complete germination of *B. megaterium* B938 by 0·04 M Ca·DPA persisted down to a_w 0·8 to 0·9 in solutions of glycerol, glycol or sucrose, although the darkening proceeded very slowly.[61]

On the other hand, control of the a_w of spores by water transfer through the gas phase is comparatively easy. Water adsorption isotherms for spore powders are similar to those for wool and other insoluble proteinaceous materials. Small piles of well-washed spores of *B. megaterium* B938 germinated during exposures of 1 day to several weeks at a_w from 0·98 to 0·88.[62] The visually "dry" spores at a_w 0·98 germinated as rapidly as pastes ($a_w = 1$) and much more rapidly than in suspensions ($a_w = 1$, or 0·98 with sucrose). Similar observations have been made with five other species of Bacillus.[63] Spores of *B. subtilis* 15u and *coagulans* 43P require heat activation for maximal counts on nutrient agar, but this was not required for germination in moist air. Spores dried onto optical cover glasses germinated more rapidly in moist air when they were contiguous than when they were well separated, but most of the isolated spores darkened finally. The germination may have been mediated by freed Ca·DPA "creeping" from the first spores to germinate (cf. p. 337).

Germination of spores dispersed in bulk solids has been discussed. [17, 64] Several response patterns are seen for storage at a_w ranges from the highest downwards; germination and outgrowth, germination and death, germination with extended survival and continued dormancy. At very low a_w the storage stability of the dormant spores may be impaired.[6]

Upon short exposure over pure water B. *megaterium* B1551 spores coated onto 0·1 mm diameter glass beads became activated for subsequent germination by 0·025 M glucose but not by 0·01 M L-alanine (0·05 M phosphate, K+, pH 7).[65] With 4 hr exposure the spores darkened spontaneously when resuspended in water. Activation was maximal in atmospheres at a_w 0·75 to 1, but some response was found at a_w as low as 0·25. Activated spores were deactivated in absolute ethanol (66°C, 18 hr), and reactivated by again exposing to water vapor, aqueous ethanol or sublethal heating. Hyatt et al.[65] proposed that the activation treatments "may permit water to reach a specific site whose hydration is essential for the functioning of germinants".

Spores attached to surfaces and exposed to high humidities may behave differently from spores exposed as powders. The effect of alkali from the glass beads of the experiments just cited is uncertain. Ordinary glass microscope slides corrode in the presence of spores germinating in humid air.[63]

4. Oxygen and oxidation-reduction conditions

The Clostridia are poisoned by oxygen and require a low oxidation–reduction potential for outgrowth and multiplication. The presence of oxygen would be expected to affect the germination of spores of bacteria with requirements for growth that vary from highly aerobic to strictly anaerobic. No strict relationship of this nature is found. Spores of some obligate anaerobes can germinate in the presence of air. The absence of oxygen does not prevent the initiation of germination of various Bacilli. Upon induction by L-alanine spores of *Cl. sporogenes* PA3679h germinate to the state of emergence.[66, 67] The germination of spores of *Cl. roseum* 6012 could be blocked at any stage by the introduction of air.[68]

Knight and Fildes[69] in 1930 reported that the limiting oxidation–reduction potential (O/R) for the germination of spores of *Cl. tetani* is 110 mv at pH 7·0. Wolf[221] has recently distinguished the effects of O/R and O_2 concentration in the germination of spores of *Cl. bifermentans* and *sporo-*

genes. Variations of O/R between −390 and +589 mv in a N_2 atmosphere were without effect on the germination rate. Kaufmann and Marshall[70] have reviewed more recent literature and discussed O/R and the development of *Cl. botulinum* in milk. Sterile milk spontaneously becomes more reducing as does meat, which is often added to media for the culture of anaerobes. Purified reducing agents such as thioglycolate are added for the same purpose. The use of reducing agents under more extreme conditions in the elucidation of the nature of the restraints on germination will be treated later.

Spores of *B. cereus* T, whether heat shocked or not, are germinated by nitrite (0·5 to 2 M, pH 6·9, 43°C) with production of nitrate.[71] H_2O_2 accelerates the germination of spores of *B. megaterium* Texas by L-alanine or inosine individually, but reduces the effectiveness of the combination.[72] De-activation of heat-activated spores of *C. bifermentans* CN 1617 is promoted by thioglycolate and prevented by continuous aeration.[14]

5. Protein reagents

The prominent role of structural protein and murein[73] in the maintenance of dormancy, and of enzymes in the initiation of germination—and probably in the sensing of germination stimuli—suggests that agencies that interact with proteins and mureins would have profound effects on dormancy and germination. Indeed, the effects of temperature, pH and ions are of this nature. Proteins bind uncharged molecules, cations and anions both non-specifically and with high specificity. Highly specific germinants presumably act through binding to specific sites on proteins. The germination promoting and inhibiting actions of alkane gases[46] probably reflect the stimulation or inhibition of particular enzymes by gas molecules dissolved within the minute non-polar ("oily") droplets that constitute the interiors of protein molecules. Finally, reagents that break or make covalent bonds on proteins may be expected to affect dormancy.

Hg^{2+} and complex mercuric ions restrain germination effectively but reversibly, presumably by reaction with protein —SH. As little as 0·0001 M Cr^{2+}, Cu^{2+}, Fe^{2+}, or Hg^{2+} inhibit germination of spores of *B. cereus* T.[74] The effects are reversed by washing with phosphate or EDTA or by the addition of glutathione or thioglycolate. Germination which is in progress can be stopped by 0·02 M $HgCl_2$.[75]

The carboxylate, amine and other groups of proteins are modified by a wide variety of reagents, some of which are quite specific. Apart from the reduction of —SS— groups or the bonding of Hg^{2+} to —SH little use has been made of such reagents with spores. They might prove useful in modifying germinative dormancy by blocking or facilitating particular activation and germination paths, in mapping the "free space" of spores, and in localizing particular proteins. Diazomethane, an esterifying agent, blocks both germination and the uptake of Ca^{2+} by mature dormant spores.[76] Preincubation of spores of *B. subtilis* ATCC 6051 with diisopropylfluorophosphate completely blocked their germination in nutrient broth[77] but not in L-alanine.[78]

Cross-linking reactions stabilize protein configurations in such diverse processes as the tanning of leather and the natural maturation of collagen. They may play a role in the sporicidal (inducing extreme germinative dormancy?) effects of dialdehydes, such as glutaraldehyde in alkaline 70% isopropanol[79] (and see Chapter 16). Formaldehyde forms cross-links in its role as a tanning agent.

Amine groups are sensitive to alkylation by epoxides and a wide variety of other reagents. Phillips[80] noted that such agents were much more active on spores, relative to their activity on sporeforming and non-sporeforming bacterial cells, than hypochlorite, Ag^+ phenols or quaternary nitrogen compounds.

Disulphide bond-breaking reagents facilitate germination of spores by lytic enzymes (see Chapter 11). Pre-exposure to such reagents increases the susceptibility of spores to H_2O_2 at alkaline pH. Buffer ion effects strongly suggest the involvement of an endogenous lytic enzyme.[81]

Disruption of protein configuration by detergents and by hydrogen-bond replacing solutes might be expected to affect dormancy. Germination is initiated for a wide variety of spores by many different ionic surface active compounds, but non-ionic surfactants are ineffective.[55] Long-chain alkyl primary amines, such as n-dodecylamine, are non-toxic during the initiation of germination.[82] Dilution (by plating) allows colony formation by almost all of the spores during the first 2 min of exposure at 40°C. The optical density and other changes of germination are arrested reversibly by acidification or by lowering the temperature to 0°C. The continued presence of n-dodecylamine is not necessary. An exposure of 1 min at 42°C or 5 sec at 60 to 70°C sufficed. Refractile DPA-free boiled spores did not respond to n-dodecylamine. The toxic quaternary nitrogen compounds released DPA but did not give phase darkening.[55] The results are consistent with a mediation of the fall in turbidity by lytic enzymes, but the involvement of enzymes in the initial action of the surfactant is uncertain.

Dimethylformamide (90%; 4°C) and dimethylsulfoxide (room temperature) rapidly activated spores of *B. pantothenticus* NCTC 8162 for growth on peptone agar.[83] Urea (8 M) gave only a slight activation at 20°C. Aqueous ethanol at 30°C activated spores of *B. megaterium* B1551, and absolute ethanol deactivated them.[15, 218] Other solvents also were effective.

Protein denaturing concentrations of urea or guanidinium chloride also arrest the drop in O.D. that occurs after germination has been initiated. Dilution or washing enables the change in optical density to be resumed.

The initiating event, whether in response to L-alanine, Ca·DPA, or n-dodecylamine, does not take place in strong urea solutions, but the capacity to be initiated is unaffected after reduction of the urea concentration. Thus, the enzymes necessary for the lytic phase must be released by the initiating event.

The triggering reaction in spores of B. licheniformis NTCC 9945 is inhibited reversibly by octanol or ethyl pyruvate.[50] Octanol is the more effective in stopping the O.D. changes once germination of the spores has been triggered. The germination of spores of B. subtilis NCTC 3610 was prevented by the presence of 0·1% p-chloro-m-cresol; the repression was reversed by the addition of 2% non-ionic polysorbate surfactant or by washing.[222]

6. Metabolic inhibitors, analogs, and antibiotics

Some typical inhibitors of metabolic enzymes—arsenate, arsenite, azide, cyanide, dinitrophenol, fluoride and fluoracetic acid—have no effect on the initiation of germination,[84] whereas a wide variety of other compounds inhibit at least certain strains. Practical incentives have led to the trial of many substances. Some common food preservatives do inhibit germination, although usually only at concentrations higher than are required to inhibit some later stage of development.[85] Diethylpyrocarbonate, $NaNO_2$ and sorbate are the more effective in preventing phase-darkening. Various substances when present during heat treatment have the effect of reducing the severity of heat processing that just suffices to keep a product from spoiling. The effect often arises from inhibition of outgrowth of heat-injured spores rather than from sporostasis or from an augmented killing during the heat treatment. In this category are subtilin and its esters, nisin, tylosin, Na-palmitoyl-L-arginyl-L-arginyl ethyl ester and related compounds, fatty acids and their oxidized and free radical derivatives, and many others.

Demonstration of a truly sporostatic action requires that normal delayed germination be differentiated from the delay induced by the sporostatic chemical. Filtration, washing or dilution may fail to remove a small but significant amount of the agent firmly adsorbed at germination-inhibiting sites on the spore. An alternative is to inactivate the agent in situ. Thus, Sacks[86] showed that either penicillin or subtilin prevents initiation or germination by a fraction of the population of B. macerans strain 7 × 1. Addition at any time of the appropriate inactivator—penicillinase or trypsin—allowed the inhibited spores to germinate and form colonies. The fractional behavior of the population persisted through a selection of colonies that originated in repressed spores, and production of a new crop of spores. The unrepressed majority of the spores were killed by the penicillin or subtilin during germination or outgrowth.

The sporostatic action of butane and other low molecular weight alkane gases for certain strains and the germination promoting action of the same gases on other strains[46] is noteworthy. Isobutyl vanillate, the most effective of a series of vanillate esters, prevented the germination of a wide variety of spores in milk when added at 0·15%.[87] Veronal (5,5-diethylbarbituric acid) prevents the O.D. drop of spores of *B. subtilis* ATCC 6051 at 0·01 M, pH 7·3, in nutrient broth or in L-alanine + phosphate.[223] Outgrowth was not inhibited, and the effect on dormant spores was eliminated by washing. Barbituric acid was ineffective.

Slepecky[88] found the O.D. drop by spores of *B. megaterium* to be inhibited reversibly by 0·05% 2-phenylethanol, a concentration well below that required to inhibit growth or DNA synthesis.

Long-chain unsaturated fatty acids have long been recognized to inhibit the germination of spores, and soluble starch to counteract this inhibition. It has been disputed whether the fatty acids as such, or derived peroxides, are the effective inhibitors. A recent study[89] shows that outgrowth is inhibited by fatty acids, by oxidized fatty acids and by free radicals derived from fatty acids. Thus, *B. licheniformis* spores were inhibited by 0·01% chlorophyll in the presence of light, a condition that allows the formation of hydroperoxides. The effect was more pronounced when 0·005% linoleic acid was added; and the effect was reduced in the presence of nitrite, a free radical quencher.

Sporostasis is demonstrated readily with structural analogs of physiological germinants. The competitive inhibition of L-alanine-induced germination by D-alanine had been thought to involve L-alanine dehydrogenase,[90] but the production of L-alanine germinable mutants from which L-alanine dehydrogenase is absent, or nearly so, suggests that a different site is involved in L-alanine-induced germination and its antagonism by D-alanine.[91, 92] The growing recognition that allosteric effects may be anticipated in the triggering of germination opens many possibilities.

The interconversion of L- and D-alanine by alanine racemase has complicated the interpretation of germinative dormancy. An auto-inhibition by exudate from the first spores to germinate in thick suspensions of *B. globigii* was traced to the formation of D-alanine from exogenous L-alanine. Inhibitors of alanine racemase such as D-cysteine, D-cycloserine and o-carbamyl-D-serine stimulate L-alanine-induced germination.[93] The racemase may control the frequency of germination in natural populations as conditions in the micro-environment of the spores change.

III. Development and Maintenance of Dormancy

The entire sporulation process, detailed in preceding chapters, is directed toward the development and maintenance of cryptobiotic dormancy. The decision of an individual vegetative cell to form or not to form a spore is unequivocal like other differentiation decisions, but the choice can vary from one cell to another,[94] just as the capability of completing the differentiation can vary. In many ways short of marked morphological aberrations and biochemical deficiencies, mature spores may fail to achieve their maximal potential dormancy. The sporulation frequency and the quality of the spores produced[95] are functions of the *sporulation conditions* (Chapter 3). The degree of germinative and presumably of metabolic dormancy is conditioned by biochemical and physical factors during and after sporulation. With some strains dormancy is so poorly achieved or so delicately balanced as to give recycling of some or most of the newly formed spores. Finally, normal mature spores can be modified reversibly in their competence to undergo germination under particular circumstances.

Levinson and Hyatt[96] found the germinative dormancy of *B. megaterium* B1551 to vary with supplementation of the sporulation medium with $CaCl_2$, phosphate and glutamate. The addition of 0·05 M glucose, lactose or D-ribose gave spores for which a heat shock repressed germinability by L-alanine, but not by glucose.[97] Spores of *Cl. sporogenes* PA3679 produced in unstirred culture germinated spontaneously, whereas when the medium was stirred continuously (anaerobically) the spore yield and germinative dormancy were increased markedly.[98] The addition of 0·5% glucose increased spore yield several fold, but subsequent germination was rapid if the spores were not harvested at the "proper time".[99] It is interesting and largely unexplained that spores of the anaerobes should tend to recycle readily, and yet when harvested they often present difficulties for obtaining high germination proportions.

The heat resistance of spores is affected by many variables operative during the sporulation process (cf. Chapters 3 and 16). Although direct comparisons are infrequent it seems likely that the same variables would affect germinative (and metabolic) dormancy. *B. coagulans* 43P grown in fermentors gave a high sporulation frequency but a low heat resistance and a low proportion of spores that required heat activation for germination on nutrient agar. Spores produced on nutrient agar plates had high heat resistance and only 5 to 10% gave colonies without heat activation.[63] Further examples of the dependence of germinative dormancy on sporulation conditions will be discussed in the following sections.

A. THE ROLE OF ENZYMES

The occurrence, localization and properties of spore enzymes have been discussed in Chapters 7 and 8. Since protein synthesis is not involved in the earliest stages of germination, the essential proteins of the dormant spore must share the spore's inactivity and resistance. Probably they are not all stabilized by the same mechanism. Cross-linking may be dominant in the coat proteins. Some amino acid sequences give inherently heat stable proteins. The enzymes of the core and core membrane may be stabilized by an overall mechanism involving reduced water content and close packing. Glucose dehydrogenase is particularly interesting in that its localization in a narrow layer peripheral to the cortex[100] and its stabilization as enzymatically inactive subunits by high concentrations of monovalent cations[101] may have been determined by an early function in germination.

1. Resistance of spores toward enzymes

Dormant spores are resistant to enzymatic attack; to exogenous enzymes in nature and in the laboratory; to the lytic enzymes of the sporangium that free the mature spore from the remains of the vegetative cell; and to the endogenous enzymes of the spore that result in loosening or dissolution of the coat and cortex. Attempts to disrupt untreated dormant spores with pure enzymes or concentrates have been largely unsuccessful,[102] as have attempts to obtain such enzymes by microbial enrichment methods. Kim and Naylor[103] obtained cleared zones on plates heavily loaded with viable spores of B. stearothermophilus around yellow colonies of an unidentified myxococcus. Germination of spores of several species by subtilisin[78] appears to have been induced by L-alanine released by self-digestion of the enzyme.[104]

A variety of spores of Bacillus and Clostridium are sensitized reversibly to lysozyme, $\beta-1,4$-muramidase, by —SS— reducing reagents under acid conditions.[81] The specificity of the enzyme as well as direct localization by fluorescent-antibody labeling[105] shows the cortex to be the site of action. Thus, the sensitization appears to increase permeability of the coat (see p. 428). B. megaterium 9885 is highly sensitive to lysozyme without a pretreatment.[106] A germination enzyme extracted from dormant spores of B. cereus and tested on sensitized spores behaves like lysozyme.[107] These experiments are described in Chapter 11. If the explanation of the effect of lysozyme on sensitized spores is correct, the effect should be paralleled by many other enzymes that have been isolated for their actions on vegetative cell wall mureins and hydrolytic products. It would also be of considerable interest to test the effects on sensitized spores of enzymes active on phosphate-containing polymers of the vegetative cell wall that supposedly are absent from bacterial spores.

2. Germination enzymes

The very low concentrations at which some organic physiological germinants act implies that the germinant receptor sites are on proteins, presumably on enzymes. The nature and function of the postulated enzymes has been of considerable interest. A search for early events in the breaking of dormancy by physiological germinants has been directed at energy-yielding systems. Although of unquestioned importance for germination, these studies have not attempted to explain the transformation of the peripheral structures of the dormant spore into a less confining assemblage of layers that permit the development of the outgrowing cell. Neither has the involvement of specific germinants with particular metabolic enzymes clarified the way in which these enzymes are kept in an inactive state within the dormant spore. The alternative view, that enzymes that alter the confining structures may exert the initial step of germination, was proposed long ago by Strange and Dark.[108] The germination enzyme of Gould et al.,[107] which presumably acts on a sensitive linkage in the cortical murein, may play such a role after release from its electrostatic binding in the dormant spore.

Following the observation of Vinter[109] that the marked conversion of —SS— to —SH during germination may indicate an active role for the —SS— groups, Gould and coworkers suggested that endogenous "enzymatic rupture of spore disulphide bonds, followed by activation of a lytic system, could invoke changes in permeability and swelling of the spore typical of germination",[110] possibly with "enzymes similar to protein disulphide reductases perhaps using reducing power from the metabolism of substances which initiate germination".[81] "However, the spore enzymes which metabolize germinants cannot be dormant; they must be capable of acting on their substrates in the otherwise dormant spore in order to initiate germination."[107] A view that germinants need not be metabolized to exert their primary function is discussed in the next section.

A germination initiator protein effective for heat-activated spores of B. cereus T has been isolated from ground dormant spores.[111] It requires L-alanine and NAD as cofactors but at lower concentrations than are required to effect germination without the initiator protein. The molecular function or substrate of the protein was not reported.

3. The allosteric effector hypothesis for the action of physiological germinants

The discovery of allosteric control of metabolic branch point enzymes by endproducts structurally unrelated, or related distantly, to the substrate of the branch point enzyme has opened the way to a recognition of the wide biological value of the allosteric control principle. It provides a mechanism by which the specific germinants could activate the lytic processes. It is possible and perhaps likely that a specific germinant activates by allosteric mechanism more than one kind of enzyme in the initiation of germination. A model based on allosteric control of a germination enzyme[112, 113, 219] gives the McCormick–Vary–Halvorson kinetics of germination[27, 114] (see Chapter 11, Section II B). Some further possibilities will be outlined next.

Metabolic dormancy is maintained by the spore structure, by close packing of components, by caulking with Ca·DPA and by the pressure exerted by the loading of the cation exchange sites and the resultant volume changes within and inside of the cortex. This dormancy is released by structural breakdown that follows solubilization of Ca·DPA and the action of lytic enzymes—probably on the cortex. The lytic enzyme(s) are inactive in dormant spores, but they are activated by a suitable combination of organic germinants and ionic environment. Proteins that function in the control of permeability and enzymes involved in the initiation of metabolism also may be activated by the same germinants. The organic germinants act on allosteric sites; thus, they need bear no structural relationship to the substrates of the enzymes. This allows for a wide variety of strain specificities for germinants. The germinants may act at low concentrations by reason of a low K_m and they may be inhibited by analogs of the germinant. Agents that affect the tertiary structure of proteins may either inhibit or potentiate the germinative processes. Suitable ion concentrations and, in some strains, specific ions are necessary to permit active configurations of the enzymes. Involvement of a variety of lytic enzymes is anticipated in view of the complexity of cortex and cell wall, the controlled expansion of the germinating spore and the advantages in re-utilization of cortical components. Independent allosteric activation allows for the effectiveness and the differing germination kinetics of a multiplicity of germinants. The actions of the lytic enzymes are cumulative, giving irreversible changes characteristic of germination. The release of metabolic dormancy initiates the structural and biochemical reconstruction of the hitherto dormant core and restraining layers to form the outgrowing vegetative cell.

The spore's lytic enzymes may have several functions: by the specific splitting of appropriate bonds they may allow a controlled expansion of the spore-structure; they may provide for a controlled release and re-utilization of murein components; and they may sense the environment through the allosteric effects of specific germinants and by their activity initiate the germinative process. A relative lability of the effector site, as has been demonstrated for allosteric enzymes of endogenous metabolism, could account for the relaxation of germination requirements that accompanies the ageing of spores; and through this the broader function of timing of germination could be accomplished.

A remarkable and illuminating aberration of a lytic control mechanism has been noted[18] in *B. megaterium* B9885. The spores germinate normally in a conventional germination medium, but in the presence of certain amino acids (0·7 mM L-leucine is outstanding) the coats are shed prematurely and the cortex and core dissolve with a precipitous clearing of turbidity of the spore suspension. Heat activation is not necessary.[224]

DPA inhibits the germination and phosphate accelerates it. The lysis does not occur in the presence of 0.05 M Ca^{2+}.

It will be of particular interest to relate the composition of the germination exudate—especially the early exudate—to the specific germinants used to obtain the exudate. Some fragmentary observations are available. Exudate from the germination of *B. licheniformis* A by L-alanine contained free D-alanine, but none was found on germination by L-cysteine or L-valine, and radioactive L-alanine or L-valine incorporated during sporulation was released from a coat fraction in substantial degree during germination induced by L-valine but not by L-alanine.[115] The concentrations and proportions of amino acids soluble in cold trichloroacetic acid differed markedly in exudate obtained on germination of *B. cereus* T by L-alanine or by inosine.[116] The release of Mn^{2+} from spores of *B. megaterium* KM and QM varied with the germinant used.[33] With 0.04 M Ca·DPA Mn^{2+} was released ahead of phase-darkening in both strains. Mn^{2+} was released more in parallel with phase darkening of KM by 0.5 mM n-dodecylamine, and Mn^{2+} lagged behind darkening of QM in 0.05 M glucose. The amino acids and other components of exudates of *B. megaterium* KM activated by exposure to Ca·DPA at a low temperature varied with several conditions for phase darkening.[9] Some exudates resembled the impoverished extracts of broken or boiled spores.

Wax *et al.*,[92, 225] isolated mutants of *B. subtilis* that differed from the parent in the greater complexity of the germination requirements. In this respect, they resembled the germination mutants found among the survivors of severe heat treatments.[41] Fractional germination was observed with the *B. subtilis* mutants. Two major routes of germination were postulated for the parent strain, of which one route was blocked in the mutants. The germination patterns were not correlated with the presence or absence of L-alanine dehydrogenase. It would be of interest to look with these mutants for a relationship between growth in the presence of a high concentration of L-alanine and germinative dormancy toward L-alanine such as was found for spores of *B. cereus* T.[117]

Fractional germination promises advantageous materials for elucidation of the roles of specific germination enzymes inasmuch as the populations as grown should be separable, biologically or physically, into more homogeneous fractions that are phenotypically distinct but genetically alike.

B. THE ROLE OF PERMEABILITY

The rapid large increase in permeability early in germination implies that impermeability is an essential aspect of the state of dormancy. Elucidation of the structural basis of impermeability is necessary for an understanding of the development and maintenance of dormancy. Unfortunately, we do not know where the permeability barriers lie or what their selectivities are. The dormant spore in its entirety is permeable to water, and possibly to somewhat larger molecules. On the other hand, some small molecules are excluded by peripheral layers. Fixatives and embedding mixtures penetrate poorly beneath the outermost coats. Unfixed spores show a sharp boundary for the penetration of mathacrylate near the juncture of the inner coat and cortex.[118] The unfixed cortex is not infiltrated by methacrylate.[119] OsO_4 penetrates poorly beneath the outer coat of dormant *B. coagulans*, whereas details of inner coat, cortex and core membrane are shown clearly with heat-activated, germinated or broken spores.[120] A short OsO_4 fixation neither extracts DPA from *B. megaterium* nor gives good internal detail, whereas $KMnO_4$ fixation extracts DPA, gives stainable spores and gives good electron micrographic detail.[118]

Dyes stain only the outer regions of dormant spores unless pretreatments or staining conditions are such as to destroy the impermeability. Germinating spores acquire stainability and a phase dark appearance and lose soluble components approximately simultaneously. Acquisition of resistance to staining by basic dyes during sporulation coincides with the formation of the cortex and expansion of the developing spore to fill the loose outermost coats in *Cl. sporogenes*.[121]

Entry of the small enzyme lysozyme (MW 14,600) to the cortex is dependent on splitting of —SS— bonds that are the natural state of cystine sulphur in the spore coat. Possibly these bonds form an important part of the inter-protein linkages of the coat inasmuch as reduction is essential for the isolation of monomeric coat protein,[122] whereas hydrogen bond-breaking solutes such as urea or guanidinium chloride are ineffective by themselves, showing little effect on dormant spores.

Free space measurements[123] suggest a permeable region for neutral solutes of MW 500 that would encompass most of the spore—certainly, of much more than the coat volume as estimated from electron micrographs of spore sections or from compositional data. This has led to the suggestion that the core of the spore is highly permeable.[124] Unfortunately, interpretation of the free space measurements rests on the assumption that the partition ratio for the test solute, the equilibrium concentration within and outside the spore, is equal to 1. Helfferich[125] in his discussion of the much simpler synthetic resins lists the many reasons whereby it is "hardly surprizing that such a behavior [molal distribution coefficient = 1] is rarely found". If the partition ratio in spores is much larger than 1 it is possible that the free space studied by Black and Gerhardt is restricted to the coat or coat + outer cortex. It also is likely that the relationship of permeable volume to the chemical nature of the test solute is not solely a function of MW in a chemically similar series such as the polyethyleneglycols or sugars, but also will reflect a changing partition ratio. Thus, a morphological entity such as the cortex may present a much sharper permeability barrier than is suggested by the relationship of Black–Gerhardt free space to MW. Pores of diminishing or distributed sizes could account for such a relationship but attempts to observe pores in dormant spores have failed.

High values of free space of dormant spores were found with methylene blue and Orange G, and a value in the inert solute range for Eosin Y, by Gerhardt and Black,[123] who write: "Contrary to common notion, several dyes were found to penetrate spores, a fact which also was observed microscopically in spores stained but left unwashed. The uniform distribution and intensity of staining provided evidence that permeability was equal for all the spores in the population." The visibility of the spores in the dye solution strongly suggests a partition of dye markedly in favor of the interior water of the spore. The fluorescigenic dye n-tolyl-α-naphthylamine-8-sulfonate gave a sharp fluorescence only with germinated spores. In dormant spores the dye did not conjugate with the negatively charged groups necessary for the fluorescence.[126]

It should be possible to delimit the penetration of dormant spores by protein reagents by observing the location in spores of components to which the reagent is covalently bonded, or of components recognizably modified by the reagent. Modification of unique components of the cortex may establish whether or not the cortex is permeated by low MW solutes.

Various mechanisms may control permeability of the dormant spore: for example, electrostatic expansion under the influence of ion concentrations and cation load, or protein configurational changes in response to ion concentrations and more specific effectors. As germination is initiated the lytic and exudative processes produce marked changes in permeability. The release of endogenous germinants and inhibitors through permeability changes may be significant in the activation or germination initiation of dormant spores.

The dramatic effects of mechanical germination—by puncture or crushing—are most readily interpreted as resulting from disruption of a permeability barrier and the activation of enzymes inhibited by the absence of substrate, by the absence of an effector, or by close-packing. Abraded spores[127] quickly become stainable and susceptible at 4°C to 80% ethanol, to 15% H_2O_2, or to 3% phenol. They release their DPA at 65°C. The composition of the solubilized material released on exposure of the abraded spores to outgrowth conditions resembles that of physiologically germinated spore exudate, and differs from the composition of the solubilized material of disintegrated dormant spores.

C. The Role of Water Content

Although the dormant spore is freely permeable to water, it clearly binds much less water than the fractionated components of the spore can bind. Grecz and Smith[128] found a marked increase in equilibrium water content of spores of *Cl. botulinum* types E and A after heat killing. The materials released absorbed considerably higher proportions of water than the dormant spores. They concluded that all or most of the polar groups of spores are masked with respect to water binding, probably on a molecular level; and that such masking is directly related to dormancy, high refractility and heat resistance. They based the conclusion on the belief that the "total spore structure is extensively permeable to relatively large molecules", a supposition that cannot be justified solely by the observations of Black and Gerhardt discussed above.

The speed of water uptake by dormant spores must be limited by the rate of configurational changes in the polymeric materials of the spore rather than by impermeable barriers, which are not reasonable on the necessary dimensional scale.[129] Thus, the a_w within the spore must parallel the a_w of the spore's environment. Readjustment is rapid. Spores equilibrated at $a_w = 0.3$ and then placed at $a_w = 0.9$ at a sterilizing temperature begin to die at the rate characteristic for $a_w = 0.9$ within 20 to 30 sec.[130]

The heat resistance of spores of widely varying resistance at a_w near 1 becomes much less variable at a_w near 0.3. Thus, the spores respond as though the most sensitive vital part of the spore were at a_w near 0.3.[131] If one is unwilling to postulate water impermeable barriers, all parts of the spore must be at the a_w of the environment. The dilemma was resolved by recognition[129] that pressure in a system at high a_w could give a reduced water content that corresponds to a low a_w in a pressure-free system. It was suggested that the cortex (the *contractile cortex*) might have this function. A lowered water content of the magnitude proposed as well as the close-packing induced by the necessary pressure—perhaps of the order of 100 kg/cm²—would maintain the metabolic dormancy of the affected regions without the need of postulating further restraining mechanisms. Release of the pressure would allow the immediate initiation of metabolic activity. It is possible that the increases of metabolic activity on activation are controlled through a change in internal pressure.

D. The Role of Metals

Two metal ions, Ca^{2+} and Mn^{2+}, play a special role in sporulation; requirements for sporulation greatly exceed those for vegetative growth. Mn^{2+} must be supplied early in sporulation, and a deficiency is expressed as a low sporulation frequency. Sporangium formation in Mn^{2+}-deficient

B. megaterium synchronous cultures was stimulated by an amino acid supplement, but the spores formed germinated reluctantly.[132] Ca^{2+} is required for maturation, and may be supplied late as in endotrophic sporulation. It is deposited in the spore after the developing spore becomes visibly refractile, and more or less coincident with the deposition of DPA and the acquisition of heat resistance.[109, 133] A Ca^{2+} deficiency is expressed as defective spores.

Both Ca^{2+} and Mn^{2+} are suspected of being associated with DPA in the dormant spore. Ca^{2+} normally is the major divalent metal ion in spores and the molar ratio of non-exchangeable divalent metals to DPA approximates unity. The electron paramagnetic resonance spectrum of Mn(II) in dormant spores is unique among biological materials and is characteristic of a chelationally bound state.[4] Both Ca^{2+} and Mn^{2+} are exuded during germination.

Incorporation of Ca^{2+} and DPA coincides with the formation of the cortex. Penicillin, an inhibitor of muropeptide synthesis, caused a slowing of Ca^{2+} and DPA deposition and a subsequent release even of that Ca^{2+} and DPA taken up before addition of the penicillin.[134] Ca^{2+} was released before DPA, and more rapidly on washing with 0.001 N HCl than with water or dilute salt solution. Fixation of diaminopimelic acid was not blocked by penicillin. Tetracycline antibiotics tend to be bound at sites of Ca^{2+} mobilization in many biological tissues; e.g., in bone fractures. The tetracyclines are bound reversibly with Ca^{2+} to sporulating cells, blocking the irreversible incorporation of Ca^{2+}, the synthesis of DPA and the normal phase brightening of the spore.[135]

There is a critical period during which the Ca^{2+} needed for sporulation must be supplied. Ca^{2+} uptake preceded DPA synthesis by 1 hr in synchronously sporulating *B. cereus* var *alesti*.[136] The addition of Ca^{2+} promptly when the Ca^{2+} deficient spores became phase-bright permitted normal DPA development, but at the time of maximum phase brightness 2.5 hr later Ca^{2+} was not taken up and DPA was not made.

By removing vegetative cells committed to sporulation, from their nutritive medium, washing and resuspending them in water or solutions of known composition (*endotrophic sporulation* or *replacement technique*) more or less well-matured spores are obtained. Ca^{2+} is found to be the only essential nutrient that has not already been accumulated in amounts adequate for spore formation.[137, 138] Ca^{2+} is not replaced effectively by other metals, but Sr^{2+}, Ba^{2+}, Ni^{2+} and Zn^{2+} gave a better degree of heat resistance than no metal at all. Replacement by Mn^{2+} gave spore crops that germinated extensively during harvesting.[139] Ca^{2+}, Sr^{2+} and Ba^{2+}-enriched spores produced endotrophically[140] differed strikingly in germinability. Sr^{2+}-enriched and especially the Ba^{2+}-enriched spores were slow to germinate. Prompt germination occurred with spores enriched in both Ca^{2+} and Sr^{2+} or Ca^{2+} and Ba^{2+}. The spores enriched endotrophically in divalent metals did not require heat activation for physiological germination as did spores grown in the presence of Ca^{2+} in the usual way. The authors suggest that Sr^{2+} and, less effectively, Ba^{2+} can serve for the morphological development and the maintenance of dormancy, but that Ca^{2+} is required for full germinative competence.

The inorganic makeup of the sporulation medium, and of the washes used in harvesting, must affect the exchangeable cation load of dormant spores and thereby affect their germinative behavior. Approximately 35% of the $^{45}Ca^{2+}$, Sr^{2+} or Ba^{2+} of spores of B. cereus T grown in the presence of these metals were replaced by soaking the harvested spores in 1·45 mM $CaCl_2$[141] The non-exchangeable Ca^{2+} presumably is that portion of the total divalent metal ion content that is associated stoichiometrically with the DPA content. The exchangeable Ca^{2+} of spores of B. megaterium Texas, the removal of which had profound effects on germinability, amounted to only about 5% of the total Ca^{2+} of the native spores.[11] The exchangeable Ca^{2+} content of these spores presumably was considerably less than the total exchange capacity of the spores.

Attempts have been made to relate the high metal ion content of dormant spores to the maintenance of dormancy. Harrell[142] found a stimulation of glucose oxidation by cell-free extracts of dormant spores of B. cereus T upon addition of DPA. He suggested that "at least one of the functions of DPA in the activated or germinating spore is the removal of some excess metallic ion, possibly calcium, which otherwise would inhibit the enzymatic activity of the resting spore".

Spores produced in the presence of a high Mn^{2+} supply may tend to germinate readily or spontaneously.[139] Levinson and Hyatt[96] found a more complex situation with B. megaterium B1551 wherein spores grown with 0·1 mM $MnCl_2$ germinated optimally in response to L-alanine but poorly in response to glucose or L-leucine. It has been suggested that Mn^{2+} acts by stimulating proteinases and pyrophosphatase.[143, 144]

The importance of metals in the maintenance of dormancy is indicated by the occasional effectiveness of metal-binding (chelating) agents in inducing germination. The important case of 0·04 M Ca·DPA (which is treated later) is atypical since in this case the chelated metal complex is the effective germinant. Removal of adventitious inhibitive metals is one mode of action,[74] but most spores do not exhibit this sort of contamination. It is curious that chelating agents are not general germinants in view of the importance of Ca^{2+} for the maintenance of dormancy. It is possible that the anionic charge of Na_2·EDTA prevents its penetration to active sites, but a neutral chelate complex of a moderately bound metal (which would exchange readily for Ca^{2+}) also is ineffective. An example is Mg·2-hydroxyethyliminodiacetate.[61]

A germinative action of EDTA on spores of *Cl. sporogenes* PA3679 was reported by Brown[145] and extended by Riemann.[146] Concentrations as low as 0·01 M were effective; 0·1 to 0·25 M were inhibitive for the derived strain 3679h but not for 3679.[146] Germination took place at pH 5·5 to 9·0. It went slowly at 4°C but rapidly as high as 65°C. DPA (0·033 M) and tripolyphosphate (0·01 M) behaved like EDTA on one batch of PA3679 spores. Other anaerobic and aerobic spores did not respond to EDTA. Loss of heat resistance preceded phase darkening by EDTA. At low temperature the separation of the processes was very marked, but upon warming the phase darkening of the heat-sensitive spores took place. EDTA, DPA and Ca(DPA)$_2$ inhibited the further release of ninhydrin-reactive materials at 37°C by ruptured spores but not the initial release during Mickle disintegration in the cold.

Spores of *Cl. thermosaccharolyticum* germinate slowly and incompletely. The highest proportion (67%) with strain TA37 was observed in 0·1% EDTA after 36 hr at 58°C under anaerobic conditions and with heat shocking.[147]

E. THE ROLE OF DPA

Much circumstantial evidence suggests a central role for DPA in the development and maintenance of dormancy, but the nature of the role remains unknown. DPA is absent from the vegetative cell. It appears in the sporulating cell during the maturation of the cortex, the uptake of Ca^{2+}, and the enhancement of refractility. The amount in matured spores varies with species and cultural conditions from less than 2% to about 15%. The amount is correlated with the amount of divalent metals, usually Ca^{2+}, in a ratio that usually approximates 1. Spores (pressed into KBr discs) show the u.v. spectrum of Ca·DPA. Upon germination part of the DPA is released quickly but the release of the final portions may be delayed. The location of DPA and its state of chemical binding within the spore are not established. Correlation of formation of DPA with the formation of the cortex in spore development, and of the solubilization of DPA with the dissolution of the cortex in germination or mechanical disruption, makes the idea of a cortical location attractive.

1. Biosynthesis and content of DPA

The biochemical route to DPA in sporulating cultures and in cell-free systems is described in Chapter 3, Section III D. Spores of low DPA content produced by mutant strains deficient for DPA synthesis, by Ca^{2+} deficiency during endotrophic sporulation, by cultures inhibited by diethyl pimelate, ethyl oxamate, or by lysozyme, or by strains with nutritional idiosyncrasies—all these have low heat resistance and germinative dormancy. The restoration of heat resistance, and presumably of dormancy, by the addition of DPA to the sporulating cultures in certain cases suggests that DPA, like Ca^{2+}, is translocated into the developing spore from the sporangium. The possibilities that DPA arises *in situ* by cyclization from precursors incorporated into a macromolecular matrix or that DPA is formed during the process of germination from a reduced precursor seems unlikely now.

Variations of DPA content induced during sporulation profoundly affect dormancy. Only a small fraction of the total DPA content of the dormant spore is released by mild heat treatments such as are used for heat activation. Changing the concentration of yeast extract and phenylalanine[25] in the sporulation medium caused the DPA content of spores to vary from 2·2 to 15%; the duration of treatments at 65°C required for maximal germination in L-alanine or L-cysteine was proportional to the DPA content. On the other hand, the O.D. changes in response to 0·04 M Ca·DPA were independent of DPA content or duration or omission of heat activation. Low DPA spores (2·3 and 3%) were damaged more by extended heat activation than normal spores (7% DPA) in that they tended to germinate spontaneously in 0·02 M tris, pH 8·5, 30°C; the O.D. changes were relatively small, and the response to L-alanine with or without adenosine was sluggish. The response to L-alanine was repressed by D-alanine, and stimulated greatly by a low concentration (0·004 M) of DPA.[148] The weak response to L alanine was attributed to a NAD deficiency for the action of L-alanine dehydrogenase, and its relief by DPA to a stimulation of NADH oxidase.

2. Release of endogenous DPA

DPA does not seem to be fixed in the dormant spore by covalent bonds, but rather by hydrophobic and hydrogen bonds. It is solubilized completely, though at variable rates, by germination, by severe heat or radiation, or by mechanical damage. The kinetics of release are complicated by the familiar problem of distinguishing between the response of parts of the population and a uniform response of the whole population. Breakage in the cold has given quantitative liberation of DPA in numerous trials.[130] The DPA is diffusible through cellophane. Young[149] found an association of the released DPA with Ca^{2+} and amino acids during migration on paper chromatograms that could be simulated partially with pure components. It is not clear whether enzymatic actions have been avoided entirely by cold disruption of spores. Rode and Foster[127] obtained soluble dialyzable DPA when spores were ground with Ballotini beads in 80% ethanol. On the other hand, Fleming[150] found that the DPA from physiological germination of *B. subtilis* was not retarded on columns of Sephadex G-50 (NH_4HCO_3, pH 8; excludes MW 10,000) or Sephadex G 75 (NH_4OOCH, pH 3; excludes MW 50,000), suggesting association of the DPA with high MW components.

Only a small fraction of the DPA is released during mild heat treatments such as are used for heat activation. The optimum heat shock released about 10% of the DPA from spores of *B. cereus* T.[25] The release from *B. megaterium* B1551 at 75° or 85°C was less marked at pH 8 than at lower pH values.[151] Rowley and Levinson[58] found that acid thioglycolate at 50°C caused spores of B1551 to lose DPA and viability, and to become stainable and sensitive to lysozyme although they remained phase bright. Spores of *Cl. sporogenes* lost progressively up to 30% of their DPA when they were heated at 75°C in water or in a buffered (0·004 M phosphate, K[+], pH 7·5) solution of caramelized glucose.[152] The latter condition gave loss of heat resistance but the former did not. Upon heating for 60 min at 65°C in pH 7·2 phosphate aged lyophilized spores of *B. cereus* T gradually released up to 12% of the DPA, and glucose oxidation was stimulated. Spores that had been stored frozen for 6 months neither released DPA nor oxidized glucose.[153]

The release of even a small fraction of the DPA might deplete some local site involved in the maintenance of dormancy.[25] Halvorson[90] commented that the idea of a critical threshold concentration of DPA for the maintenance of dormancy appears unlikely in view of the lack of a relationship, in spores of various DPA contents, of the "amount of DPA released by heat activation and germination; however, the possibility is not excluded that several reservoirs of DPA exist in the spore, only a minor fraction of which controls dormancy. The breaking of dormancy may involve the removal of a DPA reservoir, or a compound associated with DPA from some endogenous site(s)."

Lethal heat treatments cause structural disintegration and release of DPA, metal ions and other components (Chapter 16). An example with a possible relationship to the structural basis of dormancy was reported by Lund,[154] who noted an effect of physiological germinants on the release of DPA from heat-killed spores of *B. subtilis* BB2C. At the time that 99% were dead only 30 to 40% of the DPA had been labilized to the extent that it appeared in the soluble fraction after storage overnight at 20°C. When the heated spores were stored at 35°C a total of 85% of the DPA was released within 4 hr. The rate of release was speeded greatly by the presence of a germination mixture (0·28 mM adenosine + 15·5 mM L-alanine) during the exposure at 35°C. The release from heat-killed spores without germinants was much less sensitive to pH 4·5 or to a change of temperature (above 10°C) than was the release from viable heat-activated spores in the germination mixture. Unfortunately, the pH-sensitivity of the germinant-stimulated release from the heat-killed spores was not reported. The release from both heat-killed spores and from physiologically germinated spores was prevented by 1·8 M acetic acid. The presence or removal of supernate from heat-killed spores neither repressed nor stimulated the release of

DPA upon incubation of unheated or of killed spores at 35°C. The release of DPA from killed spores (without germinants) was optimal at pH 6·5 to 8 and at 50°C. The release at 65°C was markedly slower but the initial rate (0 to 3 min) was maintained for at least 1 hr, and when after 30 min at 65°C the temperature was dropped to 25°C the rate increased at once to the rate characteristic for incubation solely at 25°C. Lund thought these results inconsistent with a conventional view that the decreased rate at the higher temperature was caused by the destruction of enzymes. He observed that ethyl pyruvate completely stopped physiological germination but was without effect on the release of DPA from the heat-killed spores. Muramate and hexosamine were not released at once by the lethal heat, but incubation of the killed spores for 1 hr at 30°C without the germination mixture released about 40 % of the total content of both components. In the presence of the germination mixture about the same amounts of muramate and hexosamine were released from the heat-killed spores as from the uninjured heat-activated spores.

The results can be interpreted as arising from allosteric and other effects on lytic enzymes; especially if it is assumed that DPA-containing layers are exposed successively, that exposure of a fraction of the layers suffices for a lethal increase in heat sensitivity, and that at least part of the enzymes involved in the germinative release of DPA remains available for DPA release in the heat-killed spores. Murrell[226] has suggested a compositional basis for compartmentalization of DPA in the cortex.

Severe irradiation of various types, in dosage well beyond that required to kill spores (Chapter 16), gives release of DPA. Prior heat treatment often renders spores more sensitive to damage by irradiation. X-Radiation at a dosage of 0·5 to 1 Mr in water at 25 to 30°C produced an exudate from *B. cereus* spores that resembled that obtained through physiological germination.[155] Pure Ca·DPA, and the ionic form H·DPA$^-$ to an even greater extent, were degraded by this dosage, but DPA within the spores was more stable.

DPA is released from a variety of spores by ionic surfactants; by anionics below and by cationics above about pH4.[55] Non-ionic surfactants and polypeptidic antibiotics were not effective. Most of the surfactants were toxic, and the extent of the gains in stainability and phase darkening were variable. Cationic surfactants potentiate the action of heat on dormant spores, consistent with the sterilizing action of various quaternary nitrogen and related compounds. Cetyltrimethylammonium bromide (CTAB) is adsorbed rapidly by spores of *B. megaterium* even at 4°C. DPA is released quantitatively by 20 μg of CTAB per 10^8 spores in 1 ml at 50°C. A momentary exposure suffices. The excess CTAB may be removed by centrifuging before much DPA has been removed, and the spores resuspended at 50°C. At 4°C DPA is neither released nor activated for release after removal of surfactant and subsequent heating of the spores to 50°C. The material released at 50°C resembles that released in physiological germination. Alderton[156] has noted that cation-stripped H-spores are disrupted by CTAB in the same way that the H-form beads of polyacrylic resin (IRC 50) are disrupted.

Spores are killed upon exposure to most surfactants that cause them to lose their DPA, but n-dodecylamine at 0·06 mM is relatively nontoxic.[82] If *B. megaterium* spores are plated (i.e. diluted) soon after the DPA is released, the spores remain viable. The optical density drop was maximal at pH 7·4 to 7·8, and was terminated by acidification or stopped temporarily by lowering the temperature to 0°C. Spores from which DPA had been removed by boiling did not undergo the optical density changes with n-dodecylamine. Spores exposed to n-dodecylamine just long enough for heat resistance to be lost retained most of their DPA, but this was lost on further incubation even though the n-dodecylamine was diluted or neutralized by the addition of 0·1% Na lauryl sulphate. The secondary release of DPA was retarded by low pH or low temperature but not by a variety of poisons active on various metabolic enzymes. The secondary release of DPA resembled that observed by Lund[154] for heat-killed spores (see p. 334).

Rode[157] found substantial amounts of DPA to be released from spores of *B. megaterium* suspended in the anode compartment of an electrodialysis cell operated at 20 to 28 milliamperes. The spores collected on the cellophane membrane as a tough pellicle, and thus behaved as though negatively charged. The bulk pH of the anode compartment was 11·7 and of the cathode compartment 2·9 after 7 hrs, when the temperature had risen to 39°C. Control suspensions at these pH values did not give DPA release. A reinvestigation of electrodialysis of spores of *B. cereus, stearothermophilus* and *subtilis* in a cell in which the spores were isolated from extreme pH shifts or temperature rise failed to give any loss of DPA.[158] Here also the spores migrated toward the anode.

The release of DPA from spores of *B. stearothermophilus* at pH 4 in the range 80 to 100°C corresponded to an activation energy of 46,000 calories.[227]

The release of DPA during physiological germination follows the loss of heat resistance but precedes somewhat the lytic release of muropeptide, amino acids and peptides. Ca^{2+} was released from germinating spores of *Cl. roseum* somewhat sooner than DPA.[159]

3. Ca·DPA germination

A DPA content of 15% in dormant spores of *B. megaterium* (buoyant density 1·4) corresponds to 1·25 M overall in the spore solids. Presumably much higher local concentrations exist. Yet a 0·04 M concentration of Ca·DPA in the suspending medium suffices to initiate germination and to release spore DPA in the remarkable process discovered by Riemann and Ordal.[146, 160, 161] Ca·DPA is a relatively general germinant which acts on spores of *Bacillus* sp., of *Clostridium* sp. and of *Sarcina ureae*,[162] bypassing the specificities of the physiological germinants that presumably developed in response to specific ecological situations. One may infer that it acts on a structure common to all of these spores. The existence of the DPA moiety, if not of Ca·DPA itself, as a universal component of these spores may even imply a role for Ca·DPA germination as a part of the natural germination initiated by strain-specific germinants.

The following factors control the balance between dormancy and the initiation of germination by Ca·DPA or closely related compounds:

(*a*) *The existence of a threshold.* Maximal activity is found for spores of *B. megaterium* and other aerobic species with 0·04 M Ca·DPA; 0·02 M is inactive. The activity correlates with the calculated concentration of [Ca·DPA]°, the uncharged ionic form, when the concentrations of metal ion and ligand are varied separately.

(*b*) *The restricted replaceability of the chelated metal ion.* Sr^{2+} can replace Ca^{2+} but 0·08 M Sr·DPA is required; 0·04 M is ineffective.[163] Mg·DPA is partially effective but an even higher concentration is required. Ba^{2+} and other divalent metal ions form DPA chelates that are too insoluble for evaluation. The alkali metal salts are not effective.

(*c*) *The restricted replaceability of the chelating ligand.* Although chelating agents related only distantly to DPA may initiate the germination of certain spores, the requirements for a Ca·DPA-like germination include a molecular shape essentially unaltered from that of Ca·DPA and a metal associating strength somewhat similar to that of DPA. 4H-pyran-2,6-dicarboxylate, with a pK_{Ca} (log of the dissociation constant for the Ca^{2+} chelate) of 3·6 vs. 4·2 for DPA and near-identical geometry, is fully effective.[61] 4-Methyl-DPA, with pK_{Ca} 4·3, is ineffective. Isophthalate and pyridine-3,5-dicarboxylate, with shapes nearly identical to that of DPA but with low Ca^{2+} binding strength, are ineffective. Smaller molecules with suitable association constants, such as diglycolate, also are ineffective.

(*d*) *The ionic concentration and specific ion effects.* Germination can occur in the pH range 5 to 9 but the nature of the buffer ion is important. Tris is reasonably good at pH 8 and imidazole or o-aminopyridine at pH 7. Alkali metal cations are tolerated at higher concentrations than alkaline earth cations. Cations and anions can be ranked in series according to the inhibitive concentrations, and these rank orders vary with the strain of spore in a manner reminiscent of the ionic requirements for physiological germination. Very low ion concentrations inhibit, presumably because a certain ionic strength is necessary.

(*e*) *The solubility of the metal chelate complex.* The solubility of Ca·DPA at 25°C is 0·015 M, so a high degree of supersaturation is necessary to

o

secure the optimal concentration, 0·04 M. A low nucleation probability makes this feasible. The effective Ca·pyran dicarboxylate is not super-saturated at 0·04 M, so it seems unlikely that Ca·DPA acts by the disruptive growth of Ca·DPA micro-crystals at nucleation sites within the spore.

(f) *Temperature*. The optimal temperature may be strain dependent. Riemann and Ordal reported about 45°C for spores of PA3679 but about 25°C for spores of B. *megaterium*.[160] Keynan and Halvorson[164] found the rate of O.D. change for spores of B. *cereus* T to be relatively independent of temperature but the duration of the lag phase was strongly dependent on temperature with a ΔH^* of 14,500 cal/mole between 2° and 20°C. The lag period was minimal around 25°C, but it became very long at a slightly higher temperature. Jaye and Ordal[165] noted that exposure of spores of B. *megaterium* to 0·04 M Ca·DPA at 60°C selectively blocked their susceptibility to 0·04 M Ca·DPA at 25°C but that their susceptibility to L-alanine, glucose, NaCl or n-dodecylamine was unaltered.

Lee and Ordal[9] separated an activation phase in spores of B. *megaterium* from an O.D.-dropping phase which proceeded relatively much more slowly at a low temperature. The activated spores were stabilized by washing in ice water and storing at 3°C. They germinated spontaneously and rapidly in water at 24°C. They could be deactivated by exposure to a low pH, and they retained a competence to be reactivated. Only small amounts of materials were released by the activation or deactivation treatments. Various inhibitors of the germination of the activated spores allowed the exudation of materials that resembled more closely those released by broken dormant spores than those released during the uninhibited germination of the activated spores. The Ca·DPA activated spores did not show a lag in the response to lethal heat (84°C, pH 7, barbital buffer) as did the untreated stock spores.

(g) *Inhibitors*. Whereas L-alanine- and L-cysteine-induced germination of spores of B. *cereus* T were inhibited by related amino acids, 0·04 M Ca·DPA-induced germination was not. All three germinants were inhibited by ethyl pyruvate, atabrine or $HgCl_2$.[164] Reducing solvents such as 2-mercaptoethanol are inhibitive at much lower concentrations than are ethanol or glycols.[61]

The effects of temperature, ions and inhibitors imply an enzymatic involvement not only during the phase darkening characterized by release of structural components of the cortex but also during the initial activation or lag phase. Neither phase-darkening nor activation occurred when urea (at 2 M but not at 1 M) or guanidinium chloride (at 1·5 M but not at 0·75 M) was present during exposure of dormant B. *megaterium* spores to 0·04 M Ca·DPA. Pretreatment or simultaneous treatment with much more severe denaturing conditions (e.g. 8 M urea at an elevated temperature) left the spores unaltered in susceptibility to 0·04 M Ca·DPA after washing.[61]

The results suggest that 0·04 M Ca·DPA releases an enzyme that is inactivated by urea, by moderately elevated temperatures or at pH 3, and that the continued activity of this enzyme is essential for activation of the spore by Ca·DPA. Transfer of spores in this initial activation phase to pH 3 or to 37°C presumably inactivates the enzyme irreversibly since upon restoration of the original condition there is no cumulative effect of the previous treatment on the duration of the lag period. Successive Ca·DPA treatments may activate enzymes located at successively deeper layers within the spore. At 60°C, on the other hand, all of the enzyme within reach of succeeding Ca·DPA treatments may be destroyed and the route of germination via 0·04 M Ca·DPA barred permanently. Phase-darkening is not caused by removal of DPA alone since complete extraction of DPA by lethal heating of spores in water or aqueous alcohol leaves the spores phase bright.

If the early events in the lag phase of Ca·DPA germination are lytic, presumably they occur at specific sites in polymeric structures since only small amounts of components are released.[9] No muropeptide, free reducing sugars, or free amino acids were found although they are released in large amounts during the phase-darkening stage. Two preparations of *B. megaterium* spores with high (1%) Mn^{2+} contents released Mn^{2+} before the O.D. changes were apparent.[165] Upon treatment with 0·04 M Ca·DPA at 60°C 30 to 40% of the Mn^{2+} was released. Up to 15% of the $^{45}Ca^{2+}$ incorporated during sporulation was released continuously during the lag phase although the rate of release increased during the phase darkening phase. The $^{45}Ca^{2+}$ of the dormant spores was not exchangeable with $^{40}Ca^{2+}$.[164]

Attempts have been made to suggest a mode of attachment of Ca·DPA to high MW spore components, and a mode of release, on the basis of simple chemical equilibria. Riemann[161, 146], and Fleming and Ordal[43] have suggested an electrostatic attachment of Ca·DPA and its mobilization via the Ca·DPA$_2^=$ complex under the influence of exogenous ions (see Chapter 11, Fig. 11). In the process macromolecular structures including enzymes would be released from their dormant condition. The discovery of a dimer, [DPA·Ca·Ca·DPA]°, in the crystal of Ca·DPA·$3H_2O$,[166] and its absence from aqueous solution at spectrometrically detectible concentrations[167] offers more possibilities for the mobilization of DPA during Ca·DPA-induced germination.[168]

Thus, Ca·DPA may be embedded in the spore structure through the pyridine ring, and as a result of this association it may dimerize relatively readily with a Ca·DPA molecule in solution. If the dimerization weakens the bonding of the pyridine ring to the spore structure the dimer Ca·DPA$_{structure}$:Ca·DPA$_{solution}$ may be released into solution where it would dissociate at once into two Ca·DPA monomers. A cyclic process thus is provided for the removal of Ca·DPA from critical sites, the further actions on or of which may be the rate-limiting steps of germination. Correlation of intermolecular associations in crystals of metal chelates with the ability of solutions of the chelates to effect a Ca·DPA-like germination is underway.[61] The active Ca·4H = pyran-2,6-dicarboxylate has a crystalline trihydrate that is isomorphous with the Ca·DPA·3H$_2$O crystal.[169] Presumably it could form a mixed dimer

Spore structure: DPA·Ca$\diagup^O_{\diagdown O'}\diagdown$Ca'·pyran-dicarboxylate

corresponding to that postulated for Ca·DPA germination. The tetrahydrate crystal of the active Sr·DPA has a polymeric linkage of Sr and carboxy

O atoms with the relatively close spacing Sr—Sr' = 4·21 Å corresponding to Ca—Ca' = 3·99 Å in Ca·DPA·3H$_2$O and the pyran analog[170].

Ca·DPA is thought to induce the spontaneous germination of spores often seen in centrifuge packs, in heavy suspensions and in humid atmospheres. In the latter circumstance well-separated individual spores darkened but the lag periods were extended and variable.[63] In view of the very slow nucleation in bulk supersaturated solutions of Ca·DPA it seems likely that within the microenvironment of the spores or within the individual spores themselves supersaturated germinative concentrations might form by slow leakage from the very high concentrations of Ca^{2+} and DPA immobilized within the dormant spore.

Ca·DPA-supplemented media allowed higher counts on plates or dilution tubes than ordinary media with non-heat-activated spores of B. subtilis 5230.[171] Unfortunately, some other strains and species of spores failed to give as good recoveries.

One may ask whether most germinants activate a Ca·DPA type germination within the spore.[164] Against this idea are the findings that exposure to 0·04 M Ca·DPA at 60°C specifically inactivates the Ca·DPA germination pathway, and that the O.D. falls after longer lags for Ca·DPA germination than with some other agents such as L-alanine. Viable dormant DPA-free spores of a mutant of B. cereus T have been produced by Halvorson et al. [228, 229]

4. The functions of DPA

Although various functions for DPA have been suggested, as yet none has been established well. In the absence of any evidence that DPA is metabolized by the dormant or the germinating spore, one looks for a structural role for DPA. However, except for traces of DPA esters, no covalently bonded DPA has been found in spores. A substantial DPA content is not essential for a normal appearance of sections. Spores of *B. popilliae* have a prominent cortex[172] although the DPA content has been reported to be low.[173]

An association of DPA with Ca^{2+} and proteins has usually been suspected to be important in the maintenance of heat resistance (see Chapter 7, Section VII). "The most direct explanation [of the appearance of heat sensitive enzymes upon germination] is that the binding of these enzymes to large structures (protein, bound DPA, etc.) renders them heat resistant and inactive and thereby maintains the dormant state"—Halvorson and Church.[174]

DPA at pH 6 and 7 stabilizes the soluble heat-labile glucose dehydrogenase from *B. subtilis* spores, although not as effectively as the isomeric pyridine-2,4- and -2,5-dicarboxylates.[175]

Various proteins significant in dormancy and germination may be expected to have evolved adaptations to high concentrations of DPA whatever evolutionary pressures led to the production and deposition of DPA. Stimulation or repression of the activity of particular enzymes need not imply a primary role for DPA. The electron transport system may be regulated during germination by DPA.[90] DPA and FMN each stimulate a 20X purified NADH oxidase.[176] The compounds appear to compete for the same site and each is inhibited by atabrine. The presence of the pyridine ring in each may be significant. The authors suggest that DPA may substitute for O_2 as an electron acceptor in anaerobic germination. The labile dihydro-DPA was not observed. Keynan *et al.*[148] noted that L-alanine-induced germination of low (2·3%) DPA *B. cereus* T spores was stimulated by 0·004 M DPA, a concentration far below that required for Ca·DPA germination.

Rode and Foster[72] found that DPA and related compounds, including α-picolinate, pipecolate and nicotinate, stimulated the L-alanine-induced germination of spores of *B. megaterium* Texas even without the presence of $CaCl_2$. The supersaturated germinative concentration 0·04 M Ca·DPA was not reached in these trials. The high activity of the DPA analog pyridine-2-carbinol-6-carboxylate + $CaCl_2$, but without L-alanine or inosine, is noteworthy.

Halvorson *et al.*[177] have suggested that DPA is not fully complexed to metal in the dormant spore; that "activation leads to an increase in chelation potential by releasing bound forms of DPA. The liberated free DPA promotes the flow of electrons to molecular O_2 by (1) the removal of an inhibitory metal directly or indirectly affecting the system and by (2) a direct stimulation of DPNH cytochrome c reductase."

Ca·DPA may serve as a readily loosened and solubilized caulking material,[146, 161] with possible roles of inhibiting permeation, or of fixing a compressed state of the cortex or core, or of inhibiting enzymes designed for the initial attack on the cortical structures in germination. The spore structures may be poised on the verge of germination so that the release of minute amounts of Ca·DPA under the proper conditions gives an auto-catalytically accelerated release of Ca·DPA and breakdown of spore structures. The findings with Ca·DPA germination support this last view of the function of DPA.

F. THE ROLE OF STRUCTURE

The microscopically distinguishable structures of the dormant spore must be adapted to two complementary roles: (1) A stabilizing of individual and associations of macromolecules under adverse environmental conditions and a repression of the activities of enzymes and nucleic acids. (2) A poising of structures for the clearing operations of germination and the metamorphosis of the core into the vegetative cell. The sequential appearance of distinguishable layers from the inside outwards during spore formation suggests a packaging process, just as the steps in germination suggest a loosening and a dissolution or opening of a package. However, we are only on the threshhold of an understanding of the functions of the various peripheral layers of the dormant spore. Instructive examples of abortive sporulation and unmatured spores are discussed in Chapters 2 and 3. Thus, blockage of protein synthesis in sporulating *B. subtilis* by the addition of chloramphenicol just before the coat was to be laid down gave coatless spores that had well-developed cortexes, contained DPA and were thermoresistant.[178] The coat-deficient spores lysed spontaneously soon after formation.

Spores of *Cl. thermosaccharolyticum* 3814 and TA37 have a phase-dark Sudan Black-staining (lipid) periphery,[147] and yield about 15% extractable lipid after acid hydrolysis as compared to less than 1% without hydrolysis.[179] A relationship to the high germinative dormancy of these spores is uncertain.

1. Role of the exosporium

An exosporium is not found on all strains of spores, and spores from which the exosporium has been stripped retain both dormancy and germinability. The exosporium blocks the penetration of Congo Red into spores of *B. polymyxa*.[180] From the relative rates of solubilization under sonication the exosporium appears to carry the alanine racemase and adenosine deaminase in spores of *B. cereus* T.[181] These enzymes may have roles in restraining and promoting germination.

Hodgkiss *et al.*[182, 183] have observed microtubules coiled tightly within the exosporium of spores of some species of anaerobes. Upon sporangial lysis the exosporium absorbs fluid and enlarges, the appendages uncoil, and may eventually break through the exosporuim. The authors conjectured that the microtubules may be chemo-sensory organelles for germination.

2. Role of the spore coats

Spore coats contribute structural rigidity. They preserve their shape through germination (in those species where the coat does not dissolve), through mechanical breakage and through chemical extractions. Specific modes of rupture suggest the presence of planes of weakness or of special sensitivity to germinative processes. Spores of different species vary greatly in morphological complexity. Among the anaerobes sporangial materials tend to form the outermost layer of the spore coat and remarkable appendages, filaments, or microtubules tend to be present. *B. laterosporus* is characterized by a one-sided deposit of wall material. Even in the box-shaped *B. apiarus* the ovoid cortex + core are enclosed, and this seems to be the invariable rule among the bacterial spores.

Dry spores of *B. globigii* are plastic under high pressure, but regain their shape when pressure and adherence to surfaces are reduced.[184] Viability was not reduced greatly by squeezing the spore thickness to one-third or less, until tearing became evident. Cracking of the outer coat did not prevent outgrowth.

Microincineration disclosed a parallel fibrillar ash pattern with 10 mμ periodicity in the middle one of the three principal layers of the coat of *B. megaterium* B-938.[119] Since the pattern was shown by shed coats of germinated spores as well as by the intact coats of dormant spores, it seems unlikely to represent the metal correlated stoichiometrically with the DPA. A similar 10 mμ fibrillar structure has been disclosed in *B. megaterium* B1551 spore coats by treatment with dilute $HClO_3$.[185] Phosphate polymers in the coat may be involved in the maintenance of dormancy.[143] Isolated coats exchange divalent cations rapidly.[59]

The spore coat is the site of deposition of —SS— during sporulation (Chapter 3, Section III B). Evidence for a keratin-like structure is discussed in Chapter 7, Section III B). Isolated coats of spores of *B. cereus* and *megaterium* gave 80% solubilization of the protein on extraction with dithiothreitol but left an insoluble fraction that preserved the morphology.[122] The soluble fraction gave a single band during disc electrophoresis. Late in sporulation L-cystine goes preferentially into the alkali-insoluble fraction of the coat, presumably by an —SS— exchange reaction since the introduction of cystine is inhibited selectively by sulphite (or L-cystinyl-L-alanine) and sulphite is incorporated instead to give heat-resistant spores that contain DPA.

The coat becomes elastic early in germination to permit a rapid expansion by about 20% of the dormant volume.[186] Subsequently, the coats behave in distinctive ways in different species. They tend to dissolve in large spores such as *B. cereus* and to split and persist in small spores such as *B. subtilis*.[187] Spores of *B. subtilis* 6051 in an early stage of germination with subtilopeptidase showed diffuse borders in electron micrographs that were not seen on germination with nutrient broth.[78] The behavior of coats following germination is discussed further in Chapter 12.

The coat of *B. stearothermophilus* is cracked and split by heat to a degree that suggested a role in heat activation.[188] Heat-activated *B. anthracis* spores showed a mottled appearance of the coat and a more opaque less granular core than dormant spores.[189] In *B. coagulans* the coat and core are swelled and the cortex is disrupted by lethal autoclaving.[190] After prolonged heating only the coat retains some structural integrity.

The state of reduction of —SS— linkages in the coat protein affects the entrance of large molecules such as lytic enzymes to the cortex, as was discussed in connection with permeability (p. 327). A similar relationship of —SS— reduction and the tertiary structure of the coat protein to the state of activation has been proposed.[12, 13] Activation by acid or heat was not promoted by 6 M urea alone, showing that more than hydrogen bonding is involved. Suzuki and Rode[106] have found that a prior treatment of *B. megaterium* 9885 is not necessary for germination by minute traces (2 p.p.m) of lysozyme. The strain appears to offer an unusual opportunity for investigation of the coat (?) defect. Vinter[109] suggested that cystine-containing protein may play a dual role of stabilizing the structure of the dormant spore and of contributing to a burst of physiological activity during germination through the action of a disulphide reductase. The amount of —SH increases 2 to 14× during germination, although the amount reduced is no more than a quarter of the total —SS—.[191]

The extent of the available surface of spores as measured by gas adsorption is strongly dependent on drying history. Presumably this reflects alterations in the coat protein. Berlin *et al.*[192] found 3 to 5 m^2/g available for monolayer adsorption of N_2 at —195°C by vacuum-dried spores. This approximates the smooth geometrical surface and is far too small an area to accommodate the cation exchange sites found by Alderton and Snell[57] for spores in aqueous suspension. Neihof *et al.*[193] found the N_2 area to be ten-fold higher for spores that had the water first replaced by ethanol and the ethanol replaced by pentane before drying. The sporocidal action of ethylene oxide on *B. subtilis* spores at the optimal a_w 0·33 depended strongly on drying history.[194] The very high resistance of extremely dry spores was reversed at once by exposure to liquid water, but only slowly by exposure to humid atmospheres. *B. cereus* spores with high contents of

Ca^{2+} and DPA appeared to have a more hydrophobic surface in flotation trials than spores with low contents.[195]

The condition of the spore coat may well be involved in aberrant germinative and metabolic behavior of lyophilized spores as well as of spores that exhibit extreme germinative dormancy, through effects on permeability.

3. Role of the cortex

Whereas the coat appears to have the primary function of protecting the cortex, the cortex may play its greatest role in protecting the core and rendering it dormant. As Joan Powell wrote in 1957: " . . . the collapse of the heat stable structure of the spore may be brought about by changes in . . . that part of the spore integument which contains the hexosamine peptide. . . . We think it likely that a hydration and depolymerization of this structure occurs during germination."[32]

The cortex is seen as a region between the coat and the core in all spores, although the proportionate thickness of the cortical shell varies widely among species. Visualization of its fine structure requires special staining. Its chemical composition is described fully in Chapter 7. It contains the muropeptide. Diaminopimelic acid is unique to the cortex in the dormant spore, and is invariably present. The cortex is susceptible to muramidase and to autolytic spore enzymes of unknown specificity. Coat-deficient spores obtained by blocking protein synthesis at a late stage are not released from the sporangium because of lack of the sporangium-lysing enzyme.[178] However, the spores break down within the sporangium by dissolution of the cortex.

Mechanical damage activates the cortex-lysing enzymes. The rapidity of the phase darkening may be observed very simply by racking down a phase microscope objective on to a sparse suspension of spores with just the right pressure, and then quickly refocusing and scanning the nearby field.[129] Cortical components are solubilized and the distinctive cortical structures are lost on mechanical breakage unless it is done under appropriate conditions such as low or high pH, or high temperatures that inactivate the lytic enzymes contained with the dormant spores.[56, 130] Boiled spores lose DPA but not hexosamine.[118]

The development of the cortex, its correlation with the development of phase brightness and deposition of Ca^{2+} and DPA, and the effects of Ca^{2+} deficiency upon morphology and stability of *B. cereus* var *alesti* have been depicted beautifully by Young and Fitz-James.[136] Cortex-deficient mutants have been found (Chapter 2), and spores with defective cortexes have been produced by growth in cycloserine.[196] These spores lacked stability, although the synthesis of DPA and the uptake of Ca^{2+} were normal. Mutant A$^{(-)1}$ of *B. cereus* var *alesti*[197] forms a spore with an uncompleted cortex. Maximum refractility is not achieved. Although Ca^{2+} uptake and DAP and DPA synthesis are normal, these components are freed by a precocious lysis during the terminal lysis of the sporangium.

The cortex must have vital roles in germination as well as in the maintenance of dormancy; as a possible site for germination signal receptors, as a site for enzymes evolved for control of the lytic processes, as a device for restraint of osmotic forces on the core pending elaboration of the new cell wall, and as a foundation and a source of materials for the new cell wall. Uehara and Frank[198] have suggested that two permeability barriers are disrupted in the germination of spores of *C. sporogenes* 3679h. Penetration of the first barrier outside of the cortex is reflected in a phase darkening of the periphery of the spore, a release of DPA and a labilization of the cortex. Penetration of the inner barrier is shown by a phase darkening of the central portion of the spore, a swelling of the core and by differential staining with an acidic dye (mercurochrome) as a counterstain. The latter set of effects was not seen with autoclaved spores, or with spores exposed to physiological germination conditions at 60°C.

The basal layer of the cortex becomes the wall of the germinating cell. DAP is laid down in two periods during spore formation; that laid down early in the basal layer of the cortex persists in germinating *B. cereus*, that laid down late is lost with the dissolving cortex.[199] Some reutilization of cortical materials for new cell wall synthesis is indicated. Rode and Foster[28] have speculated that "molecules of Ca·DPA and mucopeptide are present in orderly array in alternating layers . . . under germination conditions this structure, whose existence depends on charged groups, becomes disorganized by exogenous ions of the right size and characteristics". It has been suggested that the primary event in germination is a dissolution or alteration of the cortex[200, 201] or the action of a lytic enzyme.[202]

The cortex behaves differently during germination of spores of different species. As an example, the spore of *B. cereus* loses its cortex, whereas the spore of *B. polymyxa* retains it.[203] In contrast to the morphological picture, the losses of Ca^{2+} (63 and 67%, resp.) and hexosamine (32 and 41%, resp.) are similar in both species. These components and the associated peptide and DPA represent only about half of the solids released during germination. The authors suggest that only about 30% of the muropeptide is involved in the maintenance of spore dormancy. It seems more likely that the extent of the release of hexosamine is related more to the autolytic complement of the spore than to the extent of participation of the hexosamine in the maintenance of dormancy.

As has been argued above (see also p. 327), the cortex may be a permeability barrier. It has also been suggested that the cortex may function to secure and maintain a compression of the core, and thereby a lowered water content in the core[129]—the *contractile cortex* hypothesis. Such a condition would be expected to contribute to heat resistance and to dormancy. The possibility of a cortex shrunken onto the core is suggested vividly by the highly folded inner margin of the cortex released from disrupted spores;[56] (see also Chapter 2, Fig. 8b). In an extension of the contractile cortex hypothesis Alderton and Snell[57] suggested that the cation exchange system of the dormant spore could act as a source of pressure. If cation exchange occurs upon carboxyl groups, replacement of the closely bonded protons by other cations would give an expansion of the system through the electrostatic repulsions of the ionized carboxyl groups, analogous to the effect in weak cation exchange resins. Isotropic expansion against an inextensible shell or, more likely, anisotropic expansion in a more complex structure could exert pressure upon the core. Warth *et al.*[56] gave an example of a polymer, CM-Sephadex C-25, that contracts under the influence of 1 M $CaCl_2$ at pH 7 and thus might simulate a contracting cortex. Obviously, the behavior of relatively intact and fractionated cortex needs investigation. The cortical muropeptide obtained by autolysis of *B. coagulans* spore integuments[230] appears to have a suitable structure. Intact vegetative cells and isolated cell walls of *B. megaterium* undergo substantial volume changes as the ionic strength is altered.[231]

Vinter and Šťastná[204] observed a rapid partial restoration of phase-brightness and a 25 to 30% contraction of cell volume in germinating spores of *B. cereus* upon treatment with 50 p.p.m. polymyxin B. Protein synthesis and cell development were halted. Other basic compounds (clupein and polylysine) gave similar effects. The effects were inducible from the beginning of germination up to the elongation stage when the capacity disappeared quickly. The presence of 0·17 M NaCl or other salts inhibited the effects on microscopic appearance but not on protein synthesis. The effects were interpreted as arising from an electrostatic cross-linking in the residual portion of the cortex.

Hitchins and Gould[102] observed *cores* released from dormant spores by shaking with glass Ballotini beads in media adjusted to below pH 4·2 to inhibit the cortex-lysing enzymes. The cores retained muropeptide (by staining with fluorochrome-labeled lysozyme) and responded to pH and Ca^{2+} by changes in refractility and volume. Chambon *et al.*[232] caused the core contents of *B. megaterium* spores to be extruded by the action of lysozyme at pH 7·8 after activation with 8 M urea in 10% 2-mercapto-ethanol at pH 3·0.

4. Role of the core

The central portion of the spore is functionally a vegetative bud. It contains the hereditary charter, a repressed protein synthesizing system, the enzymes necessary to initiate the synthesis of new enzymes and structural materials and, presumably, reserves for the supply of energy and intermediates. The control of the nucleic acids and the protein synthesizing system during sporulation and germination are under intensive investigation (see Chapter 4). The DNA, the ribosomes, the transfer RNA and the associated enzymes are packaged in the earliest phases of the sporulation process, presumably in non-functional states. Questions of the alterations of components during extraction from dormant spores have not been resolved.[205] Components have been isolated from dormant spores and tested in reconstituted systems. Those obtained by 10 minutes of lysozyme treatment at pH 7·8 showed only minor differences in composition or activity from those obtained from vegetative cells.[232, 233, 234] A low level of m-RNA was found, in contrast to other reports of its absence.[206, 208] The low level could have been caused by the high nuclease activity of the spores, but the possibilities that this activity represented remnants of sporulation messenger or that it was formed during the lytic period were not excluded.[234]

On the other hand, components isolated from mechanically broken dormant spores have given conflicting results.[206, 208] Blockage of m-RNA synthesis by actinomycin D allows germination but prevents outgrowth.[206] Thus, transcription of the DNA must occur early in the process of germination. m-RNA formed in germination does not contain the m-RNA characteristic of the sporulation stage.[209] Both ribosomes and transfer enzyme from dormant *B. cereus* were defective but the phenylalanyl-s-RNA synthetase was active.[206, 207] The defect in the ribosomes of the dormant spore may be related to a configurational state since the activation of the ribosomes does not require protein synthesis.[210] Bishop and Doi[211] have written that there is "insufficient evidence at the moment to decide whether dormancy is controlled primarily by regulating the transcription process or whether the regulation of the transcription process is a secondary factor in the overall process of converting a vegetative cell to a dormant spore".

The energy reserves of spores of *B. megaterium* and other species appear to be glutamic acid, the dominant free amino acid,[9, 235] and 3-phospho-D-glyceric acid.[235] Each occurs in relatively large amounts, approx. 1% of the dry weight. *B. subtilis* spores contain up to 6% of free L-3-sulpho-lactic acid; it is absent from vegetative cells and from the spores of *B. cereus, megaterium* and *thuringiensis*, it is not metabolized, and its function is not known.[235]

Relatively little structure is seen in the core region of spore sections (cf. Chapter 2). The core is surrounded by the core membrane. In *B. popilliae* an interrupted membrane has been seen parallel to and close inside the core membrane.[172] The nucleoplasm and densely packed ribosomes also were seen. *B. cereus* grown with insufficient Ca^{2+} gave labile spores with submaximal refractility.[136] The ribosomes in the core of these spores retained their affinity for lead, unlike normal dormant spores. The authors suggested the possibility of binding of Ca·DPA to the ribosomes with elimination of their affinity for lead.

The chromatin of dormant spores but not of germinated, Ca^{2+}-deficient or sectioned spores is extruded by the Feulgen staining procedure as a result of the exposure to strong acid. Young and Fitz-James[136] suggested that the bursting in dormant spores arises from osmotic pressure generated by the products of acid hydrolysis. Spores fixed only after the acid treatment showed bursting of the cortex, whereas spores fixed before acid treatment showed bursting of the coat as well. A pressure-generating cortex gives another possibility for the origin of the bursting force. The extent of retention of nucleic acids within cores released by mechanical disintegration of dormant spores at acidic pH[102] has not been reported. It may be that these particles are more representative of cortical basal layer and core membrane than of intact cores. All of the nucleic acid is extruded as a result of cortical lysis at pH 7·8.[232]

The core, as well as the cortex and coat, has been suggested for the location of DPA (Chapter 7, Section VI C) on the basis of photographs of intact spores by 270 mμ light.[212] The presence of u.v. absorbing nucleic acids and aromatic amino acids, as well as possible artifacts from the severe mismatch of the refractive indices of spore and water, make interpretation uncertain. However, germinating spores showed a uniform absorption, whereas dormant spores showed an intense absorption in the outermost and central regions with relatively little absorption between.[213] Autoclaved spores, devoid of DPA, had a similar appearance, but the central absorbing region seemed more contracted and the low-absorbing annulus seemed wider. The outer and inner regions could also be seen in untreated spores photographed with 300 mμ monochromatic light where Ca·DPA and other ionic states of DPA have negligible absorbance.[167] Isolated spore coats, which lack DPA, also could be photographed by u.v. It is clear that substances other than DPA contribute markedly to the u.v. micrographs. DPA-deficient endotrophic spores did show much less absorbance at 270 mμ than normal spores.[137]

The micrographs can also be interpreted as a DPA-containing cortical region contracted onto the core (see p. 329). A space between the cortex and coat often is seen in sectioned dormant spores (cf. Ref. 212, Fig. 5). Usually it is considered to be an artifact of methacrylate embedding. It may disclose a region of relative weakness, as is also suggested by cleavage between the coat and cortex in Ca^{2+}-deficient spores.[136]

In order to account for the magnitudes of their free space values (see p. 327), Black and Gerhardt[124] have proposed that "the core of the dormant spore exists as an insoluble and heat-stable gel, in which cross-linking between macromolecules occurs through stable but reversible bonds so as to form a high-polymer matrix with entrapped free water". They suggest that the core is the location of the heat-protective mechanism.

IV. Evolutionary Aspects

Here we will speculate on factors that may have shaped the evolution of the bacterial spore, that may have led to this architecturally incomparable housing for the dormant vegetative bud in comparison with which the resting stages of most common bacterial species appear quite undistinguished. Pre-eminent among the great influences on microbial life are the needs imposed by the fluctuating environment for alternating periods of competitive growth and of husbandry of marginal reserves. We may choose to believe with Foster[214] and Lamanna and Mallette[215] that the advantage of a reduced metabolic rate guided the selection of those features of the primitive ancestors that led to the present-day sporeformers, but this need is general and sporeformers comprise but a minority of the bacterial world. There must be disadvantages as well as advantages to the formation of a highly specialized dormant cell. Non-sporeformers compete very successfully with sporeformers in various ecological circumstances; only minor groups such as Azotobacter have developed cysts or analogous morphologically recognizable resting forms. Furthermore, sporeformers need to forgo multiplication during the extended metamorphosis, and are hypersensitive in the germination stage.

Whereas the cortex arises from membranes homologous to those that give rise to the vegetative cell wall, and the muropeptidic components of the cortex are identical or homologous to those present in the walls of most if not all Eubacteriales, the protein coat of the spore appears to be unique. A protein coat is lacking completely in vegetative Gram-positive bacteria, which instead are covered principally by a thick murein layer related compositionally to the cortex. The complex lipoprotein coat of the Gram-negative bacteria and the lipoidal coats of the acid-fast bacteria show small resemblance to the spore coat. The extraordinary resistance of the spore coat to attack by other microorganisms and by free enzymes does not seem to have been developed as successfully elsewhere in the bacterial world. Perhaps the development of an enzyme-resistant protein outer coat came early on the evolutionary road to the spore as we know it.

The evolution of the spore has included a condensation or packing of materials that is similar to the dormant structures of fungi and higher plants, although the composition of the reserves and of the protective layers varies widely. Selection pressures of economy in packaging materials and of efficiency in controlling permeability, such as the entrance of deleterious enzymes and the loss of cellular components, may have dominated. By whatever route, the spore has attained an extreme condensation illustrated by its high buoyant density and by a reduction of water content far below the water-binding capacity of broken or severely heated spores at high water activities.[128]

It is highly unlikely that within the dimensions of a spore completely water impermeable layers of the necessary thickness exist. This has led to the idea of pressure as a factor controlling the reduced water content of the dormant spore.[129] This possibility seems to be the only alternative to a necessity for masking individual polar groups on a molecular level, an unattractive prospect in view of the wide variety of hydrophilic molecules within the dormant spore and their rapid hydration on mechanical damage or germination. Moreover, the hypothesis allowed the assignment of a unique function to the cortex—a singular organelle obviously essential for formation of the dormant spore and superfluous to the outgrowing cell, but for which no distinctive function has been evident. Unfortunately, no decisive evidence for or against the contractile cortex has appeared although circumstantial observations support it. Probably, pressure (if it turns out to be significant) will be so well integrated with other protective factors that the relative importance will be hard to evaluate. The suggestion, long ago,[216] that the condensation of protoplasm into the spore resembled the syneresis of a gel (Entquellung) hints at the incidental molecular behavior that may have been significant in the initial evolution of a complex pressure-generating organelle. Reselin, the remarkable proteinoid rubber of the insect wing-hinge,[217] illustrates the potentialities for evolution from more familiar molecules of macromolecular networks which perform a specialized function.

The recent discovery of the cation exchange property of dormant spores and the recognition that it has an active role in dormancy and heat resistance open the question of the principle through which these effects are exerted. The variation of volume of synthetic and natural polyelectrolytes with ionic environment is well established. One may visualize an evolution of an incidental cation exchange property of a passive murein cell wall[231] into a functional pressure-generating property of the cortex as the evolving spore coat took over the enzyme-resisting function of the cell wall of the ancestral form.

DPA lies just off the biosynthetic route to DAP, the latter being a component of the vegetative cell wall and the cortical muropeptide. Its formation from dihydro-DPA, a DAP intermediate, involves only an easy oxidation. Its formation for a metal-binding enzyme-restraining initial role in proximity to a Ca^{2+}-accumulating network may then have led to a role for Ca·DPA as a quickly removable network-fixing component of the cortex.

The wide variety of germination signals even in closely related species and strains suggests that this is a relatively later event in the evolution of the spore-formers, presumably contemporaneous with the spreading of the ancestral forms into the niches that they occupy today. A recognition of assimilable nutrients by actual assimilation that presumably is characteristic of dormant vegetative cells has been sharpened in the evolution of spores into an instinctive recognition of suitable microenvironments. It will not be surprizing, once the details of the germination signal mechanisms are known, to find a combination of protein subunits in a manner illustrated by certain of the endproduct (feedback) inhibitions in the endproduct control of branch points in biosynthetic pathways—one protein sequence carrying out the primary germinative (lytic?) reaction, the other protein sequence carrying the germinant-recognizing control function and evolved from a metabolic or metabolic control role.

Under artificial conditions, i.e. under conditions not likely to have been effectual in the evolution of the signalling mechanism, spore germination may be triggered in circumstances unsatisfactory for outgrowth. The pressure for economy or simplicity in the signal receptors has led to simple germination patterns with respect to the specific germinants, which usually are organic molecules. The phenomena of fractional germination and of a reduction in germination requirements with age or exposure may have evolved to minimize the effect of damage from an infelicitous signal. The effect of these adaptations is a spread-out germination, a scattering of the bacterial seed in time.

REFERENCES

1. Crook, P. G. (1952). *J. Bact.* **63**, 193.
2. Church, B. D. and Halvorson, H. O. (1957). *J. Bact.* **73**, 470.
3. Desser, H. and Broda, E. (1965). *Nature, Lond.* **206**, 1270, and personal communication.
4. Windle, J. J. and Sacks, L. E. (1963). *Biochim. biophys. Acta* **66**, 173.
5. Sussman, A. S. and Halvorson, H. O. (1966). "Spores. Their Dormancy and Germination". Harper and Row, New York.
6. Marshall, B. J., Murrell, W. G. and Scott, W. J. (1963). *J. gen. Microbiol.* **31**, 451.
7. Wilson, G. S. and Shipp, H. L. (1938). *Chemy Ind.* **57**, 834.
8. Keynan, A. and Halvorson, H. (1965). *In* "Spores III" (L. L. Campbell and H. O. Halvorson, eds), p. 174. Amer. Soc. Microbiol., Ann Arbor, Michigan, U.S.A.
9. Lee, W. H. and Ordal, Z. J. (1963). *J. Bact.* **85**, 207.
10. Lewis, J. C., Snell, N. S., and Alderton, G. (1965). *In* "Spores III" (L. L. Campbell and H. O. Halvorson, eds), p. 47. Amer. Soc. Microbiol., Ann Arbor, Michigan, U.S.A.
11. Rode, L. J. and Foster, J. W. (1966). *J. Bact.* **91**, 1582.
12. Keynan, A., Evenchik, Z., Halvorson, H. O. and Hastings, J. W. (1964). *J. Bact.* **88**, 313.

13. Keynan, A., Issahary-Brand, G. and Evenchik, Z. (1965). *In* "Spores III" (L. L. Campbell and H. O. Halvorson, eds), p. 180. Amer. Soc. Microbiol., Ann Arbor, Michigan, U.S.A.
14. Gibbs, P. A. (1967). *J. gen. Microbiol.* **46**, 285.
15. Holmes, P. K. and Levinson, H. S. (1967). *Currents in Modern Biology* **1**, 256.
16. Kvaratskheliya, M. T. and Dorosinskii, L. M. (1965). *Microbiology* **34**, 907 (translation from *Mikrobiologiya* **34**, 1034).
17. Murrell, W. G. (1961). *Symp. Soc. gen. Microbiol.* **11**, 100.
18. Foerster, H. F. and Foster, J. W. (1966). *J. Bact.* **91**, 1168.
19. Sacks, L. E. (1967). *J. Bact.* **94**, 1789.
20. Knaysi, G. (1948). *Bact. Rev.* **12**, 19.
21. Freese, E., Park, S. W. and Cashel, M. (1964). *Proc. natn. Acad. Sci., U.S.A.* **51**, 1164. Ibid. **52**, 516.
22. Foster, J. W. and Wynne, E. S. (1948). *J. Bact.* **55**, 623.
23. Morrison, E. W. and Rettger, L. F. (1930). *J. Bact.* **20**, 313.
24. Snell, N. S. (1968). *Appl. Microbiol.* **16**, 436.
25. Keynan, A., Murrell, W. G. and Halvorson, H. O. (1961). *Nature, Lond.* **192**, 1211.
26. Dickson, E. C. (1927). *Proc. Soc. exp. Biol. Med.* **25**, 426.
27. Vary, J. C. and Halvorson, H. O. (1965). *J. Bact.* **89**, 1340.
28. Rode, L. J. and Foster, J. W. (1962). *Arch. Mikrobiol.* **43**, 183.
29. Lewis, J. C. (1968). U.S. patent no. 3,385,766.
30. Tamir, H. and Gilvarg, C. (1966). *J. biol. Chem.* **241**, 1085.
31. Pheil, C. G., Lee, C. K. and Ordal, Z. J. (1966). *Bact. Proc.* 17.
32. Powell, J. F. (1957). *J. appl. Bact.* **20**, 349.
33. Jaye, M. (1964). *Thesis*, University of Illinois, Urbana.
34. Curran, H. R. and Evans, F. R. (1947). *J. Bact.* **53**, 103.
35. Finley, N. and Fields, M. L. (1962). *Appl. Microbiol.* **10**, 231.
36. Fields, M. L. (1964). *Appl. Microbiol.* **12**, 407.
37. Fields, M. L. (1963). *Appl. Microbiol.* **11**, 100.
38. Edwards, J. L., Busta, F. F. and Speck, M. L. (1965). *Appl. Microbiol.* **13**, 858.
39. Alderton, G., Thompson, P. A. and Snell, N. (1964). *Science, N.Y.* **143**, 141.
40. Hachisuka, Y. (1964). *Jap. J. Bact.* **19**, 162.
41. Campbell, L. L., Richards, C. M. and Sniff, E. E. (1965). *In* "Spores III" (L. L. Campbell and H. O. Halvorson, eds), p. 55. Amer. Soc. Microbiol., Ann Arbor, Michigan, U.S.A.
42. Halvorson, H. O. (1955). *Annls Inst. Pasteur, Lille* **7**, 53.
43. Fleming, H. P. and Ordal, Z. J. (1964). *J. Bact.* **88**, 1529.
44. Hyatt, M. T. and Levinson, H. S. (1961). *J. Bact.* **81**, 204.
45. Levinson, H. S. and Sevag, M. G. (1953). *J. gen. Physiol.* **36**, 617.
46. Rode, L. J. and Foster, J. W. (1965). *Proc. natn. Acad. Sci., U.S.A.* **53**, 31.
47. Rode, L. J. and Foster, J. W. (1962). *Nature, Lond.* **194**, 1300.
48. Krask, B. J. (1961). *In* "Spores II" (H. O. Halvorson, ed.), p. 89. Burgess Publishing Co., Minneapolis, Minn., U.S.A.
49. Hachisuka, Y., Kato, N. and Asano, N. (1956). *J. Bact.* **71**, 250.
50. Halmann, M. and Keynan, A. (1962). *J. Bact.* **84**, 1187.
51. Splittstoesser, D. F. and Farkas, D. F. (1966). *J. Bact.* **92**, 995.
52. Becker, M. E. and Pederson, C. S. (1950). *J. Bact.* **59**, 717.
53. Wiley, W. R. and Stokes, J. L. (1962). *J. Bact.* **84**, 730.
54. Chislett, M. E. and Kushner, D. J. (1961). *J. gen. Microbiol.* **25**, 151.

P

55. Rode, L. J. and Foster, J. W. (1960). *Arch. Mikrobiol.* **36**, 67.
56. Warth, A. D., Ohye, D. F. and Murrell, W. G. (1963). *J. Cell Biol.* **16**, 579, 593.
57. Alderton, G. and Snell, N. S. (1963). *Biochem. biophys. Res. Commun.* **10**, 139.
58. Rowley, D. B. and Levinson, H. S. (1967). *J. Bact.* **93**, 1017.
59. Rode, L. J. and Foster, J. W. (1966). *J. Bact.* **91**, 1589.
60. Beers, R. J. (1957). *In* "Spores" (H. O. Halvorson, ed.), p. 45. Amer. Inst. Biol. Sciences, Washington, U.S.A.
61. Lewis, J. C. (1967). *Bact. Proc.* p. 23, and unpublished observations.
62. Lewis, J. C. (1961). *In* "Spores II" (H. O. Halvorson, ed.), p. 165. Burgess Publishing Co., Minneapolis, Minn., U.S.A.
63. Snell, N. S. and Lewis, J. C., unpublished data.
64. Rayman, M. M. (1957). *In* "Spores" (H. O. Halvorson, ed.), p. 51. Amer. Inst. Biol. Sciences, Washington, U.S.A.
65. Hyatt, M. T., Holmes, P. K. and Levinson, H. S. (1966). *Biochem. biophys. Res. Commun.* **24**, 701. (1967). *Bact. Proc.* 23.
66. Uehara, M. and Frank, H. A. (1965). *In* "Spores III" (L. L. Campbell and H. O. Halvorson, eds), p. 38. Amer. Soc. Microbiol., Ann Arbor, Michigan.
67. Fujioka, R. S. and Frank, H. A. (1966). *J. Bact.* **92**, 1515.
68. Hitzman, D. O., Halvorson, H. O. and Ukita, T. (1957). *J. Bact.* **74**, 1.
69. Knight, B. C. J. G. and Fildes, P. (1930). *Biochem. J.* **24**, 1496.
70. Kaufmann, O. W. and Marshall, R. S. (1965). *J. Dairy Sci.* **48**, 670.
71. Black, S. H. (1964). *Bact. Proc.* 36; *Tex. Rep. Biol. Med.* **22**, 654.
72. Rode, L. J. and Foster, J. W. (1961). *Z. allg. Mikrobiol.* **1**, 307.
73. Weidel, W. and Pelzer, H. (1964). *Adv. Enzymol.* **26**, 193.
74. Krishna Murty, G. G. and Halvorson, H. O. (1957). *J. Bact.* **73**, 230.
75. Levinson, H. S. and Hyatt, M. T. (1965). *In* "Spores III" (L. L. Campbell and H. O. Halvorson, eds), p. 198. Amer. Soc. Microbiol., Ann Arbor, Michigan, U.S.A.
76. Rode, L. J. and Foster, J. W. (1966). *Bact. Proc.* 17.
77. Sierra, G. (1964). *Can. J. Microbiol.* **10**, 929.
78. Sierra, G. (1967). *Can. J. Microbiol.* **13**, 489.
79. Pepper, R. E. and Chandler, V. L. (1963). *Appl. Microbiol.* **11**, 384.
80. Phillips, C. R. (1952). *Bact. Rev.* **16**, 135.
81. Gould, G. W. and Hitchins, A. D. (1963). *J. gen. Microbiol.* **33**, 413.
82. Rode, L. J. and Foster, J. W. (1961). *J. Bact.* **81**, 768.
83. Widdowson, J. P. (1967). *Nature, Lond.* **214**, 812.
84. Halvorson, H. O. (1959). *Bact. Rev.* **23**, 267.
85. Gould, G. W. (1964). *In* "Microbial Inhibitions in Food" (N. Molin, ed.), *4th Int. Symp. Food Microbiol., SIK, Göteborg* p. 17. Almqvist and Wiksell, Uppsala.
86. Sacks, L. E. (1955). *J. Bact.* **70**, 491.
87. Evans, F. R. and Curran, H. R. (1948). *Food Res.* **13**, 66.
88. Slepecky, R. A. (1963). *Biochem. biophys. Res. Commun.* **12**, 369.
89. Tonge, R. J. (1964). *In* "Microbial Inhibitions in Food" (N. Molin, ed.), *4th Int. Symp. Food Microbiol., SIK, Göteborg* p. 35. Almqvist and Wiksell, Uppsala.
90. Halvorson, H. O. (1963). *In* "Control mechanisms in respiration and fermentation" (B. Wright, ed.), p. 3. Ronald Press, New York, U.S.A.
91. Freese, E. and Cashel, M. (1965). *In* "Spores III" (L. L. Campbell and H. O.

Halvorson, eds), p. 144. Amer. Soc. Microbiol., Ann Arbor, Michigan, U.S.A.

92. Wax, R., Freese, E. and Cashel, M. (1967). *J. Bact.* **94**, 522.
93. Gould, G. W. (1966). *J. Bact.* **92**, 1261.
94. Schaeffer, P., Millet, J. and Aubert, J.-P. (1965). *Proc. natn. Acad. Sci., U.S.A.* **54**, 704.
95. Grelet, N. (1957). *J. appl. Bact.* **20**, 315.
96. Levinson, H. S. and Hyatt, M. T. (1964). *J. Bact.* **87**, 876.
97. Holmes, P. K., Nags, E. H. and Levinson, H. S. (1965). *J. Bact.* **90**, 827.
98. Zoha, S. M. S. and Sadoff, H. L. (1958). *J. Bact.* **76**, 203.
99. Brown, W. L., Ordal, Z. J. and Halvorson, H. O. (1957). *Appl. Microbiol.* **5**, 156.
100. Engelbrecht, H. L. and Sadoff, H. L. (1966). *Bact. Proc.* 32.
101. Sadoff, H. L., Bach, J. A. and Kools, J. W. (1965). *In* "Spores III" (L. L. Campbell and H. O. Halvorson, eds), p. 97. Amer. Soc. Microbiol., Ann Arbor, Michigan, U.S.A.
102. Hitchins, A. D. and Gould, G. W. (1964). *Nature, Lond.* **203**, 895.
103. Kim, J. and Naylor, H. B. (1966). *Bact. Proc.* 31.
104. Gould, G. W. and King, W. L. (1966). *Nature, Lond.* **211**, 1431.
105. Gould, G. W., Georgala, D. L. and Hitchins, A. D. (1963). *Nature, Lond.* **200**, 385.
106. Suzuki, Y. and Rode, L. J. (1967). *Bact. Proc.* 22.
107. Gould, G. W., Hitchins, A. D. and King, W. L. (1966). *J. gen. Microbiol.* **44**, 293.
108. Strange, R. E. and Dark, F. A. (1956). *Biochem. J.* **62**, 459.
109. Vinter, V. (1960). *Folia microbiol., Praha* **5**, 217.
110. Gould, G. W. and Hitchins, A. D. (1963). *Nature, Lond.* **197**, 622.
111. Vary, J. C. and Halvorson, H. O. (1968). *J. Bact.* **95**, 1327.
112. Woese, C. R. (1965). *Spore Newsletter* **2**, 3.
113. Halvorson, H. O., Vary, J. C. and Steinberg, W. (1966). *A. Rev. Microbiol.* **20**, 169.
114. McCormick, N. G. (1965). *J. Bact.* **89**, 1180.
115. Martin, J. H. (1963). Thesis, Ohio State University, Columbus.
116. Hsu, W.-T. (1963). Thesis, University of Illinois, Urbana.
117. McCormick, N. G. and Halvorson, H. O. (1963). *Ann. N.Y. Acad. Sci.* **102**, 763.
118. Rode, L. J., Lewis, C. W. and Foster, J. W. (1962). *J. Cell Biol.* **13**, 423.
119. Thomas, R. S. (1964). *J. Cell Biol.* **23**, 113.
120. Ohye, D. F. and Murrell, W. G. (1962). *J. Cell Biol.* **14**, 111.
121. Hashimoto, T. and Naylor, H. B. (1958). *J. Bact.* **75**, 647.
122. Aronson, A. I. and Fitz-James, P. C. (1968). *J. molec. Biol.* **33**, 199.
123. Gerhardt, P. and Black, S. H. (1961). *J. Bact.* **82**, 750.
124. Black, S. H. and Gerhardt, P. (1962). *J. Bact.* **83**, 960.
125. Helfferich, F. (1962). "Ion exchange", p. 127. McGraw-Hill, New York, U.S.A.
126. Black, S. H. and Gerhardt, P. (1962). *J. Bact.* **83**, 301.
127. Rode, L. J. and Foster, J. W. (1960). *Proc. natn. Acad. Sci., U.S.A.* **46**, 118.
128. Grecz, N. and Smith, R. F. (1966). *Bact. Proc.* 32.
129. Lewis, J. C., Snell, N. S. and Burr, H. K. (1960). *Science, N.Y.* **132**, 544.
130. Murrell, W. G. (1967). *Adv. Microb. Physiol.* **1**, 133.
131. Murrell, W. G. and Scott, W. J. (1966). *J. gen. Microbiol.* **43**, 411.

132. Gruft, H., Buckman, J. and Slepecky, R. A. (1965). *Bact. Proc.* 37.
133. Vinter, V. (1962). *Folia microbiol., Praha* **7,** 115.
134. Vinter, V. (1964). *Folia microbiol., Praha* **9,** 58.
135. Vinter, V. (1962). *Folia microbiol., Praha* **7,** 275.
136. Young, I. E. and Fitz-James, P. C. (1962). *J. Cell Biol.* **12,** 115.
137. Black, S. H., Hashimoto, T. and Gerhardt, P. (1960). *Can. J. Microbiol.* **6,** 213.
138. Black, S. H. and Gerhardt, P. (1963). *Ann. N.Y. Acad. Sci.* **102,** 755.
139. Pelcher, E. A., Fleming, H. P. and Ordal, Z. J. (1963). *Can. J. Microbiol.* **9,** 251.
140. Foerster, H. F. and Foster, J. W. (1966). *J. Bact.* **91,** 1333.
141. Halvorson, H. and Howitt, C. (1961). *In* "Spores II" (H. O. Halvorson, ed.), p. 149. Burgess Publishing Co., Minneapolis, Minn., U.S.A.
142. Harrell, W. K. (1958). *Can. J. Microbiol.* **4,** 393.
143. Levinson, H. S., Sloan, J. D. and Hyatt, M. T. (1958). *J. Bact.* **75,** 291.
144. Levinson, H. S. (1957). *In* "Spores" (H. O. Halvorson, ed.), p. 120. Amer. Inst. Biol. Sci., Washington.
145. Brown, W. L. (1956). Thesis, Univ. of Illinois, Urbana.
146. Riemann, H. (1963). Thesis, Copenhagen University.
147. Pheil, C. G. (1967). Thesis, Univ. of Illinois, Urbana.
148. Keynan, A., Murrell, W. G. and Halvorson, H. O. (1962). *J. Bact.* **83,** 395.
149. Young, I. E. (1959). *Can. J. Microbiol.* **5,** 197.
150. Fleming, H. P. (1963). Thesis, Univ. of Illinois, Urbana.
151. Levinson, H. S. and Hyatt, M. T. (1960). *J. Bact.* **80,** 441.
152. Roberts, T. L. and Wynne, E. S. (1962). *J. Bact.* **83,** 1161.
153. Harrell, W. K. and Mantini, E. (1957). *Can. J. Microbiol.* **3,** 735.
154. Lund, A. J. (1961). *In* "Spores II" (H. O. Halvorson, ed.), p. 49. Burgess Publishing Co., Minneapolis, Minn., U.S.A.
155. Farkas, J. and Kiss. I. (1965). *Acta microbiol. Acad. Sci., hung.* **12,** 15.
156. Alderton, G., personal communication.
157. Rode, L. J. and Foster, J. W. (1960). *J. Bact.* **79,** 650.
158. Harper, M. K., Curran, H. R. and Pallansch, M. J. (1964). *J. Bact.* **88,** 1338.
159. Wooley, B. C. and Collier, R. E. (1965). *Can. J. Microbiol.* **11,** 279.
160. Riemann, H. and Ordal, Z. J. (1961). *Science, N.Y.* **133,** 1703.
161. Riemann, H. (1961). *In* "Spores II" (H. O. Halvorson, ed.), p. 24. Burgess Publishing Co., Minneapolis, Minn., U.S.A.
162. Iandolo, J. J. and Ordal, Z. J. (1964). *J. Bact.* **87,** 235.
163. Jaye, M. and Ordal, Z. J. (1965). *J. Bact.* **89,** 1617.
164. Keynan, A. and Halvorson, H. O. (1962). *J. Bact.* **83,** 100.
165. Jaye, M. and Ordal, Z. J. (1966). *Can. J. Microbiol.* **12,** 199.
166. Strahs, G. and Dickerson, R. E. (1968). *Acta crystallogr.* **B24,** 571.
167. Lewis, J. C. (1967). *Analyt. Biochem.* **19,** 327.
168. Lewis, J. C. (1967). *Spore Newsletter* **2,** 117.
169. Palmer, K. J. and Lee, K. S., personal communication.
170. Palmer, K. J. and Young, R. S. F., personal communication.
171. Busta, F. F. and Ordal, Z. J. (1964). *Appl. Microbiol.* **12,** 106.
172. Black, S. H. and Arredondo, M. I. (1966). *Experientia* **22,** 77.
173. Mitruka, B. M., Costilow, R. N., Black, S. H. and Pepper, R. E. (1967). *J. Bact.* **94,** 759.
174. Halvorson, H. and Church, B. (1957). *Bact. Rev.* **21,** 112.
175. Hachisuka, Y., Tochikubo, K., Yokoi, Y. and Murachi, T. (1967). *J. Biochem., Tokyo* **61,** 659.

176. Halvorson, H., O'Connor, R. and Doi, R. (1961). *In* "Cryptobiotic Stages in Biological Systems" (N. Grossowicz, S. Hestrin and A. Keynan, eds), p. 71. Elsevier Publishing Co., Amsterdam.

177. Halvorson, H., Doi, R. and Church, B. (1958). *Proc. natn. Acad. Sci., U.S.A.* **44,** 1171.

178. Ryter, A. and Szulmajster, J. (1965). *Annls Inst. Pasteur, Paris* **108,** 640.

179. Pheil, C. G. and Ordal, Z. J. (1967). *J. Bact.* **93,** 1727.

180. Dondero, N. C. and Holbert, P. E. (1957). *J. Bact.* **74,** 43.

181. Berger, J. A. and Marr, A. G. (1960). *J. gen. Microbiol.* **22,** 147.

182. Hodgkiss, W., Ordal, Z. J. and Cann, D. C. (1967). *J. gen. Microbiol.* **47,** 213.

183. Hodgkiss, W. and Ordal, Z. J. (1966). *J. Bact.* **91,** 2031.

184. Monk, G. W., Hess, G. E. and Schenk, H. L. (1957). *J. Bact.* **74,** 292.

185. Rode, L. J. and Williams, M. G. (1966). *J. Bact.* **92,** 1772.

186. Hitchins, A. D., Gould, G. W. and Hurst, A. (1963). *J. gen. Microbiol.* **30,** 445.

187. Gould, G. W. (1962). *J. appl. Bact.* **25,** 35.

188. Franklin, J. G. and Bradley, D. E. (1957). *J. appl. Bact.* **20,** 467.

189. Moberly, B. J., Shafa, F. and Gerhardt, P. (1966). *J. Bact.* **92,** 220.

190. Hunnell, J. W. and Ordal, Z. J. (1961). *In* "Spores II" (H. O. Halvorson, ed), p. 101. Burgess Publishing Co., Minneapolis, Minn., U.S.A.

191. Blankenship, L. C. and Pallansch, M. J. (1966). *J. Bact.* **92,** 1615.

192. Berlin, E., Curran, H. R. and Pallansch, M. J. (1963). *J. Bact.* **86,** 1030.

193. Neihof, R., Thompson, J. K. and Deitz, V. R. (1967). *Nature, Lond.* **216,** 1304.

194. Gilbert, G. L., Gambill, V. M., Spiner, D. R., Hoffman, R. K. and Phillips, C. R. (1964). *Appl. Microbiol.* **12,** 496.

195. Dobiáš, B. and Vinter, V. (1966). *Folia microbiol., Praha* **11,** 314.

196. Murrell, W. G. and Warth, A. D. (1965). *In* "Spores III" (L. L. Campbell and H. O. Halvorson, eds), p. 1. Amer. Soc. Microbiol., Ann Arbor, Michigan, U.S.A.

197. Fitz-James, P. C. (1965). *In* "Mécanismes de régulation des activités cellulaires chez les microorganismes" (J. Senez, ed), p. 529. Colloques Internationaux du Centre National de la Recherche Scientifique No. 124, Marseille.

198. Uehara, M. and Frank, H. A. (1967). *J. Bact.* **94,** 506.

199. Vinter, V. (1965). *Folia microbiol., Praha* **10,** 280.

200. Mayall, B. H. and Robinow, C. F. (1957). *J. appl. Bact.* **20,** 333.

201. Kawata, T., Inoue, T. and Takagi, A. (1963). *Jap. J. Microbiol.* **7,** 23.

202. Gould, G. W. and Hitchins, A. D. (1965). *In* "Spores III" (L. L. Campbell and H. O. Halvorson, eds), p. 213. Amer. Soc. Microbiol., Ann Arbor, Michigan, U.S.A.

203. Hamilton, W. A. and Stubbs, J. M. (1967). *J. gen. Microbiol.* **47,** 121.

204. Vinter, V. and Šťastná, J. (1967). *Folia microbiol., Praha* **12,** 301.

205. Kornberg, A., Spudich, J. A., Nelson, D. L. and Deutscher, M. P. (1968). *A. Rev. Biochem.* **37,** 51.

206. Kobayashi, Y., Steinberg, W., Higa, A., Halvorson, H. O. and Levinthal, C. (1965). *In* "Spores III" (L. L. Campbell and H. O. Halvorson, eds), p. 200. Amer. Soc. Microbiol., Ann Arbor, Michigan, U.S.A.

207. Kobayashi, Y. and Halvorson, H. O. (1968). *Archs. Biochem. Biophys.* **123,** 622.

208. Doi, R. H. and Igarashi, R. T. (1964). *J. Bact.* **87,** 323.

209. Doi, R. H. (1965). *In* "Spores III" (L. L. Campbell and H. O. Halvorson, eds), p. 111. Amer. Soc. Microbiol., Ann Arbor, Michigan, U.S.A.
210. Halvorson, H. O. (1967). Symposium on molecular and cellular aspects of differentiation and morphogenesis. 67th Annual Meeting, Amer. Soc. Microbiol., New York.
211. Bishop, H. L. and Doi, R. H. (1966). *J. Bact.* **91**, 695.
212. Hashimoto, T. (1960). *Tokushima J. exp. Med.* **7**, 36.
213. Hashimoto, T. and Gerhardt, P. (1960). *J. biophys. biochem. Cytol.* **7**, 195.
214. Foster, J. W. (1956). *Q. Rev. Biol.* **31**, 102.
215. Lamanna, C. and Mallette, M. F. (1965). "Basic bacteriology. Its biological and chemical background", 3rd ed. Williams and Wilkins Co., Baltimore, U.S.A.
216. Darányi, J. v. (1927). *Zentbl. Bakt. Parasit., II Abt.* **71**, 353.
217. Weis-Fogh, T. (1961). *J. molec. Biol.* **3**, 648.
218. Hyatt, M. T. and Levinson, H. S. (1968). *J. Bact.* **95**, 2090.
219. Woese, C. R., Vary, J. C. and Halvorson, H. O. (1968). *Proc. natn. Acad. Sci. U.S.A.* **59**, 869.
220. Levinson, H. S. and Hyatt, M. T. (1969). *In* "Spores IV". (In press.)
221. Wolf, J. (1969). *In* "Spores IV". (In press.)
222. Parker, M. S. and Bradley, T. J. (1968). *Can. J. Microbiol.* **14**, 745.
223. Sierra, G. (1968). *Appl. Microbiol.* **16**, 801.
224. Holdom, R. S. and Foster, J. W. (1967). *Antonie van Leeuwenhoek* **33**, 413.
225. Wax, R. and Freese, E. (1968). *J. Bact.* **95**, 433.
226. Murrell, W. G. (1969). *In* "Spores IV". (In press.)
227. Brown, M. R. W. and Melling, J. (1968). *Biochem. J.* **106**, 44p.
228. Wise, J., Swanson, A. and Halvorson, H. O. (1967). *J. Bact.* **94**, 2075.
229. Halvorson, H. O. (1969). *In* "Spores IV". (In press.)
230. Warth, A. D. (1965). *Biochim. biophys. Acta* **101**, 315.
231. Marquis, R. E. (1968). *J. Bact.* **95**, 775.
232. Chambon, P., Wutraw, E. J. and Kornberg, A. (1968). *J. biol. Chem.* **243**, 5101.
233. Chambon, P., Deutscher, M. P. and Kornberg, A. (1968). *J. biol. Chem.* **243**, 5110.
234. Deutscher, M. P., Chambon, P. and Kornberg, A. (1968). *J. biol. Chem.* **243**, 5117.
235. Nelson, D., Spudich, J., Bonsen, P., Bertsch, L. and Kornberg, A. (1969). *In* "Spores IV". (In press.)
236. Gould, G. W., Jones, A. and Wrighton, C. (1968). *J. appl. Bact.* **31**, 357.

CHAPTER 10

Activation

ALEX KEYNAN AND ZIGMUND EVENCHIK

Institute of Life Sciences, Hebrew University, Jerusalem, and Institute for Biological Research, Ness Ziona, Israel.

I. Introduction

THREE sequential processes are known to be responsible for the changing of a dormant bacterial spore into a vegetative cell: activation, germination and outgrowth. It can be shown that these three processes are fundamentally different from each other. They are caused by different treatments, they differ in the kind of changes brought about in the spore and they differ biochemically.[1]

The following is a short description of these three stages: fresh spores exposed to optimal germination conditions will not germinate in most cases unless preheated or "aged", they are therefore in the state of biological dormancy. The process of "conditioning the spore to germination" has been called "activation". The activated spore retains most of the spore properties, and the process of activation is in most cases reversible. When activated spores are exposed to the appropriate "germination environment" an irreversible process called "germination" occurs, which involves complete loss of typical spore properties. But the germinated spore still differs from the vegetative cell in its physiology and chemical composition. After germination is completed "outgrowth" occurs which changes a germinated spore into a vegetative cell. This is a process of biological growth and differentiation and takes place only in a growth medium.

The evidence accumulated by many workers (which will be given later in this chapter) indicates that activation does not involve metabolism and consists apparently in changes in the configuration of macromolecules. Germination, the process following activation, occurs also in the presence of inhibitors of macromolecular synthesis, and it is therefore a process during which no new macromolecules are being synthesized. During germination some spore components are broken down and excreted into the medium; germination is therefore considered to be a process of degradation, during which spore metabolism is initiated. During the process of outgrowth some new kinds of protein not present in the spore stage have been found, as well as new structures typical of the vegetative stage only.

This chapter will deal with the process of activation, the first of the three sequential steps which terminate the cryptobiotic stage. It will give a description of the way in which a bacterial spore reluctant to germinate is transformed into a state which enables it to germinate readily. Some of the proposed mechanisms responsible for these changes will be discussed.

A. Nomenclature and its Limitations

In this review we shall call the native unactivated spore which is reluctant to germinate a "dormant" spore, as proposed by Murrell,[2] who stated: "Viable spores failing to germinate in apparently favourable conditions are said to be dormant." The term "dormancy" will therefore be used for the non-response of the spore to germinating conditions. "Dormancy" in this sense is commonly used by botanists and mycologists to describe the same phenomenon. For instance, Goddard[3] stated: "It has been clearly established that the ascospores of the fungi *Ascobolus* and *Neurospora* are clearly dormant and will germinate only after they have been heated." On the other hand, one can find in the literature names, like "sporostasis", "delayed dormancy" or "germinative dormancy" for the same phenomenon.

Unfortunately the term "dormancy" is also used sometimes to describe a state of low, or no metabolism of cells. In this chapter, however, the state of arrested metabolism will be called "cryptobiosis" as proposed by Keilin.[4]

According to the terminology proposed by us, the resting bacterial endospore is therefore in the state of cryptobiosis as long as it is a spore, even when the spore is activated. A spore is "dormant" as long as it is in a physiological state of reluctance to germinate. Activation, therefore, terminates the state of dormancy temporarily, without terminating cryptobiosis in the spore. During ageing dormancy is lost irreversibly. The aged spore is still cryptobiotic but is no longer dormant. When activated or aged spore suspensions are exposed to germinating agents, germination occurs and the cryptobiotic state is irreversibly lost.

It should be pointed out that there are shortcomings to these definitions, as with most definitions which describe biological phenomena not yet completely understood. According to our definition, for example, a spore is dormant if it requires activation to germinate. In some cases however, spores will require activation to germinate by one agent but not by another. Therefore, "dormancy" has to be considered not as an absolute property of the spore, but as one which is somehow also related to a definite germination-inducing condition. The reluctance to germinate which is defined here as dormancy is not always necessarily dependent on the internal physiological state of the spore, but might also result from environmental conditions inhibiting germination. Lewis (Chapter 9, p. 306) has therefore differentiated between autogenous germinative dormancy and environmental germinative dormancy. It is the autogenous germinative dormancy which will be described in this chapter.

The definition of activation is further complicated by the fact that different investigators have applied different methods for the quantitation of activation; therefore the question might be raised: have they been dealing with the same phenomena?

This problem is of special importance when experiments dealing with mesophilic bacteria are compared with those dealing with thermophilic bacteria. Most workers dealing with thermophilic or thermotolerant spores take advantage of the fact that these strains will not germinate at all unless activated. Therefore, they are measuring activation as the ability of a spore suspension to form colonies on nutrient plates. Investigators dealing with mesophilic bacteria measure activation usually as an increased rate of germination. Recently it has been suggested,[5] therefore, that one should differentiate between "activation for germination" and "activation for growth". As it will be seen in this review, we believe that in all cases activation does not influence growth but influences the germinative ability of the spore only. We are dealing, therefore, in both cases with the same phenomenon for which the same mechanism seems to be responsible. In both cases activation consists of breaking the dormancy of the spore and overcomes its inability to germinate.

The activated spore germinates if exposed to the right conditions, even if growth is inhibited by irradiation or by addition of growth-inhibiting substances. In those cases where the kinetics of activation were measured by both germination rates and colony counts, identical results were obtained. Therefore, in our opinion, activation has no direct connection with growth and activation for growth should not be differentiated from activation for germination.

B. The Discovery of Activation

Activation of mould spores was discovered before activation of bacterial spores. Several mycologists reported that after forest fires "red moulds" appeared in great quantity on carbonized trees and these were later shown to be different species of *Neurospora*.[6-9] On the basis of these observations, Shear and Dodge,[9] were the first to apply heat to a suspension of *Neurospora* spores as a method to activate them for germination.

The first use of activation of bacterial spores was in 1919 by Weizmann.[10] When developing the technology of acetone fermentation by *Clostridium acetobutylicum* he noted that fermentation did not start readily from spore suspensions, proceeding at a slow and sluggish rate or even failing to start. He observed that by exposing spore suspensions to heat, he could always get active fermentation at a predictable time. The following is his description of this procedure: "The selected tubes I now heat up to from 90–100°C for a period of one to two minutes. Many of the bacteria are destroyed, but the desired resistant spores remain. I next inoculate a sterilized maize mash with the culture which has been heated as said, and so obtain a subculture." Weizmann interpreted the effect of heating as eliminating contamination with vegetative cells from the spore suspension which, if not heat-killed, start to grow before spore germination. He assumed that the growth of these vegetative cells might inhibit spore germination by changing the medium, therefore leading to an uneven fermentation. He apparently did not realize that there is a direct effect of heat on the spore.

The first to note the direct activating effect of heat on spores were Curran and Evans,[11] who in 1945 noticed an increase in the colony count of milk after heat treatment. They stated: "A mild heating of the spores of mesophilic aerobes has been shown to hasten their subsequent germination." They also found that this sublethal heating influences the number of spores which will subsequently germinate and form colonies.

When reinvestigating this phenomenon in 1955 Powell and Hunter[12] showed that heat-activated spores revert to their dormant state after storage at room temperature and that heat activation is therefore a reversible process.

C. Methods of Activation

The simplest method of activation of spores is the exposure to sublethal heat, but other treatments have been shown to replace the heat effect. For example, exposure to low pH (1–1·5), exposure to thiol compounds and to strong oxidizing agents activates spores.

Heat treatment is the most studied method of activation and therefore most of this chapter deals with this; the other methods differ from heat activation in being less efficient.

D. Extent of Spore Dormancy and Degree of Activation

Dormancy, defined as the inability of a spore to germinate, is difficult to express quantitatively. One has to assume that dormancy is broken completely when the germination rate of the spore suspension is maximal. It has been suggested that one way of measuring the degree of dormancy of a spore suspension is by determining the energy necessary to break dormancy. For different spore suspensions this might mean either the length of time of exposure to a given temperature, or alternatively, the temperature needed to reach optimal germination rate at a given time. This approach was suggested by Keynan et al.,[13] who correlated dormancy with DPA content of spores.

II. Changes Occurring in Spores as a Result of Activation

A. Properties of Activated Spores

Activated spores retain most of the important properties of the dormant state such as resistance to heat and radiation, non-stainability and refractility. On the other hand, activation also causes several measurable changes which are described overleaf.

1. Changes in germination rate

The most easily recognizable changes which result from activation of spores are those affecting their germination characteristics. Spores from freshly harvested cultures usually do not germinate or only very slowly. Often some germination can be caused by placing the spores into rich media, but a very long lag period will precede germination, which will be slow and unsynchronized. Germination in such unactivated spores will usually be incomplete in the sense that many ungerminated spores will

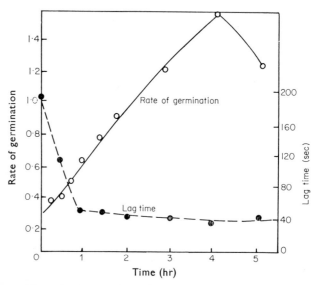

FIG. 1. Effect of heat shock at 65°C on the rate of germination of spores of *B. cereus* strain T. The rate of germination of freshly harvested spores was determined by measuring the exponential decrease in optical density as a function of time by the method of Woese and Morowitz.[82] (Reproduced with permission of O'Connor.[83])

remain in the suspension. The degree to which germination can be stimulated in unactivated spores varies. In many cases highly dormant unactivated spores will not germinate at all, and therefore do not form colonies when plated even on a very rich medium. When heat-activated, the same spore suspension germinates completely in a quite synchronized manner and forms normal colonies. In this case, germination and colony formation is completely dependent on activation. In less dormant spores some germination will occur even if not activated, but activation increases the germination rates dramatically.

Powell and Hunter[12] observed that "the extent of germination depended on the degree of activation, i.e. the temperature and duration of heating". Pretreatment at various high temperatures for various periods alters the kinetics of germination of a spore suspension. Both the time required for the first changes in the spore and the rate of germination (which measures the number of spores germinating per unit time) are obviously affected. As shown in Fig. 1, the lag-time decreases and the rate of germination increases after activation. In all cases, when the activated suspension of spores is exposed to proper amounts of a germinating agent, germination will occur within a very short time.

The degree of dormancy of individual spores in a suspension is quite variable and not all of the spores germinate even after severe activation. Although the majority of activated spores will germinate immediately, some remain dormant as long as 90 days.[14] Using a single cell method, it was observed in *Cl. acetobutylicum* by McCoy and Hastings[15] that germination of at least 5% of freshly harvested spores was delayed from 11–117 days, and that a spore isolated from a year-old culture required 222 days to germinate. The finding of different germination rates in activated spore populations indicated that a phenotypic heterogeneity existed among spores.

2. *Changes in germination requirements*

In addition to changes in rate of germination, activation influences also the qualitative and quantitative requirements of spores for agents which cause germination. Specific nutrients are needed for spore germination as was demonstrated for the first time by Hills in 1949.[16] Glucose, adenosine, inosine, L-alanine and certain other amino acids, each of these compounds either separately or in various combinations, will cause germination in over 90% of individuals in several strains of bacilli. Activation might modify this germination requirement as observed first by Powell and Hunter.[12] They demonstrated that dormant spores of *B. cereus* required either inosine or a mixture of alanine + tyrosine + adenosine for germination. The same spores after prolonged ageing or heat-treatment germinated easily in the presence of adenosine alone. An even more pronounced effect of heat activation was observed in spores of *B. megaterium*, which germinated spontaneously after activation. The rate and extent of germination was dependent on the temperature and time of the heat treatment. Hyatt and Levinson[17] found that heat activated spores of *B. megaterium* germinated when glucose was the sole germinating agent. With a rise in the preheating temperature the requirement for glucose to attain maximum germination rate decreased. They found that whilst 25 mM glucose was needed after heat treatment at 52·5°C for 10 min, maximal germination rate was reached in the presence of 2·5 mM of glucose after treatment at 60°C for the same time.

Investigating the L-alanine dehydrogenase system in spores of *B. cereus*, O'Connor and Halvorson[18] found that the heat-treatment decreased the quantitative requirement for L-alanine-induced germination (Fig. 2). They suggested that heat treatment makes the enzyme, which triggers germination, more accessible to the amino acid.

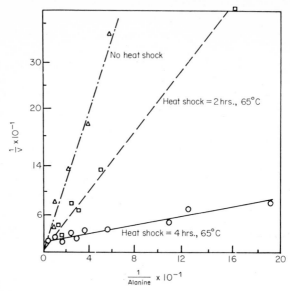

F I G. 2. Saturation of L-alanine-induced germination of dormant and heat shocked spores. The germination rate of *B. cereus* spores was measured before and after heat shock for 2 hr and 4 hr at 65°C with different L-alanine concentrations. (Reproduced with permission from O'Connoi.[83])

3. Changes in morphology, permeability and spore composition

Some changes in the morphology of the spores were noted when heat activation was carried out at low pH. Acid-activated spores tended to aggregate during heating and their optical density decreased, which has never been observed in spores activated in water. Acid treated spores examined under the phase microscope had darker contours than usual,[88] and showed an irregular side swelling (Issahary, unpublished results). Additional evidence of morphological and structural changes in activated spores were reported by Moberly *et al.*[19] They observed under the electron microscope that the spore coat of *B. anthracis* was altered. Spores after heating for 15 min at 65°C became more mottled and the cytoplasm was less granuiar and more opaque than in dormant spores (Fig. 3).

It has also been reported that activation influences the permeability of the spore. Falcone and Bresciani[20] reported that in their experiments spores of *B. subtilis* germinated in the presence of pyruvate after acid treatment. Water activated spores never germinated in the presence of pyruvate, but require L-alanine instead. The authors assumed that changes in permeability allowed pyruvate to enter through the spore cortex and this could explain the triggering of germination.

The chemical composition of spores has been reported to change after activation. Harrell and Mantini[21] demonstrated in spores of *B. cereus* T

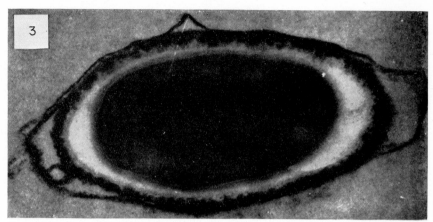

FIG. 3. View of a typical heat activated spore of *B. anthracis* showing the mottled appearance of the coat and changed appearance of the cytoplasm. × 98000. (Reproduced with the permission of Moberly *et al.*[19])

TABLE I

Effect of heat on glucose metabolism and the release of dipicolinic acid in ungerminated spores

Time heated 65 °C (min)	μ1 O_2 per hr[a]	% of total DPA released[b]
0	10	2·0
15	102	5·45
30	179	8·3
45	206	10·0
60	230	12·5

[a] Glucose (0·5 mg) was used as the substrate.
[b] Dipicolinic acid (DPA) was measured as the calcium salt at 278 mμ in a Beckman spectrophotometer.
(Reproduced with the permission of Harrell and Mantini[21].)

the release of DPA as a function of time and temperature and suggested a correlation between excretion of DPA and degree of activation. Their results, which are summarized in Table I, might indicate a correlation between the amount of DPA released and an increase in metabolism of glucose. The release of DPA during heat activation could result from a change in the permeability of the spore coat and the opening of a chemical bond between DPA and the spore enzyme. Keynan et al.[13] observed DPA release during and after activation of B. cereus T spores which contained varying amounts of endogenous DPA. The sensitivity of activated spores to certain chemicals is also altered. The resistance of heat-activated spores to phenol, formaldehyde and other chemical agents is diminished.[22] A decrease in resistance to compounds like penta-chlorophenol and vitamin K₃ was observed.[23]

4. Changes in metabolic activity of the spores

No metabolic activity was apparent in dormant spores of B. subtilis and B. anthracis as reported by Crook,[24] who could not detect any respiration with or without glucose. Activation induces some measurable metabolism in spores. Church and Halvorson[25-27] found that aged spores started to oxidize glucose when activated, although no germination was apparent. The activation of glucose metabolism by heat was reversible and diminished during storage at 5°C. At this temperature, after 24 hr, the glucose oxidizing capacity dropped to 15% of the initial rate. When the spores were stored for a longer period than 24 hr the metabolic activity was completely lost. It was found by Murrell[28] that glucose oxidation in spores of B. subtilis could be initiated also by incubating them in the presence of small amounts of L-alanine or adenosine in concentrations at which these substances do not induce germination. Following this observation Murty and Halvorson[29] demonstrated that different concentrations of alanine or adenosine added during preincubation controlled the rate of glucose oxidation. In contrast to B. subtilis spores of B. cereus T required heating for activation of glucose oxidation in addition to L-alanine. No heat treatment was needed when the same spores were incubated in the presence of adenosine and low concentrations of glucose.[27]

In addition to the glucose oxidizing enzymes other enzymatic systems are apparently also activated by heating. It has been shown that preheated spores of B. cereus T oxidize gluconate as well as 2-ketogluconate and pyruvate.

Bishop and Doi,[30] investigating ribosomes from spores of B. subtilis, found that enzymatic activity of proteases in intact spores was increased after heat shock of the spore suspension for 15 min at 60°C.

The activity of enzymes extracted from spores was increased when activation preceded extraction. When NAD-linked glucose dehydrogenase, for instance, was isolated from extracts of non-heated spores,[31] its activity was only 25% of the enzyme isolated from spores after heat treatment. In this case activity of the enzyme was a function of the degree of activation of the spore suspension before extraction. A similar observation was made by Krask,[32] who reported that enzymatic activity of acetokinase and CoA-kinase, which is found in spore extracts only, was increased when spores were heated before rupture.

B. QUANTITATIVE MEASUREMENT OF ACTIVATION

Several methods have been employed for the measurement of activation of spore suspensions. Most of them follow the kinetics of germination as a function of activation of the spore suspension. The kinetics of germination is a function of the degree of activation when germination is carried out under standard conditions. Germination can be followed by decrease in optical density of the suspension, or direct microscopic examination of loss in refractility and darkening of spores. These methods are sensitive and can show slight differences in heat activations carried out at various temperatures and with different times of heating.

Another method measures activation as the percentage of spores which form colonies when plated on an appropriate medium. This method is frequently used with highly dormant spores. These spores require high temperature activation which at the same time may kill a certain percentage of the spores. The germination of such spore suspensions will be unsynchronized and cannot represent the degree of activation. Moreover, this method is insensitive because the same number of colonies may develop from two different spore suspensions which had different germination rates. On the other hand, the differences in activation of two spore suspensions can easily be detected when germination rate is measured. Therefore, for quantitative experiments, especially with mesophilic bacteria, the method of measuring activation by following the germination rate is recommended.

III. Heat Activation

A. FACTORS AFFECTING THE TEMPERATURE REQUIREMENTS OF SPORES FOR ACTIVATION

Different species vary in their temperature requirements for activation. This variation exists also among different strains and even among different batches of the same strain. Some strains of *B. megaterium* require only a few minutes at 60°C for optimal germination rates, whereas *B. stearothermophilus* and other thermophilic and thermotolerant bacteria require from 105–115°C for optimal activation.[11] The findings indicate that the exact time and temperature of activation depends on the species, the composition of the media on which spores have been grown, and the age of the spore suspension. Although not all the factors responsible for this variation are known, there are some which have been quite well described.

Q

One of these is the composition of the medium on which the organism was grown. Keynan *et al.*[13] induced sporulation in the presence of different amounts of Ca^{2+} and phenylalanine, conditions which have been demonstrated to affect spore composition.[33] They produced spores containing different amounts of Ca–DPA. The times necessary for optimal activation of these spores correlated well with the content of Ca–DPA (Fig. 4), but this finding does not necessarily mean that heat activation

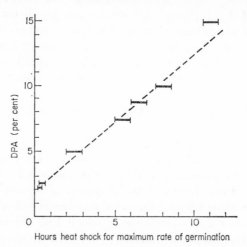

FIG. 4. Dependence of the duration of heat shock on intrasporal DPA concentration. Spores containing various concentrations of DPA were heat shocked for various periods of 65°C, centrifuged, washed and resuspended in M/50 tris buffer pH 8 to an optical density of 0·40. Germination at 30°C was initiated by the addition of L-alanine (15 mg/ml). The rate of germination was followed spectrophotometrically, as described by Woese *et al.*[82] The bars indicate the approximate time of heat shock required to obtain maximum germination rates. (Reproduced with the permission of Keynan *et al.*[13])

requirements are dependent only on Ca and DPA concentration in spores. Similar correlation between heat resistance and DPA content had been shown previously by Halvorson and Howitt,[34] but Warth *et al.*,[35] on reinvestigating this problem, found a closer correlation between heat resistance and the murein composition of the spore cortex. Another factor which influences the temperature requirements of spores for activation is the temperature of growth at which sporulation occurs.[11]

B. Conditions Influencing Activation of Spores during Heating

Powell and Hunter[12] were the first to report that heat activation must be carried out in the presence of water. Later it was observed that it was impossible to activate dry lyophilized spores, or spores resuspended in a high concentration of glycerol.[36] These results indicate the importance of the role of water for activation.

Heat activation is also pH dependent. Although heat activation is mostly carried out at neutral pH, at 60°C for 1 hr, optimal activation under the same conditions in *B. cereus* T could be also obtained at a pH between 2 and 3. A lower pH than 2 activates spores at 60°C when heated for 10 min; longer exposure times injure spores and prevent germination.[37] At pH above 8·5 spores were not activated when heated for an hour. The same spores when washed and resuspended at neutral pH and then heated for 1 hr at 60°C behaved as activated spores. This indicates that in *B. cereus* activation is inhibited at a high pH, without injury to the spores. This does not seem to be the case in other species, since it has been shown recently by Gibbs[38] that in spores of *Clostridium bifermentans* activation is optimal at high pH and inhibited at a low pH.

A number of reports exist indicating that the composition of the medium in which spores are suspended during heating influences the degree of activation. This was first reported by Curran and Evans,[11] who showed that the suspending medium determined the degree of activation of their strains. The following media used by them could be arranged in order of decreasing efficiency upon heat activation: glucose, lactose, peptone, skim milk, glucose nutrient agar, beef extract, glucose nutrient broth, distilled water and sodium chloride. From this work it appeared that some substances could interfere with activation when added during heating to the spore suspension. This finding was reinvestigated by Powell and Hunter,[12] who reported that although some substances had no effect when added to the spore suspension during activation, others like sodium fluoride inhibited activation strongly in spores of *B. megaterium*.

Ordal (personal communication) reported recently that some surface active agents in low concentrations interfered with heat activation. It was also found by several workers that certain cations may interfere with activation when present during heating. Heavy metals such as iron and copper inhibited the activation of spores of *B. cereus* T[29] and of *B. megaterium*.[39] Inhibition of the activation in *B. popilliae* spores heated for 10 min at 80°C in the presence of potassium ions has been reported.[40] The potassium effect diminished with ageing of spores and suggested that a slow activation occurred during ageing of systems blocked by potassium.

Salts have been shown to interfere with germination when present in the spore suspension during heating.[41, 52] When the spores were washed thoroughly after heat activation, they behaved like fully activated spores. It seemed that the spores absorbed salts during heat activation which interfered with germination. When washed, the salts were eluted from the spores and normal germination occurred. This indicated that spores heated in the presence of salts were activated and that the salts interfered with germination. One has therefore to be rather careful in interpretation of experiments in which a substance seems to prevent activation. The possibility has always to be considered, that the substance is being absorbed during activation and will interfere with germination later, thereby imitating an apparent inhibition of activation.

C. Heat Effects Other than Activation on Bacterial Spores

Spores react in a variety of ways when they are exposed to heat, activation being only one of these effects. These effects are especially marked in highly dormant spores which need high temperatures of activation.

Therefore before discussing the kinetics of heat activation the following effects of heat on spores have to be considered.

(1) Heat killing of spores
(2) heat damage to some specific enzymes involved in germination
(3) heat-induced dormancy
(4) quantitative population changes in heterogeneous spore suspensions
(5) permanent changes in germination requirements as a result of heating.

1. Heat killing of spores

Heat resistance varies among different species and in different spore suspensions of the same strain. Typical D values for spores of several mesophilic strains at 100°C are around one min, but spores of strains of B. megaterium survive up to 270 min at this temperature, and spores of B. coagulans and B. stearothermophilus survive up to 714 min.[42]

Heat killing does not seem to be an interfering factor during activation at 60–70°C for short periods of time in mesophilic spores. Spores of mesophilic bacteria, grown under special conditions to contain low Ca–DPA may be less heat resistant and more vulnerable at this temperature.[33] Heat killing may be an important complicating factor during activation of spores of B. stearothermophilus, at 115°C, and when the germination is determined by colony count only. In this assay it is impossible to differentiate between heat-killed and non-activated spores, as neither produce colonies. The introduction of a medium containing Ca and DPA in which every viable spore, regardless of its state of activation, produces a colony, enables a precise determination to be made of the percentage of spores activated or killed as a function of heating time.[43]

2. Heat damage to specific enzymes involved in germination

When high temperatures were applied for very short times to a strain of B. subtilis[44] some of the spores failed to germinate on regular nutrient media. When the Ca–DPA media, mentioned in the previous section, were used all the spores germinated and formed colonies. Normal heat shock failed to reactivate the spores which were previously heated for a short time at high temperature, and then failed to germinate. The non-response of spores to nutrients to which the same spore suspension responded before heating seems to indicate that some of the specific enzymes involved in physiological germination had been injured during this treatment. Similar observations were made when *B. cereus* was heat activated at low pH.[45] This strain, which germinates in low concentrations of L and D-alanine, failed to germinate after this treatment, in D-alanine. It was further shown that the alanine racemase of the spore, which is apparently needed for D-alanine germination, was inactivated during heating at low pH.

3. Heat-induced dormancy

It was found that short exposures to 80, 90 and 100°C increased the dormancy of *B. stearothermophilus* which is usually activated when exposed to 115°C for a few minutes.[46] Heat-induced dormancy can be differentiated from heat injury, since these spores can be reactivated when reexposed to high temperatures. In these studies heat-induced dormancy was measured by colony count, but this phenomenon could also be demonstrated by following the rate of germination by counting the number of "darkened" spores under the phase contrast microscope. Both these methods correlated well and demonstrated heat induced dormancy.

4. Quantitative population changes in heterogeneous spore suspensions

When the spore population is genetically heterogeneous and consists of more than one mutant, it has been shown that heating may affect one mutant more than the other, thereby complicating the interpretation of the kinetics of heat activation. One such case has been well documented by Fields,[47] who showed that the rough variant was less heat resistant, and more easily heat activated, than the smooth variant of *B. stearothermophilus* after heating for a few minutes to 110°C. Temperatures needed to activate the smooth variants killed the rough variant so that plate counts showed only the smooth variant colonies. Fields also demonstrated a difference in optimal temperatures for heat induced dormancy between spores of rough and smooth mutants.

5. Permanent changes in germination requirements as a result of heating

Some spores which survived severe heating had altered germination requirements and did not respond well to those germination media which caused germination in the parent strain.[48] It was also shown by Zamenhof[49] that heat sometimes induced in spores a high rate of permanent mutations.

D. The Effects of Heat Activation on Individual Spores

Heat activation is usually measured as a population phenomenon of a spore suspension. In order to understand this phenomenon one has to ask what the effect of heat activation is on the individual spore. Vary and McCormick[50] have investigated this problem by correlating changes occurring in a population of germinating spores with changes in the individual spores.

When germination was followed by darkening of individual spores under the phase microscope, two parameters could be measured: one was the time interval from the exposure of the spore to the germination agent to the beginning of germination. This time interval has been called the "microlag". The second event, which was called "microgermination", was the time interval needed for the individual spore to germinate. This was the period from the start of phase darkening to the completely dark spore.

The distribution of the microlag of the individual spores in a heat-shocked suspension was between 100–400 sec, with a sharp peak at 200 sec, and the distribution of microgermination was between 5–40 sec, with a sharp peak around 10 sec. The distribution of the microlag was a function of heat activation. Whereas the peak in inactivated spore populations was around 500 sec or more, the peak of the microlag dropped to about 200 sec after heat activation. No great changes occurred in microgermination times as a result of heat activation. This seems to indicate that the spore starts to germinate faster as a result of activation, but once germination starts the time it takes from the beginning until the conclusion of this process is not influenced by activation.

This observation indicates that the changes of germination kinetics induced by heat activation are due to changes in mean microlag time of individual spores rather than changes in the time it takes a spore to germinate.

E. The Kinetics of Heat Activation

The first attempt to investigate the kinetics of heat activation in bacterial spores was carried out by Powell and Hunter.[12] It was pointed out that when at a given temperature (between 56–68°C) the percentage of spontaneous germination of *B. megaterium* spores was plotted against time of heating, a sigmoid curve was obtained (Fig. 5). Plotting the temperature against the time of heating required to give 50% of spontaneous germination, the curve in Fig. 6 was obtained. The authors suggested that a very long heating period would be necessary for activation at temperatures below 55°C, and that there might be a "critical temperature" of activation. This was shown in one of their experiments in which complete germination occurred after 18 hr of heating at 45°C. When the spores were held for the same time at 44°C practically no germination could be observed. This illustrated that changes in heating time necessary to achieve maximal activation at a given temperature are not linear. The existence of a "critical temperature" (when heating for a constant time) or a critical time (when heating at a constant temperature) is usually assumed to denote the occurrence of a sudden and abrupt process. This kind of kinetics was first

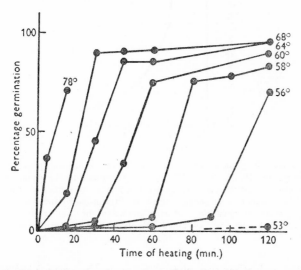

FIG. 5. Heat-activated spontaneous germination of *B. megaterium* spores. *B. megaterium* spores were heated for 15, 30, 60, 90 and 120 min at temperatures between 50–78°C. After centrifugation spores were resuspended in distilled water and remained at room temperature for 1 hr. Percentage of germination was determined by the staining method and plotted against duration of heat treatment. (Reproduced with permission of Powell and Hunter.[12])

demonstrated by Goddard in 1936 during studies with Neurospora spores.[3] It was found that the activation process at a given time has a critical temperature and this time temperature relationship could be compared to that of reactions typical for a "phase change", as for instance from a solid to a liquid state, which exhibit the same phenomena. Murrell[2] calculated the Q_{10} of this process for *B. coagulans* spores and found it to be 5·5.

In the work just cited the notion of "time temperature" relationship was applied to a narrow range of temperatures in which activation occurred during short times. Recently it was shown that activation occurs over a very wide range of temperatures and that the activation curve follows the same pattern.[43] Using as an indicator for activation the number of colonies

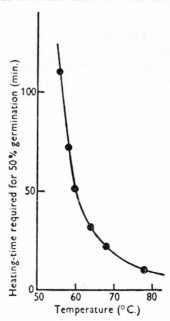

FIG. 6. Heat-activated spontaneous germination of *B. megaterium* spores. *B. megaterium* spores were heated for 15, 30, 60, 90 and 120 min at temperatures between 50–78°C. After centrifugation spores were resuspended in distilled water and remained at room temperature for 1 hr. The percentage of germination was determined by the staining method. By plotting temperature of activation against a period of heating required to give 50% germination, a smooth curve was obtained. (Reproduced with permission of Powell and Hunter.[12])

of *B. subtilis*, activation was shown to occur between 5–84°C (Fig. 7). In another experiment *B. cereus* spores were activated between 20–30°C, and between 50–60°C as shown by an increase in germination rates (Fig. 8). At low temperatures activation is very slow; for instance, 48 hr are needed to heat activate spores of *B. cereus* at 30°C, while the same spores would be activated in 45 min at 60°C. It took 525 days to activate *B. subtilis* at 5°C, while only 90 min were required for activation at 94°C.

As stated before, the exact quantitative measurement of heat activation involves other effects of heat on spores. It is important to know how many spores are viable in the suspension at every moment during the experiment. Heat may kill spores or induce dormancy in some strains and this is the main difficulty in estimating the correct number of viable spores. One has also to consider the fact that if activation is measured by colony count, some additional activation might occur during the incubation of plates, especially when the experiment is carried out at high temperatures.

All these considerations have been taken into account in the detailed investigation carried out by Busta and Ordal[43] on the kinetics and the thermodynamic properties of heat activation. Following earlier observations by Keynan and Halvorson[51] on rapid germination occurring in the presence of Ca^{2+} and DPA and unaffected by heat activation, Busta and Ordal incorporated Ca^{2+} and DPA into the assay medium in order to obtain a total viable spore count and to rule out heat induced dormancy or heat killing.[43] The number of spores obtained by plating a spore suspension on the Ca–DPA

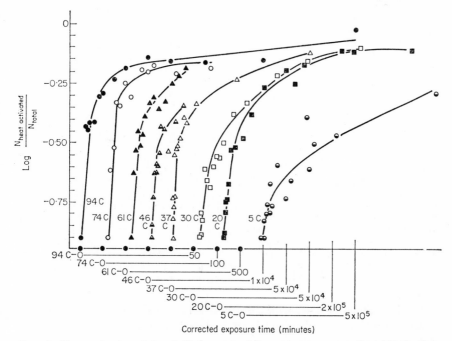

FIG. 7. Heat activation of *B. subtilis* spores at eight temperatures; semilogarithmic plot. During heat treatment a concentration of spores of ca. 1×10^8 ml were suspended in sterile distilled water. The pH of the suspensions ranged between pH 6·5–7·0. The amount of heat activation is reported as a decimal fraction (i.e. the number of heat-activated spores divided by the total number of spores present). (Reproduced with permission of Busta and Ordal.[43])

medium was equal to the number calculated by direct microscopic observation. They calculated initial velocities of activation as a function of temperature, in order to correct the rate of activation occurring in spores after plating. Using this new method they collected data describing the kinetics of heat activation at eight different temperatures which are given in Fig. 7. In this figure the logarithm of the decimal fraction of the heat-activated spores is plotted on the ordinate against corrected exposure time on the abscissa. The time axis has been manipulated in order to demonstrate the similar nature of the curves. From these data the investigators

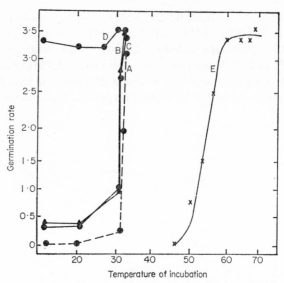

F IG. 8. Critical temperatures for heat activation. The experiments illustrated at the left of the figure (A, B, C and D) were carried out by incubating spore suspensions for 48 hr, whereas in curve E a 45-min incubation time was used. The spores were suspended during incubation in 0·2 M succinate buffer (pH 5·6). They were then washed twice in distilled water, resuspended in phosphate buffer at pH 7·8, and the rate of germination over the first 20 min was determined. (A) Incubation for 48 hr followed by germination assay. (B) As in A but with 0·02 M mercaptoethanol during incubation. (C) As in A but with 0·02 M thioglycolic acid during incubation. (D) As in A but with heat activation (65°C for 45 min) after incubation; this illustrates the potential of spores. (E) Incubation for 45 min followed by germination assay. (Reproduced with permission of Keynan et al.[37])

calculated the thermodynamic properties of the heat-activation process.

Constructing an Arrhenius plot from reaction velocities, they found that the energy of activation is very high, about 28 kcal. The fact that the activation energy is the same for 5°C and 94°C suggests that we are dealing with the same process at all temperatures. They also estimated that the change in entropy (ΔS^+) was 4·6–8·1 cal/deg, which is relatively small. Their results are consistent with the assumption that the formation of the activated state does not involve the disruption of several weak bonds but requires the breakage of a few strong bonds.

F. Reversibility of Heat Activation

When activated endospores are stored for a few days activation is lost and the spores revert to their original dormant state. This loss of activation in spores of a micro-organism was first noted by Goddard in 1936, who showed that if activated spores of Neurospora are kept under conditions which do not support germination, activation is lost.[3]

Powell[53] was the first to note the same phenomenon in bacterial spores. She showed that heat activation is necessary to obtain germination in *B. subtilis*. When activated spores were stored for 48 hr their germination rates decreased rapidly. She therefore concluded that a "temporary increase in the rate of germination was produced by heating". She also pointed out that this "temporary" state is different from the process of ageing which is irreversible and permanent.

Activation can also be shown to be reversible when measured by other techniques and not only by measuring the rate of germination. For instance, Church and Halvorson[27] showed that heat-activated ungerminated spores respired and took up oxygen in the presence of glucose. Respiration was lost several hours after heat activation, but could be recovered by reactivation.

When spores were activated by means other than heat they returned to their original dormant state after storage. Keynan *et al.*[37] showed that spores activated with mercaptoethanol deactivated on storage. Deactivation occurs also in *Clostridium bifermentans*, as was shown recently;[38] however, deactivation was prevented by aeration of the spore suspension. It might be of interest to note that under certain conditions, as activation of *B. cereus* at low pH[52] or *Cl. bifermentans* at high pH[38] deactivation does not occur. Busta and Ordal[43] could not demonstrate deactivation in their strain of *B. subtilis* when measuring activation and deactivation by plate count.

Deactivation depends on environmental conditions;[37] deactivation was temperature dependent, at 28°C activation was complete in *B. cereus* after 96 hr, no deactivation occurred at −20°C, and the rate of deactivation was slow at 4°C (Fig. 9). Gould and Hitchins[54] showed that exposure to disulphide-rupturing agents sensitized spores to lysozyme. Such sensitized spores returned to the normal lysozyme insensitive condition after being dialyzed against water for two weeks. They interpreted this finding as a reoxidation of reduced disulphide bonds. The same reducing agents also cause activation, so that it could be inferred that activation consists of rupturing disulphide bonds and deactivation consists of reoxidation of reduced disulphide bonds (Fig. 10).

While investigating deactivation one has to remember that a certain slow rate of activation, which is temperature dependent, may occur during storage. As deactivation is also temperature dependent, it seems that we are dealing here with an equilibrium of two processes. This might account for the fact that not all investigators have observed the deactivation process.

Deactivation discussed above deals with "spontaneous deactivation", but there are several investigations in which deactivation has been induced by chemical means. Spores activated by Ca–DPA for spontaneous germination can be deactivated by exposure to acetic acid for a day.[55] After this

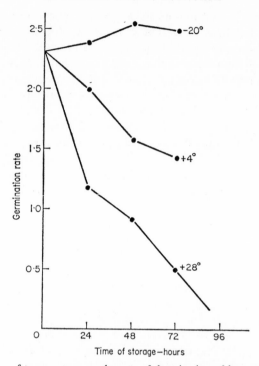

Fig. 9. Influence of temperature on the rate of deactivation of heat-activated spores of *B. cereus* T. Suspensions of freshly heat-activated spores were suspended in distilled water and stored at the temperature indicated. After 24 and 48 hr, the spores were centrifuged and resuspended in phosphate buffer (pH 7·8) and the rates of germination calculated as the percentage decrease in optical density per minute during the first 20 min, The rate at zero time was measured immediately after the heat shock. (Reproduced with the permission of Keynan et al.[37])

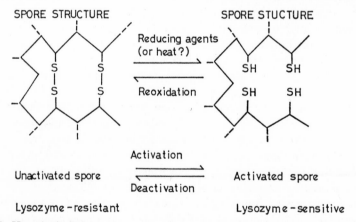

Fig. 10. Hypothetical changes accompanying activation of spores with reducing agents and deactivation with air.

treatment the spores did not germinate when washed and resuspended in neutral buffer. The deactivated spores were physiologically similar to dormant spores in their germination requirements, and they could be easily reactivated by heat or DPA. However, the deactivated spores were less heat resistant than spores before activation. There seems to be a paradox in the fact that low pH itself can sometimes activate spores under certain conditions while deactivating them under certain other conditions. This discrepancy might be explained by the role of specific acids in deactivation (acetic acid, phosphoric acid) rather than the action of low pH in general.

Another case of chemically induced deactivation was recently reported by Holmes et al.[56] and Hyatt et al.,[57] who showed that activated spores of B. megaterium were deactivated by absolute alcohol at 60°C for 18 hr and then could be reactivated by heat.

IV. Activation by Other Agents

A. Low pH and Reducing Agents

Normally spores are heat activated and germinated at a neutral pH. The degree of activation is pH dependent and several reports deal with this problem. In addition, exposure to low pH can by itself induce activation in spores which are subsequently washed and resuspended in a suitable germination medium. Keynan et al. reported that spores of B. cereus T germinated spontaneously after prolonged incubation in a buffer at low pH.[37] A case of reversible activation by low pH occurs in spores of B. stearothermophilus.[58] The activation was carried out at pH 1·5 and dormancy was restored at pH 9·7. The changes of spore states as a result of pH treatment are summarized in Chapter 9.

Another method of activation of bacterial spores consists of using reducing agents which imitate or spare the heat treatment.[37] Compounds which reduce disulphide bonds do not act as germinating agents but activate spores during prolonged incubation. When freshly harvested spores of B. cereus T were incubated in the presence of 0·20 M thioglycolate or 0·2 M mercaptoethanol, at 28°C partial germination could be obtained after washing and addition of L-alanine and adenosine. A prolonged time was required for activation with reducing agents since the effect appeared only after exposure of 12 hr. On the other hand, spores of the anaerobe Cl. bifermentans could not be activated with thioglycolate and this treatment inhibited activation.[38] Activation by reducing agents was pH dependent and it correlated well with that of spontaneous activation at low pH. It was found that optimal activation in presence of a reducing agent was around pH 6. Although activation after exposure to a solution of 6 M urea could not be observed, Ordal (personal communication) recently reported partial activation of spores using the same concentration of urea.

When the pH dependence of heat activation of *Cl. bifermentans* spores was investigated increased activation was obtained in the extremes of the pH ranges. Activation occurred at pH 10·0 or above and also at pH 2·0. Spores activated by alkali at pH 10·0 did not deactivate. Spores of *Cl. bifermentans* activated at 30°C at pH 7·4 for 2 hr completely lost the ability to germinate after storage for 6–7 days at 16–20°C or 4°C. After heating to 85°C for 10 min these spores were reactivated. On the other hand, heat-activated spores, when stored as above, were only 50% deactivated and could not be reactivated by a second heating. It was shown that the deactivation process occurs only in conditions of partial anaerobiosis. Aeration of stored activated spore suspensions inhibited their deactivation. It was suggested that in the case of a strict anaerobic bacterium, an enzymatic reaction rather than oxidation of sulphydryl groups was responsible for the deactivation.

The activation of thermophilic spores by furfural has been reported. The mode of action of furfural is not known, but it may act as a reducing agent.[59]

B. Ca–DPA

Riemann and Ordal[60] and Keynan *et al.*[13] showed that spores of *B. subtilis* and *B. cereus* T germinated without heat activation in solutions of equimolar concentrations of calcium and dipicolinic acid. On the other hand, Freeze and Cashel[61] reported that spores of *B. subtilis* 168 did not germinate in these concentrations of Ca–DPA, but were activated only. After the addition of Ca–DPA a slow decrease in optical density of the spore suspension could be observed. The decrease was caused presumably by the excretion of Ca^{2+} from the spores, since 98% of the spores remained refractile when observed microscopically. In these experiments Ca–DPA did not induce germination in spores after 200 min, but did activate the spores after 160 min to L-alanine germination. After exposure of 160 min to Ca–DPA, L-alanine was added, a rapid decrease in optical density with loss of refractility occurred. The rate of this germination was a function of the duration of pretreatment with Ca–DPA. At the same time using spores labelled with Ca^{45} rapid excretion of Ca^{2+} into the medium could be measured.

Activation of spores of *B. megaterium* by Ca–DPA occurred at 7–10°C.[55] Such spores were refractile, non-stainable and showed intermediate heat resistance, and germinated spontaneously in distilled water. By treatment with acid it was possible to deactivate these spores. The spontaneous germination of Ca–DPA-activated spores was inhibited by various cations like Cu^{2+}, Fe^{2+}, Ag^+ or Hg^+, by a chelating agent o-phenanthroline and by low pH solutions. These inhibitions could be reversed by addition of Na–DPA which suggested a competitive action of the Na ion.

C. Ionizing Radiation

Levinson and Hyatt were the first to report in 1960 the activation of spores of B. megaterium by ionizing radiation.[62] Irradiated spores were able to germinate even when radiation resulted in the inability of the germinated spores to divide. In their experiments irradiated spores showed a rise in oxygen consumption rate when exposed to appropriate germinating agents. The spores germinated in the presence of glucose + K_2SO_4 + ammonium acetate in phosphate buffer at pH 7·0. Their results suggested certain similarities in the action of heat and radiation especially at the higher doses of exposure.

Recently Gould and Ordal[63] reported activation of B. cereus PX by exposure to gamma radiation. Irradiated spores germinated at a high rate in the presence of a germinating agent, like L-alanine, inosine, n-dodecyla-mine and Ca–DPA. Increasing doses of radiation resulted in increased germination. Spores irradiated with killing doses of gamma rays (as determined by plating and counting the surviving cells) germinated rapidly. Similar treatments in the same range of radiation doses had no activation effect on spores of B. cereus T. or B. subtilis. In many respects activation by radiation resembles that by heat. Spores activated by both methods were deactivated on storage and had an elevated thiol group content, when titrated with Ag^+. In contrast to heat, irradiated spores germinated more rapidly in n-dodecylamine.

D. Activation by Various Chemicals

Gould[64] reported activation of B. cereus T spores germinating in L-alanine with D-cycloserine (oxamycin) and with o-carbamyl-D-serine. Gould suggested that both these antibiotics acted by being analogues of D-alanine and being able to inhibit the enzymatic activity of alanine-race-mase, thus preventing the formation of D-alanine and consequently allow-ing more efficient utilization of L-alanine during germination.

Spores of B. pantothenticus could be activated by pretreatment with very polar solvents like di-methylformamide.[65] By comparing the effect of heat with that of the chemical treatment, it was found that heating for 80 min at 60°C increased the capacity of producing colonies from 27·4% in an untreated suspension to 70%. Pretreatment with 90% v/v aqueous solu-tion of dimethylformamide at temperatures between 4–60°C produced rapid activation. Even a short exposure sufficed to produce maximum activation (70–80% colony formation). A similar effect was found with another solvent, dimethylsulphoxide at room temperature. An 8 M solution of urea had only a slight activation effect at 20°C.

Spores of *B. coagulans* lost their heat resistance rapidly when exposed to chloral hydrate solutions in water or ethylene glycol. Longer exposure to these conditions caused darkening of the spores. This treatment released DPA and other spore components irreversibly; the changes produced by chloral hydrate resemble germination rather than activation.[66]

Spores of *B. megaterium* germinate in the presence of manganese salts.[39] By preincubating unheated spores of *B. megaterium* for 3 hr at 30°C, in the presence of 0·1 mM manganous sulphate, respiration and germination of spores could be observed. Iron and copper in the same concentrations inhibited germination. It appears that manganese, activated proteolytic enzymes in the dormant spore causing germination and an increase in respiration. The fact that this activation was partially inhibited by D-alanine suggests that the action of manganese salts may be involved in general metabolic activities of the spore.

Lyophilized spores of *B. megaterium* were activated for germination with glucose (25 mM) or with L-alanine (1 mM) after prior exposure to water vapour (a_w, 1·0; 30 min at 25°C).[57] Exposure of lyophilized spores to water vapour increased both the percentage of germinating spores and the rate of germination. Exposure to water vapour for 2–4 hr resulted in spontaneous germination. Suitable time and temperature exposure to water activity as low as 0·6 gave maximal germination. Activation was also achieved by treatment with aqueous ethanol 60% v/v at 30°C for 5 min, but not with absolute alcohol. Activated spores were deactivated with absolute alcohol at 66°C for 18 hr and could be reactivated.[56] The authors suggested that activation by aqueous ethanol similar to that by heat shock may permit water to reach a specific site, hydration of which is essential for the functioning of germination. On the other hand, removal of firmly bound water by means of absolute alcohol deactivated already activated spores.

E. Ageing

It was mentioned before that freshly harvested spores do not germinate without some activation treatment. On the other hand, it is known that spore populations undergo marked changes upon storage even at low temperatures and under relatively dry conditions, and that the rate of these changes is temperature as well as moisture dependent. The altered spores are considered to have undergone the process of ageing. When investigating the germination and viability of aged spore populations at various times of storage one finds a decrease in viability and alterations in germination requirements. These spores behave as if they have been activated. The ageing phenomenon was first described by Powell in 1950, who observed that fresh spores germinated slowly but had an accelerated rate of germination after 20 days of storage in water at 20°C.[53] Similar changes resulting from storage were observed by Murty and Halvorson.[29] Spores of *B. cereus* T respired when incubated with low concentrations of L-alanine, but will also respire without L-alanine when stored in distilled water at 0°C. As described in Chapter 9 spores of *B. cereus* respired on glucose when activated. Spores stored for 4 months at −20°C had an endogenous and glucose oxidative activity five times higher than that of fresh spores.[27]

Ageing and heat activation seem to be similar phenomena. Long exposure of spores to low but stable temperatures causes activation; for instance, exposure at 5°C for 525 days activates spores of *B. subtilis*.[43] This can be described on the one hand as an example of the ageing phenomenon, and on the other hand this observation indicates that by all thermodynamic standards there is no difference in the activation kinetics at 5°C during 525 days and exposure to 60°C for 30 min. Therefore thermodynamically there seems to be no difference between ageing and activation. Both, activating and ageing, terminate the state of dormancy and the only difference which can be shown to exist is that during activation dormancy is lost temporarily, whereas during ageing dormancy is lost irreversibly. One way of interpreting these results is to assume that whatever the mechanism responsible for the reversion of activation might be is also lost during ageing.

V. Germination Systems not Requiring Activation

Most of the information on the first steps of germination were obtained in studies using compounds metabolized by spores, e.g. L-alanine. However, several workers showed that certain chemicals which are not metabolized can also induce germination. "Physiological germination" is induced by specific nutrients required for germination, and the kinetics of the germination is activation dependent. There is also a kind of "non-physiological germination" which does not require activation. For example, germination can be caused by surface active substances, chelating agents, or by mechanical means. They are referred to as chemical and mechanical germination and these germination kinetics do not depend on and are not changed by heat activation.

Rode and Foster were the first to note chemical germination of spores exposed to dodecylamine and several other n-alkyl primary amines.[67, 90] Spores exposed to these agents lost heat resistance and refractility, became stainable, metabolically active and remained viable. This germination seemed to proceed much faster than physiological germination.

When spores were shaken in a Mickle tissue disintegrator for 12–14 min, they germinated and produced colonies.[67] The germination kinetics was not dependent on activation.

Germination of spores of a putrefactive anaerobe (PA 3679), the National Canner Association strain as well as its mutant (h), was observed in the presence of EDTA.[68] In this experiment the loss of heat resistance was considered as a criterion of germination. 99% of spores germinated in less than 5 min when incubated at pH 5·5–9·0 in 5·5 mM EDTA at 37°C. On the other hand, spores of *B. subtilis*, *B. mycoides*, *B megaterium* and *B. thermoacidurans* did not germinate when exposed to EDTA and showed no sign of reduced refractility even after 5 hr at 20–35°C.

Spores exposed to Ca–DPA in Tris buffer germinated after a relatively long lag period.[69, 60] In some bacterial spores germination induced by chelators is dependent on duration of incubation and on temperature.[13, 51] Heat activation does not influence the kinetics of Ca–DPA-induced germination; incubation with Ca–DPA at 37°C activates spores for L-alanine germination.[61] Germination rates with L-alanine increased by a factor of four after exposure to Ca–DPA. It seems therefore that Ca–DPA added externally to spores has at least two effects: it activates some spores for L-alanine germination and in other cases causes germination.

Ca–DPA-induced germination is usually classified as "chemical germination", but there is some evidence that its effect is in some way metabolically mediated. For instance, it is diminished by some inhibitors of metabolism, and its lag period is temperature-dependent in a way which agrees with the notion that some metabolic reaction has to occur before germination starts.

It is difficult to understand how chelating agents cause germination; chelation of metal ions is the most likely explanation (but see p. 336). Theoretically they could chelate an inhibitory ion which is present in the dormant spore and which represses germination. Chelation results in the activation of an enzyme essential for initiation of germination. It was found that Ca and DPA are present in spores in a chelated form and that some of the DPA is released during heat treatment and ageing. This fact, together with the finding of increased metabolic activity of spores during germination, supports the assumption that breakage of the enzyme Ca–DPA chelate can be the first signal in a chain of reactions leading to germination.[69]

There is no universally accepted theory on the nature of chemical and mechanical germination, but many workers assume that in these cases the so-called lytic system is activated directly and is not mediated through a metabolic stage.[85, 86, 87] Physiological germination, on the other hand, involves induction of metabolism by a trigger. It seems therefore that activation is necessary for triggering, and those germination systems which circumvent the metabolic trigger are not activation-dependent.

VI. Activation Phenomena in Other Organisms

Cryptobiosis is not confined to bacteria only, and exists in many diverse organisms. Cryptobiotic stages are sometimes found in unicellular structures like the spores of moulds, and cysts of protozoa, and sometimes in multicellular structures like seeds and buds of plants. In all these forms the cryptobiotic process is not terminated by just exposing the organism to optimal growth conditions—definite activation occurs before metabolism and growth can start. Few systems have been studied in such detail as bacterial spores, and the sequence of biological events responsible for the termination of the cryptobiotic states, and their different stages, are not well defined.

In some systems the cryptobiotic stage can be produced under the direct influence of environment. There are organisms which dehydrate and remain inactive for many months until resuspended in water. After rehydration a normal life cycle starts again. In this case cryptobiosis occurs accidentally and is only a transient and resistant form. Yet it is not necessarily a part of the life cycle of the organism.

In such systems in which the cryptobiotic stage is part of the normal life cycle a cell or group of cells differentiate into a structure which is designed for cryptobiosis. In such a structure cell division stops for an indefinite time and metabolism is low or absent.

In some cases the factor responsible for activation is heat, like in many species of fungal spores. But in other organisms different factors can be responsible for activation. They may be physical, such as light of a specific wavelength, or mechanical damage to the seed or spore coat, or chemical activation as in the case of Neurospora, where furans and other heterocyclic compounds substitute for heat as activating agents.[70]

In some cases a variety of different external triggers terminate the cryptiobiotic state. Germination of mature lettuce seeds which have imbibed water depends on light and temperature. These seeds germinate in darkness at temperatures below 10°C, but they require light for germination at temperatures between 20–28°C. The light which induces germination is red light (6500–6800 Å) and its effects may be reversed by far-red light (7200–7500 Å).[71] Light requirement is strongest in fresh seeds and decreases in older ones. Mechanical treatment such as pricking the coat, or chemical treatment such as exposure to thiourea in the dark, induces germination even in young seeds. It is interesting to note that at temperatures above 30°C only mechanical treatment or exposure to atmospheres containing 40% CO_2 cause germination. Little is known about the biochemical nature of the reactions responsible for breaking the cryptobiotic state. In some cases one can show that the primary signal triggers a chain of events and is therefore not the direct cause of the termination of the cryptobiotic state. For instance, the red light-induced germination in lettuce seeds could not be reversed by far-red light when the latter was applied more than 10 min after the former. Apparently the chain of events had already started and the whole system was irreversibly committed to germination.[72]

Things seem to be even more complicated in some cryptobiotic stages of insects where the cryptobiotic state can be only terminated by a combined programme of light-dark periods with changes in temperatures at the same time.[73] An extensive and more detailed account and analysis of activation in different cryptobiotic systems has been published[74] and is also reviewed in Chapter 1.

It is evident that a mechanism exists in nature which is quite distinct from other physiological processes which control growth and differentiation. This mechanism, responsible for the dormant state, must be activated before the normal life cycle starts.

VII. The Nature of Activation

It might be presumptuous to suggest a mechanism for the activation process before the nature of dormancy itself is better understood. Still it might be useful to advance certain general concepts on the nature of the processes which are described here. Discussions on such concepts might stimulate further experiments which could lead to a better understanding of these phenomena.

The reluctance to germinate is a phenomenon not confined to bacterial spores only but is known to occur in spores of moulds, seeds of plants and other cryptobiotic systems. Theories explaining the nature of activation which have been proposed should fit all the known facts in all the various systems. One has, of course, to consider the possibility that different mechanisms could be responsible for apparently similar phenomena. It should be pointed out that at present no single hypothesis exists which is consistent with all the known facts and explains the phenomenon of dormancy and activation at the molecular level.

Some of these hypotheses, like the one which links activation to changes in permeability, is on the level of cell physiology; others, especially those which consider dormancy as a state of repressed metabolism, and activation as a process which derepresses it, are formulated on the general metabolic level. Still other hypotheses deal with the mechanism of activation as a change in structure of macromolecules and therefore try to explain these phenomena at the molecular level.

These various approaches are not necessarily contradictory to each other or mutually exclusive, but on the contrary, they may well supplement each other, explaining similar phenomena at different levels of organization.

1. Activation considered as a change in spore permeability

The meaning of permeability is not completely clear when applied to bacterial spores. It has been shown that the dormant spore is to a large extent permeable to water.[75] The fact that some substrates can be metabolized after activation, and not before, might be interpreted to show a change in permeability or alternatively an uncovering or activation of an enzyme.

There is much information indicating that the permeability of the endospore is increased by activation, whatever the mechanism of activation may be. The evidence that permeability is altered is based on the following observations: increased uptake of oxygen by activated spores when glucose is added;[39, 76, 27, 84] need for less L-alanine for the saturation of germination rates,[18] excretion of Ca–DPA and some amino acids during activation;[55] and finally, more direct proof by Gould and Hitchins,[54] who observed that exposure of spores to mercaptoethanol, a treatment which induces activation, enables a molecule as large as lysozyme to penetrate the spore coat and reach its substrate located in the cortex.

2. Activation considered as a change at a definite spore structure

Activation can be viewed as molecular changes occurring in different proteins located in various spore structures. On the other hand, some workers suggested that the targets of activation may be at a specific spore component. Gould and Hitchins analyzing the site of action of mercaptoethanol and other disulphide reducing agents which have been shown to activate spores suggested that a disulphide-bond opens in the coat fraction of the spore as a result of activation.[37, 54, 89] This suggestion is based on the finding of high content of disulphide bonds in the spore coat.[77] Gould and Hitchins assume, therefore, that the spore coat is the first target altered during activation with reducing agents.

Keynan et al.[37] also suggested the spore coat rather than the cortex as the site of heat activation. They correlated heat activation with denaturation of the proteins located in the outer parts of the spore-coat rather than in the cortex. The spore-coat is not universally accepted as the target of activation. It could be the site of a reversible activation system as a cation exchanger which "may well control the state of activation".[66] As calcium and other cations are probably more concentrated in the cortex, other investigators assume that the cortex is most probably the site which responds to heat activation.[66]

3. Non-dependence of activation on metabolic reactions

The various treatments which induce activation in spores, like heat shock, exposure to reducing agents or high concentration of urea, all point in the direction that the prime event responsible for activation is not metabolic. The "abrupt" kinetics of this reaction resemble melting curves of macromolecules, which indicates that we are dealing with physicochemical effects. The evidence that activation is reversible in most bacterial spores adds further weight to this hypothesis. One can also argue that all these activation treatments are artificial imitations of the naturally occurring process of ageing, especially since the kinetics of ageing and heat activation do not seem very different. The general consensus of opinion of most workers in this field seems to be that the passage of the dormant state to an active one is accompanied by a change in the tertiary structure of some macromolecules of the spores, and that the process of activation does not involve metabolic reactions.

4. Activation as a possible release of a factor stimulating germination

One of the first hypotheses on the mechanism by which activation occurs was proposed by Goddard in 1936, who assumed that in *Neurospora* germination depends on an endogenous substance released during heating.[3] According to this hypothesis, when the spore is heated a stimulating factor becomes available and germination starts, but this substance is unstable and disappears if germination is prevented. A variation of this theory which assumes the formation of stimulatory substances has been suggested.[78] It was noted that heat activation reduced the glucose and L-alanine requirements for germination of spores in *B. megaterium* and it was assumed that heat activation released internal stimulants which did not by themselves induce germination but which reacted with compounds added exogenously to increase germination rates.

It should be pointed out here that activation solubilizes some substances in resting spores. Ca^{2+} and DPA are known to exist in equimolar concentration in the spore in a bound form, and are also known to be released in a soluble form during activation. The release of DPA during heat activation has been well documented.[21, 13]

When activation is induced by exposure of the spores to Ca–DPA, small amounts of amino acids, especially glutamic acid, are released in addition to Ca–DPA.[55] Spores from which Ca–DPA was removed before the induction of germination and stored at low temperatures started to germinate when the temperature was raised. One has therefore to assume that removal of Ca–DPA activated the spores for spontaneous germination.

Freeze and Cashel,[61] working with *B. subtilis*, also demonstrated that under certain conditions Ca–DPA does not induce germination, but nevertheless activates spores for germination with L-alanine.

This evidence suggests that the hypothetical stimulating substance might be either Ca^{2+} or DPA or both of them. Keynan et al.[37] therefore tried to increase the rate of activation by the external addition of DPA during heat activation. This addition had no effect on the kinetics of activation induced by heat, mercaptoethanol or thioglycolic acid. The finding that activated spores stay activated after several washings, and that activation is reversible even in unwashed spores, does not support the notion that Ca–DPA or one of these substances is necessary to maintain the activated state. On the contrary, it demonstrates that the maintenance of the state of activation does not depend on the presence of soluble small molecules, which have been released during activation.

On the other hand, one has to consider the possibility that this substance may be unimportant in maintaining the activated state, but that its release causes a transient change responsible for the active state. This would assume that some bound Ca–DPA is released in the internal environment of the spore and activates a site not easily accessible to external Ca–DPA. This release brings about a reversible change of those molecular structures of the spore which are responsible for the "temporary state of activation". This hypothesis is based on the fact that addition of Ca–DPA in equimolar concentration activates the spore suspension.[61] Nevertheless, this hypothesis seems rather unlikely because the activation constants for temperature, time and pH by Ca–DPA, are all different from their respective constants induced by other ways. In any case, this hypothesis assumes that the released substances are instrumental in inducing this change rather than being themselves involved in the changed state or its maintenance.

5. Mechanism of activation—the possibility of inactivation of an inhibitor

From time to time it has been suggested that heat or other agents inactivate intercellular inhibitory substances which prevent germination of spores. Such germination inhibiting substances exist in plant seeds or mould spores and are known in these systems to be responsible for the maintenance of the dormant state.

Although several naturally occurring substances when added to spore suspensions inhibit spore germination,[2] there is no evidence that such simple soluble inhibitory substances exist in bacterial spores. No inhibitory effects were observed when extracts from dormant spores were added to germinating spore suspensions.

The hypothesis of the existence of an inhibitor was proposed in a different way by Halvorson et al.[79] They showed that DPA, which is a potent chelating agent, stimulated the activity of the endogenous electron transport system in spores. They assumed therefore that activation might lead to the release of DPA from its bound form. The liberated DPA then removed an inhibitory metal by chelation, promoting the flow of electrons from substrate to molecular O_2 which is also stimulated by a direct effect of DPA on NADH cytochrome C reductase. This hypothesis assumes the existence of a metal which retains the dormant state, but no such metal has been isolated so far. However, one has to remember that activated spores are inhibited by various cations.[41] Several groups of workers pointed out the inhibitory effect of Cu^{2+} and Fe^{2+}.[55, 76, 80] The spontaneous reversibility of activated and washed spores to their original dormant state renders the hypothesis that the activation process involves a removal of a cation or simple soluble substance very unlikely.

Still another variation on the inhibitor hypothesis is the suggestion that DPA itself is an inhibitor of metabolism by "masking enzyme activity". In dormant spores there might be a Ca–DPA chelated structure which can be broken by heat, releasing some of the Ca–DPA into the medium.[69] The existence of such a peptide Ca–DPA complex has been shown to exist. This kind of hypothesis has to account for the reversibility of activation. One possible explanation is to assume the existence of a pool of DPA in spores. During reversal such DPA might recombine with the free sites from which DPA was released at activation.

An interesting new observation has been made recently by Gould[64] which might throw some light on the role of the inhibitory reaction on germination and activation. He observed that D-cycloserine increased the germination rate of spores when they were germinated with L-alanine. Gould showed that cycloserine inhibited the alanine racemase. Alanine racemase forms D-alanine continuously from L-alanine and thereby inhibits germination. When this enzyme is inhibited, by cycloserine or D-cysteine, germination rates increased. One could therefore assume that activation might also inactivate the racemase reversibly and stimulate the germination rate.

6. Activation—explained on the molecular level

Activation may be the result of a reversible change in the tertiary structure of some of the spore macromolecules. This suggestion is supported by the following facts. The thermodynamic and the kinetic data of heat activation are very similar to the data obtained for heat denaturation of proteins or the so-called "melting curves" of double stranded nucleic acids. A very characteristic feature of this thermodynamic relation is the existence of a critical temperature range indicating that the energy of activation for the process is very high. Such data suggest the disruption of tertiary structures in macromolecules.

It should be noted that all chemicals which bring about activation are known to cause structural changes of macromolecules. Activation at low pH occurs in the range at which proteins are denatured.[37] Disulphide bond disrupting agents and reducing agents which induce activation are also known to cause denaturation of proteins.[54] In at least one species a solution of 8 M urea—an agent known to break hydrogen bonds—will induce activation (Ordal, personal communication).

Spores contain five times more cystine than vegetative cells and this is concentrated mostly in the spore coat.[77] Disulphide bonds are typical of the cryptobiotic state and during germination they are reduced to sulphydryl groups.

Summing up these observations, it is tempting to assume that the macromolecules responsible for the maintenance of the dormant state are coat proteins rich in cystine, stabilized at a certain configuration by S—S linkages. Reduction of these bonds would change the tertiary structure resulting in partial unfolding or denaturation of these proteins. It is known that spore coats are mostly composed of protein.[35]

Remembering that activation seems to increase permeability, the most probable assumption is that such a change is brought about in the spore coat protein. This change could be responsible for increased permeability as well as for the uncovering of the active sites of enzyme or increasing accessibility of the substrate to enzymes.

This interpretation is consistent with the known reversibility of activation. The reverse reaction would then consist of the reoxidation of sulphydryl groups reforming the original disulphide bridges and the original tertiary structure of the protein (Fig. 10, p. 380).

The direct measurement of the appearance of sulphydryl groups during activation would be expected if this hypothesis is correct. There have been several attempts to demonstrate such an effect using heat-activated spores, but for technical reasons the results are inconclusive (Halvorson, personal communication).

The most direct demonstration that activation consists of a change of the spore coat is by Gould et al.[81] These workers demonstrated that fluorochrome-labelled lysozyme does not stain dormant spores, but when the spores are exposed to disulphide bonds rupturing agents they become stainable by fluorochrome-labelled lysozyme. Fluorochrome lysozyme staining is an attachment of fluorochrome lysozyme to its murein substrate, which has been shown by Warth et al. to be located in the cortex of the spore.[15] Here, therefore, it has been demonstrated that a disulphide bond breaking agent changes the spore coat in such a way as to make it permeable enough to an extracellular macromolecule.

The theories offered by different workers are not opposed to each other, but rather deal with different aspects of the same phenomenon and are not mutually exclusive. It is difficult to separate the problem of activation from the more general problems concerning spores. Activation will probably only be completely understood when we understand the mechanism imposing the cryptobiotic state and the mechanisms responsible for dormancy and germination.

ACKNOWLEDGEMENT

The authors would like to express their most sincere thanks to Mrs B. Woman for her help in preparing the manuscript.

REFERENCES

1. Keynan, A and Halvorson, H. O. (1964). *In* "Spores III" (L. L. Campbell and H. O. Halvorson, eds), p. 174. Am. Soc. Microbiol. Ann Arbor, Michigan, U.S.A.
2. Murrell, W. G. (1961). *Symp. Soc. gen. Microbiol.* **11,** 120.
3. Goddard, D. R. (1936). *J. gen. Physiol.* **19,** 45.
4. Keilin, D. (1959). *Proc. R. Soc. Ser. B.* **150,** 149.
5. Gibbs, P. A., Gould, G. W., Hamilton, W. A., Hitchins, A. D., Hurst, A., King, W. L., Roberts, T. A. and Wolf, J. (1967). *Spore Newsletter,* **2,** 152.
6. Möller, A. (1901). *Bot. Mit. Trop.* **9.**
7. Kitasima, K. (1924). *Ztsch. Forstwiss. Gesell. Japan* **21,** 285.
8. Togukawa, Y. and Emotto, Y. (1924). *Jap. J. Bot.* **2,** 175.
9. Shear, C. L. and Dodge, B. O. (1927). *J. agric. Res.* **34,** 1019.
10. Weizmann, C. (1919). U.S. Patent No. 138,978.
11. Curran, H. R. and Evans, F. R. (1945). *J. Bact.* **49,** 335.
12. Powell, J. F. and Hunter, J. R. (1955). *J. gen. Microbiol.* **13,** 59.
13. Keynan, A., Murrell, W. G. and Halvorson, H. O. (1961). *Nature, Lond.* **192,** 1211.
14. Burke, V., Spraque, A. and Burnes, L. A. (1925). *J. infect. Dis.* **36,** 555.
15. Mc.Coy, E. and Hastings, E. G. (1928). *Proc. Soc. exp. Biol. Med.* **25,** 735.
16. Hills, G. M. (1949). *Biochem. J.* **45,** 353.
17. Hyatt, M. T. and Levinson, H. S. (1962). *J. Bact.* **83,** 1231.
18. O'Connor, R. J. and Halvorson, H. O. (1961). *J. Bact.* **82,** 706.
19. Moberly, B. J., Shafa, F. and Gerhardt, P. (1966). *J. Bact.* **92,** 220.
20. Falcone, G. and Bresciani, F. (1963). *Experientia* **19,** 152.
21. Harrell, W. K. and Mantini, E. (1957). *Can. J. Microbiol.* **3,** 735.
22. Reddish, C. F. (1950). *Official Proc. 37th Ann. Meeting Chem. Sp. Manuf. Ass.* p. 99.
23. Simidu, K. and Ueono, O. (1955). *Bull. Jap. Soc. scient. Fish.* **20,** 927.
24. Crook, P. G. (1952). *J. Bact.* **63,** 193.
25. Church, B. D. and Halvorson, H. O. (1955). *Bact. Proc.* p. 41.
26. Church, B. D. and Halvorson, H. O. (1956). *Bact. Proc.* p. 45.
27. Church, B. D. and Halvorson, H. O. (1957). *J. Bact.* **73,** 470.
28. Murrell, W. G. (1955). "The bacterial endospore", Monograph, Univ. Sydney, Australia.
29. Murty, G. G. and Halvorson, H. O. (1957). *J. Bact.* **73,** 235.
30. Bishop, H. L. and Doi, R. H. (1966). *J. Bact.* **91,** 695.
31. Halvorson, H. O. and Church, B. D. (1958). *In* "Spores" (H. O. Halvorson ed.), p. 359. Burgess Publ Co., Minneapolis, Minn., U.S.A.
32. Krask, B. J. (1956). Thesis, Univ. Chicago.
33. Church, B. D. and Halvorson, H. O. (1959). *Nature, Lond.* **183,** 124.
34. Halvorson, H. O. and Howitt, C. (1961). *In* "Spores II" (H. O. Halvorson ed.), p. 149, Burgess Publishing Co., Minneapolis, Minn., U.S.A.
35. Warth, A. D., Ohye, D. F. and Murrell, W. G. (1963). *J. Cell. Biol.* **16,** 579.
36. Beers, R. J. (1958). *In* "Spores" (H. O. Halvorson, ed.), p. 45. Burgess Publishing Co., Minneapolis, Minn., U.S.A.

37. Keynan, A., Evenchik, Z., Halvorson, H. O. and Hastings, J. W. (1964). *J. Bact.* **88**, 313.
38. Gibbs, P. A. (1966). *J. gen. Microbiol.* **46**, 285.
39. Levinson, H. S. and Hyatt, M. T. (1955). *J. Bact.* **70**, 368.
40. Splittstoesser, D. F. and Steinkraus, K. H. (1962). *J. Bact.* **84**, 278.
41. Halmann, M. and Keynan, A. (1962). *J. Bact.* **84**, 1187.
42. Murrell, W. G. and Warth, A. D. (1965). *In* "Spores III" (L. L. Campbell and H. O. Halvorson, eds), p. 1. Am. Soc. Microbiol., Ann Arbor, Michigan, U.S.A.
43. Busta, F. F. and Ordal, Z. J. (1964). *J. Fd. Sci.* **29**, 345–53.
44. Edwards, J. L., Busta, F. F. and Speck, M. L. (1965). *Appl. Microbiol.* **13**, 851.
45. Issahary, G., Evenchik, Z. and Keynan, A. (1967). (In preparation.)
46. Finley, N. and Fields, M. L. (1962). *Appl. Microbiol.* **10**, 231.
47. Fields, M. L. (1963). *Appl. Microbiol.* **11**, 100.
48. Campbell, L. L., Richards, M. and Sniff, E. (1965). *In* "Spores III" (L. L. Campbell and H. O. Halvorson, eds), p. 55. Am. Soc. Microbiol., Ann Arbor, Michigan, U.S.A.
49. Zamenhof, S. (1960). *Proc. natn. Acad. Sci., U.S.A.* **46**, 101.
50. Vary, J. C. and McCormick, N. G. (1965). *In* "Spores III" (L. L. Campbell and H. O. Halvorson, eds), p. 188. Am. Soc. Microbiol., Ann Arbor, Michigan, U.S.A.
51. Keynan, A. and Halvorson, H. O. (1962). *J. Bact.* **83**, 100.
52. Keynan, A., Issahary, G. and Evenchik, Z. (1965). *In* "Spores III" (L. L. Campbell and H. O. Halvorson, eds). p. 180. Am. Soc. Microbiol., Ann Arbor, Michigan, U.S.A.
53. Powell, J. F. (1950). *J. gen. Microbiol.* **4**, 330.
54. Gould, G. W. and Hitchins, A. D. (1963). *J. gen. Microbiol.* **33**, 413.
55. Lee, W. H. and Ordal, Z. J. (1963). *J. Bact.* **85**, 207.
56. Holmes, P. K. and Levinson, M. S. (1967). *Currents Mod. Biol.* **1**, 256.
57. Hyatt, M. T., Holmes, P. K. and Levinson, H. S. (1966). *Biochem. biophys. Res. Commun.* **24**, 701.
58. Alderton, G., Thompson, P. A. and Snell, N. (1964). *Science, N.Y.* **143**, 141.
59. Mefford, R. B. and Campbell, L. L. (1951). *J. Bact.* **62**, 130.
60. Riemann, H. and Ordal, Z. J. (1961). *Science, N.Y.* **133**, 1703.
61. Freese, E. and Cashel, M. (1965). *In* "Spores III" (L. L. Campbell and H. O. Halvorson, eds), p. 144. Am. Soc. Microbiol., Ann Arbor, Michigan, U.S.A.
62. Levinson, H. S. and Hyatt, M. T. (1960). *J. Bact.* **80**, 441.
63. Gould, G. W. and Ordal, Z. J. (1968). *J. gen. Microbiol.* **50**, 77.
64. Gould, G. W. (1966). *J. Bact.* **92**, 1261.
65. Widdowson, J. P. (1967). *Nature, Lond.* **214**, 812.
66. Lewis, J. C., Snell, N. S. and Alderton, G. (1965). *In* "Spores III" (L. L. Campbell and H. O. Halvorson, eds), p. 47. Am. Soc. Microbiol., Ann Arbor, Michigan, U.S.A.
67. Rode, L. J. and Foster, J. W. (1960). *Proc. natn. Acad. Sci., U.S. A.* **46**,118.
68. Brown, W. L. (1956). Thesis, University of Illinois.
69. Riemann, H. (1961). *In* "Spores II" (H. O. Halvorson, ed.), p. 24. Burgess Publishing Co., Minneapolis, Minn., U.S.A.
70. Sussman, A. S. (1953). *J. gen. Microbiol.* **8**, 211.
71. Evenari, M. (1956). *In* "Radiation Biology" (A. Holleander, ed.), p. 518. McGraw Hill Co., New York, U.S.A.
72. Evenari, M. and Neuman, G. (1953). *Bull. Res. Coun. Israel* **3**, 136.

73. Lees, A. D. (1961). *In* "Crypotobiotic stages in biological systems". Proc. 5th Biol. Conf. Oholo. (N. Grossovicz, S. Hestrin and A. Keynan, eds) p. 120. Elsevier Publishing Co., New York, U.S.A.

74. Sussman, A. S. and Halvorson, H. O. (1966). "Spores—their dormancy and germination". Harper and Row, New York, U.S.A.

75. Black, S. H. and Gerhardt, P. (1961). *J. Bact.* **82**, 743.

76. Levinson, H. S. and Hyatt, M. T. (1956). *J. Bact.* **72**, 176.

77. Vinter, V. (1961). *In* "Spores II" (H. O. Halvorson, ed.), p. 127. Burgess Publishing Co., Minneapolis, Minn., U.S.A.

78. Levinson, H. S. (1961). *In* "Spores II" (H. O. Halvorson, ed.), p. 14. Burgess Publishing Co., Minneapolis, Minn., U.S.A.

79. Halvorson, H. O., Doi, R. H. and Church, B. D. (1958). *Proc. natn. Acad. Sci.,* U.S.A. **44**, 1171.

80. Levinson, H. S. and Sevag, M. G. (1953). *J. gen. Physiol.* **36**, 617.

81. Gould, G. W., Georgala, D. L. and Hitchins, A. D. (1964). *In Proc. 3rd Int. Symp. Fleming's Lysozyme,* Session lb : 61. Scuolo Arti Grafiche, Milan.

82. Woese, C. and Morowitz, H. (1958). *J. Bact.* **76**, 81.

83. O'Connor, R. J. (1961). *In* "Spores II" (H. O. Halvorson, ed.), p. 73. Burgess Publishing Co. Minneapolis, Minn., U.S.A.

84. Falcone, G., Salvatore, G. and Covelli, I. (1959). *Biochim. biophys. Acta* **36**, 390.

85. Powell, J. F. and Strange, R. E. (1956). *Biochem. J.* **63**, 661.

86. Gould, G. W., Hitchins, A. D. and King, W. L. (1966). *J. gen. Microbiol.* **44**, 293.

87. Suzuki, Y. and Rode, L. J. (1967). *Bact. Proc.* p. 22.

88. Robinow, C. F. (1960). *In* "The Bacteria" (I. C. Gunsalus and R. Y. Stanier, eds), p. 207. Academic Press, New York, U.S.A.

89. Hitchins, A. D., Gould, G. W. and King, W. L. (1966). *J. appl. Bact.* **29**, 505.

90. Rode, L. J. and Foster, J. W. (1962). *Arch. Mikrobiol.* **43**, 183.

CHAPTER 11

Germination

G. W. GOULD

Unilever Research Laboratory, Sharnbrook, Bedford, England

I. Introduction

GERMINATION is a well-defined stage in the development cycle of spore-forming bacteria. By germination we mean essentially the conversion of a resistant and dormant spore into a sensitive and metabolically-active form. This change can occur to individual spores of some organisms in less than a minute with apparently very little utilization of exogenous substrates, and so has been aptly termed a "trigger" reaction. Research on spore germination has been largely concerned with elucidating the biochemical and biophysical steps comprising the trigger reaction and also with describing the resultant changes recognizable as germination. In spite of considerable research effort the nature of the trigger reaction has so far not been fully explained, although there is a vast amount of data available concerning the multiplicity of ways of initiating germination and describing the subsequent changes which take place. In this chapter we will review these data and discuss the nature of the germination event. Germination has been most recently reviewed by Sussman and Halvorson[1] and Halvorson, *et al.*[2]

Although activation will not be discussed in this chapter, it is important to realize that (as we saw in the previous chapter) some form of activation may be a necessary prerequisite for germination. It must be emphasized that "germination" as we have defined it terminates with the cell we may call the "germinated spore" perhaps only some seconds, rather than minutes, after germination has been initiated. Consequently the germinated spore, although now sensitive and metabolically active, must in no sense be considered equivalent to a vegetative cell. It is, for example, cytologically distinct and lacks a full complement of typical vegetative macromolecules and enzymic activities. The sequence of changes involved in the formation of a new vegetative cell from the germinated spore is the next stage in the development cycle of sporeforming bacteria, is termed outgrowth, and is the subject of Chapter 12.

II. Description of Germination

A. Changes Occurring during Germination

1. Changes in resistance

During germination, spores lose their characteristic resistance to heat, desiccation, pressure, vacuum, ultraviolet and ionizing radiations, antibiotics and other chemicals, and extremes of pH. Loss of resistance which depends on structural integrity of the spore coats, for instance to high-voltage, direct-current pulses[3] or to enzymes (e.g. lysozyme in sensitive strains,[239]) and metabolic inhibitors, occurs later, during dissolution of the cortex or emergence of the new vegetative cell from the spore coats.[4] Loss of heat resistance, which has attracted by far the most study, was the criterion first used to measure germination and occurs nearly concurrently with other changes described below (see Section II B4). Powell heated single spores germinating in microscope slide cultures and showed that loss of heat resistance of individual spores occurred rapidly and completely.[5] The extent of fall in heat resistance during germination varies according to the heat resistance of the spores (which can cover a range of about 10^5-fold[6]) but is large and normally ranges between about 10^3-fold and 10^6-fold.

The decreases in resistance to u.v. and ionizing radiations during spore germination are less dramatic, being often less than 10-fold.[7, 8, 9, 10] Furthermore, during the first few minutes of germination of *B. subtilis* and *B. megaterium* spores u.v. resistance increased prior to decreasing again before outgrowth; it appeared that the temporary increase in resistance occurred in spores which were not yet phase dark,[9, 14] the sequence being:

ungerminated, phase-bright medium u.v. resistance → phase-bright, high u.v. resistance → germinated, phase-dark, low u.v. resistance

The reason for the peak in resistance is not clear. Analysis of the DNA photoproducts of u.v.-irradiated spores suggested that DNA in dormant spores was in a state of low hydration[11, 12] and must therefore change to the hydrated vegetative form during germination and outgrowth. Such changes may well be accompanied by changes in conformation affecting u.v.-resistance.[14] Thymine incorporation demonstrable early in germination[16] was thought to indicate repair type DNA-synthesis, but a later kinetic study with *B. subtilis*[13] suggested that repair type synthesis was insignificant prior to the onset of DNA replication during outgrowth.

Although antibiotics and most chemical food preservatives have little effect on spore germination, they do inhibit synthetic processes during outgrowth or vegetative cell growth and multiplication.[4, 15, 17, 18] Spores increase in sensitivity to chemical reagents like phenol during germination;[19, 20] this can form the basis of rapid methods for measuring germination rates in place of the more conventional heat challenge.[19]

2. Breaking of dormancy

It is clear now that germination of bacterial spores is essentially a degradative process which does not involve extensive synthesis of new macromolecules. It follows that the appearance of enzymic activity during germination results mainly from the rapid unmasking or activation of pre-existing enzymes rather than from new syntheses. The manner in which the unmasking or activation occurs is conjectural; dormancy is not necessarily maintained by the same mechanisms that maintain spore resistance, but there is evidence that calcium dipicolinate, murein in the spore cortex and the state of hydration of the spore core are all involved. Mechanisms of dormancy are discussed in detail in Chapter 9. Onset of respiration during the breaking of dormancy accompanying germination has been most studied, and as far as can be measured occurs concurrently with germination detectable by other criteria.[21] Increase in respiratory activity during germination is very great, as compared to the respiration rate of unactivated ungerminated spores, which is so low as to be undetectable using sensitive radioactive tracer techniques.[22, 23]

It is important to realize, however, that spores are not wholly dormant. Some enzymes (see Chapters 8 and 9) are readily detectable in ungerminated spores. In particular, any enzymes which are operative during germination cannot be dormant or germination could never occur; indeed, enzymes active on germinants (e.g. alanine racemase and dehydrogenase, adenosine deaminase and ribosidase, nucleoside phosphorylase) are detectable in otherwise dormant and ungerminated spores. Activation results in temporary appearance of some new enzymic activities in ungerminated spores (see Chapter 10); nevertheless, the diversity of activities which appear during the breaking of dormancy associated with germination is substantially greater and irreversible.

3. Depolymerization and excretion of spore constituents

During germination spores excrete material amounting to some 30% of their dry weight. The exudate is principally composed of calcium, a roughly equimolar amount of dipicolinic acid and fragments of depolymerized murein.[24] In addition small amounts of amino acids, small peptides and proteins are present in spore germination exudate. The exuded calcium may originate from the core, which is thought to be rich in this mineral in the dormant spores[25, 26] or from the cortex. The reasons why calcium is so strongly held in the dormant spore and yet so rapidly lost during germination is not altogether clear. The most likely explanation would be that it is simply solubilized during germination as the spore becomes hydrated. This presupposes breakdown of some permeability barrier to water during germination, an event which has never been proved to occur and against which there are theoretical objections.[27] Some of the calcium in spores is exchangeable with other ions, and exchange modifies heat resistance.[28] Whether exchangeable calcium includes calcium from the core and cortex is not certain, because spore coats also bind calcium.[25] In spores of B. megaterium Texas the coat-bound calcium is readily exchangeable with other cations,[29] but the site of the additional exchangeable calcium is not known.

Calcium-binding by electronegative murein is well known in vegetative bacteria,[30] and would suggest that the spore cortex, which is rich in murein,[31] could be a major binding site and that calcium is released during germination as a result of depolymerization and consequent diffusion out of the spore of muropeptides; however, microscopic evidence that most spore calcium is held in the cortex region is lacking.[25, 26]

There is no definitive evidence on the origin of DPA in spore exudate. Its occurence in near equivalence with calcium has suggested that both occur together in spores and probably mainly as the 1 : 1 chelate, but this has not been proved.

"Spore peptide" in germination exudate certainly originates mostly from murein in the spore cortex, which is probably depolymerized during germination by enzymes like the S-enzyme of Strange and Dark.[32] Spore peptide contains the typical murein components,[24, 33] (Chapter 7), N-acetylglucosamine, [N-acetylmuramic acid, L- and D-alanine, diaminopimelic acid and D-glutamic acid. The murein formed first during sporulation is not exuded during germination but remains, probably as the cortical membrane, to become the cell wall of the new vegetative form.[34] The different composition of this more stable layer was indicated by its resistance to lysozyme in spores of B. cereus T,[31] and by its staining in dis-

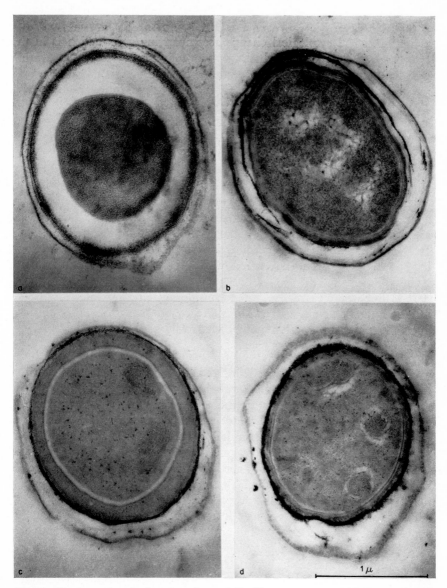

FIG. 1. Electron micrographs of sections of *Bacillus cereus* spores. (a) Ungerminated spore; (b) spore germinated by incubation with inosine (p. 417); (c and d) spores treated with thioglycollic acid, then germinated by incubation with (c) lysozyme (p. 427) or (d) spore lytic enzyme (p. 429).

R

rupted spores of *B. cereus*[35] and *B. subtilis*[36] with ferritin-labelled antisera to vegetative cells.

4. Cytological changes

The principal cytological change accompanying germination is in the structure of the cortex, which almost completely disappears during germination of spores of some organisms (e.g. *B. cereus*;[37, 38] *B. megaterium*;[39] *B. anthracis*; [40, 41]) sometimes following an initial swelling (e.g. *B. megaterium* QM B1551;[42] *Cl. tetani* and *Cl. histolyticum*;[43] *Cl. butyricum*;[44]), and becomes spongy or fibrillar without much change in volume in others (e.g. *B. subtilis*;[45, 46, 239] *B. polymyxa*;[38]). At the same time, or closely following the changes in the cortex, elements of the cytoplasm become more distinct. In particular, the nuclear areas which are indistinct in sections of dormant spores become less electron dense and more obvious (Fig. 16). In *B. subtilis* spores ribosomes become distinct and small vesicles appear in the cytoplasm but attached to the core membrane. The vesicles are thought to be precursors of mesosomes which are recognizable later on.[46] Swelling accompanies germination, so that the exosporium enveloping spores (which have this structure) may be rapidly filled although remaining intact.[41]

It is important to realize when considering electron micrographs of thin sections of spores that major spore constituents like DPA are lost during fixation and flotation methods normally used.[47]

5. Optical density changes

The optical extinction of visible light by a spore suspension decreases during germination, commonly by about 60%, but its extent differs with different organisms; furthermore, the apparent rate of fall in extinction is slightly dependent upon the type of spectrophotometer with which it is measured.[5, 20] The kinetics of the fall in extinction are dealt with below (Section II B1); typical germination curves are shown in Figs. 4 and 10. The observed fall in extinction is certainly due mostly to the excretion and solubilization of dry matter from the germinating spores and the consequent fall in refractivity and light absorption by individual spores.[5, 48] Small additional effects due to changes in spore shape and volume would be superimposed upon the major fall and would be measured differently by different instruments. Basic molecules which can cause the return of phase brightness in freshly germinated spores (see Section II A6) also cause a rise in extinction value,[49] probably by crosslinking electronegative groups in residual murein within the spore and causing contraction.

6. Phase darkening

During germination spores change from bright to dark as viewed by dark phase contrast optics (Fig. 2) at about the same time as heat resistance

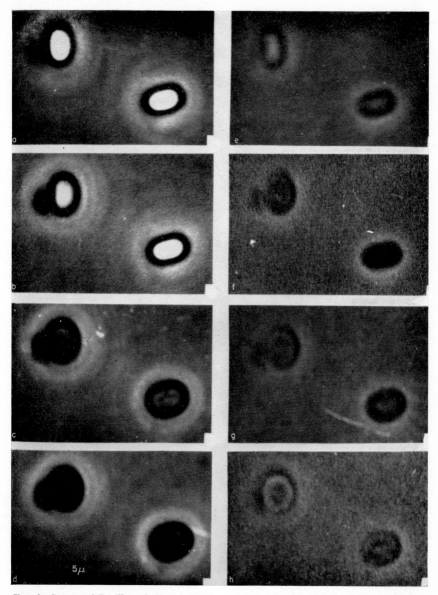

FIG. 2. Spores of *Bacillus subtilis* germinating on yeast glucose agar. (a–d) Phase contrast micrographs; (e–h) light-field micrographs; (a and e) ungerminated spores; (b and f) first signs of germination and swelling after incubation for 3 min; (c and g) core and cortex regions visible after 6 min; (d and h) spores fully germinated, phase-dark swollen and rounded after 10 min (Hitchins *et al.* [59]).

is lost.[5, 50] Phase darkening occurs because of the loss of refractive index of spores during germination, and was for long thought to be due to influx of water into essentially dry cells.[51] The proven high water content of spores makes this explanation not wholly sufficient. The fall in refractive index accompanying germination probably results from the combined effects of excretion of dry matter, slight swelling and possibly the redistribution of water within the spore during germination. For example, depolymerization of cortex murein could reduce pressure on the core allowing local hydration of this structure[27] or lead to swelling due to the osmotic influence of low molecular weight muropeptides within the spore. Spores of the thermophilic actinomycete *Micromonospora vulgaris* also show phase darkening during germination.[52]

One must avoid the assumption that the observed change from light to dark necessarily accompanies a major part of the transition from dormant to germinated spore. The observed light–dark transition is arbitrary, depending on cell size and refractivity (or density), the refractive index of the suspending fluid, the wavelength of the light, and the optical characteristics of the microscope used (particularly the absorbancy of the phase plate). The interpretation of phase-contrast images was discussed in detail by Barer.[240]

Partial phase darkening has been described for germinating spores of *Clostridium* and *Bacillus* species. Uehara and Frank[53] found that spores of *Cl. sporogenes* PA 3679h germinating in L-alanine + sodium pyrophosphate only partly phase darkened and did not become fully stainable if the temperature of incubation was above the 45°C optimum or if D-alanine was present; oddly, the typical decrease in extinction still occurred in suspensions of these "partly germinated" spores. Phase darkening of germinating *B. subtilis* SJ2 spores also occurred in two distinct stages.[54] The first stage (partial darkening) was induced by L-alanine alone, whilst phosphate was necessary for the second stage (complete darkening). Partial darkening could be due to incomplete action of cortex-lytic spore enzymes or to contraction of residual cortex material by calcium.[54] The latter explanation seems likely in view of Vinter's observations that germinated and swelling spores partially contracted and became phase-bright in the presence of various basic molecules which could crosslink charged groups in the electronegative murein.[49] Arntz[55] described phase brightening of germinated spores and vegetative cells of *Bacillus* species caused by Ca^{2+} and possibly due to precipitation of calcium phosphate within the cells. Hitchins and Gould[56] isolated structures from *B. cereus* spores which consisted of cores plus some of the surrounding cortex but free of DPA. These bodies also contracted reversibly and became phase-bright when in the presence of basic molecules and multivalent cations and at low pH values, presumably due to similar crosslinking reactions. It seems likely that phase darkening of spores during germination results from both the hydrolysis of bonds in cortex murein and the release of crosslinking cations; each of these reactions can be influenced separately.

7. Onset of stainability

Clean dormant spores are essentially unstainable except at their peripheries unless pretreated, for instance by heat or acids, whereas germinated spores are readily stained by simple stains. Stainability was first used extensively by Powell,[48, 57] and has been used since along with phase-contrast microscopy and the measurement of extinction changes as a rapid method of estimating the percentage germination in a spore suspension. Germinating spores may become stainable because of breaching of a permeability barrier[47] as discussed below. Alternatively, stainability may accompany the unmasking of reactive groups in the spore. This assumes that the majority of reactive groups in the dormant spore are unavailable for reaction with stains because they are chemically blocked. The most likely blocking agents would be dipicolinic acid (for basic groups) and calcium (for acidic groups) or perhaps the 1 : 1 calcium dipicolinate molecule.

8. Increase in permeability

The hypothesis that dormant spores have a low water content and are impermeable to water would presuppose that a large increase in permeability occurs during germination. However, the "anhydrous spore" hypothesis has been shown untenable both on theoretical grounds[27] and by actual measurement of the water content of spores. Such permeability changes that do occur in spores during germination evidently involve only certain parts of the spore (e.g. the core) and the accessibility of functional groups. Black and Gerhardt[58] measured small increases in the total space in spores accessible to water, glucose and stains which occurred during germination of spores of B. cereus T but pointed out that the small overall changes measured did not exclude the possibility of larger local changes occurring within the spore. Study of penetration of fixatives and embedding materials for electron microscopy of spores suggested that a real permeability barrier existed near the outer edge of the cortex in dormant spores[47] and was destroyed, presumably by dissolution of the cortex, during germination.

9. Increase in spore volume

Following germination in a medium which can support outgrowth (Chapter 12) spores may approximately double in volume before the new cells emerge from the spore coats. This enlargement can be seen microscopically and can be measured as an increase in packed cell volume[21, 59] and is mostly due to growth, i.e. synthesis of new cell material. The fact that swelling also actually accompanies germination was recognized by many workers.[58] Germination swelling measured microscopically is shown in Fig. 3, and can also be measured by packed cell volume techniques[59] or by use of the Coulter Counter[60, 241] The volume increases accompanying germination were about 20% in the spores studied (*B. cereus*, *B. subtilis* and *B. megaterium*); at the same time as they increased in volume the germinating spores became more spherical. This can be

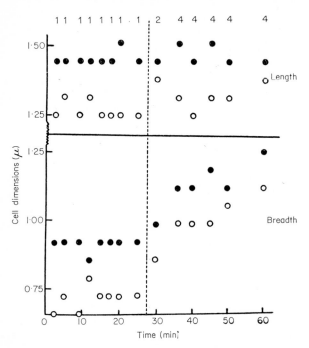

FIG. 3. Length and breadth of a typical *Bacillus subtilis* spore during germination on yeast glucose agar. Measurements from phase-contrast photomicrographs (●) and light-field photomicrographs (○). Key to stages of germination: 1, phase-bright ungerminated spore; 2, first sign of germination and phase darkening; 3, germinating but not fully phase-dark spore (not seen with this particular spore); 4, germinated, fully phase-dark spore (Hitchins, et al.[59])

seen in Fig. 3, where the increase in spore breadth during germination is relatively greater than the increase in length, and in the photographs of germinating spores in Fig. 2. Such a sudden increase in size and change in shape suggests that the spore coat is, or quickly becomes, plastic during germination, perhaps because of action of lytic enzymes or disulphide reductases.[61, 62, 63] Rehydration and expansion of the core accompanying release of pressure exerted by a contractile cortex[27] could certainly cause germinating spores to assume more spherical shapes.

B. KINETICS OF GERMINATION

1. Equations

The early time course of germination of a spore population is sigmoidal, whether measured by optical changes or by onset of heat sensitivity. Fig. 4 shows a plot of germination of *B. cereus* T spores measured

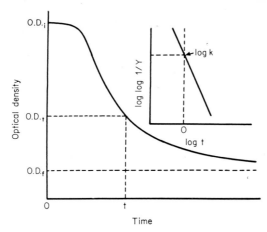

FIG. 4. Typical optical density changes in a suspension of germinating spores (*Bacillus cereus* T). The insert is a plot of the data according to equation 2 below (Vary and McCormick[64]).

spectrophotometrically.[62] Germination at any point can be expressed as a fraction Y, so that where OD_i is the initial optical density (OD), OD_f is the limiting OD as germination nears completion and OD_t is the OD after time t:-

$$Y = \frac{OD_i - OD_t}{OD_i - OD_f} \tag{1}$$

A plot of log log $1/Y$: log t is linear (Fig. 4), and of the form:

$$\log \log 1/Y = -c \log t + \log \log 1/Yo \tag{2}$$

where Yo is the value of Y at $t = 1$ and the slope is $-c$. McCormick[65, 66] developed these equations to find the solution for Y where K is $\ln 1/Yo$:

$$Y = e^{-Kt^{-e}} \tag{3}$$

so that the rate of change of the germinated fraction becomes:

$$dY/dt = \frac{KcY}{t^{c+1}} \tag{4}$$

The equation:

$$t = \left(\frac{Kc}{c+1}\right)^{\frac{1}{c}} \tag{5}$$

allows estimation of the time of the inflection point in the germination curve and, by substituting t (from equation 5) into equation 4, allows estimation of the maximal rate of germination.[66] The general germination equation below is obtained by inclusion of a further "constant", α, which is the ratio of OD_i to OD_f, i.e. a measure of the total fall in OD under the conditions pertaining:

$$OD_t = OD_i (1 - (1 - \alpha)e^{-Kt^{-c}}) \qquad (6)$$

Study of germination of *B. cereus* T spores by L-alanine[66, 67] and of *B. subtilis* SJ2 spores by L-amino acid substrates of L-alanine- and L-leucine-dehydrogenases[54] showed that the constant c (the slope) was influenced principally by temperature whereas the constants K (the intercept) and α (the amount of germination) were functions of the concentration of germinative L-amino acids and were affected by the presence of D-amino acids and the extent of heat activation.

2. Microlag and microgermination

Kinetic data on the germination of individual spores observed by phase-contrast illumination fitted the germination equations well.[67] Such microscopic observation of *B. cereus* T spores germinating in L-alanine allowed measurement of their individual lag times prior to the onset of phase darkening ("microlag"), and of the time taken for individual spores to completely germinate, or phase darken ("microgermination").[64, 67] Each

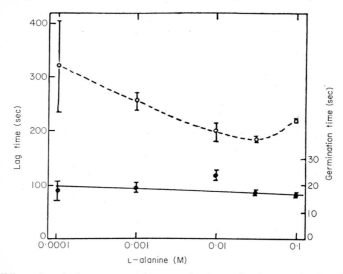

FIG. 5. Effect of L-alanine concentration on microlag and microgermination times of *Bacillus cereus* T spores. Spores were heat-activated at 65°C for 1 hr, then germinated by L-alanine at the indicated concentration. The average microlag (○) and microgermination (●) times with their respective standard deviations of the mean were calculated for each L-alanine concentration (Vary and Halvorson.[67])

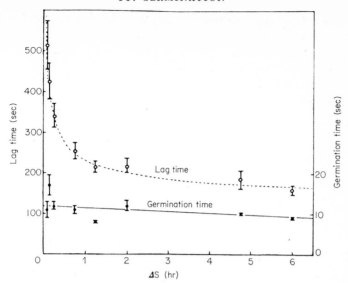

FIG. 6. Effect of heat-activation on microlag and microgermination times of *Bacillus cereus* T spores. Spores were heat-activated (ΔS) for the indicated times at 65°C. Microlag (○) and microgermination times (●) were recorded as for Fig. 5 (Vary and Halvorson.[67])

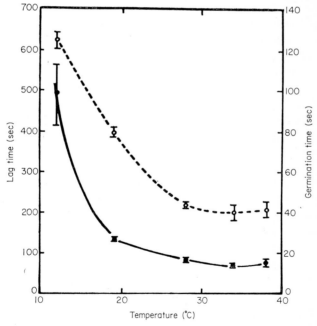

FIG. 7. Effect of temperature on microlag and microgermination times of *Bacillus cereus* T spores. Spores were heat-activated at 65°C for 1 hr, then germinated at the indicated temperatures. Microlag (○) and microgermination times (●) were recorded as for Fig. 5 (Vary and Halvorson.[67])

of these time parameters exhibited skew distribution, suggesting that the spore population was heterogeneous and could, for instance, contain small numbers of exceptionally dormant cells. The concentration of L-alanine and the extent of heat activation influenced microlag times (as they did the constants K and α) much more than microgermination times (Figs 5 and 6). Microgermination (like the constant, c) was principally influenced by temperature (Fig. 7). Sporulation conditions affected microlag and micro-germination differently.[20]

The kinetic studies suggested that during microlag, reactions occurred which depended on pretreatment of the spore, germinant concentration and temperature; for instance, reactions involving unmasking of enzyme active centres or allosteric sites during activation, or reactions dependent upon binding of germinant molecules by small numbers of acceptor sites.[2, 67] However, once a critical level of reaction had occurred (e.g. at the termination of microlag) then the germination event itself (micro-germination) was, within an individual spore, endogenous and not affected by the nature of the germinant or the pretreatment of the spore. Nevertheless, the observed temperature dependence of microgermina-tion times suggested an enzymic rather than purely physical basis for this stage.

3. Delayed germination

One aspect of the kinetics of germination which has been little studied is the occurrence of 'superdormant' spores in spore populations, i.e. those spores which exhibit delayed germination when compared with the majority in a population[249]. Their existence can be surmised from the skew nature of lag and germination times and from observations with mixed germinants (Section III A1). In any spore population, incubation[9] under conditions thought to be optimal for germination always leaves a fraction of the population ungerminated; furthermore, because it contains super-dormant individuals this fraction is difficult to detect by conventional means. Superdormant spores are characterized by 'skips', i.e. late spon-taneous germination of individual spores, and are particularly troublesome in canning practice because very long storage trials must be carried out to ensure their absence. How superdormant spores differ from their less dormant neighbours is not known. Woese, Vary and Halvorson[247] have suggested that one important factor could be the distribution within a spore population of some critical 'germination enzyme'. If molecules of the critical enzyme are Poissonally distributed in a spore population, then the number of molecules could be so low in some spores (Section III A1) as to result in a superdormant state. A heterogeneous distribution of other spore properties, like permeability, cation and dipicolinate load etc. could also conceivably contribute to superdormancy. Separation and study of superdormant individuals from spore population offers the most attractive approach to understanding the superdormant state.

4. *Sequence of events during germination*

Germination of bacterial spores is characterized by the numerous changes detailed above which are commonly observed to occur more or less concurrently. Some of the changes, such as phase darkening and fall in extinction,[68, 69] seem irreversible in that once started they are not easily arrested; Halman and Keynan showed that by suitable use of inhibitors and pH changes the "trigger" stage of L-alanine-induced germination of *B. licheniformis* spores could be separated from subsequent germination.[68]

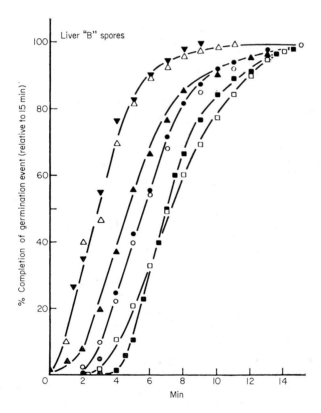

F I G. 8. Sequence of events during germination of spores of *Bacillus megaterium* QM B1551 produced on a liver-based medium. Spores were heat-activated at 60°C for 10 min, then germinated at 30°C in glucose (25 mM) plus L-alanine (1 mM). After 15 min, 88% of the spores were sensitive to heat and $HgCl_2$; 82% were stainable and phase dark; spore suspensions had lost 34% (measured by Klett spectrophotometer) and 26% (measured by Beckman DU spectrophotometer) of their original turbidity, and had shed 77% of their DPA (Levinson and Hyatt[20]). △ Heat sensitivity. ▼ Chemical sensitivity. ▲ DPA loss. ○ Turbidity loss (KLETT). □ Turbidity loss (DU). ● Stainability. ■ Phase darkening.

Knaysi described how transfer of germinating suspensions of *B. cereus* spores to environments unfavourable for initiating germination did not immediately arrest germination. Powell[5] found that heat sensitization slightly preceded fall in extinction in a suspension of germinating *B. subtilis* spores. Further careful study has shown that the changes collectively recognized as germination do proceed in sequence, and that the sequence can be influenced by sporulation conditions. Levinson and Hyatt used $HgCl_2$ (2 mM) to arrest rapidly germination of *B. megaterium* QM B1551 spores in

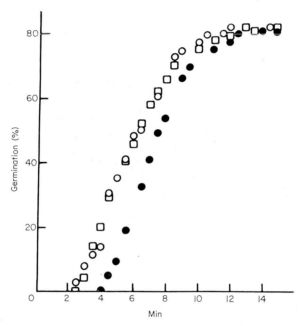

FIG. 9. Sequence of events during germination of spores of *Bacillus megaterium* QM B1551. Experimental method as in Fig. 8 with germination levels stabilized for measurement with $HgCl_2$ (2 mM). Germination was determined by stainability with 0·5% methylene blue (□); by all stages of phase darkening, from end of microlag to end of microgermination (○); and by only fully phase dark (●) (Levinson and Hyat.[70])

L-alanine plus glucose and found the sequence shown in Fig. 8 using spores grown in a liver extract broth. The onset of stainability corresponded with the termination of microlag rather than with microgermination[70] (Fig. 9). Acquisition of heat sensitivity occurred early in the sequence with a number of spore preparations, including that illustrated in Fig. 8. In contrast, in a study which relied on time of sampling to follow the sequence of events,[71] phase darkening appeared to precede heat sensitization of *B. megaterium* ATCC 8245 spores.

III. Initiation of Germination

A. METABOLIZABLE GERMINANTS

1. Amino acids

Hills was the first to discover that germinant requirements for bacterial spores could be simple and specific.[72, 73, 74] His finding that L-alanine was able to cause germination of spores of a *B. subtilis* strain, and was competitively inhibited by D-alanine, was the first of many investigations of amino acids as germinants. At suitable pH value and ionic strength L-alanine has been found germinative for most strains studied of *B. subtilis, B. cereus, B. licheniformis, B. coagulans, B. stearothermophilus, B. circulans, B. sphaericus* and some strains of *B. megaterium,*[75, 76, 77, 78, 79] but with exceptions (see below) so that germinant requirements are diagnostic neither at inter- nor intraspecies level. The apparent importance of L-alanine is probably partly due to its frequent selection for study, for a number of spores are known which respond to other amino acids and their analogues, most commonly L-α-amino butyric acid, L-cysteine, L-valine, L-leucine or its isomers.[78, 80, 81, 82, 83] The effects of different substituent groups in the alanine molecule on germinability of spores of *B. subtilis* (Marburg) were studied in detail by Woese *et al.,*[81] who showed that (1) spores bound D-alanine more strongly than L-alanine; (2) other groups could replace the methyl group of L-alanine without completely preventing its activity, but (3) D-alanine was much more sensitive to methyl substitutions with the exception that (4) addition of an electronegative element to the methyl group of L-alanine dramatically reduced binding and germination rate, but had much less effect on D-alanine. Sometimes other amino acids may only replace L-alanine at higher concentrations[82] or over a narrow concentration range,[84] and sometimes act to potentiate germination by other germinants. A well-defined instance of potentiation was that caused by L-leucine on germination of spores of some *B. megaterium* strains which were virtually non-germinable by other amino acid, riboside and glucose mixtures.[79] Heat-activated spores often germinate in a greater variety of amino acids than unheated spores. For instance, unheated spores of *B. megaterium* QM B1551 germinated in L-alanine; heated spores germinated additionally in L-cysteine, L-leucine, DL-isoleucine, DL-methionine, DL-nor-leucine, L-proline and L-valine.[80]

Germination of a spore suspension by a single amino acid is often incomplete and the same is true of germination of spores in other germinants. Such "fractional germination"[85] may occur because of heterogeneity in populations of spores so that following incubation with a germinant for some time, the remaining ungerminated spores have germination requirements quantitatively and qualitatively different to those of the initial population.[79] It is possible that spores may contain such small amounts of some enzymes, coenzymes, or substrates that single spores may statistically contain only some of them but not others.[86] Individual spores may differ in size, coat thickness and permeability, metal and DPA content, content of free amino acids or germination inhibitors and so on, so that individual spores can be considered as being in different "states of readiness" for rapid germination.[87] Populations of spores may show heterogeneity of other parameters, e.g. of resistance to chemicals;[88] of buoyant density and dormancy,[89] and temperature induced heritable changes in germination requirements of the surviving fraction of *B. stearothermophilus* spores.[90] Fractional germination may also result from changes occurring in the medium during germination. Such changes are often accentuated by using thick spore suspensions. The simplest example of this is so-called "autoinhibition" of germination of spores in L-alanine due to the formation of D-alanine catalysed by spore alanine racemase.[48, 83, 91, 92] Racemase inhibitors like D-cysteine, D-cycloserine, o-carbamyl-D-serine,[89, 93] D-α-aminobutyric acid and hydroxylamine (B. J. Krask, personal communication) can alleviate the autoinhibition, and the extent to which germination can be stimulated by inhibition of alanine racemase is sometimes surprisingly large, even with dilute spore suspensions.[93, 248]

The mechanism of action of L-alanine-induced germination is not known. Of the spore enzymes known which can act on L-alanine, the racemase can act to hinder L-alanine-induced germination by forming D-alanine, and can also allow spores to germinate slowly in D-alanine through the formation of L-alanine which then slowly initiates germination.[75] A role for racemase in driving L-alanine-induced germination, however, was proved unlikely by studies of properties of the spore enzyme and its occurrence by Church *et al.*[94] and by the lack of inhibition of L-alanine-induced spore germination by racemase inhibitors.[83, 93]

Transaminases detectable in extracts of heat-activated *B. subtilis* spores could act on oxoglutaric acid plus L-aspartic acid, L-alanine or D-alanine, but not on pyruvic acid plus L-aspartic acid or DL-phenylalanine,[95] but use of transaminase inhibitors (e.g. isonicotinic acid hydrazide) suggested that transaminase action was not associated with germination.[96]

Inhibition by Atebrin suggested involvement of a flavoprotein, perhaps L-amino acid oxidase, in L-alanine-induced germination of *B. subtilis* spores.[96] Inhibition of germination by chlorpromazine could be due to inhibition of oxidative metabolism.[97] Production of pyruvic acid and hydrogen peroxide from L-alanine by spore oxidase action has been confirmed using *B. cereus* T;[98, 99] octyl alcohol, which inhibits oxidative deamination of amino acids, also inhibited germination.[100] Initiation of germination *via* oxidative formation of pyruvate from alanine was indicated by enzyme studies of *B. cereus* T spores.[99]

NAD-linked L-amino acid dehydrogenases have been found in a number of spores[101, 102, 103, 104, 105] and spore L-alanine dehydrogenase has been purified.[101, 102, 103] *B. cereus* T spores probably contain only about 60 molecules of L-alanine dehydrogenase per spore.[106] Its properties were such that O'Connor and Halvorson[107] suggested that it was the binding site for L-alanine during L-alanine-induced germination of *B. cereus* T spores. Krask and Fulk[108] showed that L-cysteine-induced germination of *B. cereus* T spores was probably mediated by L-alanine dehydrogenase acting on the cysteine and that L- and D-cysteine desulphydrases present in these spores were not important germination enzymes. Spores of *B. subtilis* SJ2 were germinated by substrates of L-alanine and L-leucine dehydrogenase.[54] Spores of *B. cereus* T enriched in L-alanine dehydrogenase by sporulation in the presence of L-alanine unexpectedly germinated in L-alanine more slowly than controls,[106] suggesting that L-alanine dehydrogenase level alone was not the sole rate determining factor. Freese *et al.*[109] isolated an L-alanine dehydrogenaseless (AlD-) mutant of *B. subtilis*, and found its spores to germinate exceptionally slowly in L-alanine and in some other amino acids which may be acted on by alanine dehydrogenase, e.g. L-cysteine,[108] or which may give rise to L-alanine within the spore.[109] This finding seemed superficially to further implicate the dehydrogenase in L-alanine-induced germination, but it should be noted that the AlD — spores *did* germinate in L-alanine, albeit more slowly than spores of the AlD+ parent, and germinated as rapidly as AlD+ spores in complex medium. Furthermore, when first activated with calcium dipicolinate (Ca–DPA) both kinds of spores could be germinated by L-alanine, although the AlD+ spores still responded more rapidly and completely than those of the AlD— mutant:[110] D-alanine still inhibited germination of the mutant spores in L-alanine. Freese and Cashel[110] concluded that the AlD— bacteria produced spores which differed from the AlD+ ones in a number of ways, e.g. in germinability, response to Ca–DPA etc. but that since AlD—spores could still be germinated by L-alanine, and inhibited by D-alanine, a central role for alanine dehydrogenase in germination was not proved. That alternative pathways for utilization of germinant amino acids must exist was also suggested by a study of spores of *B. subtilis* (Marburg), which can be germinated by L-alanine. Wax *et al.*[86] isolated mutants which required additionally D-glucose + D-fructose + K+ or NH4+ for germination. Interestingly the parent strain also had these additional requirements when tested at high temperatures or when in the presence of D-alanine. Wax *et al.* suggested that the results indicated a dual role for L-alanine, one of which was absent in the mutants, blocked by D-alanine and inhibited by high temperatures. L-Asparagine or L-glutamine could substitute for L-alanine in the glucose-fructose system.[242] Germination mutants of *B. cereus* T have also been isolated[250] whose properties suggest at least two distinct germination pathways for this organism. The nature of the two germination pathways is not known.

It has proved difficult to follow any metabolism of L-alanine during spore germination because, once germinated, the cells metabolize amino acids by means which are not only unrelated to the initiation of germination but may be so active as to mask any initial reactions. Nevertheless, the studies that have been carried out do suggest that any uptake of alanine by spores *during* germination must be very small. For instance, *B. cereus* T spores fixed no more than several hundred molecules of ^{14}C-L-alanine or its products per spore in 45 sec, during which time 40% germination occurred.[111] A greater uptake (about 5×10^6 molecules per spore) occurred when inosine was present during a one minute lag before germination began.[112] Uchiyama *et al.*[113] found *B. subtilis* (Marburg) spores to take up only small amounts of L-alanine during germination compared with outgrowth.

Bearing in mind the possible contribution by germinated spores, most studies have suggested that L-alanine is deaminated during germination by one of the enzymes mentioned above. Possible products are therefore pyruvic acid, ammonia, $NADH_2$ and hydrogen peroxide. None of these have been found widely active as germinants, though there is some evidence implicating pyruvate in the initiation of germination.

Pyruvate and acetate caused slow germination if first pre-incubated with *B. cereus* T spores at pH 5,[98] suggesting that only the undissociated acids could sufficiently permeate spores; indeed ^{14}C-pyruvic acid was shown to enter spores of *B. subtilis* only at low pH values and to cause fractional germination.[114] Possibly activation caused by exposure to low pH could have contributed to this result. Pyruvate plus ATP caused some germination of *B. subtilis* spores,[96] and pyruvate could stimulate germination of *B. cereus* T spores by adenosine.[99] Ethyl pyruvate probably inhibited *B. licheniformis* spore germination[68] by inhibiting pyruvate-handling enzymes, but this site of action in spores has not been proven. The inhibitor of germination and thiamine antagonist, W-1435 (Bis-1,3-β ethylhexyl-5-methyl amino hexahydropyrimidine), was shown to suppress pyruvate oxidation by *B. cereus* T spores.[99, 100]

Hydrogen peroxide germinated spores of *B. subtilis*,[96] and PA 3679,[115] stimulated inosine- or L-alanine-induced germination of *B. megaterium* spores[116] and caused germination-like changes in spores treated with reagents which rupture disulphide bonds.[62] The latter effect was probably caused by metal-catalysed production of free radicals from the hydrogen peroxide rather than by its metabolism.

In conclusion, it is clear that L-alanine and certain other amino acids are important germinants for many spores and that, as evidenced by inhibitor studies, their germinative action depends on some sequence of enzyme action. This probably involves deamination, but further germination metabolism is speculative and an initial allosteric mechanism is not ruled out. Certainly only very small "trigger" amounts of L-alanine are utilized during germination compared with outgrowth. The true role of germinant amino acids may well remain obscure because of the lack of study of important adjunctive factors like activation, trace ions and other nutrients and activity of racemases.

2. Ribosides

The germinative action of ribosides was first reported by Hills[72, 73] who found that adenosine germinated spores of some *Bacillus* species. Inosine, which can be derived from adenosine by deamination mediated by spore deaminase,[117] has generally been found more active. Guanosine and xanthosine, but not the pyrimidine nucleosides cytidine or uridine or the purine nucleotides adenosine-5-phosphate or guanosine-5-phosphate, also showed some germinative activity for *B. cereus* T spores particularly when L-alanine was also present[118] but are generally relatively inactive alone.[72, 73, 79] Spores of *B. cereus*, *B. anthracis* and some *B. megaterium* strains have been found most often to be germinated by ribosides, whereas spores of *B. subtilis*, *B. coagulans*, *B. licheniformis* and *B. stearothermophilus* and also spores of *Clostridium* species have been found generally unresponsive.[236] No pyrimidine ribosidases or pyrimidine riboside deaminases were detectable in spores of *B. cereus* T until some time after germination,[119] but spores contained a number of enzymes active in metabolizing the germinative purine ribosides.

Ribosidase,[120] which catalysed formation of free base plus ribose was found in spores of *B. cereus* and *B. polymyxa*, but not in *B. globigii* which is unresponsive to inosine. However, ribosidase appeared unimportant in driving germination of *B. cereus* T spores because enzyme activity and germination differed widely in pH and temperature optima,[118] whilst products of the cleavage, whether individually or together, were not germinative.[121] That adenosine and inosine were phosphorylated during or shortly following germination was shown by incorporation of labelled phosphate into nucleoside phosphates.[122] Enzymes catalysing phosphorolytic cleavage of adenosine and inosine to form the free bases plus ribose-1-phosphate were detected in *B. cereus* T spores, as was phosphoribomutase which could further catalyse formation of ribose-5-phosphate from ribose-1-phosphate.[123] Ribose-5-phosphate could also be formed from ribose and ATP by a Mg^{2+}-activated ribokinase.[123] The apparent importance of the pentose phosphates led Krask and Fulk to suggest that ribosides may stimulate germination of some spores by acting as sources of utilizable phosphate esters obtainable with little

s

expenditure of energy by the dormant spore.[123] Such a situation may exist in stored erythrocytes, where nucleosides restore viability through resynthesis of phosphate esters and ATP.[124] This concept received support from the demonstration that aged spores of *B. cereus* T could be slowly germinated by ribose-5-phosphate.[98] Inhibition by an antagonist of pyruvate metabolism suggested that, as with L-alanine-induced germination, pyruvate was an important intermediate in germinative utilization of ribose-5-phosphate. In contrast, strong evidence against a major role for nucleoside phosphorylase in riboside-induced germination was obtained by Gardner and Kornberg[125] who isolated a mutant of *B. cereus* T whose spores, although containing only 3% as much enzyme active on inosine, germinated as rapidly and completely in inosine as the wild-type.

Sacks[243] recently made the important observation that spores of some strains of *B. macerans* could be germinated by the free bases adenine and 2,6-diaminopurine, but not by adenosine. The hypothesis that ribosides serve primarily as sources of ribose and ultimately pyruvate thus seems still less likely. Furthermore, it is likely that germination of *B. cereus* T spores is only initiated by ribosides alone when the spores contain endogenous sources of L-alanine or other amino acids.[251]

As with most metabolizable germinants the problem is not to determine which enzymes in spores can act on the ribosides, but rather to determine which, if any, of the enzymes present are essential in germination rather than playing some part in subsequent outgrowth.

3. Sugars

Germination by glucose has been mainly observed with spores of *B. subtilis*,[76,] *B. megaterium*,[77,126] and *Clostridium* species (see below) although glucose can cause slow incomplete germination of spores of some other organisms[79] and sometimes potentiates the action of other germinants. In a comprehensive study, Hyatt and Levinson[126] found that unheated spores of *B. megaterium* QM B1551 could be germinated by the following hexoses and their derivatives (25 mM); D-glucose, D-mannose, 2-deoxy-D-glucose, D-glucosamine, N-acetyl-D-glucosamine. Heated spores germinated in lower concentrations and also in a wider variety of sugars but the apparent variety was shown to reflect mostly contamination of sugar samples with traces of glucose or mannose. Manganese potentiated germination of *B. megaterium* spores by glucose,[127] possibly by stimulating manganese-dependent enzymes (Section III C 3). Glucose can usually be replaced by other germinants.[79] Hermier found spores of *B. subtilis* SJ2 to be germinated rapidly by L-alanine and more slowly by 11 sugars testedi, ncluding hexoses and pentoses.[82] Germination of *B. megaterium* spores by glucose or by glucosamine was similar in response to inhibitors and environmental conditions, but differed from germination caused by L-alanine or L-valine, by some other amino acids and by KNO_3.[128]

The pathway of glucose metabolism critical to germination and, if there is one, the key metabolic product are not known. Spores of *B. cereus* T contain the enzymes necessary for catabolism of glucose by Embden-Meyerhof (EMP) and hexose monophosphate pathways (HMP), though at lower specific activities than vegetative cells.[129, 130, 131] Spores of *B. coagulans* var. *thermoacidurans* had a complete glycolytic enzyme system which appeared from inhibitor studies to play an important part in germination, whereas outgrowth was accompanied by oxidative breakdown of glucose.[132] Although of low activity compared with vegetative cells, spores of *Cl. botulinum* were proved to have a complement of EMP enzymes.[133] It appears that complete absence of any enzymes from spores is the exception rather than the rule, however, it seems most likely that the majority of detectable spore enzymes play no part in germination but may be indispensable during the early stages of outgrowth, before synthesis of new enzymes commences. It is unlikely, for instance, that enzymes of glucose catabolism play any part in driving germination of *B. cereus* T spores initiated by L-alanine or inosine, especially since glucose itself is not an active germinant for spores of this organism. With spores germinable by glucose on the other hand, one must presume that EMP or HMP enzymes may play a catalytic role in germination.

4. *Combinations of germinants*

It will be clear from the three preceding sections that the most rapid and complete germination is often obtained by using combinations of germinants. In some instances combinations of organic germinants appear obligatory; the most effective combination studied is L-alanine + inosine, of which the inosine moiety is usually replaceable by adenosine, and which usually germinates spores of *B. cereus* and some strains of *B. megaterium*. Glucose may enhance germination of spores of some strains of *B. subtilis* and *B. megaterium* in the presence of other germinants, but is usually replaceable.[79] Spores of the *Clostridium* species which have been studied generally respond to the combinations of germinants (Section III B 5). In addition to organic germinants spores generally, but not exclusively, require a critical concentration of ions in the medium for maximal germination (Section III B 1).

The mode of action of combinations of germinants is not known. The simplest interpretation would be that metabolism of one germinant provided an intermediate necessary for metabolism of the other, or that one germinant inactivated some inhibitor so that the other germinant could be utilized. Illustrations of this latter possibility are known; for example, inhibitors of alanine racemase allow L-alanine to germinate spores more effectively by slowing the formation of inhibitory D-alanine. Germinative activity of ions may only fully be expressed following metabolism of an organic germinant that increases ion permeability of the cell (Section III B 1).

It must be remembered that some germinants are effective in extremely low concentrations, and that even well-washed spores may still have traces of germinant compounds absorbed on them. It is likely that germination of spores by single organic germinants may often be only apparent, and really occur because traces of a second germinant are present within or absorbed upon the spore.[251] An example of this was given by Black and Gerhardt.[134] Unactivated *B. cereus* T spores in suspensions as initially prepared, germinated in either L-alanine or adenosine alone; as the spores were cleaned by consecutive washes either of the germinants became less and less effective and eventually caused rapid germination only when in combination. Furthermore, combined germinants may act by each affecting a different fraction of a spore population, so that together they germinate a greater proportion of the population than singly. Such a combined effect of germinative ions is shown in Chapter 9, Fig. 1 (p. 309).

5. Germinants for spores of Clostridium species

There is no reason to believe that germination of *Clostridium* spores differs in any fundamental way from that of *Bacillus* spores; it is nevertheless true that in all phases of spore research the anaerobes have so far received far less attention than the aerobes. For instance, germinant requirements have been investigated for only few clostridia, although those which have been elucidated have sometimes proved to be strikingly different from the most common germinants for *Bacillus* spores: in contrast to some *Bacillus* species those of *Clostridium* so far studied do not require nucleosides for germination.

A simple pattern was shown by heat-activated spores of PA 3679h, which were germinated by L-alanine alone (8 mM) optimally at about 45°C in sodium pyrophosphate buffer (pH 8·5); D-alanine competitively inhibited germination, thus the germination pattern resembled closely the well-known pattern shown by spores of many *Bacillus* species.[135] However additions of L-arginine and L-phenylalanine, amino acids not generally implicated in germination of spores of *Bacillus* species, were necessary for maximal germination.[136] These amino acids have also been found germinative to various extents for spores of other clostridia, e.g. *Cl. acetobutylicum, roseum, botulinum* strains, *bifermentans* and *septicum*.[137-143] Spores of *Cl. tetani* were found to have the novel germination requirements L-methionine + L-lactate + nicotinamide + Na^+.[244] Spores of *Cl. botulinum* type E as prepared in the laboratory germinate particularly readily and germination seems to be stimulated by a variety of amino acids.[144]

Glucose initiates germination of some clostridial spores:[145, 146] caramel-ized glucose caused germination-like changes which could be potentiated by a number of carcinogens.[147] Initiation of *Cl. bifermentans* and *septicum* spore germination by lactate plus amino acids was described by Gibbs.[142, 143] Involvement of lactate as a germinant has not been pre-viously reported for spores of *Bacillus* or *Clostridium* species and its role is so far obscure.

A stimulating effect of bicarbonate on germination of spores of *Clostri-dium* species has often been described since the observation by Olsen and Scott[148] that bicarbonate increased the viable count of spore suspensions. Bicarbonate (or carbon dioxide) increased the germination rate of various *Clostridium* spores including PA 3679 and *Cl. botulinum* strains,[140, 149, 150] but has not been found essential for all species and did not stimulate germination of *Bacillus* spores.[151] Oxalacetic acid could substitute for carbon dioxide in the germination of *Cl. botulinum* spores,[151] suggesting that the carbon dioxide acted by condensing with pyruvic acid by the Wood-Werkman reaction to form the 4-carbon acid.

Like *Bacillus* spores, *Clostridium* spores can be germinated by calcium dipicolinate,[152] indeed the original observation of the germinative effects of chelating agents was made by Brown[115] using ethylene diamine tetra-acetate and PA 3679 spores. Germination of untreated *Clostridium* spores, or release of DPA from them caused by hydrogen peroxide has been described;[115] when treated with reagents which rupture disulphide bonds spores of numerous *Clostridium* species as well as *Bacillus* species can be caused to undergo germination-like changes by hydrogen peroxide.[62]

Germination of spores of aerobes is certainly not hindered by absence of oxygen,[245] but germination of spores of anaerobes can sometimes be sup-pressed by oxygen. For instance, spores of *Cl. botulinum* and *roseum* only germinated rapidly in defined medium when oxygen was excluded;[137] spores of *Cl. butylicum, acetobutylicum* and *roseum* were inhibited from germinating in a complete (trypticase) medium by oxygen.[139] In contrast, spores of *Cl. welchii, chauvei* and PA 3679 germinated well aerobically in a complex medium[153] as did spores of *Cl. welchii, chauvei,* PA 3679 and *botulinum* type A (but not type B) in defined medium.[145] Germination of *Cl. bifermentans* and *septicum* spores in complex and defined media was not inhibited by oxygen.[142, 143]

An explanation of some of these differences may be that inhibition of germination by oxygen depends not only on the species and strain, but also on the medium. Fujioka and Frank[136] gave a good example of this with the germination of PA 3679 spores, which was suppressed by oxygen in alanine-deficient medium but not in medium containing high levels of alanine. They suggested that germination *via* the alanine pathway was insensitive to oxygen but germination *via* some other pathway initiated by unidentified germinants was oxygen-sensitive. Similarly, germinants for spores of *Cl. tetani* differed according to the anaerobiosis of the medium.[244]

B. Non-metabolizable Germinants

1. *Ions*

Although spores of some bacterial species were able to germinate substantially in media containing no added ions, a study of 46 strains representing 13 *Bacillus* species showed that germination of all was increased by ions[79] though not all by the same ionic species. Generally organic ions were more effective germination adjuncts than were inorganic ions. Studies of ionic germination stem from the observations of Levinson and his colleagues[87, 127] that Mn^{2+} and a number of anions stimulated *B. megaterium* spore germination, and of Rode and Foster that in relatively ion-free media spores of *B. cereus* T,[154] *B. megaterium* QM B1551[85] and *B. megaterium* Texas[155] germinated very poorly. In particular with the

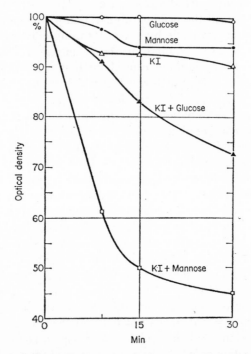

Fig. 10. Germination of spores of *Bacillus megaterium* QM B1551 by potassium iodide and sugars. Spores were heat-activated at 60°C for 1 hr before use. The germination medium consisted of deionized water plus glucose or mannose (50 μM) or potassium iodide (1 mM). Tap water could supply sufficient ions to make the sugars germinative. Incubation temperature was 40°C (Rode and Foster.[85])

B. megaterium strains (Fig. 10) one or other of the previously known organic germinants could be replaced by a variety of inorganic and organic anions,[85, 116, 155] whilst a requirement for Na^+ was clearly shown for germination of *B. cereus* T spores.[154] Although germination by strong electrolytes alone was rare, and only occurred substantially with heat-activated spores of *B. megaterium* QM B1551,[79] the general ion requirement for germination, and often the non-specificity of the requirement, have been amply confirmed in other studies with *Bacillus* and *Clostridium* species; e.g. *B. subtilis* spores germinated optimally in media of ionic concentration of about 100 mM[156] and germination of some strains was accelerated by NH_4^+ [76]; PA 3679 required salts for optimal germination,[136] for instance, about 35 mM sodium pyrophosphate,[135] and a definite requirement for Na^+ was shown for germination of heat-activated spores of *Cl. bifermentans*;[157] germination of *Cl. chauvoei* spores was accelerated by $CaCl_2$.[158] Spores of *B. popilliae* were unusual in that their germination was inhibited if K^+ was present during heat activation;[159] salts in the medium caused increased germination.[160] Nitrate and nitrite operated as germinants for heat-activated spores of *B. megaterium* QM B1551,[80] and nitrate-germination differed in response to environmental factors (like chemical inhibitors, pH, temperature) from germination caused by L-alanine or L-valine or by glucose or glucosamine.[128] High concentrations of nitrite (0·5–2·0 M sodium or potassium salts) caused germination-like changes in *B. cereus* T spores after a long lag period, and the germinating spores catalysed oxidation of the nitrite to nitrate.[161] It was suggested that germination was caused by reduction, NO_2^- serving as an electron donor, possibly mediated by spore catalase. Nitrite germination was accelerated by high spore concentrations, oxidized pyridine nucleotides, Mo^{2+}, Mn^{2+}, H_2O_2, DPA and 2,4-dinitrophenol and decelerated by O_2, reduced nucleotides and nitrate.[161] Spores of PA 3679 h were also germinated by nitrite.[184] Higher activity at pH 6 than at pH 7 suggested that undissociated nitrous acid may be the actual germinant.

The role of Mn^{2+} in germination of *B. megaterium* spores has received special attention since, in addition to stimulating germination,[127] Mn^{2+} is known to stimulate a number of spore enzymes which it is thought may be involved in germination.[32, 162, 163] This aspect is dealt with more fully below in Section III C 3.

Spores modified by partial replacement of spore calcium with other divalent cations or with hydrogen exhibited changed germination patterns. Endotrophically-formed calcium spores of *B. megaterium* QM B1551 and *B. cereus* T germinated more rapidly than barium spores whether the germinants were ions, n-dodecylamine or amino acids.[154] Hydrogen spores of *B. megaterium* Texas were not germinated by L-alanine plus inosine, which caused rapid germination of normal spores, unless a strong but non-

specific electrolyte was present (e.g. 30 mM phosphate buffer;[165] on reloading the spores with calcium, strontium, or barium, the requirement
for strong electrolyte was lost. Spores of *B. megaterium* QM B1551 treated
with thioglycollic acid at pH 2·6 became non-germinable by glucose, and
yet germinability was restored by calcium salts.[166]

The considerable data accumulated on the necessity and usual non-
specificity of ions in different germination systems led Foerster and Foster[79]
to conclude that the major event in germination was ionic; however, the
exact role of ions has not been revealed. Rode and Foster suggested that
ions could affect some structure protective to the core which depended on
associated ions for its stability;[85] the most likely candidate for this role
would be the electronegative cortex murein. The influence of cations and
chelates on spore dormancy (Chapter 9) makes such a role attractive.
Alternatively germination may involve the breaching in spores of a permeability barrier *to* ions, or involve action of ion-dependent enzymes. For
instance, the germinative action of *B. cereus* spore lytic enzyme required
an ionic strength equivalent to about 100 mM NaCl[167] and relatively non-
specific ion requirements for activity of diverse enzymes are not uncommon.
The finding of Hermier and Rousseau[54] that certain amino acids caused
only partial phase darkening of *B. subtilis* SJ2 spores unless potassium
phosphate was present has already been mentioned. They suggested that
action of amino acid dehydrogenases was necessary only during the lag
phase, and that the ions were necessary for the subsequent refractivity
change characteristic of germination.

2. *Surfactants*

Rode and Foster[168, 169, 170] invented the term 'chemical germination' to
describe germination-like changes induced in spores by surface active
agents and to distinguish this germination from 'physiological germination'
(Section A) and 'mechanical germination' (Section D). Numerous surfactants were shown to be effective germinators of spores of *B. megaterium*
and other strains.[168] Cationic surfactants were most active but some anionics
were germinative at low pH values, whilst non-ionics and antibiotic
polypeptides did not cause germination. The specially active (catonic)
alkyl primary amine n-dodecylamine (laurylamine) has received most
study, and at concentrations as low as 10 to 100 μM induces in *B. megaterium* spores the following changes: loss of refractivity, loss of resistance to
staining, chemicals and heat, swelling, fall in optical density of spore
suspensions, fall in spore dry weight by about one-third, excretion of
Ca, DPA and muropeptides.[169, 170] Furthermore, although the surfactant
killed the newly germinated spores if left in contact with them, n-dodecylamine could be effectively quenched with cephalin so that the surfactant-
germinated spores remained viable and capable of outgrowth.[170] In its

effect on the spore, therefore, chemical germination seemed to mimic completely the better known physiological germination. n-Dodecylamine germination was inhibited by high concentrations of inorganic cations; it differed from physiological germination in that it occurred at 70°C, at pH 11, in 10% ethyl alcohol, 15% hydrogen peroxide, 1% phenol and in 1% sodium cyanide[169] i.e. under conditions where one would not expect to find enzymic activity. The peculiar resistance of spores, however, makes one wary of assuming that some enzyme action is not involved. Heat-activated spores of B. megaterium[170] and of B. cereus[171] germinated no faster in n-dodecylamine than unheated spores in contrast to physiological germination; however, γ-radiation caused activation of spores of B. cereus for germination in both systems.[171] The most likely explanation of germinant activity of surfactants was thought by Rode and Foster to be that the surfactants breached a permeability barrier to solutes within the spore by acting as wetting agents or by solubilizing lipids.[170] Activity via an increase in permeability also seems likely following the work of Voss,[172] who showed that long chain alkyl amines were more effective cations than tris in making cells of Gram-negative bacteria in EDTA more permeable to surface active agents. Germination of B. megaterium spores brought about by butane may also involve spore permeability changes caused by the alkane weakening hydrophobic bonding.[173] The barrier could be a spore membrane (plasma or cortical membranes), the cortex, or a spore coat layer. Later research has not revealed this, although the observed permeability of spores to a variety of solutes[174] suggested that any such barrier must be deep within the spore rather than located near the surface. Inhibition of germination of B. subtilis spores by chlorpromazine could be caused by a decrease in permeability of some spore membrane.[97] Other roles of cationic surfactants could be to bind to electronegative spore structures and displace previously bound cations, for instance in structures containing calcium, dipicolinic acid, and murein. That spore cations influenced surfactant germination was shown by the relative non-germinability by n-dodecylamine of endotrophically formed spores of B. megaterium QM B1551 enriched in strontium or barium compared with normal spores or with endotrophically formed calcium spores.[164]

3. Chelates

In 1956 Brown reported the important observation that spores of Cl. sporogenes PA 3679 could be germinated by ethylene-diaminetetraacetic acid,[115] but intensive study of germination by chelates only began following the discovery in 1961 that the major spore constituent calcium dipicolinate (Ca–DPA) could germinate spores of a number of species of Bacillus and Clostridium.[152] Germination was most rapid when equimolar calcium and dipicolinic acid were used (optimally at about 20 to 40 mM in tris-HCl buffer). It was postulated that the effective germinant was the 1 : 1 Ca–DPA chelate, the 1 : 2 chelate being inactive. The rate of germination of clostridial spores showed temperature dependence like that shown by enzymatic reactions whereas spores of *Bacillus* species required lower temperatures for successful Ca–DPA germination than for amino acid-induced germination.[175] This reflected the rapid formation of a (non-germinative) calcium dipicolinate precipitate at the higher temperatures, and was prevented if amino acids or gelatin were present. *Clostridium* spores, which required less Ca–DPA for germination than *Bacillus* spores were less affected by the concentration drop during precipitation. Spores of *Cl. botulinum* 33 A did not germinate in Ca–DPA.[176] Ca–DPA was the only effective germinant found for spores of *Sporosarcina ureae*.[177] Relative non-specificity of the cation in chelate germination was shown by the germination of *B. megaterium* spores caused by strontium and magnesium DPA chelates.[178] Some analogues of DPA, particularly γ-pyran-2,6-dicarboxylic acid, were shown to be capable of initiating germination.[246] Ca–DPA germination appeared to involve metabolism at some stage since it was sensitive to general metabolic poisons though not to alanine analogues which inhibited L-alanine-induced germination.[179] Spores of *B. cereus* T low in DPA still germinated in Ca–DPA, though they germinated more slowly than normal in L-alanine unless exogenous DPA was added.[180] *B. megaterium* spores incubated at 60°C in Ca–DPA became non-germinable by Ca–DPA at lower temperatures, although still germinable with physiological germinants.[181]

The relationship of Ca–DPA germination to ionic germination is not clear. Rode and Foster[85] found that Ca–DPA germination of spores of *B. megaterium* QM B1551 depended on the presence of other ions (e.g. Na^+, K^+, Cl^-) and that Ca–DPA could substitute for either germinant in the L-alanine plus inosine mixture.[116] The importance of chelate concentration and the tendency of Ca–DPA to precipitate unless a stabilizing molecule (like an amino acid) is present[175] could explain the necessity for an organic molecule. The Texas strain of *B. megaterium*, in contrast, was germinated not only by Ca–DPA but also by Na–DPA particularly when other cations, not necessarily including Ca^{2+}, were present.[155] The germinative action of Ca–DPA was made the basis of a medium for obtaining full recovery when counting unheated and heat-damaged spores of *B. subtilis*.[182, 183]

Explanations of chelate germination imply that spores contain structures, SPORE STRUCTURE-Ca–DPA or ENZYME-Ca–DPA which are necessary to maintain dormancy.[156, 175] Exogenous Ca–DPA could cause germination by initially binding spore DPA to form $Ca-(DPA)_2{}^{2-}$ which would be released. Fleming and Ordal[156] pointed out that the stability constants of the complexes were such as to make this feasible, and furthermore one would expect such structures to be markedly influenced by ions in general similarly to the way normal germination is influenced (Fig. 11). Alternatively exogenous Ca–DPA may dimerise (to form $(Ca–DPA)_2$) with spore Ca–DPA more readily than with its sister molecules in solution and hence start germination by directly removing Ca–DPA from spores. The

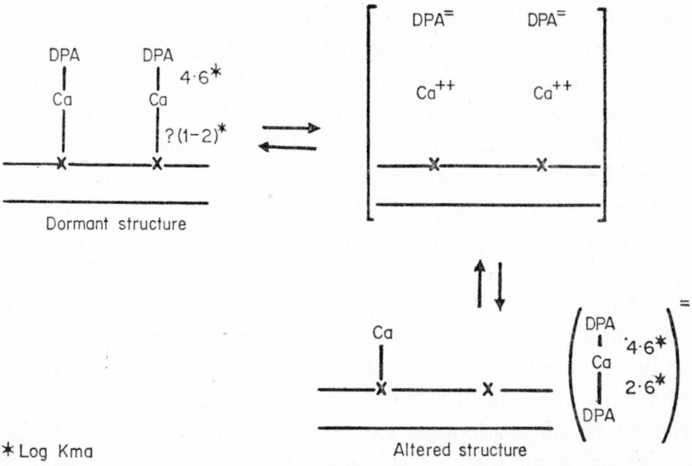

Fig. 11. Hypothetical structure of complexes of calcium and dipicolinic acid (DPA) with spore molecules and changes which may lead to germination (Fleming and Ordal.[156])

structural role of metals in *Azobacter* cysts, which resemble spores cytologically though lacking DPA, was emphasized by the ability of ethylenediaminetetraacetic acid to cause germination-like rupture of the cyst coat.[185]

C. GERMINATION OF SPORES WITH ENZYMES

1. Lysozyme

Muropeptides expelled from spores during germination, or released from spore debris incubated with a spore lytic enzyme, were shown to be sensitive to further degradation by egg white lysozyme.[186] Electron microscopy showed that the enzyme specifically removed cortex, and in some strains also cortical membrane, from spore fragments.[31] The substrate of lysozyme (N-acetylmuramide glycanohydrolase, EC 3.2.1.17) is known to be the $\beta-(1 \to 4)$ link between N-acetylmuramic acid and N-acetylglucosamine in the polysaccharide backbone of mureins,[187] so it is clear that structural integrity of the cortex, at least in those spores studied, depends on its murein component and in particular on its intact $\beta-(1 \to 4)$ glycosidic bonds.

Although containing lysozyme-sensitive bonds, spores are resistant to lysozyme[62, 188] unless first treated in some way to sensitize them to the enzyme. Many lysozyme-insensitive vegetative bacteria can be made sensitive to lysis by a variety of techniques, including treatment with acetone or chloroform or incubation at extreme pH values or with ethylenediamine tetraacetate (EDTA) in tris buffer (see review by Salton[189]); these treatments probably remove overlaying lipid or cations to expose the cell wall murein layer. The spore-like cysts of *Azotobacter* spp. could be lysed by lysozyme following treatment with EDTA.[185] Spores of *Bacillus* and *Clostridium* species, however, were not sensitized to lysozyme by the techniques which sensitize some vegetative bacteria, but were sensitized by some oxidizing (performic acid) and reducing agents (thioglycollic acid, mercaptoethanol) if treated with the reagents at low pH values.[62] The necessity for low pH may reflect the importance of spore cations in protecting disulphide bonds from rupture, since exposure of spores to acid resulted in loss of spore cations,[28] and loss of *B. megaterium* spore heat resistance and germinability caused by thioglycollic acid was partially recoverable by treating the spores with calcium salts.[166] Alternatively the low pH value, which is near the isoelectric point of whole spores,[190] may be necessary to allow adequate exposure of spore disulphide bonds to the reagent by affecting tertiary structure of spore macromolecules. Since treated spores were more permeable to lysozyme than untreated spores,[191] and the sensitizing reagents were most effective if used in conjunction with 8 M urea,[62] the reagents probably sensitized spores principally by causing sufficient rupture of disulphide bonds in the keratin-like protein component of spore coats[192, 193] to leave gaps large enough for lysozyme to pass through and reach sensitive murein in the underlying cortex.

Lysozyme caused changes in sensitized spores characteristic of the changes normally associated with germination[62] and which occurred within minutes of addition of the enzyme. Both viable sensitized spores, and sensitized spores which had been rendered non-viable by heat or by the severity of the sensitizing treatment could be 'germinated' by lysozyme, suggesting the non-involvement of spore enzymes or spore metabolism in lysozyme germination. In addition to shedding DPA and muropeptides and becoming phase-dark and heat-sensitive, suspensions of lysozyme-germinated spores became optically less dense (Fig. 12) and showed cytological changes similar to those seen in spores germinated by amino acids or ribosides (Fig. 1). Continued incubation of lysozyme-germinated spores in the presence of lysozyme led to loss of viability[62] and eventual leakage of contents, including enzymes e.g. DNA polymerase[194] and nucleic acids.[195] However, the initial changes induced in sensitized spores by lysozyme so closely resembled those which occur during normal germination as to support the hypothesis[196] that the depolymerization of

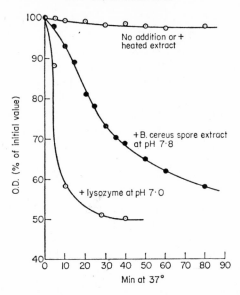

F I G. 12. Germination of *Bacillus cereus* PX spores by lysozyme (100 μg/ml) and by spore lytic enzyme. Spores were first sensitized by treatment with thioglycollic acid (10%) in urea (4 M) at 70°C for 30 min, then well washed before incubation with the enzymes at 37°C in buffer (100 mM sodium phosphate) to initiate germination. The substrate spores in this instance were non-viable and not germinable by metabolizable germinants. When milder sensitization procedures were used, viable spores germinable both by enzymes and by metabolizable germinants were obtained (Gould and Hitchins.[200])

murein in the spore cortex is an early event in normal spore germination. Spores of *B. megaterium* ATCC 9885 were unique in being sensitive to lysozyme without the necessity of pretreatment with reagents which rupture disulphide bonds.[197] It appeared that murein responsible for stability of these spores was directly accessible to lysozyme, probably because of some defect in structure of the coat layers.

2. Spore "lytic" enzyme

When Powell and Strange[24] found that germinating spores released material containing amino acids, amino sugars, DPA and calcium, it was suggested that the exudation could be the result of enzymic depolymerization of some spore constituents.[24, 196] A lytic enzyme was later extracted from spores of some *Bacillus* species,[32] and its properties were found to be compatible with its involvement in the release of germination exudate. Preparations containing spore lytic enzyme caused release of murein fragments from spore debris, and lysed the cell walls of some vegetative *Bacillus* species[32] and chloroform-treated *Escherichia coli*.[198, 199] The enzyme's substrate seemed most likely to be murein. Spores of *B. cereus* were good sources of lytic enzymes, whereas spores of strains of

B. subtilis and some other species were not. However, it was noticed that debris from broken spores of *B. sphaericus*[32] and *B. coagulans*[31] incubated in suitable buffers slowly released soluble hexosamine-containing material. It appeared therefore that these spores also contained lytic enzymes, although not readily extractable in soluble form.

Although untreated spores of *B. cereus* were resistant to attack by preparations containing spore lytic enzyme,[57] they became sensitive if first treated with reagents which ruptured disulphide bonds.[200] If the treatment was mild the spores remained viable. Presumably the reagents altered permeability of the spore coat sufficiently to allow the enzyme to enter the spore in the same way that lysozyme was able to enter such treated spores.[62, 191] There is evidence that reduction of spore coat disulphide bonds normally does occur during germination of some *Bacillus* spores.[61, 63] Action of the enzyme on sensitized spores resulted in the spores appearing to germinate,[200] the germination-like changes occurred in viable and in heat-killed spores. The enzyme could be inactivated by oxidation or by reagents for thiol groups (p-chloromercuribenzoate, N-ethylmaleimide, iodine) and partially reactivated by reducing agents.[167] It was in the active reduced form as obtained in spore extracts and, although heat-stable in the spore, was heat-labile in solution. The only major change noticed during germination was that the enzyme distribution altered; in debris made from ungerminated spores in media of low ionic strength the enzyme was mostly bound to some insoluble part of the spore, whereas in debris from the germinated spores the enzyme was mostly free; the binding site was tentatively identified as the core. Enzyme could be freed from its binding site by raising the pH value or ionic strength or by the poly-cation spermine suggesting that binding was by electrostatic bonds or salt linkages and not by covalent linkages. In addition to requiring salts for its release, the enzyme's activity as a germinant depended on the presence of a suitable ionic strength, equivalent to about 0·1 M NaCl.

Lytic enzyme may function during normal germination by being released from its binding site within the spore by some product of germinant metabolism or by DPA,[252] or by a change in spore permeability allowing ingress of ions. It would then disrupt the cortex (Fig. 1), destroying its properties, for instance as a contractile organ[27] or as a permeability barrier[47] responsible for maintaining the inert resistant state of the core. At this time however, it is not known whether action of the enzyme plays a dominant role in initiating germination (i.e. whether its action causes excretion of spore exudate and thereby directly results in loss of phase brightness, decrease in extinction, loss of heat resistance) or whether the enzyme becomes active during germination along with other spore enzymes by some as yet unknown mechanism for the breaking of dormancy.[201]

Distinct from spore lytic enzyme is a heat-activatable substance recognizable in *B. cereus* T spore extracts which can directly cause germination of the same spores when supplied with the correct cofactors.[202] The heat-activatable substance has proved to be a protein, and the effective cofactors are low levels of L-alanine and NAD. The exact mechanism involved is not yet known, but may be allosteric, involving sequential binding of L-alanine and NAD to the heat-activatable (probably enzyme) protein.†

3. Protease and other hydrolytic spore enzymes

Spores of *B. subtilis* (Marburg) and of some other *Bacillus* species were germinated by the *B. subtilis* protease subtilisin.[203] Inhibition by D-alanine suggested that subtilisin caused germination by forming L-alanine,[204] for instance by autodigestion, by hydrolysis of contaminating proteins or even by hydrolysis of some part of the spore coat.[204, 205] Indeed a *B. megaterium* spore protease was found to attack spore protein and to liberate alanine from it.[162] Germination of spores of this organism and action of the protease both increased in the presence of Mn^{2+}. The increase in germination due to Mn^{2+} could be partly reversed by D-alanine, arguing strongly for a role of the protease in producing L-alanine during Mn^{2+}-stimulated germination.[206] *B. subtilis* spore protease has been shown to be distinct from spore esterase[207] which has also been suggested as an important germination enzyme,[208, 209] which could act on lipids to increase spore permeability. High levels of spore pyrophosphatases in *B. megaterium*, *B. coagulans* and *B. subtilis*, also dependent on Mn^{2+}, have been described,[210–213] and the phosphorus content of *B. cereus* spore coats was found to decrease during germination.[214, 215] Apart from this circumstantial evidence it is not known to what extent these enzymes are implicated in germination nor to what extent thay can actually attack structures like the spore coat. An organism (probably a *Myxococcus* species) capable of lysing spores of *B. stearothermophilus* has been described[216] and presumably secretes enzymes able to attack the coats of intact spores, but the nature of the enzymes is not yet known.

D. PHYSICAL AGENTS

It was pointed out by Rode and Foster[217] that the immediate result of abrading or cracking the surface layers of spores was to induce the germination-like changes described in Section II above. They found that if such abraded spores were transferred to a nutrient medium outgrowth occurred showing that provided the abrasion was not excessive the organisms remained viable. Abrasion therefore seemed to substitute fully for other germinants, and the term "mechanical germination" was used to describe the phenomenon. That mechanical germination could occur

---

† This may involve enzymatic formation of low molecular weight germinants.[252] [253]

during a variety of routine microbiological procedures was highlighted by
Lewis *et al.*[27] who pointed out that the pressure of a coverslip on a slide
could cause phase darkening of spores, and by Knaysi and Curran[218] who
observed phase darkening and disruption of spores in suspensions on
slides during emulsification with a wire loop. Dry rupture of spores also
appeared to occur *via* a germination-like change.[219] It would appear that
methods generally used for breaking spores act by first causing germination,
so that spore extracts as generally prepared are really extracts of freshly
germinated spores.

The fact that spores can be mechanically germinated suggests that
structural integrity of the spore coats or cortex layers is an important
factor contributing to dormancy and resistance, and that during other (non-
mechanical) forms of germination a similar rupture or lysis of these layers
might occur. The rapidity of mechanical germination could be most easily
explained if one of the disrupted layers exerted some control over permeability of the core to water,[217] to ions or to other molecules. The cortex
is the most likely candidate for the role of permeability barrier because its
lysis causes germination.

Low doses of ionizing radiation may activate spores[171] and higher doses,
although killing spores may induce germination-like changes or pseudo-
germination,[110, 220] perhaps by disrupting permeability barriers either
directly or *via* the triggering or release of spore lytic enzymes.

E. ENVIRONMENTAL FACTORS

The role of oxygen in spore germination has been discussed in Section
III A 5. Briefly inhibition by oxygen of germination of spores of *Clostridium* species in some germination systems has been demonstrated but
germination of *Bacillus* spores seems unaffected by presence or absence of
oxygen. Low or high pressures have not been shown to inhibit germination
of bacterial spores.

1. pH value

The optimum pH value for spore germination varies widely and depends very much on the organism. For example, Thorley and Wolf[77] found
spores of *B. megaterium* to germinate optimally in glucose at pH 5·1;
B. cereus T germinated optimally in L-alanine at pH 8, but still germinated
slowly at pH 10;[94] alkali-resistant strains of *B. cereus* and *B. circulans*
germinated slowly at pH values as high as 10·3 and 11·0 respectively.
Furthermore, the pH range and optimum for germination of spores of
any one strain depend on the particular germinant used. For instance[128]
spores of *B. megaterium* QM B1551 germinated optimally in L-alanine at
pH 6·0, but in glucosamine or L-proline the pH optimum was above

about 8. Levinson[222] noticed that pH optima for germination of heat-activated and unheated *B. megaterium* spores differed.

The lower limiting pH value for germination, growth and toxin production by food poisoning spores has attracted much attention. Spores of *Cl. botulinum* types A and B can germinate in rich media at pH values as low as 4·6, and type E at 4·8. However, under practical conditions in food-stuffs the limiting pH value is influenced by other factors such as water activity, temperature, presence of preservatives etc. so that discussion of pH value in isolation can be misleading The interplay and practical significance of these various influences is more fully discussed in Chapters 14 and 15.

More than one optimum pH value was observed for germination of spores of some organisms,[77] possibly a reflection of different pH optima for different enzymes involved in germination. The simplest instance, is perhaps associated with different optima for L-alanine dehydrogenase or some other enzyme (catalysing germination) and alanine racemase (hindering germination) (Section III A 1).

Germination caused by calcium dipicolinate occurred optimally between pH 5 and 9 and was most rapid near pH 7.[152] The pH range over which ionic germination occurred was wide, for instance 4·1 to 10·9 for *B. megaterium* QM B1551 germinated by chloride ions, but (as with calcium dipicolinate) the sharp cut off at the extremes of pH suggested involvement of spore enzymes in these forms of germination.[85] Germination caused by cationic surfactants took place only above a critical pH value, about pH 6·0 for *B. megaterium* and although initially most rapid at pH 9·9, was most complete in the range pH 7·4 to 7·8.[168, 170] Germination by enzymes (Section III C) occurred at the pH values expected from their known optima. Germination of thioglycollic-acid treated spores by hydrogen peroxide was definitely not enzymic, and occurred more and more rapidly as the pH value was raised.[62]

Generally, germination can occur at pH values slightly higher and lower than the limits for vegetative growth; but some of the chemical germinants mentioned above are active at much more extreme values.

2. Temperature

Many of the observations which have been made on the effects of temperature on spore germination were summed up by Knaysi[69] who wrote that germination could generally occur at temperatures outside the limits for vegetative growth. For the *B. cereus* strain he studied the maximum, optimum and minimum temperatures for germination were 59°, 30° and near −1°C, and for vegetative growth 44°, 35° and 10°C respectively. The wide limits were strikingly demonstrated by Mol[223] who found spores of *B. cereus* to germinate (85%) in 6 hours at 8°C, and yet not to outgrow during 4 months at this temperature. Study of the temperature characteristics of germination of a variety of spores[224, 225] revealed that there is no one temperature which would allow germination of *all* spores and yet arrest vegetative cell multiplication. If there was such a temperature, it could be used in novel methods of food preservation.

The wide temperature ranges for germination of spores of different organisms was emphasized by Wolf and his colleagues[75, 77, 226] who found optima for some mesophilic *Bacillus* species as high as 50°C and measured slow germination of spores of many strains at 0°C. In the presence of glycerol as antifreeze spores of *B. cereus* T germinated slowly at temperatures as low as −6°C. [227] Larkin and Stokes isolated psychrophilic species of *Bacillus* that could grow, sporulate and germinate below 0°C.[228] Germination of spores of *Clostridium* species also has wide temperature limits when suitable media are used. Rapid germination of heat-activated spores of *Cl. botulinum* strains, *perfringens* and PA 3679 at 75°C in solutions of autoclaved glucose has been described,[229] whilst observation of *Cl. sporogenes* germination at 4°C (90% germination in 2 weeks, but no growth at 10°C in 6 weeks[230]) has been extended to other *Clostridium* species, some of which can germinate slowly at temperatures below 5°C, i.e. well below their minimum growth temperature.[150] Special attention has been given to the low temperature limits for germination, outgrowth and vegetative growth of food poisoning *Clostridium* species, in particular *Cl. botulinum* type E spores which can germinate and outgrow at temperatures below 4°C,[231] and is discussed in detail in Chapter 14.

Quotation of accurate temperature characteristics for spore germination can be misleading in that, as with pH value, the cardinal points depend to some extent on the germinant.[128] Similarly to pH optima, multiple temperature optima have been described for germination of some *Bacillus* spores.[77] *B. subtilis* (Marburg) has additional requirements for germination at high temperatures, and a mutant having these requirements at normal germination temperature has been obtained[86] (Section III A 1).

A spore suspension which has begun to germinate is not immediately arrested by dropping either the temperature or the pH to levels below those normally required to initiate germination.[68, 69] The reaction(s) which initiate the germination process are therefore more sensitive to low temperature (or pH) than the reactions which carry the process to completion. We saw in Section II A6 that other ways of separating initiation from completion of the germination event have been described.

Germination by calcium dipicolinate occurred optimally below 35°C for a number of *Bacillus* species.[152] The low temperature requirement is probably a function of solubility and stability of the calcium: DPA complexes rather than of any spore enzyme mechanism[175] (Section III B 3). In contrast, germination caused by the surfactant n-dodecylamine[170] began most rapidly at 60 to 70°C, although was not as extensive as at 37 to 50°C, suggesting an initial action which was not enzymic. However, the heat-resistance of spores is such that high temperature operation of the n-dode-cylamine system does not exclude the possibility of enzymes playing a part following a physicochemical trigger.

3. Water activity

In considering growth from a spore inoculum in a rich medium, the following responses to available water level are usually recognizable:[232] (a) low water activity (a_w) where no germination occurs, and the spore count remains stable; (b) medium a_w where germination alone occurs but no outgrowth or vegetative growth takes place, so the spore count decreases with or without a decrease in the total viable count; (c) high a_w where germination and subsequent outgrowth and vegetative growth can occur, so the total viable count increases. The a_w levels delineating these three areas differ with different organisms, but it has been generally found that spore germination will occur at lower a_w than required for vegetative growth. For instance *B. cereus* and *B. subtilis* spores germinated at a_w values below 0·8 but only outgrew at values above about 0·9.[232, 233] *Cl. sporogenes* spores would germinate but not grow in media containing sodium chloride (8%) or glucose (60%) added to lower the a_w.[230]

The ability of spores of *Cl. perfringens* and *Cl. botulinum* types to grow and form toxin from a spore inoculum has been extensively studied (Chapter 14). Under otherwise optimal growth conditions germination of *Cl. botulinum* spores was not completely inhibited by a_w levels as low as 0·89 (obtained with glycerol[234]) but it is important to remember that under sub-optimal growth conditions the effect of low a_w on germination may become more pronounced (or conversely, low a_w can intensify the inhibitory effects of sub-optimal conditions). For example, at low pH values germination of *Cl. botulinum* A and E spores was more sensitive to low a_w than at near optimal pH values,[234] and incubation at sub-optimal temperatures allowed low a_w levels to inhibit growth more effectively from spores of *Cl. botulinum* types.[235] Freeze-thaw cycles lowered the a_w permitting germination of *B. cereus* spores. Presumably on freezing (water plus substrate), the water tended to separate as ice which on thawing formed areas where the a_w was sufficiently high to allow germination.[232]

The mechanism of inhibition of spore germination by low a_w is not known. Spores incubated with germinants in media of low a_w (e.g. high salt content) do not subsequently germinate when incubated in germinant-free high a_w medium, suggesting that enzyme or permeability reactions involved in germination are inhibited by low a_w rather than simply the physical hydration of any part of the spore.

4. Ionic strength

The particular germinative effects of ions for some spores have been discussed in Section III B1. Germination by cationic surfactants like n-dodecylamine is hindered by cations.[169] Nevertheless, germination by the so-called metabolizable germinants normally only occurs optimally in media of suitable ionic strength. For *B. subtilis* spores an ionic strength of about 0·1 M was necessary for optimal germination in L-alanine.[156] Of various ions tested, phosphate was most effective and divalent cations least effective for *B. subtilis*, but the pattern of ionic responses is not the same for all organisms.[85] Increasing the ionic strength to 1·0 M, retarded germination. With other spores (e.g. *B. cereus* T) the ion requirement is less marked, but the general need for ions for optimal germination is well documented. The relatively non-specific ion requirement may be needed by some spore germination enzymes; for instance, spore lytic enzyme from *B. cereus* PX has a non-specific ion requirement for activity.[167] Alternatively, a particular level of ions may be necessary to accomplish some physico-chemical stage in germination; for instance stages involving stability of chelate structures in spores would be ion dependent[156] and exchangeability of some spore ions is now well known.[28] Raising the ionic strength also results in the release of previously bound spore lytic enzyme in spore homogenates, and perhaps additionally releases other spore enzymes which can catalyse germination.

IV—Discussion. The Nature of the "Trigger Reaction"

The major changes occurring during germination of bacterial spores are now known in some detail. The overall picture is of rapid depolymerization of murein, probably mostly from the outer cortex, and leakage from the spore of some of the resulting muropeptides along with calcium and dipicolinic acid. Changes in spore shape, volume, cytology, and optical properties presumably result more or less directly from the initial depolymerization reaction. What is not certain is whether depolymerization is the prime event, caused by the release or activation of lytic enzyme within the spore, or whether some other event initiates the process by activating numerous enzymes in rapid sequence, including those involved in depolymerization. If a major cause of spore dormancy and resistance is pressure exerted on the core by a contractile cortex,[27] then it is easy to understand how activation of a cortex-lytic system could rapidly release the pressure and result in the phenomena characteristic of germination. Alternatively, if the cortex is the site of a permeability barrier, to water, ions or larger molecules, then its partial dissolution would lead to a sudden increase in permeability, exposing functional groups in spore enzymes to exogenous substrates and thereby terminating dormancy. Whatever the reactions involved in initiating germination, a problem is posed by the diversity of germinants effective for the same spore and for different spores. Must one suppose that a common mode of initiation is shared by such diverse substances as L-alanine, inosine, glucose, calcium dipicolinate, n-dodecylamine and the many other germination systems described? It is difficult to imagine what the common link may be, because the pathway of metabolism of "metabolizable germinants" during germination is not known. Study of the germination of spores of mutants deficient in germinant-handling enzymes should help to reveal these pathways, but it must be borne in mind that the so-called "metabolizable germinants" could act as allosteric effectors of some spore enzyme whose allosteric activation initiates germination as indicated by work on the heat-activatable initiation system of Vary and Halvorson (Section III C2). Study of germinant metabolism would then be misleading. Chelate germination could certainly be due to destabilization of calcium dipicolinate-containing chelates within the spore, and may well be a major step in all forms of germination. However, the exact location of Ca and DPA in spores is still largely conjectural so that the effects of chelates on Ca–DPA spore structures must remain speculative. The most likely role one could ascribe to surfactant germinants would be interference with the integrity of some spore membrane, for instance by changing its permeability. It may be significant that phenethyl alcohol can inhibit spore germination[237] and is known to inhibit growth of vegetative cells by its effects on the cell permeability barriers.[238] Permeability changes during germination could be quite small and need not necessarily involve a major change in permeability of spores to water, but rather a selective change in permeability to ions or to other molecules necessary for action of some spore enzymes. The general necessity for ions in germination is now well recognized.

Spore germination is usually studied in laboratories by workers using large quantities of well cleaned spores, usually of a particular strain of *B. cereus*, *B. megaterium* or more recently *B. subtilis* grown in a particular medium. The spores chosen for such studies tend for the sake of convenience to be easy to produce and germinate rapidly. However, they may well not be typical of the spores which occur naturally and which may be, in particular, more difficult to germinate rapidly than the common "laboratory spores". A result of this unconscious selection of rapid germinators by experimentalists is that the enormous practical problem of delayed germination of spores, or superdormancy, frequently encountered in "natural" spores has been largely ignored. If *all* spores germinated as rapidly as do the first 90% in a laboratory-grown *B. cereus* T spore suspension incubated under good germination conditions, then food processors would have, in germination, an excellent procedure for preservation based on complete eradication of spores from foods. Unfortunately, populations of spores appear to be basically heterogeneous in that in any population, superdormant individuals occur which germinate much more slowly than the majority. The reason for superdormancy is obscure and study of superdormant spores is complicated because of their low frequency and difficulty of isolation. However, study of such spores should lead to a better understanding of the factors controlling the ease of germination of spores. For example the existence of superdormant spores, and even of spores so dormant as to be non-germinable (by conventional germinants) can be inferred from the hypothesis of Woese.[2, 247] This assumes that spores contain some critical "germination enzyme" (or other molecule) the numbers of molecules per spore being Poissonally distributed throughout the population. Action of this enzyme must reach a certain level before the actual irreversible degradative germination event occurs. A model based on these simple assumptions fits the kinetic data obtained on germination best when the mean number of the hypothetical germination enzyme molecules per spore is only about 9. Such low numbers suggest that in any spore population the chances of finding spores containing exceptionally low levels of "germination enzyme", even one, two or no molecules per spore, are very real. Such spores may be the superdormant ones recognized experimentally; the critical enzyme is probably not alanine dehydrogenase,[247] it may be protease, ribosidase or some other riboside-handling enzyme, spore lytic enzyme, or the activatable germination protein. Alternatively, instead of a critical enzyme, some membrane with critical permeability, or some chelate structure with critical stability differing from spore to spore could account for heterogeneity with respect to germination, and there are no doubt still further possibilities.

Superdormancy may well have evolutionary significance as a survival factor ensuring that no spore population ever *completely* germinates at any one time; it thus acts as an insurance against the potentially lethal dangers of germination in environments unfavourable for growth. Nevertheless, all spores must sooner or later pass through the rapid degradative stage of germination if they are to give rise by way of outgrowth to new vegetative cells which may grow, divide, and eventually form new spores and thus complete the development cycle of the sporeforming bacterium.

REFERENCES

1. Sussman, A. S. and Halvorson, H. O. (1966). *In* "Spores: their dormancy and germination". Harper and Row, New York.
2. Halvorson, H. O., Vary, J. C. and Steinberg, W. (1966). *A. Rev. Microbiol.* **20**, 169.
3. Hamilton, W. A. and Sale, A. J. (1968). *Biochim. biophys. Acta*, **148**, 789.

4. Gould, G. W. (1964). *In* "Microbial inhibitors in foods." 4th Int. Symp Food Microbiol. (N. Molin, ed.), p. 17, Almquist and Wiksell, Stockholm.
5. Powell, E. O. (1957). *J. appl. Bact.* **20,** 342.
6. Murrell, W. G. and Scott, W. J. (1966). *J. gen. Microbiol.* **43,** 411.
7. Kennedy, E. J. and Grecz, N. (1966). *Bact. Proc.* p. 14.
8. Briggs, A. (1966). *J. appl. Bact.* **29,** 490.
9. Irie, R., Yano, N., Morichi, T. and Kembo, H. (1965). *Biochem. biophys. Res. Commun.* **20,** 389.
10. Farkas, J. and Kiss, I. (1964). *Commun. Cent. Fd Res. Inst.* **1,** 13.
11. Donnellan, J. E. and Setlow, R. B. (1965). *Science, N.Y.* **149,** 308.
12. Smith, K. C. and Yoshikawa, H. (1966). *Photochem. Photobiol.* **5,** 777.
13. Wake, R. G. (1967). *J. molec. Biol.* **25,** 217.
14. Stafford, R. S. and Donnellan, J. E. (1967). *A. Mtg. Biophys. Soc. Abstracts* **11,** 80.
15. Duncan, C. L. and Foster, E. M. (1968). *Appl. Microbiol.* **16,** 406.
16. Yoshikawa, H. (1965). *Proc. natn. Acad. Sci., U.S.A.* **53,** 1476.
17. Kozima, T., Kawabata, T. and Okitu, T. (1965). *Nippon Suissan Gakk.* **31,** 934.
18. Parker, M. S., Barnes, M. and Bradley, T. J. (1966). *J. Pharm. Pharmac.* **18,** 103S.
19. Fernelius, A. L. (1960). *J. Bact.* **79,** 755.
20. Levinson, H. S. and Hyatt, M. T. (1966). *J. Bact.* **91,** 1811.
21. Mandels, G. R., Levinson, H. S. and Hyatt, M. T. (1956). *J. gen. Physiol.* **39,** 301.
22. Broda, E. (1967). *Angew. A. Chem.* **6,** 640.
23. Desser, H. and Broda, E. (1965). *Nature, Lond.* **206,** 1270.
24. Powell, J. F. and Strange, R. E. (1953). *Biochem. J.* **54,** 205.
25. Thomas, R. S. (1964). *J. Cell Biol.* **23,** 113.
26. Knaysi, G. (1965). *J. Bact.* **90,** 453.
27. Lewis, J. C., Snell, N. S. and Burr, H. K. (1960). *Science, N.Y.* **132,** 544.
28. Alderton, H., Thompson, P. A. and Snell, N. (1964). *Science, N.Y.* **143,** 141.
29. Rode, L. J. and Foster, J. W. (1966). *J. Bact.* **91,** 1589.
30. Vincent, J. M. and Humphrey, B. A. (1963). *Nature, Lond.* **199,** 149.
31. Warth, A. D., Ohye, D. F. and Murrell, W. G. (1963). *J. Cell. Biol.* **16,** 593.
32. Strange, R. E. and Dark, F. A. (1957). *J. gen. Microbiol.* **16,** 236.
33. Warth, A. D. (1965). *Biochim. biophys. Acta,* **101,** 315.
34. Vinter, V. (1965). *Folia Microbiol. Praha,* **10,** 280.
35. Walker, P. D., Baillie, A., Thomson, R. O. and Batty, I. (1966). *J. appl. Bact.* **29,** 512.
36. Walker, P. D., Thomson, R. O. and Baillie, A. (1967). *J. appl. Bact.* **30,** 317.
37. Chapman, G. B. and Zworykin, K. A. (1957). *J. Bact.* **74,** 126.
38. Hamilton, W. A. and Stubbs, J. M. (1967). *J. gen. Microbiol,* **47,** 121.
39. Mayall, B. H. and Robinow, C. (1957). *J. appl. Bact.* **20,** 333.
40. Hachisuka, Y., Kojima, K. and Sato, T. (1966). *J. Bact.* **91,** 2382.
41. Moberly, B. J., Shafa, F. and Gerhardt, P. (1966). *J. Bact.* **92,** 220.
42. Freer, J. H. and Levinson, H. S. (1967). *J. Bact.* **94,** 441.
43. Takagi, A. Kawata, T., Yamamoto, S., Kubo, T. and Okita, S. (1960). *Jap. J. Microbiol,* **4,** 137.
44. Takagi, A., Kawata, T. and Yamamoto, S. (1960). *J. Bact.* **80,** 37.
45. Kawata, T., Inoue, T. and Takagi, A. (1963). *Jap. J. Microbiol.* **7,** 23.
46. Rousseau, M., Flechon, J. and Hermier, J. (1966). *Annls Inst. Pasteur, Paris,* **111,** 149.

47. Rode, L. J., Lewis, C. W. and Foster, J. W. (1962). *J. Cell. Biol.* **13**, 423.
48. Powell, J. F. (1950). *J. gen. Microbiol.* **4**, 330.
49. Vinter, V. and Šťastná, J. (1967). *Folia Microbiol. Praha*, **12**, 301.
50. Pulvertaft, R. J. V. and Haynes, J. A. (1951). *J. gen. Microbiol.* **5**, 657.
51. Ross, K. F. A. and Billing, E. (1957). *J. gen. Microbiol.* **16**, 418.
52. Erikson, D. (1955). *J. gen. Microbiol.* **13**, 119.
53. Uehara, M. and Frank, H. A. (1965). *Bact. Proc.* p. 36.
54. Hermier, J. and Rousseau, M. (1967). *Annls Inst. Pasteur, Paris* **113**, 327.
55. Arntz, P. (1966). *Ant. van Leeuwenhoek*, **32**, 450.
56. Hitchins, A. D. and Gould, G. W. (1964). *Nature, Lond.* **203**, 895.
57. Powell, J. F. (1957). *J. appl. Bact.* **20**, 349.
58. Black, S. H. and Gerhardt, P. (1962). *J. Bact.* **83**, 301.
59. Hitchins, A. D., Gould, G. W. and Hurst, A. (1963). *J. gen. Microbiol.* **30**, 445.
60. Parker, M. S. and Barnes, M. (1967). *J. appl. Bact.* **30**, 299.
61. Vinter, V. (1961). *In* "Spores II" (H. O. Halvorson, ed.), p. 127, Burgess Publishing Co., Minneapolis, Minn., U.S.A.
62. Gould, G. W. and Hitchins, A. D. (1963). *J. gen. Microbiol.* **33**, 413.
63. Blankenship, L. C. and Pallansch, M. J. (1966). *J. Bact.* **92**, 1615.
64. Vary, J. C. and McCormick, N. G. (1965). *In* "Spores III" (L. L. Campbell and H. O. Halvorson, eds), p. 188. Am. Soc. Microbiol., Ann Arbor, Michigan, U.S.A.
65. McCormick, N. G. (1964). *Biochem. biophys. Res. Commun.* **14**, 443.
66. McCormick, N. G. (1965). *J. Bact.* **89**, 1180.
67. Vary, J. C. and Halvorson, H. O. (1965). *J. Bact.* **89**, 1340.
68. Halmann, M. and Keynan, A. (1962). *J. Bact.* **84**, 1187.
69. Knaysi, G. (1964). *J. Bact.* **87**, 619.
70. Levinson, H. S. and Hyatt, M. T. (1965). *In* "Spores III" (L. L. Campbell and H. O. Halvorson, eds), p. 198. Am. Soc. Microbiol., Ann Arbor, Michigan, U.S.A.
71. Hambleton, R. and Rigby, G. J. (1966). *J. Pharm. Pharmac.* **18**, 30S.
72. Hills, G. M. (1949). *Biochem. J.* **45**, 353.
73. Hills, G. M. (1949). *Biochem. J.* **45**, 363.
74. Hills, G. M. (1950). *J. gen. Microbiol.* **4**, 38.
75. Wolf, J. and Mahmoud, S. A. Z. (1957). *J. appl. Bact.* **20**, 373.
76. Wolf, J. and Thorley, C. M. (1957). *J. appl. Bact.* **20**, 384.
77. Thorley, C. M. and Wolf, J. (1961). *In* "Spores II" (H.O. Halvorson, ed.), p. 1. Burgess Publishing Co., Minneapolis, Minn., U.S.A.
78. Martin, J. H. and Harper, W. J. (1963). *J. Dairy Sci.* **44**, 663.
79. Foerster, H. F. and Foster, J. W. (1966). *J. Bact.* **91**, 1168.
80. Levinson, H. S. and Hyatt, M. T. (1962). *J. Bact.* **83**, 1224.
81. Woese, C. R., Morowitz, H. J. and Hutchinson, C. A. (1958). *J. Bact.* **76**, 578.
82. Hermier, J. (1962). *Annls Inst. Pasteur, Paris*, **102**, 629.
83. Krask, B. J. (1961). *In* "Spores II" (H. O. Halvorson, ed.), p. 89. Burgess Publishing Co., Minneapolis, Minn., U.S.A.
84. Caracò, A., Falcone, G. and Salvatore, G. (1958). *G. Microbiol.* **5**, 127.
85. Rode, L. J. and Foster, J. W. (1962). *Arch. Mikrobiol.* **43**, 183.
86. Wax, R., Freese, E. and Cashel, M. (1967). *J. Bact.* **94**, 522.
87. Hyatt, M. T. and Levinson, H. S. (1961). *J. Bact.* **81**, 204.
88. Church, B. D., Halvorson, H., Ramsey, D. S. and Hartman, R. S. (1956). *J. Bact.* **72**, 242.

89. Lewis, J. C., Snell, N. S. and Alderton, G. (1965). *In* "Spores III" (L. L. Campbell and H. O. Halvorson, eds), p. 47, Am. Soc. Microbiol., Ann Arbor, Michigan, U.S.A.
90. Campbell, L. L., Richards, C. M. and Sniff, E. E. (1965). *In* "Spores III" (L. L. Campbell and H. O. Halvorson, eds), p. 55, Am. Soc. Microbiol., Ann Arbor, Michigan, U.S.A.
91. Stewart, B. T. and Halvorson, H. O. (1953). *J. Bact.* **65**, 160.
92. Fey, G., Gould, G. W. and Hitchins, A. D. (1964). *J. gen. Microbiol.* **35**, 229.
93. Gould, G. W. (1966). *J. Bact.* **92**, 1261.
94. Church, B. D., Halvorson, H. and Halvorson, H. O. (1954). *J. Bact.* **68**, 393.
95. Falcone, G. and Caracò, A. (1958). *G. Microbiol.* **5**, 80.
96. Falcone, G., Salvatore, G. and Covelli, I. (1959). *Biochim. biophys. Acta*, **36**, 390.
97. Galdiero, G. (1967). *Farmacista*, **22**, 85.
98. Halvorson, H. and Church, B. D. (1957). *J. appl. Bact.* **20**, 359.
99. Halvorson, H. (1957). *In* "Spores" (H.O. Halvorson, ed.), p. 144, Am. Inst. Biol. Sci., Washington, U.S.A.
100. Halvorson, H. O. (1959). *Bact. Rev.* **23**, 267.
101. O'Connor, R. and Halvorson, H. (1960). *Archs. Biochem. Biophys.* **91**, 290.
102. O'Connor, R. and Halvorson, H. (1961). *Biochim. biophys. Acta*, **48**, 47.
103. Hermier, J. (1962). *C.R. hebd séance, Acad. Sci., Paris* **254**, 2865.
104. Yoshida, A. and Freese, E. (1964). *Biochim. biophys. Acta*, **92**, 33.
105. Hermier, J. (1965). *Annls Biol. anim. Biochim. Biophys.* **5**, 483.
106. McCormick, N. G. and Halvorson, H. O. (1963). *Ann. N.Y. Acad. Sci.* **102**, 763.
107. O'Connor, R. J. and Halvorson, H. O. (1961). *J. Bact.* **82**, 648.
108. Krask, B. J. and Fulk, G. E. (1966). *Bact. Proc.* p. 32.
109. Freese, E., Park, S. W. and Cashel, M. (1964). *Proc. natn. Acad. Sci., U.S.A.* **51**, 1164, and erratum in ibid. **52**, 516.
110. Freese, E. and Cashel, M. (1965). *In* "Spores III" (L. L. Campbell and H. O. Halvorson, eds), p. 144. Am. Soc. Microbiol., Ann Arbor, Michigan, U.S.A.
111. Harrell, W. K. and Halvorson, H. (1955). *J. Bact.* **69**, 275.
112. Hsu, W-T. (1963). *Thesis*, University of Illinois.
113. Uchiyama, H., Tanaka, K. and Yanagita, T. (1965). *J. gen. Appl. Microbiol. Tokyo*, **11**, 233.
114. Falcone, G. and Bresciani, F. (1963). *Experientia*, **19**, 152.
115. Brown, W. L. (1956). *Thesis*, University of Illinois.
116. Rode, L. J. and Foster, J. W. (1961). *Z. allg. Mikrobiol*, **1**, 307.
117. Powell, J. F. and Hunter, J. R. (1955). *J. gen. Microbiol.* **13**, 59.
118. Lawrence, N. L. (1955). *J. Bact.* **70**, 583.
119. Lawrence, N. L. and Tsan, Y-C. (1962). *J. Bact.* **83**, 228.
120. Lawrence, N. L. (1955). *J. Bact.* **70**, 577.
121. Powell, J. F. and Hunter, J. R. (1956). *Biochem. J.* **62**, 381.
122. Srinivasan, V. R. and Halvorson, H. O. (1961). *Biochem. biophys. Res. Commun.* **4**, 409.
123. Krask, B. J. and Fulk, G. E. (1959). *Archs. Biochem. Biophys.* **79**, 86.
124. Simon, E. R., Chapman, R. G. and Finch, C. A. (1962). *J. clin. Invest.* **41**, 351.
125. Gardner, R. and Kornberg, A. (1967). *J. biol. Chem.* **242**, 2383.
126. Hyatt, M. T. and Levinson, H. S. (1964). *J. Bact.* **88**, 1403.

127. Levinson, H. S. and Sevag, M. G. (1953). *J. gen. Physiol.* **36**, 617.
128. Hyatt, M. T. and Levinson, H. S. (1962). *J. Bact.* **83**, 1231.
129. Halvorson, H. O. and Church, B. D. (1957). *Bact. Rev.* **21**, 112.
130. Goldman, M. and Blumenthal, H. J. (1964). *J. Bact.* **87**, 377.
131. Blumenthal, H. J. (1965). *In* "Spores III" (L. L. Campbell and H. O. Halvorson, eds), p. 222, Am. Soc. Microbiol., Ann Arbor, Michigan, U.S.A.
132. Amaha, M. and Nakahara, T. (1959). *Nature, Lond.* **184**, 1255.
133. Simmons, R. J. and Costilow, R. N. (1962). *J. Bact.* **84**, 1274.
134. Black, S. H. and Gerhardt, P. (1961). *J. Bact.* **82**, 743.
135. Uehara, M. and Frank, H. A. (1965). *In* "Spores III". (L. L. Campbell and H. O. Halvorson, eds), p. 38. Am. Soc. Microbiol., Ann Arbor, Michigan, U.S.A.
136. Fujioka, R. S. and Frank, H. A. (1966). *J. Bact.* **92**, 1515.
137. Halvorson, H. O. (1957). *In* Proc. 9th Res. Conf. Am. Meat. Inst., p. 1. Chicago, Ill.
138. Hitzman, D. O., Zoha, S. M. S. and Halvorson, H. O. (1955). *Bact. Proc.* p. 40.
139. Hitzman, D. O., Halvorson, H. O. and Ukita, T. (1957). *J. Bact.* **74**, 1.
140. Treadwell, P. E., Jann, G. J. and Salle, A. J. (1958). *J. Bact.* **76**, 549.
141. Reimann, H. (1963). *Thesis*, University of Copenhagen, Denmark.
142. Gibbs, P. A. (1964). *J. gen. Microbiol.* **37**, 41.
143. Gibbs, P. A. (1968). *J. appl. Bact.* (In press.)
144. Ward, B. Q. and Carroll, B. J. (1966). *Can. J. Microbiol.* **12**, 1145.
145. Wynne, E. S., Mehl, D. A. and Schmeiding, W. R. (1954). *J. Bact.* **67**, 435.
146. Kan, B., Goldblith, S. A. and Procter, B. E. (1958). *Fd Res.* **23**, 41.
147. Hachisuka, Y., Wynne, E. S., Galyen, L. I. and Jenkins, L. L. (1959). *Bact. Proc.* p. 40.
148. Olsen, A. N. and Scott, W. J. (1946). *Nature, Lond.* **157**, 337.
149. Lund, A. J. (1956). *Rep. Hormel Inst. Univ. Minn.* p. 88.
150. Roberts, T. A. and Hobbs, G. (1968). *J. appl. Bact.* **31**, 75.
151. Wynne, E. S. and Foster, J. W. (1948). *J. Bact.* **55**, 331.
152. Riemann, H. and Ordal, Z. J. (1961). *Science, N.Y.* **133**, 1703.
153. Wynne, E. S., Collier, R. E. and Mehl, D.A. (1952). *J. Bact.* **64**, 883.
154. Rode, L. J. and Foster, J. W. (1962). *Nature, Lond.* **194**, 1300.
155. Rode, L. J. and Foster, J. W. (1962). *Arch. Mikrobiol.* **43**, 201.
156. Fleming, H. P. and Ordal, Z. J. (1964). *J. Bact.* **88**, 1529.
157. Gibbs, P. A. (1967). *J. gen. Microbiol.* **46**, 285.
158. Princewill, T. J. T. (1965). *J. comp. Pathol.* **75**, 343.
159. Splittstoesser, D. F. and Steinkraus, K. H. (1962). *J. Bact.* **84**, 278.
160. St Julian, G., Pridham, T. G. and Hall, H. H. (1967). *Can. J. Microbiol.* **13**, 279.
161. Black, S. H. (1964). *Bact. Proc.* p. 36.
162. Levinson, H. S. and Sevag, M. G. (1954). *J. Bact.* **67**, 615.
163. Levinson, H. S. and Hyatt, M. T. (1963). *Ann. N.Y. Acad. Sci.* **102**, 773.
164. Foerster, H. F. and Foster, J. W. (1966). *J. Bact.* **91**, 1333.
165. Rode, L. J. and Foster, J. W. (1966). *J. Bact.* **91**, 1582.
166. Rowley, D. B. and Levinson, H. S. (1967). *J. Bact.* **93**, 1017.
167. Gould, G. W., Hitchins, A. D. and King, W. L. (1966). *J. gen. Microbiol.* **44**, 293.
168. Rode, L. J. and Foster, J. W. (1960). *Arch. Mikrobiol.* **36**, 67.
169. Rode, L. J. and Foster, J. W. (1960). *Nature, Lond.* **188**, 1132.
170. Rode, L. J. and Foster, J. W. (1961). *J. Bact.* **81**, 768.

171. Gould, G. W. and Ordal, Z. J. (1967). *J. gen. Microbiol.* **50,** 77.
172. Voss, J. G. (1967). *J. gen. Microbiol.* **48,** 391.
173. Rode, L. J. and Foster, J. W. (1965). *Proc. natn. Acad. Sci., U.S.A.* **53,** 31.
174. Gerhardt, P. and Black, S. H. (1961). *J. Bact.* **82,** 750.
175. Riemann, H. (1961). *In* "Spores II." (H. O. Halvorson, ed.), p. 24. Burgess Publishing Co., Minneapolis, Minn., U.S.A.
176. Kuritza-Olejnikow, H. and Grecz, N. (1965). *Bact. Proc.* p. 7.
177. Iandolo, J. J. and Ordal, Z. J. (1964). *J. Bact.* **87,** 235.
178. Jaye, M. and Ordal, Z. J. (1964). *J. Bact.* **89,** 1617.
179. Keynan, A. and Halvorson, H. O. (1962). *J. Bact.* **83,** 100.
180. Keynan, A., Murrell, W. G. and Halvorson, H. O. (1962). *J. Bact.* **83,** 395.
181. Jaye, M. and Ordal, Z. J. (1966). *Can. J. Microbiol.* **12,** 199.
182. Busta, F. F. and Ordal, Z. J. (1964). *Appl. Microbiol.* **12,** 106.
183. Edwards, J. L., Busta, F. F. and Speck, M. L. (1965). *Appl. Microbiol.* **13,** 851.
184. Duncan, C. L. and Foster, E. M. (1968). *Appl. Microbiol.* **16,** 412.
185. Socolofsky, M. D. and Wyss, O. (1961). *J. Bact.* **81,** 946.
186. Strange, R. E. and Dark, F. A. (1956). *Biochem. J.* **62,** 459.
187. Salton, M. R. J. and Ghuysen, J. M. (1959). *Biochim. biophys. Acta,* **36,** 552.
188. Douglas, H. W. and Parker, F. (1958). *Biochem. J.* **68,** 94.
189. Salton, M. R. J. (1958). *J. gen. Microbiol.* **18,** 481.
190. Douglas, H. W. (1957). *J. appl. Bact.* **20,** 390.
191. Gould, G. W., Georgala, D. L. and Hitchins, A. D. (1963). *Nature, Lond.* **200,** 385.
192. Vinter, V. (1962). *Folia Microbiol. Praha,* **7,** 115.
193. Kadota, H., Iijima, K. and Uchida, A. (1965). *Agric. Biol. Chem.* **29,** 870.
194. Falaschi, A., Spudich, J. and Kornberg, A. (1965). *In* "Spores III" (L. L. Campbell and H. O. Halvorson, eds), p. 88, Am. Soc. Microbiol., Ann Arbor, Michigan, U.S.A.
195. Tanooka, H. and Hutchinson, F. (1965). *Radiat. Res.* **24,** 43.
196. Powell, J. F. and Strange, R. E. (1956). *Biochem. J.* **63,** 661.
197. Suzuki, Y. and Rode, L. J. (1967). *Bact. Proc.* p. 22.
198. Work, E. (1959). *Annls Inst. Pasteur, Paris,* **96,** 468.
199. Gould, G. W. (1962). *J. appl. Bact.* **25,** 35.
200. Gould, G. W. and Hitchins, A. D. (1965). *In* "Spores III" (L. L. Campbell and H. O. Halvorson, eds), p. 213. Am. Soc. Microbiol., Ann Arbor, Michigan, U.S.A.
201. Falcone, G. (1961). *Ann. Microbiol. Enzimol.* **11,** 133.
202. Vary, J. C. and Halvorson, H. O. (1968). *J. Bact.* **95,** 1327.
203. Sierra, G. (1964). *Can. J. Microbiol.* **10,** 929.
204. Gould, G. W. and King, W. L. (1966). *Nature, Lond.* **211,** 1431.
205. Sierra, G. (1967). *Can. J. Microbiol.* **13,** 489.
206. Levinson, H. S. and Hyatt, M. T. (1955). *J. Bact.* **70,** 368.
207. Sierra, G. (1967). *Can. J. Microbiol.* **13,** 673.
208. Sierra, G. (1963). *Can. J. Microbiol.* **9,** 643.
209. Roberts, T. L. (1965). *Thesis,* Clark University.
210. Murrell, W. G. (1955). *The Bacterial Endospore,* Mimeo, The University, Sidney.
211. Levinson, H. S., Sloan, J. D. and Hyatt, M. T. (1958). *J. Bact.* **75,** 291.
212. Tono, H. and Kornberg, A. (1967). *J. biol. Chem.* **242,** 2375.
213. Tono, H. and Kornberg, A. (1967). *J. Bact.* **93,** 1819.

214. Fitz-James, P. C. (1955). *Can. J. Microbiol.* **1**, 502.
215. Fitz-James, P. C. (1955). *Can. J. Microbiol.* **1**, 525.
216. Kim, J. and Naylor, H. B. (1966). *Bact. Proc.* p. 31.
217. Rode, L. J. and Foster, J. W. (1960). *Proc. natn. Acad. Sci., U.S.A.* **46**, 118.
218. Knaysi, G. and Curran, H. R. (1961). *J. Bact.* **82**, 691.
219. Sacks, L. E., Percell, P. B., Thomas, R. S. and Bailey, G. S. (1964). *J. Bact.* **87**, 952.
220. Farkas, J. and Kiss, I. (1965). *Acta microbiol. hung.* **12**, 15.
221. Chislett, M. E. and Kushner, D. J. (1961). *J. gen. Microbiol.* **25**, 151.
222. Levinson, H. S. (1961). *In* "Spores II" (H. O. Halvorson, ed.), p. 14. Burgess Publishing Co., Minneapolis, Minn., U.S.A.
223. Mol, J. H. H. (1957). *J. appl. Bact.* **20**, 454.
224. Knaysi, G. (1965). *Appl. Microbiol.* **13**, 500.
225. Lundgren, L. (1966). *Physiologia. Pl.* **19**, 403.
226. Williams, D. J., Clegg, L. F. L. and Wolf, J. (1957). *J. appl. Bact.* **20**, 167.
227. Halvorson, H. O., Wolf, J. and Srinivasan, V. R. (1961). *In* "Low Temperature Microbiology Symposium", p. 27. Campbell Soup Company, Camden, New Jersey, U.S.A.
228. Larkin, J. M. and Stokes, J. L. (1966). *J. Bact.* **91**, 1667.
229. Wynne, E. S., Galyen, L. I. and Mehl, D. A. (1955). *Bact. Proc.* p. 39.
230. Mundt, J. O., Mayhew, C. J. and Stewart, G. (1954). *Fd Technol.* **8**, 435.
231. Schmidt, C. F., Lechowich, R. V. and Folinazzo, J. F. (1961). *J. Fd Sci.* **26**, 626.
232. Hagen, C. A., Hawrylewicz, E. J. and Ehrlich, R. (1967). *Appl. Microbiol.* **15**, 285.
233. Bullock, K. and Tallentire, A. (1952). *J. Pharmac.* **4**, 917.
234. Baird-Parker, A. C. and Freame, B. (1967). *J. appl. Bact.* **30**, 420.
235. Ohye, D. F., Christian, J. H. B. and Scott, W. J. (1966). *In* "Botulism" (M. Ingram and T. A. Roberts, eds), p. 136. Chapman and Hall, London.
236. Murrell, W. G. (1961). *Symp. Soc. gen. Microbiol.* **11**, 100.
237. Slepecky, R. A. (1963). *Biochem. biophys. Res. Commun.* **12**, 369.
238. Silver, S. and Wendt, L. (1967). *J. Bact.* **93**, 560.
239. Balassa, G. and Contesse, G. (1965). *Annls Inst. Pasteur, Paris* **109**, 683.
240. Barer, R. (1959). *In* "Analytical Cytology" (R. C. Mellor, ed.), 2nd ed., p. 169. McGraw-Hill Book Co. Inc., London.
241. Hibbert, H. R. (1967). *Thesis*, University of Manchester.
242. Wax, R. and Freese, E. (1968). *J. Bact.* **95**, 433.
243. Sacks, L. E. (1967). *J. Bact.* **94**, 1789.
244. Shoemith, J. G. and Holland, K. T. (1968). *Biochem. J.* **106**, 38P.
245. Roth, N. G. and Lively, D. H. (1956). *J. Bact.* **17**, 162.
246. Lewis, J. C. (1967). *Bact. Proc.* p. 23.
247. Woese, C. R., Vary, J. C. and Halvorson, H. O. (1968), *Proc. natn. Acad. Sci., U.S.A.* **59**, 869.
248. Jones, A. and Gould, G. W. (1969). *J. gen. Microbiol.* **53**, 383.
249. Gould, G. W., Jones, A. and Wrighton, C. (1968). *J. appl. Bact.* **31**, 357.
250. Warren, S. C. (1968). *J. gen. Microbiol.* **54**, (In press.)
251. Warren, S. C. and Gould, G. W. (1968). *Biochim. biophys. Acta,* **170**, 341.
252. Gould, G. W. (1969). *In* "Spores IV" (L. L. Campbell, ed.), *Am. Soc. Microbiol., Ann Arbor, Mich.* (in press).
253. Vary, J. C. (1969). *In* "Spores IV" (L. L. Campbell, ed) *Am. Soc. Microbiol., Ann Arbor, Mich.* (In press.)

CHAPTER 12

Outgrowth and the Synthesis of Macromolecules

R. E. STRANGE AND J. R. HUNTER

Microbiological Research Establishment, Porton, Salisbury,
Wiltshire, England

I. Introduction

With the approval of many interested microbiologists, the events leading to the formation of a fully grown vegetative cell from a dormant bacterial spore are divided into the following series of stages: germination, pre-emergent swelling, emergence, and growth of the new cell (Fig. 1). That part of the growth cycle following germination and terminating in the formation of the first mature vegetative cell is referred to as outgrowth. A brief but adequate discussion of the evolution of this terminology is given by Campbell.[1] In earlier studies "germination" referred to the entire transition from a resting spore to a fully grown vegetative cell but is now generally accepted as referring only to the change from a heat-resistant, refractile spore to a heat-labile non-refractive cell. One result of dividing the complete process into separate stages is that criteria defining the initial stages of germination and pre-emergent swelling take no account of the ability of the spore to complete the subsequent stages. Thus, changes characteristic of germination, as it is now defined, can be induced by various means, some of which are lethal to the cell (see Chapter 16). The terminology used for the stages in spore outgrowth is self-explanatory: pre-emergent swelling invokes and ends with rupture of the spore cases and then, provided that conditions are favourable, the embryo vegetative cell emerges

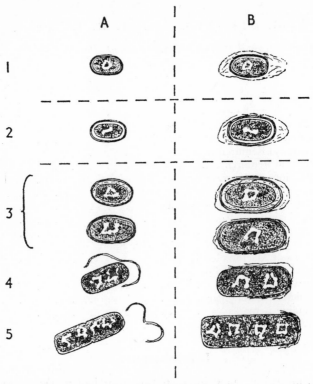

FIG. 1. Diagrammatic representation of the development from spore to dividing vegetative cell. (A) Small-celled, (B) Large-celled species. Stage 1, resting; 2, germination; 3, swelling; 4, emergence and elongation; 5, division.

and develops into a fully grown vegetative cell. A spore suspension does not necessarily germinate synchronously so that there may be no sharp separation of the stages during outgrowth but in individual spores there is probably no overlap in time of the different processes.[2] While the obvious criteria for defining these stages and determining their duration are morphological changes, measurable physiological changes are also evident; for example, during germination and outgrowth of aerobic spores of *Bacillus megaterium*, characteristic patterns of oxygen uptake are associated with pre-emergent swelling, spore rupture and elongation of the emergent vegetative cell.[3] It has also been established that the chemical and physical environmental requirements necessary for each stage of outgrowth may differ. Consideration of the physical and biochemical differences between spores and vegetative cells makes it evident that outgrowth, above all else, is a process of synthesis of new macromolecules and a highly ordered sequence of biosynthetic events occurs resulting in the differentiation of structures characteristic of the vegetative cell.[4, 125]

In this chapter, we are concerned with the sequence of physical, physiological and biochemical events associated with the change from a germinated spore into a mature vegetative cell.

II. Technical Considerations

Selection of an organism for studying outgrowth involves a number of considerations. Many aerobic Bacilli have simple growth requirements and lend themselves more easily to studies of certain aspects of outgrowth, for example, those concerned with the essential nutrients. Anaerobic Clostridia are more difficult to deal with and require various undefined complex media for growth; broth and digests are generally unsuitable media for studies concerned with nutritional requirements. Some organisms form clean spores easily whereas others produce spores with attached vegetative cell remnants which are difficult to remove. This residual vegetative cell material will persist during germination and outgrowth, and interfere, in particular, with chemical analyses. Various methods have been used to clean spores including prolonged storage of washed suspensions followed by repeated further washing, treatment with lysozyme[5] and/or other enzymes, centrifugation in sucrose density gradients,[6] biphasic systems[7] and foam flotation[8] (see p. 216).

The progress of outgrowth is monitored by means of light or electron microscopy and measured by determination of the extinction values of samples at a selected wavelength in a spectrophotometer and/or by determination of dry weight.

There may be a variation in the time taken by spores to germinate because populations are not homogeneous: some spores will therefore begin outgrowth before others have completed germination. A fair degree of sychronous development during outgrowth can often be achieved by completely germinating heat-activated spores in a simple medium in which further development cannot occur (because of lack of essential nutrients) and then transferring the germinated spores into the appropriate growth medium; synchronous development will then usually occur for two or three division cycles.

A method has been reported for arresting suspensions of spores in any stage of outgrowth without the use of inhibitors or the need to determine the exact nutritional requirements of the organisms.[9] Heated (65°C: 1 h) spores of *B. cereus* were germinated at 30°C in buffer containing L-alanine + adenosine, then transferred to Trypticase Soy broth in which they were incubated at 30°C with aeration. At various stages of outgrowth, samples of the suspension were chilled and the organisms separated and washed twice in buffer by centrifugation. The washed organisms, resuspended in buffer containing 3 mM glucose, did not change morphologically during 4 h at 30°C, although they metabolized glucose, remained viable and immediately continued the process of outgrowth if resuspended in Trypticase Soy broth. This technique would appear to be extremely useful in cases where comparative studies during outgrowth involving many samples are being made.

III. The Resting Spore and the Vegetative Cell

The physical and chemical differences between the mother spore and the mature vegetative cell provide a basis for establishing the biosynthetic events and structural changes that occur during outgrowth.

In Chapter 2, it is seen that the spore has an outer envelope structure that may include an exosporium surrounding a coat of two or more layers, a cortex and a core consisting of a germ cell membrane enclosing the germ cell itself; in contrast, the mature vegetative cell may have an outer capsule (in Bacilli this is usually composed of polyglutamyl peptide), a cell wall of different composition from that of the spore envelope, a cytoplasmic membrane with which is associated a number of enzymes and the cytochromes, and the protoplast. The main events during germination (see Chapter 11) are degradative processes leading to the release of various spore constituents (muropeptides, peptides, amino acids, calcium and dipicolinic acid)[10] and, with the possible exception of amino acids, there is no evidence that these released constituents are utilized by the developing cell during outgrowth.

The spore coat which is shed intact in some species[11] but absorbed[11] or fragmented[12] in others during outgrowth, contains keratin-like disulphide-rich proteins, a phosphomuramyl polymer and other proteins.[13, 14] Apparently bound sulphur in spore coats is not utilized during outgrowth[15] and S-amino acids are apparently absent from soluble germination exudates.[16] The vegetative cell wall of Gram positive spore-bearing bacteria is structurally and chemically different from the spore coat but it is of interest that cell walls contain a "basal" structure[17] of murein with a composition similar to that of spore murein.[18] There is biochemical evidence that some spore muropeptides may provide the wall of the emerging cell[19] but surprisingly, the location of murein in the resting spore is not generally clear. More recent work[22] has shown that the cortex[20, 21] may not be the major site of murein in spores of all species.

Components and structures contributing to the dormancy of resting spores are discarded during germination and outgrowth, and appear to be of no nutrient value to the developing cell. Therefore, spore cores are of primary interest in the context of outgrowth; these have been isolated[23] and their structure and chemical composition is described in Chapter 7.

Qualitative and quantitative differences in the inorganic, organic and macromolecular compositions of spores and vegetative cells are discussed in previous chapters of this book. Compared with vegetative cells, spores are rich in calcium and deficient in phosphorus, sodium and potassium.[24, 25, 26] In fact, an exogenous source of phosphorus is required for outgrowth[27] and alkali metals are usually present in excess in growth media.

Spores and vegetative cells contain a considerable amount of bound hexosamine and muramic acid which are released on acid hydrolysis. These acetylated amino sugars together with D-alanine, L-alanine, D-glutamic acid and αε-diaminopimelic acid (DAP) or lysine are present in the vegetative cell wall and spore muropeptides. Although the monomers in the spore polymer and cell polymer are usually the same, in *Bacillus sphaericus* the vegetative cell murein contained lysine but no DAP, whereas the spore murein contained DAP.[28] The presence of lysine-containing murein in these spores was not rigorously investigated but this is of some interest since its absence would indicate that the cell wall of the outgrowing cell is synthesized completely *de novo*.

Nucleic acid patterns in spores and vegetative cells differ quantitatively,[29, 32, 33] but DNA, transfer RNA and ribosomal RNA with similar physical properties were found in both types of cell (Fig. 2). Of importance in the context of outgrowth is that messenger RNA[4, 31] and polysomes[30] appear to be absent from spores.

A major difference in the proteins of spores and vegetative cells has already been mentioned i.e. the much higher proportion of cystine-rich proteins in spores. More important for outgrowth are qualitative and quantitative differences in the enzyme contents of spores and cells (see Chapter 8 and ref. 34). The initial events during outgrowth will depend on enzymes pre-existent in the spore and later events on the development of protein-synthesizing capacity to produce required enzymes that are absent or present in only limiting amounts.

Compared with vegetative cells, activated aerobic spores have a low cytochrome content[35] and their respiration is cyanide resistant. Apparently, activated aerobic spores use mainly a soluble flavoprotein system for terminal oxidation rather than a particulate cytochrome system.[34] Functional electron transport particles are normally attached to the cytoplasmic membrane of the vegetative cell[36] and the absence of these in the inner

T

membrane of the spore indicates another deficiency to be repaired during outgrowth.

It is evident that during spore outgrowth, changes in structures, the pattern of RNA and classes of ribosomes, the relative concentrations of

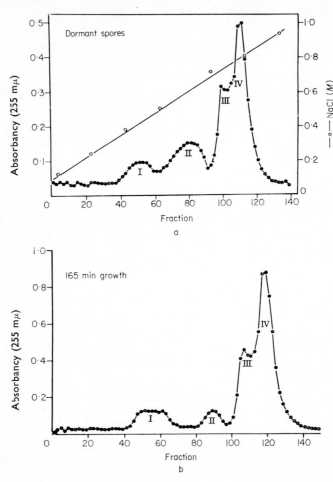

FIG. 2. Elution pattern (from a methylated albumin–kieselguhr column) of nucleic acids extracted from (a) resting spores and (b) cells after growth of the spores for 165 min in defined medium. Peaks I, III and IV contained RNA and peak II, DNA. Figure by courtesy of the American Society for Microbiology.[29]

various enzymes and overall metabolic activity, occur. Some of these changes are unique to spore outgrowth, but probably not all. During the lag phase of vegetative growth, changes in the pattern of RNA and classes of ribosomes, concentrations of enzymes and overall metabolic activity also occur.

IV. Environmental Requirements for Spore Outgrowth

A. CHEMICAL REQUIREMENTS

The spore germ cell does not contain reserve substances and is probably completely dependent on exogenous nutrients for growth and survival until the next sporulation cycle.

Re-sporulation may occur without completion of the normal vegetative growth and division cycle if nutrients are limited. The phenomenon of "microcycle sporogenesis"[126] that has been observed with *B. cereus*[126] and *B. megaterium*,[127] involves the direct transition of the primary cell after germination into a new sporangium without intermediate division stages (Chapter 3).

At some stage during outgrowth, the physical and chemical requirements become identical to those for normal vegetative growth and these will vary with the particular organism. Some bacteria grow in simple defined media, others require one or more complex substrates. Evidence suggests that organisms are more fastidious during outgrowth compared with vegetative growth but whether this is generally true is not easily established because requirements for vegetative growth may be complex and not defined. Certain strains of *B. subtilis* and *B. megaterium* grow when supplied with sources of carbon and energy, nitrogen, phosphorus, sulphur and inorganic salts. In such cases, the need for all these nutrients during outgrowth has been established. In the case of a vegetative spore bearing organism with more sophisticated but defined growth requirements, extra nutrients are usually required for normal outgrowth of the spores.

1. Sulphur

Spores of *B. megaterium* germinated but did not outgrow in phosphate buffer (pH 6·8–7·0) containing glucose and L-alanine.[15] Outgrowth proceeded when manganese sulphate was added to the medium and substitution with other manganese salts showed that sulphate ions were responsible for the stimulation. Sulphate ions could be replaced by other inorganic or organic sources of sulphur. Evidently, sulphur present in the cystine-rich spore coats was not utilizable.[15] The block in outgrowth caused by withholding a sulphur source from *B. megaterium* spores occurred about the time of emergence; in contrast, however, outgrowth of *B. subtilis* spores occurred in the absence of an added sulphur source presumably because an endogenous source of sulphur was available.[37]

2. Phosphorus

Spores of B. megaterium germinated and outgrew in phosphate buffer (pH 7·5–8·0) containing L-alanine, glucose, potassium sulphate and ammonium acetate but germinated without outgrowth in a similar medium containing tris buffer (pH 7·5–8·0) in place of phosphate buffer.[27] Addition of phosphate to the tris buffer medium stimulated outgrowth. In sodium cacodylate buffer, normal outgrowth occurred in the presence of 0·5–1·0 mM phosphate and many inorganic and organic phosphates would replace the orthophosphate.[27] The stimulating effect of organic phosphates suggested the involvement of phosphatases but the production of orthophosphate from many of the phosphate compounds supporting outgrowth could not be demonstrated. Nevertheless, intact germinated spores produced orthophosphate from pyrophosphate and to a lesser extent from phenylphosphate and polyphosphate, supporting the view that phosphatases are active during germination and outgrowth.

3. Nitrogen

It has been shown that a nitrogen source is required for outgrowth and that nitrogen compounds able to support germination do not necessarily support outgrowth.[38] Of 48 nitrogen-containing compounds examined, only ammonium sulphate, potassium nitrate, D-alanine, L-alanine, L-arginine, L-aspartate, L-glutamate, glutamine, L-proline, adenine, adenosine and guanosine supported outgrowth of B. megaterium spores in phosphate buffer (pH 7·0) containing glucose and potassium sulphate. The nitrogen content of the intact spores was 11% but apparently none of this was utilizable during outgrowth.

4. Carbon

It has been unequivocally shown that an exogenous carbon energy source is required during outgrowth. Experiments in which combinations of glucose and L-alanine in the medium were varied showed that glucose supported outgrowth of B. megaterium whereas L-alanine alone did not, although spores germinated in L-alanine were viable and capable of outgrowth if glucose was added.[39] Of 77 carbon compounds tested, 21 supported outgrowth but not germination and 5 supported germination but not outgrowth.[40] During outgrowth of an indole-requiring mutant of B. subtilis (Marburg strain), glucose and 9 amino acids were required; citrate or glycerol was not equivalent to glucose even in the presence of 18 amino acids.[41]

5. Amino acids

Outgrowth of spores of certain organisms does not occur unless amino acids are provided. Development of young vegetative cells from spores of *B. subtilis* and *B. mycoides* occurred in the absence of added amino acids whereas outgrowing *B. natto* spores needed isoleucine and *B. cereus* spores, valine.[42] The minimum amino acid requirements for outgrowth of *B. stearothermophilus* spores at 55°C were isoleucine, leucine, valine, methionine, histidine, arginine (in addition, thiamin, biotin and nicotinic acid were required), whereas these amino acids except for leucine supported vegetative growth (in the presence of thiamin and biotin).[43] *B. coagulans* spores required glutamic acid, histidine, isoleucine and valine for outgrowth in addition to the nutrients necessary for vegetative growth.[43] *B. cereus* T organisms required isoleucine, valine and methionine for vegetative growth but leucine was required in addition to these for outgrowth.[43] *B. subtilis* (Marburg strain) spores germinated in a defined medium containing L-alanine or L-asparagine but outgrowth occurred only when L-glutamate was present in addition to either of these components.[44, 48] Incorporation of ^{32}P into the acid-insoluble fraction of the spore occurred in the absence of glutamine and asparagine but the synthesized material played no part in outgrowth.[37] An indole-requiring mutant of *B. subtilis* (Marburg strain) required glucose, L-alanine and tryptophan for germination but normal outgrowth did not occur unless at least 9 amino acids were provided.[41] In the presence of groups of 3 or 4 amino acids, outgrowth was extremely slow except with arginine, glutamate and aspartate. A mixture of aspartate, glutamate, arginine, asparagine and glutamine was more effective. Spores of a mutant of *B. subtilis* requiring phenylalanine for vegetative growth, germinated but did not develop in the absence of this amino acid.[46]

Some of these examples of amino acid requirements during outgrowth support the view that nutrient demands are greater during this process than during normal vegetative growth. It should be emphasized, however, that the effect obtained with one or more amino acids is frequently an acceleration of a process which will still occur, albeit at a slower rate, in the absence of the compound(s). Nevertheless, the fact that in general the developing cell has more exacting growth requirements indicates that, compared with the homologous vegetative cell, its capacity to synthesize certain substances is low.

6. Metal ions

Manganese ions stimulated outgrowth of *B. megaterium* spores.[15] Cobalt and nickel ions inhibited outgrowth but their effects were evident at different stages. Cobalt ions added to suspensions of intact spores allowed germination but delayed outgrowth; however, when cobalt was added at the beginning of the emergence stage, no effect was evident. Nickel ions delayed outgrowth including the emergence stage but had no effect on elongated cells about to divide.[15]

7. Miscellaneous substances

Attempts have been made to devise defined media supporting rapid synchronous germination and outgrowth. In a medium containing glucose, potassium phosphate, L-alanine, glutamic acid, asparagine, lysine, methionine and magnesium chloride, asynchronous emergence occurred with *B. subtilis* spores. In a medium containing 8 amino acids, glucose, magnesium, potassium phosphate, sodium sulphate, ammonium nitrate, calcium, iron, manganese, outgrowth proceeded similarly to that in the "best" medium containing Casamino acids and salts.[47]

Outgrowth of spores of *B. subtilis* mutants requiring uracil, phenylalanine or nicotinic acid would not occur in the absence of these substances.[46]

Thioglycollate was necessary for the outgrowth of *Clostridium bifermentans*[49] and *Cl. botulinum* spores[50] although this substance inhibited germination.

B. PHYSICAL REQUIREMENTS

1. Temperature

The optimum temperature for outgrowth of aerobic spores is usually different from that for germination. At 8°C, 85% of *B. cereus* spores germinated in yeast extract phosphate broth but underwent no further development.[51] A study of the effect of temperature on the rate of germination and outgrowth (increase in volume) of *B. cereus* spores showed that the maximum, optimum and minimum temperatures for germination were 59°, 30° and −1°C, respectively, and for swelling they were 44°, 35°, and 10°C respectively.[52] A lower minimum temperature for germination compared with outgrowth was also observed with *B. subtilis* spores.[53] However, whereas *Cl. botulinum* spores germinated at 75°C, outgrowth and vegetative growth were arrested at 48°C.[54] The effect of temperature on germination and outgrowth is discussed in detail elsewhere;[34, 55] there is no evidence to show that the temperature range for outgrowth differs from that for normal vegetative growth.

2. pH

The optimum pH for outgrowth may differ from that for germination; for example, optimal outgrowth and germination of *B. megaterium* spores occurred at pH values of 7·5–8·0 and 6–7 respectively, and in a medium nutritionally adequate for outgrowth at pH values >6·4, no outgrowth occurred at pH 6·0.[15]

3. Solute concentration

There is not much information concerning the effect of solute concentration on spore outgrowth. A greater resistance to saturated sodium chloride (2 min exposure) was observed during outgrowth of spores of *Cl. botulinum* and a putrefactive anaerobe than was the case with the homologous vegetative bacteria.[56] A similar effect was observed with *B. subtilis* where hypertonic sodium chloride allowed some elongation of outgrowing cells to occur but prevented their division.[57]

C. INHIBITORY SUBSTANCES

In general, substances that inhibit normal vegetative growth also inhibit outgrowth but do not necessarily affect germination. However, certain substances at concentrations just insufficient to affect germination, inhibit various stages of outgrowth and vegetative growth. Examples of such inhibitory substances that have either an effect on any form of bacterial

TABLE I

Substances inhibiting spore outgrowth

Inhibitor	Organism	Stage affected	Ref.
L-serine	*B. subtilis*	outgrowth	61
Co²⁺ (0·2 mM)	*B. megaterium*	swelling	15
Ni²⁺ (0·2 mM)	*B. megaterium*	swelling, emergence, elongation	15
Cu²⁺, Cr²⁺, Hg²⁺	*B. cereus*	outgrowth	62
Arsenate, arsenite (10 mM)	*B. cereus*	outgrowth	62
Monoiodoacetate (10 mM)	*B. cereus*	outgrowth	62
Borate, cyanide	*B. cereus*	outgrowth (partial effect)	62
Chloramphenicol Puromycin Actinomycin D	*B. cereus*	outgrowth	63
Nisin (2– > 100 RU/ml) Subtilin (5–20 μg/ml) Diethylpyrocarbonate (0·125–0·25%) Nitrite (0·03%)	Bacilli	initiation of outgrowth	57
Benzoic acid (0·01–0·03%)	Bacilli	elongation	57
Tylosin (0·5–1·0 μg/ml) Sorbic acid (0.015–0·05%) Metabisulphite (0·005–0·02% SO₂)	Bacilli	cell division	57
Phenethyl alcohol	*B. megaterium*	outgrowth	58
Unsaturated fatty acids	Bacilli	outgrowth	59
Chloroquin	*B. subtilis*	outgrowth	60

growth or a specific effect at a certain stage of growth, are given in Table I. In all cases, the extent of inhibition depends on the concentration of the substance, composition of the suspending medium, pH value and other factors, details of which are given in the cited reports. In some cases, the mechanism of inhibition is at least partially clear, for example, heavy metal respiratory poisons and inhibitors of protein or nucleic acid synthesis. Of particular interest are those substances which preferentially affect certain stages of outgrowth and the mechanisms involved in these cases are not clear.

V. Mass and Volume Changes

The second stage in the evolution from spore to vegetative cell is characterized by further swelling of the germinated spore. The spore coats have become more permeable to water and nutrients[64] and although there is certainly a considerable increase in the water content,[65, 66] the swelling is not due solely to the imbibition of water, but also to uptake of nutrients such as glucose, with resulting synthesis of new materials.[39, 67, 68]

The amount of pre-emergent swelling varies with the type of spore.[11, 68] Spores of the small-celled aerobic species, exemplified by *B. subtilis*, do not expand as much in volume as the large-celled species, e.g. *B. megaterium*. In the first case the swelling is limited to about 100% increase, but in the second case three times or more increase in volume has been found to occur. Associated with this difference in the degree of expansion is the parallel difference in the disposal of the spore coats. The small-celled species shed a firm, possibly less elastic coat which remains for some time, while the coats of the large-celled species appear to be much more flimsy and often appear to be absorbed or dissolved[11, 22, 69] (Fig. 1). Few studies have been made on spores of anaerobes, but here again there is a difference. Propst and Möse[70] attach great importance to the swelling of the cortex and its subsequent liquefaction in *Cl. butyricum*, while on the other hand the spore of *Cl. botulinum* appears to undergo very little secondary swelling.[50]

An increase in dry weight takes place concurrently with the expansion in volume, and in suitable conditions this can occur very rapidly,[71] as early as ten minutes after resting spores have been placed in the appropriate medium. The conditions and requirements for outgrowth have been discussed in Section III above, but it is worth re-iterating that the main demarcation between germinated spore and vegetative cell occurs at the beginning of the phase of swelling, and that if conditions are not satisfactory, the cell can remain in a state of suspended animation for a considerable time[9, 37, 46, 64, 65, 66, 72] even up to a few days, and will quickly resume development when suitable nutrients are again made available.

Further development of the swollen organism involves the emergence of the young bacterium, and this topic will be dealt with in the next section.

VI. Cytological and Antigenic Studies

A. CYTOLOGICAL

Gross features of spore anatomy, such as the presence of spore coats and an exosporium have been known since studies were first begun on spores, but the inadequacy of the conventional light microscope for the demonstration of the finest details of bacteria delayed until comparatively recently any precise knowledge of the intimate structure of spores. The advent of the electron microscope overcame this deficiency, and although later developments in technique have questioned some of the earlier interpretations, there is now fairly general agreement on, and nomenclature for, the architectural details of both spores and vegetative cells. However, there is still dissension as to the fate of some spore structures and the origin of some vegetative cell structures, though the discrepancies reported may be due to differences in processing the organisms for electron microscopy, or perhaps in the behaviour of the various species studied.

1. Cell wall and cell membrane development

The earliest investigations using the electron microscope, on whole spores of *B. cereus* var. *mycoides*[73] appeared to show that the wall of the vegetative cell was developed anew by the germinated spore. However, later studies employing ultrathin sections of *B. cereus*,[22, 74, 75, 76] *B. cereus* var. *anthracis*,[72, 77] *B. megaterium*,[78, 128] *B. coagulans*,[75] *B. subtilis*,[75, 80, 81, 82] *B. stearothermophilus*[75] and *Cl. butyricum*[70, 83] demonstrate that at least the foundation of the wall of the future vegetative cell is present in the resting spore. It seems quite evident in all these species that cell wall is built on, or actually is, the membrane which lies at the base of the cortex and surrounds the spore core or cytoplasm. A plasma or cytoplasmic membrane lying interior to the cortical membrane has also been observed and it is thought that this becomes the cytoplasmic membrane of the vegetative cell.

2. Cortex

The cortex, which in the resting spore is featureless and electron transparent, becomes swollen and spongy in the germinated spore and usually diminishes and disappears as outgrowth proceeds. Various hypotheses have been put forward to account for its function during outgrowth. Ohye and Murrell[20] have shown that the cortex contains at least some material similar in composition to that of the vegetative cell wall, and suggest that this is utilized for assembly into the growing wall of the germ cell as the cortex is broken down. This suggestion appears to have been confirmed in radioactive studies by Vinter.[19] Propst and Möse[70] working with *Cl. butyricum* have conjectured that the swollen and liquefied cortex, which apparently does not diminish during outgrowth in this organism, exerts turgor pressure on the expanded spore coats and exosporium to rupture them and expel the young vegetative cell. The prior observation that emergence in *Cl. pectinovorum* occurs with explosive swiftness[84] adds weight to the evidence that, among the clostridia at least, this may be an active mechanism. In *B. polymyxa* also, Hamilton and Stubbs[22] have found that the cortex does not undergo diminution in volume and appears to retain most of its structure until the vegetative cell has emerged. In their comparison between this organism and *B. cereus* they make the interesting point that the solubilization of the dipicolinic acid, calcium and murein which occurs to a similar extent in both organisms, is in *B. cereus* associated with the disappearance of the cortex.

3. Emergence

As has been mentioned in Section V, p. 456, the coats, and exosporium if present, of spores of large-celled species are apparently dissolved or absorbed when seen under the light microscope,[11, 85] but studies with the electron microscope have shown that fragments of these integuments are still present after the bacillus has emerged, and shed remnants of spores could also be found in the surrounding medium.[22, 72, 73, 74] The young vegetative cells of the small-celled species emerge after the spore coats have ruptured, and the firm coats remain easily visible for up to several generations of bacilli. The emergence can occur at right angles to either axis of the spore giving equatorial or polar modes but twisting of the spore case or "levering off" of a still-attached hemisphere of a spore case can give the impression that the young bacillus is emerging at some other angle (Fig. 3). The mode of emergence was thought originally to have some taxonomic significance, but it has been found to vary not only in different strains of a species, but also to a small extent in a given strain with the particular environmental conditions.[69]

4. Development of flagella

There are very few reports in the literature concerning the development of flagella on the young bacterium of motile strains emerging from the germinated spore, and only two or three definite studies on this topic appear to have been made. Leifson[86] quotes the observation of Klein[87] that, with an organism called *B. leptosporus* in hanging drop preparations, motility did not appear until an hour after germination was complete, at a total time of nearly seven hours, when many four-membered chains of cells could be seen. Leifson illustrated the development of flagella on cells emerging from germinated spores of three species, *B. vulgatus*, *B. cereus* and *B. flavus*, and found that motility can be observed at $1\frac{1}{2}$ hr in the last two species when only two or three short flagella may be present. It appears that in one Clostridium species—*Cl. pectinovorum*—flagella may even be present on emergence, since the liberated cell moves about very actively

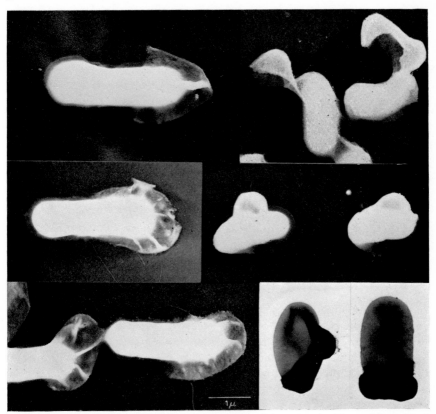

FIG. 3. Electron micrographs showing emergence of the young bacillus and the type of spore coat. Left-hand column: *B. cereus*, polar emergence, with flimsy, collapsed and attached spore coats. Right-hand column: *B. subtilis*, equatorial emergence, with firm empty spore coats. Bottom right-hand pair, negatively stained with phosphotungstic acid, all others gold-palladium shadowed.

from the start.[84] The observation of Klein has been paralleled with *B. subtilis*, when spores were watched forming colonies over several hours on an agar surface.[88] Motility became evident after about 4–5 hr incubation when the colony contained a few hundred organisms and was surrounded by a moat of free liquid. With the exception of one comment to the effect that the germ cell of *B. mycoides* does not bear flagella,[73] the use of the electron microscope for studying outgrowth has not yet added much information about the time of development of flagella. Two explanations can be advanced for this: (a) that observations have not often been made on whole spores and cells, but on sections with other cytological objectives in mind, and flagella would be much less likely to be seen in thin sections, especially if only two or three are present; and (b) that observations have been terminated too soon after the cell has emerged, even though they may have gone as far as the first division of the vegetative cell. However, in the course of preparing illustrations for this chapter, spores of three species were germinated and taken through their development to the first division of the vegetative cells. Since all the clean spore suspensions had been stored for ten years at 4°C, and were heat-shocked for 30 min at 60°C, we achieved almost 100% synchrony. Spores of *B. subtilis*, *B. cereus* and *B. megaterium* (all laboratory strains) were inoculated to give spore population densities of about 0·25 mg/ml ($1-2 \times 10^8$ spores/ml) in pre-warmed nutrient broth + 1% glucose, 2 mM inosine and 2 mM L-alanine and incubated at 37°C in a water-bath with aeration. Progress was followed by phase contrast microscopy and samples were taken at appropriate stages for electron microscopy. *B. cereus* spores germinated extremely rapidly and completely, within 5 min under these conditions, and emergence had occurred from all the spores at 45 to 50 min. A small number of cells were actively motile by 60 min and some are illustrated in Fig. 4. All the cells were motile and had gone through their first division by 90 min. *B. subtilis* was somewhat slower in development but most of the emerged cells were motile at 90 min.

5. Nuclear material

The nuclear material of plant and animal cells undergoes a periodic break-up from a "resting" virtually structureless body surrounded by a nuclear membrane into the constituent chromosomes and these duplicate during mitotic division. In contrast, the nuclear apparatus in bacteria is present at all times in a variable and varying configuration and apparently undergoes an almost continuous duplication while the bacteria are growing, this process being amitotic in the classical sense.[89] That nuclear material consisting of chromatin is present in bacteria is confirmed by staining reactions—Feulgen positive after acid hydrolysis, Giemsa etc. which are exactly the same as those found in classical cytological genetics. It is interesting to note that when seen by both phase contrast and electron microscopy, the material is less dense than the surrounding cytoplasm. There is precise correspondence between the position and shape of these less dense areas seen under phase contrast in the living bacterium and the Feulgen-positive areas in the *same* bacterium when it has been fixed and stained.[89] There is also an exact parallel between the visible increase in the chromatin and the increase in deoxyribonucleic acid in the cell as shown by chemical analysis,[79] demonstrating that this is indeed the genetic material of the cell.

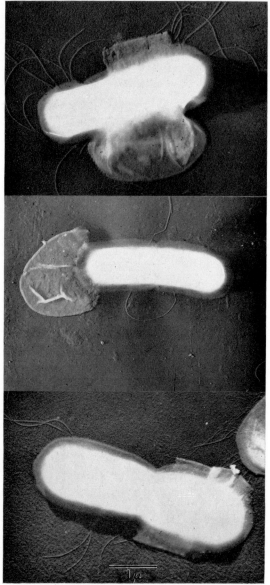

FIG. 4. Electron micrographs of *B. cereus*, 60 min sample (see text), showing flagella, and the commencement of the first division of the vegetative cell. Gold-palladium shadowed.

Since the resting spore is highly refractile and not permeable to stains, the chromatin structure cannot be seen by phase contrast or in stained preparations, except when the spore has been mechanically cracked open or hydrolysed by acid treatment, after which it shows as a ring usually

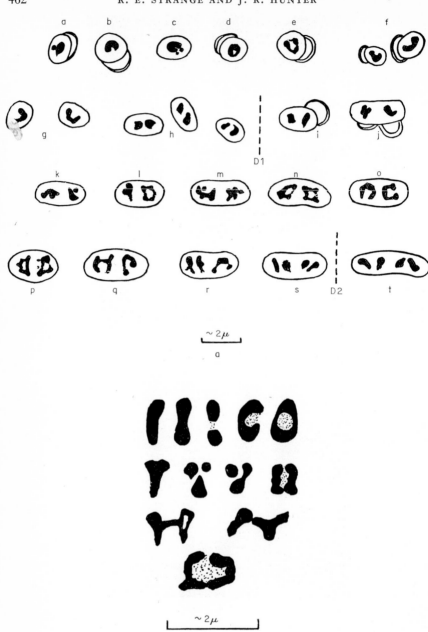

FIG. 5. a. Stages in the multiplication of the chromatin bodies during the development of a germinating spore of *B. megaterium*. b. Diagrams of types of chromatin bodies commonly seen in *Bacillus* cells. Top row: recently divided bodies; remainder: division figures. (Figures by courtesy of the Society for General Microbiology and the author.[90])

with at least three segments (Fig. 5). Following germination, however, the chromatin bodies can be seen and stained and then are no longer found as rings, but instead have become a single solid bar, round, or dumb-bell shaped. As development continues, these rapidly change into open annular rings or three- or four-cornered clusters of interconnected granules[90] which divide into two, and in *B. megaterium* divide again as the vegetative cell elongates to give four nuclear bodies per cell (Fig 5). *B. cereus* appears to contain only two chromatin structures in the fully outgrown vegetative cell.[79]

B. ANTIGENS

Spores and vegetative cells have long been known to possess serologically distinct as well as common antigenic properties, which have been used not only for taxonomic studies to reveal interspecies relationships in both Bacilli and Clostridia[91, 92, 93] but also recently to confirm that the vegetative cell wall is pre-existent as the cortical membrane in the spore.[76] As one might expect, the distinctive antigens are those which occur on the surface, as agglutinogens or precipitinogens. By coupling the respective spore and vegetative cell antibodies with colour-contrasted fluorescent dyes, Walker and Batty[94, 95] have most elegantly demonstrated the emergence of young bacilli from germinated spores of *Cl. sporogenes* and *B. cereus*. Since both "H" and "O" antigens can be obtained from vegetative cells, it might be thought that the use of fluorescent H antibody could elucidate the question of the time of appearance of flagella on emerged cells. It has been suggested, however, that the implication by some workers that "H" antigens in the genus *Bacillus* are necessarily associated with flagella is not supported by the evidence, and that gel diffusion studies have indicated that it is probably incorrect to regard these "H" antigens as flagellar in nature.[91]

VII. Respiration: Oxygen Uptake by Aerobic Spores During Outgrowth

The various cytological changes that can be seen occurring during outgrowth from a germinated spore are a reflection of the profound alterations in metabolic activity taking place during the transition into a vegetative cell. In any single organism the process appears to follow a smooth continuous progression, which means that when using the large populations necessary in respirometric methods for determining metabolic activity, a high degree of synchrony is required so that changes occurring at any particular stage are not obscured by earlier or later differences. With many species this synchrony can be achieved by suitable pretreatment such as ageing by storage[96, 97] and/or heat activation[96, 98, 99] of the resting spore suspension to increase the speed and completeness of germination of the population, by withholding some metabolite essential for outgrowth[9, 27, 100] or by germination at a low pH[100] so as to allow all the spores that do germinate, however different their rate of germination might be, to reach the same physiological state. Using these methods, at least 80% and often over 90% of the organisms are in synchrony, and it is possible to evaluate the effects of various metabolites on their respiration.

Free oxygen apparently is not measurably utilized for respiration by clean resting spores of *B. subtilis* and *B. megaterium*[101] or *B. cereus*[71] though a very small oxygen uptake is detectable when phosphate or glucose is also present,[101, 102] nor is oxygen required for the germination "trigger" reaction.[27, 103] However, as soon as germination has taken place, uptake of oxygen begins, the rate of uptake depending not only upon the conditions and available substrates in the respirometer but also upon the stage of development during outgrowth (Fig. 6). Thus Fitz-James[71] found

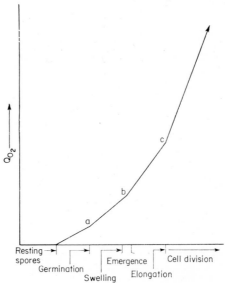

FIG. 6. Idealized representation of the oxygen uptake during germination and outgrowth, after Refs 38, 71 and 97. Development can be halted at points a, b and c (see text).

that the Qo_2 (μl O_2/mg resting spore wt/hr) of *B. cereus* spores in 35 μM adenosine plus 5 mM L-alanine was 50 at 25 min, but supplementation of this mixture with 100 mM glucose, 1% lactalbumin hydrolyzate, 0·1% yeast hydrolyzate and phosphate to 30–40 μg P/ml increased the Qo_2 to 280 by the same time.

Hyatt and Levinson in an extensive series of experiments on the nutritional requirements and respiratory activity of germinating and outgrowing spores of a strain of *B. megaterium* have shown that not only are sulphur,[15] phosphate,[27] nitrogenous compounds[38] and a carbon energy source[40] necessary for development of the germinated spore into a dividing vegetative cell, but that the changes in the rate of consumption of oxygen correspond to the morphological stages in outgrowth. As might be expected, the different sulphur, phosphate and metabolically active nitrogen and carbon sources had different oxygen uptake rates, the last two reflecting the different pathways by which they were utilized.

1. Sulphur

Of the nine inorganic sources of sulphur which were tested only two sulphates—cobalt and nickel—depressed the respiratory rate and delayed outgrowth, but if these cations were chelated with ethylenediaminetetraaceticacid normal outgrowth and typical respiratory changes occurred. Among the remaining seven sources was sulphuric acid which could supply the necessary sulphur. Only three (thiamin, biotin and coenzyme A) of the eight organic sulphur sources did not promote outgrowth, perhaps for reasons of permeability or configuration. Unlike fungal spores, bacterial spores could not utilize selanate in place of sulphate.[15]

Increase of the sulphur concentration above about 3 ppm (as either potassium sulphate or cysteine) had no effect on respiration or on the duration of any morphological phase, but concentrations lower than this adversely affected the changes in respiratory rate. It is of interest that the sulphur (0·3% dry weight of *B. megaterium*) contained in spores themselves apparently cannot be utilized for outgrowth.

2. Phosphorus

The rate of uptake of oxygen varies much more with the phosphate concentration than is the case with sulphur. There is a three-fold increase in oxygen utilization when the phosphate is increased five-hundred-fold (Q_{O_2} 80 with 0·1 mM, 240 with 50 mM). Again, the 2·5% phosphorus in the dry weight of the spore is not available for outgrowth or has been used up in germination. Since the vegetative cells contain about twice as much phosphorus as the spores[24] it is obvious that this element must be supplied for outgrowth and vegetative growth.

3. Nitrogen

Both amino and ammonium nitrogen could be utilized for outgrowth and gave different rates of oxygen uptake and outgrowth. Among the 48 compounds tested, for example, ammonium sulphate and the amino acids L-alanine, L-asparagine, glutamine, L-arginine and L-aspartic and L-glutamic acids promoted the most rapid and complete development, while with L-proline there were lower oxygen consumption rates during

U

elongation and cell division. With adenosine and potassium nitrate the oxygen uptake rate changes associated with elongation and also cell division were delayed until about twice the normal time of development. Where cell division did not take place, as with L-leucine and with L-phenylalanine, the rate of oxygen consumption changes accompanying swelling and elongation occurred after some delay but then the rate of oxygen uptake levelled off. As with phosphate, it was not necessary that the nitrogen source be present from the beginning of incubation, and provided that it was added at the appropriate time there was no delay in the initiation of respiratory changes; if addition was delayed, emergence or elongation and the concomitant increases in oxygen uptake rate started immediately. The only effect of varying the concentration of nitrogen was to limit the amount available for completion of development. Under the experimental conditions used, with 1 mg spores/ml, as little as 2·5 mM allowed swelling, emergence and elongation to occur, but was then exhausted so that division could not take place. With higher concentrations, greater proportions of the young vegetative cells divided and the slope of the oxygen uptake rate curve increased, and about 10 mM was required for at least one cell division of 5×10^8 bacilli.

4. Carbon

As with nitrogen sources, the oxygen consumption rates reflected the progress and extent of outgrowth in the various carbon compounds, of which 77 were tested. Those that supported outgrowth included some pentoses and hexoses, some sugar derivatives, many disaccharides and oligosaccharides, some alcohols and some of the more readily permeable tricarboxylic acid cycle intermediates. Phosphorylated sugars were neither oxidized nor utilized. The rate of oxygen uptake did not depend upon the class of compound, D-glucose, malate, maltose, sucrose, trehalose and gentobiose all giving similar rapid rates. With gluconate, D-fructose, glycerol, mannitol, N-acetyl-D-glucosamine, 2-D-ketogluconate and fumarate there was slight delay. In D-ribose, D-xylose, lactose, melibiose, turanose, raffinose, D-galactose, and L-aspartate, cell division was delayed about 60 min beyond the normal 130 min achieved with D-glucose, while with D-glucosamine, L-glutamate, β-methyl-D-glucoside, cellobiose and succinate it was delayed a further 40 min. Swelling, emergence and elongation occurred but with low and delayed oxygen consumption rates in α-methyl-D-glucoside, melezitose, inositol, glucuronate and citrate but only a few cells started to divide within 6 hours. Outgrowth in 25 mM D-glucose, gluconate or fumarate was inhibited by 1 mM potassium cyanide, but with 10 mM sodium fluoride there was a 10 min delay in oxygen consumption rates at all stages in D-glucose, and a 30 min delay in gluconate and fumarate. In a similar manner to nitrogen, the concentration of D-glucose determined the extent of development during outgrowth and not the rate at which oxygen was taken up. Here the concentration which limited the emerged cells to completion of elongation, and which was exhausted before their division was 5 mM.

VIII. Synthesis of Macromolecules

A. NUCLEIC ACIDS AND PROTEINS

During outgrowth, the embryo cell develops the capacity to synthesize various macromolecules in an ordered manner and at the rate required for differentiation into a mature vegetative cell capable of multiplication. To some extent, an analogous situation exists during the lag phase of vegetative growth during which the bacterial cell equips itself with the necessary synthesizing machinery to grow exponentially at a rate dependent on the physico-chemical conditions of the environment. Additional nutrients may be required for normal outgrowth to occur but this does not undermine the analogy since the lag growth phase of vegetative bacteria with extremely simple growth requirements is often drastically shortened by the addition to the medium of certain amino acids and/or growth factors. It has been established that outgrowth is dependent upon a repair in the protein synthesizing system and an ordered synthesis of proteins.[4]

It is emphasized that the macromolecular composition of the vegetative cell, i.e. the product of outgrowth, is dependent upon the growth rate imposed by the environment. A large proportion of the biomass of all bacteria is protein which may account for more than 70% of the bacterial dry weight, although 50–60% is more usual in *Bacillus* species. The protein content of vegetative bacteria varies slightly with growth rate in a given medium but is significantly decreased when endogenous reserve materials are accumulated. In contrast, the RNA content of vegetative bacteria varies 3–4 fold with growth rate and RNA may account for 25% of the dry weight of rapidly growing bacteria but only 8% of slow growing bacteria. The relationship between growth rate and cellular RNA content arises because, the faster the bacteria grow, the faster they must synthesize protein, and RNA is involved in protein synthesis (see below). The DNA content of vegetative organisms varies less and in *B. subtilis* vegetative cells, for example, DNA accounted for about 2% of the dry weight at all growth rates from 0·1 to 0·6 hr^{-1} in magnesium-limited chemostat cultures.[104] The amount of protein per unit of DNA varies little in vegetative bacteria, whereas the amount of RNA per unit of DNA varies by a factor of 3–4 at different growth rates.[105] To a large extent then the pattern of macromolecular synthesis during outgrowth will depend on the physico-chemical conditions of the environment.

Proteins are synthesized on ribosomal particles in response to a specific mRNA with a base composition complementary to that of the DNA template. Activated amino acids are carried to the ribosome/mRNA complex by sRNA where they are joined in the correct sequence according to the "message" relayed from DNA by mRNA,[106] to form polypeptides. Since spores are apparently devoid of mRNA[4, 31] and polysomes,[30] and certain of the enzymes required are present in limiting amounts,[4] protein synthesis during germination and/or outgrowth is delayed until these deficiencies are repaired. mRNA has been characterized as a polynucleotide with an average molecular weight not less than 3×10^5 and with a high turnover rate.[107] Actinomycin D has been used to elicit information concerning the synthesis and turnover of mRNA; the antibiotic specifically attaches to DNA and prevents DNA-dependent RNA synthesis without interfering with the functioning of mRNA once it is formed, i.e. protein synthesis. Thus, no protein synthesis will occur if actinomycin D is present during germination of bacterial spores because mRNA cannot be synthesized. If the antibiotic is added at a later stage when protein synthesis has begun, synthesis continues for a short period during which pre-formed mRNA rapidly decays into small molecular weight acid-soluble fragments. The formation and decay of mRNA is studied by exposing germinated spores to short pulses of radio tagged RNA precursors (e.g. [14C] uracil)

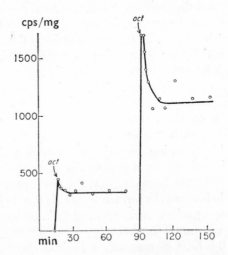

Fig. 7. Degradation of rapidly labelled RNA during outgrowth of spores of *B. subtilis* (ind⁻ mutant of Marburg strain). Spore suspensions (extinction value 0·2–1·0 at 420 mμ) were germinated in nutrient medium at 37°C; 0·04mM [14C] uracil (1 μc/ml) was added at 15 and 90 min, followed by actinomycin (100 μg/ml) 3 min later in each case. Samples (0·5 ml) of the suspensions were removed at intervals, trichloroacetic acid-insoluble fractions prepared and their radioactivity determined. (Figure by courtesy of *Ann. Inst. Pasteur* and the authors.[45])

when mRNA is rapidly and preferentially labelled compared with ribo-somal RNA (Fig. 7); addition of actinomycin D stops mRNA synthesis and the rate of decay of the pre-formed mRNA (Fig. 7) is determined by measuring the decay of labelled RNA or the release of acid-soluble radio-labelled fragments.[45, 108]

In addition to the lack of mRNA in spores, other parts of the protein synthesizing system are apparently defective.[4] For example, in the presence of 19 amino acids and other additives required for protein synthesis, extracts of *B. cereus* strain T spores had virtually no ability to incorporate [^{14}C] phenylalanine into protein.[4] Germinated spore extracts had some incorporating ability but vegetative cell extracts were about 10 times more active than extracts from spores germinated for 20 min. Even in the pre-sence of polyuridylic acid and saturating concentrations of soluble RNA from vegetative *B. cereus*, the incorporating activity of resting spore extracts was only about 2·5% that of vegetative cell extracts and this low activity was apparently not due to an inhibitor in the spore extract. Addition of the soluble fraction of disrupted vegetative cells increased incorporating activity of spore extracts about 5-fold suggesting that the defect in the protein synthesizing machinery arises from a defi-ciency of soluble enzymes involved in protein synthesis. Thus, delay in protein synthesis during outgrowth is probably due to the time required to synthesize new mRNA and certain enzymes. Extracts of *B. cereus* spores at various stages of outgrowth were tested for protein synthesizing ability and it was found that the rate of amino acid incorporation increased markedly 20 min after the initiation of germination.[4]

Synthesis of RNA starts immediately after germination and proceeds during the swelling and elongation stages.[108] Incorporation of [^{14}C] uracil into RNA of a "synchronously" germinated population of *B. cereus* (NCIB 8122) spores was maximal during the period of maximum swelling and then decreased; a second minor increase in the rate of incorporation occurred at the start of emergence of the germ cell. During elongation and cell division, incorporation into newly synthesized RNA was steady. Long (15 min) pulse labelling of RNA with [^{14}C] uracil led to relatively stable labelled RNA (ribosomal) whereas short (3–5 min) pulse labelling followed by the addition of actinomycin D revealed a rapidly labelled RNA fraction which was degraded rapidly and corresponded to mRNA. By the addition of [^{14}C] uracil at different times during outgrowth followed by actinomycin D after exposure for 3 min, it was shown that the rate of degradation of this labelled RNA fraction differed, being somewhat lower during swelling than during elongation. During swelling and the start of elongation, the content of [^{14}C] uracil in the cold acid-soluble nucleotide pool increased in parallel with incorporation of nucleotide into RNA.

The timing of germination and outgrowth in suspensions of *B. subtilis*

(Marburg strain) spores at 35°C is shown in Fig. 8 and under the conditions used, increases in RNA occurred after about 20 min and the RNA content doubled after about 120 min. Synthesis of soluble RNA and ribosomal RNA began at about the same time and their ratio indicated a

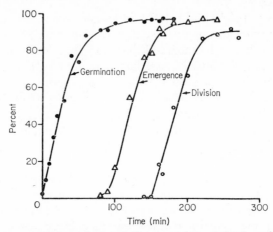

FIG. 8. Development of cleaned *B. subtilis* spores in a defined medium containing buffered salts, glucose and several amino acids at 35°C. Phase-contrast microscopy was used to score for germination, emergence and cell division. Each point represents the average of from two to nine determinations. (Figure by courtesy of the American Society for Microbiology.[29])

relatively increased synthesis of ribosomal RNA during outgrowth. At about the time the total RNA had doubled, DNA synthesis began and a dramatic increase in RNA synthesis was observed at or shortly before the onset of DNA synthesis (Fig. 9). Thus, there is a differential synthesis of nucleic acids during germination and outgrowth.[29] During synchronous outgrowth of strains of *B. subtilis* (prototroph, SB19; ind-meth-auxotroph, SB26) no increase in DNA occurred for 160 min and then synthesis started abruptly at about the time of emergence, after which the DNA content doubled in about 90 min.[109] Examination of the transforming activity of DNA from lysates prepared from the organisms taken at 12 min intervals beginning at 170 min with SB26 as the recipient cell showed clearly that the indole marker was copied before the methionine marker during new DNA synthesis.

Synthesis of ribosomal RNA during outgrowth leads to changes in the relative and absolute amounts of each class of ribosome initially present in the resting spore.[30] A relatively small proportion of the total RNA in *B. subtilis* spores is present as ribosomes, the major fraction being non-ribosomal; the ribosomes are mainly of the 50S and 70S classes. Early during outgrowth, there was a rapid increase in 30S and 50S classes, the relative increase in the 30S class being greater than that in the 50S class.

Ribosomal particles in the 70S class did not start to increase in number until 30 min following germination and 100S particles (polysomes) began to appear after about 60 min.[30]

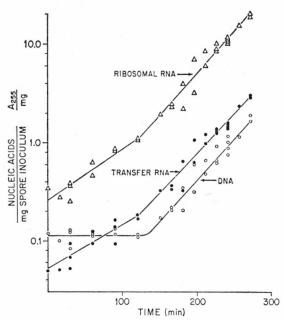

FIG. 9. Total of ribosomal RNA, transfer RNA and DNA per mg *B. subtilis* spore inoculum during outgrowth in a defined medium at 35°C. Samples from the culture were taken at intervals and poured over a frozen buffered salts solution to stop nucleic acid synthesis. Cells were collected and washed by centrifugation at 4°C, and ground in a mortar with acid-washed sand. Nucleic acids were isolated and separated on methylated albumin-kieselguhr columns. The relative amounts of transfer RNA, ribosomal RNA and DNA were determined by adding the extinction values at 255 mμ of each fraction comprising the appropriate peaks. (Figure by courtesy of the American Society for Microbiology.[29])

RNA synthesis starts very soon after germination but there is a discernible lag before protein synthesis begins. During outgrowth of spores of *B. cereus* strain T, RNA synthesis started about 5 min after the initiation of germination whereas protein synthesis started about 8 min after and from then on lagged behind RNA synthesis by about 3 to 5 min.[4] This result suggests that there are other deficiencies besides the lack of messenger RNA in spores.

Several kinetic studies have been made of the synthesis of RNA, protein and DNA during outgrowth and the results obtained are in general agreement. One such investigation was made with resting spores of an indoleless mutant of the Marburg strain of *B. subtilis* that were germinated and allowed to outgrow in a complex medium containing Difco nutrient broth, potassium chloride, magnesium sulphate and manganese chloride.

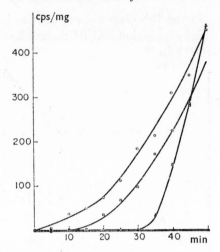

FIG. 10. Synthesis of macromolecules during outgrowth of spores of a mutant of the Marburg strain of *B. subtilis*. Spores were germinated in nutrient medium and syntheses of RNA (○), protein (□) and DNA (△) at 37°C were followed by measuring the incorporation of [14C] uracil, [14C] valine and 32P, respectively, into the appropriate fractions. (Figure by courtesy of *Ann. Inst. Pasteur* and the authors.[45])

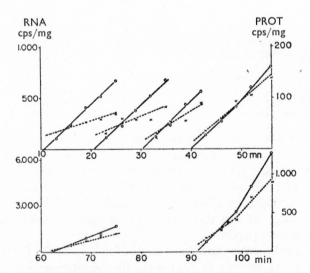

FIG. 11. Synthesis of RNA and protein during outgrowth of spores of a mutant of the Marburg strain of *B. subtilis*. [14C] uracil and [14C] valine were added to suspensions of organisms in nutrient medium at 37°C at various times during outgrowth and incorporation into RNA (○) and protein (×), respectively, was measured. (Figure by courtesy of *Ann. Inst. Pasteur* and the authors.[45])

Emergence took place after about 100 min and vegetative cells appeared after 180–200 min. Following the short initial phase of germination, RNA synthesis began and rapidly reached an exponential rate whereas, although protein synthesis followed similar kinetics, there was a lag of 10 min before it commenced; DNA synthesis started 30–40 min after the onset of germination and then continued at its maximum rate (Fig. 10). When [¹⁴C] uracil and [¹⁴C] valine were added to spore suspensions at various times during outgrowth, incorporation started almost immediately on addition of the labelled precursor and was linear for at least 20 min. The rate of synthesis of RNA was constant up to about 30 min after the onset of germination and then increased progressively (Fig. 11). The rate of synthesis of protein increased gradually during outgrowth so that the ratio of synthesis of RNA and protein diminished progressively to become constant after about 60 min (Fig. 11). If chloramphenicol was added to suspensions before macromolecular synthesis was due to begin, no DNA synthesis occurred; if, however, chloramphenicol was added after the start of DNA synthesis, a residual synthesis of DNA occurred, the extent of which depended on the time of addition of chloramphenicol. This suggests that DNA synthesis depends on the prior synthesis of protein and that this protein is absent from the resting spore. The absence in the spores of stable messenger RNA was confirmed by showing that protein

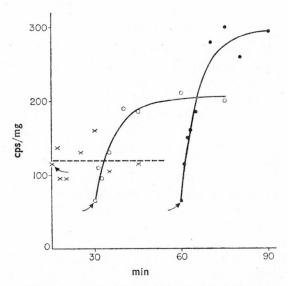

FIG. 12. Effect of actinomycin on the synthesis of protein during outgrowth of spores of a mutant of *B. subtilis* in nutrient medium. [¹⁴C] valine and actinomycin (100 μg/ml) were added together at 15 (×), 30 (○) and 60 (●) min after the onset of germination and incorporation of radioactivity into protein was measured. (Figure by courtesy of *Ann. Inst. Pasteur* and the authors.[45])

synthesis was completely inhibited when actinomycin was added at an early stage during outgrowth; however, when the antibiotic was added at a later stage, protein synthesis continued for several minutes and the extent of the synthesis increased with the time of its addition (Fig. 12). This indicates that there is an enrichment of the spore messenger RNA during outgrowth. It is interesting that, whereas during normal vegetative growth DNA duplication is correlated with the formation of a division system, this investigation showed that DNA duplication occurs during outgrowth without detectable modification of the cytoplasmic membrane.

The effects of actinomycin D, chloramphenicol and puromycin on both $[2-{}^{14}C]$ L-leucine incorporation into proteins and the rates of glucose oxidation during outgrowth of B. cereus strain T have been studied.[63] In uninhibited spore populations there was a rise in respiratory activity during germination and a second rise, beginning 5 min after germination was complete, which continued for 11 min leading to a 50% overall increase in respiratory activity. Chloramphenicol and actinomycin D inhibited both the second rise in respiratory activity and amino acid incorporation into protein suggesting that synthesis of protein components of the respiratory system occurs during this stage of outgrowth. This synthesis of elements of the respiratory system is initiated 10 min after the onset of protein synthesis and lasts only a brief period.

B. ENZYMES

A considerable proportion of the protein synthesized during outgrowth consists of enzymes produced in an ordered manner so that the structural and metabolic requirements of the changing germ cell are catered for at the appropriate time. Studies of enzymes in spores and vegetative cells[34] have shown deficiencies in spores but the necessary genetic information for formation of the missing enzymes is presumed to exist.

The glycolytic (EM) and hexosemonophosphate (HMP) pathways function in vegetative cells of Bacillus species but there is some doubt regarding their relative concentrations in resting and germinating spores.[34] Hexokinase, phosphofructose kinase and aldolase have been reported as absent from resting spores of B. cereus strain T,[110, 111, 112] but these enzymes were found to be present according to the results obtained with more sensitive assay procedures.[113, 114] Measurements of the recovery of $^{14}CO_2$ from $[1-{}^{14}C]$ glucose and $[6-{}^{14}C]$ glucose[115] indicated that during germination, elongation and division, the relative proportions of glucose utilized by the EM and HMP pathways were 80 and 20%, 90 and 10%, 98 and 2%, respectively.[116] It has been pointed out that difficulties in interpretation of the data may arise with the technique used.[34]

Intact germinated spores of B. cereus strain T apparently do not have an operative tricarboxylic acid cycle and an active cycle may not develop until after several generations following germination.[114, 117] When [^{14}C] glucose was used as the substrate for germinated spores, [^{14}C] acetate accumulated: little or no $^{14}CO_2$ was evolved from [6—^{14}C]-glucose. Fumarase was absent from spore extracts and this enzyme was not synthesized until sometime late in the elongation stage; after the 6th division of the resulting vegetative cells, fumarase activity was 80 times that present during the elongation stage.[114] However, a particulate tricarboxylic acid cycle for the oxidation of pyruvate was reported in fortified cell-free extracts of activated spores of the same organism.[112] This apparent contradiction has been discussed but not resolved.[34] If a functional tricarboxylic acid cycle is absent in germinated spores the question arises as to how the α-ketoacids required for amino acid and protein synthesis during outgrowth are provided. However this debate is resolved, there is no doubt that changes do occur in the pathways of glucose metabolism during outgrowth and that these changes are influenced by the composition of the growth medium. For example, the rate of formation of oxidative enzymes involved in the metabolism of glucose, gluconate, pyruvate and succinate by B. subtilis spores during outgrowth differed in defined and complex media.[118] Also, when B. cereus strain T spores developed to the swollen stage in an undialysed pancreatic extract of casein (Casitone) containing glucose, only weak tricarboxylic acid cycle activity was evident and even when glucose was omitted there was practically no increase in activity. In contrast, during swelling and elongation in dialysed Casitone medium, a very active tricarboxylic acid cycle developed and this was repressed by the presence of glucose or lactate; at the late elongation stage, over 50% of the added [6—^{14}C] glucose was oxidized to $^{14}CO_2$[116]. Thus, the development of the tricarboxylic acid cycle activity in outgrowing spores may be repressed by diffusible substances in growth media.

Evidence for the ordered synthesis of enzymes during outgrowth of B. cereus strain T spores was provided by the study of three enzymes, L-alanine dehydrogenase, α-glucosidase and alkaline phosphatase.[4] A low level of each of these enzymes was maintained during synchronous outgrowth until the first division when the activity of each increased to a new level. The time when renewed synthesis occurred was different for each enzyme and this relative time difference was repeated over several division cycles while synchronous growth persisted (Fig. 13). Thus, the timing of the synthesis of specific macromolecules during outgrowth is different and sequential transcription and translation may continue after outgrowth into normal vegetative growth.[4]

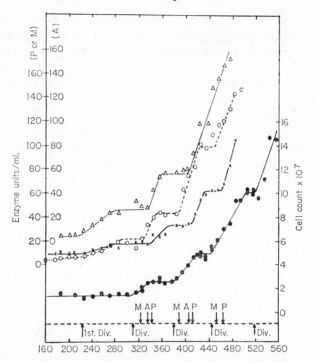

Fɪɢ. 13. Timing of enzyme synthesis during synchronous growth of *B. cereus* strain T spores. Spores were germinated and allowed to outgrow in a defined medium. Samples of the suspension were removed at intervals and the separated cells washed and repeatedly frozen and thawed; the resulting suspensions were used for assays of α-glucosidase (M) (○), alkaline phosphatase (P) (×) and L-alanine dehydrogenase (A) (△). Cell counts (●) were determined in duplicate with a Petroff–Hauser bacteria counter. In the lower bar graph, the time of initiation of cell division (Div) and of enzyme synthesis (M, A, P) are indicated by arrows. (Figure by courtesy of the American Society for Microbiology; for details see Ref. 4.)

C. Cell Wall Components

The cell wall of a vegetative bacterium which protects the osmotically fragile protoplast in environments of unfavourable tonicity, contains a variable amount of murein in a layer that apparently provides rigidity and maintains the shape of the cell. Studies of the biosynthesis of cell wall murein have established that the pathways are distinct from those for the synthesis of protein since chloramphenicol inhibits protein but not murein synthesis.[119, 120] In *Bacillus* species, muropeptides (glycosamino-peptides) containing muramic acid, D-alanyl-D-alanine, L-alanine, D-glutamic acid and αε-diaminopimelic acid (or lysine) are formed by sequential addition to uridine diphosphate-acetylmuramic acid and these muropeptides are transferred with glucosamine and other components into the bacterial cell wall. A comprehensive account of bacterial cell wall synthesis should be consulted for details.[121]

The emerging germ cell enters into its new environment in a relatively unprotected state compared with the previous conditions where it was surrounded by a tough outer envelope system and an inner cortex. In general, information concerning the origin and composition of the cell wall of the emerging germ cell is sparse and suggests that the situation varies in different organisms. Morphological evidence is available showing that either the inner spore coat[74] or the innermost layer of the cortex[78, 80] may become the cell wall of the emerging cell.

Recent work[22] has shown that only about 30–40% of the spore murein with a composition similar to that of the vegetative cell wall murein[18] is solubilized during germination of spores of large celled (e.g. *B cereus*) and small celled (e.g. *B. polymyxa*) species. The remainder (60–70%) is retained by the organism during at least the initial stages of outgrowth and therefore seems to have a function quite separate from that of preserving dormancy. The retained murein appears to reside in the exosporium of *B. cereus* but in the core wall and cortex of *B. polymyxa*. Measurements of the hexosamine contents of *B. cereus* and *B. polymyxa* spores during outgrowth in buffered peptone-yeast extract-meat extract-glucose medium at 37°C, established that re-incorporation and resynthesis of muropeptides associated with vegetative cell wall synthesis began about 10 min after the start of germination of *B. cereus* and after 30 min in the case of *B. polymyxa*.

Other evidence is available supporting the view that part of the spore muropeptide is utilized to provide the outgrowing germ cell with wall material.[19] A structure containing αε-diaminopimelic acid (DAP), probably the spore cortical membrane, was found to persist during germination and outgrowth of *B. cereus* (NCIB 8122), and may be the primary wall structure of the new vegetative cell. *B. cereus* cells were labelled with [14C] DAP during two different stages (prespore and maturation) at the end of vegetative growth and development was then continued until the spore release stage. With spores from organisms labelled during the prespore stage, [14C] DAP was mainly present in a hot acid-precipitable fraction whereas with labelling during the maturation stage, the amount of [14C] DAP incorporated into this fraction was much lower in relation to the total incorporated. During outgrowth, most of the [14C] DAP incorporated during the prespore stage remained in the cells in the hot acid-insoluble fraction and only a small proportion was released in the germination exudate. Of particular interest was the fact that during swelling and outgrowth of the spores labelled during the prespore stage, there was a small but significant increase in the radioactivity of the hot acid-insoluble fraction indicating transfer of components of hot acid labile [14C] DAP-containing material into this fraction. In contrast, the [14C] DAP-containing structures of spores labelled during maturation were largely released during germination and only a small proportion was retained in the outgrowing cells. Evidence for turnover of muropeptides during outgrowth was obtained by adding penicillin to the medium; under these conditions, the radioactivity of the hot acid-insoluble fraction decreased. Turnover may explain why the increase in the amount of muropeptide in this fraction during outgrowth in the absence of antibiotic appeared to be small.

An attempt has been made to relate cell wall synthesis during outgrowth to the synthesis of RNA and protein.[122] During swelling of *B. cereus* (NCIB 8122) there was an increasingly rapid synthesis of RNA and protein, and incorporation of [14C] DAP into the cellular hot acid-insoluble fraction started in the late phase of swelling. Sensitivity of [14C] DAP incorporation to chloramphenicol decreased during outgrowth suggesting that synthesis of the cell wall of the developing germ cell is dependent on the prior formation of certain proteins, which may be enzymes required for cell wall production. A comparison was made of the ability of outgrowing cells of *B. cereus* in divided synchronized cultures to incorporate [14C] DAP i.e. to synthesize new cell wall. In one case synthesis of new RNA was inhibited by actinomycin D (2 μg/ml) and in the other synthesis of new protein was inhibited by chloramphenicol (50 μg/ml), the inhibitors and [14C] DAP being added at different times during germination and outgrowth. The results confirmed that the effect of chloramphenicol on cell wall synthesis decreased during outgrowth and showed that new mRNA synthesis was required for the synthesis of the protein(s) involved. The necessary mRNA was formed from about the 10th minute after the start of germination, and after 20 min cell wall synthesis was much less dependent on the synthesis of new protein.

The effect of antibiotics interfering with cell wall synthesis has also been studied during spore outgrowth. The antibiotics used were penicillin which prevents transfer of muropeptides into the cell wall (i.e. the final stage), and cycloserine which inhibits the initial conversion of L-alanine to D-alanine and the formation of D-alanyl-D-alanine (Fig. 14). Neither antibiotic affected the swelling phase indicating that interference with cell wall formation did not lead to osmotic destruction of the swollen cell deprived of the protection of a large part of its envelope. Presumably, therefore, preexisting stable spore murein structures are present that protect the emerging cell (see above). However, in the presence of these antibiotics there was an increase in sensitivity at the end of the elongation phase.[123] When penicillin (500 units/ml) was present from the start of germination, incorporation of [14C] DAP was inhibited until the antibiotic had been degraded and then normal swelling occurred but at the onset of and during elongation, the density of the culture decreased markedly. The production of penicillinase abolished the effect of penicillin at the later stages of outgrowth but with increased amounts of antibiotic, elongated cells were highly sensitive. Attempts to form protoplasts or spheroplasts by adding sucrose (10, 15 or 20%, w/v) to the cultures during outgrowth were unsuccessful although swollen cells were sometimes observed. In

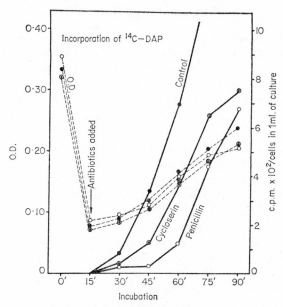

FIG. 14. Synthesis of new cell wall in outgrowing spores of *B. cereus*. Spores (1–2 × 10⁸/ml) were germinated and allowed to develop in peptone medium containing [¹⁴C] DAP at 30°C. Incorporation of [¹⁴C] DAP into the trichloroacetic acid-insoluble fraction of the cells and extinction values of the cultures were measured in the absence (●) and presence of penicillin (○) and cycloserine (⊙). Solid lines show incorporation of [¹⁴C] DAP and dashed lines extinction values of the cultures. (Figure by courtesy of the American Society for Microbiology.[124])

the presence of cycloserine (mM) spores swelled, germ cells elongated and indeed some cells divided after a lag period; [¹⁴C] DAP incorporation was decreased but after the lag period before division there was a new increase in incorporation into murein. The division of the cycloserine-treated culture was abnormal and the pairs bulged atypically, one of the cells in the pair often being thicker than the other, or club-shaped or spherical. This suggests that the cell wall of one of the pair was incomplete and therefore the shape of the cell changed as a result of osmotic effects. There was a slight decrease in the density of the culture during the later stages indicating that some of the resulting daughter cells were destroyed.

It appears from these experiments that a relatively stable pre-existing murein structure in spores persists during outgrowth and this may provide at least the foundation of the future vegetative cell wall. The spore envelopes lose their protective function during germination and outgrowth but there is no evidence to show that the exposed germ cell is osmotically sensitive, suggesting that an outer protective layer is present.

IX. Concluding Remarks

The material in this chapter represents an attempt to provide a coherent picture of the process of spore outgrowth. We hope the material presented fairly represents the work of the numerous investigators in a field that has expanded rapidly over the last 10 years. We found that a large proportion of the reported work concerned aerobic spores and this is no doubt due to the greater technical difficulties associated with investigations of anaerobic organisms.

Among the problems that are not yet satisfactorily resolved are: (a) Location of all the spore murein in different spores and the fate of that retained in the spore structures after germination; there is cytological, immunological and biochemical evidence that at least the foundation of the future vegetative cell wall is present in the spore but the sequence of events during outgrowth is not yet generally clear. (b) A similar situation exists with respect to the cytoplasmic membrane of the emerging cell; there is cytological evidence that a cytoplasmic membrane pre-exists in the resting spore, situated below the inner cortical membrane (which becomes the cell wall?). However, the absence from aerobic spores of cytochromes and certain enzymes that are involved in oxidative metabolism in the vegetative cell and that are usually present in particles associated with the cytoplasmic membrane, suggests that a new cytoplasmic membrane may be formed during outgrowth.[34]

Many elegant studies of macromolecular synthesis during outgrowth have been reported but, in particular, the conclusion of Kobayashi and his associates is pertinent:[4] "Outgrowth is dependent upon the repair in the protein-synthesizing system and an ordered synthesis of protein. Studies on the conversion of a dormant spore to a vegetative cell therefore provide a well-defined model system for understanding the mechanisms whereby irreversible, sequential events are controlled during intracellular differentiation."

REFERENCES

1. Campbell, L. L., Jr. (1957). *In* "Spores" (H. O. Halvorson, ed.), p. 33. American Institute of Biological Sciences, Washington, U.S.A.
2. Keynan, A. and Halvorson, H. O. (1965). *In* "Spores III" (L. L. Campbell and H. O. Halvorson eds.), p. 174. Am. Soc. Microbiol., Ann Arbor, Michigan, U.S.A.
3. Mandels, G. R., Levinson, H. S. and Hyatt, M. T. (1956). *J. gen. Physiol.* **39**, 301.

4. Kobayashi, Y., Steinberg, W., Higa, A., Halvorson, H. O. and Levinthal, C. (1965). *In* "Spores III" (L. L. Campbell and H. O. Halvorson eds), p. 200. Am. Soc., Microbiol. Ann Arbor, Michigan, U.S.A.

5. Brown, W. L., Ordal, Z. J. and Halvorson, H. O. (1957). *Appl. Microbiol.* **5,** 156.

6. Krask, B. J. and Fulk, G. E. (1959). *Archs Biochem. Biophys.* **79,** 86.

7. Sacks, L. E. and Alderton, G. (1961). *J. Bact.* **82,** 331.

8. Gaudin, A. M., Mular, A. L. and O'Connor, R. F. (1960). *Appl. Microbiol.* **8,** 84.

9. Goldman, M. and Blumenthal, H. J. (1961). *Can. J. Microbiol.* **7,** 677.

10. Powell, J. F. and Strange, R. E. (1953). *Biochem. J.* **54,** 205.

11. Lamanna, C. (1940). *J. Bact.* **40,** 347.

12. Knaysi, G. and Hillier, J. (1949). *J. Bact.* **57,** 23.

13. Vinter, V. (1961). *In* "Spores II" (H. O. Halvorson, ed.), p. 127. Burgess Publishing Co., Minneapolis, Minn., U.S.A.

14. Kondo, M. and Foster, J. W. (1967). *J. gen. Microbiol.* **47,** 257.

15. Hyatt, M. T. and Levinson, H. S. (1957). *J. Bact.* **74,** 87.

16. Vinter, V. (1960). *Folia microbiol., Praha* **5,** 217.

17. Work, E. (1957). *Nature, Lond.* **179,** 841.

18. Strange, R. E. and Dark, F. A. (1957). *J. gen. Microbiol.* **17,** 525.

19. Vinter, V. (1965). *Folia microbiol., Praha* **10,** 280.

20. Ohye, D. F. and Murrell, W. G. (1962). *J. Cell Biol.* **14,** 111.

21. Warth, A. D., Ohye, D. F. and Murrell, W. G. (1963). *J. Cell Biol.* **16,** 593.

22. Hamilton, W. A. and Stubbs, J. M. (1967). *J. gen. Microbiol.* **47,** 121.

23. Hitchins, A. D. and Gould, G. W. (1964). *Nature, Lond.* **203,** 895.

24. Curran, H. R., Brunstetter, B. C. and Myers, A. T. (1943). *J. Bact.* **45,** 485.

25. Powell, J. F. and Strange, R. E. (1956). *Biochem. J.* **63,** 661.

26. Rouf, M. A. (1964). *J. Bact.* **88,** 1545.

27. Hyatt, M. T. and Levinson, H. S. (1959). *J. Bact.* **77,** 487.

28. Powell, J. F. and Strange, R. E. (1957). *Biochem. J.* **65,** 700.

29. Donellan, J. E., Nags, E. H. and Levinson, H. S. (1965). *In* "Spores III" (L. L. Campbell and H. O. Halvorson, eds), p. 152. Am. Soc. Microbiol., Ann Arbor, Michigan, U.S.A.

30. Woese, C. R., Langridge, R. and Morowitz, H. J. (1960). *J. Bact.* **79,** 777.

31. Higa, A. (1964). *Thesis*, Mass. Inst. Technol., Cambridge, U.S.A.

32. Balassa, G. (1963). *Biochim. biophys. Acta* **72,** 497.

33. Doi, R. H. and Igarashi, R. T. (1964). *J. Bact.* **87,** 323.

34. Sussman, A. S. and Halvorson, H. O. (1966). "Spores, Their Dormancy and Germination". Harper and Row (Publishers), New York and London.

35. Keilin, D. and Hartree, E. F. (1949). *Nature, Lond.* **164,** 254.

36. Marr, A. G. (1960). *In* "The Bacteria", Vol. I (I. C. Gunsalus and R. Y. Stanier, eds), p. 443. Academic Press, New York and London.

37. Woese, C. R. (1959). *J. Bact.* **77,** 690.

38. Levinson, H. S. and Hyatt, M. T. (1962). *J. Bact.* **83,** 1224.

39. Hyatt, M. T. and Levinson, H. S. (1961). *J. Bact.* **81,** 204.

40. Hyatt, M. T. and Levinson, H. S. (1964). *J. Bact.* **88,** 1403.

41. Balassa, G. and Contesse, G. (1966). *Annls Inst. Pasteur, Paris* **110,** 25.

42. Amaha, M. and Sakaguchi, K. (1952). *J. agric. Chem. Soc. Japan* **26,** 353.

43. O'Brien, R. T. and Campbell, L. L. (1957). *J. Bact.* **73,** 522.

44. Uchiyama, H., Tanaka, K. and Yanagita, T. (1965). *J. gen. appl. Microbiol., Tokyo* **11,** 233.

45. Balassa, G. and Contesse, G. (1965). *Annls Inst. Pasteur, Paris* **109,** 683.

46. Demain, A. L. and Newkirk, J. F. (1960). *J. Bact.* **79**, 783.
47. Donellan, J. E., Nags, E. H. and Levinson, H. S. (1964). *J. Bact.* **87**, 332.
48. Demain, A. L. (1958). *J. Bact.* **75**, 517.
49. Gibbs, P. A. (1964). *J. gen. Microbiol.* **37**, 41.
50. Treadwell, P. E., Jann, G. J. and Salle, A. J. (1958). *J. Bact.* **76**, 549.
51. Mol, J. H. H. (1957). *J. appl. Bact.* **20**, 454.
52. Knaysi, G. (1964). *J. Bact.* **87**, 619.
53. Wolf, J. and Mahmoud, S. A. Z. (1957). *J. appl. Bact.* **20**, 124.
54. Wynne, E. S., Schmeiding, W. R. and Daye, G. T. (1955). *Fd Res.* **20**, 9.
55. Murrell, W. G. (1961). *Symp. Soc. gen. Microbiol.* **11**, 100.
56. Reimann, H. (1957). *J. appl Bact.* **20**, 404.
57. Gould, G. W. (1964). *4th Int. Symp. Food Microbiol.*, SIK, Göteborg, p. 17. Almqvist and Wiksell, Uppsala.
58. Slepecky, R. A. (1963). *Biochem. biophys. Res. Commun.* **12**, 369.
59. Tonge, R. J. (1964). *4th Int. Symp. Food Microbiol.*, SIK, Göteborg, p. 35. Almqvist and Wiksell, Uppsala.
60. McDonald, W. C. (1967). *Can. J. Microbiol.* **13**, 611.
61. Hachisuka, Y., Sugai, K. and Asano, N. (1958). *Jap. J. Microbiol.* **2**, 317.
62. Krishna Murty, G. G. and Halvorson, H. O. (1957). *J. Bact.* **73**, 230.
63. Steinberg, W., Halvorson, H. O., Keynan, A. and Weinberg, E. (1965). *Nature, Lond.* **208**, 710.
64. Black, S. H. and Gerhardt, P. (1962). *J. Bact.* **83**, 301.
65. Stuy, J. H. (1956). *Biochim. biophys. Acta* **22**, 241.
66. Black, S. H. and Gerhardt, P. (1962). *J. Bact.* **83**, 960.
67. Stuy, J. H. (1958). *J. Bact.* **76**, 179.
68. Hitchins, A. D., Gould, G. W. and Hurst, A. (1963). *J. gen. Microbiol.* **30**, 445.
69. Gould, G. W. (1962). *J. appl. Bact.* **25**, 35.
70. Propst, A. and Möse, J. R. (1965). *Zentbl. Bakt. Parasitenk. Orig. Abt.* 1. **195**, 500.
71. Fitz-James, P. C. (1955). *Can. J. Microbiol.* **1**, 525.
72. Moberly, B. J., Shafa, F. and Gerhardt, P. (1966). *J. Bact.* **92**, 220.
73. Knaysi, G., Baker, R. F. and Hillier, J. (1947). *J. Bact.* **53**, 525.
74. Chapman, G. B. and Zworykin, K. A. (1957). *J. Bact.* **74**, 126.
75. Warth, A. D., Ohye, D. F. and Murrell, W. G. (1963). *J. Cell Biol.* **16**, 579.
76. Walker, P. D., Baillie, A., Thomson, R. O. and Batty, I. (1966). *J. appl. Bact.* **29**, 512.
77. Hachisuka, Y., Kojima, K. and Sato, T. (1966). *J. Bact.* **91**, 2382.
78. Mayall, B. H. and Robinow, C. F. (1957). *J. appl. Bact.* **20**, 333.
79. Fitz-James, P. C. and Young, I. E. (1959). *Nature, Lond.* **183**, 372.
80. Kawata, T., Inoue, T. and Takagi, A. (1963). *Jap. J. Microbiol.* **7**, 23.
81. Ryter, A. (1965). *Annls Inst. Pasteur, Paris* **108**, 40.
82. Rousseau, M., Flechon, J. and Hermier, J. (1966). *Annls Inst. Pasteur, Paris* **111**, 149.
83. Takagi, A., Kawata, T. and Yamamoto, S. (1960). *J. Bact.* **80**, 37.
84. Robinow, C. F. (1960). *In* "The Bacteria", Vol. I (I. C. Gunsalus and R. Y. Stanier, eds), p. 207. Academic Press, New York and London.
85. Pulvertaft, R. J. V. and Haynes, J. A. (1951). *J. gen. Microbiol.* **5**, 657.
86. Leifson, E. (1931). *J. Bact.* **21**, 357.
87. Klein, L. (1889). *Zentbl. Bakt. Parasitenk. Orig. Abt.* **1**, 6, 313.
88. Pearce, T. W. and Powell, E. O. (1951). *J. gen. Microbiol.* **5**, 91.
89. Mason, D. A. and Powelson, D. M. (1956). *J. Bact.* **71**, 474.

90. Robinow, C. F. (1956). *Symp. Soc. gen. Microbiol.* **6**, 181.
91. Norris, J. R. and Wolf, J. (1961). *J. appl. Bact.* **24**, 42.
92. Moussa, R. S. (1959). *J. Path. Bact.* **77**, 341.
93. Walker, P. D. (1963). *J. Path. Bact.* **85**, 41.
94. Walker, P. D. and Batty, I. (1964). *J. appl. Bact.* **27**, 137.
95. Walker, P. D. and Batty, I. (1965). *J. appl. Bact.* **28**, 194.
96. Powell, J. F. and Hunter, J. R. (1955). *J. gen. Microbiol.* **13**, 59.
97. Amaha, M. and Nakahara, T. (1959). *Nature, Lond.* **184**, 1255.
98. Evans, F. R. and Curran, H. R. (1943). *J. Bact.* **46**, 513.
99. Levinson, H. S. and Hyatt, M.T. (1955). *J. Bact.* **70**, 368.
100. Levinson, H. S. and Hyatt, M. T. (1956). *J. Bact.* **72**, 176.
101. Spencer, R. E. J. and Powell, J. F. (1952). *Biochem. J.* **51**, 239.
102. Crook, P. G. (1952). *J. Bact.* **63**, 193.
103. Roth, N. G. and Lively, D. H. (1956). *J. Bact.* **71**, 162.
104. Tempest, D. W., Dicks, J. W. and Meers, J. L. (1967). *J. gen. Microbiol.* **49**, 139.
105. Neidhart, F. C. (1963). *A. Rev. Microbiol.* **17**, 61.
106. Arnstein, H. R. V. (1963). *Rep. Prog. Chem.* **60**, 512.
107. Jacob, F. and Monod, J. (1961). *J. molec. Biol.* **3**, 318.
108. Vinter, V. (1966). *Folia microbiol., Praha* **11**, 392.
109. Wake, R. G. (1963). *Biochem. biophys. Res. Commun.* **13**, 67.
110. Church, B. D. and Halvorson, H. (1957). *J. Bact.* **73**, 470.
111. Doi, R. H., Halvorson, H. and Church, B. D. (1959). *J. Bact.* **77**, 43.
112. Halvorson, H. and Church, B. D. (1957). *J. appl. Bact.* **20**, 359.
113. Goldman, M. (1961). *Thesis*, University of Michigan, Ann Arbor, Michigan, U.S.A.
114. Goldman, M. and Blumenthal, H. J. (1964). *J. Bact.* **87**, 377.
115. Wang, C. H., Stern, I., Gilmour, C. M., Klungsoyr, S., Reed, D. J., Bialy, J. J., Christensen, B. E. and Cheldelin, V. H. (1958). *J. Bact.* **76**, 207.
116. Blumenthal, H. J. (1965). *In* "Spores III" (L. L. Campbell and H. O. Halvorson, eds), p. 222. Am. Soc. Microbiol., Ann Arbor, Michigan, U.S.A.
117. Goldman, M. and Blumenthal, H. J. (1964). *J. Bact.* **87**, 387.
118. Hachisuka, Y., Asano, N. and Sugai, K. (1958). *Jap. J. Microbiol.* **2**, 79.
119. Mandelstam, J. and Rogers, H. J. (1958). *Nature, Lond.* **181**, 956.
120. Hancock, R. and Park, J. T. (1958). *Nature, Lond.* **181**, 1050.
121. Salton, M. R. J. (1964). "The Bacterial Cell Wall". Elsevier Publishing Co., Amsterdam, London and New York.
122. Vinter, V. (1965). *Folia microbiol., Praha* **10**, 288.
123. Vinter, V. (1963). Int. Symp. on the Physiology, Ecology and Biochemistry of Germination. Greifswald, Germany.
124. Vinter, V. (1965). *In* "Spores III" (L. L. Campbell and H. O. Halvorson, eds), p. 25. Am. Soc. Microbiol., Ann Arbor, Michigan, U.S.A.
125. Torriani, A. and Levinthal, C. (1967). *J. Bact.* **94**, 176.
126. Vinter, V. and Slepecky, R. A. (1965). *J. Bact.* **90**, 803.
127. Holmes, P. K. and Levinson, H. S. (1967). *J. Bact.* **94**, 434.
128. Freer, J. H. and Levinson, H. S. (1967). *J. Bact.* **94**, 441.

CHAPTER 13

Sporeformers as Insecticides

JOHN R. NORRIS

"Shell" Research Ltd, Milstead Laboratory of Chemical Enzymology, Sittingbourne, Kent, England

I. Introduction

A. Sporeformers as Insect Pathogens

THE existence of infectious disease in insect populations has been recognized for a very long time. Indeed, it is fair to say that the recorded history of insect pathology begins with Aristotle's description of diseases of the honey bee in his *Historia Animalium*. As an experimental science insect pathology dates from the demonstration, in 1834, by Agostino Bassi, that the fungus *Beauveria bassiana* caused an infection in the silkworm and it is in the writings of Bassi that we find the first suggestion that micro-organisms might be used to control insect pests. Thus although the basic concept of a microbial insecticide was suggested when Pasteur was still a boy, it is only today, some 130 years later, that microbial methods are beginning to play a significant role in the control of pest insects.

To be effective a microbial insecticide must possess various characteristics of which stability during prolonged storage and persistence in an infective form after dispersal are of prime importance. It is not surprising therefore that particular attention has been focused on bacteria and fungi producing resistant forms which will persist outside the host insect. In the bacteria it has been the ability to form spores that are resistant to most of the detrimental effects of the environment which has attracted the attention of workers attempting to achieve microbial control and the species which today come closest to providing effective insecticide preparations are members of the family *Bacillaceae*.

Over 100 named species of spore-forming bacteria have been isolated from, or found associated with, insects,[1] but few of these are valid species and many are not true pathogens in the sense that they will initiate infection via a normal portal of entry to the insect body. The few spore-forming bacteria that are capable of producing pathogenic effects in insect populations are, however, of great importance in the field of insect pathology since work on them and on their relationships to the host insects has added much to our understanding of the basic principles of the subject. Although this chapter is concerned primarily with those species which show potential as microbial insecticides, the others will be mentioned in the interests of completeness.

B. AEROBIC SPOREFORMERS PATHOGENIC FOR INSECTS

1. Bacillus cereus

Bacillus cereus has been isolated from diseased insects on numerous occasions.[2, 3] The bacterium is capable of infecting larvae via the gut, but the details of its pathogenicity are not fully understood. *B. cereus* produces an active phospholipase and there is some correlation between pathogenicity and ability to produce this enzyme. Heimpel,[4] working with a relatively virulent isolate, Pr 1017, showed that toxicity was dependent on the bacterium actually growing in the larval gut and suggested that phospholipase synthesized during growth was responsible for initial damage to the gut wall.

Although attempts have been made to use *B. cereus* for pest control, the results have not been encouraging.

2. Bacillus thuringiensis

This bacterium is closely related to *B. cereus* but is vigorously pathogenic for lepidopterous larvae. It occupies an important place in microbial control and will be discussed in detail later in this chapter.

3. Bacillus alvei

Bacillus alvei was originally isolated from bees suffering from a rather ill-defined condition known as European Foulbrood.[5] The relationship of the bacterium to the disease condition is not fully understood, several different species of bacteria having from time to time been implicated as causative agents. In recent years an anaerobic streptococcus, *Streptococcus pluton*, has been recognized as the primary pathogen in European Foulbrood, but *Bacillus alvei* and *Streptococcus faecalis* are often found as major components of the flora of unhealthy bee larvae and these species may play a secondary role in the disease. For a fuller account of this subject, the reader is referred to an article by Heimpel and Angus.[6]

4. Bacillus larvae

American Foulbrood is a condition distinct from European Foulbrood and has been recognized as a disease of young bee larvae since about 1900. *Bacillus larvae* is the causative organism and its resistant spores remain viable for years in contaminated material under natural conditions. Bee larvae are most susceptible to infection during the first 50 hr of life. It has been suggested that the peritrophic membrane, which is laid down about two days after hatching, protects the larval gut epithelium from bacterial attack after that time. The reader is referred to publications by Patel and Cutkomp[7] and Patel and Gochnauer[8] for a more detailed discussion of the pathology and toxigenicity of the disease.

5. Bacillus popilliae *and* Bacillus lentimorbus

These two bacteria grow poorly on routine bacteriological media but are of considerable importance as microbial insecticides. They will be described in detail later.

C. Anaerobic Species Pathogenic for Insects

Until the early 1950s no clostridia were known to be associated with disease in insects. Since then, however, publications by Bucher[9, 10, 11] have shown that anaerobic sporeformers are not uncommon as insect pathogens. Bucher demonstrated extensive multiplication of a sporeforming bacterium in the gut of larvae of *Malacosoma pluviale* suffering from a condition known as brachyosis.[9] Although he was able to demonstrate the pathogenicity of the bacterium for *Malacosoma sp.* by infection experiments using contaminated gut contents, he experienced great difficulty in cultivating the pathogen. This was finally achieved using a complex medium containing leaf extract and reducing substances. Two bacteria, *Clostridium brevifaciens* and *Cl. malacosomae* were involved and both were capable of causing disease in young larvae.[11] A bacterium similar to *Cl. brevifaciens* was isolated from diseased larvae of the Essex skipper (*Thymelicus lineola*).[10]

II. Bacterial Insecticides

A new impetus has been given to the search for biological methods of controlling insect populations by concern over possible hazards arising from the persistence of some chemical insecticides and their ubiquitous distribution in the environment. Considerable effort has been directed to the production of bacterial and viral insect pathogens and such preparations have been tested against a wide range of pest insects in many parts of the world. The main attraction of these pathogens is their inability to affect man, animals or plants and in this respect they offer considerable advantages over many of the chemical insecticides with which they must compete.

Safety alone will not sell an insecticide however. To compete successfully on the insecticide market a microbial preparation must compare in cost and efficiency with more conventional products; it must remain effective during storage and after dissemination; it must kill the insect in such a way as to minimize damage to the infested crop or foodstuff; it must present no formulation problems and, ideally, it should possess a sufficiently broad spectrum of activity against different insect species to render it useful in attacking the pest complexes which are found on many crops. Of the many species of bacteria that have been examined as potential insecticides, those that have so far come closest to success are two aerobic spore-forming bacteria, *Bacillus popilliae* used in the control of larvae of the Japanese Beetle and *Bacillus thuringiensis* used against various lepidopterous larvae. Both of these species have reached the stage of commercial production and have been used in the field for a number of years. They present an interesting contrast in that *B. popilliae* depends for its effectiveness on active colonization of the pest species over a period of a few years, while it is the toxins of *B. thuringiensis* that are the effective control agents and preparations of this organism must be used repeatedly to control pest populations in much the same way as conventional chemical insecticides are used. This distinction underlies the whole approach to commercial exploitation of the bacteria and should be borne in mind during the following discussion of the two organisms and their use as insecticides.

III. *Bacillus popilliae* and *Bacillus lentimorbus*

A. The Milky Diseases

The milky diseases are infectious diseases of scarabaeid grubs of which the disease of the Japanese Beetle (*Popillia japonica*) is by far the best known.

The beetle was introduced into the United States from Japan in 1916 and spread from a limited area in Burlington County, New Jersey, to cover a large part of the New England States and Canada. It is responsible for extensive damage to lawns, pastures, shrubberies and other plants. Eggs are laid during the late summer at a depth of 2–4 inches below the soil surface where the first instar larvae emerge after 2–3 weeks. The various larval instars feed actively on the roots of grasses and other plants. Over-wintering usually occurs at a depth of about 6 inches in the third larval instar. Feeding recommences the following April and pupation takes place at a depth of 2–4 inches in May. Adult beetles emerge about the second week in June and live for 30–45 days, during which time the female will lay 40–60 eggs.

Various diseases were recognized in Japanese beetle larvae shortly after the insect began to increase its territory in the United States, and the different micro-organisms concerned were studied at the U.S.D.A. Bureau of Entomology, Japanese Beetle Laboratory. Bacterial diseases were of two types, the black group and the white group, of which the latter became known as the milky diseases. Larvae were frequently found infected in the field and presented a characteristic chalky-white appearance. Microscopic examination showed the body cavities of infected larvae to contain large numbers of bacteria which could not be cultivated on ordinary laboratory media.

The milky diseases are caused principally by two closely similar bacteria, *Bacillus popilliae* and *Bacillus lentimorbus*.[12] The milky disease organisms are able to infect only certain closely related beetles of the family *Scarabaeidae*. In some cases infected larvae of species other than Japanese Beetle have been found in the field. These have usually been carrying *B. popilliae*, but other, similar, bacteria have occasionally been isolated.[13, 14, 15, 16, 17] For a detailed list of insects suceptible to infection with *B. popilliae* the reader is referred to a recent review by Dutky.[18]

Although the milky disease bacteria are closely similar, they can be distinguished from one another by cultural and immunological methods. Whole cell antigens of *B. popilliae* and *B. lentimorbus* do not cross-react with one another or with antigens of other aerobic sporeformers. *B. popilliae* is the organism which has been studied most extensively because it has shown most potential as a biological control agent. The other species differ in detail from *B. popilliae*, but their main cultural and pathological characteristics are similar.

B. The Development of *B. Popilliae* in Larvae of the Japanese Beetle

In nature, infection of Japanese beetle larvae normally follows ingestion of viable spores of the bacterium as the insects feed on grass roots or other plant materials. Spores germinate in the larval gut and vegetative cells penetrate through the wall of the gut and multiply in the haemocoel. Infection can also be induced by direct injection of spores into the body cavity. The infected insect will usually live for a considerable time and growth of vegetative bacteria in the haemolymph produces visible turbidity which increases until the internal organs are obscured and the larva is virtually opaque and moribund.

Early in the disease, the vegetative cells appear as uniform rods measuring 0·9 μ by 5·2 μ (Fig. 1). As the infection progresses the cells begin to

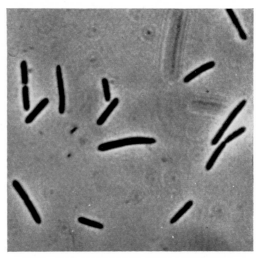

FIG. 1. Vegetative cells of *Bacillus popilliae*. Phase contrast × 3,600.

sporulate, becoming first oval and then distinctly pyriform. Each bacterial cell develops a central oval spore and, later, a refractile parasporal body situated at one end of the sporangium which increases in width to accommodate it. The mature sporangium measures roughly 1·6 μ by 5·5 μ and the spore is not released but remains, together with the parasporal body, inside the sporangial wall (Fig. 2). Growth and sporulation proceed side by side as long as the larval body provides sufficient nutrient and the cadaver is finally packed with millions of mature sporangia.

The rate at which the disease develops depends to a large extent on the prevailing temperature. Development of the organisms occurs over a range of about 16°C to 36°C. Both the time to onset of symptoms and the longevity of diseased larvae are increased at lower temperatures within this range. These factors compensate one another and the final yield of spores per larva is reasonably constant over much of the range. Other factors such as the state of health of larvae and the availability of food, by varying the longevity of infected larvae, may postpone or advance death following infection and so increase or decrease spore yields.[18]

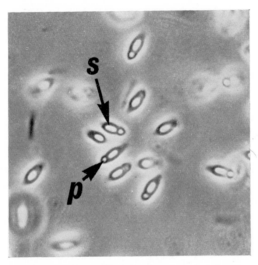

FIG. 2. Mature sporangia of *B. popilliae* showing spores and parasporal bodies. Phase contrast × 3,600. **s**—spore; **p**—parasporal body.

The pathogenicity of milky disease bacteria is poorly understood at the present time. Certainly much of the effect on the host may be attributed to the removal from the blood of nutrients and essential growth factors which become locked away in the growing microbial cells. Toxins are also apparently involved since cell-free filtrates of cultures of *B. popilliae* are lethal when injected in small amounts into larvae. Toxicity is lost on heating the preparations to 50°C for 10 mins. The refractile parasporal body, which is synthesized by *B. popilliae* and *B. fribourgensis* but not by *B. lenti-morbus*, may play a role in the disease, but if this is so there is as yet no indication of its mode of action. The granule is apparently crystalline in nature, like the parasporal body of *B. thuringiensis*, and it is also soluble only at extremely high pH (12·5). Unlike the *B. thuringiensis* crystal, however, the granule of *B. fribourgensis*—which alone has been examined in detail—contains nucleic acids and there is no evidence that it dissolves, or is in any way activated when taken into the larval gut.[19]

C. Growth of *B. popilliae* in Culture Media

Although the vegetative cells of *B. popilliae* grow vigorously in the haemolymph of infected larvae, yielding as many as 2×10^{10} spores per ml of haemolymph during a period of disease development which may extend from 7 to 21 days, attempts to grow the organism *in vitro* failed, for a long time, to produce consistent growth or cell counts approaching the *in vivo* figures. More recently, media and growth conditions have been devised which allow vegetative cells to be grown to reasonably high densities, but sporulation under artificial cultural conditions still presents problems.

Using a medium containing yeast extract, dipotassium hydrogen phosphate and glucose, counts up to 2×10^9 cells/ml may be achieved in incubation periods of 18–24 hr, but maximum viable populations persist for very short periods of time, cessation of growth being followed by rapid death of the cultures which are often sterile after 72 hr.[20] Spores are not formed in this medium. Hydrogen peroxide produced during growth may be responsible for the rapid death of cells and more stable cultures have been obtained by decreasing aeration towards the end of the growth phase.[21] On a medium containing trypticase and barbituric acid a high proportion of the cells produce refractile bodies visible under the phase contrast microscope (Fig. 3). These bodies are not heat resistant and lack

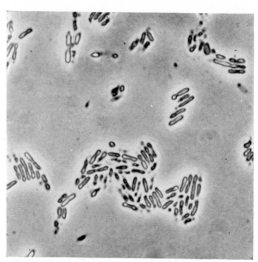

Fig. 3. Refractile bodies of *B. popilliae*. Phase contrast $\times$ 3,600.

other characteristics of endospores but they do confer fairly high viability on cultures for extended periods. Refractile bodies probably represent abortive spore-formation in cells committed to sporulation.[21] Certain

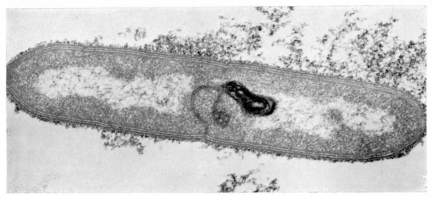

Fig. 4. Vegetative cell of *B. popilliae*. Electron micrograph section × 45,000.

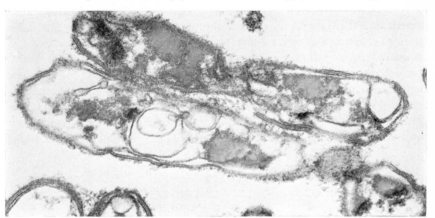

Fig. 5. Refractile bodies of *B. popilliae*. Electron micrograph section × 45,000.

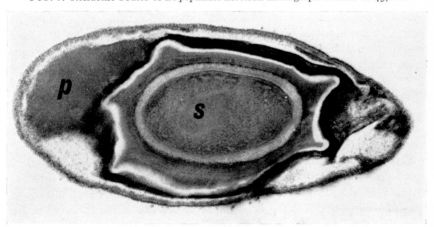

Fig. 6. Mature sporangium of *B. popilliae*. Electron micrograph section × 58,280. s—spore; p—parasporal body.

strains of *B. popilliae* are reported to show as much as 20% authentic sporulation on solid media.[22] Vegetative cells do not contain catalase, but sporulating cells, spores and refractile bodies have catalase activity, and cells from late phases of growth show increased peroxidase activity. These observations support the thesis that accumulation of toxic peroxides may result in cell death and possibly prevent spore formation.

Spores from infected larvae and refractile bodies produced in trypticase-barbiturate medium resemble one another in protein, nucleic acid and enzyme composition. Spores contain dipicolinic acid but the refractile bodies do not. Refractile bodies contain higher levels of poly-β-hydroxy-butyrate than do spores. Electron microscopy shows that the refractile bodies are structurally quite different from vegetative cells and spores (Figs 4, 5 and 6). Spores are structurally complex. The sporangial wall is derived from the wall of the vegetative cell and the spore itself has prominent lateral ridges on the surface running parallel to the long axis. The spore coat consists of several layers of varied density with an extensively laminated inner structure bounding the cortex. The parasporal body, connected to the spore by the delicate wall of the sporangium, is regular in outline and crystalline structure is sometimes seen on its surface. Refractile bodies, by contrast, are smaller than either spores or vegetative cells, possess no structures suggestive of spores but contain numerous internal membranes and electron dense objects resembling parasporal bodies.[23]

D. Use of *B. popilliae* for control of the Japanese Beetle

A spore-forming bacterium capable of remaining viable during storage away from its host for long periods of time and causing a spreading infection in populations of Japanese beetle larvae was obviously attractive as a potential biological control agent. Although inability to produce spore crops on artificial media was a major obstacle to the development of an effective control programme, workers at the U.S.D.A. Bureau of Entomology Laboratories at Morestown, New Jersey, tackled the problem of mass producing spores almost as soon as the nature of the milky disease bacteria became known and by 1939 had worked out a method which remains substantially unchanged to the present time.

Viable spores from diseased larvae are preserved as dried films of blood on microscope slides, or are lyophilized, and resuspended in water to produce inocula for healthy larvae. Doses of about 0·03 ml (1 million spores) are injected into the body cavity of healthy larvae collected in the field. The larvae are then reared in soil containing grass seed until disease is well developed. When sporulation is complete the grubs are refrigerated until required for use. Spores are released from the larvae by crushing and the suspension standardized by direct counting of the spores under a microscope. The suspension of spores is added to calcium carbonate and the resulting concentrate mixed with a dry powder carrier to produce a dust containing 100 million spores per gramme.

Use of the spore dust in the field depends on the fact that disease will spread in populations of Japanese beetle larvae from initially localized infection sites. Spore powder is usually applied to the soil in 2 gramme amounts at intervals of about 10 feet and spread occurs by various means including wandering of infected larvae, dispersal by water and wind, passive transfer by birds and animals and human activity. Such a treatment programme enables the disease to spread throughout the Japanese beetle population in three seasons.

Starting in 1939 the U.S.D.A , in co-operation with several Federal and State agencies, has introduced milky disease into ever-increasing areas of the Eastern United States. Preparations made by the U.S.D.A. were soon augmented by commercial materials available to farmers under the trade names "Doom" (*B. popilliae*) and "Japidemic" (*B. lentimorbus*). In treated areas there has often been a marked decrease in the population of Japanese beetle, but it is not always easy to attribute this directly to the pathogen. The bacteria have usually provided only one arm of an integrated programme involving the combined use of biological agents, chemical insecticides and improved agricultural practices. Nevertheless, the bacteria clearly become established in grub populations in which the pathogen persists for many years. Its efficiency as a long-term control agent is beyond doubt and the pest is now of relatively little economic significance over much of the treated area. Spores have been known to persist in soils for as long as 22 years after the grub has been eliminated.

Production of spore crops by growing the bacteria in living insects is necessarily a laborious and expensive process and the shortage of viable spores, coupled with high production costs, has been the main factor limiting the field use of these preparations. It is now clear that artificial media will soon be available for spore production and the resulting increase in infective material at lower cost should give a big impetus to the colonization programme. The reduction of Japanese beetle to a level where it is no longer an economically important pest, over its entire range in the New World, appears to be a possibility for the immediate future.

IV. *Bacillus thuringiensis*

A. History and Classification of *B. thuringiensis*

In 1902 the Japanese bacteriologist, Ishiwata, isolated from diseased silkworms an aerobic spore-forming bacterium which he showed to be the cause of the disease. One of the most striking characteristics of infection with the bacterium was the speed with which disease would sweep through a colony of silkworms, reducing them to blackened, spore-packed cadavers. Ishiwata called his isolate "Sotto-Bacillen" ("sudden collapse bacillus"). Two other Japanese, Aoki and Chigasaki,[24, 25] studied the bacillus and

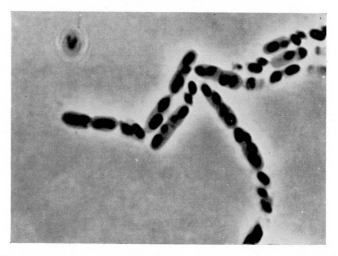

F I G. 7. Phase contrast photomicrograph of mature sporangia of *B. thuringiensis* (Serotype **g**) showing spores and protein crystals × 3,500.

demonstrated that its pathogenicity was due to a toxin present in sporulated cultures but absent from young vegetative cell cultures.

A similar bacterium was isolated in Germany on two occasions from diseased larvae of the Flour moth (*Anagasta kühniella*) and called *Bacillus thuringiensis*.[26, 27] The second isolate, the so-called Mattes isolate, was widely distributed to laboratories in various parts of the world. Most of the attempts to use *B. thuringiensis* for insect control purposes have been centred on this Mattes isolate. Since the early 1950s many more isolates of *Bacillus thuringiensis* have been obtained from diseased caterpillars in various parts of the world and there are today some 150 different isolates available for study.

"*Bacillus thuringiensis*" is, in fact, a blanket term for a group of closely related bacteria. They resemble *Bacillus cereus* in their general morphology and cultural characteristics. Some strains of *B. thuringiensis* share spore antigens with some *B. cereus* strains and there exist bacteriophages which cross-react between the two organisms.[28]

The German workers noted an important physical difference between the two species; as cells of *B. thuringiensis*, growing in or on nutrient media, reached maturity they produced not only a typical endospore, as did *B. cereus*, but also synthesized a second body, the so-called Restkörper or Parasporal body which came to lie, together with the spore, inside the sporangium (Fig. 7). Hannay[29, 30] rediscovered this parasporal body while studying spore formation in *B. thuringiensis* using the electron microscope and showed that it was a small rhomboidal crystal which was soon demonstrated to consist of protein.[31] Angus[32, 33] discovered a direct connection between the parasporal body and insect disease; the protein crystal was a potent toxin which, when consumed by a silkworm, paralysed the caterpillar and led to its death. Angus' conclusion that "paralysis is not caused by the growth of the micro-organism in the host tissue" laid the foundation for the large volume of work on the toxic products of *B. thuringiensis*, which was to follow.

As more strains of crystal-forming insect pathogens have been isolated, it has become possible to classify the bacteria into a number of groups differing from one another in their biochemical reactions, flagella antigens and in the characteristic molecular forms of esterase enzymes revealed when extracts of vegetative cells are examined by starch gel electrophoresis. Nine serotypes based on H antigens can be distinguished and one of them (Serotype 4) is divided into two sub-types, depending on the presence of one or other secondary antigens. There is a close correlation between serotypes and esterase groupings.[34, 35] As with many groups of bacteria, different workers have emphasized different characters when studying the classification of *B. thuringiensis* and the present picture of the group is a confusing one. One of the main arguments centres on a proposal that ability to produce different toxins should be reflected in schemes of classification,[36] a concept which has led to the recent establishment of several new varieties of *B. thuringiensis*.

The difference between *B. thuringiensis* and the other large-celled species in Morphological Group 1 of the genus *Bacillus*[37] is evident enough; the protein parasporal body is found only in *Bacillus thuringiensis* and appears to be a remarkably stable character. However, even this distinction has been questioned by Lysenko,[38] and it is clearly premature to try to finalize the classification of these organisms. Table I attempts to summarize present ideas about the classification of the crystal-forming aerobic sporeformers.

Y

TABLE I

Classification of Bacillus thuringiensis

Serotype	H Antigens	Acetylmethyl Carbinol Prodⁿ	Lecithinase	Acid from Salicin	Acid from Sucrose	Acid from Mannose	Hydrolysis of Starch	Urease Prodⁿ	Proteolysis	Pellicle formation	Esterase type	Exotoxin	Variety (after Heimpel[36])
1	I	+	+	+	+	+	+	−	+	+	Berliner	+	*thuringiensis*
1	I	+	+	+	+	+	+	−	+	+	Berliner	−	*amuscatoxicus*
2	II	+	+	+	+	−	−	−	+	+	Finitimus	−	*finitimus*
3	III	+	+	−	−	−	+	−	+	−	Alesti	−	*alesti*
4A	IV, a	+	+	−	+	−	+	−	+	−	Sotto	−	*sotto*
4A	IV, a	+	+	−	−	−	+	−	+	−	Sotto	−	*dendrolimus*
4B	IV, b	+	+	−	−	−	+	−	+	−	Kenya	−	*—*
5	V	−	−	+	−	−	+	+	+	+	Galleriae	±	*galleriae*
6	VI	+	−	+	+	+	+	+	+	−	Entomocidus	−	*entomocidus*
6	VI	−	−	−	+	+	+	−	+	−	Entomocidus	−	*entomocidus*
6	VI	+	−	−	−	−	+	−	+	+	Entomocidus	−	*subtoxicus*
7	VII	+	+	+	−	−	+	+	+	−	Galleriae	+	*aizawai*
7	VII	+	+	+	+	−	+	+	+	−	Galleriae	−	*pacificus*
8	VIII	−	−	−	+	−	+	−	+	+	Morrison	−	*anagastae*
9	IX	+	+	+	−	−	+	−	+	+	Tolworth	+	*—*

Biochemical Reactions

B. The Natural Disease

B. thuringiensis is normally isolated from populations of lepidopterous larvae which may be showing obvious signs of disease or which may appear healthy. Disease sometimes spreads in insect populations with remarkable speed, as in the case of silkworm colonies or the "population collapse" which occurred in a Kenya grain store heavily infested with *Cadra cautella*.[39] In this latter case, masses of diseased larvae appeared in the caterpillar population and yielded a crystal-former of Serotype 4B. Frank disease is not always evident in a population shown to be carrying *B. thuringiensis*. Environmental stress predisposes an insect population to epizootics of *B. thuringiensis* disease and it is not unusual to find that a population of caterpillars, apparently quite healthy when collected in the field, succumbs to infection when brought into the laboratory and held for some time under abnormal conditions. A similar situation develops in populations of stored products caterpillars infesting grain shipments. Intense congestion can develop at infestation sites in the holds of ships and several strains of *B. thuringiensis* have been isolated from caterpillars, frass and webbing collected from such sites when grain ships have docked in the United Kingdom after lengthy journeys from various parts of the world.[34] Once established in a population, the bacillus can persist for a long time, as is demonstrated by recurring outbreaks of infection in laboratories rearing caterpillars[39] and by the repeated isolation of particular strains from flour mills and grain warehouses over a number of years (Norris, unpublished observations).

In nature, caterpillars become infected by ingestion of spores and protein crystals released from diseased larvae, either as contaminants of faecal pellets or by the disintegration of cadavers packed with spores and crystals. Little is known about factors influencing the initiation and spread of infection under natural conditions. Efforts to induce persisting infection in insect populations by treating them with bacterial preparations have met with almost complete failure.

C. The Distribution of *B. thuringiensis* Strains

With the isolation of new strains of *B. thuringiensis* and the development of more refined methods for their identification it is becoming possible to build up patterns of world-wide distribution for some of the types. Information is as yet fragmentary, and some of the observations certainly reflect the distribution of bacteriologists rather than of bacteria, but even so some conclusions appear to be justified. Serotype 1 is world-wide in its distribution and is found infecting many different species of lepidoptera. Perhaps this picture has been influenced by man's efforts, since this serotype has been widely used in field trials over many parts of the world. Serotype 3 occurs mainly in the silk farms of France and in French flour mills. Serotype 4A (sotto) is found mainly in the Japanese silk farms and the closely similar Serotype 4A (dendrolimus) was isolated in the Eastern U.S.S.R. Serotype 4B is found all over the world infecting stored products insects in grain stores, warehouses, flour mills and in ships carrying grain.

A recent incident illustrates the persistence of *B. thuringiensis* in certain environments and the kind of ecological study possible. Specimens of larvae from an infested grain cargo were examined when the ship concerned docked in the United Kingdom and were found to be carrying a typical Serotype 4B strain of *B. thuringiensis*. Enquiries revealed that the grain had been loaded at Caotzacoalcos in Mexico and further samples of grain, insects and infestation site material were obtained from the warehouse in which the grain had been stored prior to shipment. These specimens were also contaminated with an apparently identical Serotype 4B organism. A similar case is documented by Norris and Burges.[39]

D. The Toxins of *B. thuringiensis*

Bacillus thuringiensis has been implicated as the aetiological agent of disease in larvae of many species of lepidoptera as well as in a few sawflies and one or two other insects. Over 130 species of lepidoptera are known to be susceptible and it is possible that all species may be susceptible to some degree.[36] Various strains of the bacillus produce toxins of different types, but the chemical nature and mode of action of the various toxins and the part they play in the disease process are not fully understood.

Three main toxins have been well documented: the protein crystal or endotoxin; a thermostable, water-soluble exotoxin active against the housefly and other lepidopterous and non-lepidopterous species; and a water-soluble exotoxin active against sawfly larvae. In addition to these toxins certain enzymes produced by growing vegetative cells of the bacterium, such as lecithinase C and hyaluronidase, have been implicated as playing some part in the disease process.

1. Protein crystal toxin (protein toxin; endotoxin; B.t.—δ endotoxin)

The protein crystal is certainly the most important toxin produced by *B. thuringiensis*. Protein crystals are produced by all types of *B. thuringiensis* and, although the crystals vary from strain to strain in shape, size and antigenic composition, they appear to play similar roles in pathogenicity. Angus[32, 33] showed that the protein crystal could be dissolved in alkaline solutions and that the extract, free from viable spores, was toxic for the silkworm *Bombyx mori*, causing gut paralysis followed by general body paralysis in as little as 60 min. Neither the intact, spore-free crystal nor a spore-free solution of crystal protein was toxic on injection into the haemocoel of the silkworm. Subsequent work has shown that the crystal dissolves under the alkaline, reducing conditions encountered in the midgut of most lepidopterous larvae and that the protein is digested by the complex of proteolytic enzymes in the caterpillar gut with the release of one or more toxic fragments.

Crystal formation is a remarkably stable characteristic of *B. thuringiensis* cultures, some of which have been cultivated on laboratory media for

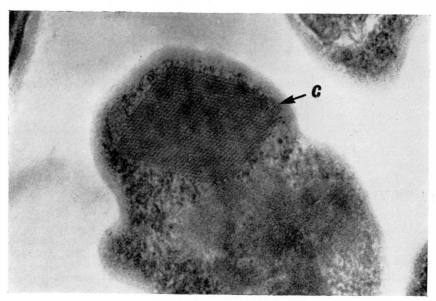

FIG. 8. Electron micrograph of thin section of a sporulating cell of *B. thuringiensis* (Serotype 9) showing a protein crystal × 112,000. c—protein crystal.

almost 70 years without loss of crystal-forming ability or toxicity. Crystal synthesis and spore-formation proceed at the same time in *B. thuringiensis* cells and the two processes appear to be intimately associated with each other. Asporogenic mutants fail to produce crystals, although loss of crystal-forming ability without loss of spore-formation has been noted.[40] Crystal formation in the absence of spore production can occur when cultures are grown at low temperatures.[41] Antigens characteristic of the crystal protein may be detected at an early stage in spore-formation when the vegetative cell is slightly granular, but there are no obvious signs of spore or crystal (Norris, unpublished observation). Electron micrographs of the surface of the crystal show regular structure,[42] and Holmes and Monro[43] have put forward a theory of structure of the crystal based on a cubic close-packing arrangement of protein sub-units. The evidence from electron microscopy suggests that the protein sub-unit is rod shaped or ellipsoidal measuring approximately 47Å by 120Å in Serotype 9 strains. Fig. 8 shows an electron micrograph of a thin section through a crystal *in situ* inside a developing sporangium.

The protein of the crystal is synthesized from amino-acids resulting from breakdown of intracellular protein during a massive "protein turnover" in the cell during the early phases of sporulation. The toxic inclusions are not formed by simple crystallization of a protein which exists in solution in the vegetative cell prior to sporulation. An interesting idea has recently been advanced to explain the "crystallization" of the protein as it is being formed; silicon in quantities of the order of 0·3–0·5% (w/w) can be detected in pure crystals, and it is suggested that protein may be laid down on a silicon skeleton. This suggestion, if correct, could account for the requirement for highly alkaline conditions for solution of the protein, as well as the crystallization phenomenon.[44] Similar silicon contents have been detected in the crystalline inclusion bodies of polyhedrosis viruses, organelles which show other similarities in structure, chemical characteristics and solubility to the *B. thuringiensis* crystals.

The protein of the crystal has an amino acid composition which is not in any way unusual.[36] When the crystals are dissolved in alkali (0·01–0·05 N NaOH), or in buffers containing reducing agents and/or urea,[45, 46, 47, 48] the resulting solution contains several proteins which differ in their molecular size and antigenicity and are not all toxic for caterpillars. The picture emerging from these studies is a complex one and the reader should consult the review by Heimpel[36] and the paper by Cooksey[48] for detailed discussions.

An enzyme complex obtained by methanol precipitation from regurgitated gut juice of *Pieris brassicae* may be used to digest *B. thuringiensis* crystals.[40] The protein is broken down into several fragments, one at least of which is dialyzable and toxic for *Pieris* by injection. Angus[50] worked with the silkworm, *Bombyx mori*, and was unable to demonstrate release of a dialyzable toxin active by injection when crystals of a Serotype 4A (sotto) organism were digested with silkworm gut juice.

The chemical structure of the crystal toxin is complex and solutions of the protein contain several different components as judged by physical separation or by antigenic composition.[51, 52, 53, 48] Toxic fragments with molecular weights of the order of 40,000 have been obtained by several workers using proteolytic digestion with caterpillar gut juice and a variety of pure proteases. Smaller toxic molecules have occasionally been reported.[48, 49]

(*a*) *Mode of action of the crystal toxin.* When a mixture of spores and crystals is ingested by a silkworm the crystals pass to the midgut where they dissolve in the alkaline gut fluid and are digested by proteolytic enzymes. The gut is rapidly paralysed and changes occur in the permeability of the gut wall. These permeability changes are associated with breakdown of the gut epithelium the cells of which loosen from the basement membrane and from one another. Changed gut permeability results in a leakage of alkaline gut contents into the haemocoel and the rise in blood pH causes a general body paralysis. Spores germinate in the gut after its pH has fallen and vegetative cells invade the haemocoel liberating enzymes which attack the host tissues and soon reduce it to a blackened cadaver. The vegetative cells produce spores and crystals which are released when the cadaver disintegrates. This sequence of events is characteristic of a few, highly susceptible caterpillars, including the silkworm, the Chinese Oak silkworm, the Tobacco hornworm and the Tomato hornworm, which have been labelled Type I insects.[54]

The majority of susceptible lepidoptera fall into a second group, the so-called Type II insects. These larvae develop gut paralysis shortly after feeding on spore/crystal mixtures, but there is no gut leakage, no change in blood pH and no general paralysis.

The caterpillars die more slowly than is the case with Type I larvae and the processes leading to death are complex and incompletely understood. In many cases there appears not to be an extensive proliferation of *B. thuringiensis* in the tissues of the moribund larvae although there is often extensive growth of other bacterial species (usually derived from the gut microflora) during the terminal stages. The relative importance of toxins, starvation and bacterial growth in the disease process, are not at present known.

Both Type I and Type II insects are susceptible to pure crystal preparations or to spore-free solutions of crystal protein. A further pattern of behaviour is shown by two Type III insects, *Anagasta kühniella* and *Porthetria dispar*, which are only killed by spores and crystals acting together.

A fourth class of insects was described by Martouret,[55] who reported certain noctuids as not susceptible to the crystal toxin. There is, however, considerable variation in susceptibility of different insect species for various strains of *B. thuringiensis* and it is probable that some degree of susceptibility can be demonstrated with all lepidoptera.[36]

Studies of the mode of action of the toxin have been largely concerned with the disintegration of the epithelium lining the gut. Heimpel and Angus[54] proposed that the cell-cementing substances might be a substrate for the toxin and it is certainly true that N-acetylglucosamine, probably released from the mucopolysaccharide cell-cementing substance, increases in the blood of silkworms 60 min after feeding. Changes in gut permeability following intoxication are more complex than was at first supposed.[56] The transfer of carbonate from gut lumen to haemocoel increases during the development of general paralysis in silkworms, but glucose travels in the reverse direction. There is, however, no clear indication as yet of the mode of action or location of receptor sites for the protein toxin. No explanation has been proposed for the initial paralysis of the gut which is a common feature of intoxication in widely different lepidopterous species. Recent observations in the laboratories of "Shell" Research Ltd suggest that toxin released by digestion of crystal protein by *Pieris* gut juice has a blocking action for cholinergic synapses in the central nervous system of the cockroach. Whether this pharmacological action, which is not shown by undigested protein, has any relationship to the mode of action of the toxin *in vivo* it is not, as yet, possible to say.

2. Thermostable exotoxin (fly factor; fly toxin; exotoxin; B.t.—β exotoxin)

The presence of a heat-stable toxin in the supernatants of cultures of certain strains of *B. thuringiensis* was first demonstrated by McConnell and Richards,[57] who showed that supernatants of autoclaved cultures killed larvae of the Wax moth (*Galleria mellonella*) when injected into the body cavity. Further studies have shown that supernatants of cultures of Serotype 1 and 9 organisms contain a thermostable toxin active against a wide range of species including Lepidoptera, Diptera, Hymenoptera, Coleoptera and Orthoptera.[58, 59, 60, 61, 62, 63, 64]

Although there is a great deal of confusion concerning the species activity spectra of the various preparations described and their activity when administered by different routes, it is probable that all of these reports concern the same toxin.[47]

Chemical studies[64, 65, 66, 67] of exotoxin from *B. thuringiensis* and from the non-crystalliferous *B. gelechiae* show that the toxin molecule contains adenine, phosphate (in a 1 : 1 ratio) and an as yet unidentified sugar moiety.

The effects of feeding small doses of exotoxin to larvae are only seen at moulting or during metamorphosis. Intoxicated larvae of *Musca domestica* either fail to pupate properly, or adults fail to emerge from the pupae, or emerge deformed, depending on the amount of toxin consumed. It is suggested that the toxin acts by interfering with hormonal control of these physiological changes or possibly with protein synthesis. Purified exotoxin kills housefly larvae when fed to them at doses as low as 0·5 μg/ml food and the LD_{50} for *Galleria mellonella* by injection is of the order of 0·005 μg/larva (Boyce, personal communication). The exotoxin exhibits toxicity for mice at oral dosages of 400–800 mg/kg body weight and 100–200 mg/kg by injection when administered in single doses. Higher toxicity is observed when exotoxin is administered as repeated doses.

Exotoxin is synthesized by vegetative cells during the logarithmic growth phase and accumulates in the culture fluid prior to spore and crystal formation. Cultural conditions profoundly influence the amount of exotoxin synthesized, production being maximal when the bacterium is grown on complex growth media containing, for instance, casein hydrolysate. In defined media valine stimulates production of exotoxin and leucine and isoleucine depress it.[68] Glucose in the medium has a markedly depressant effect on exotoxin production. In complex growth media of the type normally used to grow *B. thuringiensis* commercially, exotoxin is produced in concentrations of the order of 20 mg per litre.

3. Thermolabile exotoxin

When certain batches of "Thuricide", a wettable powder formulation of *B. thuringiensis* (Serotype I) prepared for use as a field insecticide, are extracted with water, the sterile-filtered extracts contain a toxin which is rapidly inactivated by heat, oxidation and u.v. irradiation. The toxin is lethal, on feeding, for 19 species of sawflies, but its lability will probably prevent its use in the field for control purposes. Toxic extracts contain free amino acids and polypeptides, but the nature of the toxin itself is not yet known.[69, 70]

4. Exocellular enzymes

Exocellular enzymes, especially phospholipase and hyaluronidase, have been implicated in the disease process,[71] but the evidence seems slight. Certainly histological examination of bacillus-infected larvae shows extensive damage to midgut cells and the damaged tissues are known to contain phospholipids and hyaluronic acid cell-cementing substances. It is clear, however, that purified crystal toxin alone can produce similar tissue damage and there is no evidence suggesting that the pathogenicity of a particular bacterial strain correlates with its ability to produce exocellular enzymes. Again, most of the toxic effects are seen before the spores germinate, but enzymes are only synthesized by actively growing vegetative cells. Exocellular enzymes produced by *B. thuringiensis* will certainly play a part in the terminal destruction of body tissues of infected larvae, but it is unlikely that they are to any extent involved in the events preceding that stage.[47]

E. *BACILLUS THURINGIENSIS* AS AN INSECTICIDE

When Berliner[26] and Mattes[27] isolated and described *B. thuringiensis* it was immediately clear that this spore-forming bacterium, growing readily on normal bacteriological media, highly infectious for lepidopterous larvae, non-pathogenic for humans and animals and exhibiting a long storage life, might have some potential as an insecticide. Extensive field trials of spore dusts, particularly against the European Corn Borer, during the 1930s were encouraging[72] and the first commercial preparation (Sporeine) was produced by a French firm just before the Second World War. Sporeine was a dust consisting of dried bacterial spores (and presumably protein crystals, although their presence was not recognized) and an inert carrier (bentonite).[73]

Shortly after the war Steinhaus[74] achieved impressive control of Alfalfa caterpillar in California, using a spore dust produced in the laboratory, and this work, more than any other, stimulated widespread interest in the potential of *B. thuringiensis* for pest control. Several companies in the United States, and subsequently in other parts of the world, began to explore the possibility of mass producing the bacterium and formulating it for use against a variety of lepidopterous pests.

Unlike *Bacillus popilliae*, *B. thuringiensis* grows vigorously on bacteriological media and sporulates with ease under normal fermentation conditions in periods of 28–36 hours after inoculation. Growth media used on the commercial scale are based on complex semi-soluble natural materials such as cottonseed meal, soybean meal and fish meal. Although, at first, bacteria were grown on solid media, submerged culture in liquid medium soon became the production method of choice. When the cells reach sporulation, they have converted part of the medium components into bacterial cells, spores, crystal protein, soluble toxins, exocellular enzymes and other products. When commercial media are used much insoluble material originating from the medium also remains in suspension. The properties of the final product will be influenced by the nature of the medium and by the method used to harvest the spores and crystals from the spent fermentation liquid.

One fermentation process is carried out in a semi-solid medium based on bran soaked in growth medium and aerated with moist air in a tower fermenter. At the end of fermentation, the mass is dried by passing dry air through it and the product ground to a fine powder. Such a preparation contains viable spores, protein crystals, undigested medium components and also any substances, such as thermostable exotoxin, that are dissolved in the growth medium and resist the drying process. Fluid cultures are normally harvested by centrifugation followed by drying or direct formulation of the cell paste as a flowable liquid. Depending on the bacterial strain used and the amount of culture fluid components present in the final product, the preparations will vary widely in their content of exotoxin. To the harvested fermentation products are added diluents, carriers, stickers, ultraviolet light absorbents and so forth as required by the individual product to ensure good performance in the field. A wide range of such formulation additives has been found compatible with *B. thuringiensis*.[75]

The first commercial materials containing *B. thuringiensis* became available for testing in the United States in 1958 as wettable powder and dust formulations. The dust formulations gave good coverage and proved effective in controlling Cabbage looper on lettuce, cabbage and cauliflower. Wettable powders were at first unsuccessful, but improvements in formulation have led to their more extensive use, especially against Alfalfa caterpillar. *B. thuringiensis* insecticide is also available as a stabilized spore-crystal suspension (flowable liquid) and has been used in granular formulations against European cornborer, as encapsulated formulations, and as corn-meal baits for Tobacco budworm. The flowable liquid is a very effective preparation but proves to be less stable than dry powders which can be stored for ten years without loss of virulence. Nevertheless, stability of the liquid formulations is quite adequate for normal storage requirements.[36] Many preparations are today available for field tests or for sale from industry or governmental agencies. The trade names include: Agritrol, Bakthane, Biotrol, Parasporin and Thuricide from the United States; Bactospeine and Sporeine from France; Bathurin from Czechoslovakia; Biospor from Germany; Dendrobacillin and Entobakterin from the U.S.S.R.

During the last decade many field trials and laboratory assays of *B. thuringiensis*-based insecticides against a wide range of insect pests have been reported and many more trials by commercial firms have never appeared in the literature. These tests have achieved several objectives; they have defined the range of insect species susceptible to the various preparations, they have demonstrated that, under suitable conditions, *B. thuringiensis* preparations will effectively control pest insects in the field, and they have shown that these insecticides are devoid of toxicity for higher animals, plants, economically valuable insects and insect predators. In addition, they have emphasized the importance of carefully selecting and developing strains of *B. thuringiensis* with high toxicities and spectra of activity against different insect species relevant to their intended use in the field. There is considerable variation in the toxicity of different bacterial strains when tested against one particular insect[76, 77, 78] or when individual strains are tested against a range of insects.[79] The careful selection of bacteria for high toxicity to chosen test insects has led to marked improvements in the dose/effectivity characteristics of formulations over the past few years, but the potential for tailor-making insecticides to fit particular field outlets by strain selection and improvement has scarcely been exploited. An interesting and instructive study along these lines was recently reported.[80] Because of possible introduction of disease organisms into the silk industry, work with *B. thuringiensis* is severely restricted in Japan. In order to overcome these restrictions, Japanese workers have selected bacterial strains for low toxicity to the silkworm but high toxicity for pest lepidoptera. They have produced a strain which has an LD_{50} (spores/larva) of 10^6–10^7 for the silkworm but only 10^4 for the Cabbage worm (*Pieris rapae*), 10^4–10^5 for the Fall webworm (*Hyphantria cunea*) and less than 10^5 for the Rice stem borer (*Chilo suppressalis*).

The factors influencing the relative toxicities of different bacterial strains for different insect species are poorly understood. An attempt to select a strain of a Serotype 9 *B. thuringiensis* for high toxicity to *Pieris brassicae* yielded an organism highly pathogenic for that species and for the Flour moth (*Anagasta kühniella*) but with low toxicity for the Wax moth (*Galleria mellonella*). By comparison a preparation based on a Sero-

TABLE II

The activity of three B. thuringiensis *preparations against three species of lepidoptera*

| Serotype | Toxic Units/mg dry weight against: | | |
	Pieris brassicae	*Anagasta kühniella*	*Galleria mellonella*
1 (standard)	1,000	1,000	1,000
9	10,150	27,450	847
5	2,950	7,180	121,000

type 5 organism was highly active against *Galleria* and relatively inactive against the other two species (Table II) (Norris and Burges, unpublished).

It is beyond the scope of this chapter to review the mass of field assay and toxicological data which exists on *B. thuringiensis* insecticides. The reader is referred to recent reviews by Franz,[81] Martouret and Milaire,[82] Jaques,[83] Vankova[84] and Herfs.[75] The safety of these preparations for man and a range of animals, birds and fish has received exhaustive examination and readers are referred to the review by Heimpel[36] and to the paper by Fisher and Rosner.[85]

F. STANDARDIZATION OF *B. THURINGIENSIS* INSECTICIDES

For small-scale laboratory work, field experimentation and production of insecticides for the market, methods of comparing toxicities and for standardization of preparations are essential. With chemical insecticides the active principles can often be measured by chemical analysis and standardization is relatively simple, being expressed as a statement of percentage of active ingredient in a formulation. With *B. thuringiensis*, however, the problem is more difficult since there are at least three active principles (spores, crystal toxin and exotoxin) produced in differing amounts during manufacture. None of these can be determined chemically at the present time.

Pure spores alone have little insecticidal activity. The protein toxin is a potent stomach poison for susceptible lepidoptera having an LD_{50} as low as 0·25 μg, pure crystals/gm of body weight for *Pieris brassicae* larvae.[86] The exotoxin attacks a wide range of insects and is also highly active in the pure form, 50% inhibition of emergence of pupae of *Galleria mellonella* being seen after administration of approximately 0·005 μg pure exotoxin/larvae (Boyce, personal communication). Although these components cannot be measured chemically they can be determined by biological methods; viable spore counts and bioassay of the toxins against suitably susceptible insects.

At an early stage in the development of *B. thuringiensis* insecticides it was assumed that insecticidal activity was proportional to the content of viable spores. This is not necessarily the case, however, since the amounts of toxins present vary greatly, depending on the manufacturing process, and the choice of bacterial strain. There is also considerable batch to batch variation for any particular process. Although the viable spore count is still used as a standardization method in the United States, its inadequacy for the purpose is now widely recognized and it is being replaced by bioassy methods. Different insects, however, vary widely in their susceptibility to different bacterial preparations and the choice of assay insect is critical. Activity may be measured in several ways. Crystal toxin may be indicated by direct mortality of an assay caterpillar or indirectly by inhibition of feeding since intoxicated larvae cease to feed, although they may remain alive for some time. Exotoxin is conveniently indicated by failure of adult flies to emerge from pupae after larvae have been fed on treated food. Recommended assay species include, for crystal toxin, *Anagasta kühniella*, *Pieris brassicae*, *Estigmene acrea*, *Trichoplusia ni*, *Galleria mellonella* and *Bombyx mori*, and for exotoxin, *Musca domestica* and *Drosophila melanogaster*.

Bioassay techniques, especially where the situation is a complex one like this, require the use of standard reference materials. The Pasteur Institute in Paris holds reference preparations for *B. thuringiensis* crystal toxin and exotoxin. In addition commercial firms use internal standards of their own. The crystal toxin standard is defined as having 1,000 insect toxin units/mg and the number of units in a test material is then calculated:

$$\frac{\text{Measure of effect of the standard}}{\text{Measure of effect of the test}} \times 1,000 \text{ units of toxicity}$$

The assay insect must, of course, be specified.

e.g. $\dfrac{\text{L.D.}_{50} \text{ standard}}{\text{L.D.}_{50} \text{ test}} \times 1,000$ *Pieris* toxicity units

The bioassay and standardization of *B. thuringiensis* preparations has

been the subject of an extensive exploratory programme organized inter-
nationally among laboratories working in this field; the results of this study
have been fully presented and the various issues discussed in detail by
Burges.[87]

G. THE PRESENT STATUS OF *B. THURINGIENSIS* INSECTICIDES

Bacillus thuringiensis-based products are now produced at the rate of
hundreds of tons per year in several countries and have been certified by the
appropriate licensing authorities for use on a number of crops. Products
registered with the U.S. Department of Agriculture are exempt from
residue tolerance requirements and are registered for use on the following
crops: alfalfa, artichokes, banana, broccoli, cabbage, cauliflower, celery,
forest trees, lettuce, melons, potatoes, tomatoes and tobacco.

Laboratory tests show that *B. thuringiensis* is active against over 120
insects including many pests of agricultural crops, forests, orchards, vine-
yards and stored foods. Control of the following species in the field is a
commercial reality or at the development stage of extensive field trials:
Cabbage looper (*Trichoplusia ni*), Cabbage worm (*Pieris rapae*), Tobacco
hornworm (*Protoparce sexta*), Tobacco budworm (*Heliothis virescens*),
Tomato hornworm (*Protoparce quinquemaculata*), Alfalfa caterpillar (*Colias
eurytheme*), Gypsy moth (*Porthetria dispar*), Spring cankerworm (*Palae-
crita vernata*), Fall cankerworm (*Alsophila pometria*), Tent caterpillars
(*Malacasoma* sp.), Plume moth (*Platyptilia pusillodactyla*), general defolia-
ting and peel scarring caterpillars (*Ceramidia* sp., *Platynota* sp., *Opsiphanes*
sp.) and the Wax moth (*Galleria mellonella*).

Control of pests on crops, forest and orchard trees and in stored products
is largely dependent on the activity of the protein crystal toxin and the
target species are lepidoptera. Another possible application concerns the
exotoxin. When preparations of *B. thuringiensis* containing exotoxin are
fed to chickens or calves the toxin survives passage through the gut and
persists in the faeces—or in some cases bacterial growth derived from
spores in the faeces may actually cause it to be produced there. In either
case the development of fly larvae in the faeces is inhibited.

Perhaps the best way to assess the present position of *B. thuringiensis*
is to summarize the advantages and disadvantages of this bacterium in
insect control.

1. Advantages

(*a*) The products are safe. They show no toxicity for man, mammals,
     birds, bees, earthworms, fish or plants. Even specificity among the
     insects themselves is quite striking. Other than the target lepidop-
     tera only a few Diptera and sawflies are known to be susceptible. No

useful predatory or parasitic insects have been found to be susceptible.

(b) The use of the preparations leads to little detrimental effect on the biocoenosis.

(c) Development of resistance in insect populations is so far unknown.

(d) Formulations show good resistance to adverse conditions. Even ultraviolet light, which kills spores on leaves exposed to sunlight, often has little effect on the toxicity of the preparation, since the toxins are much more resistant to ultraviolet irradiation.

(e) The bacterium and its toxins mix readily with a variety of formulation additives, chemical insecticides, bait materials and insect hormones, without significant loss of virulence.

(f) The cost of preparations, which at first was high, has been reduced by developments in production methods and is now comparable to that of chemical insecticides on a cost-effectivity basis.

2. Disadvantages

(a) The specificity of the preparations for lepidoptera, an advantage from the point of view of limited interference with the environment, is at the same time a disadvantage when the treatment of a crop with several pests of different types is proposed.

(b) Strict timing of field application is sometimes necessary.

(c) The effective period in the field may in some cases be as short as two weeks.

(d) The preparations are stomach poisons and will only be effective against actively feeding insects. This necessitates a high degree of efficiency in coverage which is not always easy to attain.

(e) Larvae cease to feed very quickly after they have ingested B thuringiensis preparations but they may not die for several days. It has proved difficult to explain to farmers that living insects are not causing economic damage to his crops.

Consideration of these characteristics leads to the conclusion that B. thuringiensis insecticides are likely to find their first applications in situations where crops are damaged principally by lepidoptera and the use of chemical insecticides is restricted by problems of toxicity or developing insect resistance. This appears to be the pattern of development at the present time. In the south-eastern United States, government regulations on insecticide residues on tobacco placed the grower in a position where he almost had to try the bacillus. The stimulus of a potentially large market led to the development of a bait-dust technique leading to effective control of the two major pests, Tobacco hornworm and Tobacco budworm. In California, state restrictions on the use of chemicals and developing resistance in the insect populations have created a broad demand for *B. thuringiensis* preparations on lettuce and cole crops, the primary target being the Cabbage looper. A recent State recommendation has been made for the use of the bacterial agents for control of the Grape leaf folder.[88] Slightly different considerations have led to the development of a method of using *B. thuringiensis* to control the Wax moth in infested beehives.[89] Here the difficulties of using conventional insecticides are toxicity for bee larvae and toxic residues contaminating honey.

V. The Future of Spore-forming Bacteria as Insecticides

In comparing *B. popilliae* and *B. thuringiensis* it is important to recognize that they represent two quite different approaches to insect control. *B. popilliae* is introduced in relatively small amounts into the Japanese beetle population which it colonizes. It then spreads by natural processes and becomes established in the environment where it persists for many years, reducing the pest to numbers which are no longer economically significant. In only one isolated case[90] has *B. thuringiensis* been found to persist in an insect population after use in the field. Control applications use *B. thuringiensis* as though it were a simple chemical agent toxic for the target pest.

The effectiveness of *B. popilliae* is well demonstrated. The factor limiting its use at the present time is the cost of manufacture arising from the need to grow the bacterium *in vivo* in infected grubs. When growth media and fermentation methods are developed which allow of large-scale and economic production of infective spores, the rapid extension of the present control programme, with dramatic effects on the Japanese beetle infestations in the United States and Canada, can be expected.

The future of *B. thuringiensis* as an insecticide is less easy to forecast. The efforts of insect pathologists have provided a range of formulations which are safe to use and reasonably effective in the field against a range of lepidopterous pests without killing useful insects such as bees, hymenopterous pollinators, and predators of insects. They have the advantage of a broad spectrum of activity as compared with the other prominent biological control agents, the viruses, which are usually species specific. On the face of it they should be highly successful, but their growth on the market has been slow, suggesting that they are not particularly attractive to growers. This is undoubtedly due to the variability exhibited by the preparations in the field and the fact that there are relatively few outlets available where lepidoptera are the only economically important pests.

Variability of performance is a technological problem. Improvements in the selected bacterial strains, their fermentation, formulation, and distribution and a better understanding of the factors influencing susceptibility of an insect in the field have already done much to improve the efficiency and reliability of *B. thuringiensis* insecticides and this trend will certainly continue. The exploitation of wider outlets depends on a more thorough understanding of insect complexes in the field. Experience with the Cabbage looper in Arizona and California indicates probable lines of development in this field. Control of the Cabbage looper is complicated by the fact that it has developed resistance to some insecticides and by the necessity of avoiding residues at harvest. By using a carefully selected chemical insecticide to control other pests such as thrips and aphids in combination with *B. thuringiensis*, an acceptable control of the pest complex can be achieved without residue problems being encountered.

It can be expected that with the developing pattern of agriculture towards intensive, large-scale, single crop operation the needs of specific pest situations—especially where resistance problems occur or populations of different susceptibilities and resistance exist on plants—will necessitate expansion of integrated control procedures, not only microbial–chemical but also microbial–microbial, where different pathogens will be applied to suppress the total pest population. The future of microbial insecticides probably lies not in replacing chemical control agents but in playing a part with them in integrated programmes.

There is another aspect to the future development of microbial insecticides. The search for chemicals which are highly toxic for insects but which have no effect on other forms of life is a notoriously difficult one. It implies the discovery and exploitation of features in the structure, biochemistry or behaviour of insects which differentiate them from other living things. The micro-organisms which infect insects are usually highly specific for their hosts—in other words, they have found and exploited just those features for which man is looking in his search for new insecticides. When we understand fully the mechanisms by which micro-organisms and their toxins kill insects we shall have a much deeper understanding of the differences between insects and other life forms, and the knowledge may well open up entirely new approaches to the development and use of insecticidal chemicals.

Attempts to use micro-organisms to control insect pests have, so far, been made only at the crudest of levels; the blanket introduction of the organism and its toxins into the environment with little understanding of the way in which the agent causes disease or spreads and persists in the insect population. Nor have attempts to produce in the laboratory new pathogens of enhanced virulence received anything like the attention they merit. As more is learned of the biology of the processes involved, more refined products and approaches will undoubtedly become available. The horizons of microbial control which are evident today may well appear in retrospect to be remarkably limited.

z

514   J. R. NORRIS

ACKNOWLEDGEMENTS

I am grateful to Dr S. H. Black of Baylor University College of Medicine and to the Editors of the *Journal of Bacteriology* for permission to reproduce Figs. 1–6 and to "Shell" Research Ltd. for permission to quote unpublished work.

REFERENCES

1. Steinhaus, E. A. (1946). *Bact. Rev.*, **10**, 51.
2. Heimpel, A. M. and Angus, T. A. (1958). *Proc. 10th Int. Congr. Entomol. Montreal, 1956*, **4**, 711.
3. Heimpel, A. M. (1961). *J. Insect Path.* **3**, 271.
4. Heimpel, A. M. (1955). *Can. J. Zool.* **33**, 311.
5. Cheshire, F. R. and Cheyne, W. W. (1885). *J. roy. Microscop. Soc.* **5**, 581.
6. Heimpel, A. M. and Angus, T. A. (1963). *In* "Insect Pathology", Vol. 2, p .21. Academic Press Inc., New York.
7. Patel, N. G. and Cutkomp, L. K. (1961). *J. econ. Entomol.* **54**, 773.
8. Patel, N. G. and Gochnauer, T. A. (1959). *J. Insect Path.* **1**, 190.
9. Bucher, C. E. (1957). *Can. J. microbiol.* **3**, 695.
10. Bucher, G. E. (1960). *J. Insect Path.* **2**, 172.
11. Bucher, G. E. (1961). *Can. J. Microbiol.* **7**, 641.
12. Dutky, S. R. (1940). *J. agric. Res.* **61**, 57.
13. White, R. T. (1947). *J. econ. Ent.* **40**, 912.
14. Harris, E. D. (1959). *Fla. Ent.* **42**, 181.
15. Wille, H. (1956). *Mitt. schweiz. ent. Ges.* **29**, 271.
16. Hurpin, B. (1959). *Entomophaga* **4**, 233.
17. Beard, R. L. (1956). *Can. Ent.* **88**, 640.
18. Dutky, S. R. (1963). *In* "Insect Pathology", Vol. 2, p. 98. Academic Press Inc., New York.
19. Lüthy, P. and Ettlinger, L. (1967). *Proc. Int. Colloq. Insect Pathol. Microbial Control, Wageningen, 1966*, p. 54.
20. Rhodes, R. A., Sharpe, E. S., Hall, H. H. and Jackson, R. W. (1966). *Appl. Microbiol.* **14**, 189.
21. Costilow, R. N., Sylvester, C. J. and Pepper, R. W. (1966). *Appl. Microbiol.* **14**, 161.
22. Rhodes, R. A. (1967). *Microbial Control of Insect Pests* (Abstr.), p. 31. United States–Japan Committee on Scientific Co-operation Seminar, Fukuoka.
23. Mitruka, B. M., Costilow, R. N., Black, S. H. and Petter, R. E. (1967). *J. Bact.* **94**, 759.
24. Aoki, K. and Chigasaki, Y. (1915a). *Mitt. med. Fak. K. Univ.* **13**, 419.
25. Aoki, K. and Chigasaki, Y. (1915b). *Mitt. med. Fak. K. Univ.* **14**, 59.
26. Berliner, E. (1915). *Z. angew. Ent.* **2**, 29.
27. Mattes, O. (1927). *Sber. Ges. Beförd ges. Naturw. Marburg* **62**, 381.
28. Norris, J. R. (1967). *J. gen. Microbiol.* **26**, 167.
29. Hannay, C. L. (1953). *Nature*, **172**, 1004.
30. Hannay, C. L. (1956). *Symp. Soc. gen. Microbiol.*, **6**, 318.
31. Hannay, C. L. and Fitz-James, P. C. (1955). *Can. J. Microbiol.* **1**, 694.
32. Angus, T. A. (1954). *Nature* **173**, 545.
33. Angus, T. A. (1956). *Can. J. Microbiol.* **2**, 416.
34. Norris, J. R. (1964). *J. appl. Bact.* **27**, 439.
35. DeBarjac, H. and Bonnefoi, A. (1967). *C.r. hebd séanc. Acad. Sci., Paris* **264**, 1811.

36. Heimpel, A. M. (1967). *A. Rev. Ent.* **12**, 287.
37. Smith, N. R., Gordon, R. E. and Clark, F. E. (1952). "Aerobic Sporeforming Bacteria". *U.S. Dept. Agr. Misc. Pub. No. 16.*
38. Lysenko, O. (1963). In "Insect Pathology", Vol. 2, p. 1. Academic Press Inc., New York.
39. Norris, J. R. and Burges, H. D. (1963). *J. Insect Path.* **5**, 460.
40. Fitz-James, P. C. and Young, I. E. (1959). *J. Bact.* **78**, 743.
41. Smirnoff, W. A. (1963). *J. Insect Path.* **5**, 242.
42. Labaw, L. W. (1964). *J. Ultrastruct. Res.* **10**, 66.
43. Holmes, K. C. and Monro, R. E. (1965). *J. mol. Biol.* **14**, 572.
44. Estes, Z. E. and Faust, R. M. (1966). *J. Invert. Path.* **8**, 145.
45. Young, I. E. and Fitz-James, P. C. (1959). *J. biophys. biochem. Cytol.* **6**, 483.
46. Lecadet, M. (1966). *C.r. hebd. séanc. Acad. Sci., Paris* **262**, 195.
47. Rogoff, M. H. (1966) *Adv. appl. Microbiol.* **8**, 291.
48. Cooksey, K. E. (1968). *Biochemical J.* **106**, 445.
49. Lecadet, M. and Martouret, D. (1962). *C.r. hebd. séanc. Acad. Sci., Paris* **254**, 2457.
50. Angus, T. A. (1964). *J. Insect Path.* **6**, 254.
51. Krywienczyk, J. and Angus, T. A. (1965). *J. Invert. Path.* **7**, 175.
52. Pendleton, I. R. and Morrison, R. B. (1966). *J. appl. Bact.* **29**, 519.
53. Pendleton, I. R. and Morrison, R. B. (1967). *J. appl. Bact.* **30**, 402.
54. Heimpel, A. M. and Angus, T. A. (1959). *J. Insect. Path.* **1**, 152.
55. Martouret, D. (1961). *Symp. Phytopharm. Phytiat. 12th, Ghent, Belgium* **8**, 1.
56. Fast, P. G. and Angus, T. A. (1965). *J. Invert. Path.* **7**, 29.
57. McConnell, E. and Richards, A. G. (1959). *Can. J. Microbiol.* **5**, 161.
58. Burgerjon, A. and DeBarjac, H. (1964). *Entomophaga* **2**, 121.
59. Hall, J. M. and Arkawa, K. Y. (1959). *J. Insect Path.* **1**, 351.
60. Liles, J. N. and Dunn, P. H. (1959). *J. Insect Path.* **1**, 309.
61. Briggs, J. D. (1960) *J. Insect Path.* **2**, 418.
62. Herfs, W. (1963). *Entomophaga* **8**, 163.
63. Kreig, A. and Herfs, W. (1963). *Z. PflKrankh. PflPath. PflSchutz* **70**, 11.
64. Cantwell, G. E., Heimpel, A. M. and Thompson, M. J. (1964), *J. Insect Path.* **6**, 466.
65. DeBarjac, H. and Dedonder, M. R. (1965). *C.r. hebd. séanc. Acad. Sci., Paris* **260**, 7050.
66. Benz, G. (1966) *Experientia* **22**, 81.
67. Sebesta, K., Horská, K. and Vanková, J. (1967). *Proc. Int. Colloq. Insect Pathol. Microbial Control, Wageningen, p.* 238.
68. Conner, R. M. and Hansen, P. A. (1967). *J. Invert. Path.* **9**, 114.
69. Smirnoff, W. A. (1964). *Entomophaga* **2**, 249.
70. Smirnoff, W. A. and Berlinquet, L. (1966). *J. Invert. Path.* **8**, 376.
71. Heimpel, A. M. (1955). *Can. J. Zool.* **33**, 311.
72. Husz, B. (1931). *Sci. Rept. Intern. Corn Borer Invest.* **4**, 22.
73. Jacobs, S. E. (1950). *Proc. Soc. appl. Bact.* **13**, 83.
74. Steinhaus, E. A. (1951). *Hilgardia* **23**, 1.
75. Herfs, W. (1965), *Z.PflKrankh. PflPath. PflSchutz* **72**, 584.
76. Broersma, D. B. and Buxton, J. A. (1967). *J. Invert. Path.* **9**, 58.
77. Angus, T. A. (1967). *J. Invert. Path.* **9**, 256.
78. Angus, T. Z. and Norris, J. R. (1968). *J. Invert. Path.* (In press.)
79. Burgerjon, A. and Biache, G. (1967). *Entomologia exp. appl.* **10**, 211.
80. Aizawa, K. and Fujiyoshi, N. (1967). *Microbial Control of Insect Pests,*

(Abstr.) p. 29. U.S.–Japan Committee on Scientific Co-operation Seminar, Fukuoka.

81. Franz, J. (1961). *A. Rev. Ent.* **6**, 183.
82. Martouret, D. and Milaire, H. (1963). *Phytiat-Phytopharm.* **12**, 71.
83. Jaques, R. P. (1964). *Ann. entomol. Soc. Queb.* **9**, 17.
84. Vanková, J. (1964). *Entomophaga* **2**, 271.
85. Fisher, R. and Rosner, L. (1959). *J. agric. Fd Chem.* **7**, 686.
86. Martouret, D., Lhoste, J. and Roche, A. (1965). *Entomophaga* **10**, 349.
87. Burges, H. D. (1967). *Proc. Int. Colloq. Insect. Pathol. Microbial Control*, p. 293, *Wageningen.* North Holland Pub. Co., Amsterdam.
88. Dulmage, H. T. (1967). *Proc. Int. Colloq. Insect Pathol. Microbial Control*, p. 122, *Wageningen.* North Holland Pub. Co., Amsterdam.
89. Burges, H. D. (1966). *Am. Bee J.* **106**, 48.
90. Talalaev, E. V. (1959). *Trans. 1st Int. Conf. Insect Pathol. Biol. Control, Praha*, p. 51.
91. de Barjac, H. and Dedonder, R. (1965). *C.R. Acad. Sci., Paris* **260**, 7050.
92. Bond, R. P. M., Boyce, C. B. C. and French, S. J., submitted to *Biochem J.*
93. Šebesta, K., Horská, K. and Vankova, J. (1967). *"Proc. Int. Colloq. on Insect Pathol. and Microbial Control"*, p. 238, *Wageningen*, North Holland Pub. Co., Amsterdam.
94. Benz, G. (1966). *Experientia* **22**, 81.
95. Bond, R. P. M., submitted to *Chem. Commun.*
96. Farkas, J., Šebesta, K., Horská, K., Samek, Z., Dolejs, L. and Šorm, F. *Coll. Czech, Chem. Comm.* (In press.)
97. Šebesta, K. and Horská, K. (1968). *Biochim. Biophys. Acta.* **169**, 281.

NOTE ADDED IN PROOF

Purification or partial purification of the exotoxin(s) of *B. thuringiensis* has now been reported frm four labooratories.[91, 92, 93, 94] It is not yet established that the purified materials so obtained are identical. It is clear, however, that if not identical, all represent a closely related class of high molecular weight (800–900) adenine nucleotides containing one phosphate group/adenine. That the phosphate group is present as a mono-ester is clear from the results of Bond et al.,[95] who showed that the phosphate can be removed by acid or alkaline phosphatase or by chemical hydrolysis in buffer at pH 4. Removal of the phosphate with alkaline phosphatase has also been achieved by Šebesta et al. Both Bond and Šebesta et al. have found that allomucic acid is a constituent of their toxin [95, 96] and the Czech group have, in addition, presented evidence for the presence of glucose and ribose, linked by an unusual ether bond.[96] Their proposed structure as shown in structure on left. Although the mode of action of exotoxin in the insect is not known Šebesta and Horská[98] have shown that their toxin inhibits DNA dependent RNA polymerase when tested *in vitro* against purified enzymes of bacterial origin. Inhibition of RNA synthesis would be compatible with the effects observed when the toxin is administered to insect larvae.

CHAPTER 14

Medical and Veterinary Significance of Spore-forming Bacteria and Their Spores

A. C. BAIRD-PARKER

Unilever Research Laboratory, Sharnbrook, Bedford, England

I. Introduction

ALTHOUGH members of the genera *Bacillus* and *Clostridium* are mainly soil saprophytes or harmless inhabitants of the intestinal tract, certain species are highly pathogenic to members of the animal kingdom. This pathogenicity is frequently associated with a high degree of mortality resulting from the elaboration of extremely potent toxins either in the infected animal, in its food or in the associated environment of its habitat. In this chapter we are concerned with the many diseases and intoxications caused by spore-forming bacteria that are of either medical, public health or veterinary importance. Examples of these are the large range of diseases caused by the different types of *Cl. perfringens*, anthrax caused by *B. anthracis*, botulism caused by *Cl. botulinum* and tetanus caused by *Cl. tetani*. It is difficult to estimate the economic losses caused by these diseases and intoxications. Some idea of their importance can perhaps be gauged from a recent estimate[1] that more than 50,000 human deaths are caused every year by tetanus alone and that many times this number of deaths are caused in animals through botulism and anthrax. From the point of view of human disease, tetanus causes more deaths each year than any of the more publicized diseases such as small-pox, plague, leptospirosis or poliomyelitis.[1]

II. Clostridial Diseases and Intoxications

Clostridia are seldom able to infect healthy tissue. They cause infection and disease only when conditions permit growth and the formation of the powerful array of enzymes and toxins that are responsible for their invasion into healthy tissues and for their pathogenicity. Clostridia are infective and usually cause disease by one or other of two routes. The first of these results from the ingestion of spores or vegetative cells, which under rather ill-defined conditions in the gut are able to grow and release the toxins that are responsible for the disease symptoms. Examples of diseases caused by this route of infection are the various enteric diseases of animals such as lamb-dysentery caused by *Cl. perfringens* Type B, struck in sheep and lambs caused by *Cl. perfringens* Type C and infectious enterotoxaemia in sheep and calves caused by *Cl. perfringens* Type D. Some workers believe that *Cl. perfringens* Type A may cause a fatal toxaemia following obstruction of the intestine or acute appendicitis; the evidence for this is, however, rather controversial.[2] The infective route of *Cl. septicum* which causes braxy in sheep (a disease characterized by inflammation of the fourth stomach), remains obscure although evidence of infection by the intestinal route is strong.[3] The second main route of infection is via damaged tissue in which the conditions favourable to growth, i.e. a low Eh and/or oxygen tension, resulting from the growth of aerobic or facultative anaerobic micro-organisms or from the interruption of the blood supply to the damaged tissue. Examples of diseases caused by this route of infection are the various histotoxic clostridial diseases of man and animals such as oedematous myonecrosis (gas gangrene) which in man is caused by a group of toxigenic clostridia, of which the most important species are *Cl. perfringens* Type A, *Cl. oedematiens* (*Cl. novyi*) and *Cl. septicum*.[4] *Clostridium tetani*, which causes tetanus, is infective solely by the damaged tissue route, but *Cl. chauvoei*, which causes Black leg, a myonecrotic disease similar to gas gangrene, in cattle, sheep and pigs, is usually infective by the intestinal route in cattle and via damaged tissue in sheep.[2] In Black disease, which is an infectious necrotic hepatitis of sheep caused by *Cl. oedematiens*, the conditions favourable for the growth of this organism occur in necrotic areas in the liver resulting from the invasion of the liver with immature liver flukes.[2]

At least two species of clostridia can cause food poisoning in man. The most important of these is *Cl. botulinum*, which is able to grow in a variety of foods and form a neurotoxic protein which when ingested causes botulism. This toxin causes paralysis of the cholinergic and certain adrenergic nerves of the efferent autonomic nervous system supplying muscles and glands and the somatic nerves supplying skeletal muscles;[2, 5] it has been estimated that *c* 28 g of the toxin could kill 200 million people.[6] Botulism is a typical example of the intoxication type of food poisoning. The only other *Clostridium* species definitely capable of causing food poisoning in man is *Cl. perfringens*. *Cl. perfringens* Type A causes a relatively mild food poisoning; heat-resistant strains of Type C have also been implicated in a more serious type of food poisoning (*enteritis necroticans*), but the involvement of *Cl. perfringens* in this type of food poisoning has been disputed recently by several workers.[7, 8] *Cl. perfringens* Type A food poisoning is an example of the infective type of food poisoning. *Cl. bifermentans* was reported by Duncan[213] in a food-poisoning incident involving a meat and potato pie, but the evidence incriminating this organism as a food-poisoning agent is inconclusive.

A. CLOSTRIDIUM PERFRINGENS

1. Isolation and enumeration

Dilutions of suitably dispersed samples can be plated directly, or after enrichment in cooked meat medium or Differential Reinforced Clostridial Medium (DRCM),[9] on to a neomycin-blood agar[10, 11] or sulphite–polymyxin–sulphadiazine agar;[12, 13] further enumeration methods are described by Green and Litsky[15] and Marshall *et al.*[16] Suspect colonies of *Cl. perfringens* can be confirmed by streaking across half-α-antitoxin plates of Willis and Hobbs' medium[14] or by stab inoculation into nitrate-motility medium.[12]

Clostridium perfringens forms spores poorly in most foods and bacteriological media, so attempts at isolation should be made without pasteurization. If this fails, or if a spore count is required, samples should be heated at 60°C for 30–60 min. Heating at higher temperatures should be avoided unless only heat-resistant spores are to be detected or counted. In a recent publication, Duncan and Strong[208] found that heating at 60°C for 20 min was optimal for germination of heat-sensitive spores of *Cl. perfringens* and heating at 80°C for 20 min was optimal for heat-resistant spores.

2. Classification

Clostridium perfringens is divided into five toxigenic types (Types A to E) according to the types of "toxins" produced by the various strains. The pattern of toxins produced by a strain is usually characteristic of a particular toxigenic type and each of these types produces specific diseases in man and other animals.[2] The θ toxin-negative, heat-resistant strains of *Cl. perfringens* Type A can be divided into at least thirteen serotypes by slide agglutination.[10]

3. Distribution

The different toxigenic types of *Clostridium perfringens* are widely distributed in nature and are commonly found in human and animal intestines (particularly carnivora)[20] and in soils, dust and water and in milk, meat and vegetable products.

(*a*) *Soils*. Opinions are divided as to whether *Cl. perfringens* is primarily a soil saprophyte or an inhabitant of the intestinal tract.[29] It is well established that this organism is more frequently isolated from well-manured, cultivated soils than from clean desert soil,[17] but it is apparently not known whether this difference is because conditions in the soil permit growth and survival in one but not the other.[18]

(*b*) *Intestinal tract*. The human and animal intestinal tract is usually contaminated with *Cl. perfringens* during the first days of life.[19] Maximum numbers are present during the first week and thereafter decline in numbers. A study of the distribution of *Cl. perfringens* in the animal intestine[20] showed that large numbers of organisms were present in the anterior part of the stomach, lower numbers in the posterior part of the stomach and that numbers increased between the duodenum and the ileum. There is a pronounced influence of diet on the incidence of *Cl. perfringens* in the animal intestine. In general *Cl. perfringens* is absent or present only in small numbers in intestines of animals fed on cereal diets.[20] Feeding with meat results in an increase in the incidence of this organism. This was explained[20] as an effect of the growth-supporting abilities of the different diets. Further factors are probably the increased risk of infection from feeding a meat diet and also the effect of diet on conditions in the intestine. In a recent study[21] lysolecithin released by the hydrolytic cleavage of the bile lecithin and other lipids and phospholipids was suggested to be responsible for inhibiting the growth of *Cl. perfringens* in the small intestine of the pig. Sidorenko and Borovik[22] found that whereas only 6·2% of faeces from new-born babies contained *Cl. perfringens* Type A the incidence of this organism increased with age, until in persons aged 61 years and over the incidence was 78%. Previous studies by Sidorenko[23] showed that almost all strains isolated from human faeces belong to Type A; Type B, C and D strains were isolated only from persons who worked with animals or lived in cattle-breeding districts. Type A is usually found in the intestine of pigs[21] but other types may also be common in other animals.[2]

(*c*) *Heat-resistant strains*. The frequent involvement in food poisoning of weakly haemolytic or non-haemolytic strains of *Cl. perfringens* Type A that are able to produce spores of higher heat resistance than classical Type A strains has resulted in a large number of studies concerned with the distribution of these particular organisms. The incidence of heat-resistant spores able to survive 100°C for 1 hr[10] in faeces of persons not suffering from food poisoning caused by such strains is generally low. For example, an incidence of 1·5–6·0% was found for different communities in Australia[11] and 4·9% for food handlers in the United States.[24] Persons in institutions and hospitals generally have a higher carrier rate (15–30%),[11] The higher carrier rate in closed communities is probably the result of an exposure to a higher risk of infection. There is no evidence that heat-resistant strains can establish themselves as part of the gut flora; the carriage of these strains is transient and one serological type can be readily replaced by another.[11] The distribution of heat-resistant spores in meat and meat products has been studied in a number of different countries. Hall and Angelotti[25] found that although 43% of a variety of meat and meat products contained *Cl. perfringens* Type A, less than 1% of isolates were able to produce spores surviving heating for 30 min at 100°C. Studies in the British Isles have shown a variable but generally higher incidence of heat-resistant spores in carcass meat produced in this country. Thus Sylvester and Green[26] reported that 28% of abattoir carcasses were contaminated with heat-resistant spores of *Cl. perfringens* and that pig carcasses were most contaminated; Hobbs *et al.*[27] also found that the heat-resistant spores of this organism were more common in pig faeces than cattle faeces. Heat-resistant strains are probably uncommon in soil.[28]

4. The role of the spore in infection

In natural infections the spore is probably mainly involved in the initiation of infection. Lowbury and Lilly[184] showed that vegetative cells of *Cl. perfringens* die rapidly when dried on glass slides. However, it appears possible that vegetative cells of *Cl. perfringens* in faeces would be sufficiently protected to survive for some time in the vegetative state. Experimentally, infections have been induced by the use of both vegetative cells and spores.

(a) *Wound infection.* Neither spores nor vegetative cells can grow in healthy tissue as the Eh of live muscle (pH 6·0–6·5) is generally > + 250 mv,[30] which is too high for growth. Hanke and Bailey[31] reported that at the pH of 6·4 the growth of *Cl. perfringens* was inhibited at an Eh of + 160 mv. More recent observations[32] showed that vegetative cells of some strains of *Cl. perfringens* may grow after a long lag period at an Eh of about + 250 mv in meat digest broth at pH 6·0; most, however, grow only at a much lower Eh. Although no studies appear to have been done of the limiting Eh for growth from spores, there would appear to be no reason to assume that the limiting Eh would be very different to that of vegetative cells.

The most important condition for the initiation of infection by *Cl. perfringens* and other clostridia is the lowering of Eh and/or oxygen tension in tissue surrounding the infective spores or vegetative cells. The ways in which this is brought about in wound infections are listed by MacLennan[4] and others as follows:

(1) Failure of blood supply to the infected area caused by trauma, overtight torniquets or dressings, cold, sho.. ..r local oedema.
(2) The presence of foreign bodies, e.g. pieces of clothing, particles of soil and in experimentally induced infections, agar.
(3) The occurrence of necrotic tissue or haemorrhage as a direct result of trauma or the action of necrotizing agents, such as calcium chloride or medicants such as adrenalin or epinephrine.
(4) The presence of aerobic and facultative anaerobic bacteria.

The importance of oxygen tension in controlling the growth of *Cl. perfringens* in muscle has been recently demonstrated in *in vivo* experiments. The use of hyperbaric oxygen[33] or hyperoxygenation of tissue by local or regional injection of hydrogen peroxide[34] was shown to be effective in preventing certain experimental infections.

The importance of calcium chloride as an activator in *Cl. perfringens* infections has been known for a long time.[4] Recent studies by Princewill[35] using *Cl. chauvoei*, which causes similar histotoxic infections to *Cl. perfringens* in animals, have shown that calcium chloride may also play a direct role in infection by breaking spore dormancy.

(b) *Oral infection.* The conditions required to cause infection via the mouth are far from understood. Most information relates to infectious enterotoxaemia in sheep caused by *Cl. perfringens* Type D.

This disease generally results when an animal is transferred from a poor to a rich diet and is believed to be the result of a change in the conditions in the intestinal tract which results in an environment more favourable for the survival and growth of the organism. Bullen and colleagues[36, 37] observed that spores are usually destroyed in the rumen but that overeating results in a reduction of rumen acidity, allowing spores to reach the lower intestine. The reduction of rumen acidity also permits undigested food to reach the intestine and supply the necessary nutrients for growth and toxin formation. The previously mentioned studies of Smith[20] also demonstrate the importance of diet in affecting the growth of *Cl. perfringens* in the intestinal tract.

5. Clostridium perfringens *Type A food poisoning*

(a) *Symptoms.* The main symptom of *Cl. perfringens* Type A food poisoning is generally mild gastroenteritis usually occurring 8–16 hr after ingestion of food contaminated with this organism.[27] The clinical disease is characterized mainly by diarrhoea and abdominal cramp. Some patients also suffer nausea and headaches, but vomiting and fever are uncommon.[38]

(b) *Conditions for food poisoning.* Food poisoning is caused by the ingestion of a food, generally a meat product, containing at least 10^7–10^8/g of viable vegetative cells of *Cl. perfringens* Type A. Spores, dead vegetative cells or sterile culture filtrates are unable to cause food poisoning.[27, 39] Recent studies by Hauschild *et al.*[40] have shown that the age of the vegetative cell is important in causing food-poisoning symptoms in man. Early log phase cells are sensitive to gastric juices and only cause mild abdominal discomfort when fed to human volunteers, stationary phase cells are more resistant to gastric juices and cause typical food-poisoning symptoms. Recent studies by Duncan *et al.*[214] and by Hauschild *et al.*[221] have shown that by the use of ligated ileal loops it is possible to study *Cl. perfringens* food poisoning in experimental animals.

(c) *Strains involved in food poisoning.* Until recently it was accepted that *Cl. perfringens* Type A food poisoning was caused only by atypical strains of *Cl. perfringens* Type A[10, 27] that lacked θ toxin and produced only small amounts of α toxin and variable amounts of κ toxin; such strains form spores that are generally more heat resistant than the classical type and many survive heating for at least 1 hr at 100°C.[27, 41] It would now appear that food poisoning may also be caused by strains which have normal heat resistance and which are indistinguishable from typical fully toxigenic Type A strains[42, 215] and by heat-sensitive strains which lack θ toxin and which are therefore similar to the heat-resistant, spore-forming strains.[43, 44] In Europe and Asia most *Cl. perfringens* Type A food poisoning outbreaks have been reported to have been caused by heat-resistant strains, whereas in America heat sensitive strains are much more commonly incriminated.[42] However, this may be a reflection of the fact that until recently, in Europe and Asia, heat-sensitive forms of *Cl. perfringens* have tended to be ignored as a possible cause of food poisoning. For example, in a recent study in the British Isles, heat-sensitive *Cl. perfringens* Type A was incriminated in a number of cases.[209]

(d) *Food-poisoning toxins.* An extensive search for a food-poisoning toxin has so far yielded inconclusive results. Nygren,[45] after a very detailed investigation using animals, concluded that phosphorycholine released by the action of the α toxin on lecithin was responsible for the food-poisoning symptoms. However, subsequent workers[46, 47] were unable to confirm this conclusion. The involvement of phosphory-choline would appear unlikely as strains forming only traces of α toxin under optimal conditions appear to be equally able to cause food poisoning as fully toxigenic strains.[43] The mechanism of *Cl. perfringens* food poisoning therefore remains obscure. The viable vegetative cells reaching the lower intestine may perhaps form a toxic compound, perhaps from some substance present in the gut, or purely affect the gut by upsetting its normal microbial balance.

(e) *Types of foods involved in food poisoning.* Foods mainly involved are large joints of meat or poultry that have been cooked and left for some time at room temperature and either eaten cold or after only mild reheating, or boiled, braised, steamed or stewed meat cooked on the day prior to eating.[10] The common factor in all cases is that the meat has been insufficiently heated to destroy all the spores and that during cooling these have germinated, initiated vegetative cell growth in the food which has been eaten without further cooking to destroy the vegetative cells. In America where heat-sensitive strains of *Cl. perfringens* have usually been implicated in food poisoning[42] it has been suggested that contamination with spores or vegetative cells occurs after cooking. However, heat-sensitive spores may survive cooking in foods,[48] so there is probably no need to postulate that heat-sensitive strains have contaminated the food after cooking.

(*f*) *Growth of* Cl. perfringens *in food*. Spores of heat-resistant strains of *Cl. perfringens*, unlike those of most classical Type A strains, generally require heat activation for germination.[212] Barnes *et al.*[49] found that under optimal conditions less than 3% of the spores of a heat-resistant strain that they studied were able to germinate without heat shock. In fresh meat spores were unable to germinate at temperatures up to 37°C unless they were heat activated either by heating in the meat (70°C for 30 min) or heat activated prior to inoculation. Spore germination is extremely rapid after heat activation. Heat shocked spores and vegetative cells were able to commence growth at 20°C but not 15°C in both fresh and cooked meat at pH 5·7–6·0. Optimum temperature is 43 to 47°C, maximum growth temperature is about 30°C.[50] In artificial media at a neutral pH, strains will grow in the presence of up to 8% NaCl, 0·04% $NaNO_2$ or 1% $NaNO_3$.[51]

B. CLOSTRIDIUM TETAN

1. Isolation and identification

Although a strict anaerobe, *Clostridium tetani* is usually easy to isolate and identify. The swarming ability of this organism on an agar surface is traditionally used for isolation.[52] A suitably dispersed sample is divided into two parts and one heated at 65°C for 30 min. The heated and unheated parts of the sample are then inoculated into bottles of freshly steamed and cooled cooked meat medium or an equivalent medium such as RCM[85] or VL medium[53] and incubated for up to 7 days at 37°C. At the same time a loopful of the sample is inoculated at the outer edge of a blood agar plate and the plate incubated anaerobically for 24 hr; after incubation, the enrichment cultures are similarly plated. Fluorescent labelled antibodies can be used to detect the presence of *Cl. tetani* at this stage.[65] The spreading growth of *Cl. tetani* is readily visible by the use of a suitably angled light. Growth from the leading edge is plated out anaerobically and aerobically for purity. This technique is only suitable for isolating motile strains of *Cl. tetani*. Where it is desired to isolate non-motile strains, or when the above method fails, it is suggested that selective enrichment and plating should be tried.[54] The only certain way of confirming that an isolate is a toxigenic strain of *Cl. tetani* is by mouse toxicity tests. A pair of mice, one protected by intraperitoneal injection of 500 units of tetanus antitoxin at least 1 hr prior to injection of the test culture, each receive 0·3 ml of a mixture of equal volumes of 2·5% calcium chloride and a 10 day old Robertson's cooked meat culture.[54] The mixture is injected into the muscles of the left thighs of the protected and unprotected animals. Death of the unprotected animal, with the characteristic symptoms of tetanus and survival of the protected animal, confirms the isolate as *Cl. tetani*. Non-toxic strains may be isolated which colonially and microscopically resemble *Cl. tetani*. In order to identify these strains biochemical tests should be done.[52, 55]

2. Classification

Ten types of *Clostridium tetani* can be recognized by use of the flagellar antigens.[2]

3. Distribution

Clostridium tetani occurs with variable frequency in soils throughout the world. It has also been isolated from the faeces of man and a wide variety of animals, but the frequency of isolation from any one species is extremely variable.[54] It has been suggested that *Cl. tetani* is only a transient inhabitant of the intestine and that it is unable to establish itself in the intestinal tract;[2] contamination of the intestine results from the ingestion of spores. However, the importance of wild and domestic animals in the contamination of soil has been expressed by a number of workers and it is well established that *Cl. tetani* can be isolated most frequently from well-manured soils. The most favourable conditions for the survival of this organism are a hot, damp climate in conjunction with a soil rich in organic matter.[1] The ecological relationship of *Cl. tetani* to other soil micro-organisms and nutritional factors affecting its survival in soil have been little studied. Recently some evidence has been obtained that certain plants may influence the *Cl. tetani* population of the soil. For example, Bychenko[1] cites evidence that the associated rhizosphere of wheat stimulates the growth of this organism. In a recent survey of the distribution of *Cl. botulinum* and *Cl. tetani* in Russian soils[57] it was found that *Cl. tetani* occurred more frequently than *Cl. botulinum* but that, like *Cl. botulinum*, the organism was most frequently isolated from soils in the southern territories. In contrast to *Cl. botulinum*, *Cl. tetani* was more commonly present in sandy soil than in silt soil. Both organisms were commonly isolated from the rich black earth of the Primorye territory.

4. The relationship of spores to disease

Many more cases of tetanus occur in tropical and subtropical regions of the world than in the temperate regions and more cases occur in rural than urban areas.[1] Therefore, there is a good correlation between the incidence of *Cl. tetani* spores in the environment and the number of tetanus cases. Spores of *Cl. tetani* are unable to grow in healthy tissue, as the Eh is too high. Knight and Fildes[58] established that the highest Eh at which spores of *Cl. tetani* could initiate vegetative growth was +110 mv; the Eh of healthy tissues is > + 250 mv.[30] The recent studies of Shoesmith and Holland[59] on the germination requirements for *Cl. tetani* spores indicate that methionine, lactate, nicotinamide and sodium ions are required for germination. They also found that although the *Cl. tetani* spores germinated in the presence of air, outgrowth of germinated spores and vegetative cell growth takes place only under strictly anaerobic conditions; it has not yet been established whether spores of *Cl. tetani* can germinate in healthy tissue and remain viable. It is known, however, that spores of this organism can remain dormant in healthy tissue for up to 14 years.[2] It is unlikely that a metabolically active germinated spore could survive for such a long time and it is more probable that such spores remain ungerminated in the tissue. The requirement of lactate for spore germination is of interest as reduction of Eh required for the growth of *Cl. tetani* in tissue also stimulates the production of lactate from the muscle glycogen; also under the conditions of low Eh and pH endogenous tissue enzymes release amino acids into the damaged tissue.[60]

5. Tetanus

Tetanus results from the contamination of a wound or broken skin with spores of *Cl. tetani*. Under suitable conditions[2] the spores germinate and multiply in the area of the infection and elaborate the powerful neurotoxin that is responsible for the 30–80% mortality of tetanus cases. The clinical symptoms are characterized by a series of muscular spasms usually commencing at the site of infection. The frequent involvement of the masseter muscles of the jaw has resulted in the popular term of lockjaw for this disease. Although there is very good evidence that tetanus toxin acts at the level of the central nervous system after it is first bound by specific sphingolipids in the brain,[61] the routes whereby the toxin reaches the brain and central nervous system are not completely known. For further information on tetanus it is suggested that the reader refers to the reviews by Oakley,[62] Wright,[63] van Heyningen and Arseculeratne.[64]

C. CLOSTRIDIUM BOTULINUM

1. Isolation and detection

It is usually easy to detect the presence of *Clostridium botulinum* by mouse toxicity tests (see below). However, in the absence of toxin some difficulty may be experienced in detecting it in fish viscera and certain soil and mud samples which contain organisms either inhibiting spore germination[81, 82] or destroying the toxin.[118, 158] The isolation of this organism is much more difficult and it is recommended that this should not be attempted by an inexperienced worker; for a recent review of the problems see Ref. 83. The isolation or detection of *Cl. botulinum* should not be attempted without special laboratory facilities and without prior immunization of workers with polyvalent toxoid.[84]

(*a*) *Detection.* If the sample is a fish or meat product with a pH about 5·0 this can be used for a direct enrichment procedure. Such samples are divided into 3 parts; one part is heated at 60°C for 30 min, one at 80°C for 15 min and the last unheated. After these treatments the samples are vacuum-packed in gas-impermeable bags or packed in 1–2 oz glass bottles leaving as small an air space as possible but taking care to release gas once growth has commenced. The samples are incubated at 30°C for 5 days, and then if sufficient free liquor is present for toxicity tests this is poured off into a sterile container. If insufficient liquor is present the sample is washed with a small quantity of 0·85% saline or macerated in an equal weight of saline. In materials not supporting the growth of *Cl. botulinum* or as an alternative procedure to the above, samples are inoculated into one or more of the following media: Reinforced Clostridial Medium (RCM),[85] cooked meat medium or Trypticase-peptone glucose broth[87] with or without the addition of trypsin;[88] samples are heated and incubated as described for the direct enrichment procedure. Fluorescent antibody techniques can be used to detect the presence of *Cl. botulinum* cells in the enrichment cultures[89, 90, 206] and in pathological specimens.[119]

In order to be certain that failure to detect *Cl. botulinum* in a sample is not caused by inhibition of growth, or inactivation of the toxin, it has been recommended that low levels of *Cl. botulinum* are introduced into a separate sample of the material under test and treated as outlined above (Dr L. N. Christiansen, personal communication). If *Cl. botulinum* is not detected in the inoculated sample it is not safe to assume that *Cl. botulinum* is absent from the test sample.

For toxin detection, the washings, macerates or cultures are centrifuged and the supernatants sterilized by passing them through a membrane filter; 0·25 ml of the sterile filtrates are injected intraperitoneally (IP) into each of a pair of mice. To save sterilization by filtration, a useful alternative procedure is to mix the supernatants with equal volumes of 2% streptomycin sulphate and inject 0·5 ml of the mixture IP into each of a pair of mice. The mice are observed for up to 3 days. If they die with symptoms of botulism, i.e. laboured breathing, with indrawn flanks, followed usually by stretching of the front and hind legs at death, protection tests are done using a polyvalent mixture of antitoxins to *Cl. botulinum* Types A–F and *Cl. tetani*. The polyvalent antitoxin mixture, suitably diluted in 2% streptomycin or distilled water, is mixed with an equal volume of supernatant and left for at least 1 hr at room temperature; 0·5 ml is injected intraperitioneally into each of a pair of mice; controls of boiled and unheated supernatants are also injected. If the mice given the injection of the boiled supernatant and the supernatant-antitoxin mixture survive, the type of *Cl. botulinum* present is found by specific neutralization tests using monovalent antitoxins. It is important to use dilutions of the toxic supernatants and to dilute the monovalent antitoxins to their recommended titres as cross-reactions may occur between *Cl. botulinum* Type C and D antitoxins[100] and high titres of E antitoxin may neutralize Type F toxin;[56, 203] also low levels of tetanus antitoxin may be present in undiluted sera as the result of tetanus immunization of horses used for serum production. Toxins of non-proteolytic strains of *Cl. botulinum* are activated by trypsin (0·25% trypsin 1 : 250, Difco) at pH 6·0[205] and some workers routinely use trypsin-treated suspensions for toxicity tests. This is a sound practice for detecting toxin in pure cultures but may result in the destruction of activated toxin in a mixed culture such as a stored food. Therefore if trypsinization is used both trypsin-treated and untreated supernatants should be tested for toxicity.[99] If more than one toxic type is present mixtures of monovalent antitoxins should be used to resolve them.

Samples from food suspected to be involved in a botulism outbreak should be tested for mouse toxicity before and after incubation.

Passive haemagglutination,[91] immunoelectrophoresis,[92] and acetylcholine esterase inhibition[93, 216] have been used for the *in vitro* detection of *Cl. botulinum* toxins. A warning note on the use of serological techniques for the detection of *botulinum* toxins is indicated by the observations of Sugiyama *et al.*[211] on Type E toxins. These workers have confirmed that the sites on the toxin molecule for serological specificity and toxicity are distinct and that when the Type E toxin is trypsin activated, with 100-fold increase in lethality, this is not accompanied by an increase in immunological reactivity. A method for concentrating toxins in aqueous solutions is described by Pigoury *et al.*[120]

(*b*) *Isolation.* Samples of toxic enrichment cultures are sealed in ampoules and heated at either 80°C for 15 min if heat-resistant spores of Types A, B, C, D or proteolytic Type F are believed to be present, or at 60°C for 30 min for heat-sensitive spores of Type E and C and non-proteolytic Types B and F.[94, 203] Type E spores have also been isolated by mixing culture sediments or food macerates with an equal volume of ethyl alcohol and leaving for 1 hr at room temperature with occasional mixing.[95] After one or the other of the above procedures the samples are plated out onto blood agar or Willis and Hobbs' medium[14] and incubated anaerobically for 3 days at 30°C. Suspect colonies are inoculated into RCM, incubated for 3–5 days at 30°C and tested for mouse toxicity. If the isolates prove to be toxic the previously described procedures should be used to identify the toxigenic type(s). Purity should be checked by repeated plating on blood agar plates incubated both aerobically and anaerobically.

2. Classification

Clostridium botulinum is divided into 6 main toxigenic types (Types A–F) according to the production of 6 antigenically distinct toxins (A–F) by the different strains; naturally occurring strains produce only one of these toxins: *Clostridium botulinum* Type A is usually saccharolytic and proteolytic, Type B and F strains are either saccharolytic and proteolytic or only saccharolytic, Types C, D and E are saccharolytic. Type A, proteolytic Type B strains and Type C and D strains are typical mesophiles and fail to grow below 10–15°C.[96] Non-proteolytic Type B and Type E and F strains are psychotrophs and will grow at 3–5°C.[97]

Spores of toxigenic Types A, B, C and D are heat resistant; Type A spores are generally more heat resistant than Type B spores,[207] Type E and non-proteolytic Type F spores are heat sensitive.[203] Spores of non-proteolytic Type B strains may be only moderately heat resistant.[217]

3. Distribution

The ecology and distribution of *Cl. botulinum* in natural habitats has been studied extensively. From these studies it can be concluded that these organisms are primarily soil saprophytes,[6] and although they are not infrequently isolated from the gut contents of land, marine and fresh water animals, there is no experimental evidence that they multiply in the intestinal tract other than that of certain birds.[86]

The distribution of *Cl. botulinum* appears to depend on a number of geographical, climatic and geological features.[57] The erratic distribution of spores of *Cl. botulinum* in the soil has been noted by Dolman[98] and many other workers. This distribution is probably genuine, but may be a reflection of the fact that extensive surveys have been done only in certain areas and that frequently these surveys have not included all *Cl. botulinum* types. A further fact is that certain soils and muds contain organisms that inhibit the growth and hence detection of *Cl. botulinum* spores; such inhibitory organisms have been demonstrated for Type E[81] and Type F[82] spores. Other organisms may inhibit toxin production.[118, 151]

Types A, B and C appear to be the most uniformly distributed and have been isolated from soils and marine and fresh-water muds in North America, South America, Europe, Asia, Africa and Australia.[2, 6, 99, 204] In the U.S.A. Type A apparently predominates in the eastern states, particularly in the cultivated soils of the Mississippi valley and in the Great Lakes areas.[6] Non-proteolytic Type B strains have until recently been rarely found in North America,[98] but now a number of isolations have been made from marine muds;[99] they apparently predominate in European soils[98] and have recently been isolated from marine muds along the south and west coasts of the British Isles.[92] Type C, which is commonly isolated from soils in most parts of the world, multiplies in the mud and vegetation bordering alkaline marshes.[100] Type D is rarely isolated in Europe and North America but is common in certain parts of Africa[100] and Australia.[6]

Type E spores and vegetative cells are commonly isolated in some parts of the world from lake mud, soils and mud from rivers, coastal muds and from marine and fresh-water animals.[101-109] Although it is usually isolated from water and land near water, it is generally accepted to be of soil origin.[102] It has so far only been found in the Northern Hemisphere,[6] and although quite extensive surveys have been made in the Southern Hemisphere only certain areas have been examined and it is very probable that eventually this organism will be found.[104] Type E would appear to have rather catholic tastes and certain areas are very much more contaminated than others. For example, it is common in Swedish soils[110] and in the mud and fish of the Baltic and fish caught in the Norwegian Sea.[102, 109, 111] However, it is only rarely found in the nearby British Isles and only after extensive surveys have two Type E strains been isolated from mud and herring sampled in the Moray Firth in Scotland.[92] Difference between the degree of contamination of neighbouring areas is also shown by the recent surveys of the Great Lakes of North America. The incidence of Type E in the Great Lakes is generally low, with the exception of Green Bay of Lake Michigan in which the incidence is extremely high.[106] The reason for the high incidence in the Green Bay area is not understood.[218] It was thought that spores from the surrounding land mass of the bay were being washed into the bay from the soil and from the water of the complex Fox river system that drains into the bay. However, the incidence of the organism in the soil decreases with increasing distance from the shore line and the incidence of spores in mud samples from the Fox river is also low.[107] Therefore some factors must be allowing Type E spores to concentrate in this area. One suggestion has been that the waste products from the paper mills surrounding the bay render the water relatively anaerobic and supply nutrients for growth.[112]

Cl. botulinum Type F has so far only been isolated in Denmark and in North and South America and recently in Scotland.[92] The original Danish strain[113] is proteolytic, but the North American strains are non-proteolytic. It has been isolated from marine mud from the coast of Oregon and California,[99] from a salmon caught in the Columbia river,[114] from a sewage lagoon and a small stream in North Dakota,[82] and from sand collected on a beach in the Brazilian State of Ceara.[204] The isolation of this organism from sewage lagoons was shown to be influenced by the presence of *Bacillus licheniformis* which at certain times of the year occurs in large numbers and suppresses the growth of *Cl. botulinum* Type F.

The recently published study of Kravchenko and Shishulina[57] of the incidence and distribution of *Clostridium botulinum* in the soils, rivers and lakes of the U.S.S.R. is one of the most complete and extensive surveys for *Cl. botulinum* that has been published and deserves special mention. They examined a total of 4,345 soil and water samples and found an overall incidence of 10·5% positive samples. The incidence was markedly higher in the southern territories of U.S.S.R. than the northern territories. Type E was isolated most frequently (62·2% samples positive), Type B occurred in 28·1% samples, Type A in 8·3% samples and Type C in 2·1%, Type D was only once isolated. The distribution of Type A and B spores was influenced by a complex collection of factors such as type of soil, geographical position and possibly vegetation. Type E was only isolated from soil samples obtained from along rivers and lakes and their associated meadows. Moist soils rich in organic substances appear to favour the growth and sporulation of Type E.

4. The role of the spores in botulism

Both human and animal botulism usually results from the ingestion of toxin preformed in a food;[98, 115] the different types of botulism caused by the 6 toxigenic types are shown in Table 1. Types A, B and E are usually involved in human botulism although Types C, D and F have also been implicated.[100]

(a) *Human botulism*. Although there is no conclusive experimental data that spores are infective there are reports of human botulism apparently resulting from wound infection.[2, 115, 219] Experiments in animals in which botulism was caused by the injection or feeding of large numbers of washed spores, heated to destroy toxin, are inconclusive as it is probable that sufficient toxin remained after the heat treatment to cause botulism without growth and *de novo* synthesis of toxin. Recently Grecz and co-workers[116, 210] have reported the presence of small amounts of a heat-resistant botulinum toxin in spores of *Cl. botulinum* Type A. When such spores were stored at 2–4°C in phosphate buffer the toxin was released and became as heat labile as vegetative cell toxin. This work may explain botulism resulting from an experimental inoculum of large numbers of spores. A number of Russian workers[115] believe that botulism is both an intoxication and an

infection and that the symptoms of botulism are caused first by the ingestion of preformed toxin and that the spores present in the ingested food
are able to initiate vegetative growth in the intestinal tract and release
further toxin. Although the evidence for this in human botulism is mainly
circumstantial it should not be ignored.

TABLE I

Diseases and toxigenic types of Clostridium botulinum

Toxigenic type	Species mainly affected	Commonest vehicle	Highest geographic incidence[a]
A	Man; chickens ("Limberneck")	Home-canned vege-tables and fruits; meat and fish	Parts of North America and U.S.S.R.
B	Man, horses and cattle	Prepared meats, especially pork products	North America, U.S.S.R., Europe (non-proteolytic strains)
C_a	Aquatic wild birds (Western duck sickness)	Rotting vegetation of alkaline marshes	North and South America, South Africa, Australia
C_β	Cattle ("Midland cattle disease"); horses ("forage poisoning") and mink	Toxic food; carrion; pork liver	Australia, South Africa, Europe, North America
D	Cattle ("Lanziekte")	Carrion	Australia, South Africa
E	Man	Uncooked products of fish and marine mammals	Northern Japan, British Columbia, Labrador, Alaska, Great Lakes, Sweden, Denmark, U.S.S.R.
F	Man	Liver paste; "venison jerky"	Denmark, North and South America, Scotland

(Modified from Dolman and Murakami[100])

[a] Based on published survey and disease incidences.

(b) *Animal botulism.* In animal botulism the spore acts as a reservoir of
infection in the soil. Botulism of cattle and sheep is widespread in certain
parts of South Africa[121] and Australia.[122] It results mainly from the
animal feeding on putrefying carcasses contaminated with *Cl. botulinum*
from the soil. In domestic birds and animals, botulism usually results
from the feeding of toxin-containing food. Botulism of aquatic birds is
believed to result when conditions favourable for growth and toxin production by Type C occur in the rotting vegetation of alkaline marshes and

mud flats;[123, 124] recent work in America[125] has indicated that Type E toxin may be important in increasing Type C toxicity in birds. Boroff and Reilly[86] have demonstrated that both intoxication and infection may play a role in botulism of quail. Although there is no evidence that *Cl. botulinum* can grow in the intestine of live fish, the recent report[126] that rainbow trout are killed by Type E toxin would indicate that fish may die as a result of the feeding of contaminated food. For further information on botulism and botulinum toxin the reader is referred to two recent excellent symposia on botulism.[5, 84]

5. Growth of Clostridium botulinum *in food*

A vast variety of foods have been involved in human botulism.[117] The common feature of such foods is that they have been subjected to some form of preservation process to extend their keepability; spores of *Cl. botulinum* either survive the process or contaminate the food after processing. Further features are that foods have been stored under conditions allowing growth and toxin formation and have been eaten without further cooking or adequate heating to destroy the heat labile toxins.

The kinds of foods responsible for botulism outbreaks in various parts of the world are to some extent reflected by the eating habits of different nations. Most European outbreaks have been caused by liver sausage, ham, preserved meats, pâtés and brawn; various fish products have also been involved in outbreaks in Scandinavian countries. In America most outbreaks have been caused by home-preserved vegetables and vegetable-containing products, but more recently they have also involved imported produce, canned and smoked fish and canned liver paste. In Russia pork and home-processed fish (smoked, dried, salted or pickled) are commonly implicated and in Northern Japan a fermented fish delicacy (Izushi) is a common cause of botulism. *Clostridium botulinum* Type E is usually involved in botulism of fish products and Types A and B of meat products.

The factors controlling the growth of *Cl. botulinum* in food are numerous and for simplicity these will be discussed separately.

(a) *Redox potential (Eh) and oxygen tension.* The redox potential of a food depends both on the redox potential of the food itself, its poising capacity, i.e. ability of the oxidizing and reducing systems in the food to maintain the redox potential in the presence of oxygen, and on the oxygen tension of the atmosphere and its availability to the food.[144] For growth, *Cl. botulinum* requires an Eh of $< + 150$ mv[127] and an oxygen tension of not more than 4–8% of atmospheric pressure.[128] Such conditions occur a few millimetres below the surface of meat and fish muscle after death, in heat-sterilized foods such as milk[127] or on the surface of foods contaminated with aerobic and facultative anaerobic bacteria and in vacuum-packed foods. The introduction of vacuum packing and the involvement of vacuum-packed fish in several food-poisoning outbreaks in the United States[117] has resulted in a large number of studies on the effect of vacuum packing on the growth and toxin production of *Cl. botulinum*.

Vacuum packing was introduced primarily as a convenient and attractive way of displaying a product but it was also found to have a number of other advantages to the producer, namely suppression of the aerobic spoilage flora and increased shelf life (particularly at refrigerator temperatures), maintenance of cured-meat colour, decrease of oxidative rancidity and dehydration, and improvement of general microbiological quality by avoiding post-factory contamination.[129] However, despite these advantages to the producer, the consumer does run an increased botulism risk with certain vacuum-packed foods unless they are stored under conditions that prevent growth or are processed in such a way that all spores of *Cl. botulinum* are killed. Recent studies on the effect of vacuum packing on the toxin production by *Cl. botulinum* Type E in plastic film wrapped, fresh, smoked and irradiated fish show that growth and toxin production occur in both aerobic and vacuum packs[130-132] but toxins may be produced more rapidly or in larger amounts in vacuum packs.[131, 132] In cured meat products toxin again develops in both aerobic and vacuum packs,[133-135] but in the vacuum pack the aerobic spoilage flora is depressed and toxin may develop without the warning signs of spoilage.[133, 136] The delay in obvious signs of spoilage, and toxicity developing without spoilage in both vacuum packed fish[131] and meat products, has caused great concern to public health authorities throughout the world. Their great concern is emphasized by the recent Canadian regulations on vacuum-packed meat[137] and the previous similar regulations concerning smoked fish.[198]

(*b*) *Nutritional requirements*. The nutritional requirements for the growth of *Cl. botulinum* are complex. Defined media have been devised for Types A and B[139] and Type E[140-142] and Type F;[143] the nutritional requirements for Types A, B and F are similar. Although the rate and amount of growth in defined media may be similar to that of the complex meat digest media usually used for growing these organisms, the production of toxin in defined media may be poor or absent.[139, 142] Kindler and his colleagues[145] found that toxin formation by *Cl. botulinum* Type A was stimulated by the addition of tryptophan and inhibited by EDTA at concentrations that had no effect on growth; the inhibitory effect of EDTA was removed by Mg^{++}. More recently Boroff and DasGupta[146, 147] showed the importance of tryptophan both as a constituent of the toxin molecule and for its biological activity and Gullmar and Molin[142] showed that the concentration of tryptophan that was optimal for Type E toxin production was six times the concentration that was optimal for growth; these workers also noted that small amounts of linoleic acid also stimulated toxin production. Ward and Carroll[140] were only able to obtain toxin production when citric acid or lactic acid were added to their synthetic medium. However, further work by these workers[148] established that neither of the compounds was an absolute requirement for toxin production.

It is difficult to relate nutritional requirements to the growth and toxin production by *Cl. botulinum* in a particular food, as growth depends on a variety of factors of which the chemical composition of the food is only one. The observations made by several workers[130, 149, 150] that fatty fish, such as herring, mackerel and salmon generally support better toxin production by Type E than plaice or cod may be an example of an effect of food constituents on growth and/or toxin production. Also the observation of Johannsen[130] that *Cl. botulinum* Type E was unable to grow in spinach, although spinach has been involved in a number of food-poisoning outbreaks involving *Cl. botulinum* Type A or B,[117] may be a further example.

(c) *Influence of food micro-organisms.* Food micro-organisms may either stimulate or inhibit the growth of *Cl. botulinum*.[152] Their effects are not understood precisely, but their importance has been shown by a number of workers. In a well-buffered food, lactic acid bacteria may stimulate growth of *Cl. botulinum* by the supply of growth factors or by making the Eh more favourable to growth;[153] yeasts may also aid growth by oxidizing an acid, causing the pH to rise to a value favourable to growth.[154] In a poorly buffered and easily fermented food, lactic acid bacteria may prevent growth by the production of acid conditions.[155] Growth of *Cl. botulinum* may also be prevented by nisin-producing strains of *S. lactis*,[155] and fatty acids[156] and possibly antibiotics[157] produced by *Brevibacterium linens* in surface-ripened cheese and bacitracin[82] produced by strains of *Bacillus licheniformis*. Bacteriocins produced by clostridia closely related to *Cl. botulinum* Type E may play an important role in preventing the growth of this organism in fish products.[159] Proteolytic bacteria may either stimulate or inhibit toxin formation.[158, 160]

(d) *Temperature.* Non-proteolytic Type B strains,[161] Type E[162, 163] and Type F[164] strains are psychrotrophs and are able to grow and form toxin from both spore and vegetative cell inocula at temperatures down to 3–5°C; the minimum growth temperature is, however, very dependent on the strain, inoculum size and the growth medium. Spores and vegetative cells of *Cl. botulinum* Types A and B are unable to initiate growth at 10°C,[165] but grow from vegetative cell inocular at 12·5°C and from spore inocula at 15°C.[166] The minimum growth temperatures for Types C and D strains do not appear to have been determined.[97] Optimum growth temperature for Types A, B and E is between 35–40°C;[162, 166] the optimum temperature for toxin production is between 25 and 30°C.[97, 167] Maximum growth temperatures for vegetative cell inocula are between 45 and 48°C; for spore inocula the maximum growth temperature tends to be lower.[162, 166]

(e) *pH*. The limiting pH for growth and toxin production by *Cl. botulinum* Types A, B and E is between pH 4·8 and 5·0.[168, 169, 170] The validity of reports of strains growing in certain foods at pH values below 4·5 is questionable.[169] It appears probable that growth and toxin production occurred in such foods at a higher pH than that recorded for the food at the time of determining growth or toxicity; an alternative explanation is that the pH value of the food was not uniform. The effect of pH on spore germination varies with toxigenic type and may also vary between strains. Using single strains of Types A, B and E, Baird-Parker and Freame observed[170] that the germination of Type B spores occurred at about the same rate at pH values between 7·0 and 5·3 but that the germination of Type A spores was markedly delayed at below pH 6·0 and was completely prevented at pH 5·0; vegetative cells of Type A grew at this pH. Type E spores were able to germinate at pH 5·0, but neither spores nor vegetative cells initiated growth at this pH. The maximum pH for growth is about 8·5 and, although good toxin production occurs at all pH values permitting good growth, the toxin is unstable above pH 7·0 and therefore cultures and foods at pH values below 7·0 are more likely to be toxic.[165]

(f) *Water activity*. Measurement of the water activity (a_w) of a food is a convenient means of determining the available moisture for microbial growth. Numerically it is expressed as the water vapour pressure of a solution or food relative to the vapour pressure of pure water at the same temperature and pressure. In order to study the effect of a_w on microbial growth in laboratory media, a variety of solutes have been used such as sugars, inorganic salts and alcohols and the results obtained have been compared with growth obtained in foods of equivalent water activities. As a result of such studies Scott and his co-workers[171] concluded that the ability of a number of bacteria to grow in foods containing differing amounts of water, salts or carbohydrates could generally be explained in terms of the a_w of the particular food. In a recent study Ohye *et al.*[172] showed that the minimum a_w for the growth of vegetative cells of *Cl. botulinum* Type E was the same when a mixture of salts (NaCl, KCl and Na_2SO_4) or NaCl alone was used to adjust the growth medium to the required a_w; similar results were obtained for spores of Types A, B and E.[173] The minimum a_w at which spores of *Cl. botulinum* initiated growth in salt-containing media incubated under optimum conditions of pH and temperature was 0·95 for Type A, 0·94 for Type B and 0·97 for Type E spores.[173] However, when growth of *Cl. botulinum* was compared in RCM adjusted to different water activities by the addition of glycerol or NaCl similar results to the above were obtained for the salt-containing media, but quite different results for the glycerol media.[170] For example, spores of *Cl. botulinum* Types A and B were able to initiate growth at the minimum a_w of 0·96 in the RCM containing NaCl (pH 7·0) when incubated at 30°C, whereas under the same conditions, with glycerol present the minimum a_w for growth was 0·92; similar results were obtained with vegetative cells. Therefore, although the effect of NaCl in preventing or delaying growth of *Cl. botulinum* in a food may be partly explainable in terms of its effect on a_w, some additional inhibitory effect must be caused by the NaCl molecule. The minimum a_w for growth increases markedly at temperatures and pH values below and above the optima[170] and is also affected by inoculum concentration. A good example of the effect a_w on the growth of *Cl. botulinum* in a food is the observation[188] that liver sausage packed in natural casing supports poor growth of *Cl. botulinum*, whereas packing in less water-permeable Saran, which prevents dehydration of the sausage, results in good growth and toxin production.

(*g*) *Curing ingredients.* The sodium chloride concentration that is inhibitory to growth depends on the toxigenic type,[173] the strain,[169, 172] growth conditions (particularly pH and temperature),[167, 169] inoculum level[169, 174] and degree of heat injury.[175] The maximum NaCl concentration at which *Cl. botulinum* Types A and B can grow under optimal growth conditions is about 9% w/w;[173, 176] the maximum salt concentration tolerated by Type E is 5 to 6% w/w.[169, 170, 172] Spore germination of *Cl. botulinum* Types A, B and E occurs at much higher NaCl concentrations than outgrowth.[169, 170, 177]

Nitrate, although it has been used to inhibit the growth of clostridia in cheese,[190] probably has no inhibitory properties at concentrations used in cured meats;[178] there is some evidence that it may in fact stimulate the growth of certain micro-organisms.[179] Nitrite, however, has been shown to be a powerful inhibitor of spore outgrowth and of vegetative cell growth of *Cl. botulinum*; spore germination is not inhibited by concentrations permitted in food. The role played by nitrite in inhibiting microbial growth is complex.[180, 181] The minimum inhibitory concentration of nitrite for a particular organism inoculated in a food or bacteriological media depends on pH,[180] additives,[102, 183] heat treatment[179, 185, 186, 191] and the numbers of organisms present.[179, 185, 187–189] The concentration of nitrite required to inhibit a given concentration of organisms at different pH values corresponds closely to the undissociated nitrous acid present in nitrite solutions of different pH;[192] the optimum pH for nitrite inhibition and for nitrous acid formation is about pH 5·5.[180] The addition of ascorbate to improve colour stability in cured meats increases the inhibitory effect of nitrite at a neutral pH,[182] although at an acid pH it will reverse the inhibitory effect of nitrite by reducing the nitrite to nitric oxide.[183] Although it has been demonstrated by a number of workers[185, 186] that heat treatment makes the surviving organisms more sensitive to nitrite inhibition, there is now evidence that nitrite when heated in a bacteriological medium forms an unknown compound which is extremely toxic to vegetative cells of

Cl. sporogenes[191] to spores and vegetative cells of other clostridia including *Cl. botulinum*.[193]

(*h*) *Smoke constituents*. The growth of *Cl. botulinum* in certain smoked fish is delayed or prevented by smoking;[174, 194, 195] the amount of toxin produced is small compared with the unsmoked fish.[196] It has been claimed that either phenols[174] or formaldehyde[194] are responsible for the growth inhibition in smoked fish. However, smoking cannot be claimed to give good protection against *Cl. botulinum* as the most serious botulism outbreak in the United States for a number of years involved vacuum-packed smoked white fish chub.[197] A number of experiments have demonstrated that Type E is able to grow and form toxin in this product and it has been recommended that such smoked fish products should be frozen[138] or stored for not longer than 6 days at 4°C.[198]

III. Bacillus Diseases and Intoxications

Few diseases are caused by *Bacillus* species. Almost all species are harmless soil saprophytes with the exception of the specific insect pathogens discussed in Chapter 13, and *Bacillus anthracis* which is an animal parasite. There are a number of reports that *Bacillus subtilis* may give rise to eye infections and cause secondary infections[201] and that *B. cereus* may be pathogenic to mice, guinea pigs and rabbits, but such disease would appear to be rare.[2] *B. cereus* also causes food poisoning,[200] and has been incriminated in ear infections. *B. subtilis* has been incriminated in food poisoning.[220]

A. BACILLUS ANTHRACIS

1. Isolation and identification

Vegetative cells of *B. anthracis* can usually be identified in fresh or preserved pathological specimens by their appearance when stained with methylene blue[66] or by fluorescent labelled antibodies.[67] Although easily isolated from cases of anthrax its isolation from such sources as soil, wool, hides and feedstuffs is more difficult and care must be taken to distinguish it from closely related *Bacillus* species. Of the various selective media that have been proposed for isolating *B. anthracis* the most effective is probably Morris' medium[68] which contains propamidine isethionate as a selective agent. The medium is, however, only suitable for detecting spores, as vegetative cell inocula are inhibited by propamidine. For total count purposes the polymyxin medium of Gillissen and Scholz[69] is more suitable. Various tests have been suggested for distinguishing *B. anthracis* from related bacilli.[2, 56] The two simplest remain the absence of motility determined by the Craigie tube technique and pathogenicity tests in mice or guinea pigs. Young colonies can be identified by use of the species specific γ phage.[70]

2. Distribution

Bacillus anthracis is an animal parasite causing diseases mainly amongst herbivora. Spores of this organism have been isolated from soils in most parts of the world. There is some evidence that the organism is able to survive and multiply in the soils of certain areas, in particular low-lying marshy areas in subtropical regions. The persistence of this organism in soil would appear to depend on its ability to form spores. Davis[71] has shown that at relative humidities above 80% *Bacillus anthracis* spores are able to initiate vegetative cell growth between 20° and 44°C. However, vegetative cells are unable to compete with soil flora and rapidly die out unless the temperature of the soil is sufficiently high for rapid sporulation. Rapid sporulation in soil depends on both temperature and humidity.[71] Optimum sporulation occurs at 37°C and at an equilibrium relative humidity (ERH) of 100%. At temperatures below 30°C sporulation is slow and therefore in temperate countries this organism is believed to be unable to survive for long in the soil. For example, in Britain, spores are not usually present in soils except when bone-meal containing fertilizers contaminated with *B. anthracis* spores are used as a winter soil dressing. It has been assumed that such spores remain dormant during the winter, germinate when the soil warms during the summer months and that the resulting vegetative cells are then destroyed by the soil flora before resporulation occurs.[71] As well as soil and fertilizers the organism is commonly found in dust, animal hair and hides and in animal feeds.

3. The role of the spore in the initiation and spread of infection

Anthrax in animals usually results from the ingestion of spores. It is common therefore in areas where spores persist in the soil and where they are present on the vegetation on which the animal feeds. In such areas anthrax occurs most commonly during the late summer months when the ambient temperature is most favourable for sporulation and when the spore population of the soil reaches a maximum.[2] In temperate countries the organism is believed to be unable to survive in the soil, and therefore most cases of anthrax occur during the winter months as a result of the use of artificial feeds contaminated with spores of *B. anthracis*.[72] The means whereby the ingested spores initiate infection is uncertain, but it is believed that some abrasion of the pharyngeal or intestinal mucosa must be involved.[2] During the later stages of the disease the organism is excreted in the urine, faeces and saliva. Before death the organism invades the blood stream, causing septicaemia. At death, bloody fluids exude from the body openings. At this stage the organism exists in its vegetative form and it is only after the exudates are released from the body that the vegetative cells form spores. Sporulation occurs only when temperatures are favourable. Minett[73] found that in the infected blood from open carcases sporulation was optimum between 32° and 38°C and that at ambient temperatures of between 15–21°C the bacilli were destroyed by the saprophytic contaminants before sporulation had taken place. As spores are the means whereby further infection occurs, in the absence of conditions favouring sporulation, the disease rapidly dies out.

In man the mode of infection by anthrax spores is quite different to that in animals. In the first place, anthrax is primarily a disease of animals and man is infected only through the handling of animals or animal products.[2] Secondly, the alimentary route of infection is rare in man, and the two main routes of infection are by inhalation of spores present in the air "inhalation or pulmonary anthrax" or a cutaneous anthrax caused by contamination of an abrasion or wound. Cutaneous anthrax is much more common than pulmonary anthrax, which is much the more fatal form of the disease. Pulmonary anthrax has been studied extensively experimentally.[74, 75] It has been established that infection results from the inhalation of spores only in particles of $< 5\,\mu$; spore clumps or larger spore containing particles are unable to reach the lung alveoli which are the primary infection sites. In a study of pulmonary anthrax in guinea pigs, Ross[76] found that spores on reaching the alveoli are rapidly phagocytosed by macrophages and transported to the draining lymphatics. In the sinusoids of the lymph nodes the spores germinate and initiate vegetative cell growth. Following multiplication, the bacilli invade the lymphatic system and are disseminated throughout the blood stream, causing the characteristic symptoms of anthrax. In external or cutaneous anthrax spores initiate growth in the region of the infection and cause the formation of the characteristic malignant pustule. The bacilli are present in largest number in the central necrotic area of the pustule and from this invasion of the blood stream may occur.[2]

4. Anthrax as a disease

The epidemiology of anthrax infection is complex and for information on this subject it is suggested that the reader refers to Refs. 74, 75 and 77. It is now well established that death from anthrax results from the elaboration of a complex toxin which is composed of at least three components that have been called factors I, II, III by the British workers[78] and respectively as the oedema factor, protective antigen and lethal factor by the American workers.[79] Whether these toxins form a single complex toxin or act as a mixture of toxins is not known.[80]

B. *BACILLUS CEREUS*

1. Food poisoning

Bacillus cereus has been involved in a number of food-poisoning cases and has been particularly studied by European workers. The symptoms of this relatively mild food poisoning are very similar to that of *Cl. perfringens* and without a bacteriological examination it may be impossible to decide the cause; a medium for isolating *B. cereus* from foods has recently been described by Mossel *et al.*[199] Another similarity with *Cl. perfringens* food poisoning is that at least 10^7/g viable vegetative cells must be present in the food to cause food poisoning.[200] As with *Cl. perfringens* food poisoning, Nygren (1962) claims that phosphoryl-choline released by the action of the lecithinase on lecithin was the cause of food poisoning, although this was not confirmed by Dack and his colleagues.[46] Nikodemusz claims to be able to reproduce the food-poisoning symptoms in cats.[202] Types of food involved in *B. cereus* food poisoning are vegetable products, minced meat, liver sausages, puddings and sauces.[199]

REFERENCES

1. Bytchenko, B. (1966). *Bull. W.H.O.* **34**, 71.
2. Wilson, G. S. and Miles, A. A. (1964). Topley and Wilson's "Principles of Bacteriology and Immunity". 5th Ed. Arnold, London.
3. Borthwick, G. R. (1934). *Brit. J. exp. Path.* **15**, 153.
4. MacLennan, J. D. (1962). *Bact. Rev.* **26**, 177.
5. Whaler, B. C. (1967). *In* "Botulism 1966". Proc. 5th Int. Symp. Fd Microbiol, Moscow, July 1966 (M. Ingram and T. A. Roberts, eds), p. 377. Chapman and Hall, London.
6. Gordon, R. A. and Murrell, W. G. (1967). *CSIRO Fd Preserv. Q.* **27**, 6.
7. Wright, D. H. (1967). *East Africa Med. J.* **43**, 544 (abstr. *Bull. Hyg. Lond.* **42**, 1268).
8. Murrell, T. G. C., Egerton, J. R., Rampling, A., Samels, J. and Walker, P. D. (1966). *J. Hyg. Camb.* **64**, 375.
9. Gibbs, B. M. and Freame, B. (1965). *J. appl. Bact.* **28**, 95.
10. Hobbs, B. C. (1965). *J. appl. Bact.* **28**, 74.
11. Sutton, R. G. A. (1966). *J. Hyg. Camb.* **64**, 65.
12. Angelotti, R., Hall, H. E., Foter, M. J. and Lewis, K. H. (1962). *Appl. Microbiol.* **10**, 193.
13. Hauschild, A. H. W., Erdman, I. E., Hilsheimer, R. and Thatcher, F. S. (1967). *J. Fd Sci.* **32**, 469.
14. Willis, A. T. and Hobbs, G. (1958). *J. Path. Bact.* **75**, 299.
15. Green, J. H. and Litsky, W. (1966). *J. Fd Sci.* **31**, 610.
16. Marshall, R. S., Steenbergen, J. F. and McClung, L. S. (1965). *Appl. Microbiol.* **13**, 559.
17. MacLennan, J. D. (1962). *Bact. Rev.* **26**, 177.
18. MacLennan, J. D. (1943). *Lancet* **II**, 63, 94, 123.
19. Smith, H. W. and Crabb, W. E. (1961). *J. Path. Bact.* **82**, 53.
20. Smith, H. W. (1965). *J. Path. Bact.* **89**, 95.

21. Fuller, R. and Moore, J. H. (1967). *J. gen. Microbiol.* **46**, 23.
22. Sidorenko, G. I. and Borovik, E. B. (1967). Abstr. *Bull. Hyg. Lond.* **42**, 465.
23. Sidorenko, G. I. (1966). Abstr. *Bull. Hyg. Lond.* **41**, 346.
24. Hall, H. E. and Hauser, G. H. (1966). *Appl. Microbiol.* **14**, 928.
25. Hall, H. E. and Angelotti, R. (1965). *Appl. Microbiol.* **13**, 352.
26. Sylvester, P. K. and Green, J. (1961). *Medical Officer* **105**, 289.
27. Hobbs, B. C., Smith, M. E., Oakley, C. L., Warrack, G. H. and Cruickshank, J. C. (1953). *J. Hyg. Camb.* **51**, 75.
28. Nakamura, M. and Converse, J. D. (1967). *J. Hyg. Camb.* **65**, 359.
29. Willis, A. T. (1956). *J. appl. Bact.* **19**, 105.
30. Barnes, E. M. and Ingram, M. (1955). *J. Sci. Fd Agric.* **6**, 448.
31. Hanke, M. E. and Bailey, J. H. (1945). *Proc. Soc. exp. Biol. N.Y.* **59**, 163.
32. Barnes, E. M. and Ingram, M. (1956). *J. appl. Bact.* **19**, 117.
33. Kaye, D. (1967). *Proc. Soc. exp. Biol. Med.* **124**, 360.
34. Finney, J. W., Haberman, S., Race, G. J., Balla, G. A. and Mallams, J. T. (1967). *J. Bact.* **93**, 1430.
35. Princewill, T. J. T. (1965). *J. comp. Path.* **75**, 343.
36. Bullen, J. J., Scarisbrick, R. and Maddock, A. (1953). *J. Path. Bact.* **65**, 209.
37. Bullen, J. J. and Scarisbrick, R. (1957). *J. Path. Bact.* **73**, 495.
38. Despaul, J. E. (1966). *J. Am. diet. Ass.* **49**, 185.
39. Weiss, K. F., Strong, D. H. and Groom, R. A. (1966). *Appl. Microbiol.* **14**, 479.
40. Hauschild, A. H. W., Hilsheimer, R. and Thatcher, F. S. (1967). *Can. J. Microbiol.* **13**, 1041.
41. Weiss, K. F. and Strong, D. H. (1967). *J. Bact.* **93**, 21.
42. Hall, H. E., Angelotti, R., Lewis, K. H. and Foter, M. J. (1963). *J. Bact.* **85**, 1094.
43. Hauschild, A. H. W. and Thatcher, F. S. (1967). *J. Fd Sci.* **32**, 467.
44. Taylor, C. E. D. and Coetzee, E. F. C. (1966). *Mon. Bull. Minist. Hlth* **25**, 142.
45. Nygren, B. (1962). *Acta path. microbiol. scand. Suppl.* 160.
46. Dack, G. M., Sugiyama, H., Owens, F. J. and Kirsner, J. B. (1954). *J. infect. Dis.* **94**, 34.
47. Weiss, K. F., Strong, D. H. and Groom, R. A. (1966). *Appl. Microbiol.* **14**, 479.
48. Woodburn, M. and Kim, C. H. (1966). *Appl. Microbiol.* **14**, 914.
49. Barnes, E. M., Despaul, J. E. and Ingram, M. (1963). *J. appl. Bact.* **26**, 415.
50. Collee, J. G., Knowlden, J. A. and Hobbs, B. C. (1961). *J. appl. Bact.* **24**, 326.
51. Gough, B. J. and Alford, J. A. (1965). *J. Fd Sci.* **30**, 1025.
52. Willis, A. T. (1960). "Anaerobic Bacteriology in Clinical Medicine". Butterworth, London.
53. Buttiaux, R., Beerens, H. and Tacquet, A. (1962). "Manual de Techniques Bacteriologiques". E.M.F., Paris.
54. Byrd, T. R. and Ley, H. L. Jr. (1966). *Appl. Microbiol.* **14**, 993.
55. Sanada, I. and Nishida, S. (1965). *J. Bact.* **89**, 626.
56. Wolf, J. and Barker, A. N. (1968). *In* "Identification Methods for Microbiologists", Part B (B. M. Gibbs and D. A. Shapton, eds), p. 93. Academic Press, London.
57. Kravchenko, A. T. and Shishulina, L. M. (1966). *In* "Botulism 1966". Proc. 5th Int. Symp. Fd Microbiol., Moscow, July 1966 (M. Ingram and T. A. Roberts, eds), p. 13. Chapman and Hall, London.

58. Knight, B. C. J. G. and Fildes, P. (1930). *Biochem. J.* **24**, 1496.
59. Shoesmith, J. G. and Holland, K. T. (1968). *Biochem. J.* **106**, 38P.
60. Oakley, C. L. (1954). *Br. med. Bull.* **10**, 52.
61. van Heyningen, W. E. (1959). *J. gen. Microbiol.* **20**, 310.
62. Oakley, C. L. (1954). *A. Rev. Microbiol.* **8**, 411.
63. Wright, G. P. (1955). *Pharmacol. Rev.* **7**, 413.
64. van Heyningen, W. E. and Arseculeratne, S. N. (1964). *A. Rev. Microbiol.* **18**, 195.
65. Batty, I. and Walker, P.D. (1965). *J. appl. Bact.* **28**, 112.
66. Cruickshank, R. (1965). "Medical Microbiology". Livingstone, London.
67. Cherry, W. B. and Moody, M. D. (1965). *Bact. Rev.* **29**, 222.
68. Morris, E. J. (1955). *J. gen. Microbiol.* **13**, 456.
69. Gillissen, G. and Scholz, H. G. (1961). *Zentbl. Bakt. Parasitenkunde Abt. I Orig.* **182**, 232.
70. Chadwick, P. (1959). *J. gen. Microbiol.* **21**, 631.
71. Davies, D. G. (1960). *J. Hyg. Camb.* **58**, 177.
72. Report (1959) *Mon. Bull, Minist. Hlth.* **18**, 16.
73. Minett, F. C. (1950). *J. comp. Path.* **60**, 161.
74. Albrink, W. S. (1961). *Bact. Rev.* **25**, 268.
75. Brachman, P. S., Kaufmann, A. F. and Dalldorf, F. G. (1966). *Bact. Rev.* **30**, 646.
76. Ross, J. M. (1957). *J. Path. Bact.* **73**, 485.
77. Klein, F., Walker, J. S., Fitzpatrick, D. F., Lincoln, R. E., Mahlandt, B. G., Jones, W. I. Jr., Dobbs, J. P. and Hendrix, K. J. (1966). *J. infect. Dis.* **116**, 123.
78. Stanley, J. L. and Smith, H. (1961). *J. gen. Microbiol.* **26**, 49.
79. Beall, F. A., Taylor, M. J. and Thorne, C. B. (1962). *J. Bact.* **83**, 1274.
80. Bonventre, P. F., Lincoln, R. E. and Lamanna, C. (1967). *Bact. Rev.* **31**, 95.
81. Kautter, D. A., Harmon, S. M., Lynt, R. K. Jr. and Lilly, T. Jr. (1966). *Appl. Microbiol.* **14**, 616.
82. Wentz, M. W., Scott, R. A. and Vennes, J. W. (1967). *Science* **155**, 89.
83. McClung, L. S. (1967). *In* "Botulism 1966". Proc. 5th Int. Symp. Fd Microbiol., Moscow, July 1966 (M. Ingram and T. A. Roberts eds), p. 431. Chapman and Hall, London.
84. Cardella, M. (1964). *In* "Botulism. Proceedings of a Symposium" (K. H. Lewis and K. Cassel, eds), p. 113. P.H.S. Publ. No. 999-FP-1, Public Health Service, Cincinnati.
85. Gibbs, B. M. and Hirsch, A. (1956). *J. appl. Bact.* **19**, 129.
86. Boroff, D. A. and Reilly, J. R. (1962). *Int. Archs Allergy Appl. Immun.* **20**, 306.
87. Schmidt, C. F., Nank, W. K. and Lechowich, R. V. (1962). *J. Fd Sci.* **27**, 77.
88. Harmon, S. M. and Kautter, D. A. (1967). *Bact. Proc.*, p. 5.
89. Georgala D. L., and Boothroyd, M. (1967). *In* "Botulism 1966". Proc. 5th Int. Symp. Fd Microbiol., Moscow, July 1966. (M. Ingram and T. A. Roberts, eds), p. 494. Chapman and Hall, London.
90. Walker, P. D. and Batty, I. (1967). *In* "Botulism 1966". Proc. 5th Intern. Symp. Fd Microbiol., Moscow, July 1966 (M. Ingram and T. A. Roberts, eds), p. 482. Chapman and Hall, London.
91. Johnson, H. M., Brenner, K., Angelotti, R. and Hall, H. E. (1966). *J. Bact.* **91**, 967.
92. Hobbs, G. (1967). Personal communication.

93. Marshall, R. and Quinn, Y. (1967). *J. Bact.* **94**, 812.
94. Greenberg, R. A., Bladel, B. O. and Zingelmann, W. J. (1966). *Appl. Microbiol.* **14**, 223.
95. Johnston, R., Harmon, S. M. and Kautter, D. A. (1964). *J. Bact.* **88**, 1521.
96. Ohye, D. F. and Scott, W. J. (1953). *Aust. J. biol. Sci.* **6**, 178.
97. Roberts, T. A. and Hobbs, G. (1968). *J. appl. Bact.* **31**, 75.
98. Dolman, C. E. *In* "Botulism. Proceedings of a Symposium" (K. H. Lewis and K. Cassel, eds), p. 5. P.H.S. Publ. No. 999-FP-1, Public Health Service, Cincinatti.
99. Eklund, M. W. and Poysky, F. (1967). *In* "Botulism 1966". Proc. 5th Int. Symp. Fd Microbiol., Moscow, July 1966 (M. Ingram and T.A. Roberts, eds), p. 49. Chapman and Hall, London.
100. Dolman, C. E. and Murakami, L. (1961). *J. infect. Dis.* **109**, 107.
101. Ward, B. Q., Carroll, B. J., Garrett, E. S. and Reese, G. B. (1967). *Appl. Microbiol.* **15**, 629.
102. Johannsen, A. (1965). *J. appl. Bact.* **28**, 90.
103. Chapman, H. M. and Naylor, H. B. (1966). *Appl. Microbiol.* **14**, 301.
104. Rhodes, D. N., Roberts, T. A. and Hobbs, G. (1966). *In* "Technical Report Series" No. 54. Intern. Atomic Energy Agency, Vienna p. 39.
105. Nakamura, Y. (1963). *Japan J. med. Sci. Biol.* **16**, 304.
106. Bott, T. L., Deffner, J. S., McCoy, E. and Foster, E. M. (1966). *J. Bact.* **91**, 919.
107. Prévot, A. R. and Huet, N. (1951). *Bull. Acad. natn. Méd., Paris* **135**, 432.
108. Craig, J. M. and Pilcher, K. S. (1967). *In* "Botulism 1966". Proc. 5th Int. Symp. Fd Microbiol., Moscow, July 1966 (M. Ingram and T. A. Roberts, eds), p. 56. Chapman and Hall, London.
109. Cann, D. C., Wilson, B. B., Hobbs, G., Shewan, J. M. and Johannsen, A. (1965). *J. appl. Bact.* **28**, 426.
110. Johannsen, A. (1963). *J. appl. Bact.* **26**, 43.
111. Cann, D. C., Wilson, B. B., Hobbs, G. and Shewan, J. M. (1967). *In* "Botulism 1966". Proc. 5th Int. Symp. Fd Microbiol., Moscow, July 1966 (M. Ingram and T. A. Roberts, eds), p. 62. Chapman and Hall, London.
112. Bott, T. L., Deffner, J. S. and Foster, E. M. (1967). *In* "Botulism 1966". Proc. 5th Int. Symp. Fd Microbiol., Moscow, July 1966 (M. Ingram and T. A. Roberts, eds), p. 21. Chapman and Hall, London.
113. Moller, V. and Scheibel, I. (1960). *Acta path. microbiol. scand.* **48**, 80.
114. Craig, J. M. and Pilcher, K. S. (1966). *Science* **153**, 311.
115. Petty, C. S. (1965). *Am. J. med. Sci.* **249**, 345.
116. Grecz, N. and Lin, C. A. (1967). *In* "Botulism 1966". Proc. 5th Intern. Symp. Fd Microbiol., Moscow, July 1966 (M. Ingram and T. A. Roberts, eds), p. 302. Chapman and Hall, London.
117. Meyer, K. F. and Eddie, B. (1965). "Sixty-five years of human botulism in the United States and Canada. Epidemiology and Tabulation of reported cases 1899 through 1964". University of California, U.S.A.
118. Jordan, E. O. and Dack, G. M. (1924). *J. infect. Dis.* **35**, 576.
119. Hunter, B. F. and Rosen, M. N. (1967). *Avian Dis.* **11**, 345.
120. Pigoury, L., Chabassol, C. and Stellman, C. (1965). *Bull. Acad. vét. Fr.* **38**-319.
121. Sterne, M. and Wentzel, L. M. (1953). *Rep. 14th Int. Vet. Congr.* **3**, 329.
122. Meyer, K. F. (1956). *Bull. W.H.O.* **15**, 281.
123. Quortrup, E. R. and Holt, A. L. (1941). *J. Bact.* **41**, 363.

124. Sherman, J. M., Stark, C. N. and Stark, P. (1927). *Proc. Soc. exp. Biol. N.Y.* **24,** 546.
125. Jensen, W. I. and Gritman, R. B. (1967). *In* "Botulism 1966". Proc. 5th Int. Symp. Fd Microbiol., Moscow, July 1966 (M. Ingram and T. A. Roberts, eds), p. 407. Chapman and Hall, London.
126. Skulberg, A. and Grande, M. (1967). *Trans. Am. Fish. Soc.* **96,** 67.
127. Kaufmann, O. W. and Marshall, R. S. (1965). *Appl. Microbiol.* **13,** 521.
128. Meyer, K. F. (1929). *J. infect. Dis.* **44,** 408.
129. Cavett, J. J. (1968). *Progress in industrial Microbiology* 7 (D. J. D. Hockenhull, ed.). Heywood, London.
130. Johannsen, A. (1961). Svenska Institutet förr Konserveringsforskning, Göteborg, Sweden, Report, No. 100.
131. Kautter, D. A. (1964). *J. Fd Sci.* **29,** 843.
132. Abrahamsson, K., De Silva, N. N. and Molin, N. (1965). *Can. J. Microbiol.* **11,** 523.
133. Pivnick, H. and Bird, H. (1965). *Fd Technol., Champaign* **19,** 132.
134. Christiansen, L. N. and Foster, E. M. (1965). *Appl. Microbiol.* **13,** 1023.
135. Warnecke, M. O., Carpenter, J. A. and Saffle, R. L. (1967). *Fd Technol., Champaign* **21,** 433.
136. Pivnick, H. and Barnett, H. (1967). *In* "Botulism 1966". Proc. 5th Int. Symp. Fd Microbiol., Moscow, July 1966 (M. Ingram and T. A. Roberts, eds), p. 208. Chapman and Hall, London.
137. Canadian Food and Drugs Act, PC 1967–1209, Sections B.14.013 and B.14.014.
138. Thatcher, F. S. (1964). *In* "Botulism. Proceedings of a Symposium" (K. H. Lewis and K. Cassel, eds), p. 240. P.H.S. Publ. No. 999-FP-1. Public Health Service, Cincinnati.
139. Kindler, S. H., Mager, J. and Grossowicz, N. (1956). *J. gen. Microbiol.* **15,** 386.
140. Ward, B. Q. and Carroll, B. J. (1966). *Can. J. Microbiol.* **12,** 1145.
141. Snudden, B. H. and Lechowich, R. V. (1967). *In* "Botulism 1966". Proc. 5th Int. Symp. Fd Microbiol., Moscow, July 1966 (M. Ingram and T. A. Roberts, eds) p. 144. Chapman and Hall, London.
142. Gullmar, B. and Molin, N. (1967). *In* "Botulism 1966". Proc. 5th. Int. Symp. Fd Microbiol., Moscow, July 1966 (M. Ingram and T. A. Roberts, eds), p. 185. Chapman and Hall, London.
143. Holdeman, L. V. and Smith, L. D. (1965). *Can. J. Microbiol.* **11,** 1009.
144. Mossel, D. A. A. and Ingram, M. (1955). *J. appl. Bact.* **18,** 232.
145. Kindler, S. H., Mager, J. and Grossowicz, N. (1956). *J. gen. Microbiol.* **15,** 394.
146. Boroff, D. A. and DasGupta, B. R. (1964). *J. Biol. Chem.* **239,** 3694.
147. Boroff, D. A., DasGupta, B. R. and Fleck, U. (1967). *In* "Botulism 1966". Proc. 5th Int. Symp. Fd Microbiol., Moscow, July 1966 (M. Ingram and T. A. Roberts, eds), p. 278. Chapman and Hall, London.
148. Ward, B. Q. and Carroll, B. J. (1967). *Can. J. Microbiol.* **13,** 108.
149. Cann, D. C., Wilson, B. B., Hobbs, G. and Shewan, J. M. (1965). *J. appl. Bact.* **28,** 431.
150. Shewan, J. M. British Food Manufacturing Industries Research Assoc. Tech. Circ. 303, Leatherhead, England.
151. Hall, I. C. and Peterson, E. (1923). *J. Bact.* **8,** 319.
152. Riemann, H. (1967). *Fd Technol., Champaign* **21,** 759.

153. Benjamin, M. I. W., Wheater, D. M. and Shepherd, P. A. (1956). *J. appl. Bact.* **19**, 159.
154. Meyer, K. F. and Gunnison, J. B. (1929). *J. infect. Dis.* **45**, 135.
155. Saleh, M. A. and Ordal, Z. J. (1955). *Fd Res.* **20**, 340.
156. Grecz, N., Wagenaar, R. O. and Dack, G. M. (1959). *Appl. Microbiol.* **7**, 33.
157. Grecz, N. (1964). *In* "Microbiol Inhibitors in Food". Proc. 4th Int. Symp. Fd Microbiol. (N. Molin, ed.). Almquist and Wiksell, Stockholm.
158. Skulberg, A. (1964). "Studies on the formation of toxin by *Clostridium botulinum*". Monograph A/S Kaare Grytting. Orkanger, Norway.
159. Kautter, D. A. and Bartram, M. (1967). *In* "Botulism 1966". Proc. 5th Int. Symp. Fd Microbiol., Moscow, July 1966 (M. Ingram and T. A. Roberts, eds), p. 448. Chapman and Hall, London.
160. Sagaguchi, G. and Tohyama, Y. (1955). *Jap. J. med. Sci. Biol.* **8**, 247.
161. Eklund, M. W., Weiler, D. I. and Poysky, F. T. (1967). *J. Bact.* **93**, 1461.
162. Ohye, D. F. and Scott, W. J. (1957). *Aust. J. biol. Sci.* **10**, 85.
163. Schmidt, C. F., Lechowich, R. V. and Folinazzo, J. F. (1961). *J. Fd Sci.* **26**, 626.
164. Walls, N. W. (1967). *In* "Botulism 1966". Proc. 5th Int. Symp. Fd Microbiol. Moscow, July 1966 (M. Ingram and T. A. Roberts, eds), p. 158. Chapman and Hall, London.
165. Bonventre, P. F. and Kempe, L. L. (1959). *Appl. Microbiol.* **7**, 374.
166. Ohye, D. F. and Scott, W. J. (1953). *Aust. J. biol. Sci.* **6**, 178.
167. Abrahamsson, K., Gullmar, B. and Molin, N. (1966). *Can. J. Microbiol.* **12**, 385.
168. Ingram, M. and Robinson, R. H. M. (1951). *Proc. Soc. appl. Bact.* **14**, 73.
169. Segner, W. P., Schmidt, C. P. and Boltz, J. K. (1966). *Appl. Microbiol.* **14**, 49.
170. Baird-Parker, A. C. and Freame, B. (1967). *J. appl. Bact.* **30**, 420.
171. Scott, W. J. (1957). *Adv. Fd Res.* **7**, 83.
172. Ohye, D. F., Christian, J. H. B. and Scott, W. J. (1967). *In* "Botulism 1966". Proc. 5th Int. Symp. Fd Microbiol. Moscow, July 1966 (M. Ingram and T. A. Roberts, eds), p. 136, Chapman and Hall, London.
173. Ohye, D. F. and Christian, J. H. B. (1967). *In* "Botulism 1966". Proc. 5th Int. Symp. Fd Microbiol. Moscow, July 1961 (M. Ingram and T. A. Roberts, eds), p. 217, Chapman and Hall, London.
174. Spencer, R. (1967) *In.* "Botulism 1966". Proc. 5th Intern. Symp. Fd. Microbiol. Moscow, July 1966 (M. Ingram and T. A. Roberts, eds), p. 123 Chapman and Hall, London.
175. Baird-Parker, A. C. (1967). Unpublished observations.
176. Greenberg, R. A., Silliker, J. H. and Fatta, L. D. (1959). *Fd Technol., Champaign* **13**, 509.
177. Mundt, J. O., Mayhew, C. J. and Stewart, G. (1954). *Fd Technol., Champaign* **8**, 435.
178. Scott, W. J. (1955). *Ann. Inst. Pasteur Lille* **7**, 68.
179. Silliker, J. H., Greenberg, R. A. and Schack, W. R. (1958). *Fd Technol., Champaign* **12**, 551.
180. Shank, J. L., Silliker, J. H. and Harper, R. H. (1962). *Appl. Microbiol.* **10**, 185.
181. Ingram, M. (1959). *Chem. Ind.* **552**.
182. Henry, M., Joubert, L. and Goret, P. (1954). *C.r. Séanc. Soc. Biol.* **148**, 819.
183. Greenberg, R. A. and Silliker, J. H. (1961). *J. Fd Sci.* **16**, 622.
184. Lowbury, E. J. L. and Lilley, H. A. (1958), *J. Hyg. Camb.* **56**, 169.

185. Riemann, H. (1963). *Fd Technol. Champaign* **17**, 39.
186. Roberts, T. A. and Ingram, M. (1966). *J. Fd Technol.* **1**, 147.
187. Bulman, C. and Ayres, J. C. (1952). *Fd Technol. Champaign* **6**, 255.
188. Steinke, P. K. W. and Foster, E. M. (1951). *Fd Res.* **16**, 477.
189. Spencer, R. (1966). *Food Manufacture* **41**, 39.
190. Riemann, H. (1963). *In* "Chemical and Biological Hazards in Foods" (J. C. Ayres, A. A. Kraft, H. E. Snyder and H. W. Walker, eds), p. 279. Iowa State University Press, Ames, Iowa, U.S.A.
191. Perigo, J. A., Whiting, E. and Bashford, T. E. (1967). *J. Fd Technol.* **2**, 377.
192. Castellani, A. G. and Niven, C. F. Jr. (1955). *Appl. Microbiol.* **3**, 154.
193. Perigo, J. A. (1967). Personal communication.
194. Nielsen, S. F., Pedersen, H. O. (1967). *In* "Botulism 1966". Proc. 5th Int. Symp. Fd Microbiol. Moscow, July 1966 (M. Ingram and T. A. Roberts, eds), p. 66. Chapman and Hall, London.
195. Abrahamsson, K. (1967). *In* "Botulism 1966". Proc. 5th Int. Symp. Fd Microbiol, Moscow, July 1966 (M. Ingram and T. A. Roberts, eds), p. 73. Chapman and Hall, London.
196. Cann, D. C., Wilson, B. B., Hobbs, G. and Shewan, J. N. (1967). *In* "Botulism 1966". Proc. 5th Int. Symp. Fd Microbiol. Moscow, July 1966 (M. Ingram and T. A. Roberts, eds), p. 202. Chapman and Hall, London.
197. Dack, G. M. (1964). *In* "Botulism. Proceedings of a Symposium" (K. H. Lewis and K. Cassel, eds), p. 33. P.H.S. Publ. No. 999-FP-1. Public Health Service, Cincinnati.
198. Pace, P. J., Krumbiegel, E. R., Wisniewski, H. J. and Angelotti, R. (1967). *In* "Botulism 1966". Proc. 5th Int. Symp. Fd Microbiol. Moscow, July 1966 (M. Ingram and T. A. Roberts, eds), p. 40. Chapman and Hall, London.
199. Mossel, D. A. A., Koopman, M. J. and Jongerius, E. (1967). *Appl. Microbiol.* **15**, 650.
200. Hauge, S. (1955). *J. appl. Bact.* **18**, 591.
201. Lázár, J. and Jurcsak, L. (1966). *Zentbl. Bakt. Parasitenbunden Abt. I Orig.* **199**, 59.
202. Nikodemusz, I. (1965). *Zentbl. Bakt. Parasitenbunden Abt. I Orig.* **196**, 81.
203. Eklund, M. W., Poysky, F. T. and Wieler, D. I. (1967). *Appl. Microbiol.* **15**, 1316.
204. Ward, B. Q., Garrett, E. S. and Reese, G. B. (1967). *Appl. Microbiol.* **15**, 1509.
205. Duff, J. T., Wright, G. G. and Yarinsky, A. (1956). *J. Bact.* **72**, 455.
206. Midura, T. F., Inouye, Y. and Bodily, H. L. (1967). *Public Health Reports* **82**, 275.
207. Esty, J. R. and Meyer, K. F. (1922). *J. infect. Dis.* **31**, 650.
208. Duncan, C. L. and Strong, D. H. (1968). *Appl. Microbiol.* **16**, 82.
209. Sutton, R. G. A. and Hobbs, B. C. (1968). *J. hyg. Camb.* **66**, 135.
210. Grecz, N., Lin, C. A., Tang, T., So, W. L. and Sehgal, L. R. (1967). *Jap. J. Microbiol.* **11**, 384.
211. Sugiyama, H., von Mayerauser, B., Gogat, G. and Heimsch, R. C. (1967). *Proc. Soc. exp. biol Med.* **126**, 690.
212. Roberts, T. A. (1968). *J. appl. Bact.* **31**, 133.
213. Duncan, J. T. (1944). *Mon. Bull. Minist. Hlth.* **3**, 61.
214. Duncan, C. L., Sugiyama, H. and Strong, D. H. (1968). *J. Bact.* **95**, 1560.
215. Hauschild, A. H. W. and Thatcher, F. S. (1968). *Can. J. Microbiol.* **14**, 705.
216. Sumyk, G. B. and Yocum, C. F. (1968). *J. Bact.* **95**, 1970.

548   A. C. BAIRD-PARKER

217. Dolman, C. E., Tomsick, M., Campbell, C. C. R. and Laing, W. B. (1960). *J. infect. Dis.* **106,** 5.
218. Bott, T. L., Johnson, J. Jr., Foster, E. M. and Sugiyama, H. (1968). *J. Bact.* **95,** 1542.
219. Condit, P., Cosand, M. and Scott, M. (1968). *Morbidity and Mortality,* **17** 199.
220. Tong, J. L., Engle, H. M., Cullyford, J. S., Shimp, D. J. and Love, C. E. (1962). *Am. J. Public Hlth.* **52,** 976.
221. Hauschild, A. H. W., Niilo, L. and Dorward, W. J. (1968). *Appl. Microbiol.* **16,** 1235.

CHAPTER 15

Sporeformers as Food Spoilage Organisms

M. INGRAM

Agricultural Research Council, Meat Research Institute,
Langford, Bristol, England

I. Introduction

THE aim of this chapter is to illustrate, in a comparatively general way, what is the practical importance of some of those properties of spores which are discussed in detail in other chapters. In any attempt at disinfection, bacterial spores are apt to present a special problem because of their unusual powers of resistance; and the preparation for example of canned foods is governed by rules largely aimed to control the activities of heat-resistant spores which would otherwise lead to spoilage of the food. Because the treatment of the subject in this chapter is general, references are to books, reviews or leading articles, more than to individual papers of restricted scope.

To write of the sporeforming bacteria as food spoilage organisms, in a book entirely devoted to spores, involves strict limitation in choice of material. The spores are biochemically inert, and therefore not themselves responsible for spoilage. Spoilage is actually effected by vegetative cells; and one of the points to be remembered is that the various significant environmental factors may have distinctly different effects on spores and on vegetative cells. It would seem out of place to write at length about the biochemistry of spoilage, or about many of the effects of environmental factors on its development, these being topics not strictly relevant to our subject of spores. Similarly, simple description of methods seems inappropriate. This chapter must, rather, try to show how the peculiar characters of bacterial spores, described at length in other chapters, determine the part sporeforming bacteria play in food spoilage.

The capacity of the almost inert spores to spoil food, should they survive, depends on their ability to germinate, outgrow, and achieve extensive vegetative multiplication in the food. Interruption of this chain of events at any point will prevent spoilage. Further, the seriousness of the spoilage will depend to some extent upon its nature, which in turn depends on the biochemical powers of the species and the nature of the food as substrate.

A. Important Physiological Characteristics

A considerable limitation follows from the physiological characteristics of the sporeforming species.

1. Temperature relations

Take, for example, the relations to temperature. It is a fact, though entirely unexplained, that very few sporeforming bacteria are capable of growing at all at chill temperatures below about 10°C; and the few which can, only do so slowly in comparison with non-sporing psychrotrophic species. Consequently, sporeformers are seldom important in spoilage of refrigerated foods. Indeed, refrigeration is the principal safeguard against the development of sporeforming anaerobes in foods like meat which, if they remain warm, are likely to be spoiled in a few hours by massive growth of these bacteria. Most of the sporeforming species are mesophiles, active only under more or less warm conditions; but, even under these conditions, they often represent only a subordinate element in the spoilage flora. This is because, compared with similar mesophilic species which do not form spores, the sporeformers are not outstanding in characters such as rate of growth, or the secretion of metabolic products like acid or alcohol which exclude competing species. Only among species which grow under warmer conditions, from about 40°C upwards, are the sporeformers apparently predominant; but this impression may be partly illusory, arising from the circumstances in which thermophilic spoilage occurs. On the whole, therefore, sporeformers are very important in spoilage under warm conditions, negligible under chill conditions, and at intermediate temperatures they become important when some selective agency which spores resist (such as heat) eliminates the competition of otherwise similar non-sporing species.

2. Tolerance of acidity

A further important limitation is for example that due to acidity. Foods cover the range from neutrality down to about pH 2·5, but only yeasts and moulds can spoil foods more acid than about pH 3·7. Even at acidities from pH 4·5 to 3·7, only two or three types of sporing bacteria can grow, and these are not dangerous species. Some of the species with the most heat-resistant spores are inhibited under conditions more acid than pH 5·3. The really dangerous species among the sporeforming bacteria, *Clostridium botulinum*, will not grow in foods more acid than pH 4·5. Hence it is convenient to divide canned foods into four groups on a pH basis, separated by the boundaries indicated, pH 5·3, 4·5 and 3·7.[1] In the fourth group there are no spore-forming bacteria, and in the third, only a very limited variety; hence it is possible to restrict processing requirements accordingly.

These examples serve to show that it is only in particular cases that sporeforming bacteria are important in food spoilage, and that these cases occur only when circumstances favour the physiological peculiarities of the sporeformers. Their main peculiarity being the possession of resistant spores, the usual decisive element is some factor which spores resist and vegetative organisms do not.

The same resistance makes the spores themselves (as distinct from the derived vegetative cells) extremely difficult to destroy in a food, without damaging the food also. Orthodox chemical agents are either ineffective or too poisonous, and most physical agents either unreliable or too damaging.

3. Resistance to heat

Almost the only procedure for killing spores which is both effective and tolerable in foods has, hitherto, been heating. But even here, the most heat-resistant spores require so severe a treatment as commonly has an undesirable effect on the food, so that one uses as little heating as possible. This leads to three broad categories of treatment:

(i) In food whose quality is not ruined by high degrees of heating, a "nearly sterile" condition can be achieved (i.e. a predominant pro-portion of units—e.g. cans or bottles—contain no viable spores). Even here, quality is usually better when heating is minimized. Foods in this class will be devoid of enzyme activity and, when packed in hermetic containers needed to maintain sterility, are called "conserves".

(ii) In foods which tolerate only a rather lower degree of heating, some of the more resistant spores will remain alive though they may not develop. If the spores remain dormant, they cause no change in the food. This is as if the spores were not present, and the food is consequently described as "commercially sterile".

(iii) In foods sensitive to heating, it is only possible to destroy vegetative cells and the more sensitive bacterial spores, and some auxiliary factors are needed to ensure preservation even for limited periods. This type of treatment is commonly called "pasteurization" (e.g. pasteurized canned ham) and foods in this class are called "semi-conserves" or "semi-preserved". It will be evident that, with in-creasing degrees of heat treatment, spores become increasingly important, because they are likely to survive while competing species are eliminated. Among bacterial spores, only species of the highest heat resistance will be important in conserves, while the most lightly processed semi-conserves may contain heat-sensitive spores, besides vegetative cells of relatively heat-resistant types such as those of faecal streptococci.

Similar general considerations apply with agencies other than heat; though the same species may not necessarily be the most resistant to different agents.

B. Obnoxious Properties

None of the sporeforming bacteria significant in foods are infective pathogens, so that their activities only become important when large numbers of vegetative cells develop in the food. The biochemical activity of the dormant spores is negligible, nor have they themselves appreciable pathogenic activity. It follows therefore that obnoxious conditions arise only when viable spores are exposed to conditions where they can germin-ate, outgrow, and multiply extensively, within the food.

1. Spoilage

The species of bacteria which form spores are diverse in their bio-chemical activities. Some ferment carbohydrates with formation of acid and/or gas. Others cause breakdown of proteins, with liberation of ammonia and more complex amino compounds, and may or may not produce gas in the process. A few—but not uncommon—versatile species do both. Some can reduce nitrates or nitrites, present in cured meats, to nitrogen or oxides of nitrogen. The type of spoilage produced may therefore vary, from stinking putrefaction, to acid production, or to production of gas without much obvious change otherwise; and the nature of this spoilage will obviously depend, in considerable measure, on the nature of the food, as well as on the nature of the spore-forming organisms which happen to be in it.

2. Food poisoning

Even more serious than obnoxious spoilage is the prospect that the food might be rendered harmful, for certain sporeforming bacteria are capable of causing more or less dangerous food poisoning. Some (but by no means all) strains of *Bacillus cereus* and *Cl. perfringens* (syn. *Cl. welchii*) may cause rather severe but brief (a few hours) diarrhoea. While these species are similar in secreting a lecithinase enzyme, this is not responsible for their ill-effects. *Cl. botulinum*, on the other hand, secretes a most powerful toxin causing nervous paralysis, leading to a form of poisoning called "botulism" which is commonly fatal, so that a high proportion of those eating toxin-containing food die. The social and economic consequences of such an incident are intolerable; so that, where circumstances are such as to make this risk conceivable, all the operations of the food manufacturer and handler must be aimed to prevent it.

Cl. botulinum is a typical putrefactive anaerobe, resembling *Cl. sporogenes* and related species which do not cause poisoning, with very similar growth requirements. Therefore, if there has been large multiplication of these species with consequent putrefaction and gas formation, this is an indication that *Cl. botulinum* might have grown and made the food dangerous, had it been present; this is the basic reason for the fear to eat putrid food. In fact, this is an unreliable guide. On the one hand, *Cl. botulinum* is fortunately rare, so that most putrid food is in reality almost harmless, even if unpalatable; the old idea that putrid food contained poisonous "histamines" has been found to be largely erroneous. On the other hand, there is a conceivable possibility of toxin-formation without putrefaction. Of the four immunologically distinct types of *Cl. botulinum*, designated A, B, E and F, which have been implicated in human botulism, type E is non-proteolytic. Further, though types A, B and F are most commonly proteolytic, non-proteolytic strains of both type B and type F have recently been isolated; and these latter, together with type E, are capable of growing and producing toxin at domestic refrigeration temperatures (*c.* 5 °C). The absence of proteolysis is particularly sinister because the toxic foods are apparently normal. Fortunately *Cl. botulinum* toxins are readily inactivated by heating, so that this form of poisoning is almost always due to food not cooked just before eating.

In the case of *Cl. perfringens (welchii)*, there has been disagreement about the nature of the particular strains which cause food poisoning. It was at first thought that they were characterized by the possession of spores of unusually high heat resistance, resisting hours of boiling, roughly equal to that of *Cl. botulinum* (cf. Table IV, p. 577). It now seems that such strains were preferentially selected by the circumstances, both of occurrence and isolation, and that poisoning may also be caused, when circumstances are favourable, by strains of normal heat resistance for this species.

Similarly with *B. cereus*, there is at present no indication that the strains involved in food poisoning have unusual spores.

C. General Considerations

The nature of the bacteria involved and of the food spoilage they cause, and hence the procedures required to prevent spoilage, are determined by a marriage of the requirements of the bacteria with the properties and condition of the food. There are thus three broad questions resembling those of seed, soil and climate in agricultural practice.

(i) The first is the nature of the micro-organisms, for present purposes bacterial spores, which occur in the food. This leads in turn to consideration of the occurrence and distribution of these spores in the environment, and the means whereby they gain access to the food.

(ii) The second point is whether the properties of the food are such as will permit development of the spores in question—acidity, as just indicated, is a very important property in this connection, as is the water activity (a_w).

(iii) Finally, the other environmental conditions must be appropriate, and the limiting effects of temperature have already been mentioned.

More extensive discussion of each of these topics is the subject of Section II of this chapter.

II. Spoilage of Foods by Sporeformers

Because of the importance of these questions—the occurrence of spores in the food, the relevant properties of the food and the environmental circumstances—they will first be considered individually in some detail. It is then easier to consider how they interact, to select the various species of spore-forming bacteria which are important in different circumstances.

A. Occurrence of Spores in Foods and the Environment

The problems arising from the presence of bacterial spores in a food will obviously depend on the nature and numbers of the spores present, in relation to the nature of the food and its treatment. The nature and number present depend, in turn, on the general occurrence and distribution of spores and the likelihood that they will gain access to the food during handling. The subject is large, complex in detail, and can only be illustrated here with a few examples.

It is of course the spores which are of paramount importance in this connection. Vegetative cells, of clostridia especially, are readily killed by treatments like exposure to air, or chilling or drying; and consequently, they are seldom important in spreading contamination, especially from ubiquitous natural sources like soil and water which—apart from direct pollution from the bodies of animals—are the common sources of contamination. The importance of spores springs from two main causes: (i) their remarkable powers of dormancy and resistance, which lead to (ii) their almost ubiquitous distribution. The former can be dismissed briefly here, the latter calls for more extended description.

For details of the resistance of spores, see Chapter 16. It suffices to recall that they are practically unaffected by drying or freezing, salts and acids, and mostly survive long exposure to heating or to solar radiations including ultraviolet light. For these reasons, they can probably persist more or less indefinitely under natural conditions, and there are recorded examples of survival for about fifty years.

1. Natural distribution of spores

Soil normally contains spores in considerable numbers; though there are few systematic data.

Spores of aerobic *Bacillus* species are the most frequent in number and variety. Waksman[2] surveyed the frequency of isolation of different strains. Most common were those of *B. petasites* now included in *B. megaterium* or *B. subtilis*, and of *B. cereus*; these together made up roughly 350 out of 500 isolations. Occurring less frequently were a wide variety of species.

The occurrence of *Clostridium* strains in soil has received comparatively little attention, apart from the nitrogen-fixing species. Waksman[2] says that more than 100,000 nitrogen-fixing clostridia have been found per g soil, and quotes other work showing a range of 100–1,000,000 total anaerobes per g. Clostridia considerably outnumber the cells of non-sporing nitrogen-fixing *Azotobacter* species, by two to tenfold in the work quoted. A limited survey showed relatively high numbers of clostridia,[3] including many *Cl. perfringens*. Most surveys have been concerned mainly with *Cl. botulinum*, and these are reviewed in Chapter 14. According to

Burges[4] 31 species of *Clostridium* have been isolated from soil, and 27 species of *Bacillus*, and it seems that spores of almost any species might be present occasionally.

The vague impression is that in general *Bacillus* spores outnumber *Clostridium* spores by a factor of about 100. For instance, in Broadbalk (Rothamsted, U.K.) soil the clostridial spore count was about 10^3/g and the general anaerobic spore count about 10^5, roughly three-quarters of the latter representing strains of *B. mycoides*, which are known to be able to grow anaerobically (summarized from F. A. Skinner, personal communication).

With this vast reserve of bacterial spores in the soil, it is not surprising that these same spores form one of the major groups of contaminants in the air and in dust, particularly as they are extremely resistant to desiccation. Once again data are scarce, particularly on the anaerobes; there is some information about aerobic *Bacillus* spp.[5]

2. Contamination of food from the environment

The presence of spores throughout the environment leads to infection of any portions of nutritious material wiped or spilled onto utensils or left lying in imperfectly cleaned equipment. Accordingly, growth of sporeforming species is likely in these places, the more so as the spores provide a means of surviving unfavourable conditions associated with stoppages or cleaning operations. As a result, it is common to find spore-formers as contaminants of food-processing equipment, and the species involved are naturally those adapted to develop on the nutritional basis of the foodstuff in question and hence well able to spoil it.

Avoidance of casual contamination, and cleaning schedules which remove spores as well as vegetative organisms, are necessary to minimize numbers of spores in foods. Discussion of such matters of factory housekeeping and hygiene is not now appropriate; information is available in standard textbooks.[6, 7] The important point here is that the resistance of spores to sanitation procedures makes it very difficult to ensure their elimination. The ubiquitous reservoirs of spores ensure that all foods are themselves likely to carry some spores. This is especially so with foods like vegetables or spices which are apt to suffer contamination with soil, and hence arises the unusual difficulty of processing foods containing vegetables or spices.

This contamination is, however, at first largely confined to external surfaces, for living tissues, whether of plants or animals, contain few microorganisms internally. It becomes much more troublesome if the nature of the food is such that, as a result of chopping or mixing, this surface contamination becomes distributed throughout the mass, for the contaminating spores are then protected from the direct action of external influences.

3. Intrinsic contamination of foods

The question whether living tissues are sterile internally is one of great practical importance; because if it were so, complete removal of surface contamination would in principle suffice to ensure preservation. Textbooks indeed often state that normal healthy tissues are sterile, which encourages the belief that surface sterilization (e.g. with radiation of low penetration) is a promising food-preserving process; but these views do not seem to be justified in actual practice.

With plants, the soil normally provides such a large surface contamination with spores that there is a high risk of penetration into the tissues as a result of accidental damage, mechanical or by insects. The surface of the plant itself carries a characteristic microflora, but bacterial spores are not common in it, so far as is known. The seed may often be sterile inside, so that sterile plants can fairly readily be grown from surface-sterilized seeds if suitable precautions are taken against re-contamination. Hence it appears at first sight that, for example, aseptic handling and de-podding of leguminous crops might be a useful process. But in practice sufficient contamination occurs (presumably through invasion by insects etc.) to make it impossible to achieve a sterile product.

The skin surface in animals and humans is usually heavily loaded with bacteria and, though the intrinsic flora of the skin contains few sporeformers, bacteria of this type are always present among the non-resident contaminants from dirt. Bacteria from the skin penetrate into any lesions (e.g. the spores of gangrene, tetanus and anthrax), and the body normally clears them through the lymphatic system, so that they are especially likely to be found in the lymph nodes. The gut contains sporeformers and spores aerobic and anaerobic, in vast numbers, attaining millions per g of faeces in appropriate circumstances; and these too are capable of leaking into the blood and lymphatic system under various conditions of stress to which animals are commonly subjected in practice.[8, 9] Moreover, it is practically impossible, in the commercial slaughter, evisceration and butchering of meat animals, to avoid contaminating the rest of the carcass from the above sources. The consequence is that, under practical conditions, meat tissue is not sterile, though the numbers of spores it contains is small even in lymph nodes, with good abattoir practice. For example, in a survey of freshly slaughtered beef, spores of *Clostridium* were recovered from 8% of mesenteric lymph nodes in cattle carefully slaughtered, but in 60% of lymph and muscle samples from cattle less carefully slaughtered, though the numbers were usually less than 10 per g.[10] More widespread surveys suggest that (though there are wide variations) the average number of aerobic spores in commercial meat is of the order 100 per g and of anaerobic spores 1 per g[11]—a ratio roughly corresponding to that perhaps existing in soil. Such estimates of spore populations are based on examination of samples which have been heated, usually at 80°C for upwards of 10 min. The number of spores which remain capable of surviving such treatment may be much reduced, however, because they can germinate in meat if conditions are appropriate, even at cool temperatures where outgrowth is not possible.

The exact numbers of spores which remain in sound tissues are necessarily difficult to determine, because they are so few that a high level of technique is necessary to avoid accidental contamination which confuses the issue. Consequently, the subject has always been open to argument and interpretation. In the writer's opinion there is now ample reliable evidence to show that "sound healthy tissues" *as commonly understood in practice* are not completely sterile, though they contain very few bacteria and even fewer spores. The question is still one of great interest as it decisively affects conceivable process requirements.

4. *Importance of nature and number of spores*

The reader may well feel that the above text lacks data to illustrate it. The reasons are first that such data are scarce, and second that numbers of spores are so much a matter of circumstances that data have little permanent value. From the practical standpoint the important conclusions are that all foods may be expected to contain spores, and that the spores will be very numerous if conditions are dirty, while special precautions will be needed if their numbers are to be kept at low levels. If the source of contamination is soil, directly or indirectly, a wide range of species may occur, many of doubtful significance for the food. If the contamination is from uncleaned processing equipment, the nature of the organism is likely to be more specific and troublesome. The precautions to be taken to limit contamination, and the processing needed to control it, depend directly on these considerations.

The difficulties will obviously be greater if the contaminating spores happen to have a high resistance to the influence being applied to control them. Thus, for example, heat-resistant spores are especially troublesome in heat-processed foods, a topic resumed in Section III.

Equally important is the number of spores, for it is well established that a process which controls small numbers satisfactorily will fail if the number of spores is increased, as illustrated in Table I.

There are several conceivable ways of explaining the phenomenon.

(i) It will be shown later (Fig. 1) that the heating needed to kill spores decreases regularly with the numbers of spores present; but this decrease is relatively small.

(ii) Vas and Proszt[13] showed clearly that a small proportion, about

TABLE I

Effect of curing salts and number of spores, on spoilage of pork, uninoculated, or inoculated with spores of Cl. sporogenes strain 3679, and heated 20 min at 80°C[a]

Curing salt mixture			% spoilage after storage at 37°C for (weeks):			
3·5% NaCl	0·12% NaNO₃	0·01% NaNO₂	1	2	4	16
Inoculated meat: *c*. 50 spores per g						
−	−	−	100	100	100	100
−	+	−	100	100	100	100
−	−	+	0	80	100	100
−	+	+	100	100	100	100
+	−	−	100	100	100	100
+	+	−	50	100	100	100
+	−	+	0	0	0	100
+	+	+	0	0	0	20
Uninoculated meat: < 1 spore per g						
+	+	−	100	100	100	100
+	−	+	0	0	0	0
+	+	+	0	0	0	0

[a] Summarized from Bulman and Ayres.[12]

10^{-7}, of spores of *B. cereus* were unusually resistant to heat. In such a case, samples of even 100 spores would be very unlikely to contain a resistant one, the probability being about 10^{-5}; whereas samples of 10^8 spores would almost certainly contain several resistant ones, and accordingly exhibit greater heat resistance, much more than corresponding to the simple increase in number of cells. However, while this is a plausible explanation in principle, it is doubtful how far it may be generally valid. The classical experiments of Esty and Meyer[14] with *Cl. botulinum* failed to reveal any abnormally resistant spores down to proportions well below one per 10^7 spores.

(iii) Recently, too, it has been pointed out that low concentrations of curing ingredients are capable of inhibiting the vegetative cells of spore-formers when their number is small, and it has been justly remarked that misleading impressions may well arise if experiments are made with excessive numbers of spores.[15]

(iv) Probably, similar relations hold for limiting factors like acidity, or minimum growth temperature, though no relevant data are known to the writer.

The impression nevertheless remains that these explanations may be inadequate from the quantitative viewpoint. The spore numbers in practical conditions are very low, and it is a moderate increase at these levels which may dramatically increase the rate of spoilage, i.e. the number of spores developing after a given process. Mere increase in initial number seems insufficient to explain the difference. If it is to be explained in terms of a proportion of resistant spores, that proportion must be a substantial fraction of the whole, say 10^{-1} instead of 10^{-7}. It seems improbable that this is generally the case, from the frequent observation of regular exponential destruction of spores by heat; though it is arguable that "laboratory" spores behave differently from "natural" spores; and indeed, a mixed natural contamination of spores, with diverse heat resistances, might well produce a population with a substantial proportion of unusual resistance. Further, in Spencer's experiments[15] the observed diminution in limiting NaCl concentration was small; and, in fact, the numbers of (vegetative!) cells he used (10^2–10^5) were still some thousandfold greater than levels of real interest. The practical difficulties of experimentation at spore concentrations of orders around 1 per g are of course great, and have so far prevented satisfactory investigation of these problems.

B. PROPERTIES OF THE FOOD

The nature of the food itself is important in several ways which are well understood, and in others less well understood.

1. Acidity

The importance of this has already been mentioned, in permitting a practically useful classification of foods into three broad groups (a) "high-acid", (b) "acid", and (c) "low-acid"; because, as will shortly be illustrated, different degrees of acidity prevent the development of different species which are important in food spoilage.

Acidity has a further effect, in considerably diminishing the heat resistance of bacterial spores. For example, with Cl. botulinum type A, the heat resistance at 115°C was at pH 4 only about one-third that at pH 7.[16] This effect is, however, of less importance than the effect of acidity in controlling the development from viable spores of particular species.

(a) High-acid foods. These are foods with pH < 3·7. No sporeforming bacteria grow in them, so they are not considered further here.

(b) Acid foods. These are defined as those in the pH range 3·7–4·5, the limiting pH being that below which it is certain that Cl. botulinum will not grow. Accordingly, foods in this class need not be given the process necessary to control this dangerous species. This being so, it is clearly important that the limiting pH of 4·5, below which it is possible to relax processing requirements, should be firmly established, and there have been several careful investigations, besides much experience, to confirm the validity of this limit.[17] It has, however, yet to be shown clearly whether the inhibition is of germination, outgrowth or vegetative multiplication.

The species important in acid foods are the following. The most heat resistant is the partial thermophile *B. thermoacidurans* (syn. *B. coagulans*). It is aciduric down to pH 4·2, and is a spoilage hazard in tomato juice especially, producing souring but no gas (the "flat-sour" type of spoilage); it can be controlled by acidification to pH 4·1. Strains of the mesophilic anaerobes *Cl. butyricum* and *Cl. pasteurianum* have somewhat lower heat resistance but grow over the range pH 4·0–4·5; they produce gas from carbohydrates in the food, and consequently distend cans, a phenomenon called "blowing". *B. macerans* and *B. polymyxa* are facultative anaerobes with similar pH tolerance which have been isolated from spoiled canned fruits.

As reference to Table IV below will indicate, those spore-bearing bacteria which can develop in acid foods have on the whole lower heat resistance than those growing in less acid foods. Consequently, it is both safe and practicable to give lower heat processes to "acid" than to "low-acid" foods; and the same would apparently apply to analogous processes involving radiation instead of heat.

(c) *Low-acid foods.* These support the growth of the widest range of species, among which *Cl. botulinum* is much the most important. However, though it is among the more heat-resistant species, the spores of some strains of *Cl. sporogenes* are still more resistant, and those of the thermophiles *B. stearothermophilus*, *Cl. thermosaccharolyticum* and "*Cl. nigrificans*" (syn. *Desulphovibrio desulphuricans*) are much more resistant still.

These thermophilic species differ, not only in being aerobic or anaerobic, but also in that *Cl. nigrificans* does not spoil foods more acid than about pH 5·3, whereas *Cl. thermosaccharolyticum* affects foods down to pH 4·5. This has been the basis for separation of a "medium-" or "semi-acid" class of foods. None of the three species grows at 30 °C; but if cans containing *B. stearothermophilus* are incubated for long periods near this temperature, a progressively smaller proportion spoil when subsequently placed at 55°, which does not happen with *Cl. thermosaccharolyticum*. This is a different phenomenon from "auto-sterilization", where spores develop into vegetative cells, which multiply and then die away, leaving no viable cells. The present phenomenon suggests that the spores of *B. stearothermophilus* may germinate and gradually die at temperatures below that at which vegetative development is possible; this has been reported with other species, as noted later.

Because spores of these thermophiles will not develop except at temperatures well above 30 °C, to which foods in temperate climates are seldom exposed for any length of time (except in hot-vending machines, or unless they are cooled too slowly after heat processing or are subjected to quite wrong conditions of storage), it is often unnecessary to give the severe heat treatments which would be needed to eliminate thermophilic spores; and, because these treatments are so severe (e.g. autoclaving for some time at 120°C) that they greatly damage the quality of most foods, they are avoided whenever possible. In this case, viable thermophilic spores remain in the product, but do not develop so long as the temperature limitations are respected, and we have an example of the state termed "commercial sterility", the pack behaving for practical purposes as if sterile.

When reduced heat treatments are given to low-acid foods, the most likely survivors next in order of heat resistance are the group of putrefactive anaerobes including *Cl. sporogenes* and *Cl. botulinum*; though, to give due credit to the control exercised by modern canning industry, such occurrences are now almost invariably due to the use of altogether inadequate heat treatment by some ill-informed operator, commonly the housewife making home preserves.

It happens, fortunately, that spores of *Cl. botulinum* are as a rule very uncommon; examination of 2,000 samples of different meats, for example, recovered some 20,000 spores of putrefactive anaerobes—the group which includes *Cl. botulinum* and *Cl. sporogenes*, of which only one proved to be *Cl. botulinum*. Experience has shown that there occur much more commonly strains of *Cl. sporogenes* which have similar properties to *Cl. botulinum*, except that they produce no toxin and are even more heat resistant: such strains are extremely useful as "indicators" in processing to control *Cl. botulinum*. In foods of the low-acid class which permit development of putrefactive anaerobes, it is in any case necessary to control heat-resistant strains of *Cl. sporogenes* like PA 3679, otherwise they cause putrefactive spoilage of the product. The heat process must therefore be adequate to destroy spores of strain 3679 and, because they are both more numerous and more heat resistant than those of *Cl. botulinum*, a process adequate for this purpose will also be more than adequate to control *Cl. botulinum*. A process which allows development of appreciable numbers of strain 3679 spores will be suspect as regards *Cl. botulinum*.

2. Water activity

There are, however, other ways than acidity to control the putrefactive anaerobes, including *Cl. botulinum*, ways which consequently permit the use of less severe processing treatments than would otherwise be regarded as safe. For example, it has long been known that cured meat can be protected from spoilage with a heat treatment which is only a small part of that which is necessary to protect fresh meat. Experiments with inoculated packs show that this is because curing salts act as if they made the spores of bacteria more sensitive to heat. Data like those of Table I indicate that the effect is due to the common salt (sodium chloride) and the nitrite which are the essential ingredients of curing salts. The separate role of nitrate is uncertain, save as a precursor for nitrite.

Experiments (Table I) in which the spores, salts and the growth medium are all heated together, do not reveal whether the effect of the curing salts is directly on the heat resistance of the spores, or is exerted after the spores germinate, or whether it arises by some interaction between the salts and the medium. This question is, however, of secondary practical importance, and its consideration is deferred to the Section on Heat Resistance (p. 548). What is of immediate concern is to distinguish the effects of sodium chloride and nitrite.

As regards sodium chloride used alone, numerous experiments have indicated that concentrations of 6–10% are usually needed to inhibit *Cl. botulinum*, and that this concentration must be calculated on the quantity of water in the system to give an index called the "brine-concentration". With many *Bacillus* species, salt has been found to allow germination of the spores, but to prevent the later transition from outgrowth to vegetative multiplication. Detailed investigations with salted systems have revealed inhibition at aqueous activities below 0·97–0·95. This corresponds with the inhibitory salt concentrations well enough to suggest that the effect of sodium chloride is exerted mainly via a change in aqueous activity; and that the primary effect of other salts like sodium nitrate should be similar, for which there is in fact direct evidence, though recent work indicates some minor specific effects.

These relations make it possible, for example, to pack sliced pre-cooked salt beef into cans, and preserve it by heating for only a few minutes at 100°C if it is salted so that the brine concentration exceeds about 12%; a mild heat treatment of only this order is essential, if the slices are to retain their coherence. Such a product is, however, much too salty for general use; and if lower salt concentrations are desired, nitrite must be present also.

3. Chemical inhibitors

Nitrite acts as a chemical preservative, and concentrations much less than 0·1% are effective. Its action is complex and imperfectly understood.

In part, it can probably be related to the concentration of undissociated HNO_2, since the effectiveness of nitrite in preventing development of bacterial spores is increased roughly tenfold for a fall of 1 unit of pH in the region pH 5–6 (as also happens with non-sporing bacteria e.g. *Staphylococcus aureus*). Such relations have been demonstrated with strains of both *Bacillus* and *Clostridium*.

However the concentrations of nitrite which act in the above manner in laboratory experiments are on the whole distinctly larger (exceeding 100 mg/kg), than those which are often effective in practice (below 10 mg/kg), and examination of this discrepancy has recently revealed that a more powerful antimicrobial substance is produced when nitrite is heated in certain culture media. The nature of this substance is still unknown, but it is clearly different from nitrite because its activity is not much affected by pH.[18] It prevents development of large inocula of vegetative cells of various *Clostridium* species, when obtained by heating nitrite at pH 6 in concentrations of the order of 10 mg/kg[19] and preliminary experiments suggest that it is effective against *Clostridium* spores at similar concentrations (T. A. Roberts, personal communication). It has, however, yet to be demonstrated in a food product, and its stability in storage is unknown.

Many other chemical preservatives (e.g. sulphur dioxide) have been tested for the same purpose,[46] but none have proved satisfactory (cf. Section V D (p. 567).

4. Synergistic effects

The factors described above interact in cured meat. The interaction between pH and nitrite has just been mentioned. There is a similar interaction between pH and salt or water activity: less salt (i.e. a higher water activity) suffices to inhibit if the pH is low; or, put differently, increase of salt concentration narrows the pH optimum for development. This has long been known in canning practice.

Recent work has explored the relation between these factors singly or combined, and the minimum temperature at which spores can develop and grow.[20, 21] The effect of lower pH, or increased salt concentration, or still more of both together, is to limit development at the lower end of the temperature range. The result is apparently to raise the minimum temperature for growth to such a degree that even not very heat-resistant species become unlikely to develop at normal temperatures.

In cured meats with not less than 50 g sodium chloride and 50 mg nitrite per kg water, and of normal pH near 6, it is in fact found to be safe to give heat treatments only about one-tenth of those for uncured meats and similar low-acid foods. When this is done, the important spoilage organisms are a group of *Bacillus* species characterized by their ability to reduce nitrate and by an unusually high tolerance for nitrite. Some of these strains also tolerate up to 20% sodium chloride and slow growth in 14% is quite common. These concentrations are much greater than are tolerable on grounds of palatability, and the spores are common in soil and therefore widely distributed. Fortunately they are not particularly heat resistant, and can be controlled with moderate heat treatments. If spores of putrefactive anerobes with higher heat resistance are present, they are likely to survive, providing another example of commercial sterility, for they do not multiply if the composition of the food as regards curing salts is appropriate.

C. ENVIRONMENTAL FACTORS

These are determined by the conditions of packing and storing the food, and much the most important factor is the temperature which prevails in the food.

1. Temperature

At a high temperature, only thermophilic species can spoil food; at a low temperature, only cold-tolerant species; at intermediate warm temperatures, spoilage is likely to be caused by mesophiles. Although the foregoing is obvious, precisely what temperature relations are implied by the above terminology is a matter for discussion.

The following may serve as a general guide, in relation to food spoilage. Thermophiles, in the strict sense, grow best at temperatures above 50°C, and not at all at temperatures below about 35°C. Mesophiles have their optimum near 35°C, with an upper limit at about 45°C and a lower at 15–10°C. The common cold-tolerant bacteria have a maximum near 35°C, their optimum below 30°C, and grow relatively vigorously down to 5°C, or below. Among sporing bacteria, *B. stearothermophilus* is a typical thermophile, *Cl. botulinum* types A and B and *Cl. sporogenes* are typical mesophiles, whereas *Cl. botulinum* type E is rather cold-tolerant.

There is a broad correlation between the growth temperatures of bacteria and the heat resistance of their spores, as already noted. Those growing at lower temperatures (e.g. *Cl. botulinum* type E) have heat-sensitive spores; those growing at high temperatures, the thermophiles, tend to have spores of high heat resistance, and the mesophiles fall in between. This rough correlation will become evident on scrutiny of Table IV. Because of this relation, the spores most likely to survive heat processing are those of more or less thermophilic species (p. 577).

In fact, in different foods, heat processing selects a variety of "facultative" thermophiles most of which do not fit satisfactorily into the restricted definition of thermophile given above, falling between that and the typical mesophile. These intermediate types have been divided into two, according to whether their optimum temperature inclines to that of the thermophile or that of the mesophile.[22] The former were called "thermophile-facultative mesophile", the important example being *B. thermoacidurans*; though the optimum temperature is near 55°C, growth occurs readily down to 30°C and slowly at 25°C. The latter were called "mesophile-facultative thermophile", and a variety of these exist, both aerobes and anaerobes. Among the aerobes are various strains in the *B. subtilis-licheniformis* group, which can grow up to 55°C though the optimum is usually 40–45°C; and some strains of *Cl. sporogenes* and *Cl. perfringens* behave similarly though to a lesser degree.

Experience indicates that the unusually heat-resistant spores of the above-mentioned species are much less common than those of ordinary mesophiles, though there are no systematic data to support this. The precautions taken in manufacture will usually reduce them to an order well below 1 per container. Consequently, when a food is only lightly heated, for example 30 min at 70°C, they are heavily outnumbered by mesophilic spores, among which those of *Bacillus* species strongly predominate. There is consequently a relation between the severity of the heating which a food has received and the temperature at which it may be safely stored. With less than about 30 min at 70°C, which will not kill any save the most sensitive spores, vegetative organisms which can grow in cool condition (e.g. faecal streptococci) may survive to spoil the food slowly even at chill temperatures near 0°C. 30 min at 100°C eliminates these, and suffices to destroy all the more sensitive spores; but many mesophilic spores will remain; so that spoilage is likely if the temperature strays into the mesophilic range, roughly above 10°C. If storage temperatures beyond the mesophilic range are to be encountered, then heating for several hours at 100°C, or several minutes at 120°C will be necessary to control thermophiles.

2. *Access of air*

Strictly speaking, it is the access of oxygen rather than air which is significant, and mainly in so far as it influences the redox status of the food in which the spores are to grow.

Fresh foods, consisting of still-respiring tissue, have a low redox potential internally even if exposed to air. The attempt to control bacterial spores, however, usually involves heating, to a degree more than sufficient to destroy the respiratory activity of the food itself; and then, especially if the food is cut into small pieces and exposed over long periods, the access of oxygen would have a greater influence. But foods intended for heat processing are usually sealed into hermetic containers while still raw, and then heated without breaking the seal, so that the conditions in the container usually become and remain more or less anaerobic. Where oxygen might remain in the container, it is common practice to take special steps to remove it (e.g. vacuum filling) because it usually has a deleterious effect on quality during storage. Accordingly, in heat-processed foods one is usually concerned with anaerobic species of *Clostridium*, or with facultatively anaerobic species of *Bacillus* of which there are several which are widespread.

The opposite may be true, however, of cured meats containing nitrate. When these are packed in oxygen-proof containers, they may nevertheless provide the equivalent of aerobic conditions for nominally aerobic bacteria which are capable of getting oxygen by reduction of nitrate. There are numerous strains of *Bacillus* capable of spoiling canned cured meats for this reason; and a more restricted number are capable of using nitrite similarly, producing gas which "blows" the container.

If, as seems likely, plastic containers come into use which are capable of withstanding the heat treatments needed to sterilize their contents, slow diffusion of oxygen through the plastic will have to be prevented, or the above familiar relations will no longer hold.

D. SELECTIVE INTERACTION OF DIFFERENT FACTORS

We can now take a broad view of the combined effects of all the above-mentioned factors, to produce a synoptic table (Table II) showing the important organisms in the different sets of circumstances.

TABLE II

Sporeforming bacteria important in spoilage of heated foods

| Approx. temperature range for vigorous growth °C | Acidity status of food | |
	"Acid" pH 4·0–4·5	"Low-acid" pH > 4·6
Thermophilic (55–35°)	*B. thermoacidurans*	*Cl. thermosaccharolyticum* "*Cl. nigrificans*" *B. stearothermophilus*
Mesophilic (40–10°)	*Cl. butyricum* *Cl. pasteurianum* *B. macerans* *B. polymyxa*	*Cl. botulinum A and B* *Cl. sporogenes* *B. licheniformis*[a] *B. subtilis*[a]
Cold-tolerant (35– < 5°)		*Cl. botulinum E*

[a] Particularly in cured meats.

In Table II the species at the top are those which grow only in warm conditions and have spores of high heat resistance. They are, accordingly, those likely to be encountered in "commercially sterile" packs stored under temperate conditions. Conversely, the species at the bottom are more likely to be important in foods stored under cool conditions: the *B. licheniformis-subtilis* group in pasteurized canned ham held in temperate climates; or *Cl. botulinum* type E in chilled vacuum-packed fish, if such a product were marketed.

The species on the left are, obviously, relatively tolerant of acidity and hence are those which are troublesome in fruit products and the more acid vegetables; while those on the right are characteristic of neutral foods like fish and meat. The *B. licheniformis-subtilis* group are the nitrate-reducing, salt- and nitrite-tolerant strains of *Bacillus* which typically spoil cured meats.

Since Table II refers to heated foods, all the species represented are those which are most heat resistant although in the relevant circumstances the selection capable of growing is wider, and this is true even though there is a wide range of heat resistance among the species in Table II. Type E *Cl. botulinum*, for example, is comparatively sensitive to heat; but since its spores are nevertheless more heat resistant than the cells of non-sporing cold-tolerant species, and since it is one of the few sporeformers capable of growing vigorously at chill temperatures, it claims attention in lightly heated but cold-stored foods because it might make them toxic. Similarly, spores of the *B. licheniformis-subtilis* group are not specially heat resistant, but these are the only spore-formers whose resistance to curing salts permits them to grow in adequately cured meat.

It will be evident that, given a particular set of circumstances, the characteristic spoilage can occur only if spores of the causative species are present; and in practice the heat treatments which would otherwise be necessary are usually minimized by taking steps to limit the numbers which gain access to the food. For this reason, for example, there are bacteriological standards for canners' sugar which limit the number of thermophilic spores it may contain, and spices (which frequently contain 10^9 spores per g) are often subjected to partial sterilization before use.

These examples should suffice to illustrate the principle stated at the outset, that the nature of the species causing spoilage is determined by marriage of the properties of the spore-former with the properties and treatment of the food. The subject cannot, however, well be considered further without a better understanding of the nature and measurement of heat resistance, to which the next section of this chapter is accordingly devoted.

III. Resistance to Heat or Radiation

The resistance of spores is discussed at length in Chapter 16. Here, however, it is necessary to give some consideration to this topic, especially as regards heat resistance; enough to show how this property of spores is of practical importance in the control of food spoilage.

Heat or irradiation kills bacterial spores in a systematic manner: the longer the period of exposure, the more are killed; and the higher the temperature or the greater the intensity of radiation, the sooner this occurs. Resistance is the name given to the ability of spores to "survive" such treatment, which in practice usually includes subsequent development, because tests for survival almost always involve cultivation of the surviving cells.

Resistance effectively depends, therefore, on three things:

(i) The inherent resistance of the spores, which depends both on the species and on environmental influences which act during spore formation (Chapter 3).

(ii) Environmental influences active during the time of exposure of the spores (Chapter 16), among which may be noted the acidity and chemical composition of the supporting medium.

(iii) Environmental influences active during germination (Chapter 11) and outgrowth (Chapter 12), which will decide whether a potentially viable spore will begin to develop, or whether this development may be arrested at some stage short of vegetative multiplication; and this aspect of the matter is similar whether the question is that of ability of the organism to "be recovered" on an experimental culture medium or that of its ability to grow in, and spoil, a foodstuff.

Finally, in most practical situations heat treatment is not likely to be uniform, and it is necessary to have a way of comparing treatments made under different conditions.

It therefore seems necessary to explain as briefly as possible:

(A) How resistance is measured; by which is meant the way in which it is determined and expressed, not the details of bacteriological technique involved (for which see ref.[23]).

(B) The effect of practical conditions on resistance, and

(C) The way in which the spore-killing effect of a practical treatment is evaluated.

A. MEASUREMENT OF RESISTANCE

A brief statement on methods will be followed by discussion of the indices used as measures of resistance.

1. Methods

There are two ways of measuring resistance to heat or radiation, rather different in principle though similar in practice.

(a) "End-point" method. In the first, one determines the time of treatment needed to destroy all of a known number of spores in a container. In practice, this means preparing a number of replicate containers each containing the known number of spores, treating them individually for successively longer periods of time, then culturing the whole contents of each container to see if any spore remains viable, and noting the shortest time of treatment which leaves not one survivor. Table III presents results from classical experiments of Esty and Meyer[14] using this procedure.

TABLE III

Minutes of heating required to kill all spores of Cl. botulinum, *in sealed tubes containing 2 cm³ lots of a suspension in veal-infusion peptic digest gelatin, containing 19,200 million spores per cm³* [a]

Heating temp. (°C)	100	105	110	115	120
Survivors	315	90	28	9	4
No survivors	330	100	30	10	5

[a] Data refer to strain 97 in Table 3 of Esty and Meyer.[14]

(*b*) "*Multiple-point*" *method.* In the alternative procedure, a spore suspension is treated continuously, aliquots are withdrawn at suitable time intervals, the numbers of surviving spores in these aliquots are determined by viable counting and, from these, the numbers surviving in the suspension after different times are calculated. In practice, the aliquots are usually ready-prepared in individual containers, the containers are treated together, and a few are withdrawn at each interval for viable counting.

In either procedure, it is important to make arrangements and use a container such that its whole contents reach the desired temperature almost instantly, otherwise complicated corrections of the heating time become necessary. Also, it is essential that the containers be sealed so that they can be completely immersed in the heating medium; when this is not done, it is common experience to find occasional survivors after unduly long periods of heating (this happens with vegetative cells, as well as spores).

In order to make definite comparisons between resistances, for example between species, it is necessary to find some regularity of behaviour which can be expressed in numerical terms. Such regularity of behaviour actually exists, and can be expressed in terms of parameters called respectively D for heat or radiation, and Z and F for heat.

2. D value

The results of three experiments of Esty and Meyer,[14] using the end-point method, are collected in Fig. 1. The slope of the straight line in Fig. 1 gives the diminution in the time of heating which is needed when the initial number of spores is diminished by a factor of 10; and it appears

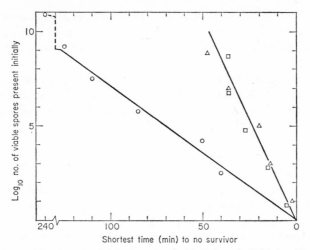

FIG. 1. Relation between the initial number of viable spores (*Cl. botulinum*) and the time of heating needed to kill them all. (From data of Esty and Meyer,[14] their Table 4: circles—strains 97 at 100°C; triangles, squares—strain 90 at 105°C two experiments.)

that this time is independent of the absolute number of spores involved, over a very wide range. (There was an exception with strain 97 at the enormous spore concentration of 7×10^{10} per cm³.) Using the second, multiple-point, method a similar result is obtained, as Fig. 2 shows. A tenfold reduction (called "decimal reduction" or, more ambiguously, "log

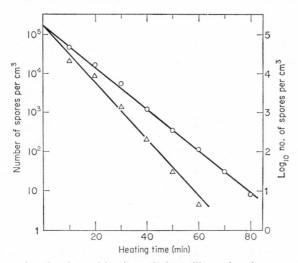

FIG. 2. Observations by the multi-point technique, illustrating the exponential form of the relation between number of surviving spores and time of heating, in the presence as well as the absence of salt. (Data of Viljoen,[24] Thermophile spores heated at 110°C suspended in pea liquor: triangles—no NaCl; circles—with 1·5% NaCl.)

cycle") in the number of surviving viable spores is always caused by the same number of minutes heating, whatever the number of spores. Evidently the two methods can be regarded as equivalent, if it be assumed that spores already killed have no effect on the survivors.

This means that if time of heating at a particular temperature is plotted against $\log_{10}n$ (n being the number of spores surviving at each particular time), a straight line results (as in Figs 1 and 2), representing a relation described as "exponential".

$$(\log_{10}n_0 - \log_{10}n) \times D = t_n$$

where n_0 is the initial number of spores per g, and n that after t_n min heating. The characteristic number of minutes, to change the viable spore number by a factor of 10^{-1} (i.e. $\log_{10}$ n by one unit), is a convenient index of the resistance to heating, and is called the D value: it follows that killing to a surviving fraction of 10^{-n} follows heating for n $\times$ D minutes at the temperature to which D refers. It is also obvious that, to obtain a given end result, more heating is required the larger the number of spores initially present.

There may be two kinds of complication, which usually do not affect the main part of the relation. The first is when, with many viable spores at the start of heating, the D value is high and falls rapidly, producing a line with the upper portion convex upwards, called a "shoulder" (see Chapter 16, Fig. 4). It may be caused by "activation" (see Chapter 10). Usually it can be neglected when very small proportions of survivors are in question, which is common in process calculations.

The second is the converse phenomenon, of a rising value of D as the numbers of surviving spores become small, causing the line to become concave upwards at its foot, producing what is called a "tail" (see Chapter 16, Fig. 3). Such a phenomenon could conceivably be caused by the production, during treatment, of protective substances or resistant mutants. It is, however, usual to regard it as indicating the presence at the outset of a small proportion of specially resistant spores. If so, the residual number of resistant survivors should vary in proportion to the initial number of spores, and should still diminish gradually with the treatment applied, as was convincingly demonstrated in the experiments of Vas and Proszt.[13] But this sort of demonstration is rarely given, and a common observation is, on the contrary, that small practically constant numbers remain, more or less independent of the initial number or increasing treatment; as was the case, for example, in the experiments of Wheaton and Pratt.[25] Such a phenomenon could conceivably be explained by a low but fairly regular contamination of samples in the bacteriological technique, and as it is not readily reproducible it is best regarded with caution. If tailing is a real phenomenon, it is extremely important, as will appear shortly.

Similar relations hold for treatment with ionizing radiation. Exposed to a source of fixed intensity (e.g. a radioactive isotope), a tenfold reduction in number of viable spores, regardless of absolute numbers, again follows during a particular time interval corresponding to the application of a particular quantity ("dose") of radiation, and it is this radiation dose which is called the D-value.

3. Z value

The D value, in heating, refers only to the particular temperature at which it is determined; and general experience, showing that shorter times of heating are effective at higher temperatures, indicates that D must decrease as temperature rises. To be able to compare the effects of different temperatures of heating it is necessary to measure this relationship.

Investigation again reveals a remarkable regularity of behaviour. D value is found to vary with temperature according to the relation

$$\frac{T_1 - T_2}{Z} = \log D_2 - \log D_1$$

(where T_1, T_2 are temperatures D_1, D_2 are corresponding D values and Z is a characteristic constant). This means that the Z value is that change of temperature which changes the D value by a factor of 10. The numerical value of Z will of course depend on the scale in which the temperatures T_1 and T_2 have been measured; in the original definition of Z, by Ball, it was measured in degrees Fahrenheit. Fig. 2 Chapter 16 illustrates this relationship.

It should be understood that it is D and not Z which is an absolute measure of heat resistance. Z only measures a temperature coefficient, which is the rate at which D varies with temperature; it is possible for species highly resistant at some particular temperature to have either high or low values of Z, with the implication that their placing in order of heat resistance is different in different temperature ranges.

The Z relation holds only over a limited range of temperature, such as is relevant to practical heat processing. Over a wider range, Z is not constant but tends to have lower values at lower temperatures, as would be expected from the relation between the temperature coefficient and the energy of activation in a chemical reaction.

4. F value

Whereas the Z value permits comparison of time/temperature treatments at different temperatures, it is desirable to choose some standard temperature as a basis to which treatments at other temperatures can be referred. Ball, who was first to appreciate this, chose 250°F as the temperature at which the F value was defined as the number of minutes heating which would produce the same spore-killing effect as the process requiring to be specified. The concepts of D, Z and F are illustrated in diagrams like Fig. 3.

Fig. 3 presents the classical data of Esty and Meyer,[14] which formed the original basis for present methods of calculating heat processes to control *Cl. botulinum*. The circular symbols represent data from their actual experiment which revealed the highest heat resistance recorded in detail. The triangles represent an idealized curve summarizing the results of many of their experiments. The continuous straight line represents what Olson and Stevens[26] considered to be the best representation of the latter data, corresponding to the widely quoted values $Z = 18°F$ and $F = 2.78$ min: Z represents the slope of the line in degrees per tenfold diminution in time

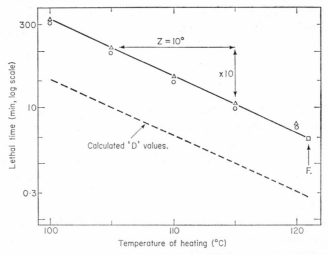

FIG. 3. Time/temperature relations in killing *Cl. botulinum* spores (suspended in phosphate buffer) by heating. (Data from Esty and Meyer[14]: circles—strain 97 in their Table 3; triangles—"ideal" data in their Chart 2.)

to kill, here about $10°C$ ($18°F$); F registers the point at which the line crosses the $250°F$ ($121°C$) ordinate, denoted by the square symbol. A line like this is called a "thermal death time curve" (TDT curve).

The corresponding D values depend on the numbers of spores involved. Consider first the experiment represented by the circular symbols, which appears to refer to tubes containing 38,400 million spores (see Table III). Since they were heated to the point of not one survivor, this represents roughly $10^{10.6}$ to 10^0, i.e. 10.6 decimal reductions. Since the corresponding F value was about 2.3, this leads to a D value of 0.22 min at $250°F$. In the case of the ideal data, Townsend, *et al.*[27] say "while the number of spores used by Esty and Meyer in their *TDT* runs was 6×10^{10} their thermal death times were based on complete destruction. It is now *customary to assume* [writers' italics!] their classical TDT curve resulted in reducing 1×10^{12} spores to one spore or that $F = 12D$. On this basis, if $F = 2.78$, $D_{250} = 0.23$ min." The writer, however, sees no good ground for this assumption. The total number of spores involved is stated explicitly by Esty and Meyer[14] as "60 billion" in the legend to their Chart 2; it is indeed almost the only one of their sets of experiments in which they leave no possible doubt about the number of spores involved. Moreover, they state "the ideal destruction curve shown in Chart 2 . . . includes the maximum resistance observed on 1,804 spore suspensions and shows the negative results obtained at the temperatures stated". It seems certain that the maximum resistances recorded must have depended on the most resistant spore suspensions, which (to judge from the rest of their data) almost certainly formed a minority of the 1,804 suspensions surveyed. This means that the effective number of spores contributing to the recorded resistances, far from being 10^{12}, must have been *less* than 6×10^{10}, possibly much less; with the implication that the maximum D values should be higher than those calculated.

Since the experiments used the same number of spores at each temperature, corresponding calculations at different temperatures lead to a corresponding set of D values, represented by the dotted line in Fig. 3. Such a line, which illustrates the relation between D value and temperature independent of spore number, has been called the *thermal resistance curve* by Schmidt;[28] and the value of D at the reference temperature of 250°F is a standard measure of heat resistance.

The estimates of Esty and Meyer for the heat resistance of *Cl. botulinum* were thoroughly confirmed in much later work by Townsend, *et al.*[29] who further pointed out that the previous F value of 2.78 needed revision to 2.45 to allow for heat penetration, and found $Z = 17.8°F$. Their corresponding values for *Cl. sporogenes* strain 3679 were $F = 4$ and $Z = 16.5$ in phosphate buffer and, since they used 1,200,000 spores per container, this represents 6.08 decimal reductions and corresponds to a D_{250} about 0.65 min, a value considerably higher than that for *Cl. botulinum* under the same circumstances. However, Schmidt[28] gives values $F = 9.5$ and $Z = 20$ for lots of 10,000 spores of strain 3679 suspended in beef and gravy, leading to a value of D_{250} about 2.4 min, a heat resistance some fourfold greater than in buffer. It is usual to observe this kind of "protective" effect with spores suspended in organic media.

These examples illustrate several notable points. Different species have different values of D at the same temperature (i.e. different "heat resistance" in the terminology previously used). Different species have different Z values, though the range is much less than in D values. High values of D are not necessarily associated with high values of Z. The values of D (and therefore F), and of Z, vary with the suspending medium, the former especially; and heat resistance in foods is much higher than in mineral solutions.

5. Resistance of different species

With this background, it is possible to give more precise meaning to the statements hitherto made, for example, about the "high" heat resistance of spores of thermophiles and the "low" resistance of those of cold-tolerant strains. Table IV lists the species of interest in the previous discussion, together with the relevant D values at particular temperatures and, where available, Z values which permit comparison between measurements at different temperatures.

As a matter of convenience, D values for radiation are included; they refer to measurements at room temperature, being little affected by temperature change in the range 5–40°C.

The striking feature of Table IV is the broad relation between temperature relations for growth—thermophile/mesophile/cold-tolerant—and heat resistance of the spores. Though there are individual variations, D values of several minutes at 120°C or hours at 100°C, characterize the thermophiles; whereas values in minutes at 100°C are typical of mesophiles; while the cold-tolerant species has similar values at only 80°. It is equally evident from Table IV that, apart from this, heat resistance of spores is a character not systematically distributed; it occurs equally in the genera Bacillus and Clostridium, and among species of diverse physiological character. The large difference in Z value between B. stearothermophilus and Cl. thermosaccharolyticum, though they have similar heat resistance at 120°C, illustrates a point just mentioned above.

Comparison of the corresponding D values for radiation with those for heat shows that there is no correlation between resistance to heat and to radiation, the value for radiation being much the same with the thermophile B. stearothermophilus as with the cold-tolerant Cl. botulinum type E. Noteworthy are the high radiation D values for the putrefractive anaerobes, which are preeminent in resistance to radiation, but not in resistance to heat; for this reason, they are even more important in considerations of radiation processing than in heat processing.

6. General comment

The foregoing discussion clearly rests on the presumption that the relation between number of viable spores and time of treatment is exponential, as indicated in Figs 1 and 2. Some minor irregularities have been mentioned already. There are also reports of entirely non-exponential relations with various species, though there is seldom sufficient statistical information to decide whether they depart significantly from the exponential model; the question is, indeed, a difficult one. However, resist-

TABLE IV

Approximate resistance[a] of some bacterial spores to heat or to γ-radiation

Bacterial species	Food class[b]	Heat resistance D Value (min) at:			Z value °C	Radn. res. D_γ (Mrad)
		120	100	80°C		
Thermophilic anaerobes						
Clostridium thermosac- *charolyticum*	LA	3–4			12–18	
Cl. nigrificans	LA	2–3				
Thermophilic aerobes						
Bacillus stearothermo- *philus*	LA	4–5	3000		7	0·22
B. thermoacidurans	A	0·1				
Mesophilic anaerobes						
Cl. sporogenes (incl. PA 3679)	LA	0·1–1·5			9–13	0·15–0·22
Cl. botulinum A and B	LA	0·1–0·2	50		10	0·1–0·35
Cl. perfringens	LA		0·3–20		10–30	0·1–0·35
Cl. caloritolerans	LA		3			0·15
Cl. histolyticum	LA		1	115	10	0·18–0·22
Cl. butyricum	A⎫		0·1–0·5			
Cl. pasteurianum	A⎭					
Cl. sordellii	LA			40		0·15
Cl. subterminale	LA			30		0·16
Mesophilic aerobes						
B. cereus	LA		5		10	0·16
B. licheniformis	LA		13		6	0·22
B. megaterium	LA		1		9	0·12
B. subtilis	LA		11		7	0·22
B. macerans	A⎫		0·1–0·5			
B. polymyxa	A⎭					
Cold-tolerant anaerobes						
Cl. botulinum E	LA			0·3–3	9–14	0·1–0·2
Cold-tolerant aerobes **absent from foods**						
Vegetative bacteria	LA⎫ A⎭			<1		most <0·1

[a] These values are approximations from a diverse literature and, not being all obtained under the same circumstances, are not strictly equivalent, though useful for broad comparative purposes.

[b] Low-acid food (pH > 4·5)—LA; acid food (pH 4·0–4·5)—A.

ance to heat, and radiation, and other influences, develop gradually after the spore is formed and not simultaneously. If a culture is harvested before all its spores have had time to mature in this sense, the resulting spore crop is likely to exhibit anomalous behaviour; and since sporulation is in general non-synchronous and re-cycling may occur (see Chapter 3), a standard pattern of resistance may well not be attained. How far such phenomena are relevant to "natural" spores is still a matter for conjecture.

The existence of the exponential relation poses some problems about the distribution of resistance among the spore population to which it applies as noted by Ball and Olson[30]. The fact that more heat is needed if more spores are present implies either (i) that the maximum resistance of individual spores should increase regularly with numbers of spores, or (ii) that the increasing resistance of spores in higher concentrations is due to a mutually protective effect. Hypothesis (i) is inherently improbable, and the data of Vas and Proszt[13] provide evidence to the contrary. Hypothesis (ii), which refers to viable spores, may be doubted because (as noted earlier) the same relation indicates that spores already killed have no effect on those remaining viable; moreover, it is not easy to imagine interactions between inert spores at distances of 10^2–10^4 spore radii. The suggestion that the relation reflects the monomolecular destruction of one vital enzyme (or gene) seems to require the conditions (i) that all spores contain the same quantity of this, and (ii) that it is the only factor deciding rate of death. Similar arguments apply in relation to radiation.

These questions are still undecided. But, at least, the exponential assumption seems well supported in the important case of *Cl. botulinum*.

B. Effect of Environmental Conditions on Resistance

Because this subject is discussed in detail in Chapter 16, it suffices here to bring out only points of general and practical interest.

1. Environmental influences during spore formation

Physical conditions, like temperature of the medium in which the spores are developed, and its chemical composition as regards for example minerals and particular organic compounds, are known to affect the heat resistance of the spores formed, though in ways not properly understood. It is thus to be expected that the heat resistance even of a particular type of spore will be different according to the food in which it occurs; for instance, it has even been suggested that spores produced in fresh meat may have different resistance from those produced in cooked meat.

This leads to the proposition that spores produced in the laboratory cannot be the same as "natural" spores; with the corollary that experiments with the former are an uncertain guide to resistance under practical conditions. A thorough understanding of the effect on the resistance of spores of environmental influences during their formation is evidently necessary to bridge this gap. In the meantime, realistic experiments with spores in the relevant food are essential in developing any practical process (see "Inoculated Packs", p. 586).

2. Environmental influences during actual treatment

Similarly, during the actual heating, various environmental influences besides temperature affect the heat resistance of the spores. The effect of acidity of the suspending medium or food has already been mentioned. The water activity of the system also has a marked effect, heat resistance being much greater if this is low. It is also greater in the presence of organic matter, so that higher values are recorded with spores suspended in foodstuffs than in water or buffer. The nature of the foodstuff is important. For example, resistance is greatly raised in the presence of high concentrations of soluble carbohydrate, partly because this raises the content of organic matter and partly due to lower water activity. Again, the presence of fat may raise greatly the apparent heat resistance, a phenomenon called "fat protection".

There are broadly similar responses in resistance to radiation which is lower in acid foods, greater in dry ones, and higher in foods than in water or buffer. Freezing increases the radiation resistance of spores much less than it increases the resistance of foods to the deleterious effects of radiation;[31] hence irradiation below $-20\,°C$ is a way to minimize the degree of quality-damage associated with a given degree of spore inactivation.

The general consequence is, again, that laboratory experiments with spores suspended in artificial media give a misleading impression of resistance under practical conditions. There is good reason to experiment with spores suspended in the food in question.

3. Environmental influences during germination and outgrowth

In the period between heating and the establishment of normal vegetative growth, a spore is in an unusually sensitive state; and influences which do not affect unheated spores may decide whether such "damaged" spores will survive and grow, or on the other hand die. Temperature is important. Starch (in solution) is a substance known to promote recovery of heat-damaged spores; sodium chloride, or antibiotics like nisin, are examples of substances which promote their death. If such substances are present in a food in which spores are heated, they appear to affect the heat resistance of the spores. A substance which promotes recovery, like starch, increases the apparent heat resistance; one like salt (in sufficient concentration) apparently diminishes it.

Of the substances which act in this way, sodium chloride is much the most important in practice, and almost the only one extensively investigated. There are plenty of data similar to those in Table I; but experiments like these in which the spores, salts, and the growth medium are all heated together, do not reveal whether the effect of the salts is directly on the heat resistance of the spores, or is exerted after the spores germinate, or whether it arises by some interaction between the salts and the medium. This is not clear even from data like those in Fig. 2. It appeared, from the early work of Esty on spores of *Cl. botulinum*, that the direct effect of salts on heat resistance must be small. This has recently been confirmed by Roberts and Ingram.[32] They heated spores in the presence or absence of salt, and transferred the heated spores to media with and without salts. The extra inhibition only occurred when heated spores were transferred to a medium containing the curing salts (Fig. 4). Experiments

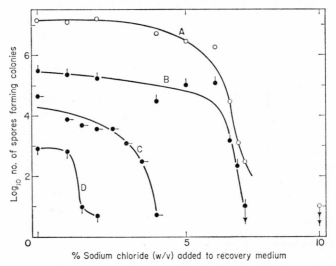

FIG. 4. The effect of sodium chloride in inhibiting heat-damaged spores. (From data of Roberts and Ingram,[32] their Table 1. Replicate samples of spore suspension of *Cl. sporogenes* strain 3679, heated in distilled water:

A at	80°C	for 20 min
B	100°	20 min
C	105°	5 min
D	105°	6 min

and plated on medium with indicated concentrations of NaCl.)

of this kind with spores of different species of *Bacillus* and *Clostridium* have clearly shown that there is little direct effect of curing salts on heat resistance, their influence being on recovery of the heated spores.

Laboratory investigations by Gould,[33] so far only with *Bacillus* spores, have helped to clarify the situation, by showing that curing salts, at the practically relevant concentrations, act at different stages of outgrowth between germination and vegetative multiplication. With 4–7% sodium chloride, germination occurred normally, the spore wall might be shed and the developing cell elongate slowly, but multiplication was prevented. With nitrite concentrations not exceeding 300 mg $NaNO_2$ per litre, development was arrested immediately after germination, before rupture of the spore wall. Analogous studies on clostridia are needed.

From curves like those of Fig. 4, it is possible to infer a far from "normal distribution" of salt resistance among the spore population. A small proportion of the spores resist somewhat greater salt concentrations than the majority; and this proportion, and the maximum tolerable salt concentration, diminish markedly with increasing heat treatment of the suspension. Is there any consistent mathematical relation between the degree of heating and the proportion of spores which resist a particular salt concentration? Nobody has analysed this question; and, when one adds similar questions about nitrite and nitrate, it would seem that prediction of processes for cured meats might be impossible.

4. Evaluating the effect of environmental influences

As has been said, however, "survival" in the practical sense only implies the ability of the spore to develop in and alter the medium, or spoil the food. Simply from this viewpoint, therefore, the stage at which the above influences act—whether germination, outgrowth, or multiplication—is not important. This being so, it would appear that if an influence does not destroy the exponential form of relation between treatment time and number of surviving spores, so that a characteristic D value can still be derived, its effect can be directly evaluated through the change in D value. If this is halved, the indicated process requirement is halved, etc.

This assumption is often implicitly accepted as valid. "Differences in heat resistance", as for example between spores of different species, or of one species grown under diverse conditions, are expressed through exponential concepts and derived D values, and these D values are then manipulated freely to evaluate corresponding processes; as has indeed been done above. Probably, few would question such a procedure, in relation to any influence acting during spore-formation. The same probably applies to influences acting during treatment: for example, when it is demonstrated that lower D values are observed in acid solutions, it is generally accepted that correspondingly lower process requirements will apply.

When one turns to curing salts acting after treatment, the problem looks less simple. With sodium chloride alone, there are occasional records of exponential death/time relations for spores heated in foods containing salt (e.g. Fig. 2), and data like those of Anderson et al.[34] seem to suggest that sodium chloride would not seriously interfere with process calculations by orthodox methods. But there is no substanial body of observations to support this view; and the writer knows of no corresponding evidence for nitrate or nitrite. Moreover, the observations with sodium chloride refer to temperatures higher than those commonly used in processing cured meats, where it seems likely that one would be operating on the "shoulder" of the death/time curve and that the Z-relation might fail. At present there is need for a way of evaluating processes for cured meats, to decide which of the many possible alternative combinations of treatment might be equivalent.

C. CALCULATION OF HEAT PROCESSES

This is a subject in itself, to which entire textbooks have been largely devoted,[30, 23] and reference should be made to them for details. Nevertheless, certain points of general microbiological interest require mention here.

From the simple relation, that killing to a surviving fraction of 10^{-n} follows heating for $n \times D$ minutes (at the particular temperature to which D refers) it is in principle simple to decide how much heating will be required to produce a given degree of survival. In practice there are two serious difficulties. The first is to decide what degree of survival is acceptable. The second arises from the fact that, in any practical process of heat sterilization, the temperature does not remain constant, but changes both as the heat penetrates and as the pack cools after heating ends, and is different in different parts of the pack. These problems will be considered separately.

1. The tolerable proportion of survivors

It is convenient to discuss this aspect of the subject with special reference to *Cl. botulinum*; for the reason that, because of the exceptional hazard which this species represents, the subject has been most carefully considered in that connection. Indeed, until recently the subject had scarcely received public discussion in relation to any other species; though food processers must use analogous arguments in private, when trying to rationalize the processes they use, for example, to control thermophilic spoilage.

In referring to pathogens in foods, it is usual to specify that they 'shall be absent" (cf. Table VI, p. 591). But absolute absence is a wholly ideal concept, which in practice can be neither attained nor verified. What can be done is to reduce the probability of presence to some small, even negligible, but in principle definite, level. What should this be?

When the problem of controlling *Cl. botulinum* was investigated by Esty and Meyer,[14] they experimented with the largest number of *Cl. botu-*

linum spores that they could manipulate, a larger number than anybody has used since in similar experiments; and, as indicated earlier, it has been assumed that they used the equivalent of 10^{12} *Cl. botulinum* spores. By other approximations, the indicated thermal death time of 2·78 min at 250°F was rounded off to 3 min. The canning industry accepted these as the best available indications of the heat process required for safety: the indicated process $F = 3$ is familiarly termed "the bot. cook", and it is supposed that this corresponds to 12 decimal reductions. Quoting Perkins,[35] "These thermal resistance values are still used as a basis for minimum heat processes for low-acid canned foods. That they have stood the test of time is attested to by the fact that not one outbreak of botulism attributable to underprocessing has occurred in commercially canned foods since the National Canners Association (NCA) began issuing bulletins recommending heat processes based on Esty and Meyer's classical *Cl. botulinum* resistance values."

This requirement implies in turn the application of a treatment $12 \times D$, the value of D being chosen appropriately to the processing conditions envisaged; and these considerations are accordingly referred to by the term "12-D concept". The same arguments apply if the food is to be treated with radiation instead of heat: 12 times the D value of about 0·37 Mrad for the most resistant strains of *Cl. botulinum* leads to an indicated processing requirement of about 4·5 Mrad.

Hindsight has justified all this by reflections such as the following. The largest canning factories process some 10^9 cans of food per year, each containing some 10^2–10^3 grammes of food; a substantial proportion of this output is capable of supporting *Cl. botulinum*, and total protection is required over periods certainly of the order of 10 years. It is necessary therefore to protect something of the order 10^{12} g of food. Assuming an infection of the order of 1 (i.e. 10^0) *Cl. botulinum* spore per g, this means the destruction of 10^{12} spores.

It is only in recent years that extensive surveys have shown that spores of *Cl. botulinum* are on the whole much less common than 1 per g. The evidence from the spore surveys on meat (mentioned in the previous section) indicates an average incidence of the order only 1 in 10^4 g, which suggests that the general requirement for an inactivation factor of 10^{-12} will usually contain a large margin of safety. Correspondingly, there have recently appeared suggestions that this factor is perhaps excessive.

With particular reference to the properties of spores, the above arguments conceal two important points which are different aspects of the same question.

The first is the assumption that the D value is constant, even at levels of spore population which are too low to be accessible to experimental investigation. Experimenters often begin with spore populations, for example, of the order $10^7/g$ and use counting methods which are reliable only down to $10/g$ (e.g. Fig. 2); in which case they actually observe only 6 decimal reductions, and are obliged to extrapolate over the lower 6 to calculate the treatment needed for 12 decimal reductions. In these circumstances, serious error will arise if D value is different in the lower part of the range.

The second point is implicit in comparing experiments like those of Esty and Meyer with the situation in cans of food. Can it be assumed that the survival (including outgrowth!) of spores in a dense suspension is quantitatively the same as that of isolated spores at high dilution? We know, indeed, that such population differences are very important in deciding development of inocula of vegetative cells.

The appearance of "tails" on survivor curves calls both these assumptions into question; especially when the tails have been demonstrated down to population levels of the order 1 per 100 g of substrate. If such tails are not caused by technical error, they mean that the D value is greater at low spore populations, so that process estimates based on downward extrapolation are likely to be too low—a dangerous miscalculation. Moreover, if the D value becomes greater when as a result of the treatment the numbers of viable spores has been reduced to perhaps only one per container, one may well ask if it would not be greater still if there were only a similar number as a result of natural contamination, again implying too low an estimate of D, and possible danger. Consequently, until such matters are better understood, it is consoling to think that calculations of process requirement on the above-indicated basis may contain a substantial margin of safety.

The situation is rather different with thermophilic spoilage, though similar principles can be applied. Assuming that one spoiled can per thousand can be tolerated (compare "Process Control", p. 589), and that each can contains 100 g of food, the tolerable level is 1 spore per 10^5 g. With a D_{250} of 2–3 min (cf. Table IV), the greatest heat treatment that almost any food will satisfactorily withstand, $F = 10$, will still leave a surviving fraction of 10^{-4}, which means 1 spore per 10^5 g only if the initial infection is 1 per 10 g. Consequently, the control of thermophilic spoilage, with acceptable thermal processes, depends mainly on limiting the access of spores to the product, if the product is to be kept at a vulnerable temperature.

2. Computation of the effect of fluctuating temperature

To calculate the overall effect of a practical heat process, in which the temperature rises gradually during heating, is then held for a longer or shorter time, and falls gradually during cooling, it is necessary to integrate the lethal effect over the whole of the temperature/time programme. The rates of change of temperature themselves depend on the penetration of heat into the containers; this topic will be mentioned later in the discussion.

The Z-relation relating D value to temperature makes the integration of lethal effect at a point possible when the temperature/time relation there is known; but rigorous mathematical treatment over the whole contents of a container is complex.[23, 30] The usual simple practice is to assume that the centre of the container will be heated least and is therefore the critical point (an assumption questioned in Ref. 23), to observe at that point the actual variation of temperature with time, to plot this relation on "lethality" graph paper, and to obtain total integrated lethality by measuring the area under the curve so obtained.

The principle of the "lethality paper" can be most simply illustrated by thinking of spores supposed to have a D value of 1 min at 121 °C and a Z value of 10°, in which case the D value at 111° would be 10 min, and at 101° would be 100 min. Accordingly, if we regard the lethal effect of 1 min at 121° as unity, 1 min at 111° will give 0·1 of this unit and 1 min at 101° gives 0·01. The graph paper, with time as abscissa, is prepared with an ordinate marked in temperature but scaled according to the corresponding lethal proportions thus 121°: 1, 111°: 0·1, 101°: 0·01 etc. The respective lethal proportions at temperatures other than 121° would obviously be different for spores with different Z values; and they could be derived by direct observation of D values at the desired temperatures even if there were no constant Z value. It will be evident that the scale is greatly expanded in the higher temperature range and that this is the decisive part of the curve. If the temperature/time curve is now plotted on this special graph paper, the ordinate at any moment measures the lethal effect at that particular temperature, and the area under the curve therefore measures the total lethal effect of the process, in units equal to the effect of 1 min heating at 121 °C. Such a calculation (more strictly, that corresponding to 250 °F) thus measures the F value of the process.

We may recall the conclusion derived from the experiments of Esty and Meyer and Townsend et al. (see above), that an F value of 3 is the safe process to control Cl. botulinum spores, and that processes with F values of 10 may be necessary to control thermophilic spores. By contrast, as little as $F = 0·3$ may be adequate to control bacterial spores in cured meats.

There is a further non-microbiological question of the external heat treatment to be applied, to a container of a particular size and shape, filled with a particular kind of food, in order to produce a particular temperature/time treatment within. It will be obvious, on general grounds, that penetration of heat to the centre will take longer if the container is large, and regular in shape; that penetration will be slower if it is filled with a solid ("conduction pack") than with a mobile liquid ("convection pack"); and that it can be accelerated in the latter situation by devices like agitating the can. The details are, however, complex, and out of place here; and the cited specialist textbooks should be consulted.

In treatment by radiation, these problems are simpler. It is usually possible to arrange a nearly uniform distribution of dose throughout the interior of a container of appropriate size, so that the same D value is applicable throughout (though it too should have been determined in the relevant food). In this case, as D values for radiation are little affected by temperature between $c.$ $5°–40°$, the whole process is simply controlled at normal temperatures by regulating the time of application.

D. Use of Inoculated Packs

In the final stages of developing a process for controlling development of spores in a food, it is usual to proceed to experiments with realistic containers of the food (e.g. cans) artificially inoculated with known numbers of spores of the bacterial species of interest.

For this there are several reasons, some of which have already been considered.

(i) Resistance is observed directly in the foodstuff (though the spores are usually not "natural").

(ii) It is possible to explore low levels of spore survival down perhaps to the order of 1 spore per 10^4 g of food.

(iii) The validity of the assumptions used above in process calculations can be subjected to direct experimental verification.

(iv) There is opportunity to examine other possible difficulties in practice, besides questions of spore resistance, e.g. leakage of containers.

In short, the whole operation can be made into a direct simulation of real conditions.

Since these procedures may involve search for one surviving spore per many containers, it has to be detected by the spoilage which it causes, which leads to the questions of routine control discussed in the next Section. This automatically takes account of inhibitory actions of the foodstuff on surviving spores, because one only detects those capable of spoiling the foodstuff.

In order to get a satisfactory viable count of the "untreated" spores being inoculated, it may be necessary to "activate" them by suitable preliminary heating. If this is done, the equivalent of the treatment giving maximum activation should be deducted from the actual times of treatment, to indicate the rest which is effective in killing, but the correction is usually negligible (cf. "shoulders", Section III A2).

Comparison with the previous method of process calculation requires a common basis of comparison, in the assumption that spore destruction follows an exponential course in inoculated packs also. In that case, the relation

$$(\log_{10} N_0 - \log_{10} N) \times D = t$$

applies (cf. page 572), where N_0 and N are now the initial and final number of spores *per container* (all containers being the same, with equal amounts of food and numbers of spores initially). N_0 being known, the crux of the matter is to determine N in circumstances where only an occasional spore may survive in many containers. Fortunately, this can be done, through analysis of the probability of survival.

Three relatively simple ways of calculating the final number have been suggested. All rely on analysis of "partial spoilage" data. This means heating replicate groups of inoculated containers for successively larger time intervals, from an initial treatment time when all spoil and have at least one surviving spore, to a final treatment time when none spoil, i.e. where no spores survive. The actual numbers spoiling or not spoiling, after successively greater treatments, can then be used in one of the following ways.

(i) Among a group of containers heated for a particular time, *when some do not spoil*, it is supposed that each container which does spoil will contain a single spore; hence, the number of spoiled containers gives the total number of surviving spores in all the containers of the group, and division gives the average number per container.

(ii) A more accurate estimate of this number can be derived by application of "most probable number" calculations. But in this procedure, as in the preceding one, a particular estimate of N, and hence of D, is obtained for each time of heating; and these values are usually averaged for all the heating times employed.

(iii) A third procedure uses all the observations to calculate the most probable treatment time (t_{50}) which would cause half the total number of containers to spoil. When half the containers spoil, the most probable number of surviving spores is 0.69 per container and hence

$$D = \frac{t_{50}}{\log_{10} N_0 - \log_{10} 0.69} = \frac{t_{50}}{\log N_0 + 0.16}$$

Schmidt[28] gives an excellent discussion of these methods with numerical illustrations, and tabulates estimates of D by all three methods of calculation based on the same set of observations. A few of his figures are reproduced in Table V, which shows that agreement between the different methods is good. Though more complex procedures exist, they scarcely seem worthwhile.[23]

Nevertheless, there appear to be almost no published systematic comparisons of D values obtained by these inoculated pack procedures with those derived from both "end-point" and "multiple-point" laboratory studies. It is disconcerting that, when using the first two of the above

TABLE V

D_{250} *values for spores of* Cl. sporogenes *strain 3679, obtained from a single set of inoculated can experiments,[a] calculated by three different methods[b]*

Spores Suspended in	D Values by method[b]		
	(i)	(ii)	(iii)
Phosphate buffer solution	1·30	1·31	1·32
Whole kernel corn	1·69	1·82	1·85
Cream style corn	2·44	2·47	2·53

[a] Summarized from Schmidt.[28]
[b] Methods described in text (p. 587).

methods of calculation for inoculated packs, the estimated values of D are commonly found to increase with time of heating, as though the line relating log N to t were concave upwards. Using 100 spores per container containing 100 g of food, one would be exploring approximately the range 10_0 to 10^{-2} spores per g of food; so such behaviour is highly suggestive of "tailing". On the other hand, Schmidt[36] states: ". . . examination of a large amount of data on thermal resistance in our laboratories suggested an empirical correlation between the F values derived from TDT curves and the D value calculated from partial spoilage data. . . . This empirical correlation may be expressed as follows . . .

$$F = D \, (\log M + 1)"$$

where M = the number of spores per replicate (N_0) times the number of replicates, i.e. M = the total number of spores involved in the particular treatment. Since $F = D \log M$, if death follows the exponential relation (see above), this means that the estimate of D from inoculated packs must be less than that expected on an exponential basis, much less if M is small. This implies the reverse of "tailing". Either discrepancy calls into question the initial assumption of exponential destruction at low spore concentrations. Considering its importance, this part of the subject does not seem to be well worked out; and one cannot escape the suspicion that the satisfactory outcome of process calculations against Cl. botulinum results from the fact that they include large margins of safety. In view of current pressure to relax processing requirements, it would seem desirable to understand these matters more precisely.

IV. Process Control and the Problem of Spore Dormancy

Any manufacturing process requires some means of routine control, on whose efficacy the quality and reliability of the product depend. But the case of manufactured foods is different from that of, say, engineering parts, in that the defects leading to spoilage or food poisoning are not immediately apparent, yet the manufacturer cannot afford to wait till their obvious manifestations develop. Consequently, routine control of food processing always involves an element of forecasting on the basis of advance or partial observations. The implication is, that whatever methods are employed will be capable of predicting the behaviour of spores which may be damaged and in a dormant state: hence the need to consider these two topics together.

A. METHODS OF CONTROL

The statement, that a particular species (whether of spore-forming bacteria or not) is the agent of some particular spoilage in a food, implies the ability to isolate and identify that species from the food and, moreover, the ability to reproduce the spoilage by re-inoculating the organism into sound food. In any attempt at control of a process aimed to prevent spoilage these procedures must accordingly be involved at the outset. But they are unsuitable for routine control of a manufacturing process, for which simpler and more expeditious, though not much less reliable methods are used.

1. Direct bacteriological examination

To isolate and identify a sporeformer from a spoiled food, suitable conditions of oxygenation, incubation temperature and culture medium must be selected; and, since the species in question are diverse, the necessary conditions are correspondingly diverse and no general procedure is applicable. Description of these matters of technique seems out of place here, and the reader is referred to textbooks giving appropriate bacteriological methods.[23, 37] Instead, discussion will be devoted to general matters particularly related to the sporing habit.

When trying to detect sporing bacteria in foods, it has been normal practice to eliminate non-sporing species by a preliminary heating of the sample, classically for 30 min at 80°C. Use of this procedure alone is not recommended for two reasons. First, the spores of many species are destroyed by such treatment, for example those of *Cl. botulinum* type E; though the more heat-resistant species, which are important in heat processed foods, are not so affected. For this reason, recent spore surveys have inclined to use a preliminary heating at only 60°C, though this may be insufficient to eliminate the most resistant non-sporing species; alcohol might be used to eliminate vegetative contamination. Second, however, it cannot safely be assumed that a species with the power to form spores will actually have done so; for example, *Cl. perfringens* forms few spores in most circumstances and, though it is one of the common clostridia in fresh meat, spores are seldom formed in that food.

When examining a heat-processed food, the problem is simplified by the elimination of all except heat-resistant species, and by the protection of the food from re-contamination by means of a sealed package, commonly a can or bottle. Direct microscopic examination is then useful. The heat-resistant spore-forming bacteria, whether aerobic or anaerobic, are Gram-positive rods visible in Gram-stained smears of the spoiled food. The presence of cocci, Gram-negative rods, yeasts or fungi, indicates either serious under-heating or leakage of the container. Microscopic examination will not, however, reveal surviving spores directly, if only because they are far too few; and its value in the diagnosis of post-processing spoilage is obviously nullified if the foodstuff was significantly contaminated before processing.

Any commercial process aimed to control the development of spoilage organisms in a food requires a routine control of its efficacy; a control which, for reasons of economy, cannot be allowed to spoil more than a very small proportion of the food.

Though casual consideration might suggest that bacteriological examination could be used, any such procedure has in fact very serious limitations.

(i) If checking for sterility, the first limitation is statistical. Even if a few spores are present, one expects to find nothing, in a high proportion of examinations; while conversely, if an occasional positive is found, a high level of bacteriological technique must be assured before it can be presumed that the organism detected actually originated in the food.

(ii) The necessity to open the containers to sample the food will normally render them unsaleable, even if the contents prove sound.

(iii) Because a pack may well contain dormant spores, it does not follow that all organisms isolated by ordinary methods of culture would necessarily be detrimental; and further investigation may consequently be needed to try to decide whether this would be so. To take a simple case, if checking a food likely to contain only an occasional thermophilic spore, whenever a positive is found it is necessary to check that it is not of mesophilic type—which should present little difficulty but adds to the time and labour of the examination. But the situation may be far more difficult than this. For instance, in examining a commercially stable pack of cured meat, a variety of spores might be encountered, as Table VI illustrates;

and their mere recovery would give little information about the probable stability of the pack, since it would be necessary to decide whether these spores would have been "capable of changing the quality". Any attempt to answer this question usually involves transfer to a medium simulating the food—e.g. a "micro-ham" test for isolates from canned ham; but we know so little about the factors deciding dormancy in such a situation that these tests cannot be wholly reliable.

TABLE VI

Extracts relating to sporeformers from "Suggestions regarding bacteriological norms for non-sterile hams preserved in cans"[a]

A. Pathogenic bacteria

(1) *Cl. botulinum* should be absent. Or, if present, the conditions in the ham should be sufficient to prevent toxin formation.

(2) Toxigenic clostridia such as *Cl. perfringens* should be absent.

B. Bacteria indicative of inadequate processing[b]

C. Bacteria related to the stability of the pack

(1) Gas-forming clostridia should be absent, except those which cannot develop in canned ham.

(2) Gas-forming denitrifying bacilli may be present. If present, the cans may blow if they are not kept in cool storage.

D. Bacteria likely to be present, but not known to affect the quality of the pack if present in reasonable numbers

The spores of inert clostridia and of inert bacilli may be present. (In this context, "inert" means non-pathogenic and not able to change the quality of canned hams at temperatures below 5°C.)

---

[a] Recommendations of a Scientific Committee.[38]

[b] No sporeformers were mentioned in this category, it being recognized that a great variety of spores might conceivably occur.

(iv) The labour involved in such exercises greatly curtails the number of containers which can be examined, which leads to large uncertainty because of statistical limitations.

For these reasons, routine control is seldom based on bacteriological examination.

2. Indirect methods

Instead, the common custom is to observe whether any cans of a large sample number blow during an accelerated incubation test, and to reject if the proportion blowing reaches some predetermined level, say 1 per 1,000. Besides the statistical weakness in such a procedure, there are bacteriological problems also.

The procedure rests on the assumption that spoilage involves production of gas; the assumption commonly holds, however, because the most heat-resistant spore-forming bacteria—the likely spoilage organisms—usually have the ability to produce abundant gas from the foods in which they grow.

This procedure is, however, clearly valueless with species which have heat-resistant spores but are not gas formers, such as those which cause "flat-souring" of vegetable products. But in this case also, bacteriological examination is avoided as far as possible, though control must involve opening cans. Routine measurement of pH in the contents of incubated cans is used to reveal those developing acidity, a reduction of pH by as little as 0·1 unit below the average being significant in this connection. Where spoilage has occurred, with a shift of perhaps 1 pH unit, conditions become unfavourable to the causative organism and the vegetative cells may die without the opportunity to form spores, in which case no viable organisms may be recoverable ("auto-sterilization"). In this situation, control by seeking viable cells would evidently be unreliable; examination of stained smears reveals numerous vegetative cells, often in the form of poorly stained "ghosts" due to autolysis.

Safety is of course even more important than spoilage, in the philosophy of production control. The presumption is, above all, that gas will be produced by *Cl. botulinum* and by those species, notably strains of *Cl. sporogenes* like no. 3679, which act as its "indicators" (see above), in the foods in which they are likely to develop. Fortunately these species are vigorous gas formers, so that this presumption usually holds.

There are, however, situations in which *Cl. botulinum* does not form gas or plainly alter a food, while making it toxic, and such situations obviously have a special interest. When this is suspected, the best recourse is a direct test for the botulism toxin, and a serious present difficulty lies in the unfortunate fact that the only satisfactory tests at present available involve the use of experimental animals, usually mice, and up to 10 may be required for each food sample tested. Such a procedure is clearly quite unsuitable for routine control, hence the reliance on indirect methods to control *Cl. botulinum*.

All these methods of control, where one looks for the indirect effect of the surviving spore instead of trying to observe it directly, suffer from a fundamental weakness. They assume that all the spores which could eventually develop will indeed have done so by the time that the examination has ended; that is to say, that all the surviving spores in the pack will, under the conditions during incubation, break dormancy and develop within the limited period of incubation. The procedure depends, in essence, on an attempt to accelerate the breaking of dormancy.

B. DORMANCY

Previous chapters have described the dormant state of bacterial spores, and the processes of germination and outgrowth which mark the breaking of this dormant state.[39] Treatments so severe as to leave only a few surviving "damaged" spores may have the effect of extending the dormancy of those survivors almost indefinitely, especially when applied in foodstuffs which exert some degree of inhibitory influence on the spores (such as the presence of curing salts in cured meats). These phenomena are fundamental to the stability of processed foods, since food which contains only dormant spores will remain stable so long as the spores remain dormant, but is likely to spoil shortly after this dormancy is broken at temperatures where multiplication can occur.

1. General

The general pattern is of a shorter or longer period of dormancy. terminated by germination. The dormant period may be long and extend over many months in canned foods. This is followed, in suitable circumstances, by outgrowth and vegetative multiplication; in a period probably short compared with the dormant period, in practical conditions, as will be seen later. It may thus happen that there is a long period in which the pack appears perfectly stable, and then suddenly one can after another will begin to spoil. Such events are of course extremely embarrassing to the manufacturer, the more so because, by their very nature, they cannot be controlled by routine checking at the time of manufacture, unless some satisfactorily reliable means of greatly accelerating the phenomena can be devised. Schmidt[28] notes: "It is quite common to observe spoilage to occur after 75 to 100 days incubation, and in some cases swells have occurred after 200 days incubation. Product incubation for thermophilic anaerobes is continued for 60 days . . . incubation in thermal death-time cans for mesophilic clostridia is continued for at least 120 days." For these reasons, the factors which control the maintenance and breaking of dormancy in spores are of very great interest in the microbiology of canned foods.

It is evidently a far from simple matter to decide whether a bacterial spore has been "killed". A long period of reliable dormancy may, for practical purposes, be equivalent to death. The point was emphasized in an authoritative statement by Cameron and Bohrer,[40] which particularly referred to Cl. botulinum but which is equally applicable to other sporeformers: a satisfactory process must either kill spores, or alternatively it must permanently inhibit spores and it must be decided whether permanent inhibition is the practical equivalent of spore destruction. Because of this ambiguity, it is common practice to talk of "inactivating" rather than "killing" spores (though inactivation in this sense has misleading connotations vis-à-vis "activation" in the sense of Chapter 10).

2. The effect of substrate

The first facet of the problem, that of the nature of the culture medium or food in which the spores are suspended, may be illustrated by reference to Table VII, where the following points are noteworthy.

TABLE VII

The effects of heat, irradiation, and recovery medium[a] on viable counts[b] of spores of Cl. sporogenes strain 3679

Treatment[d]			Treated in buffer and counted in media[c]:			Treated and counted in ham
	°C	min	I	II	III	
Preliminary activation	75°	20 m	880	11,000	3,100	—
Then either heated F = 0·6	100°	90 m	280	230	1,600	160
Or irradiated 1 Mr	—	—	95	20	330	0·05

[a] Summarized from Riemann[41]: aliquots of suspension distributed in (i) buffer or (ii) chopped ham; treated with heat or γ-radiation; and counted in (i) different media or (ii) in the ham.

[b] Counts determined by Most Probable Number technique, and calculated to thousands per ml of original suspension.

[c] Media: I, skim-milk + neopeptone; II, ditto + 0·06% $NaHCO_3$; III, pork-pea infusion; all in tubes incubated at 30°C. Ham: chopped cured ham inoculated with different spore concentrations, packed in sealed cans for treatment, then incubated at 30°C for 3 months, and swelled cans checked for clostridia.

[d] The heat and irradiation treatments were chosen to give about the same degree of kill of the spores.

(i) Different substrates record different numbers of spores as viable, even after a mild heating quite insufficient to kill any of them. This means that one can have no certain knowledge of the initial number of spores; and that the number which can develop in and spoil a food will vary with the nature of the food, in a manner not understood.

(ii) Similar differences between substrates appear when heated or irradiated spores are counted. Medium III was obviously the best for these damaged spores, whether heated or irradiated; but it was not the best for lightly heated spores. Consequently, the estimate of the damage caused by a particular treatment depends on the medium used, and there may be no clear indication which is the most correct.

(iii) In a partially inhibitory substrate like cured meat, much smaller proportions of damaged spores are capable of developing. At least $1,600 \times 10^3$ spores (equivalent) remained viable after 90 min at 100°C—i.e. were capable of growing on Medium III; but only

160×10^3 developed in ham, even during incubation for 3 months at 30°C, though independent evidence suggests that the immediate effect of heating would be not so greatly different in ham from that in buffer. It appears, then, that only about 10% of the spores potentially viable after heating them in ham were capable of developing in the partially inhibitory medium represented by ham.

(iv) These results do not support the widely publicized view, based mainly on a restricted range of experiments with *Escherichia coli*, that damaged cells recover best on poor media.

Consideration shows that these phenomena are probably a reflection of the dormancy behaviour of the spores. In certain media they break dormancy, develop more readily than in other media and are recorded as "viable"; and use of the latter will give unduly low counts which are revealed as erroneous only when comparison is made with one of the former media. Difficulties of this kind are more troublesome with *Clostridium* than with *Bacillus* spores, the latter having received far more attention. A medium devised by Busta and Ordal,[42] containing calcium and dipicolinic acid, is generally excellent for *Bacillus* spores, though its value with spores damaged other than by heat is not well documented; it has not yet been tried with clostridia.

By the same token, the behaviour of spores in artificial media is an imperfect guide to what will happen in a food. From the practical viewpoint, it is clearly important to know, for example in the ham to which Table VII refers, what prevented 90% of the potentially viable heated spores or 99·99% of viable irradiated spores (compare ham with Medium III), from developing from the dormant state. It is interesting, further, to infer that the storage stability of irradiated cured meats might be greater than that of equivalently heated cured meats.

3. The effect of time

To turn now to the influence of time, it will first be noted that the cans of ham in Table VII had been incubated for 3 months, and it is conceivable that more of the spores might have developed if incubation had gone on longer (much longer periods may be required in practice). While strict terminology regards dormancy as broken when spores begin to germinate, in experiments like those of Table VII, what one actually observes is the culmination of germination, outgrowth and vegetative multiplication; a spore might break dormancy, in the strict sense, but fail to be recorded as viable if it were inhibited at some later stage. Both of these considerations involve the element of time.

As regards the overall effect of incubation time, in any group of spores derived from a single source, as in a culture, the times before the different spores develop are log-normally distributed, according to Riemann; and with spores "damaged" by heating or irradiation, though the dormant period may be greatly extended, the nature of the distribution is apparently otherwise similar as shown in Fig. 5.

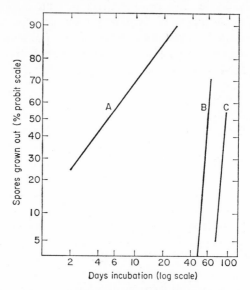

FIG. 5. To illustrate the normal distribution of the logarithms of the times required for germination and outgrowth of spores. (Representing calculations of Riemann.[41, 43]) Spores of *Cl. sporogenes* strain 3679, incubated at 30°C:

    A. In Hartley's broth + thioglycollate
    B. Heated and incubated in ham—cf. Table VII
    C. Irradiated and incubated in ham—cf. Table VII

Inspection of Fig. 5 indicates an inconvenient characteristic of this time distribution: viz. that the last few spores to develop take a relatively long time to do so. If spores are distributed sparsely among a large number of packages (bottles or cans) of food, the spoilage of these packages should follow a similar time pattern (no practical data are available to the writer). This means that among a batch of packages undergoing incubation test it will seldom be possible, from those which have spoiled within a limited time, to establish confidently the proportion which will eventually spoil. Accordingly, the actual duration of the test will tend to represent an uneasy compromise between practical duration and certainty in result. One might suppose that the regularity of behaviour could permit prediction of the eventual development of spoilage. Unfortunately, the spore contamination in canned food under practical conditions may not be homogeneous in the relevant sense, though it might perhaps be so if it happened to be derived from a single source of contamination; there is no information. In the practical case, therefore, it cannot safely be presumed that spoilage will develop in the manner here indicated, and many more actual observations are desirable.

It follows from the same characteristic that viable spore counts in artificial media increase with time of incubation. Under highly favourable circumstances, the proceedings may be so much accelerated that all spores able to do so have already germinated and multiplied within a day or two. It is however necessary, in developing any spore counting procedure, to check that longer incubation periods do not reveal more spores (cf. line A, Fig. 5). Long incubation periods can be used less conveniently with plate counts than with deep tubes as in the Most Probable Number technique.

As regards the difference between the breaking of dormancy and the development of spoilage, the distinction cannot be readily observed in a packaged food, but the importance of the question has stimulated much recent laboratory research. The indications with cured meats are that the inhibition occurs after germination (cf. p. 581). However, spores whose development is thus interrupted are believed not to remain long in a state of suspended animation, but to either overcome the inhibition and develop, or die. It is thus reasonable to suppose that the vegetative development of spores after long periods of dormancy during incubation takes place relatively rapidly, and that accordingly the induction period substantially represents the time required to break dormancy in the strict sense, as was stated in Section IV B1. Again, more observations in packaged foods seem desirable.

4. The effect of temperature

Incubation tests are urgent by nature, and consequently conditions are arranged so as to accelerate them as much as possible, notably by incubating at what is believed to be the optimum temperature for the spoilage in question. This is a temperature which is usually distinctly higher than that likely to prevail in normal handling. It is a basic presumption that such tests do not seriously misrepresent the real situation.

In the present context, this implies the assumption that, when the temperature is raised, the *rate* at which spores break dormancy is increased while the *number* which do so is not affected. It is especially important, from the practical viewpoint, that the number should not be falsely diminished. The writer is not aware of any published evidence to confirm this assumption, but it must presumably be supported by the unwritten experience of countless incubation tests carried out under commercial conditions.

V. Future Trends

Future developments in this field will certainly be directed by the commercial pressure of a persistent trend in public preference, for foods less and less damaged by over-cooking, and for foods which are less acid or less salty. In addition, the growing control of chains of food distribution, and the increasing use of refrigeration, encourage the production of non-sterile foods with correspondingly limited storage life. All this points to attempts to reduce processing treatments, while maintaining accepted standards of security from spoilage and from risk of food-poisoning.

This section must accordingly consider the possibilities which exist, or might conceivably exist, for reducing processing treatments and using milder curing treatments. It should be evident, from what has gone before, that in following this road with non-acid foods we must sooner or later expose ourselves to the risk that *Cl. botulinum* may develop. And since, with cured foods especially, we cannot precisely evaluate the influences which keep *Cl. botulinum* and related anaerobes under control, it is not at present possible to say precisely at what point that danger will arise. For these reasons, those in authority have an understandable reluctance to relax the existing standards of practice, which long experience has shown to be safe.

As processing and curing treatments are reduced, the tendency is to rely more and more on control by refrigeration below the minimum temperature for growth of the surviving sporeformers; hence for instance the recommendation "keep below 40°F" on pasteurized cans of ham. This accounts for the present commercial interest in interactions between salts, acidity and minimum growth temperature[20]; much more information of this kind is needed, relative to species likely to cause spoilage. Moreover, if processing treatments for canned hams are to be relaxed below the present effective minimum level of about 30 min at 70°C, spores of type E *Cl. botulinum* might survive and expose us to hazards, only lately appreciated, owing to the ability of this type to grow and form toxin readily at temperatures down to 5°C. This is why much attention has recently been directed to the occurrence, heat resistance and temperature relations of type E; fortunately it seems to be extremely rare, save in marine products.

A. REDUCTION IN SPORE LOAD

The most obvious immediate steps are in the direction of better housekeeping: the exclusion from food of spores of the especially troublesome kinds, and a reduction in the numbers of spores of all kinds which gain access to it. The principles should be plain from what has been written earlier.

Such steps are already being taken by firms with a sufficient degree of technical control, but they seldom receive publicity. Brown,[44] however, described a systematic diminution over several years of the content of very heat-resistant spores of *Cl. sporogenes* in meat being processed in a meat packing plant.

As already mentioned, a limit to progress in this direction is set by the intrinsic contamination within the food, when the problem shifts from relatively simple matters of plant hygiene to the treatment and properties of the raw food itself.

B. CHANGES IN GROSS COMPOSITION OF FOOD

Useful diminution in processing requirements would clearly be possible if the composition of the food could be appropriately modified. Whether such modification might be unacceptable for other reasons is a question not discussed here.

1. Acidity

The previously mentioned investigations of Baird-Parker and Freame[20] showed clearly that the limiting pH for development of spores of *Cl. botulinum* is substantially increased if salt is present in the medium, or if the temperature is far below the optimum. This implies that the division of foods into "acid" and "non-acid" categories, with a general boundary at pH 4·5, is probably unnecessarily rigid. In foods of appropriate composition, or stored under prescribed temperatures, it should be possible to raise this limit.

Though this argument is based on the requirement to control *Cl. botulinum*, it seems that similar relations would apply to other spore-forming bacteria important in spoilage, though these relations mostly remain to be worked out. Reference was earlier made to the possibility of controlling *B. thermoacidurans* by acidifying tomato to pH 4·1.

2. Curing

In the same way, the pH interacts with the amount of salt which can be tolerated. Indeed, as indicated already, the stability of heat-processed cured meats depends on the interaction of several factors. Altogether, these include: the number of spores, the pH of the meat, the concentrations of sodium chloride, nitrate and nitrite, and the severity of heating. Experience shows empirically that different combinations of these can give satisfactory control of spore-forming bacteria.

In aiming to diminish both the heat treatment and the concentration of sodium chloride, there is nevertheless not much room for manœuvre Increase of acidity leads to increased cooking-out losses which are commercially unacceptable; indeed, to diminish such loss it is customary to add polyphosphates which have the effect of raising the pH to the possible detriment of microbiological stability.

The point at which progress seems conceivable is in relation to nitrite. The indication from recent work, that only a very small fraction of the added nitrite is highly effective, implies that if means could be found to increase this active fraction without increasing the total quantity of nitrite present, it might be possible to relax other factors correspondingly. However, this is a question for the future involving some difficult experimentation. Because of the complexity of the inter-relations involved, there is reluctance to modify the practices which experience has shown to be satisfactory.

C. DIMINUTION OF HEAT TREATMENT

The loss of quality which many foods suffer, when sterilized by heat, stimulates the search for means of amelioration. Broadly, these are the following:

(i) One may seek to avoid the need to heat, by trying to exclude spores altogether, through the use of aseptic techniques.

(ii) The spores may be accepted but their heat resistance avoided by allowing or encouraging the spores to germinate, making use of the fact that heat resistance is lost at an early stage of germination. A procedure of this kind, familiar to most bacteriologists, is that called "tyndallization" which involves successive steamings to sterilize culture tubes.

(iii) One may accept the spores but seek to modify their effective heat resistance; that is to say, one tries to change the relative rates of destruction of spores and of destruction of food quality, to the advantage of the former. This may be done by using unusually high temperatures (UHT process), or by sensitizing the spores to heat by addition of some appropriate substance.

(iv) Finally, of course, one may try processes quite different from heating, such as the use of radiation, or of antibiotics, or both.

Each of these possibilities is discussed at more length below.

1. Aseptic techniques

This type of procedure is not successful, for reasons already outlined. Few foods, in practical conditions, are sterile internally and, when using asepsis as sole defence, vegetative organisms are just as damaging as spores. Consequently, it is possible to get only a proportion of sterile samples even on a small experimental scale, and to do this requires precautions far too elaborate for commercial food processing. Given a previously sterilized food, aseptic packing is already a commercial possibility.

2. Germination before heating

In time, spores germinate under suitable circumstances and lose their resistance to heat; and when the circumstances are most favourable, the time required may be short—less than an hour (cf. Chapter 11). Since many foods are good media for growth of bacteria, they might be expected (though precise information is scanty) to favour spore germination if brought to an appropriate temperature. The idea is accordingly to warm, allow the spores to germinate, and then kill them with a comparatively mild heat treatment. The same notions are applicable to irradiation.

If all the spores cannot be relied on to break dormancy and germinate at once (and we have seen that a few will only do so after a relatively long time), it is necessary to repeat the procedure; indeed, two repetitions are usually recommended to ensure success when steaming culture media to sterilize by "tyndallization".

Success in such a procedure evidently depends entirely on the time relations in breaking dormancy, for heat must be applied well before the spores which first germinated have had time to develop and cause spoilage, and one aims at circumstances in which *all* the previously surviving spores will have germinated within that time.

Special possibilities therefore follow from the discovery that spores may germinate under conditions where they cannot develop further. It was noted earlier that spores of *B. stearothermophilus* might germinate at temperatures below the minimum for vegetative growth, and analogous behaviour is reported for several mesophilic species of *Bacillus* and of *Clostridium*. For example, Mundt, *et al.*[45] observed that spores of *Cl. sporogenes* could germinate in ham at a temperature of 5 °C, which is too low to permit outgrowth and multiplication; it appeared, accordingly, that a period of holding at 5° before heating, to allow spores to germinate, might remove the need for severe heat treatment. Unfortunately, it proves that germination of various species is not sufficiently rapid or complete in these circumstances.

This leads one to consider the possibility of stimulating germination in various ways, for example by the addition of chemicals. This subject has been extensively investigated in the laboratory (see Chapters 10 and 11) and very many substances have been found to have this property— Sussman and Halvorson[39] list about a hundred. From the practical viewpoint, several comments may be made on such a list. First, the systems studied are often highly artificial, using specially treated spores and very short periods of observation. Second, it is often not clear (e.g. in nephelometric observations) whether *all* the spores have germinated; whereas, in the practical case, completeness of germination would be as important as speed. Again, it seems unlikely that any one or two substances alone would be generally effective in practice, as quite different compounds are active on spores of different species. Finally, many of the substances in question (e.g. particular sugars and amino acids) would be naturally present in appreciable concentrations in various foods, so that the benefit of addition appears doubtful at first sight. In fact, at the practical level this approach is still largely unexplored.

An alternative way of stimulating germination is by physical stimuli. Of these, heat shock would seem to be the method of choice as a preliminary to a further heating process. The effect of heat shock in stimulating germination, called "heat activation", has been discussed in detail elsewhere (Chapter 10). The only special points to be made here are that the phenomenon is more frequent and pronounced with spores of the genus *Clostridium* than of *Bacillus*, and particularly with species which are specially significant in food processing, like food-poisoning strains of *Cl. perfringens* and *Cl. botulinum*. The idea is, accordingly, to heat the food sufficiently to break the dormancy of the spores, then to cool to a temperature permitting germination, for a time long enough for germination to occur, and then to heat again to a temperature high enough to kill vegetative cells. In practice this means successive exposure to steam or boiling water, with an at present ill-defined interval between for cooling and germination, as in "tyndallizing" culture media.

Recently, unpublished attempts were made to use this procedure for "cooked-in-the-bag" poultry. If cooked poultry carcasses are packed in plastic bags, with air or without, and are exposed to warm conditions, they suffer catastrophic spoilage from clostridia (especially *Cl. perfringens*) and it was demonstrated that *Cl. botulinum* could grow and form toxin in the same circumstances. The rapid spoilage was supposed to arise from spores already heat shocked, but not killed, in the act of cooking; so it was reasoned that a further heat treatment after packing should suffice to prevent this development of clostridia. The process proved to be insufficiently reliable, however. Possible reasons are that the cooking was inadequate to heat shock all the spores into germination, or that subsequent conditions were not appropriate for complete germination to occur in time; or, of course, that contamination occurred in packing the cooked carcasses into the bags, with extraneous spores not subject to heat shock.

3. Alteration of effective heat resistance

Two practical possibilities exist here:

(*a*) to diminish heat resistance, relative to damage to the food, by using high treatment temperatures, or

(*b*) to add substances which make it possible to inactivate the spores, with a lesser heat treatment, which is—in effect—to make the spores more heat sensitive.

(a) *Ultra-high temperature processing.* The possibility of gaining advantage from the use of higher temperatures arises because of the unusually high temperature-coefficient for the destruction of bacterial spores by heat.[1] The Z value used in heat processing, which represents the temperature rise which accelerates killing tenfold, is simply related to the more familiar temperature-coefficient Q_{10}, which represents the ratio of rates at temperatures differing by 10°C. It happens (cf. Table IV) that Z values for spore destruction are usually about 10°C, i.e. the Q_{10} for this process is about 10 over the relevant temperature range. By contrast, the Q_{10} coefficients for processes of heat damage to foods are much lower, near the value 2 usual for chemical reactions; for example, for the destruction of thiamin 2·1, and of riboflavin 2·3; for the inactivation of peroxidase 2·5; for the browning of milk, about 3. This means that it is possible to kill, for instance, most of the spores in milk with little change in colour and flavour if a sufficiently high temperature of treatment is used. Suppose that a certain bactericidal effect and a certain tolerable browning of milk are caused by a certain treatment at 100°C, and regard the relation between the two effects as representing a ratio 1 : 1. If the temperature of treatment is raised by 10°C and the time reduced to $\frac{1}{3}$, the colour change will be the same at 3/3, while the effect on the spores will have increased by about 10/3. That is, raising the treatment temperature to 110° improves the kill : damage ratio from 1/1 to 3·3/1. Similarly, at 120°C, the ratio would be $(3·3)^2/1$ or approximately 1/0·1; that is, for a given sterilizing effect, the heat damage at 120°C would be only of the order one-tenth that for processing at 100°C. There is thus advantage in operating with as high a sterilizing temperature as possible, which is the basis of the so-called UHT processing. At 135–140°C, for example, the treatment time corresponding to $F = 3$ is only about 3 sec. Such a process can be generated by rapid flow through a heated tube; but the liquid in the centre of the tube tends to flow fastest, which leads to a risk of under-processing there. Direct injection of superheated steam ("uperization") is an alternative, which minimizes the problems of heat transfer and permits use of temperatures up to 150°C, but involves the addition of water; the water can, however, readily be boiled off again by passing through a pressure reduction valve, and the consequent cooling sharply brings to an end the period of effective heat treatment.

It is a fortunate circumstance that in the UHT treatment of milk a substance is produced which is inhibitory to heated spores of *B. stearothermophilus.* These spores are not uncommon in milk, and their high heat resistance would necessitate more severe heating than is actually effective, if this inhibitor were not present.

In practice, there are definite limits to the exploitation of this principle.

With too high a temperature, sterility may be achieved before the most heat-resistant enzymes like peroxidase are completely inactivated; also, the correspondingly very short times involved become more difficult to control accurately. Further, with present techniques, very rapid heat transfer is practicable only with non-viscous liquids; hence the main interest of this process lies in attempts to improve the quality of sterilized milk. Finally, of course, the sterilized liquid has to be packed hermetically with virtually complete avoidance of re-contamination.

(b) *Heat-sensitizing agents*. Many substances have been tried which effectively diminish heat resistance. That certain additives will do this should already be clear: for example, sufficient acid to bring the pH near to 4·5, or sufficient curing salts. But these are excluded from consideration now, because they have the effect of transforming the treated food into something different; here we consider substances which have specific effects, in concentrations so small as to be inappreciable.

Two very large surveys were made in the 1950s in the U.S.A.[46] which covered most of the then available substances which appeared likely both to have specific effects and to be useful and acceptable. The results, while interesting in detail, were in general disappointing. Few substances were significantly effective (some even increased heat resistance), and the most effective were two related antibiotics, subtilin and nisin, which have serious shortcomings in other directions (see p. 607). The prospects of success by this approach therefore do not seem good, though it remains one of the most attractive in principle.

Empirically, it is well established that smoking makes a substantial contribution to the preservation of canned cured meat from spoilage by sporeformers.[47, 48] The reasons are, however, not properly understood. The increased acidity which follows smoking could be one reason; but this may not be a sufficient explanation, as the pH change is largely confined to surfaces exposed to the smoke. Chemical substances (phenols?) included in the smoke might be the effective agents, and the particular substances in question might have been overlooked in the surveys hitherto made. The essential oils of certain spices have similar effects.

D. NOVEL METHODS

It always remains possible that entirely new principles might offer new scope, and two deserve special mention. These are the use of irradiation or of antibiotics, both possibilities uncovered and investigated only during the last twenty years.

It should be understood that such procedures are only at the experimental stage. There still exist technological problems, such as changes in flavour or appearance resulting from the treatment; and medico-legal questions, related to the safety of such treatments and means of controlling them. These matters are quite beyond the scope of this discussion, which must be confined to that part of the microbiological aspects of these processes which concern spores.

1. Irradiation

The use of ionizing radiation, X-, β- or γ-rays, to kill micro-organisms in foods has attracted attention because it has a number of special character-istics which offer technical advantages in processing foods. It does not involve any considerable rise in temperature, so that it is possible to sterilize foods while keeping them raw, or even frozen. Further, such radiation penetrates instantly; and it can be arranged for X- and γ-rays to be distributed almost uniformly within food units of diameter not more than about 20 cm, which makes it much easier to achieve uniformity of treatment than with heat processing.

We have already seen that the kinetics of spore destruction by these radiations resembles that by heat; and in particular, that a characteristic D value can be derived, which permits predictions of effectiveness analo-gous to those used in heat processing. Experience so far indicates that, with rare exceptions and perhaps none among spore-forming bacteria, the spores of certain strains of *Cl. botulinum* type A have the highest resistance among micro-organisms, having D values approaching 0·4 Mrad under normal circumstances; so that, in order to achieve a 12-D treatment, radiation doses of about 4·5 Mrad are needed. Because of this outstanding radiation resistance of *Cl. botulinum*, the earlier classification of foods, into those which will and those which will not support development of that species, seems just as significant in radiation processing as it is with heat, though in fact these relations are as yet imperfectly explored in the radia-tion context.[49]

Acid foods of pH less than 4·5 in which *Cl. botulinum* cannot grow should be treatable with lower doses of radiation than those in which *Cl. botulinum* spores must be killed. Especially is this so for irradiation, because the thermophilic sporeformers which are so troublesome' when processing acid foods by heat prove not to have unusual resistance to radiation, as was shown in Table IV. Further, it has recently been shown that with greater acidity there is somewhat greater sensitivity to radiation, as to heat. But it is too soon to decide whether exactly the same pH limit as used in heat treated foods (4·5) will be appropriate in radiation process-ing to divide the "acid" from the "non-acid" foods.

With cured foods, also, there are evidently close similarities between radiation and heat processing. The activities of *Cl. sporogenes* strain 3679 spores in causing spoilage, or of *Cl. botulinum* spores in leading to toxicity, have been controlled in cured meats with radiation doses about one-half those needed for uncured meats (cf. Table VIII). This has, moreover, been shown to be due to a similar basic mechanism: sodium chloride is without direct effect on radiation resistance, but previous irradiation makes surviving spores unusually sensitive to salt.[51] As with heat, the overall effect is as

TABLE VIII

The effect of curing salts on the survival of spores in beef when it is irradiated[a]

| Curing salt mixture | | | Total | | No. of cans | |
2·5% NaCl	0·1% NaNO₃	0·2% NaNO₂	no. of cans	Swollen	not swollen but toxic	Sound
—	—	—	25	25	0	0
+	—	—	25	15	10	0
+	+	—	125	0	0	125
+	—	+	125	1	5	119

[a] Summarized from Anderson *et al.*[50] Cans inoculated with 10^6 spores *Cl. botulinum* strain 33-A per g, irradiation dose 2·0 Mrad, incubation at 30°C for up to 120 days.

if the radiation resistance of the spores were diminished by sodium chloride, to a degree increasing with the salt concentration. As yet, nothing is known about the separate effects of nitrate and nitrite, though they are apparently important in radiation processing too (cf. Table VIII). In the light of these relations, caution is evidently necessary in regarding current suggestions that cured meats in general can be "sterilized" with a radiation dose of 2·5 Mrad. The required dose will apparently vary with the concentrations of curing salts in the meat, as in the case of heating. It is disturbing to see, in Table VIII, high proportions of cans toxic but not swollen, the significance of which was mentioned earlier; nitrate appears to have prevented this phenomenon.

As with heat, a search for chemical sensitizing agents to radiation has so far failed to reveal any which appear promising.

We may conclude that there is some prospect that irradiation could replace heat, in various connections where its peculiar characters are an advantage. However, further scientific, medical and legal questions need resolving at present.

2. Antibiotics

In general, antibiotics are inadequate to control spoilage, as their antimicrobial spectra are too restricted. But things seem more hopeful when the range of species involved has been limited by previous heating to a few spore-forming bacteria,[52] for some antibiotics are relatively effective inhibitors of spore-forming bacteria.

Once again, attention centres first on *Cl. botulinum*, and tylosin is an antibiotic shown in late years to be highly effective in inhibiting spores of that species, the more so as heating sensitizes the spores to tylosin without destroying it. The antibiotic is, further, effective against the spores of thermophilic bacteria, so that it can control thermophilic spoilage besides *Cl. botulinum*. The difficulty is that this antibiotic inhibits the spores only after germination, and that in stored foods the antibiotic gradually vanishes in a manner at present unpredictable. It is thus conceivable that the antibiotic might disappear before *Cl. botulinum* spores surviving the heating (presumed to be less severe than usual) had germinated, which might take several months. In this case germination could be followed by outgrowth and toxin production. That this is a real possibility was shown in inoculated pack experiments by Denny et al.[53] Tylosin has the further disadvantage of being used in veterinary therapy, and of readily developing cross resistance to ethryromycin used in human therapy.

Nisin does not have the two latter disadvantages, but it is a much less effective inhibitor of *Cl. botulinum* than tylosin, though it is effective against other clostridia and thermophiles.[54] Like tylosin, it appears to diminish heat resistance, by killing surviving spores after they have germinated. Nisin is accordingly used to control thermophilic spoilage with diminished heat processes, but it is generally agreed that processes should not be diminished below that which is in itself safe against *Cl. botulinum*.[52] The greatest value of nisin is with acid foods, in which there is no need that the processing should control *Cl. botulinum*.

E. CONCLUDING COMMENTS

To a large extent, the present high degree of control of sporeformers in processed foods is the result of a century or more of empirical experience without any detailed understanding of the mechanism of spore death or dormancy. It is only comparatively lately that extensive investigations, mainly in industrial laboratories in U.S.A., have laid the foundation for what have become sophisticated and on the whole very satisfactory methods of calculating the effectiveness of heat processes. But, now accustomed to this achievement, when facing a new process, authority seeks some similar alternative to the long and perhaps bitter experience which is too costly under modern commercial conditions of the mass market.

This means an accepted methodology for demonstrating that spores are killed or indefinitely inhibited. It means, moreover, an accepted system of quantitative evaluation, which will give confidence in predictions of effectiveness beyond the range of circumstances which can be encompassed in laboratory experiments; if one is seeking to control a phenomenon with a probability of say 10^{-12}, some extrapolation can hardly be avoided.

Accordingly, among the procedures surveyed above, it is an advantage for example of the ultra-high-temperature process that its effectiveness depends on principles which are well established and are amenable to calculation on a generally accepted basis. Similarly, it is an advantage of the irradiation process that its effects resemble those of heating, to such an extent that the philosophy and methods of calculating effectiveness are largely transferable from the established process to the new one. By contrast, it is a defect of proposed antibiotic-based processes that they have never been subjected to any corresponding systematization; and similarly, it is evident that any attempt to base practical processes on control of dormancy would require much better factual information about that phenomenon than is at present available.

The problems of estimating effectiveness are far greater with processes which depend on a combination of several factors, and such processes seem likely to proliferate in future. Reference was earlier made to the present impossibility of evaluating precisely the effectiveness of heat treatments given to cured meats, where a number of ill-understood factors are involved synergistically. Even for the possible double combination of heat plus irradiation, where there exist accepted and similar methods of process calculation for either treatment individually, there is at present no basis in prospect for calculating the total effect of a combination of the two: indeed, it is scarcely understood that there is need for it. In future, to an increasing degree, new processes are likely to be acceptable commercially only when such gaps in knowledge have been made good.

REFERENCES

1. Gillespy, T. G. (1962). *In* "Recent Advances in Food Science" (J. Hawthorn and J. M. Leitch, eds), Vol. 2. Butterworths, London.
2. Waksman, S. A. (1952). "Soil Microbiology". Wiley, New York.
3. Gibbs, B. M. and Freame, B. (1965). *J. appl. Bact.* **28**, 95.
4. Burges, A. (1958). "Micro-organisms in the Soil". Hutchinson, London.
5. Gregory, P. H. (1961). "The Microbiology of the Atmosphere". Interscience Publ., New York.
6. Tanner, F. W. (1944). "Microbiology of Foods" (2nd ed.). Garrard Press, Champaign, Illinois, U.S.A.
7. Frazier, W. C. (1958). "Food Microbiology". McGraw Hill, New York.
8. Haines, R. B. (1937). "Microbiology in the preservation of animal tissues." *Fd Invest. spec. Rep.* No. 45. Dep. Scient. Ind. Res., London.
9. Jensen, L. B. (1954). "The Microbiology of Meats" (3rd ed.). Garrard Press, Champaign, Illinois, U.S.A.
10. Narayan, K. G. (1966). *Acta vet. Acad. Sci. Hung.* **16**, 65.
11. Riemann, H. (1963). *Fd Technol. Champaign* **17**, 39.
12. Bulman, C. and Ayres, J. C. (1952). *Fd Technol. Champaign* **6**, 255.
13. Vas, K. and Proszt, G. (1957). *J. appl. Bact.* **20**, 431.
14. Esty, J. R. and Meyer, K. F. (1922). *J. infect. Dis.* **31**, 650.

15. Spencer, R. (1967). *In* "Botulism 1966" (M. Ingram and T. A. Roberts, eds), p. 123 *Proc. 5th Int. Symp. Fd Microbiol.* Chapman and Hall, London.
16. Xezones, H. and Hutchings, I. J. (1965). *Fd Technol., Champaign* **19**, 1003.
17. Segner, W. P., Schmidt, C. P. and Boltz, J. K. (1966). *Appl. Microbiol.* **14**, 49.
18. Perigo, J. A., Whiting, E. and Bashford, T. E. (1967). *J. Fd Technol., Lond.* **2**, 377.
19. Perigo, J. A. and Roberts, T. A. (1968). *J. Fd Technol., Lond.* **3**, 91.
20. Baird-Parker, A. C. and Freame, B. (1967). *J. appl. Bact.* **30**, 420.
21. Ohye, D. F. and Christian, J. H. B. (1967). *In* "Botulism 1966" (M. Ingram and T. A. Roberts, eds), p. 217. *Proc. 5th Int. Symp. Fd Microbiol.* Chapman and Hall London.
22. Gillespy, T. G. and Thorpe, R. H. (1968). *J. appl. Bact.* **31**, 59.
23. Stumbo, C. R. (1965). "Thermobacteriology in Food Processing". Academic Press, New York.
24. Viljoen, J. A. (1926). *J. infect. Dis.* **39**, 286.
25. Wheaton, E. and Pratt, G. B. (1962). *J. Food Sci.* **27**, 327.
26. Olson, F. C. W. and Stevens, H. P. (1939). *Fd Res.* **4**, 1.
27. Townsend, C. T., Somers, I. I., Lamb, F. C. and Olson, N. A. (1954). "A Laboratory Manual for the Canning Industry". National Canners' Association, Washington, D.C.
28. Schmidt, C. F. (1954). *In* "Antiseptics, Disinfectants, Fungicides and Sterilization" (G. F. Reddish, ed.), p. 724. Lea and Febiger, Philadelphia.
29. Townsend, C. T., Esty, J. R. and Baselt, F. C. (1938). *Fd Res.* **3**, 323.
30. Ball, C. O. and Olson, F. C. W. (1957). "Sterilization in Food Technology". McGraw Hill, New York.
31. Ingram, M., Coleby, B., Thornley, M. J. and Wilson, G. W. (1959). *In Proc. Int. Conf. Preservation of Foods by Ionizing Radiations* (Cambridge, Mass., July 1959). M.I.T. Press, Cambridge, Mass.
32. Roberts, T. A. and Ingram, M. (1966). *J. Fd Technol. Lond.* **1**, 147.
33. Gould, G. W. (1964). *In* "Microbial Inhibitors in Food" (N. Molin and I. Erichsen, eds), p. 17 *Proc. 4th Int. Symp. Fd Microbiol.* Almqvist and Wiksell, Stockholm.
34. Anderson, E. E., Esselen, W. B. and Fellers, C. R. (1949). *Fd Res.* **14**, 499.
35. Perkins, W. E. (1964). *In* "Botulism, Proceedings of a Symposium" (K. H. Lewis and K. Cassel, Jnr., eds), p. 187. Publ. Hlth. Serv. Publ. No. 999-FP-1. U.S. Dept. Health, Education and Welfare, Cincinnati, Ohio.
36. Schmidt, C. F. (1961). *In Report of the European Meeting on the Microbiology of Irradiated Foods, Appendix II.* FAO/IAMS (Paris, April 1960). FAO., Rome.
37. Baumgartner, J. G. and Hersom, A. C. (1956). "Canned Foods, an Introduction to their Microbiology" (4th ed.). J. and A. Churchill, London.
38. Scientific Committee (1955). *In* "Bacteriologie des Semi-conserves de Viande". *Proc. 1st Int. Symp. Fd Microbiol. Annls Inst. Pasteur, Lille* **7**, 254.
39. Sussman, A. S. and Halvorson, H. O. (1966). "Spores: their Dormancy and Germination". Harper and Row, London.
40. Cameron, E. J. and Bohrer, C. W. (1951). *Fd Technol. Champaign* **5**, 340.
41. Riemann, H. (1960). *Nord. VetMed.* **12**, 86.
42. Busta, F. F. and Ordal, Z. J. (1964). *Appl. Microbiol.* **12**, 106.
43. Riemann, H. (1957). *J. appl. Bact.* **20**, 404.
44. Brown, W. L. and Vinton, C. (1964). Pers. comm. to 10th European Conference of Meat Research Workers.
45. Mundt, J. O., Mayhew, S. J. and Stewart, A. (1954). *Fd Technol. Champaign* **8**, 435.

cc

46. Michener, H. D., Thompson, P. A. and Lewis, J. C. (1959). *Appl. Microbiol.* **7**, 166.
47. Jensen, M. (1951). *Konserves* **4**, 45.
48. Handford, P. M. and Gibbs, B. M. (1964). *In* "Microbial Inhibitors in Food" (N. Molin and I. Erichsen, eds), p. 333 *Proc. 4th Int. Symp. Fd Microbiol.* Almqvist and Wiksell, Stockholm.
49. Ingram, M. and Roberts, T. A. (1966). *In* "Food Irradiation", *Proc. Int. Symp. Fd Irrad.* IAEA/FAO (Karlsruhe, June 1966), p. 267. I.A.E.A., Vienna.
50. Krabbenhoft, K. L., Corlett, D. A. and Anderson, A. W. (1967). *In* "Botulism 1966" (M. Ingram and T. A. Roberts, eds), p. 76 *Proc. 5th Int. Symp. Fd Microbiol.* Chapman and Hall, London.
51. Roberts, T. A., Ditchett, P. J. and Ingram, M. (1965). *J. appl. Bact.* **28**, 336.
52. Ingram, M. and Roberts, T. A. (1968). *Annls Inst. Pasteur, Lille* **19** (in press).
53. Denny, C. B., Reed, J. M. and Bohrer, C. W. (1961). *Fd Technol. Champaign* **15**, 338.
54. Hawley, H. B. (1957). *Fd Mf.* **32**, 370.

CHAPTER 16

Resistance of Spores

T. A. ROBERTS AND A. D. HITCHINS*

Meat Research Institute (Agricultural Research Council), Langford, Bristol, England, and Unilever Research Laboratory, Sharnbrook, Bedford, England

I. Introduction

A. SCOPE AND LIMITS OF DISCUSSION

AN attempt has been made in this chapter to summarize the literature contributing to the understanding of the resistance of bacterial spores, and to place before the reader some of the difficulties encountered in resistance studies on spores. Only representative references have been chosen, as the reader will, in reading the references listed, be referred to much of the earlier work.

The consideration of radiation has been limited to ionizing radiations, although significant contributions on the mechanisms of radiation damage have been made using ultraviolet light.[1, 2]

* Present address: Department of Microbiology and Public Health, Michigan State University, East Lansing, Michigan, U.S.A.

Resistance has been arbitrarily defined as the ability of a treated spore to germinate and outgrow to form a vegetative cell capable of dividing and forming a macro-colony. Hence loss of viability is the result of the inactivation of any part of this sequence of events. In fact, after damaging treatments, the ability to reproduce is usually lost first, and in some instances spores which are "non-viable" after heating or irradiating may still be germinated under suitable conditions.

Factors affecting germination are omitted entirely from this chapter and are considered in detail in Chapter 11, but it must be appreciated that if germination is inhibited, outgrowth and subsequent division of vegetative cells cannot occur since these events normally follow germination.

B. TERMINOLOGY, SURVIVOR CURVES AND KINETICS

The resistance of bacterial spores to deleterious agents is commonly determined in the laboratory by counting the number of viable survivors (i.e. colony-forming units) remaining after a series of increasingly severe treatments. In this way a measure of the rate of inactivation is obtained, and rates of inactivation under varying environmental conditions may be compared. Some of the methods commonly used to determine heat resistance are reviewed by Stumbo,[3] to which should be added the use of freeze-drying ampoules or similar containers containing spore suspensions heated totally immersed in a water or oil bath.[4] Radiation resistance may be measured by similar methods, e.g. irradiating a bulk of spore suspension under constant gas conditions and withdrawing samples for counting, or irradiating several discrete identical samples with different doses.

The inactivation of bacteria often follows an exponential order. Consequently if the logarithm of the number of viable units is plotted against the severity of treatment (duration of heating, dose of radiation) a straight line will, in many instances, be obtained. This is commonly referred to as a logarithmic order of death. In such an instance, the slope of the inactivation curve can be estimated from only two viable counts instead of calculated from a series of counts made after different treatments. The limits of error of the calculated line and of the inactivation rate may only be calculated from sufficient determinations to permit statistical analysis. For some purposes, e.g. the screening of substances for their ability to modify the resistance of spores to damaging agents, an approximation of the slope, such as that obtained from only two viable counts, is adequate, and, of course, the saving in experimental time is considerable. Conversely, it is possible to expend too great an effort attempting to determine more precisely the limits of error of measurements in a relatively poorly defined system.

An idealized situation is illustrated in Fig. 1, where three survivor curves are shown, all linear on a plot of logarithm of the percentage survivors against duration of heat treatment. The y axis may be referred to as (i) "log numbers", when actual counts are plotted; (ii) "log percentage survivors", when the initial (untreated) count is taken to be 100% and counts after treatment are taken as a percentage of this figure; or (iii) "surviving fraction", when counts after treatment are expressed as a fraction of the initial number. (i) is less satisfactory than (ii) or (iii) since initially different numbers of spores will cause curves to originate at different points on the y axis, sometimes making visual comparisons confusing, but (ii) and (iii) should only be used when the initial number of spores has no effect on the slope of the survivor curves.

These curves may be described by the general formula:

$$\frac{N}{N_0} = 1 - (1 - e^{-kd})^n \qquad (1)$$

where N_0 = initial cell population

N = number surviving treatment "d" (dose of radiation, time of heating)

n = extrapolation number equal to the intercept on the y axis

k = inactivation constant, or slope

d = dose

A computer least squares technique has been published for the rapid and simultaneous estimation of n and k and their errors.[5] Where $n = 1$, as in Fig. 1, equation (1) becomes

$$\frac{N}{N_0} = e^{-kd} \qquad (2)$$

The term "sigmoid" is sometimes used when $n > 1$; but such instances are probably more accurately considered as special cases of exponential inactivation.[6]

Although the slopes of the curves in Fig. 1 are readily calculated, comparisons are more readily made in terms of "D", the decimal reduction value, mathematically equal to the reciprocal of the slope of the survival curve. "D" is equivalent to the treatment which destroys 90% of the cells, i.e. reduces the viability by one power of ten. The term D has come to be used with both heat and radiation with the same meaning, and has almost completely superseded the confusion of D_{90}, D_{37}, D_{10}, which referred to inactivation to the percentage survival indicated in the subscript. Where a subscript is now used it refers to the temperature of heating, $D_{90°C}$ indicating the decimal reduction value at 90°C; but in order to avoid possible confusion with earlier terminology, it is desirable to specify °C.

If curve 1 in Fig. 1 is taken as the standard survival, curve 2, by comparison, has a more rapid inactivation rate, i.e. a smaller D value. This might have been achieved by heating at a higher temperature, or by heating, or irradiating, in the presence of a chemical capable of rendering spores more sensitive to the treatment. Curve 3 has a slower inactivation rate, i.e. a larger D value, possibly achieved by the inclusion of a protective agent, or by heating at a lower temperature than in 1 or 2.

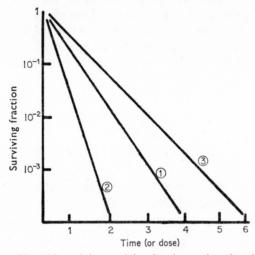

FIG. 1. Plot of logarithm of the surviving fraction against duration of treatment.

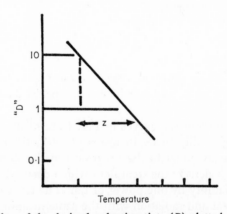

FIG. 2. The logarithm of the decimal reduction time (D) plotted against temperature.

If the curves in Fig. 1 had been achieved by heating at three different temperatures, the effect of the change in temperature on D value is evaluated by plotting the logarithm of the D value against the temperature. The number of degrees change in temperature to achieve a tenfold change in D is referred to as the Z value (Fig. 2). Without recourse to graphical methods, z may be obtained from:

$$Z = \frac{T_1 - T_2}{\log D_2 - \log D_1} \qquad (3)$$

where $D_2 =$ the D value at temperature T_2
$D_1 =$ the D value at temperature T_1

The Z value is used in thermal process calculations, and since these are frequently made by engineers, is traditionally calculated in °F. However, there is no reason why Z should not be specified in °C, although care should be taken to make the units clear. Q_{10} is a quotient indicating how much more rapidly death occurs at temperature T_2 than at temperature T_1, commonly given for a temperature increase of 10°C. (See Ref. 3.)

$$Z = \frac{18}{\log Q_{10}} \qquad (4)$$

The Z value is less dependent on the composition of the heating menstruum and other variables connected with heating than the D value. Z is essentially similar to the energy of activation of a chemical reaction. Attempts have been made to explain both activation and inactivation of spores in terms of thermodynamic functions. However, since each process may involve simultaneously several chemical reactions, and each reaction may have different thermodynamic properties, inferences are inconclusive. For further details the reader is referred to three papers considered most useful in this context.[522, 549, 557]

A plot of log D against temperature is frequently represented by a straight line. However it must be appreciated that Z cannot be constant over a wide range of temperature since upon reducing the temperature D tends to infinity, and correspondingly Z tends to zero. Considerable variation in Z with species is common, and a high heat resistance does not necessarily mean a high Z value.

There are numerous documented instances of survival curves deviating from linearity such as those illustrated in Figs 3–6. Although every effort is made to maintain the purity of bacterial cultures, even in the best run laboratories contamination occasionally occurs. It is most important that this be recognized as such. In Fig. 3, curve 1 is typical of an inactivation study on a mixed spore suspension, when the more sensitive spores are inactivated first, followed by the more resistant part of the population. Where contamination is suspected (Fig. 3, curve 1) and is not immediately revealed by plating the suspension, it is worth while repeating the inactivation at a different temperature, when curve 2 might be obtained. If the suspension is mixed, the intercept of the second component of curves 1 and 2 would occur at the same value of y, and would indicate the number of resistant units present. This would occur if a mixture of vegetative cells and spores, or a mixture of sensitive and resistant spores, were heated. Curves are occasionally obtained when the value of the intercept on y is not constant.[7, 8] Such data could be explained if the fraction of resistant units differed at different temperatures, but this seems a remote possibility. Alternatively, the constituent being inactivated could exist in two inter-convertible forms, such as appears to be the case with certain viruses.[9]

The possibility of highly resistant fractions occurring in a spore popula-tion should also be considered, i.e. perhaps 1 spore in 10^8 possesses a

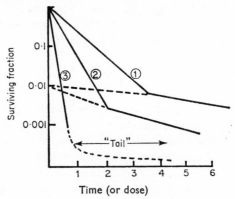

FIG. 3. Plot of logarithm of the surviving fraction against duration of treatment.

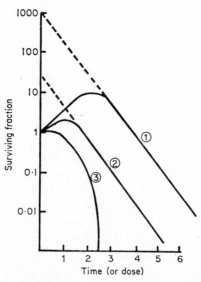

FIG. 4. Plot of logarithm of the surviving fraction against duration of treatment.

resistance appreciably greater than the rest of the population. Claims to this effect have been made for both heat[10] and radiation,[11, 12] but most workers have failed to consolidate their claims experimentally. If several different concentrations of spores are treated in the same way, the resistant

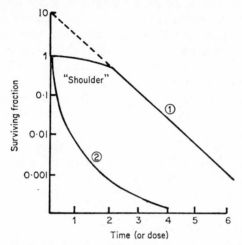

FIG. 5. Plot of logarithm of the surviving fraction against duration of treatment.

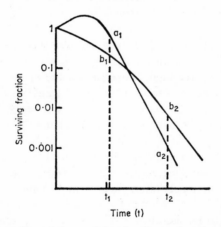

FIG. 6. Plot of logarithm of the surviving fraction against duration of treatment.

"tail" (Fig. 3, curve 3) would occur at the same surviving fraction regardless of the initial numbers. A claim of 1 resistant spore in, perhaps, 10^{10} is extremely difficult to substantiate experimentally, since c. 10^{12} spores would be required to attempt inactivation on different initial numbers and obtain a "tail" in each case. Claims for 1 resistant spore in 10^6 or less[13] should not be made without such substantiation.

Fig. 4 illustrates further curve types. In curves 1 and 2 the inactivation rates are the same, but the degree of activation differs (activation is considered in detail in Chapter 10). A simple statement of D values would be confusing and meaningless, since although the D values are the same, the curves are clearly different. The extent of the treatment before inactivation commences should be clearly indicated, either by stating, together with the D value, a lag-value, i.e. the greatest value of x resulting in no inactivation, or better, by extrapolating the linear portion of the curve to $x = 0$ and recording the value of y at this point (see Fig. 4). In radiation studies the value of y at $x = 0$ is referred to as the "extrapolation number",[14] and equals "n" in equation (1).

An ever-increasing rate of inactivation (i.e. D decreasing steadily with increasing treatment) has occasionally been reported[15, 16, 17] (Fig. 4, curve 3).

In curve 1, Fig. 5, the initial delay in inactivation could be caused by the inactivation being balanced almost exactly by the activation of dormant spores. (Dormancy is considered in Chapter 9.) If a comparison of total spores and the viable count revealed that such a situation could not exist, an alternative explanation must be sought. In radiation processing, curves of this type are not uncommon, and are used to calculate an "extrapolation number", which may indicate either the number of "hits" required for the inactivation, or the number of essential targets each requiring one hit (see III C). Curve 2 (Fig. 5) illustrates an ever-decreasing rate of inactivation.[16, 17]

Should non-linear survival curves be observed experimentally, the simplified approach of making only two viable counts (i.e. an untreated count and one after treatment) to estimate the slope must be avoided, since this fails to take into account possible changes in shape of survival curves. Fig. 6 illustrates an extreme case where the use of either treatment t_1 or t_2 alone would not have revealed the change in curve shape. At t_1 the surviving fraction $a_1 > b_1$; but after t_2, $b_2 > a_2$. Whatever the explanation of non-linear survivor curves, comparisons of inactivation rates become less simple, and workers are sometimes reduced to ignoring the first part of the curve where no inactivation occurs, the "shoulder", and considering only that part of the curve showing a logarithmic death rate.

The discussion so far assumes that resistance is a constitutive property of a spore and that a given resistance will be exhibited whenever standard conditions of sporulation, treatment and recovery are used. This assumption has been shown to be incorrect for heat resistance.[18, 19] After sporulation, and before heating and recovery in a standard manner, the heat resistance of mature isolated spores may be manipulated between relatively large limits. The altered resistance is a property of the spore, not of the reagents, as the spores are heated in an environment free of the reagent used in the treatment. The phenomenon is mentioned briefly at this point since appreciation that the resistance of spores might change during the heating (or irradiation) process should be borne in mind when attempting analyses of non-linear survival curves where the resistance (i.e. the slope of the survivor curve) appears to change during the treatment, e.g. Fig. 4, curve 3, and Fig. 5, curve 2.

Attempts have been made to devise methods for fitting the best line to certain non-linear survival curves[20, 21, 22] using high-speed electronic digital computers, but there is a very real danger of attaching a greater mathematical significance to a curve than the data warrant. A high degree of precision is required to enable a choice of mathematical model fitting such data to be made, and such precision is frequently lacking in bacteriological data. A review of kinetics of inactivation has recently been published,[7] and although it is with particular reference to virus particles, the principles are the same and are concisely presented. In a similar context hypothetical curves for consecutive first-order activation and death (illustrated in Fig. 4, curves 1 and 2) have been computed, but are reported not to agree closely with experimental curves for B. coagulans.[23]

Although much of the preceding discussion of survivor curves applies to chemical resistance studies two considerations specific to the use of chemicals need to be mentioned. First, to study the kinetics of chemical killing the concentration of the chemical must be kept constant. This can be assured by using as high a concentration as is consistent with an easily measurable rate of kill, by keeping the cell concentration as low as possible and by eliminating organic or other reactive materials from the experimental system. Theoretical aspects of the effect of chemical concentration on kill rate can be found in some of the textbooks cited at the beginning of Section II D. The second consideration is that to recover survivors efficiently the effects of chemical carryover into the viable count plates must be counteracted by dilution or by quenching (neutralizing) the test chemical with a suitable and relatively non-toxic chemical. Chemical quenching is essential at very low survivor levels when the viable count dilution is insufficient to eliminate inhibition in the recovery plates. Details of quenching methods will be found in the references given for individual chemicals in Section II D.

II. Factors Affecting Resistance

A. HEAT RESISTANCE

Spores of different species of both bacilli and clostridia exhibit a wide range of heat resistance. Among the more sensitive are B. megaterium ($D_{100°C} = 1$ min) (Ref. 24) and Cl. botulinum type E ($D_{80°C} = 0.6-3.3$ min) (Ref. 4), while among the more resistant are the thermophiles

B. stearothermophilus ($D_{115°C} = 22.6$ min) (Ref. 24) and *Cl. thermosaccharo-lyticum* ($D_{132.2°C} = 4.4$ min) (Ref. 25). Variation in resistance of spores of different strains also occurs[4, 26] even when these spores are obtained from the same batch of sporulation medium.

1. Conditions during sporulation

With any strain, certain conditions during sporulation are capable of altering the resistance of its spores to heat.

(*a*) *Temperature*. Spores of *Bacillus* species produced at higher incubation temperatures were sometimes more resistant than those produced at lower temperatures.[27, 28, 29] Although spores of *Cl. botulinum* obtained at 37°C were more resistant than those from 24°C or 29°C (Ref. 36), they were also more resistant than those obtained at 41°C. A detailed analysis of the effect of sporulation temperatures from 15–41°C on spore composition and heat resistance of *B. cereus*[30] revealed optimal resistance in spores produced at 30°C. It would be interesting to extend these observations over a greater temperature range such as might be achieved by using a non-proteolytic strain of *Cl. botulinum* type B capable of growth from *c.* 3°–45°C (Ref. 31). However, sporulation is often depressed at temperatures too far from the optimum, and the preparation of adequate numbers of spores over such a temperature range might present a difficult problem.

(*b*) *Ionic environment*. Claims that spores occurring naturally are more resistant than those produced in laboratory media[32] tend to be supported by observations that spores were more resistant if aged[33] or produced in the presence of soil[34], possibly implicating mineral salts as important factors. Omission of calcium from the sporulation medium caused a reduction in the heat resistance of *B. megaterium* spores,[35] while omission of iron or calcium reduced the resistance of *Cl. botulinum* spores.[36] Addition of manganese or calcium increased *B. coagulans* var *thermoacidurans* spore resistance, but magnesium was without effect. The most resistant spores were obtained in a medium containing 50 p.p.m. $MnSO_4$ and 45 p.p.m. $CaCl_2$.[37] Other authors claim that magnesium contributes significantly to spore heat resistance.[38] High levels of phosphate (K_2HPO_4) markedly decreased the resistance of *B. coagulans* var *thermoacidurans* spores,[39] although a more detailed study revealed spores from a medium containing 0.05% K_2HPO_4 to be more resistant than those from a medium devoid of phosphate.[37] Spores of *B. megaterium* were also more sensitive when obtained from a medium supplemented with phosphate.[40]

(*c*) *Organic compounds*. p-Aminobenzoic acid increased sporulation of *B. cereus*, but the heat resistance of the spores was not affected by any one growth factor, or by yeast nucleic acid.[41] Different concentrations of cysteine or thioproline ($1–5 \times 10^{-4}$ M) added to prespores of *B. cereus* led to the formation of less heat-resistant spores containing a reduced amount of DPA. These spores were inactivated rapidly from the outset of the treatment, exhibiting no initial "shoulder" on the survivor curve[42] (cf. radiation resistance). The role of sulphydryl compounds in spore resistance is dealt with in detail in Chapters 3 and 7. Supplementation of the sporulation medium with L-glutamic acid or L-proline resulted in a lowered heat resistance.[40]

(d) *Lipids*. The presence of low concentrations of certain saturated or unsaturated fatty acids during sporulation of *Cl. botulinum* increased heat resistance,[36] the longer the carbon chain length the greater the increase. Linoleic acid, however, decreased heat resistance.

(e) *Other factors*. The effects of pH, water activity and oxygen on sporulation are reviewed by Murrell[43] but no study of the effect of their presence during sporulation on resistance of the spores formed has been made.

2. Conditions during treatment

(a) *pH and buffer components*. Spores are more sensitive to heating at extremes of pH than near neutrality,[44] although variation between pH 5 and pH 9 has little effect.[45] Optimal resistance may be exhibited at slightly alkaline (*B. anthracis*, pH 8)[46] or slightly acid pH (*Cl. welchii*, pH 5).[47] A recent detailed study[48] indicated that the D value for *Cl. botulinum* (strain 62A) spores decreased with pH from pH 7 to 4 by a factor of about 5, the size of this factor being largely dependent upon the temperature of heating, and greater at 110°C (230°F) than at higher temperatures. Buffer constituents have also been reported to affect the heat resistance of spores. Phosphate ions reduced the resistance of *B. coagulans*[37] and *B. stearothermophilus*,[49] but increased the resistance of spores of *B. megaterium* previously irradiated with 0·8 Mrad.[50] Different phosphate salts had no influence on the destruction of spores of *B. megaterium* or *B. polymyxa*, stability being greater from 0·005–0·05 M phosphate. Citrate, phthalate or ammonium ions in the buffer usually reduced spore resistance to a level below that demonstrated in phosphate buffer.[51]

(b) *Sodium chloride*. Early claims that sodium chloride protects spores against heat damage[47, 52] were contradicted by the report of a progressive increase in the death rate of spores of *B. coagulans* var *thermoacidurans* with increasing concentrations of sodium chloride in tomato juice.[53] Other workers have been unable to detect any effect of sodium chloride on resistance[8, 54] when used in concentrations of interest to the food industry. Water activity may be adjusted by using salts such as sodium chloride generally at concentrations greater than those of interest to food processors (see (d) p. 622).

(c) *Ionic environment*. Low concentrations of magnesium[36, 50] or calcium[36] have been reported to reduce spore resistance in phosphate buffer, as does the presence of chelating agents such as M/100 ethylenediamine-tetraacetic acid (EDTA) or M/100 glycylglycine.[37] Tris buffer (M/200) had a similar effect.

(*d*) *Water activity*. Equilibration of spores to different water activities[55] showed that the resistance of the relatively sensitive spores of *Cl. botulinum* type E was increased *c.* × 30,000 by equilibration to $a_w = 0.8$. The same treatment increased the resistance of spores of *B. megaterium* and *B. stearothermophilus* by factors of *c.* × 3,000 and × 10 respectively. In dilute buffer spores of *B. stearothermophilus* were *c.* × 30,000 as resistant as those of *Cl. botulinum* type E. It was considered that the maintenance of some part of the spore in a dry state was necessary for a high degree of resistance.

(*e*) *Organic environment*. High carbohydrate concentrations increase heat resistance[53] an effect attributed to a partial dehydration, but equimolal concentrations of different sugars did not show identical protective effects.[36] Although the maximum resistance of aerobic thermophilic spores was increased three- to sevenfold by the presence of up to 70% sucrose, spores of a "putrefactive anaerobe" were unaffected by these conditions.[56] The resistance of spores of another putrefactive anaerobe, *Cl. sporogenes* (PA 3679), was unaffected by 10–50% sucrose, glucose, or glycerol, 2.5% pectin, gum arabic, glycogen, or dextrin.[57] Up to 2,500 p.p.m. L-ascorbic acid was also without effect,[57] although 100 p.p.m. had been previously reported[58] to afford protection to spores of a *Clostridium* sp. Gelatin, casein, globulin, glutelin, or fibrin did not protect at 2.5%; protection afforded by serum albumin, ovalbumin and yeast nucleic acid was concentration dependent. Many compounds, including antibiotics, have been screened for their ability to lower the heat resistance of spores.[59, 60] Chillies and mustard oil proved most effective using *B. subtilis* spores,[61] and garlic oil and allylisothiocyanate among the most effective using *Cl. sporogenes*.[62] In general, activity was frequent among substances known to be mutagenic for higher organisms, and among organic sulphur compounds, but was infrequent among antimicrobial chemicals and antibiotics.[60] Several miscellaneous compounds have been observed to reduce the resistance of spores of *B. coagulans*: extracts of edible green plants[63] synthetic plant auxins,[64] fungicides;[65] but two compounds effective against *B. coagulans*, viz. indole-3-acetic acid and β-naphthoxyacetic acid, were ineffective against two thermophilic flat-sour organisms in cream-style corn.[66]

(*f*) *Other cells*. The resistance of spores was unaffected by the presence of large numbers of inactivated spores[36] or vegetative cells[67] although some protection was afforded by viable cells of other species.[57]

(*g*) *Lipids*. Although it is well known that bacteria are more resistant to heat when suspended in lipid materials than suspended in aqueous systems[68] little systematic work has been done, probably due to experi-

mental difficulties. Soybean oil afforded greater protection to spores of
B. cereus, *B. megaterium* and *B. subtilis* than olive oil, triolein, or liquid
paraffin. Although different degrees of protection were observed in the
different species, the order of protection of the above species can be
gauged from the $D_{95°C}$ in phosphate buffer of 13, 8 and 4·5 min respectively, as compared to $D_{121°C}$ in soybean oil of 30, 108 and 6 min. In
triolein $D_{121°C}$ was 10, 8 and 6 min respectively.[69] Z values in phosphate
buffer were 8·5–9·0°C, but in soybean oil 23–31°C. The calculated $D_{90°C}$
for *B. subtilis* in soybean oil was 2,350 min compared with 55 min in
phosphate buffer. The heat resistance of *B. cereus* spores in soybean oil was
appreciably reduced by 1,000 p.p.m. β-propiolactone, sorbic acid or 2,5,9-
trimethyl-4,8-decadienoic acid. Vitamin K_3 and palmitic acid slightly
increased resistance.[69] It is believed that fats afford protection by localizing
the spore in an environment of low water activity;[68] the spores used in the
experiment reported above[69] had been freeze-dried before re-suspension
in the lipids. If this is the sole effect, spores previously equilibrated to a
certain a_w before re-suspension in lipids would show the resistance typical
of that a_w in an aqueous system. It is also conceivable that the peroxide
content of the lipid might influence resistance, and this might be associated
with the degree of unsaturation of the lipid. No systematic studies have
been made. It is also not apparent whether suspension of dried spores in
lipid causes the lipid to penetrate the spore. It might be argued that this is
more likely to happen in dried spores than in wet spores. Studies under
carefully defined conditions are clearly required.

(*h*) *Gaseous environment.* The effect of the gaseous environment on heat
sterilization has been studied by few workers. Heating in inert gases is
slightly more effective than in air or oxygen[70] and nitrogen is as effective
as air[71] (cf. radiation).

(*i*) *Other factors.* Spores of *B. subtilis* occluded in crystals of calcium
carbonate and then exposed to moist and dry heat were approximately
900 and 9 times as resistant at 121°C as unoccluded spores.[72]

(*j*) *Sensitizers.* Attempts to discover substances capable of reducing the
heat resistance of bacterial spores ("sensitizers") have generally been unsuccessful (but see 2 (*e*), p. 622). Initial observations with the antibiotic
subtilin appear to indicate that it was capable of reducing spore resistance,
but a more thorough screening of some 600 substances[60] revealed that its
effect was not to alter spore resistance, but to prevent outgrowth and cell
division after germination had occurred.[73]

3. *Recovery conditions*

(*a*) *Growth inhibitors.* Spores surviving heating are sometimes more
exacting in their growth requirements than are unheated spores.[74, 75]
Starch increased recovery of heated spores of *Cl. botulinum*[76] and *B. coag-*

ulans,[82] possibly by adsorbing long-chain fatty acids in the medium capable of inhibiting spore germination,[77] but not all workers have been able to detect such an increase in recovery.[24, 37] Charcoal and serum albumin also overcame the effect on natural inhibitors in media used for spore recovery.[78, 79] Furfural (1 p.p.m.) markedly increased the counts of thermophiles[80] and a source of glutamic acid has been shown to be important for the maximum recovery of severely heated spores of *B. subtilis*.[81] Heated spores of *Cl. sporogenes* were more sensitive to increasing concentrations of sodium chloride than unheated spores.[8]

(b) *pH value*. Spores of *B. stearothermophilus* are particularly susceptible to variation of pH value of the recovery medium.[29, 83]

(c) *Temperature*. The optimal growth temperature for heat-damaged spores is not necessarily the same as that for unheated spores. Certain strains of *Cl. botulinum*[36, 84] and of *B. subtilis*[85] gave greater recovery at incubation temperatures below the normal optimum for growth, but unpublished data indicate that it is not a general phenomenon, and further study is clearly required.

(d) *Recovery or "repair"*. Similarly, the effect of conditions (e.g. time, temperature, pH) between the cessation of heating and recovery has not been evaluated. Repair of thermal injury of *Staph. aureus* has been demonstrated[86] and it is possible that similar repair occurs in heat damaged spores under suitable conditions.

(e) *Germinants*. Recovery of suspensions containing a high percentage of dormant spores presents an awkward problem, since, ideally, incubation should be continued to permit the maximum number of survivors to grow. The time for germination of individual spores is approximately lognormally distributed[87, 88] and, therefore, most survivors will grow out within a relatively short incubation period, whereas the dormant remainder could require very prolonged incubation. Hence incubation of most probable number (MPN) counts is frequently continued for several months, and in studies of inhibitors likely to be used in foods, incubation for one year is not uncommon. In making plate, or deep agar, counts, such prolonged incubation is impracticable, and the number of colonies usually increases only very slightly after incubation for a few days. The importance of such small increases in numbers is minimized by constructing survival curves of several log cycles of destruction, and by calculating counts using two, or more, dilution levels.[89]

Assuming that a spore suspension containing predominantly dormant spores is under study, few systematic attempts have been made to induce germination and outgrowth within a relatively short time, although detailed studies have shown that heated spores of *B. megaterium* germinated in a greater range of organic nitrogen compounds than unheated spores,[90] while heated spores of *B. stearothermophilus* also have altered requirements for germination.[91] The inclusion in the recovery medium of germination stimulants should minimize problems of dormancy in simple spore inactivation studies. Ca–DPA[92] has been used in this way, but has proved difficult to use owing to its tendency to crystallize in the medium. In the case of spores of *Bacillus* species, chemical germinants such as L-alanine are effective against a large number of species, and the answer appears relatively simple particularly since heated spores of *Bacillus* species have more simple germination requirements than unheated.[90] The effects of different levels of damage on germination requirements have not been studied.

Germination requirements of clostridial spores are less well understood. Investigations on a relatively small number of species suggest that they are more complex than those of spores of bacilli;[93] hence a system applicable to the majority of clostridial species seems, at the moment, improbable.

(*f*) *Medium.* A comprehensive study of the recovery of heated *Cl. sporogenes* PA 3679 spores in seven media[94] revealed that the choice of recovery medium could greatly affect the apparent D value: the maximum and minimum $D_{121°C}$ values for wet heat were 1·4 and 0·48 min and $D_{148·9°C}$ values for dry heat 12·0 and 6·4 min; Z values were much less affected, but some variation with recovery media occurred. Changes in germination requirements produced by heating might, therefore, differ for wet and dry heat. In the same context, recovery medium supplemented with $CaCl_2$ and sodium dipicolinate gave higher recovery of *B. subtilis* spores suspended in skim-milk and heated in ultra-high temperature processing equipment, than an unsupplemented medium, resulting in different survivor curves;[95] Z values were also different in the two recovery media, and approximately doubled above 116–121°C in the basal medium from 8·9 to 18°C, and above 124–129°C in the supplemented medium from 6·7 to 13°C. The upward concavity of survivor curves (e.g. Fig. 5, curve 2) plotted from data using the basal medium was markedly reduced by the presence of Ca–DPA, thus confounding one possible explanation of such curves, viz. that resistance increases during heating.[19]

4. *Actinomycete spores*

The heat-resistance properties of actinomycete spores have been largely neglected by bacteriologists probably because these organisms were earlier thought to be more closely related to the fungi and to have properties similar to those of fungal conidia. Spores of most mesophilic species are killed after a few minutes exposure to temperatures between 65–70°C. There are, however, some reports of moderate heat resistance in mesophilic species belonging to the genera *Streptomyces*.[96] *Nocardia*[97, 98, 99] and *Micromonospora*.[100]

On the other hand, there is good evidence for the property of heat resistance in the spores of certain thermophilic actinomycetes. In 1899 Tsiklinsky[101] briefly described the monosporic species *Thermoactinomyces vulgaris* and showed that the spores were able to survive 20 min at 100°C. This property was later confirmed by Schutze[102] and Erikson.[103, 104] Cross[105] has recently examined the spores of this organism and of *Actinobifida dichotomica* in more detail and found $D_{100°C}$ values of 11 min and 71 min respectively when suspended in $M/1000$ pH 7·0 phosphate buffer in sealed tubes immersed in a mineral oil bath. The spores were also shown to contain dipicolinic acid (*T. vulgaris* 6·0%; *A. dichotomica* 3·58%, using the method of Lewis[106]) and to exhibit many of the properties of endospores, e.g. refractility, impermeability to simple stains, activation by heat, resistance to desiccation.[107]

It seems possible that similar properties may also be common to the spores of other thermophilic actinomycete species. Fergus[108] recently determined the thermal death times of spores of a range of species when suspended in 1% sucrose and found that those of *Thermomonospora curvata*, *Thermoactinomyces vulgaris*, *S. thermoviolaceus* var *pingens* and *S. rectus* were able to develop into colonies when subcultured after 2 and in some cases 4 hours exposure to 100°C.

Very little is known about the fine structure of such spores and the stages leading to spore formation, and it will be interesting to see how they compare with bacterial endospores.

B. RADIATION RESISTANCE

The sensitivity of different species of bacteria to ionizing radiations differs widely[109, 110] and the resistance of vegetative cells varies during the growth cycle.[111] In general, spores are more resistant than vegetative cells[112, 113, 114] although the most resistant bacterium known, *Micrococcus radiodurans*,[115] is not a sporeformer.

1. Conditions during sporulation

In contrast to the voluminous data on the effect of sporulation conditions on heat resistance, there are few data on their effect on radiation resistance.

(*a*) *Ionic environment.* Variation of divalent cations in the sporulation medium did not influence the resistance of spores of *B. megaterium*.[38] In the case of non-spore-forming bacteria, the radiation sensitivity of *Aerobacter aerogenes* was not affected by growing it in media deficient in magnesium, but a *Pseudomonas* sp. was slightly more sensitive.[116]

(*b*) *Organic environment.* Radiation resistance and cystine-rich structures develop simultaneously during sporulation,[117] (see Chapter 3) before the appearance of refractile spores, and finally the development of thermo-stability. The addition of cysteine or thioproline to sporing cultures of *B. cereus* slightly increased spore radiation resistance.[42] Variations in the cysteine and glucose concentrations in the sporulation medium, shown to affect the response of *Cl. botulinum* type B spores to heat,[118] had no effect on radiation resistance (Roberts, T. A., unpublished).

The pronounced effect of growth conditions on the resistance of vegetative cells will not be considered here[82, 119, 120, 121] since they have no direct bearing on spore studies.

2. Conditions during irradiation

Many conditions which have been shown to affect the radiation resistance of vegetative cells do not affect the resistance of spores. Such conditions will be mentioned briefly for the sake of completeness.

(a) *Cell concentration.* An apparent increase in radiation resistance at very high concentrations of vegetative cells[122] was shown to be due to oxygen depletion,[123] cells being more resistant to radiation in the absence of oxygen than in its presence (see (d) below). An effect of inoculum size was detected with spores of *Cl. botulinum* in ground beef, most marked at irradiation temperatures from $-196°$ to $0°C$, and gradually decreasing with increasing temperature until it was not detectable at $85-95°C$.[124]

(b) *pH value.* Differences of pH value during irradiation had little effect on the survival of spores of *B. subtilis*,[125, 126, 127] but *Cl. sporogenes* was most resistant at pH 7.[127] *B. thermoacidurans* spores were also most resistant at pH 7[128] and were equally sensitive at pH 2·2 and pH 10. This study employed isoionic buffers to obviate possible differences due to different ionic concentrations. Within the pH range 5–8, radiation resistance of spores of *B. cereus, B. coagulans* and *B. pumilis* was not affected, but below pH 5, resistance was reduced.[129]

(c) *Temperature.* Most early studies of the effect of temperature of irradiation on spore resistance were made with a view to irradiating foods at low temperatures where radiation-produced organoleptic changes are minimal. Some workers reported spores to be more resistant unfrozen than frozen,[125, 130] but not all spores responded similarly.[127] The overall difference in resistance was usually small.[131] *Cl. botulinum* spores are more resistant in phosphate buffer or ground beef at $-196°C$ ($D = 0·46$ and 0·68 Mrad) than at $0°C$ ($D = 0·29$ and 0·40 Mrad).[132] The transition of water from the liquid to the solid state markedly increased spore resistance,[133] further reduction in temperature resulting in a small but steady increase in resistance of spores in phosphate buffer, which was absent if spores were placed in pork pea broth.[133] Spores of *Cl. botulinum* in beef slices gave the same D value at $20°C$ and $5°C$, which then progressively increased with decreasing temperature from 0·286 Mrad at $5°C$ to 0·4

Mrad at −196°C (Ref. 134). All survivor curves could be described by equation (1), p. 613, n remaining constant at about 15.

Studies on dry *B. megaterium* spores in a precisely defined system[135] revealed a temperature independence in either nitrogen or helium from c. −268°C to −148°C, followed by a steady decrease in resistance with temperature from −148°C to 37°C. Above room temperature and in the absence of oxygen, there was a sharp increase in radiation resistance reaching a maximum at approximately 80°C, beyond which temperature it again decreased. In the presence of oxygen, temperature independence was again observed below −145°C, and above this temperature there was a similar dependence to that in nitrogen, excepting that the spores were more sensitive than in nitrogen (see later) and the resistance showed no sharp increase at higher temperatures. Spores of *Cl. botulinum* also exhibit an increased resistance at high temperatures[136] similar to that described above for *B. megaterium*.

(d) *Oxygen*. (see also (e) and (g) below). Most studies of the effect of the gaseous environment have been made with vegetative cells[82, 119, 120, 137–140] and indicate clearly that very low concentrations of oxygen are sufficient to confer most of the oxygen-dependent sensitivity.[119, 141] Although an oxygen effect was observed with *E. coli*, *B. subtilis* and *Cl. sporogenes*, no such effect was detectable with spores of *B. thermoacidurans*.[127, 128] However, all these studies were imprecise by comparison with those on spores initiated by Powers and discussed in more detail in Sections (c), (e), (f), (g) and (h). There is no doubt that an oxygen effect can be demonstrated with bacterial spores,[142] but not all radiation damage is oxygen dependent (see later and Ref. 135).

(e) *Nitric oxide*. Nitric oxide sometimes appears to substitute for oxygen in sensitizing cells to radiation,[143] but there are certain differences. Whereas pretreatment with oxygen caused no effect, pretreatment with nitric oxide causes a time-dependent latent irreversible change.[144] Nitric oxide can also protect against irradiation injury in both vegetative cells[144] and spores.[135, 145] Furthermore, nitric oxide may increase recovery of spores when introduced after irradiation by causing annealment of latent damage which may be revealed by the introduction of oxygen.[135] Post-irradiation heating also anneals this type of spore damage, but the effects of heat and nitric oxide are not additive.[135] Three categories of radiation damage have been demonstrated for "dry" spores:[135]

Class I, oxygen-independent and temperature-dependent;
Class II, oxygen-dependent and temperature-dependent;
Class III, oxygen-dependent and involving long-lived free radicals.

Detailed investigation of the inversion of radiation resistance at higher temperatures mentioned above revealed that it is due entirely to a post-irradiation annealment of radiation damage.[146] Provided that oxygen is excluded from dry systems the D value can be approximately doubled by exposing spores to high temperature after cessation of irradiation. This "thermorestoration" is temperature-dependent, and fails to reach the possible maximum if the delay between irradiating and heating exceeds 2 hr.

(f) *Water content.* Using *B. megaterium* spores equilibrated to different water contents, the contribution of water to radiation damage has been evaluated.[147] The probability of damage from oyxgen-independent causes (Class I damage)[135] is unchanged from about 10^{-5} torr (mm Hg) to 10^{-1} torr, but may rise as saturation (22 torr at 25°) is approached. The probability of damage from radiation-induced radicals which must combine with oxygen after formation (Class III) decreases from the very dry (10^{-5} torr) spore to 1 torr, remains constant to about 10 torr, and then decreases to zero at 22 torr. Thus water protects spores against these (Class III) radicals. The probability of damage observed only when oxygen is present during irradiation (Class II) decreases from a high value at 10^{-5} torr to 1 torr, remains constant to about 10 torr, and then increases as saturation is approached. Thus water has an overall protecting effect, and at the same time can be sensitizing in two ways.

(g) *Hydrogen sulphide.* The initial observations that hydrogen sulphide protected dried spores of *B. megaterium* when present during irradiation, or added afterwards,[148] have been extended to spores in buffered suspension.[149] Class I and Class III damage are reduced when hydrogen sulphide is present.[150] Investigations of the radioprotective effect of hydrogen sulphide on lysozyme and vegetative bacterial cells[151] indicate a possible correlation with the protection afforded to vegetative cells by L-cysteine.[152]

(h) *Glycerol.* A detailed examination of the effect of glycerol on radiation damage in dry spores of *B. megaterium*[152] showed that it protects by virtue of its ability to prevent Class III damage, and by acting also on Class I damage. The presence of glycerol, rather than the absence of water, was required. Although hydrogen sulphide acts on the same classes of damage, there is no evidence of equivalence of these two protective effects. Ethylene glycol, ethyl alcohol and glycerol protect vegetative cells equally well at equimolar concentrations in the presence of oxygen.[153]

(i) *Other protective agents.* A clear distinction should be made between compounds exerting their protective effect only when added before or during irradiation, for which the term "protective agent" should be reserved, and those which assist a cell to recover after irradiation.

Four groups of compounds have been shown to protect vegetative bacteria against radiation damage; sulphydryl compounds, reducing compounds, alcohols and glycols, and metabolites such as pyruvate[154] and citrate.[155] Protection increases with increasing concentration up to a certain plateau level.

Preincubation is required for metabolites[156] and alcohols in low concentrations to exert their effects, but not for sulphydryl or reducing compounds. The protective capacity of inorganic sulphur compounds was attributed to their ability to remove molecular oxygen from solution.[157] Sulphydryl compounds present a complex picture, some protecting and others sensitizing. Those effective in higher organisms are often without effect in bacteria. Cysteine,[158] glutathione, cystamine and cysteamine[159] are all protective, the maximum activity being associated with a free —SH group, a 2- to 3-carbon chain, and a strong basic function.[160, 161, 162] The protective activity of sulphydryl compounds has been ascribed to oxygen depletion within the cell[119] though not all work fully substantiates this view.[158] Alternative explanations are competition with oxygen at a sensitive reaction site,[137] acting as a free radical acceptor[161, 163] or the masking of —S—S— or —SH groups of cell proteins.[164] Each theory may be criticized. More recently binding to and stabilizing those parts of DNA not covered by histones has been suggested.[165] The protective effect of β-mercaptoethylamine,[166, 167] has been ascribed to a combination of inhibition of protein synthesis coupled with continuing DNA synthesis and an added effect during irradiation.[168] Ethanol[169] and glycerol,[153, 169] (see (h), p. 269) also protect against ionizing radiations.

Fewer observations have been made on bacterial spores than on vegetative cells, and it is not possible to state categorically that all the above substances protect both aerobic and anaerobic spores. Spores of *B. subtilis* freeze-dried from glucose, lactose or fructose were more resistant to gamma radiation than when freeze-dried from water.[170] This protection was closely linked with the formation of a "glass", since maltose did not form a glass and did not protect. The mechanism is unexplained, and may be simply a combination of water content and anoxia, or a stabilization of free radicals,[171] or a particularly radiation resistant molecular aggregation.[172]

(j) *Sensitizers.* Radiotherapy and the possible use of ionizing radiation in food processing have prompted numerous searches for compounds which might enable the effective dose of radiation to be lowered. Most observations have been made on vegetative cells.

Like protective agents, those compounds active in higher animals are not necessarily so in bacteria. β-homocysteine sensitized rats and mice[173] but protected *E. coli* B/r [174] in a system where oxygen depletion could occur. N-Ethylmaleimide (NEM) sensitized *E. coli* B/r in air and in nitrogen.[174] Subsequently NEM, iodoacetic acid (IAA) and phenylmercuric acetate (PMA) were all observed to sensitize a pseudomonad to X-irradiation.[175] Iodoacetamide sensitized the resistant bacteria *E. coli* B/r,

M. radiodurans and *M. sodonensis* to a greater degree than the relatively sensitive *E. coli* and *Ps. fluorescens*.[176]

Iodoacetamide added to *M. radiodurans* 3×10^{-3} sec before irradiation commenced conferred full sensitization. Added 3×10^{-3} sec after the cessation of irradiation, no sensitization was demonstrable.[177] p-Hydroxymercuribenzoate sensitized *M. radiodurans*, *Sarcina lutea* and *E. coli* B/r.[178]

Vitamin K_5 has also been reported to sensitize *E. coli*, *M. radiodurans*, *Ps. fragi* and *Torulopsis rosae*,[179] the bacteria only in a nitrogen atmosphere, and the yeast only in air. Analogues of vitamin K_5 also sensitized *E. coli* and *Strep. faecalis*, the former in the absence of oxygen, the latter in the presence or absence.[180] Vitamin K_5 sensitized *Salmonella typhimurium* under vacuum, and protected under air,[181] the sensitizing effect decreasing with increasing temperature from 32°F to 120°F.

X-ray sensitivity of *E. coli* was enhanced when purine or pyrimidine analogues were incorporated into the DNA.[182] Thioguanine (TG) was most effective, followed by 5-bromouracil deoxyriboside (BUDR), 5-bromouracil (5-BU), 5-iodouracil (5-IU) and 2,6-diaminopurine (2,6 DAP). Two other analogues, 5-fluourouracil (5-FU) and 2-aminopurine (2-AP), failed to enter the bacterial cell and were without effect. 5-BU and 5-IU exerted the same sensitizing effect on *E. coli* whether radiation was carried out in oxygen or in nitrogen, but sensitization by TG, 6-mercaptopurine (6-MP), and 2-AP completely disappeared upon anoxic irradiation.[183]

Substituted nitroxides, similar to nitric oxide (see (e), p. 628) in possessing an unpaired electron associated with the N—O bond, sensitized *E. coli* in nitrogen,[184] but the maximum sensitivity demonstrated was no greater than the sensitivity in oxygen. Folic acid antagonists,[185] 2-fluoroadenosine[186] and certain tetracyclines[187] sensitize *E. coli* to ionizing radiation.

The combined effects of radiation and halogenophenols deserve wider attention. In studies on *E. coli* and *Zygosaccharomyces soya* the effectiveness of the substituted halogen diminished from iodine through bromine to chlorine, and increasing the number of halogen atoms increased the effect.[188] The effectiveness of monosubstitution in the different positions was also investigated, there being a tendency for *p*- to be more effective than *m*- or *o*-.

The mechanism of sensitization is obscure, and generalizations are best avoided since most of the above compounds are ineffective, or have not been tested, on spores. One instance where spores had apparently been sensitized to radiation[189] was later shown to be due to irradiated spores possessing a reduced tolerance to sodium chloride.[190]

Early studies suggested that sulphydryl groups were in some way involved in radiation resistance,[175, 176] since many of the compounds known at that time to reduce radiation resistance would combine with free —SH groups. Detailed studies by Vinter (see Chapter 3) confirmed that spore-radiation resistance increased concurrently with the synthesis of cystine-rich materials.[191, 192] Spore resistance also correlated well with an increased ratio of disulphide bonds to thiol groups in *B. cereus*.[193] However, when spore disulphide bonds were ruptured by treatment with thioglycollic acid, there was no loss of viability or change in resistance to radiation (or to heat), although the resistance to lysozyme was decreased,[194] casting doubt upon the implication of thiol groups in radiation resistance. In general, attempts to sensitize spores using chemicals previously shown to sensitize vegetative bacteria (NEM, IAA, PCMB) have failed, even using concentrations ten or a hundred times higher than those effective in vegetative cells.

3. Recovery conditions

(a) *Incubation temperature.* Increased recovery of *E. coli* after X-irradiation was demonstrated at suboptimal incubation temperatures,[195] but this could not be confirmed for spores.[196] Sufficiently detailed studies have yet to be made with spores, but a reduced incubation temperature resulted in lower recovery of irradiated spores of *Cl. botulinum* type E, C, F, (Schmidt, personal communication).

(b) *pH value.* Irradiated *E. coli* B survived better on acid than on alkaline medium. Irradiated spores of *B. cereus*, *B. coagulans* and *B. pumilus* were more sensitive to extremes of pH than unirradiated spores.[129]

(c) *Growth inhibitors.* Irradiated spores are appreciably more sensitive to inhibition by sodium chloride than are unirradiated spores.[190]

(d) *Recovery, or "repair".* The definition of recovery becomes particularly critical in this context, since the number of spores recovered from a particular system may be modified by nitric oxide (see 2(e) above), by post-irradiation heating[135] or by hydrogen sulphide (2(g) p. 629). These interactions are admirably reviewed by Powers and Tallentire.[197] Excluding such treatment, it must be considered whether spores are themselves capable of "repairing" radiation damage. The answer appears to be negative since Powers and Tallentire were able to account fully for all radiation damage in terms of oxygen, water and temperature (see Section II B, 2 (c), (d), (e), (f)) thereby taking into account any normal laboratory conditions after irradiation. Whereas spore radiation damage can be repaired there is no suggestion that any part of this repair is enzymic.

(e) *Germinants.* The possible effect of germinants incorporated into recovery media has not been studied. Even the example of calcium dipicolinate[92] has not been confirmed with irradiated spores (see Section A, 3(e)).

(f) *Medium.* The recovery medium appears to play only a minor part in recovery of spores after irradiation. No effects were detectable with spores of *B. subtilis*.[198]

C. OTHER PHYSICAL AGENTS

1. Mechanical stress

Although spores are more mechanically resistant than vegetative cells,[199] with the possible exception of some cocci, both kinds of cells can be disrupted using the same mechanical devices. These devices have been described,[200-206] reviewed[207, 208] and some comparisons of their spore breakage ability made.[209, 210, 211] The most generally used methods of disruption involve rapid agitation of aqueous spore suspensions with glass beads or abrasive particles. Grinding with abrasives,[212] treatment of frozen suspensions in the Hughes' press,[213] blending with glass beads[214] or abrasives,[211] and shearing[211] can disrupt spores.

Disruption of spores initially causes loss of refractility[199] and then produces fragments of coats and other insoluble components. Spore dipicolinic acid (DPA) is solubilized.[249] There is a logarithmic decrease in viability during disruption.[199] Factors influencing the disruption rate are the type, amount and particle size of the abrasive, agitation speed and spore concentration.[199] The mode of disruption varies with the species.[215] Mild shearing removes exosporia, leaving the spores intact and refractile but depleted of DPA.[211] No correlation exists between the mechanical resistance and Ziehl–Nielson stainability of spores,[216] (see Section II D, 7).

For studies of spore structure, complete breakage can be prevented by the presence of fixatives,[217] and autolysis of the cortex which generally occurs during disruption can be inhibited.[218] Like vegetative cells, spores are less readily disrupted at low pH values. Below pH 4·2 the intact refractile centres of spores (probably core plus plasma and cortical membranes) are released during disruption.[219] The spore centres obtained are non-viable. Spore centres stain with fluorochrome-labelled lysozyme which specifically indicates the presence of murein.[220] Breakage of spores at low pH was used in locating an enzyme[221] and an antigen.[222]

Foaming during disruption can be controlled with antifoam[223] or by breaking spores dry using the techniques of grinding with alumina[212] or shaking with NaCl crystals.[224, 225] The kinetics of dry breakage are first order[226] and release of soluble protein is faster and more complete with dry breakage than with wet breakage. Dry breakage has been used to prepare glucose dehydrogenase[224] and a membranous structure from spores.[225]

Gentle mechanical abrasion results in changes typical of spore germination ("mechanical germination", Chapter 11) with no loss of viability.[227] Similar changes can be induced by pressing spores between a cover glass and a microscope slide[521] or by rubbing spores with a wire loop.[228] The rupture rate of dry spores has been considered as being a measure of the mechanical germination rate.[226] Spores seem to be plastic since they regain their shape after crushing (diameter reduced by 86%) by a pressure of

3 tons per in² (Ref. 229). Crushing is not necessarily lethal even when the spores are cracked.

2. Ultrasonic and sonic waves

Water-borne ultrasonic and sonic waves cause cavitation, and the rapid large pressure changes involved are responsible for cell disruption. Generators of such waves have been described.[207, 230, 231] Spores are so resistant that these waves can be used to free spore suspensions of vegetative cells.[232] Glass beads are sometimes added to increase the rate of spore disruption.[211] Ultrasonic and mechanical disruption have been compared.[209] Treatments of 400 Kc per sec[233] and 10 Kc per sec and 75 acoustical watts for 150 min[121] can disrupt spores, but *Bacillus thuringiensis* spores are broken only slightly by 26 Kc per sec and 500 watts input for 20 min.[230] The latter treatment broke *B. thuringiensis* vegetative cells in 5 min.

Ultrasonication produces refractile spores that are reduced in size by surface erosion and also produces non-refractile spores swollen to twice their original size, and causes shattering of spores and non-refractile spores.[234] Exosporia are destroyed faster than the spores, and at an exponential rate, but have a protective effect on the spore until they are stripped off.[121] Ultrasonication does not effect the subsequent initiation of germination but stimulates subsequent growth.[228] High-intensity airborne sonic and ultrasonic waves are sporicidal[235] being synergistic[236] with ethylene oxide (see Section II D, 5(*a*)).

3. Hydrostatic pressure

Hydrostatic pressure does not deform a cell since pressure is transmitted throughout a suspension. In contrast, mechanical pressure deforms a cell due to forces unequally applied to the different parts of its surface. The effects of hydrostatic pressures on microbial systems have been reviewed.[237] The use of pressure to cause shearing breakage of bacteria[204, 205] and spores[206, 211] has already been mentioned. Bacteria can be killed by high hydrostatic pressures, spores being more resistant than vegetative bacteria. Growth and cell division are inhibited by 1,400 to 7,000 p.s.i.,[238] while the death rate of vegetative cells is accelerated by 5,600 to 14,000 p.s.i.;[238] cells are killed by 88,000 p.s.i. for 14 hr.[239] In contrast spores have been reported as being killed by 176,000 p.s.i. for 14 hr but not by 88,000 p.s.i. for 14 hr,[239] 150,000 p.s.i. for 0·5 hr[240] or 250,000 p.s.i. for 0·75 hr.[241] Lower pressure (8,400 p.s.i.) is reported to have accelerated the rate of death of a strain of spores which lost viability easily at 25°C.[242] The reason for the instability of these spores was not clear.

Pressure resistance of spores is optimal near pH 8. Damaging effects at lower and higher pH values are inhibited by salt and glucose.[240] Pressure is believed to kill by causing an increase in ionization, with the result that proteins precipitate as complexes due to interactions between the additional charged groups.[240] Pressure is known to denature proteins,[242-244] but DNA is relatively pressure-resistant.[245] Relatively low pressures inhibit heat denaturation of enzymes and heat killing of cells and spores[242] and also inhibit the stimulating effect of urethane upon heat killing.[246]

4. Freezing and freeze-drying

(a) Freezing. Spores are more resistant than vegetative cells to freezing.[247] Rapid freezing has little effect on spore viability though slow freezing may kill a small proportion of suspensions of some spores.[248] Eight cycles of freezing to minus 80°C and thawing did not kill B. subtilis spores or cause release of DPA,[249] but up to 100 cycles to minus 10°C killed 50%.[228] Survivors of the latter treatment germinated less completely and the heat activateable portion of the population seemed to be more resistant.[228] Rapid freezing of Cl. botulinum spores to −196°C caused retardation of the early stages of growth during recovery.[132] It was not clear whether this was an effect on germination or stages of outgrowth. It was concluded that the effect is due to violent physical changes occurring during freezing since holding time in the frozen state was not important, but the effect of thawing was not evaluated. The effect of freezing on radiation resistance has been discussed in Section II B, 2c.

(b) Freeze-drying. This subject has been recently reviewed.[250, 251] Cowan (cited in Ref. 250) suggested that immature spores occurring during sporulation are probably less resistant to freeze-drying than mature spores. Freeze-drying has no effect on the resistance of spores to heat or electrodialysis.[252] Freeze-dried spores are relatively impervious to nitrogen gas.[253, 254]

5. Vacuum

Spores are very resistant to storage at ambient temperatures in ultra high vacuum of 10^{-5} to 10^{-10} torr,[255, 256, 257] though there has been one report of spores being killed.[258] At higher temperatures spores survive less well in ultra-high vacuum than in air.[259] Prolonged storage in a low vacuum (2×10^{-2} torr) increases the heat resistance of spores and vegetative cells.[260] Presumably the main effect of ultra high vacuum is desiccation. Drying in ultra-high vacuum increases the sensitivity of spores to irradiation.[261]

6. Storage

There are reports of spores surviving from 37 to 118 years in a variety of environments including alcohol and canned meat.[262]

(a) Storage in water. Although spores of some anaerobes can survive for many years in phosphate buffers at 0 to 37°C,[263] others, such as *Cl. botulinum* type E, germinate very readily, even at 5°C. Viability loss is a function of pH and temperature, but other factors are certainly involved to account for variation between crops of anaerobic spores of the same strain stored under the same conditions. Prestorage heat shock decreases the storage resistance of aerobic spores. Activation or deactivation may occur during storage depending on the temperature (see Chapter 10). Germination requirements may alter during ageing of spores.[43]

(b) *Storage of dried spores.* The major factor affecting the viability of freeze-dried spores during storage is the relative humidity.[264] Viability is little affected by storage for 6 years in air or vacuum at water activity (a_w) values between 0·2 and 0·8 (i.e. 20–80% equilibrium relative humidity), but loss of viability of some spores occurs at a_w 0·00.[265] After storage for short periods at high a_w values spores of some species may germinate upon rehydration[264, 265, 266] or be activated.[267] Spores are also activated by 60% ethyl alcohol, but absolute alcohol reversibly deactivates the spores.[268] Spores freeze-dried in crushed lava show a greater decrease of viability in air or *in vacuo* than in nitrogen;[269] spores can survive simulated Martian conditions.[270]

7. Aerosolization

The effect of aerosolization on *Bacillus globigii* spores has been studied in connection with its use as a tracer organism in aerosol experiments. Its orange colonies make this sporeformer easily recognizable on recovery plates. Although spores might be expected to be stable during aerosolization, germination of spores during such experiments lessens their tracer value.[271] Germination has been reported to be due to the stress of dissemination and collection in liquid impingers,[272] but this was not the cause in another case.[273] Like vegetative cells, spores lose viability when dispersed at high RH values[274] and also become heat-sensitive (cited in Ref. 274).

8. Electrical treatments

Electrodialysis of *B. megaterium* spores for 7 hr results in a DPA loss of 64% (Ref. 249). Changes of pH and temperature occurred in this work but were not regarded as sufficient to cause the DPA loss. However, other workers using the same and other species of spores did not detect DPA release, loss of viability or loss of resistance to heat and staining.[252] Differences in apparatus design might account for the discrepancy.[252]

Spores are insensitive to pulsed direct current electric shock (30 Kv per cm) but become progressively sensitive during germination and outgrowth.[275] The corresponding vegetative cells can be killed by shocks of 10 to 15 Kv per cm. Killing was associated with leakage of cell contents due to membrane damage. *E. coli*, spores and yeast, in order of increasing resistance, were killed by high-voltage arc discharge.[276] Survivors decreased logarithmically with increase in input energy. It was suggested that shock waves forced chemically active radicals and ions through the cell wall and that these killed the cells.

D. RESISTANCE TO CHEMICALS

Several textbooks and reviews of the effect of chemicals on bacteria have proved invaluable in preparing this section.[277-288] Special mention must be made of the textbooks of Sykes[279] and Rubbo and Gardner[280] and the reviews by Russell[281] and Hugo,[287] all published since 1965. A review on the effects of chemicals on proteins[289] was also invaluable.

1. Acids

Spores are acid resistant,[290, 291, 292, 293] resistance depending upon the species. Some strains can resist constant boiling hydrochloric acid at 20°C for up to 30 min, but for such extreme resistance to be manifest predrying of the spores is essential.[290] Lower concentrations of hydrogen ions are not sporicidal but replace spore cations, the spore behaving as a weak cation exchanger.[18, 19, 23] Replacement of cations by hydrogen ions markedly lowers the spore heat resistance, but this can be recovered by restoration of the cations, especially calcium ions. Cation removal occurs in 5 hr at pH 4 and 25°C[19] but no spore DPA is lost under these conditions[18] (but compare Ref. 467, where low pH releases DPA). During such treatments with mineral aids activation of viable counts of spore suspensions may also occur.[23] The rate of germination of spores may be stimulated by acid pretreatment.[294] Thioglycollic acid can also remove spore cations[295] and sensitize spores to lysozyme. Low pH alone does not sensitize spores to lysozyme.[296]

Low pH values increase the mechanical resistance of spores (see Section II C, 1). Acid hydrolysis causes extrusion of DNA and some protein as a sidebody[297, 298, 299] and loss of DPA. Extrusion, which may be correlated with an intact cortex, does not occur with calcium-deficient spores,[300] or with spore sections, partly disrupted spores and vegetative cells. Strong mineral acids render spores stainable by cold dyes;[299] boiling dilute mineral acid extracts calcium dipicolinate;[301] hot tricholoracetic acid is moderately effective in releasing hexosamine from preparations of coat plus cortex material.[218]

2. Alcohols and other solvents

Spores are much more resistant than vegetative cells to methanol, ethanol, isopropanol and acetone.[277, 288, 302, 303, 304, 308] Although ethanol inhibits germination,[305] ethanol pretreatment is reported to activate the rate of germination.[306] Methanol, ethanol and acetone at 56°C extract some DPA.[307] Carbon tetrachloride, benzene and petroleum are more deleterious to spores than trichlorotrifluoroethane.[308]

Spores are resistant to octyl alcohol, but germinated spores and vegetative cells are sensitive.[309] However, octyl alcohol inhibits initiation of germination.[305, 310] During sporulation octyl alcohol resistance develops at about the same time as chloroform resistance but prior to heat resistance.[311] β-Phenethyl alcohol and related compounds inhibit germination, sporulation and cell division.[312, 313] The polar solvents dimethylformamide and dimethylsulphoxide activate spores of B. pantothenticus in the sense of increasing the total count of a spore suspension[314] but had no effect on spores of Cl. botulinum type A (T. A. Roberts, unpublished results).

3. Aldehydes

(a) Formaldehyde. Formaldehyde (HCHO) is only slowly sporicidal.[304] Reported rapid activity[290] was mainly due to its sporostatic effect.[315] Although methanol can be used to prevent polymerization of HCHO it also reduces the sporicidal activity,[316] as do other alcohols.[280] Activity is hardly affected by pH in the range 3·6 to 7·8, and generally destruction lags before proceeding more rapidly.[280] HCHO is sporicidal in the vapour state if the relative humidity (RH) is greater than 50%,[317] the optimal RH being 80 to 90%.[317, 318, 319] The killing rate of the vapour is about the same as that of a solution with the same HCHO vapour pressure.[317] Rate of kill varies little between 0 and 37°C[318] but is faster at higher temperatures partly because polymerization is inhibited.[317] Spores are only 2–15 times more resistant than vegetative cells.[230, 284] Aldehydes alkylate the —SH and —NH$_2$ groups of proteins.[321]

(b) Glutaraldehyde. Glutaraldehyde is sporicidal,[322–325] being much more active than HCHO[280] and active in alcoholic solution.[322] Activity increases with pH, the optimum being about 8,[280] at which pH it is in the monomeric state, being an equilibrium mixture of the open chain (25%) and hydrated ring forms.[326] The aldehyde groups are necessary for activity since the bis-bisulphite and oxime derivatives are inactive, whereas β-methyl-glutaraldehyde is active.[326] Glutaraldehyde has been shown to react very strongly with wool, collagen and gelatin[327] and this reactivity with proteins probably accounts for its successful use as a spore fixative.[328]

(c) Other aldehydes. Glyoxal, succinaldehyde[322] and adipic dialdehyde[326] are sporicidal, but malondialdehyde and β-phenyl-glutaraldehyde are not active.[326] Octyl aldehyde, unlike octyl alcohol, does not inhibit germination.[305]

4. Alkalis

Alkalis are sporicidal,[329, 330, 331] their activity depending on the activity of the hydroxyl ions.[329, 330] Below 40°C the sporicidal efficiency of sodium hydroxide (NaOH) is unaffected by temperature,[330, 332] possibly due to the mobility of hydroxyl ions being relatively unaffected by temperature.[277] Later workers found that temperature increase (1 to 28°C) increased the sporicidal activity.[333, 334] However, their spores were initially only 10% viable by plate count, and they noted some stimulation of count after short exposures.[334] At higher temperatures (50 to 85°C) the sporicidal activity of NaOH increases.[335-338]

Resistance to NaOH varies with the species of spore.[338] The shapes of survivor curves range from concave upward[335, 339] through linear or slightly concave[338] to concave downward or sigmoidal.[334, 337] Spores killed by sodium hydroxide are still refractile and stain-resistant.[340] Hot dilute NaOH did not sensitize spores to lysozyme.[341] NaOH has been used to fractionate spore coats[342] but is not effective in differentiating spore structure[297] or in solubilizing the hexosamine of preparations of coat plus cortical material.[218] Sodium carbonate, tri-sodium phosphate and sodium sesquicarbonate are slightly sporicidal.[330, 343] Mixtures of certain alkalis are more sporicidal than their single components.[335]

5. Alkylating agents

Sporicidal effects of alkylating agents have been reviewed[279, 280, 281, 344] and there are reviews on their biochemical action.[345, 346] Table I classifies

TABLE I

Classification of alkylating agents

Class	Specific name
(A) Non-ring compounds	Formaldehyde
	Glutaraldehyde
	Methyl bromide
(B) 3-Membered ring compounds	
(1) N-substituted	Ethylene imine
	Propylene imine
(2) S-Substituted	Ethylene sulphide
	Chloromethylethylenesulphide
(3) O-Substituted (epoxy)	Ethylene oxide
	Propylene oxide
	Glycidaldehyde
	Styrene oxide
(C) 4-Membered ring compound	β-propiolactone

those tested against spores. Aldehydes have already been discussed in Section II D, 3.

(a) *Ethylene oxide*. The early bacteriological studies with ethylene oxide (ETO) gas have been reviewed by Phillips and Kaye,[347] who in 1949 published their work on its sporicidal effects.[348, 349, 350] Doubling the concentration halves the sterilizing time in the range 22 to 884 mg/1.[348] Sterilization with ETO gas is slow, taking at least 4 hr,[285] and is critically dependent on relative humidity (see below). ETO is most effective above pH 6·0 within the range 2·6 to 9·0.[352] The gas is effective against aerosolized spores.[349] ETO is soluble in water,[344] its solution being sporicidal.[288]

Reports of temperature coefficient (Q_{10}) range from 1·5 to 3·2.[348, 351–353] The Q_{10} varies with concentration at low but not at high temperatures due :o physical effects;[353] it also varies with the relative humidity.[352] RH is the most important variable controlling the sterilization rate,[350, 354] being more important than the partial water vapour pressure.[351] The optimum RH is about 33% and the activity at this RH is 10 times that at 97%.[350] Higher RH values are necessary when natural dry earth[355] or other absorbent materials[356] are present. Spores dried on hard impermeable surfaces are less susceptible to ETO[350] and pre-humidification and increased RH may be necessary for effective kills.[356] Spores encrusted in various materials are also less susceptible.[308, 357–359]

When spores or vegetative cells are dried at RH values below 30% their ETO survival curves can become non-logarithmic, as though super-resistant organisms equal to 0·001% to 0·1% of the initial population were present.[360, 361] The lower the RH the greater the super-resistant fraction; the effect is reversed by re-wetting the spores or cells. Acetone dried spores are also very difficult to kill.[362]

Spores are usually about ten times more resistant to ETO than the corresponding vegetative cells,[348, 363, 364, 365] although the resistance ratio to other chemicals may be about 10,000.[284] A coccus with ETO resistance equal to that of spores has been reported.[366] Desiccated cocci are more resistant than spores to ETO[357, 367] but not to methyl bromide.[367] Germinated spores are intermediate in resistance between spores and the corresponding vegetative cells.[368] Spores killed by ETO were unable to initiate germination.[368]

Resistance of spores to ETO varies with species[354] and with age.[369] It is not related to their moist heat resistance.[357] The resistance of the most resistant spores in a population correlates with an increased extractable lipid content, and the removal of this lipid renders these spores less resistant.[368]

ETO and methyl bromide can decompose to bactericidal compounds, but these do not account for the bactericidal action since the other active alkylating agents do not decompose to toxic products.[285] Alkylating agents are known to react with the —SH, —OH, —COOH and —NH$_2$ groups of proteins[370] and nucleic acids.[371] Proteins are the probable site of sporicidal action since mutants have never been observed among survivors of ETO treatment.[361] The marked difference in sensitivity of spores to alkylating agents compared with sensitivity to chemicals which only react with free —SH groups may reflect the low numbers of unmasked —SH groups in spores. The relative sensitivity to alkylating agents would then imply that the —NH$_2$ —OH and —COOH groups of spores were not masked but just as available for alkylation as those of vegetative cells.[284] The fact that the killing action of gaseous ETO depends on its concentration in the spore moisture[350] is understandable because alkylation involves replacement of labile protons with alkyl groups and water facilitates such proton reactions.[372]

(b) *Other alkylating agents.* The liquid β-propiolactone (BPL) is sporicidal against a variety of spores,[373] being much more active than formaldehyde (50,000 : 1), ethylene oxide (2,000 : 1) or methyl bromide (16 : 1) (Ref. 374). Spores are only about 4 times as resistant as vegetative cells.[375] BPL is sporicidal in solution[375, 376] or in the vapour state.[374] The sporicidal activity increases greatly with temperature,[376] the Q_{10} with the vapour being 2 to 3.[374] As with ETO and peracetic acid vapour the Q_{10} value increases at lower temperatures.[377] Activity of the vapour is optimal in the RH range 80 to 90%, there being little action at 45%.[374]

The sporicidal activity of propylene oxide vapour is a quarter of that of ETO.[378] The activity increases 7 to 20 times with increasing RH,[379] though a decrease with increasing RH has also been reported.[378] At low RH values some cocci were as resistant as spores.[379]

Gaseous methyl bromide is slowly sporicidal.[380] Liquid methyl bromide has been reported inactive[380] and active.[284] Its sporicidal action is due either to its alkylating ability[284] or to the fact that it can decompose to yield HBr.[380] Spores are about 5 times more resistant than vegetative cells,[284] even including desiccated cocci.[367]

Of the three-membered ring alkylaters (Table I) the N-substituted ring compounds are more sporicidal than the S-substituted ring compounds; the epoxy compounds are the least sporicidal.[349] Ethylene imine is the most effective compound. The RH for effective glycidaldehyde activity must be greater than 75%.[381] Peracetic acid is discussed on p. 646.

6. Antibiotics

The following antibiotics have been tested for effects on spores: nisin,[73, 382-386] tylosin,[73, 387] subtilin,[73, 382, 384, 388] bacitracin,[389] penicillin,[388, 390, 391] streptomycin,[392] tetracyclines,[391, 392] chloramphenicol,[392] carbomycin,[392] neomycin, celiomycin, streptin and circulin.[393] With few reported exceptions[393] these antibiotics are not sporicidal. Neither do they

D D

inhibit initiation of germination with two possible exceptions.[385, 388] o-Carbamyl D-serine and D-cycloserine stimulate the rate of germination[394] (see Chapter 11).

7. Dyes

Unlike vegetative cells, spores resist cold dyes, only their coats staining[400] unless heat plus phenol or heat alone is used to aid penetration.[395] The heat plus phenol method is similar to the Ziehl–Nielson procedure for staining acid-fast bacilli with carbol-fuchsin. The stained spores, like the stained acid-fast bacteria, do not decolourize with dilute H_2SO_4, nitric acid plus ethanol[395] or Na_2SO_3.[396] In Ashby's modification[397] of Schaeffer and Fulton's method[398] heat aids penetration by malachite green and then cold counter stain is used to displace the malachite green in vegetative cells. The cold counter stain cannot penetrate the spores.[399] The following treatments render spores stainable with cold dyes: initiation of germination (Chapter 11), heat,[299, 401] acid hydrolysis,[297, 299] ultraviolet light[402] and mechanical rupture.[217, 227, 299, 403] Stain resistance develops during formation of the cortex in sporulating cells.[403, 404] Some dyes inhibit outgrowth from spores.[281]

8. Enzymes

(a) Proteases. Trypsin, α-chymotrypsin, pepsin, papain, a peptidase mixture[296] and yeast keratinase[219, 405] have no visible effects upon the bright appearance of spores as seen by the phase contrast microscope even when the spores are pretreated with disulphide bond-rupturing reagents[223] which can sensitize spores to the action of lysozyme, spore lytic enzyme and hydrogen peroxide. Subtilisin causes spores to germinate[406] by enzymically generating germinants.[407, 408]

(b) Enzymes which degrade murein. In contrast to some of the corresponding vegetative cells,[409] spores incubated with lysozyme do not lose their viability[410] or alter in appearance[296] but their electrophoretic mobility changes.[411] Pretreatment with disulphide bond-cleaving agents renders a variety of spores susceptible to lysozyme, which partly lyses them causing loss of refractility and dipicolinic acid.[296, 412] This effect resembles germination (Chapter 11). The pretreatment agents are lethal doses of ultraviolet rays or high-speed electrons, the reducing agents thioglycollic acid, β-mercaptoethanol, cuprammonium sulphite or the oxidizing agent performic acid. Sensitization need not kill the spores but loss of viability occurs during the lysis, though it lags behind loss of refractility. The rupture of spore coat disulphide bonds during sensitization probably allows lysozyme to penetrate through the coat layers to the cortex, where it lyses its murein substrate.[215, 218, 413] The penetration can be demonstrated using fluorochrome-labelled lysozyme.[220] These results suggest that the cortical murein is an important factor in maintaining refractility.

Lysozyme has been used to extract DNA,[414, 415] transducing phage DNA,[415] DNA polymerase[416] and inorganic pyrophosphatase[417] from spores. Recently a spore strain has been discovered that is lysed by lysozyme without having to be sensitized first.[418] These spores are not lysed by proteases and they are not coatless.

Spores can also be sensitized to the cell wall-lytic enzyme of Strange and Dark[419] which occurs in *B. cereus* spores and which can release muropeptide fragments from preparations of spore coat plus cortex.[420] A similar enzyme probably occurs in other kinds of spores but is sometimes difficult to extract.[218, 420] The significance of this enzyme and these results with regard to the mechanism of initiation of germination is discussed in Chapter 11. Little has been published about other enzymes reported to lyse spores[421] or to initiate spore germination.[422, 423]

9. Halogens

(a) *Chlorine*. Vegetative cells[424] are very sensitive to chlorine water, while spores are moderately resistant.[339, 425–429] Chlorine dissolves in water, forming hypochlorous acid (HOCl) plus hydrochloric acid as in equation (5), the HOCl ionizing as in equation (6) (Ref. 425) with an ionization constant of 3.5×10^{-8} at 22°C. (Ref. 428.)

$$Cl_2 \ + H_2O \rightleftharpoons HOCl + HCl \tag{5}$$
$$HOCl \quad\quad \rightleftharpoons H^+ \ + OCl^- \tag{6}$$

The fact that pH markedly affects the sporicidal action[425, 426] and the general bactericidal action[424] of chlorine water suggests that HOCl is the major killing component. The HOCl molecule is estimated to be 100 times as sporicidal as the hypochlorite ion.[425] Sporicidal concentrations range from 2 to 50 p.p.m. chlorine, depending on the pH in the range 3 to 9.[426] At a given pH the sporicidal rate increases with chlorine concentration[425, 426] and with temperature,[425] the Q_{10} being about 2·1.[339] The survivor curve is non-logarithmic with a marked lag[339, 425, 427] which may be due to impermeability of the spores.[424] Chlorine water or sodium hypochlorite in HCl differentiates spore structures.[297]

Sodium and calcium hypochlorites are sporicidal.[284, 339, 430, 431] Their activity is markedly affected by the pH value[339, 430, 431, 432] as with chlorine water. The effects of concentration[339] and temperature[430] have been reported and survival curves have a distinct lag phase.[339, 430, 431] Spores are about 10,000 times as resistant as vegetative cells to hypochlorites.[284] Electrolysis products from chlorides (presumably hypochlorites) were sporicidal at −12°C.[433] Sodium hypochlorite (0·5%) lyses bacteria, including spores, within 20 min at 37°C and the lytic effect has been used to prepare spore integuments for ultrastructure studies.[434]

The chloramines are sporicidal, those tested being the N-chloro derivatives of ammonia, methylamine and glycine[426] and chloramine T.[431, 435] Their mechanism of action is not clear,[424] but may be due to HOCl released by hydrolysis.[435] In contrast to the hypochlorites and chlorine water there is little[432] or no lag[339, 427] in the action of chloramine (NH_2Cl). There is a lag with chloramine T.[432] The Q_{10} of NH_2Cl action is 3·1 and its action is affected by concentration but not affected by pH.[432] Chloramine-T action is markedly pH-dependent.[432] Organic compounds containing available chlorine are sporicidal.[431]

(b) *Bromine*. Aqueous bromine is sporicidal. Although less effective than chlorine below pH 8, bromine is about 4 times more effective above pH 9.[436] This is probably due to the fact that hypobromous acid (HOBr) is a weaker acid than HOCl and so is less ionized than the latter above pH 9.[437] This explanation assumes that undissociated HOBr is the sporicidal component of bromine water. The organic available-bromine carrier dibromodimethyl-hydantoin is sporicidal.[431] Bromides stimulate the bactericidal action of hypochlorites[437, 438] probably because HOBr is produced via the following reaction:

$$OCl^- + Br^- \rightleftharpoons Cl^- + OBr^-$$

Thus the sporicidal action of the organic available chlorine compound dichloroisocyanurate is enhanced by the presence of KBr at pH 9 but not at pH 7 or 8.[431]

(c) *Iodine*. The bactericidal effects of iodine have been reviewed.[439] Iodine is sporicidal[315, 439–443] and spores are more resistant than vegetative cells.[277] Activity increases with decrease in pH.[440, 442] Iodine killing is believed to be due to free iodine molecules[444] rather than hypoiodous acid which, however, may be partly responsible for killing at higher pH values[440] (cf. chlorine). Due to the low solubility of iodine in water potassium iodide is added as a solubilizer. Solubilization is due to the formation of triiodide ion which itself has a low sporicidal activity,[440] so that addition of excess iodide must be avoided.

Other iodine solubilizing agents or carriers such as polyvinyl-pyrrolidone or other surfactants are available. Such mixtures are called iodophors[279, 439, 445] because the complexes in them slowly liberate free iodine when diluted with water. Iodophor activity is proportional to the concentration of free iodine released rather than to the total iodine content.[443] Sometimes iodine may be lost by iodination of the surfactant.[446] Iodophors are sporicidal.[279, 315, 447, 448] Maximum bactericidal activity is usually exhibited in acid solutions,[279] though the sporicidal activity of one iodophor was not affected by pH in the range 2·3 to 6·5.[431] Iodine trichloride[449, 450] and bis-p-chlorophenyl-iodonium sulphate[451] are sporicidal, but diphenyl-iodonium chloride[279] is not.

10. Heavy metals

(*a*) *Mercury.* Mercury compounds are bacteriostatic,[281, 452] though prolonged bacteriostasis probably results in death,[285] the bacteriostatic effect can be reversed by sulphydryl compounds.[453] Reviews of the early literature on the effect of mercury compounds on spores indicate that their effect is not sporicidal.[277, 452] Instances of sporicidal effects were probably due to inefficient quenching. Sporostasis, which is due primarily to inhibition of germination by mercuric ions[454, 455, 456] or organic mercurials,[310, 457] is reversed by washing or by sulphydryl compounds. Inhibition of germination is not efficient with all kinds of spores.[455]

(*b*) *Silver and other heavy metals.* Silver compounds are bactericidal[458] and sporicidal.[331] Oligodynamic[285, 459] silver preparations which are of various kinds include sporicidally active[460] and inactive compounds.[461] Spores are much more resistant to colloidal silver than vegetative cells.[284] The bactericidal effects of heavy metals other than mercury and silver have been reviewed.[462] Copper sulphate has been reported both as having and not having a sporicidal effect.[277] Iron, copper, zinc, nickel, chromium and cobalt ions can inhibit germination.[457, 463, 464]

11. Oxidizing agents

Hypochlorites and other oxidizing halogen compounds have been discussed above.

(*a*) *Hydrogen peroxide.* This compound kills spores in a logarithmic fashion.[465] Killing rate increases with temperature but is affected little by pH. However instability at high temperatures or pH values lowers the peroxide concentration, resulting in lower rates of kill. Small fractions of the spore population often survive, probably because of the peroxide losses that occur. Such losses may be due to the presence of heat stable catalase in spores[246, 466] or any vegetative cells present, and to metal ions (Fe^{2+} or Cu^{2+}) which stimulate peroxide breakdown.[277] DPA release occurs during peroxide treatment[467] and its extent increases with temperature.[307]

Spores lyse in 5% peroxide at 56°C.[307] At lower temperatures there is little or no lytic effect unless the spores have been pretreated with disulphide bond-reducing agents. Lysis of such pretreated spores by peroxide is optimal at high pH values, being stimulated by metal ions, especially cupric ions.[296, 412] Sensitized spores are killed by the peroxide before lysis is apparent (King and Gould, unpublished data); lysis of untreated spores by hydrogen peroxide at low temperatures can occur providing the hydrogen bond-breaking agent lithium bromide is present. Preparations of spore coats are lysed by hydrogen peroxide to soluble proteins, peptides and amino acids (King and Gould, unpublished data). The mechanism of peroxide-lysis of spores and coats is probably similar to the free radical-catalysed depolymerization of collagen or gelatin which is stimulated by cupric ions.[468]

(b) *Inorganic acids.* The internal structure of spores can be differentiated by treatment of spores for short times with oxidizing acids or acidified oxidizing agents, but not with sulphuric acid.[297, 400] Heat-killed spores, germinated spores and vegetative cells cannot be differentiated by such treatments.

One such treatment, N/3 HCl plus 0·1% KMnO$_4$ at room temperature, causes extrusion of sidebodies containing DNA and protein.[298] The sidebodies produced this way can be solubilized with sodium acetate, whereas sidebodies produced by hydrolysis with HCl cannot. The oxidative treatment seems to affect the protein of the sidebody, rather than the DNA, in some way which allows solubilization to occur. Cold perchloric acid also causes extrusion of DNA from spores as a sidebody.[217]

(c) *Organic acids.* Spores treated with performic acid become sensitized to the lytic action of certain enzymes and hydrogen peroxide.[296, 412] Sensitization by performic acid is not reversed by oxidation as is the sensitization caused by reducing agents. Failure oxidatively to reverse sensitization is expected if the sensitization involves oxidation of coat disulphide bonds with the production of sulphonate groups which are stable to oxidation. Peracetic acid is extremely sporicidal,[291] being active in solution or the vapour state.[469] It is much more sporicidal than β-propiolactone.[377] Peracetic acid (3%) has a D value of about 2 hr at $-40°C$ and 8 min at $-30°C$. There is a lag period the length of which is temperature dependent. Oxidized ascorbic acid[470, 471] is sporicidal at 75°C, but the effect may be due to autoxidation products such as hydrogen peroxide.

(d) *Other oxidizing agents.* Osmium tetroxide is used to fix spores for electron microscopy. Fixation is not always complete due to lack of penetrating power.[472, 473] Little DPA is extracted during routine fixation with 1% at 26°C, but more vigorous fixation extracts a larger amount.

Potassium permanganate (KMnO$_4$)[320] preserves membrane structures well[474] but releases DPA from spores even at 4°C.[472] Nitrogen tetroxide is sporicidal while monomethylhydrazine and 1,1 dimethylhydrazine are slowly sporicidal to dried spores.[475] Ozone at 0·12 mg/ml kills *Cl. perfringens* cells in 5 min, but the spores take 15 min to kill even at 0·25 mg ozone/ml.[502] Temperature in the range 10° to 24°C has little effect on the rate of kill.

12. Phenols

The bactericidal[477, 478, 479] and sporicidal[281] effects of phenols have been reviewed. The bactericidal action of phenols seems to be mainly due to their effect on the plasma membrane. Spores are extremely resistant to phenol,[24, 480, 481] trichlorophenol[284] aerosolized hexylresorcinol,[482] and lysol, which contains cresols and xylenols.[277] Spores are much more resistant than vegetative cells[284] and germinated spores.[483] The sporicidal effect of phenol only becomes apparent with prolonged incubation. Thus a 99·99% kill of *B. megaterium* spores, which are relatively sensitive, takes about 2 days in 5% phenol at 37°C; *B. stearothermophilus* spores are not killed by this treatment.[24] Survival curves are concave upwards or sometimes linear.[24]

The sporicidal activity of phenol increases with decrease in pH.[480] Heat-activated spores have been reported to be more sensitive to phenol than untreated spores.[484] There is usually little correlation between heat resistance and phenol resistance,[24] but one exception has been reported.[485] High temperatures increase the sporicidal activity of phenol,[486] pentachlorophenol[487] and chlorocresol,[488] though there is doubt about the last of these reports.[489]

Phenol appears slightly to stimulate the increase in viable count observed when *B. stearothermophilus* spores are incubated at 37°C.[24] Spores may lose DPA during treatment with phenol[490] without losing their refractility[307] and dead spores remain refractile.[24] Low concentrations of phenol (about 0·2%) are sporostatic,[491] the effect depending on the spore concentration,[492] whereas 0·5% inhibits germination in slide cultures[493] and prevents oxygen uptake by spores in glucose.[491] Dinitrophenol does not inhibit initiation of germination.[310] Recovery of spores from phenol treatment is increased if germinants are added to the recovery medium.[486]

13. Reducing agents

Spores which are normally resistant to the lytic action of lysozyme and spore-lytic enzyme or hydrogen peroxide can be chemically or physically sensitized to them by agents which are known to be capable of cleaving disulphide bonds,[296, 412] including the reducing agents thioglycollic acid, 2-mercaptoethanol and cuprammonium sulphite. Sensitization by reducing agents, in contrast to that induced by performic acid, can be reversed by oxidation. The viability and some of the DPA of the spores may or may not be lost during sensitization with reducing agents, depending upon conditions.

Sensitization by reducing agents is most effective at pH 2 to 3 (near the isoelectric point of spores,[411]) and will also occur at pH 11 and 12,[296] although alkaline or acidic conditions alone do not cause sensitization. Thioglycollic acid[295] and other acids can remove cations from spores and decrease spore heat resistance.

Spore coat protein is known to have a high cystine content,[117, 191, 192, 494] and it is probable that it is rupture of the cystine disulphide bonds that allows the lytic agents to penetrate to their substrates within the spore.

Recent work has shown that most of the coat cystine sulphur is in a para-crystalline fraction which has a keratin-like X-ray diffraction pattern.[342] Incorporation of cystine into coat protein can occur by an exchange reaction.[495]

Amperometric estimation of thiol groups formed during thioglycollic acid treatment of spores suggests that at least 10 to 30% of disulphide bonds are ruptured during sensitization to lysozyme.[194] Other methods can be used to measure the disulphide–sulphydryl ratios of spores but have not been applied to sensitized spores.[193, 340] Sensitization with thiogly-collic acid does not alter radiation or heat resistance compared with spores pretreated with hydrochloric acid at the same pH value as the thioglycollic acid. Reducing agents will also activate some spores for germination.[496] Nitrite lysed spores by unknown mechanisms.[497]

14. Surfactants

Early work showed that quaternary ammonium compounds (QACs) are not sporicidal.[279, 284, 498, 499, 500] Apparent sporicidal activity is due to QACs (e.g. cetrimide) sticking to spores.[401] Repeated washing[498] or the use of a quenching agent[499, 501] allows recovery of the treated spores. Spores may survive heating at 95°C with QACs.[498] However, some QACs are still reported as being sporicidal[315] and the QAC Arquad T causes lysis of spores and release of DPA.[307] Depending on the pH value, nearly all cationic (including QACs) and anionic surfactants induce the release of DPA from spores at 30°C. Non-ionic and surfactant antibiotic peptides do not induce DPA release. Further information is available in the comprehensive study by Rode and Foster.[502] Another group of surfactants, the long-chain alkyl primary amines, also cause changes similar to those mentioned,[503, 504] resembling those occurring during germination, but the changes can occur in non-physiological conditions. This phenomenon is called "chemical germination" (Chapter 11).

III. Mechanisms of Resistance

A. INTRODUCTION

There are obvious commercial reasons for gaining an insight into the mechanisms of spore resistance, whether to heat, radiation or chemicals. For example, if some food processes could be made less severe without loss of microbiological stability, not only would time and costs be reduced, but quality would be improved and new markets opened. Additionally, the academic reasons for spore resistance are equally intriguing. Spores occur universally, although no generally acceptable theory of their role has been propounded,[505, 506] and are able to withstand extreme conditions for prolonged periods. It is probably true to say that failure to isolate clostridial spores from many natural environments reflects more upon technical inadequacies than on the absence of spores.

Any explanation of resistance must take into account the enormous range of resistance of spores to heat (about 10^5-fold between *Cl. botulinum* Type E and *B. stearothermophilus*) as well as smaller differences in resistance of spores of different species, and different strains. Yet these spores have remarkably similar structures, and very similar chemical compositions.

There are several possible approaches to the problem. One is to assume that spores contain some unique substance, or combination of substances, which renders them resistant, for example DPA was initially believed to be that substance. Alternatively, it may be assumed that all spores have the same basic mechanism of resistance, but that small chemical differences between species perhaps cause different spores to contain different amounts of water or other compound in some essential structure, thereby causing differences in resistance.

No entirely satisfactory mechanism of resistance has been proposed for heat, radiation, chemicals or enzymes, although many detailed studies have been made using the two former agencies. A summary will be presented here, and the reader is referred to other chapters for more detailed analyses.

B. HEAT

The isolation of DPA from bacterial spores[507] immediately stimulated research into its involvement in determining heat resistance. DPA has been thought to be associated with the cytoplasm[508, 556] and not the cortical zone, since spores depleted of DPA by autoclaving retain the latter,[509] but its true location is still not clear. Initial observations on DPA synthesis revealed it to coincide with, or to follow immediately, the development of thermostability[510] and to parallel closely calcium uptake.[191] In *B. cereus* T resistance varied directly with DPA content,[511] but in *Cl. roseum*[512] and *B. coagulans*[513] or between species[30, 514, 515] this simple relationship was not found. Inactivation of spores of *Cl. botulinum* coincided with the release of Ca and DPA at pH 7·2 at 100°C (in 15 min) and at pH 2·2 (2 min), the molar ratio remaining constant at 2·2–2·4.[516] Spores of *B. megaterium* treated with thioglycollate (0·4 M, pH 2·6) lost their heat resistance (but not their radiation resistance) and yet retained their DPA. Heat resistance could be restored by exogenous cations, suggesting these to be implicated in resistance.[295]

Calcium had previously been thought to play an important role in heat resistance,[35, 36, 37, 517] but again the content in different spores did not correlate well with resistance.[513, 514, 515, 518] In a study of 20 spore preparations, Murrell and Warth[30] found a significant relationship between calcium and heat resistance, while magnesium showed a significant inverse relationship. The Mg–Ca ratio was very significantly related to heat resistance, confirming a similar earlier observation.[515] Calcium and DPA occur in nearly equimolar amounts in mature spores[515, 519, 520] and in two species heat resistance correlated well with this ratio,[513, 514] but such a correlation was not evident between other species.[30]

Heat resistance is also correlated with diaminopimelic acid (DAP) but not with hexosamine content of spores.[30] The importance of DAP in heat resistance may well be to supply electronegative groups which can be cross-linked by spore Ca (or CaDPA) to form a rigid cortex. If Ca is chelated to DPA, as seems likely, some other cation might be required to decrease the electronegative repulsion of the carboxyl groups. Such a structure would possess many of the properties demanded by the contractile cortex hypothesis of Lewis et al.[521] At the same time, changes in heat resistance of spores accompanying exchange of spore Ca for other cations[18] become understandable as effects due to weakening of the cation cross-links. Detailed evidence for the roles of DAP and other spore constituents in heat resistance is described in Chapter 7, and the importance of cation exchange in Chapter 9.

The development of resistance has also been related to the appearance in the spore of cystine-rich materials,[192] although the determination of resistance in these studies was not precise owing to spores forming in chains. This is discussed fully in Chapter 3.

Other workers have approached heat resistance by studying the resistance of individual cell components. The final inactivation of cellular materials is most probably protein denaturation (even if it be contained within the contractile cortex described above). Calculations of activation energy and thermodynamic functions for spore inactivation coincide well with those for inactivation of proteins.[522] Studies on isolated enzymes under conditions closely resembling those within the cell have revealed some remarkable variations in resistance. Spores which have survived high temperature must contain enzymes which have resisted denaturation. Comparison of B. cereus spore and vegetative cell catalases showed marked differences in heat resistance at 80°C and in inactivation by guanidine.[523] Similar differences were observed with glucose dehydrogenase, and additionally, the heat-labile enzyme from germinated spores possessed identical serological specificity and electrophoretic mobility to its stable counterpart in the refractile spore.[523] A measure of the degree of change in resistance is indicated by the half-life of the resistant enzyme being 18 min at 69°C, and that of the labile enzyme 45 sec at 50°C. Subsequent studies by the same workers revealed that the million-fold change in heat resistance of glucose dehydrogenase which occurs in vivo may be mimicked in vitro by control of ionic environment, and closely parallels monomer/dimer conversion.[524]

Enzymatic activities have also been determined in a non-sporing marine pyschrophile. Initial denaturation of malic dehydrogenase was reversible, and the presence of several malic isozymes, each of differing heat lability, was suggested.[525] The presence of cells reduced enzyme activity, but appeared capable of protecting the enzyme against heat denaturation.[526] Although these studies were not aimed at elucidating the mechanism of heat resistance, the similar conclusions are evident. Heat resistance of spore enzymes is further discussed in Chapter 8.

C. RADIATION

The current unit of radiation dose is the rad, corresponding to the absorption of 100 ergs per g of irradiated materials. Reference has been limited, in the main, to ionizing electromagnetic radiation, i.e. X- and gamma radiation. There is a fundamental similarity between their effects and those of electrons, the difference being mainly one of penetration and distribution of secondary electron tracks throughout the irradiated material. Ultraviolet light is not discussed in detail since it produces only an excited state in the irradiated molecule.

The most important processes by which the energy of electromagnetic radiations may be transferred to materials through which they pass are (i) photoelectric absorption, occurring with radiations of energies up to 60 KeV, the photon disappearing and a single energetic electron being ejected; (ii) Compton scattering, predominantly in the range 60 KeV to 1 MeV, the initial photon interacting with an orbital electron, the energy being shared between the less energetic photon and an ejected electron; (iii) pair production, occurring only above 1 MeV, the photon disappearing and being replaced by a pair of energetic electrons, one positive and one negative, and photon energy in excess of 1 MeV appearing as kinetic energy of the electrons. Ionizations therefore occur along electron tracks of varying energies and randomly distributed.

The studies initiated by Powers (reviewed in Ref. 197) give detailed and precise information on spore radiation resistance, but have not yet offered any answer to the vexing question of why spores are resistant to ionizing radiations, or, indeed, whether there is one cell component constituting the sensitive site (or "target" as it was referred to in early papers.[527]) In some respects spores are more complex systems than vegetative cells, since inactivation may occur at any one of a number of stages between the mature refractile spore and the macro-colony. Spores of *B. megaterium* and *Cl. botulinum* irradiated with large doses undergo germination as rapidly as unirradiated spores[50, 528] and these spores, although non-viable in terms of colony formation, retain many fermentative activities, and part of their ability to synthesize new protein.[528]

DPA appears to play no part in radiation resistance,[50] although it can be released from radiation-inactivated spores of *B. megaterium* and *Cl. botulinum* by irradiation far beyond the lethal dose.[50, 136] One study on *B. subtilis*, using spores and vegetative cells and both u.v. light and X-rays,[529] detected a genetic locus for radiation sensitivity. Differences in resistance of spores and vegetative cells could not be attributed to differences in radiation resistance of their DNA (see also Ref. 530).

Many calculations have been made attempting to determine the size of the sensitive site[136, 531, 532, 533] without special reference to its identity, but frequently assuming that it will be shown to be DNA. There is evidence however, that the radiation-sensitive units in spores are not concentrated in a compact nucleus, but dispersed throughout the internal regions of the spore.[534] As soon as the extrapolation number exceeds unity, speculations regarding the size of the sensitive site or sites grow beyond the limits of the precision of the data. For example, assuming that $n = 4$, 4 "hits" or "events" in 1 site might inactivate the cell. Alternatively 1 hit in each of 4 sites would yield exactly the same survivor curve. Sites might be paired, and the pairs might differ in size. Finally the site may, or may not, be spherical. If it is not, the thickness becomes important. Presumably ionizations occurring within the site are effective, and those occurring more than a certain distance away are not, possibly limited by the limits of radical diffusion. However, if many such ionizations occur simultaneously close to the site, a finite, although perhaps low, probability of their being effective exists. The effect of heterogeneity on the shape of survival curves has been considered for groups of sites differing in n[531] or for groups with differing D_{37}.[535] Before embarking on protracted calculations, the reader would be well advised to consider several detailed publications.[22, 531, 535 536]

For the sake of completeness, brief mention must be made of studies on vegetative cells. The most resistant bacterium, *Micrococcus radiodurans*, will tolerate about 350 Krad without any fall in colony count occurring, a dose which would have produced measurable inactivation of most spores. A study of DNA base composition in 8 species of bacteria and their resistance to irradiation suggested an inverse relationship between radiation resistance and guanine–cytosine (GC) content,[537] i.e. resistance increased with decreasing GC content. The same authors reported preliminary experiments indicating that the reverse might be true for u.v. light. However, *M. radiodurans* is resistant to both types of radiation, and has a high GC content (67%;[538]) the same as the very sensitive Pseudomonads. The extrapolation number for *M. radiodurans* is 5,000–10,000; which could imply 5,000–10,000 single targets each requiring a single hit, or alternatively about 10 hits in a single target only, approximately twice the volume of that corresponding to each of the 10,000 targets.[539] Suggestions that radiation energy might be dissipated in a "sink" involving the pigment[540] have been shown to be incorrect by the isolation of a non-pigmented resistant strain, with an identical sigmoid curve to the wild-type, and a non-pigmented mutant possessing an exponential survivor curve of even greater resistance (i.e. D value) than the wild-type.[541, 542]

The remarkable radiation resistance of *M. radiodurans* has been shown to be due principally to a repair mechanism,[541] repair occurring in a period after cessation of radiation and before logarithmic growth begins. No such repair was evident with *Salmonella typhimurium*. *M. radiodurans* is also capable of repairing u.v. damage to its DNA.[543] Damaged bases, including thymine dimers, are enzymatically removed and the DNA molecule is reconstructed from information on the complementary strand.[1, 2, 544] In the case of *M. radiodurans* repair of damage by u.v. and ionizing radiations has been shown to be similar,[545] hence damage by ionizing radiation is also likely to involve DNA. Repair of X-ray damage of single stranded DNA has also been demonstrated *in vitro*.[546] *B. megaterium* spores labelled with tritiated thymidine and irradiated with monochromatic u.v. light yielded no thymine dimers, but three other thymine photoproducts were observed. These products were also produced *in vitro* by irradiation of *E. coli* DNA that had been dried in the presence of sodium chloride or calcium DPA.[547]

D. OTHER PHYSICAL AGENTS

The recently increased knowledge of spore structure and composition has been little applied in studies of the mechanism of spore resistance. Advances will probably depend on first increasing knowledge about the physical properties of the spore components and then comparing them with those of vegetative cell components. The relevant physical properties are size, shape, density, hardness, compressibility and elasticity. The coat is probably the mechanically strongest part of the spore. Little has been published on the effects of drying, freezing, freeze-drying and aerosolization on spores over long time periods. However, such knowledge may be important in considering the survival of sporeformers in their natural state.

E. CHEMICALS

The results in Section III D show that spores are more resistant than vegetative cells to most chemicals. However, with few exceptions our knowledge of the effects of chemicals on spores is limited. This is due in part to a corresponding lack of knowledge about the effects of these chemicals on vegetative cells. A decline in research impetus occurred in this field when it was realized that many chemicals had very restricted practical applications, due to their toxic properties. The effects of antibiotics, with their immense medical potential, inevitably received research priority.

The authors therefore decided that a comprehensive catalogue of studies on the effects of chemicals upon spores would best serve the interests of workers, especially newcomers in the field. In following this aim we do not claim to have exhausted the published literature on the subject and make no apology for the fact that a catalogue is hard to read and contains a certain amount of non-critical reporting.

Although little is known about how most chemical agents kill cells, brief mention might be made of possible mechanisms of resistance, with examples where available. Resistance to lytic enzymes, and perhaps to stains, may be due to important parts of the spore being impermeable to chemicals. Alternatively, sensitive sites or chemical groups could be masked in some way by interactions of different macromolecules. Perhaps calcium, DPA, or calcium dipicolinate play a role. This mechanism has been used to explain the relative lack of resistance of spores to alkylating agents, and the resistance of spore coats to proteolytic enzymes may be due to such a masking mechanism. A third alternative is that the absence of metabolism in the dormant spore renders it resistant. Resistance to metabolic inhibitors and antibiotics can be explained in this way, but lack of penetration cannot be ruled out in these cases. For example, spores are resistant to octyl alcohol and ethanol but germination is inhibited by these substances. Thus either the spores are permeable or the germination system is located outside the permeability barrier of the spore. A fourth, but unlikely, explanation of resistance could be that spores completely lack certain sensitive sites or molecules present in vegetative cells. There is no known example.

The few examples given above indicate that spores resist different chemicals in different ways. Once the resistance of spores is overcome, they are probably killed in the same way as vegetative cells. In determining why spores are more resistant to chemicals the choice of chemicals to be studied is important. Mechanisms of inactivation are more readily proposed for chemicals that are relatively group specific in their action. Thus there are plausible mechanistic theories of spore resistance to aldehydes, alkylating agents and enzymes but there are none for non-specific compounds such as the acids and alkalis.

In studying the mode of action of a chemical on spores certain effects in addition to loss of viability can be observed. Supra lethal concentrations of chemicals may cause gross structural changes such as loss of refractility, release of dipicolinic acid, increase in stainability and even complete lysis. Other effects of chemicals worth looking for include effects on initiation of germination and on outgrowth, heat resistance and radiation resistance.

Such additional effects may possibly give a clue as to the mechanism of action of the chemical concerned.

There have been several reports that chemicals, like heat and radiation, may activate spores by either increasing the total count of a spore suspension due to activation of apparently non-viable spores or by increasing the rate at which spores germinate. Examples of these effects have been reported for acids, ethanol, dimethylformamide and dimethylsulphoxide, sodium hydroxide, an iodophor and phenol. The use of chemicals to activate spore germination should prove to be an exciting field of research.

Aspects of resistance of spores to chemicals that have been little investigated include the effects of sporulation media, preheating and species variation. From the point of view of mechanisms it is probably best to compare spore resistance with that of the corresponding vegetative cells. This has not been done in all cases. Particularly interesting would be more comparisons between dormant spores and spores which have just initiated germination. Germinated spores have an altered or solubilized cortex and contain virtually no calcium or dipicolinic acid. Such comparisons could highlight any protective roles of the intact cortex or calcium and dipicolinic acid against the chemical being studied.

F. Relationship of Resistance to Different Agents

It would be most convenient if one mechanism of resistance were adaptable to both heat and radiation (chemicals present a more complex picture), but this is unlikely to be the case. Spores differing widely in heat resistance may differ only slightly in resistance to ionizing radiations. There are limited similarities between the two. Heat is capable of activating both germination[548] and colony-forming ability,[118, 549] as is radiation,[550, 556] but spores responding to heat do not necessarily respond to radiation.[26] Both heat and radiation render spores more sensitive to inhibition by sodium chloride,[8, 190] but such sensitizations seem quantitatively different. Pre-irradiation sensitizes spores to heating,[551, 552] but pre-heating has no effect on radiation resistance,[551] unless germination is initiated, when the germinated spores are less resistant to radiation[553, 554] than the ungerminated spores.

There are few studies on the relation of chemical resistance to the other kinds of resistance. Ethylene oxide resistance of spores does not correlate with resistance to moist heat. The correlation of phenol and heat resistance of spores is not clear due to conflicting reports. Resistance to octyl alcohol and chloroform develops prior to heat resistance during sporulation. Thus the little evidence available suggests that heat and chemical resistances are seldom related, but mentioned in Section II D there are interesting reports that some chemicals can mimic the effect of heat activation.

There also appear to be few studies on the relationships between resistance to chemicals and resistance to radiation and other physical agents. There is no correlation between mechanical resistance and stainability. Nevertheless, circumstantial evidence suggests that the germination caused by surfactants, mechanical germination and differentiation of spore structure by acids, and perhaps even physiological germination itself, have in common an effect on the cortex. More studies are needed in this area.

There is no obvious theory to account for the evolution by spores of resistance to such diverse agencies. One can only assume that resistance to desiccation and to hydrolytic enzymes, and perhaps to heat, have been selected through having real survival value in nature. The ability to resist other extreme and diverse stresses may well have evolved concurrently and dependently upon these, but have no survival role in nature at all.

ACKNOWLEDGEMENTS

The authors wish to thank Dr T. Cross, Department of Biological Sciences, University of Bradford, for the section on Actinomycete spores.

REFERENCES

1. Boyce, R. P. and Howard-Flanders, P. (1964). *Proc. natn. Acad. Sci. U.S.A.* **51**, 293.
2. Setlow, R. B. and Carrier, W. L. (1964). *Proc. natn. Acad. Sci. U.S.A.* **51**, 226.
3. Stumbo, C. R. (1965). "Thermobacteriology in food processing". Academic Press, London.
4. Roberts, T. A. and Ingram, M. (1965). *J. appl. Bact.* **28**, 125.
5. Tyler, S. A. and Dipert, M. H. (1962). *Physics Med. Biol.* **7**, 201.
6. Powers, E. L. (1962). *Physics Med. Biol.* **7**, 3.
7. Hiatt, C. W. (1964). *Bact. Rev.* **28**, 150.
8. Roberts, T. A., Gilbert, R. J. and Ingram, M. (1966). *J. appl. Bact.* **29**, 549.
9. Woese, C. (1960). *Trans. N.Y. Acad. Sci.* **83**, 741.
10. Vas, K. and Proszt, G. (1957). *J. appl. Bact.* **20**, 431.
11. Anellis, A., Grecz, N. and Berkowitz, D. (1965). *Appl. Microbiol.* **13**, 397.
12. Wheaton, E. and Pratt, G. B. (1962). *J. Food Sci.* **27**, 327.
13. Weiss, K. F. and Strong, D. A. (1967). *J. Bact.* **93**, 21.
14. Alper, T., Gillies, N. E. and Elkind, M. M. (1960). *Nature, Lond.* **186**, 1062.
15. Anand, J. C. (1961). *J. scient. ind. Res.* **20C**, 295.
16. Humphrey, A. E. and Nickerson, J. T. R. (1961). *Appl. Microbiol.* **9**, 282.
17. Davis, F. L. and Williams, O. B. (1948). *J. Bact.* **56**, 555.
18. Alderton, G. and Snell, N. (1963). *Biochem. biophys. Res. Commun.* **10**, 139.
19. Alderton, G., Thompson, P. A. and Snell, N. (1964). *Science, N.Y.* **143**, 141.
20. Bender, M. A. and Gooch, P. C. (1962). *Int. J. Radiat. Biol.* **5**, 133.
21. Porter, E. H. (1964). *Br. J. Radiol.* **37**, 610.

22. Biavati, B. J. (1965). *Trans. N.Y. Acad. Sci. Series II,* **27,** 551.
23. Lewis, J. C., Snell, N. S. and Alderton, G. (1965). *In* "Spore III" (L. L. Campbell and H. O. Halvorson, eds), p. 47. Am. Soc. Microbiol., Ann Arbor, Michigan, U.S.A.
24. Briggs, A. (1966). *J. appl. Bact.* **29,** 490.
25. Xezones, H., Segmiller, J. A. and Hutchings, I. J. (1965). *Fd Technol., Champaign* **19,** 111.
26. Roberts, T. A. (1968). *J. appl. Bact.* **31,** 133.
27. El-Bisi, H. M. and Ordal, Z. J. (1956). *J. Bact.* **71,** 10.
28. Lechowich, R. V. and Ordal, Z. J. (1960). *Bact. Proc.* p. 44.
29. Yokoya, F. and York, G. K. (1965). *Appl. Microbiol.* **13,** 993.
30. Murrell, W. G. and Warth, A. D. (1965). *In* "Spores III" (L. L. Campbell and H. O. Halvorson, eds), p. 1. Am Soc. Microbiol., Ann Arbor, Michigan, U.S.A.
31. Eklund, M. W., Wieler, D. I. and Poysky, F. T. (1967). *J. Bact.* **93,** 1461.
32. Vinton, C., Martin, S. and Gross, C. E. (1947). *Fd Res.* **12,** 173.
33. Curran, H. R. (1935). *J. infect. Dis.* **56,** 196.
34. Gillespy, T. G. (1947). *Ann. Rep. Fruit Veg. Preserv. Res. stn,* Chipping Camden, Glos.
35. Grelet, N. (1952). *Annals Inst. Pasteur, Paris* **83,** 71.
36. Sugiyama, H. (1951). *J. Bact.* **62,** 81.
37. Amaha, M. and Ordal, Z. J. (1957). *J. Bact.* **74,** 596.
38. Tallentire, A. and Chiori, C. O. (1963). *J. Pharm. Pharmac.* **15,** T, 148.
39. El-Bisi, H. M. and Ordal, Z. J. (1956). *Fd Res.* **20,** 554.
40. Levinson, H. S. and Hyatt, M. T. (1964). *J. Bact.* **87,** 876.
41. Williams, O. B. and Harper, O. F. Jr. (1951). *J. Bact.* **61,** 551.
42. Vinter, V. and Vechet, B. (1964). *Folia Microbiol., Praha* **9,** 238.
43. Murrell, W. G. (1961). *Symp. Soc. gen. Microbiol.* **11,** 100.
44. Amaha, M. (1961). *Proc. 4th Int. Con. Canned Foods, Berlin,* p. 13.
45. Murray, T. J. and Headlee, M. R. (1931). *J. infect. Dis.* **48,** 436.
46. Murray, T. J. (1931). *J. infect. Dis.* **48,** 457.
47. Headlee, M. R. (1931). *J. infect. Dis.* **48,** 468.
48. Xezones, H. and Hutchings, I. J. (1965). *Fd Technol., Champaign* **19,** 113.
49. Williams, O. B. and Hennessee, A. D. (1956). *Fd Res.* **21,** 112.
50. Levinson, H. S. and Hyatt, M. T. (1960). *J. Bact.* **80,** 441.
51. Walker, W. (1964). *J. Fd Sci.* **29,** 360.
52. Esty, J. R. and Meyer, K. F. (1922). *J. infect. Dis.* **31,** 650.
53. Anderson, A. A., Esselen, W. B. and Fellers, C. R. (1949). *Fd Res.* **14,** 499.
54. Stumbo, C. R., Gross, C. E. and Vinton, C. (1945). *Fd Res.* **10,** 283.
55. Murrell, W. G. and Scott, W. J. (1957). *Nature, Lond.* **179,** 481.
56. Braun, O. G., Hayes, G. L. and Benjamin, H. A. (1941). *Fd Inds* **13,** 64.
57. Amaha, M. and Sakaguchi, K-I. (1954). *J. Bact.* **68,** 338.
58. Reynolds, H. and Lichtenstein, H. (1950). *Bact. Proc.* p. 28.
59. Lewis, J. C., Michener, H. D., Stumbo, C. R. and Titus, D. S. (1954). *J. agric. Fd Chem.* **2,** 298.
60. Michener, H. D., Thompson, P. A. and Lewis, J. C. (1959). *Appl. Microbiol.* **7,** 166.
61. Bose, A. N. and Roy, A. K. (1960). *J. scient. ind. Res.* **19B,** 277.
62. Michener, H. D., Thompson, P. A. and Lewis, J. C. (1959). *U.S. Dept. Agric. Western Utilization Res. Dev. Division ARS* 74–11.
63. LaBaw, G. D. and Desrosier, N. W. (1953). *Fd Res.* **18,** 186.
64. LaBaw, G. D. and Desrosier, N. W. (1954). *Fd Res.* **19,** 98,

658   T. A. ROBERTS AND A. D. HITCHINS

65. Schmidt, C. F. (1957). *In* "Antiseptics, Disinfectants, Fungicides and Chemical and Physical Sterilization" (2nd ed.) (G. F. Reddish, ed.), p. 831. Lea and Fabiger, Philadelphia, U.S.A.
66. El-Bisi, H. M., Ordal, Z. J. and Nelson, A. I. (1955). *Fd Res.* **20**, 554.
67. Frank, H. A. and Campbell, L. L. (1957). *Appl. Microbiol.* **5**, 243.
68. Yesair, J., Bohrer, C. W. and Cameron, E. J. (1946). *Fd Res.* **11**, 327.
69. Molin, N. and Snygg, B. G. (1967). *Appl. Microbiol.* **15**, 1422.
70. Pheil, C. G., Pflug, I. J., Nicholas, R. C. and Augustin, A. L. (1967). *Appl. Microbiol.* **15**, 120.
71. Rowe, J. A. and Koesterer, M. G. (1965). *Bact. Proc.* p. 8.
72. Doyle, J. E. and Ernst, R. R. (1967). *Appl. Microbiol.* **15**, 726.
73. Gould, G. W. (1964). *In* "Microbial Inhibitors in Food" (N. Molin and I. Erichsen eds), p. 17. Almqvist and Wiksell, Stockholm.
74. Curran, H. R. and Evans, F. R. (1937). *J. Bact.* **34**, 169.
75. Nelson, F. E. (1944). *J. Bact.* **48**, 473.
76. Olsen, A. M. and Scott, W. J. (1946). *Nature, Lond.* **157**, 337.
77. Foster, J. W. and Wynne, E. S. (1948). *J. Bact.* **55**, 495.
78. Olsen, A. M. and Scott, W. J. (1950). *Aust. J. scient. Res. B.*, **3**, 219.
79. Murrell, W. G., Olsen, A. M. and Scott, W. J. (1950). *Aust. J. scient. Res. B* **3**, 234.
80. Mefferd, R. B. and Campbell, L. L. (1951). *J. Bact.* **62**, 130.
81. Hachisuka, Y. (1964). *Jap. J. Bact.* **19**, 162.
82. Hollaender, A., Stapleton, G. E. and Martin, F. M. (1951). *Nature, Lond.* **167**, 103.
83. Cook, A. M. and Brown, M. R. (1965). *J. appl. Bact.* **28**, 361.
84. Williams, O. B. and Reed, J. M. (1942). *J. infect. Dis.* **71**, 225.
85. Edwards, J. L. Jr., Busta, F. F. and Speck, M. L. (1965). *Appl. Microbiol.* **13**, 858.
86. Iandolo, J. J. and Ordal, Z. J. (1966). *J. Bact.* **91**, 134.
87. Vary, J. C. and Halvorson, H. O. (1965). *J. Bact.* **89**, 1340.
88. McCormick, N. G. (1965). *J. Bact.* **89**, 1180.
89. Farmiloe, T. J., Cornford, S. J., Coppock, J. B. M. and Ingram, M. (1954). *J. Sci. Fd Agric.* **65**, 292.
90. Levinson, H. S. and Hyatt, M. T. (1962). *J. Bact.* **83**, 1224.
91. Campbell, L. L., Richards, C. M. and Sniff, E. E. (1965). *In* "Spores III" (L. L. Campbell and H. O. Halvorson, eds), p. 55. Am. Soc. Microbiol., Ann Arbor, Michigan, U.S.A.
92. Busta, F. F. and Ordal, Z. J. (1964). *Appl. Microbiol.* **12**, 106.
93. Roberts, T. A. and Hobbs, G. (1968). *J. appl. Bact.* **31**, 75.
94. Augustin, J. A. L. and Pflug, I. J. (1967). *Appl. Microbiol.* **15**, 266.
95. Edwards, J. L. Jr., Busta, F. F. and Speck, M. L. (1965). *Appl. Microbiol.* **13**, 851.
96. Freyer, T. F. and Sharpe, M. E. (1965). *J. Dairy Res.* **32**, 27.
97. Silberschmidt, W. (1899). *Annls Inst. Pasteur, Paris* **13**, 841.
98. Di Marco, A. and Spalla, C. (1957). *G. Microbiol.* **4**, 24.
99. Erikson, D. (1955). *J. gen. Microbiol.* **13**, 127.
100. Bredermann, G. and Werner, W. (1932). *Zenbtl. Bakt. ParasitKde. Abt II*, **86**, 479.
101. Tsiklinsky, P. (1899). *Annls Inst. Pasteur, Paris* **13**, 500.
102. Schutze, H. (1908). *Arch. Hyg. Bakt.* **67**, 35.
103. Erikson, D. (1952). *J. gen. Microbiol.* **6**, 286.
104. Erikson, D. (1955). *J. gen. Microbiol.* **13**, 119.

105. Cross, T. (1968). *J. appl. Bact.* **31**, 36.
106. Lewis, J. C. (1967). *Analyt. Biochem.* **19**, 327.
107. Cross, T. (1968). Personal communication.
108. Fergus, C. E. (1967). *Mycopath. Mycol. appl.* **32**, 205.
109. Thornley, M. J. (1963). *J. appl. Bact.* **26**, 334.
110. Bridges, B. A. (1964). *Prog. ind. Microbiol.* **5**, 283.
111. Stapleton, G. E. (1955). *Bact. Rev.* **19**, 26.
112. Lawrence, C. A., Brownell, L. E. and Graikoski, J. T. (1953). *Nucleonics* **11**, 9.
113. Dunn, C. G., Campbell, W. L., Fram, H. and Hutchins, A. (1948). *J. appl. Phys.* **19**, 605.
114. Niven, C. F. (1958). *A. Rev. Microbiol.* **12**, 507.
115. Anderson, A. W., Nordan, H. C., Cain, R. F., Parrish, G. and Duggan, D. (1956). *Fd Technol., Champaign* **10**, 575.
116. Bridges, B. A. and Horne, T. (1959). *J. appl. Bact.* **22**, 96.
117. Vinter, V. (1961). *Nature, Lond.* **189**, 589.
118. Roberts, T. A. and Ingram, M. (1967). *In* "Botulism 1966" (M. Ingram and T. A. Roberts, eds), p. 169. Chapman and Hall, London.
119. Howard-Flanders, P. and Alper, T. (1957). *Radiat. Res.* **7**, 518.
120. Sargeant, T. (1961). *Radiat. Res.* **14**, 323.
121. Adler, H. I. and Engel, M. S. (1961) *J. cell. comp. Physiol. 58, Suppl.* **1**, 95.
122. Koh, W. Y., Morehouse, C. T. and Chandler, V. L. (1956). *Appl. Microbiol.* **4**, 143.
123. Gunter, S. E. and Kohn, H. I. (1956). *J. Bact.* **72**, 422.
124. Grecz, N., Walker, A. A., Anellis, A. and Berkowitz, D. (1965). *Bact. Proc.* p. 1.
125. Edwards, R. B., Peterson, L. J. and Cummings, D. G. (1954). *Fd Technol., Champaign* **8**, 284.
126. Pratt, G. B. and Eklund, O. F. (1954). *Quick froz. Fds Locker Pl.* **16**, 50.
127. Fuld, G. J., Proctor, B. E. and Goldblith, S. A. (1957). *Int. J. appl. Radiat. Isotopes* **2**, 35.
128. Proctor, B. E., Goldblith, S. A., Fuld, G. J. and Oberle, E. M. (1958). *Radiat. Res.* **8**, 51.
129. Farkas, J., Kiss, I. and Andrassy, E. (1967). *Proc. Symp. Radiosterilization of Medical Products, Budapest*, p. 343. Internat. Atomic Energy Authority, Vienna.
130. Pepper, R. E., Buffa, N. T. and Chandler, V. E. (1956). *Appl. Microbiol.* **4**, 149.
131. Matsuyama, A., Thornley, M. J. and Ingram, M. (1964). *J. appl. Bact.* **27**, 125.
132. Grecz, N., Snyder, O. P., Walker, A. A. and Anellis, A. (1965). *J. Bact.* **13**, 527.
133. Grecz, N., Upadhyay, J. and Tang, T. C. (1967). *Can. J. Microbiol.* **13**, 287.
134. El-Bisi, H. M., Snyder, O. P. and Levin, R. E. (1967). *In* "Botulism 1966" (M. Ingram and T. A. Roberts, eds), p. 89. Chapman and Hall, London.
135. Powers, E. L., Webb, R. B. and Ehret, C. F. (1960). *Radiat. Res. Suppl.* **2**, 94.
136. Grecz, N. (1965). *J. appl. Bact.* **28**, 17.
137. Alper, T. (1956). *Radiat. Res.* **5**, 573.
138. Alper, T. and Howard-Flanders, P. (1956). *Nature, Lond.* **178**, 979.
139. Van Soestbergen, A. A. (1962). *Int. J. Radiat. Biol.* **5**, 567.
140. Van Soestbergen, A. A. (1965). *J. Bact.* **89**, 1032.
141. Howard-Flanders, P. and Jockey, P. (1960). *Radiat. Res.* **13**, 466.

142. Proctor, B. E., Goldblith, S. A., Oberle, E. M. and Miller, W. C. Jr. (1955). *Radiat. Res.* **3**, 295.
143. Howard-Flanders, P. (1957). *Nature, Lond.* **180**, 1191.
144. Dale, W. M., Davies, J. V. and Russell, C. (1961). *Int. J. Radiat. Biol.* **4**, 1.
145. Powers, E. L., Kaleta, B. F. and Webb, R. B. (1959). *Radiat. Res.* **11**, 461.
146. Webb, R. B., Powers, E. L. and Ehret, C. F. (1960). *Radiat. Res.* **12**, 682.
147. Tallentire, A. and Powers, E. L. (1963). *Radiat. Res.* **20**, 270.
148. Powers, E. L., Webb, R. B. and Kaleta, B. F. (1960). *Proc. natn. Acad. Sci. U.S.A.* **46**, 984.
149. Russell, C. (1966). *Experientia*, **22**, 80.
150. Powers, E. L. and Kaleta, B. F. (1960). *Science, N.Y.* **132**, 959.
151. Brustad, T. (1964). *Radiat. Res.* **22**, 421.
152. Webb, R. B. and Powers, E. L. (1963). *Int. J. Radiat. Biol.* **7**, 481.
153. Dewey, D. L. (1963). *Radiat. Res.* **19**, 64.
154. Thompson, T. L., Mefferd, R. B. and Wyss, O. (1951). *J. Bact.* **62**, 39.
155. Liebecq, C. and Ostereith, P. M. (1963). *Archs. int. Physiol.* **71**, 112.
156. Stapleton, G. E., Billen, D. and Hollaender, A. (1952). *J. Bact.* **63**, 805.
157. Burnett, W. T., Morse, M. L., Burke, A. W. Jr. and Hollaender, A. (1952). *J. Bact.* **63**, 591.
158. Kohn, H. I. and Gunter, S. E. (1959). *Radiat. Res.* **11**, 732.
159. Hollaender, A. and Doudney, C. O. (1955). *In* "Radiobiology Symp., Liege" (Z. M. Bacq and P. Alexander, eds). Butterworth's Sci. Pub., London.
160. Bacq, Z. M. and Alexander, P. (1960). "Fundamentals of Radiobiology", 2nd ed. Pergamon Press, Oxford.
161. Doherty, D. G., Burnett, W. T. Jr. and Shapira, R. (1957). *Radiat. Res.* **7**, 13.
162. Langendorff, H. and Koch, R. (1956). *Strahlentherapie* **99**, 567.
163. Hannan, R. S. (1955). *Spec. Rpt Fd Invest. Bd* No. 61. H.M.S.O., London.
164. Eldjarn, L. and Pihl, A. (1960). *In* "Mechanisms in Radiobiology, II" (M. Errera and A. Forssberg, eds), p. 231. Academic Press, New York.
165. Brown, P. E. (1967). *Nature, Lond.* **213**, 363.
166. Elias, C. A. (1961). *Radiat. Res.* **15**, 632.
167. Cromroy, H. L. and Adler, H. I. (1962). *J. gen. Microbiol.* **28**, 431.
168. Ginsberg, D. M. (1966). *Radiat. Res.* **28**, 708.
169. Webb, R. B. (1963). *Radiat. Res.* **18**, 607.
170. Cook, A. M., Roberts, T. A., Widdowson, J. (1964). *J. gen. Microbiol.* **34**, 185.
171. Cloutier, J. A. R. (1961). *Can. J. Phys.* **39**, 514.
172. Phillips, G. O. and Baugh, P. (1963). *Nature, Lond.* **198**, 282.
173. Koch, R. (1957). "Advances in Radiobiology", p. 170. Oliver and Boyd, Edinburgh.
174. Bridges, B. A. (1960). *U.K.A.E.A. Res. Gp Rpt* No. R. 3263.
175. Bridges, B. A. (1962). *Radiat. Res.* **16**, 232.
176. Dean, C. J. and Alexander, P. (1962). *Nature, Lond.* **196**, 1324.
177. Dewey, D. L. and Michael, B. D. (1965). *Biochem. biophys. Res. Commun.* **21**, 392.
178. Bruce, A. K. and Malchman, W. H. (1965). *Radiat. Res.* **24**, 473.
179. El-Tabey Shehata, A. M. (1961). *Radiat. Res.* **15**, 78.
180. Silverman, G. J., El-Tabey Shehata, A. M. and Goldblith, S. A. (1962). *Radiat. Res.* **16**, 432.
181. Chandler, H. K., Licciardello, J. J. and Goldblith, S. A. (1965). *J. Fd Sci.* **30**, 893.
182. Kaplan, H. S., Smith, K. C. and Tomlin, P. (1961). *Nature, Lond.* **190**, 794.

183. Kaplan, H. S., Zavarine, R. and Earle, J. (1962). *Nature, Lond.* **194**, 662.
184. Emmerson, P. T. and Howard-Flanders, P. (1964). *Nature, Lond.* **204**, 1005.
185. Pittillo, R. F., Lucas, M., Blackwell, R. T. and Woolley, C. (1965). *J. Bact.* **90**, 1548.
186. Pittillo, R. F. and Lucas, M. (1965). *Nature, Lond.* **205**, 824.
187. Pittillo, R. F. and Lucas, M. (1967). *Radiat. Res.* **31**, 36.
188. Matsuyama, A., Namiki, M., Okazawa, Y. and Kaneko, I. (1963). *Agric. Biol. Chem.* **27**, 349.
189. Krabbenhoft, K. L., Corlett, D. A., Anderson, A. W. and Elliker, P. R. (1964). *Appl. Microbiol.* **12**, 424.
190. Roberts, T. A., Ditchett, P. J. and Ingram, M. (1965). *J. appl. Bact.* **28**, 336.
191. Vinter, V. (1962). *Folia Microbiol. Praha* **7**, 115.
192. Vinter, V. (1961). *In* "Spores II" (H. O. Halvorson, ed.), p. 127. Burgess Publishing Co., Minneapolis, Minn., U.S.A.
193. Bott, K. F. and Lundgren, D. G. (1964). *Radiat. Res.* **21**, 195.
194. Hitchins, A. D., King, W. L. and Gould, G. W. (1966). *J. appl. Bact.* **29**, 505.
195. Stapleton, G. E., Billen, D. and Hollaender, A. (1953). *J. cell. comp. Physiol.* **41**, 345.
196. Donnellan, J. E. and Morowitz, H. J. (1957). *Radiat. Res.* **7**, 71.
197. Powers, E. L. and Tallentire, A. (1968). *In* "Biological and chemical action of radiation XII" (M. Haissinsky, ed.), p. 74. Masson et cie, Paris.
198. Freeman, B. M. and Bridges, B. A. (1960). *Int. J. appl. Radiat. Isotopes* **8**, 136.
199. Curran, H. R. and Evans, F. R. (1942). *J. Bact.* **43**, 125.
200. Merkenschlager, M., Schlossman, K. and Kurt, W. (1957). *Biochem. Z.* **329**, 332.
201. Garver, J. C. and Epstein, R. L. (1959). *Appl. Microbiol.* **7**, 318.
202. Hughes, D. E. and Cunningham, V. R. (1963). *Biochem. Soc. Symp.* **23**, 8.
203. Nossal, P. M. (1953). *Aust. J. exp. Biol. med. Sci.* **31**, 583.
204. Hughes, D. E. (1951). *Brit. J. exp. Path.* **32**, 97.
205. Edebo, L. (1960). *J. biochem. microbiol. Technol. Engng* **2**, 453.
206. Ribi, E., Perrine, T., List, R., Brown, W. and Goode, G. (1959). *Proc. Soc. exp. Biol. Med.* **100**, 647.
207. Hugo, W. B. (1954). *Bact. Rev.* **18**, 87.
208. Wimpenny, J. W. T. (1967). *Process Biochem.* **2**, 41.
209. O'Connor, R. J., Doi, R. H. and Halvorson, H. O. (1960). *Can. J. Microbiol.* **6**, 233.
210. Sierra, G. (1963). *Can. J. Microbiol.* **9**, 643.
211. Gerhardt, P. and Ribi, E. (1964). *J. Bact.* **88**, 1774.
212. Lawrence, N. L. and Halvorson, H. O. (1954). *J. Bact.* **68**, 334.
213. Pollock, M. R. (1953). *J. gen. Microbiol.* **8**, 186.
214. Simmons, R. J. and Costilow, R. N. (1962). *J. Bact.* **84**, 1274.
215. Warth. A. D., Ohye, D. F. and Murrell, W. G. (1963). *J. Cell. Biol.* **16**, 579.
216. King, H. K. and Alexander, H. (1948). *J. gen. Microbiol.* **2**, 315.
217. Fitz-James, P. C. (1953). *J. Bact.* **66**, 312.
218. Warth. A. D., Ohye, D. F. and Murrell, W. G. (1963). *J. Cell. Biol.* **16**, 593.
219. Hitchins, A. D. and Gould, G. W. (1964). *Nature, Lond.* **203**, 895.
220. Gould, G. W., Georgala, D. L. and Hitchins, A. D. (1964). *Nature, Lond.* **200**, 385.
221. Gould, G. W., Hitchins, A. D. and King, W. L. (1966). *J. gen. Microbiol.* **44**, 293.

222. Walker, P. D., Baillie, A., Thomson, R. O. and Batty, I. (1966). *J. appl. Bact.* **29,** 512.
223. Spencer, R. E. J. and Powell, J. F. (1952). *Biochem. J.* **51,** 239.
224. Sacks, L. E. and Bailey, G. F. (1963). *J. Bact.* **85,** 720.
225. Sacks, L. E. and Thomas, R. S. (1965). *J. Bact.* **89,** 1615.
226. Sacks, L. E., Percell, P. B., Thomas, R. S. and Bailey, G. F. (1964). *J. Bact.* **87,** 952.
227. Rode, L. J. and Foster, J. W. (1960). *Proc. natn. Acad. Sci. U.S.A.* **46,** 118.
228. Knaysi, G. and Curran, H. R. (1961). *J. Bact.* **82,** 691.
229. Monk, G. W., Hess, G. E. and Schenk, H. L. (1957). *J. Bact.* **74,** 292.
230. Davies, R. (1959). *Biochem. biophys. Acta* **33,** 481.
231. Hughes, D. E. (1961). *J. biochim. microbiol. Technol. Engng* **3,** 405.
232. Heiligman, F., Desrosier, N. W. and Broumand, H. (1956). *Fd Res.* **21,** 63.
233. Beckwith, T. D. and Weaver, C. E. (1936). *J. Bact.* **32,** 361.
234. Sehgal, L. R. and Grecz, N. (1967). *Bact. Proc.* p. 5.
235. Pisano, M. A., Boucher, R. M. G. and Alcamo, I. E. (1966). *Appl. Microbiol.* **14,** 732.
236. Boucher, R. M. G., Pisano, M. A., Tortora, G. and Sawick, E. (1967). *Ultrasonics,* **5,** 168.
237. Hedén, C-G. (1964). *Bact. Rev.* **28,** 14.
238. Zobell, C. E. and Cobet, A. B. (1962). *J. Bact.* **84,** 1228.
239. Larsen, W. P., Hartzell, T. B. and Diehl, H. S. (1918). *J. infect. Dis.* **22,** 271.
240. Timson, W. J. and Short, A. J. (1965). *Biotech. Bioeng.* **7,** 139.
241. Basset, J. and Macheboeuf, M. A. (1932). *C.r. hebd. séanc. Acad. Sci., Paris* **196,** 1431.
242. Johnson, F. H. and Zobell, C. E. (1949). *J. Bact.* **57,** 353.
243. Suzuki, C. and Suzuki, K. (1963). *Archs Biochem. Biophys.* **102,** 367.
244. Miyagawa, K. (1965). *Archs Biochem. Biophys.* **110,** 381.
245. Hedén, C-G., Lindahl, T. and Toplin, I. (1964). *Acta chem. scand.* **18,** 1150.
246. Johnson, F. H. and Zobell, C. E. (1949). *J. Bact.* **57,** 359.
247. Hilliard, C. M. and Davis, M. A. (1918). *J. Bact.* **3,** 423.
248. Haines, R. B. (1938). *Proc. R. Soc. Ser. B* **124,** 451.
249. Rode, L. J. and Foster, J. W. (1960). *J. Bact.* **79,** 650.
250. Meryman, H. T. (1966). *In* "Cryobiology" (H. T. Meryman, ed.), p. 609. Academic Press, London.
251. Fry, R. M. (1966). *In* "Cryobiology" (H. T. Meryman, ed.), p. 665. Academic Press, London.
252. Harper, M. K., Curran, H. R. and Pallansch, M. J. (1964). *J. Bact.* **88,** 1338.
253. Neihof, R. A., Thompson, J. K. and Deitz, V. R. (1967). *Nature, Lond.* **216,** 1304.
254. Berlin, E., Curran, H. R. and Pallansch, M. J. (1963). *J. Bact.* **86,** 1030.
255. Prince, A. E. (1960). *Devs. ind. Microbiol.* **1,** 13.
256. Portner, D. M., Spiner, D. R., Hoffman, R. K. and Phillips, C. R. (1961). *Science, N.Y.* **134,** 2047.
257. Morelli, F. A., Fehlner, F. P. and Stembridge, C. H. (1962). *Nature, Lond.* **196,** 106.
258. Brueschke, E. E., Suess, R. H. and Willard, D. M. (1961). *Planet. Space Sci.* **8,** 30.
259. Davis, N. S., Silverman, G. J. and Keller, W. M. (1963). *Appl. Microbiol.* **11,** 202.
260. Zamenhof, S. (1960). *Proc. natn. Acad. Sci. U.S.A.* **46,** 101.

261. Silverman, G. J., Davis, N. S. and Beecher, N. (1967). *Appl. Microbiol.* **15,** 510.
262. Sussman, A. and Halvorson, H. O. (1966). "Spores (Their Dormancy and Germination)". Harper and Row, New York.
263. Evans, F. R. and Curran, H. R. (1960). *J. Bact.* **79,** 361.
264. Bullock, K. and Tallentire, A. (1952). *J. Pharm. Pharmac.* **4,** 917.
265. Marshall, B. J., Murrell, W. G. and Scott, W. J. (1963). *J. gen. Microbiol.* **31,** 451.
266. Lewis, J. C. (1961). *In* "Spores II" (H. O. Halvorson, ed.), p. 165. Burgess Publishing Co., Minneapolis, Minn., U.S.A.
267. Hyatt, M. T., Holmes, P. K. and Levinson, H. S. (1966). *Biochem. biophys. Res. Commun.* **24,** 701.
268. Hyatt, M. T., Holmes, P. K. and Levinson, H. S. (1967). *Bact. Proc.* p. 23.
269. Hawrylewicz, E. J., Gowdy, B. and Ehrlich, R. (1962). *Nature, Lond.* **193,** 497.
270. Hagen, C. A., Hawrylewicz, E. J. and Ehrlich, R. (1964). *Appl. Microbiol.* **12,** 215.
271. Harper, G. J., Hood, A. M. and Morton, J. D. (1958). *J. Hyg. Camb.* **56,** 364.
272. Levine, M. A. and Cabelli, V. J. (1963). *Bact. Proc.* p. 26.
273. Anderson, J. D. (1966). *J. gen. Microbiol.* **45,** 303.
274. Cox, C. S. (1966). *J. gen. Microbiol.* **43,** 383.
275. Hamilton, W. A. and Sale, A. J. H. (1967). *Biochim. biophys. Acta* **148,** 789.
276. Allen, M. and Soike, K. (1966). *Science, N.Y.* **154,** 155.
277. McCulloch, E. C. (1945). "Disinfection and Sterilization". Henry Kimpton, London.
278. Wyss, D. (1948). *A. Rev. Microbiol.* **2,** 413.
279. Sykes, G. (1965). "Disinfection and Sterilization". 2nd ed. E. and F. N. Spon, London.
280. Rubbo, S. D. and Gardner, J. S. (1965). "A Review of Sterilization and Disinfection". Lloyd-Luke Ltd, London.
281. Russell, A. D. (1965). *Mfg Chemist* **36,** 38.
282. Reddish, G. F. (1957). "Antiseptics, Disinfectants, Fungicides, and Chemical and Physical Sterilization". Henry Kimpton, London.
283. Curran, H. R. (1952). *Bact. Rev.* **16,** 111.
284. Phillips, C. R. (1952). *Bact. Rev.* **16,** 135.
285. Phillips, C. R. and Warshowsky, B. (1958). *A. Rev. Microbiol.* **12,** 525.
286. Rahn, O. (1945). "Injury and death of bacteria by chemical agents". Biodynamica Monograph No. 3. Biodynamica, Normandy, Mo.
287. Hugo, W. B. (1967). *J. appl. Bact.* **30,** 17.
288. Opfell, J. B. and Miller, C. E. (1965). *Adv. appl. Microbiol.* **7,** 81.
289. Olcott, H. S. and Fraenkel-Conrat, H. (1947). *Chem. Rev.* **41,** 151.
290. Ortenzio, L. F., Stuart, L. S. and Friedl, J. L. (1963). *J. Ass. off. agric. Chem.* **36,** 480.
291. Friedl, J. L. (1957). *J. Ass. off. agric. Chem.* **40,** 759.
292. Fredl, J. L. (1960). *J. Ass. off. agric. Chem.* **43,** 386.
293. Dozier, C. C. (1924). *J. infect. Dis.* **35,** 156.
294. Keynan, A., Issahary-Brand, G. and Evenchik, Z. (1965). *In* "Spores III" (L. L. Campbell and H. O. Halvorson, eds), p. 180. Am. Soc. Microbiol., Ann Arbor, Michigan, U.S.A.
295. Rowley, D. B. and Levinson, H. S. (1967). *J. Bact.* **93,** 1017.
296. Gould, G. W. and Hitchins, A. D. (1963). *J. gen. Microbiol.* **33,** 413.
297. Robinow, C. F. (1951). *J, gen. Microbiol.* **5,** 439.

298. Fitz-James, P. C., Robinow, C. F. and Bergold, G. H. (1954). *Biochim. biophys. Acta* **14**, 346.
299. Lechtman, M. D., Bartholomew, J. W., Phillips, A. and Russo, M. (1965). *J. Bact.* **89**, 848.
300. Young, E. I. and Fitz-James, P. C. (1962). *J. Cell Biol.* **12**, 115.
301. Perry, J. J. and Foster, J. W. (1955). *J. Bact.* **69**, 337.
302. Morton, H. E. (1957). *In* "Antiseptics, Disinfectants, Fungicides, and Chemical and Physical Sterilization" (G. F. Reddish, ed.), p. 377. Henry Kimpton, London.
303. Beeby, M. M. and Whitehouse, C. E. (1965). *J. appl. Bact.* **28**, 349.
304. Klarmann, E. G. (1959). *Am. J. Pharm.* **131**, 86.
305. Curran, H. R. and Knaysi, G. (1961). *J. Bact.* **82**, 793.
306. Hyatt, M. T., Holmes, P. K. and Levinson, H. (1967). *Bact. Proc.* p. 23.
307. Rode, L. J. and Foster, J. W. (1960). *J. Bact.* **79**, 650.
308. Doyle, J. E. and Ernst, R. R. (1967). *Appl. Microbiol.* **15**, 726.
309. Halvorson, H. O. (1961). *In* "Cryptobiotic Stages in Biological Systems" (N. Grossowicz, S. Hestrin and A. Keynan, eds), p. 32. Elsevier, Amsterdam.
310. Halvorson, H. O. (1959). *Bact. Rev.* **23**, 267.
311. Ryter, A. (1965). *Annls Inst. Pasteur, Paris*, **108**, 40.
312. Slepecky, R. A. (1963). *Biochem. biophys. Res. Commun.* **12**, 369.
313. Slepecky, R. A. and Celkis, Z. (1964). *Bact. Proc.* p. 14.
314. Widdowson, J. P. (1967). *Nature, Lond.* **214**, 812.
315. Klarmann, E. G. (1956). *Am. J. Pharm.* **128**, 4.
316. Willard, M. and Alexander, A. (1964). *Appl. Microbiol.* **12**, 229.
317. Nordgren, G. (1939). *Acta path. microbiol. scand. Suppl.* p. 40.
318. Report of the Committee on Formaldehyde Disinfection of the Public Health Laboratory Service (1958). *J. Hyg. Camb.* **56**, 488.
319. Bullock, K. and Rawlins, E. A. (1954). *J. Pharm. Pharmac.* **6**, 859.
320. Public Health Laboratory Service (1958). *Mon. Bull. Minist. Hlth*, **17**, 270.
321. Greenstein, J. P. and Winitz, M. (1961). "Chemistry of the Amino Acids". John Wiley and Sons, London.
322. Pepper, R. E. and Chandler, V. L. (1963). *Appl. Microbiol.* **11**, 384.
323. Borick, P. M., Dondershine, F. H. and Chandler, V. L. (1964). *J. Pharm. Sci.* **53**, 1273.
324. Stonehill, A. A., Krop, S. and Borick, P. M. (1963). *Am. J. Hosp. Pharm.* **20**, 458.
325. Duquette, E. and Snyder, R. (1966). *Bact. Proc.* p. 52.
326. Rubbo, S. D., Gardner, J. F. and Webb, R. L. (1967). *J. appl. Bact.* **30**, 78.
327. Filachione, E. M., Korn, A. H. and Ard, J. S. (1967). *J. Am. Leath. Chem. Ass.* **62**, 450.
328. Hamilton, W. A. and Stubbs, J. M. (1967). *J. gen. Microbiol.* **47**, 121.
329. Krönig. B. and Paul, T. (1897). *Z. Hyg. InfektKrankh.* **25**, 1.
330. Myers, R. P. (1929). *J. agric. Res.* **38**, 521.
331. Topley, W. W. C. and Wilson, G. S. (1929). "The Principles of Bacteriology and Immunity". Edward Arnold Ltd, New York.
332. McCulloch, E. C. (1933). *J. Bact.* **25**, 469.
333. Clegg, L. F. L. (1956). *J. Soc. Dairy Technol.* **9**, 95.
334. Whitehouse, R. L. and Clegg, L. F. L. (1963). *J. Dairy Res.* **30**, 315.
335. Levine, M., Toulouse, J. H. and Buchanan, J. H. (1928). *Ind. Engng Chem.* **20**, 63.
336. Hobbs, B. C. and Wilson, G. S. (1942). *J. Hyg. Camb.* **42**, 436.
337. Franklin, J. G. and Clegg, L. F. L. (1956). *Dairy Inds.* **21**, 454.

338. Van de Gehuchte, E. E. (1962). *Milchwissenschaft* **17**, 693.
339. Levine, M. (1952). *Bact. Rev.* **16**, 117.
340. Blankenship, L. C. and Pallanch, M. J. (1966). *J. Bact.* **92**, 1615.
341. Gould, G. W. and Hitchins, A. D. (1963). *J. gen. Microbiol.* **33**, 413.
342. Kondo, M. and Foster, J. W. (1967). *J. gen. Microbiol.* **47**, 257.
343. Davis, J. G. (1960). *J. Pharm. Pharmac.* **12**, 29T.
344. Phillips. C. R. (1957). *In* "Antiseptics, Disinfectants, Fungicides and Chemical and Physical Sterilization" (G. F. Reddish, ed.), p. 746. Henry Kimpton, London.
345. Bruch, C. W. (1961). *A. Rev. Microbiol.* **15**, 245.
346. Wheeler, G. P. (1962). *Cancer Res.* **22**, 651.
347. Phillips, C. R. and Kaye, S. (1949). *Am. J. Hyg.* **50**, 270.
348. Phillips, C. R. (1949). *Am. J. Hyg.* **50**, 280.
349. Kaye, S. (1949). *Am. J. Hyg.* **50**, 289.
350. Kaye, S. and Phillips, C. R. (1949). *Am. J. Hyg.* **50**, 296.
351. El-Bisi, H. M., Vondell, R. M. and Esselen, W. B. (1963). *Bact. Proc.* p. 13.
352. Liu, T. (1966). *Thesis*, University of Massachusetts.
353. Ernst, R. R. and Shull, J. J. (1962). *Appl. Microbiol.* **10**, 337.
354. Friedl, J. L., Ortenzio, L. F. and Stuart, L. S. (1956). *J. Ass. off. agric. Chem.* **39**, 480.
355. Mayr, G. (1961). *In* "Symposium on Recent Developments in the Sterilization of Surgical Materials", p. 90. Pharmaceutical Press, London.
356. Ernst, R. R. and Shull, J. J. (1962). *Appl. Microbiol.* **10**, 342.
357. El-Bisi, H. M., Vondell, R. M. and Esselen, W. G. (1963). *Bact. Proc.* p. 13.
358. El-Bisi, H. M., Vondell, R. M. and Esselen, W. G. (1963). *Bact. Proc.* p. 13.
359. Doyle, J. E. and Ernst, R. R. (1967). *Bact. Proc.* p. 13.
360. Phillips, C. R. (1961). *In* "Symposium on Recent Developments in the Sterilization of Surgical Materials", p. 59. Pharmaceutical Press, London.
361. Gilbert, G. L., Gambill, V. M., Spiner, D. R., Hoffman, R. K. and Phillips, C. R. (1964). *Appl. Microbiol.* **12**, 496.
362. Shull, J. J. (1963). *Bull. parent. Drug. Ass.* **17**, 9.
363. Thomas, C. G. A. (1960). *Guy's Hosp. Rep.* **109**, 57.
364. Toth, L. Z. (1959). *Archs Microbiol.* **32**, 409.
365. Znamirowski, R., McDonald, S. and Roy, T. E. (1960). *Can. med. Ass. J.* **83**, 1004.
366. Freeman, M. A. R. and Barwell, C. F. (1960). *J. Hyg. Camb.* **58**, 337.
367. Opfell, J. B. and Shannon, J. L. and Chan, H. (1967). *Bact. Proc.* p. 13.
368. Church, B. D., Halvorson, H., Ramsey, D. S. and Hartman, R. S. (1956). *J. Bact.* **72**, 242.
369. Bomar, M. (1962). *Folia Microbiol. Praha* **7**, 259.
370. Fraenkel-Conrat, H. L. (1944). *J. biol. Chem.* **154**, 227.
371. Alexander, P. and Stacey, K. A. (1958). *Ann. N.Y. Acad. Sci.* **68**, 1225.
372. Windmueller, H. G., Ackerman, C. J., Bakerman, H. and Mickelen, O. (1959). *J. biol. Chem.* **234**, 899.
373. Spiner, D. R. and Hoffman, R. K. (1960). *Appl. Microbiol.* **8**, 152.
374. Hoffman, R. K. and Warshowsky, B. (1958). *Appl. Microbiol.* **6**, 358.
375. Hartman, F. W., Piepes, S. C. and Wallbank, A. M. (1951). *Fedn Proc. Fedn Am. Socs exp. Biol.* **10**, 358.
376. Curran, H. R. and Evans, F. R. (1956). *J. Inf. Dis.* **99**, 212.
377. Jones, L. A., Hoffman, R. K. and Phillips, C. R. (1967). *Appl. Microbiol.* **15**, 357.
378. Bruch, C. W. and Koesterer, M. G. (1961). *J. Fd Sci.* **26**, 428.

379. Himmelfarb, P., El-Bisi, H. M., Read, R. B. and Litsky, W. (1962). *Appl. Microbiol.* **10**, 431.
380. Kolb, R. W. and Schneiter, R. (1950). *J. Bact.* **59**, 401.
381. Dawson, F. W. (1962). *Am. J. Hyg.* **76**, 209.
382. Campbell, L. L. and Sniff, E. E. (1959). *J. Bact.* **77**, 766.
383. Ramsierer, H. R. (1960). *Archs Microbiol.* **37**, 57.
384. Gould, G. W. and Hurst, A. (1962). VIIIth Int. Congr. Microbiol. *Abstracts*, A2–11.
385. Tramer, J. (1964). *In* "Microbial Inhibitors in Food" (N. Molin and I. Erichsen, eds), p. 25. Almqvist and Wiksell, Stockholm.
386. Hitchins, A. D., Gould, G. W. and Hurst, A. (1963). *J. gen. Microbiol.* **30**, 445.
387. Greenberg, R. A. and Silliker, J. H. (1962). *J. Fd Sci.* **27**, 64.
388. Sacks, L. E. (1956). *J. Bact.* **70**, 491.
389. Schiebel, I. and Lennert-Peterson, O. (1958). *Acta Path. microbiol. scand.* **44**, 222.
390. Wynne, E. S. and Harrell, K. (1951). *Antibiotics Chemother.* **1**, 198.
391. Wynne, E. S., Collier, R. E. and Mehl, D. A. (1952). *J. Bact.* **64**, 883.
392. Kosaki, N. (1959). *Jap. J. Bact.* **14**, 161.
393. Kaufman, O. W., Ordal, Z. J. and El-Bisi, H. M. (1955). *Fd Res.* **19**, 483.
394. Gould, G. W. (1966). *J. Bact.* **92**, 1261.
395. Cruickshank, R. (1960). Ed. "Mackie and McCartney's Handbook of Bacteriology". 10th ed. E. & S. Livingstone Ltd, London.
396. Meynell, G. G. and Meynell, E. (1965). "Theory and Practice in Experimental Bacteriology". University Press, Cambridge.
397. Ashby, G. K. (1938). *Science N.Y.* **87**, 443.
398. Schaeffer, A. B. and Fulton, D. (1933). *Science, N.Y.* **77**, 194.
399. Bartholomew, J. W., Roberts, M. A. and Evans, E. (1950). *Stain Technol.* **25**, 181.
400. Robinow, C. F. (1960). *In* "The Bacteria: A Treatise on Structure and Function" (I. C. Gunsalus and R. Y. Stanier, eds), Vol. 1, p. 207. Academic Press, New York.
401. Bartholomew, J. W. and Mittwer, T. (1950). *Stain Technol.* **25**, 153.
402. Bartholomew, J. W. and Mittwer, T. (1952). *J. Bact.* **63**, 779.
403. Hashimoto, T. and Naylor, H. B. (1958). *J. Bact.* **75**, 647.
404. Hashimoto, T., Black, S. H. and Gerhardt, P. (1959). *Bact. Proc.* p. 38.
405. Nickerson, W. J. (1963). *Bact. Rev.* **27**, 305.
406. Sierra, G. (1964). *Can. J. Microbiol.* **10**, 929.
407. Gould, G. W. and King, W. L. (1966). *Nature, Lond.* **211**, 1431.
408. Sierra, G. (1967). *Can. J. Microbiol.* **13**, 489.
409. Mustafa, A. A. (1964). *Thesis*, Kansas State University.
410. Tomcsik, J. and Baumann-Grace, J. B. (1959). *J. gen. Microbiol.* **21**, 666.
411. Douglas, H. W. (1957). *J. appl. Bact.* **20**, 390.
412. Gould G. W. and Hitchins, A. D. (1963). *Nature, Lond.* **197**, 622.
413. Ohye, D. F. and Murrell, W. G. (1962). *J. Cell. Biol.* **14**, 111.
414. Yoshikawa, H. (1965). *Proc. natn. Acad. Sci. U.S.A.* **59**, 1476.
415. Takahashi, I. (1964). *J. Bact.* **87**, 1499.
416. Falaschi, A., Spudich, J. and Kornberg, A. (1965). *In* "Spores III" (L. L. Campbell and H. O. Halvorson, eds), p. 88. Am. Soc. Microbiol., Ann Arbor, Michigan, U.S.A.
417. Tono, H. and Kornberg, A. (1967). *J. biol. Chem.* **242**, 2375.
418. Suzuki, Y. and Rode, L. J. (1967). *Bact. Proc.* p. 22.

419. Gould, G. W. and Hitchins, A. D. (1965). *In* "Spores III" (L. L. Campbell and H. O. Halvorson, eds), p. 213. Am. Soc. Microbiol., Ann Arbor, Michigan, U.S.A.
420. Strange, R. E. and Dark, F. A. (1957). *J. gen. Microbiol.* **16**, 236.
421. Kim, J. and Naylor, H. B. (1966). *Bact. Proc.* p. 31.
422. Vary, J. C. (1965). *Bact. Proc.* p. 37.
423. Halvorson, H. O., Vary, J. C. and Steinberg, W. (1966). *A. Rev. Microbiol.* **20**, 169.
424. Hadfield, W. A. (1957). *In* "Antiseptics, Disinfectants, Fungicides and Chemical and Physical Sterilization". (G. F. Reddish, ed.), p. 558. Henry Kimpton, London.
425. Brazis, A. R., Leslie, J. R., Kabler, R. W. and Woodward, R. L. (1958). *Appl. Microbiol.* **6**, 338.
426. Tilley, F. W. and Chapin, R. M. (1930). *J. Bact.* **19**, 295.
427. Weber, C. R. and Levine, M. (1944). *Am. J. publ. Hlth* **34**, 719.
428. Fair, G. M., Morris, J. C. and Chang, S. L. (1947). *J. New Engl. Wat. Wks Ass.* **61**, 285.
429. Moores, E. W. (1951). *Wat. Sewage Wks*, **98**, 130.
430. Rudolf, A. S. and Levine, M. (1941). *Bull. Ia. Engng Exp. Stn* **150**, No. 6.
431. Cousins, C. M. and Allan, C. D. (1967). *J. appl. Bact.* **30**, 168.
432. Charlton, D. B. and Levine, M. (1935). *J. Bact.* **30**, 163.
433. Shura-Bura, B. L. (1959). *J. Microbiol. Epidem. Immunobiol.* **30**, 58.
434. Rode, L. J. and Williams, M. G. (1966). *J. Bact.* **92**, 1772.
435. Johns, C. K. (1930). *Scient. Agric.* **10**, 553.
436. Marks, H. C. and Strandskov, F. B. (1950). *Ann. N.Y. Acad. Sci.* **53**, 163.
437. Kristoffersen, T. (1958). *J. Dairy Sci.* **41**, 942.
438. Shere, L., Kelley, M. J. and Richardson, J. H. (1962). *Appl. Microbiol.* **10**, 538.
439. Gershenfeld, L. (1957). *In* "Antiseptics, Disinfectants, Fungicides and Chemical and Physical Sterilization". (G. F. Reddish, ed.), p. 223. Henry Kimpton, London.
440. Wyss, O. and Strandskov, F. B. (1945). *Archs Biochem.* **6**, 261.
441. Gershenfeld, L. and Witlin, B. (1949). *Am. J. Pharm.* **121**, 95.
442. Gershenfeld, L. and Witlin, B. (1952). *J. Am. pharm. Ass. (Sci.)* **41**, 451.
443. Allawala, N. A. and Riegelman, S. (1953). *J. Am. pharm. Ass. (Sci.)* **42**, 396.
444. Niyiri, W. and Dubois, L. (1931). *J. Am. pharm. Ass. (Sci.)* **20**, 546.
445. Davis, J. G. (1962). *J. appl. Bact.* **25**, 195.
446. Bartlett, F. G. and Schmidt, W. (1957). *Appl. Microbiol.* **5**, 355.
447. Lawrence, C. A., Carpenter, C. M. and Naylor-Foote, A. W. C. (1957). *J. Am. pharm. Ass. (Sci.)*, **46**, 500.
448. Gershenfeld, L. (1962). *Am. J. Pharm.* **132**, 78.
449. Mackie, T. J. (1928). "An inquiry into post-operative tetanus. A report to the Scottish Board of Health". H.M.S.O., London.
450. Zinsser, H. and Bayne-Jones, S. (1939). "A Textbook of Bacteriology". 8th ed., p. 107. D. Appleton Century Co., New York.
451. Gershenfeld, L. and Witlin, B. (1948). *Am. J. Pharm.* **120**, 158.
452. Brewer, J. H. (1957). *In* "Antiseptics, Disinfectants, Fungicides and Chemical and Physical Sterilization" (G. F. Reddish, ed.), p. 278, Henry Kimpton, London.
453. Fildes, P. (1940). *Br. J. Exp. Path.* **21**, 67.
454. Keynan, A. and Halvorson, H. O. (1962). *J. Bact.* **83**, 100.
455. Hyatt, M. T. and Levinson, H. S. (1962). *J. Bact.* **83**, 1231.

456. Powell, J. F. (1950). *J. gen. Microbiol.* **4**, 330.
457. Murty, G. G. K. and Halvorson, H. O. (1957). *J. Bact.* **73**, 230.
458. Romans, I. B. (1957). *In* "Antiseptics, Disinfectants, Fungicides and Chemical and Physical Sterilization" (G. F. Reddish, ed.), p. 457. Henry Kimpton, London.
459. Romans, I. B. (1957). *In* "Antiseptics, Disinfectants, Fungicides and Chemical and Physical Sterilization" (G. F. Reddish, ed.), p. 465. Henry Kimpton, London.
460. Nachum, R. and Lechtman, M. D. (1967). *Bact. Proc.* p. 9.
461. Supfle, K. and Werner, R. (1951). *Mikrochemie mikrochem. Acta* **36/37**, 866.
462. Salle, A. J. (1957). *In* "Antiseptics, Disinfectants, Fungicides and Chemical and Physical Sterilization." (G. F. Reddish, ed.), p. 308, Henry Kimpton, London.
463. Levinson, H. S. and Sevag, M. G. (1953). *J. gen. Physiol.* **36**, 617.
464. Hyatt, M. T. and Levinson, H. S. (1957). *J. Bact.* **74**, 87.
465. Curran, H. R., Evans, F. R. and Leviton, A. (1940). *J. Bact.* **40**, 423.
466. Lawrence, N. L. and Halvorson, H. O. (1954). *J. Bact.* **68**, 334.
467. Brown, M. R. W. and Melling, J. (1968). *Biochem. J.* **106**, 44P.
468. Deasy, C. L. (1967). *J. Am. Leath. Chem. Ass.* **63**, 258.
469. Trexler, P. C. and Reynolds, L. I. (1957). *Appl. Microbiol.* **5**, 406.
470. Wynne, E. S., Edwards, F. F. and Eller, C. (1967). *Bact. Proc.* p. 14.
471. Eller, C., Edwards, F. F. and Wynne, E. S. (1967). *Bact. Proc.* p. 14.
472. Rode, L. J., Lewis, C. W. and Foster, J. W. (1962). *J. Cell. Biol.* **13**, 123.
473. Robinow, C. F. (1951). *J. Bact.* **66**, 300.
474. Bradbury, S. and Meek, G. A. (1960). *J. microsc. Sci.* **101**, 241.
475. Godding, R. M. and Lynch, V. H. (1965). *Appl. Microbiol.* **13**, 10.
476. Leiguards, R. H., Pero, O. A. and Polazzolo, A. Z. (1950). *Biol. Abstr.* No. 37082.
477. Klarmann, E. G. and Wright, E. S. (1957). *In* "Antiseptics, Disinfectants, Fungicides and Chemical and Physical Sterilization", (G. F. Reddish ed.), p. 506. Henry Kimpton, London.
478. Bennett, E. O. (1959). *Adv. appl. Microbiol.* **1**, 123.
479. Cook, A. M. (1960). *J. Pharm. Pharmac.* **12**, 19T.
480. Sykes, G. and Hooper, M. C. (1954). *J. Pharm. Pharmac.* **6**, 552.
481. Loosemore, M. and Russell, A. D. (1963). *J. Pharm. Pharmac.* **15**, 558.
482. McKay, I. (1952). *J. Hyg. Camb.* **50**, 82.
483. Fernelius, A. L. (1960). *J. Bact.* **79**, 755.
484. Reddish, C. F., cited in Schmidt, C. F. (1955). *A. Rev. Microbiol.* **9**, 387.
485. Davis, F., Wyss, O. and Williams, O. B. (1948). *J. Bact.* **56**, 561.
486. Russell, A. D. and Loosemore, M. (1964). *Appl. Microbiol.* **12**, 403.
487. Shimizu, W. and Ueno, S. (1955). *Bull. Jap. Soc. scient. Fish.* **20**, 927.
488. Berry, H., Jensen, E. and Siller, F. K. (1938). *J. Pharm. Pharmac.* **11**, 729.
489. Davies, C. E. and Davison, J. E. (1947). *Q. J. Pharm. Pharmac.* **20**, 212.
490. Hachisuka, Y., Nagaya, A., Hijikata, Y. and Fundhashi, K. (1963). *J. Nagoya City Univ. Med. Ass.* **14**, 200.
491. Loosemore, M. and Russell, A. D. (1964). *J. Pharm. Pharmac.* **16**, 817.
492. Loosemore, M. (1964). *Thesis*, University of Wales.
493. Lund, B. M. (1962). *Thesis*, University of London.
494. Vinter, V. (1960). *Folia Microbiol., Praha* **5**, 217.
495. Aronson, A. I. and Fitz-James, P. C. (1966). *Bact. Proc.* p. 31.
496. Keynan, A., Evenchik, Z., Halvorson, H. O. and Hastings, J. W. (1964). *J. Bact.* **88**, 313.

497. Black, S. H. (1964). *Bact. Proc.* p. 36.
498. Kivela, E. W., Mallman, W. L. and Churchill, E. S. (1948). *J. Bact.* **55**, 565.
499. Davies, G. E. (1949). *J. Hyg. Camb.* **47**, 271.
500. Klarman, E. G. and Wright, E. S. (1950). *Am. J. Pharm.* **122**, 330.
501. Chiori, C. O., Hambleton, R. and Rigby, G. J. (1965). *J. appl. Bact.* **28**, 322.
502. Rode, L. J. and Foster, J. W. (1960). *Arch. Mikrobiol.* **36**, 67.
503. Rode, L. J. and Foster, J. W. (1960). *Nature, Lond.* **188**, 1132.
504. Rode, L. J. and Foster, J. W. (1961). *J. Bact.* **81**, 768.
505. Foster, J. W. (1956). *Q. Rev. Biol.* **31**, 102.
506. Lamanna, C. (1952). *Bact. Rev.* **16**, 90.
507. Powell, J. F. (1953). *Biochem. J.* **54**, 210.
508. Lechowich, R. V. (1958). *Thesis*, University of Illinois.
509. Hashimoto, T., Black, S. H. and Gerhardt, P. (1960). *Can. J. Microbiol.* **6**, 203.
510. Halvorson, H. O. (1957). *J. appl. Bact.* **20**, 305.
511. Church, B. D. and Halvorson, H. O. (1959). *Nature, Lond.* **183**, 124.
512. Byrne, A. F., Burton, T. H. and Koch, R. B. (1960). *J. Bact.* **80**, 139.
513. Lechowich, R. V. and Ordal, Z. J. (1962). *Can. J. Microbiol.* **8**, 287.
514. Levinson, H. S., Hyatt, M. T. and Moore, F. E. (1961). *Biochem. biophys. Res. Commun.* **5**, 417.
515. Walker, H. W., Matches, J. R. and Ayres, J. C. (1961). *J. Bact.* **82**, 960.
516. Tang, T. C. and Grecz, N. (1966). *Bact. Proc.* p. 32.
517. Black, S. H., Hashimoto, T. and Gerhardt, S. A. (1962). *Can. J. Microbiol.* **6**, 213.
518. Curran, H. R., Brunstetter, B. C. and Myers, A. T. (1943). *J. Bact.* **45**, 485.
519. El-Bisi, H. M., Lechowich, R. V., Amaha, M. and Ordal, Z. J. (1962). *J. Fd Sci.* **27**, 219.
520. Pelcher, E. A., Fleming, H. P. and Ordal, Z. J. (1963). *Can. J. Microbiol.* **9**, 251.
521. Lewis, J. C., Snell, N. S. and Burr, H. K. (1960). *Science, N.Y.* **132**, 544.
522. Amaha, M. and Sakaguchi, K. (1957). *J. gen. appl. Microbiol.* **3**, 163.
523. Sadoff, H. L. (1961). *In* "Spores II" (H. O. Halvorson, ed.), p. 180. Burgess Publishing Co., Minneapolis, Minn., U.S.A.
524. Sadoff, H. L., Bach, J. A. and Kools, J. W. (1965). *In* "Spores III" (L. L. Campbell and H. O. Halvorson, eds). p. 97. Am. Soc. Microbiol., Ann Arbor, Michigan, U.S.A.
525. Burton, S. D. and Morita, R. Y. (1963). *J. Bact.* **86**, 1019.
526. Morita, R. Y. and Burton, S. D. (1963). *J. Bact.* **86**, 1025.
527. Lea, D. E. (1956). "Action of Radiation on Living Cells '. University Press, Cambridge.
528. Costilow, R. (1962). *J. Bact.* **84**, 1268.
529. Zamenhof, S., Bursztyn, H., Reddy, T. K. R. and Zamenhof, P. J. (1965). *J. Bact.* **90**, 108.
530. Alexander, P. and Bacq, Z. M. (1961). *In* "The initial effects of ionizing radiation on cells" (R. J. C. Harris, ed.), Academic Press Inc., New York.
531. Atwood, K. C. and Norman, A. (1949). *Proc. natn. Acad. Sci. U.S.A.* **35**, 696.
532. Ore, A. (1957). *Radiat. Res.* **6**, 27.
533. Wijsman, R. A. (1956). *Radiat. Res.* **4**, 257.
534. Davis, M. (1954). *Archs Biochem. Biophys.* **48**, 469.
535. Dewey, W. C. and Cole, A. (1962). *Nature, Lond,* **194**, 660.
536. Woese, C. (1958). *Archs Biochem. Biophys.* **74**, 28.

537. Kaplan, H. S. and Zavarine, R. (1962). *Biochem. biophys. Res. Commun.* **8,** 432.
538. Moseley, B. E. B. and Schein, A. H. (1964). *Nature, Lond.* **203,** 1298.
539. Oliver, R. and Shepstone, B. J. (1964). *Physics Med. Biol.* **9,** 167.
540. Kilburn, R. E., Bellamy, W. D. and Terni, S. A. (1958). *Radiat. Res.* **9,** 207.
541. Moseley, B. E. B. and Laser, H. (1965). *Proc. R. Soc. Ser. B.* **162,** 210.
542. Moseley, B. E. B. (1967). *J. gen. Microbiol.* **49,** 203.
543. Setlow, J. K. and Duggan, D. E. (1964). *Biochim. biophys. Acta,* **87,** 664.
544. Pettijohn, D. and Hanawalt, P. (1964). *J. molec. Biol.* **9,** 395.
545. Moseley, B. E. B. and Laser, H. (1965). *Nature, Lond.* **206,** 273.
546. McGrath, R. A. and Williams, R. W. (1966). *Nature, Lond.* **212,** 534.
547. Donnellan, J. E. Jr. and Setlow, R. B. (1965). *Science, N.Y.* **149,** 308.
548. Curran, H. R. and Evans, F. R. (1945). *J. Bact.* **49,** 335.
549. Busta, F. F. and Ordal, Z. J. (1964). *J. Fd. Sci.* **29,** 345.
550. Gould, G. W. and Ordal, Z. J. (1968). *J. gen. Microbiol.* **50,** 77.
551. Kempe, L. L. (1955). *Appl. Microbiol.* **3,** 346.
552. Licciardello, J. J. and Nickerson, J. T. R. (1962). *J. Fd Sci.* **27,** 211.
553. Farkas, J., Kiss, I. and Andrassy, E. (1966). *Acta Microbiol.* **13,** 35.
554. Kennedy, E. J. and Grecz, N. (1966). *Bact. Proc.* p. 14.
555. Berger, J. A. and Marr, A. G. (1960). *J. gen. microbiol.* **22,** 147.
556. Roberts, T. A. and Ingram, M. (1965). *J. Food Sci.* **30,** 879.
557. Ball, C. O. (1958). *J. gen. appl. Microbiol.* **4,** 312.

Author Index

The numbers in parentheses are reference numbers and are included to assist in locating references when authors' names are not mentioned in the text. The numbers in *italics* refer to the pages on which the references are listed.

A

Abraham, E. P., 167(1) 170(15) 171(34) 172(40) 177(40) *180, 181*

Abrahamsson, K., 534(132) 535(167) 537(167) 538(195) *545, 546, 547*

Acha, I. G., 28(176) *38*

Ackerman, C. J., 641(372) *665*

Ackers, G. K., 291(100), *299*

Adkisson, P. L., 17(105) *36*

Adler, H. I., 627(121) 630(167) 634(121) *659, 660*

Agnihotri, V. P., 26(162) *37*

Aida, K., 243(82) 245(82) *271*

Ainsworth, G. C., 170(28) *181*

Aizawa, K., 507(80) *515*

Akashi, S., 241(75) 242(75) *271*

Albertini, A., 176(64) 179(64) *182*

Albrink, W. S., 540(74) *543*

Alcamo, I. E., 634(235) *662*

Alderton, G., 217(5, 12) 218(5, 12) 246(5, 87, 88) 249(87) 263(87, 88) 264(87, 88) *269, 271,* 305(10) 309(10) 311(39) 314(10, 39, 57) 316(39) 336(156) 344(57) 347(57) *352, 353, 354, 356, 381*(58) 384(66) 389(66) *395,* 414(89) *441,* 447(7) *481,* 618(18, 19) 619(23) 625(19) 637(18, 19, 23) 650(18) *656, 657*

Alderton, H., 400(28) 428(28) 436(28) *439*

Alexander, A., 638(316) *664*

Alexander, H., 633(216) *661*

Alexander, P., 104(341, 342) *122,* 630(160) 631(176) 641(371) 652(530) *660, 665, 669*

Alfoldi, L., 112(361) *122*

Alford, J. A., 524(51) *542*

Allan, C. D., 643(431) 644(431) *667*

Allawala, N. A., 644(443) *667*

Allen, E. H., 156(137) *165*

Allen, M., 637(276) *663*

Allen, M. B., 31(200) *38*

Allen, P. J., 20(113, 115) 22(113) 24(113) *36*

Allison, L. E., 28(181) *38*

Alper, T., 618(14), 627(119) 628(119, 137, 138) 630(119, 137) *656, 659*

Altman, P. L., 5(26, 27) 6(27) 7(27) 8(26, 27) *34*

Amaha, M., 79(50) 80(50, 93) *115, 116,* 419(132) *442,* 453(42) 463(97) 464(97)

481, 483, 615(522) 620(37) 621(37, 44) 622(37, 57) 624(37) 649(37) 650(519, 522) *657, 669*

Ames, B. N., 140(62) *164,* 291(102) *299*

Anagnostopoulos, C., 30(198) *38,* 185(30) 186(39) 187(57, 62) *209, 210*

Anand, J. C., 262(138) *272,* 618(15) *656*

Anders, M., 129(16) *162*

Anderson, A. A., 621(53) 622(53) *657*

Anderson, A. W., 606(50) *610,* 626(115) 631(189) *659, 661*

Anderson, E. E., 581(34) *609*

Anderson, E. S., 208(191) *213*

Anderson, J. D., 636(273) *663*

Anderson, L. E., 81(108) 102(324) *116, 121,* 259(130) *272*

Andrassy, E., 627(129) 632(129) 655(553) *659, 670*

Anellis, A., 79(70) 112(360) *115, 122,* 216(3) 217(3) *269,* 617(11) 627(124, 132) 635(132) *656, 659*

Angelo, N., 153(129) *165*

Angelotti, R., 519(12) 521(25) 523(42) 524(42) 528(91) 534(198) 538(198) *541, 542, 543, 547*

Angus, T. A., 93(247, 249) *119,* 486(2) 487(6) 497(32, 33) 500(32, 33) 502(50, 51) 503(54) 504(56) 507(78) *514, 515*

Aoki, H., 157(142, 144) 158(142, 144) *165, 166*

Aoki, K., 496(24, 25) *514*

Ard, J. S., 638(327) *664*

Ard, W. B., 105(346) *122*

Arkawa, K. Y., 504(59) *515*

Armani, G., 80(100) *116*

Arnaud, M., 88(182) 92(182) *118,* 200(147) *212*

Arnstein, H. R. V., 468(106) *483*

Arntz, P., 404(55) *440*

Aronson, A., 153(129) *165*

Aronson, A. I., 50(33) 51(33) 52(33) *71,* 83(150) 94(259) 95(269, 270, 271) 98(269, 270, 271, 297) 103(336) 110(150, 269, 271) *117, 120, 121, 122,* 133(31) 141(68) 144(68, 82) 145(82) 146(91) 147(68, 82, 93, 94) 148 (82, 94, 97) 153(31) *163, 164,* 195(116, 117) 199 (137) *212,* 224(180)

EE

674

Subject Index

B. macerans—contd.
  nucleic acids and sporulation, 129, 131, 158, 184
  parasporal body formation by, 62, 63, 64
  spores, germination of, 306, 315, 320, 418
B. megaterium
  chromatin bodies of, 462
  microcycle sporogenesis by, 68, 69, 451
  spores of,
    activation,
      calcium dipicolinate, by, 382
      chemicals by, 319, 384
      heat by, 365, 369, 371, 374, 375, 376
      mechanism of, 390
      radiation, by, 383
    chemical composition, *see* 218–237, *see* 240–245, *see* 249–255, 264, 268
    deactivation, 381
    dormancy, 305, 315, 322, 343, 347, 348
    enzymes, 280
    germination, 20, 308, 309, 312, 316, 317, 318, 321, 323, 325, 326, 327, 330, 331, *see* 334–338, 341, 398–399, 400, 402, 406, 411, 412, 413, 417, 418, *see* 422–426, *see* 431–433, 464
    heat resistance, 314, 372, 577, *see* 619–624, 649
    outgrowth from, 446, *see* 451–457, 460, 465
    radiation resistance, 577, 629, 651
    survivability, 29
  sporulation,
    control of, proteases and, 199
    morphology of, 46, 49, 52, 56, 68, 69
    nucleic acid changes during, 127, 129, 130, 131, 136, 139, 141, 143, 146, 147, 158
    physiology and biochemistry of, 76, 78, 80, 81, 84, 85, 86, 89, 92, 94, 95, 98, 100, 103, 106, 113
B. mesentericus, nitrogen content of, 219
B. mycoides,
  chemical composition of, 218, 220, 221, 222
  spores,
    cysteine in, 98
    germination of, 312
    nucleic acids in, 131
    outgrowth requirements, 453
B. natto,
  DNA from, 129
  genetic transformation, 157, 186
  spore outgrowth requirements, 453
B. niger,
  DNA from, 128, 129
  genetic transformation, 157, 186

B. pantothenicus, spore activation, 319, 383, 638
B. pasteurii, spore germination and pH, 313
B. petasites, spore distribution, 555
B. polymyxa,
  antibiotics, from, 170, 174, 179
  chemical composition of, 219, 220, 242, 243, 244
  DNA from, 129, 157, 158
  food spoilage by, 561, 567, 577
  genetic transformation, 157, 158, 186
  spores of,
    germination, cortex and, 346, 402
    heat resistance, 621
    outgrowth of, 458, 477
    ribosidase of, 417
B. popilliae,
  growth in culture media, 492–494
  insect pathogen as, 487, 488–495, 490–491, 512
  parasporal body of, 63, 66
  spores of,
    activation, 371
    composition, 494
    dipicolinic acid, in, 341
    germination, 302, 308, 310, 313, 423
    membrane surrounding core, 348
    nucleic acids of, 131
  sporulation in defined media, 81
B. pumilus,
  DNA from, 129, 158
  spore radiation resistance, 632
  sporogenesis in continuous culture, 77
B. sotto, nucleic acids of, 131, 143
B. sphaericus,
  cell walls, D-amino acids, in, 225
  DAP decarboxylase of, 92
  DNA from, 129
  spores of,
    cortical muropeptide, 234, 237
    germination, 413, 430
B. stearothermophilus,
  food spoilage by, 561, 565, 567
    heat and radiation resistance and, 601, 603
  spores of,
    activation, 311, 314–315, 369, 381
    dormancy, 323, 336, 344, 373
    germination, 413, 414, 417, 431
    heat resistance, 250, 372, 577, 620, 621, 622, 624, 649
    outgrowth, 453, 457
  sporulation by,
    nucleic acids and, 78, 80, 81, 129, 141, 150, 156–157
B. subtilis,
  antibiotics of, 30, 168, 172, 173, 174, 201

W

Wall-lytic enzymes, *see under* Lytic enzymes
Wall structure, of dormant systems, 8
Water, activation of spores, and, 370–371
    activity,
        dormancy of spores and, 316–317
        food spoilage sporeformers, and growth of, 536–537, 562
        germination of spores and, 435
        resistance of spores and, 622, 629

content, dormancy and, 6–8, 11, 21, 25, 316–317, 329
Wax moth, *see Gallieria mellonella*, *Woodruffia metabolica*, 13

X

X-ray radiation, spore resistance to, 104, 105, 106

Z

Zinc, in spores, 248, 253, 256